PHYSIOLOGY AND BIOCHEMISTRY *of* ALGAE

Single coccolith from S. Pacific. Eocene. (Electron micrograph of replica, platinum/carbon shadowed at 65°. ×56,000. From B. Reimann.)

PHYSIOLOGY AND BIOCHEMISTRY *of* ALGAE

EDITED BY

RALPH A. LEWIN

Scripps Institution of Oceanography
University of California
La Jolla, California

ACADEMIC PRESS • 1962
New York and London

ACADEMIC PRESS INC.
111 Fifth Avenue, New York 3, New York

United Kingdom Edition published by
ACADEMIC PRESS INC. (LONDON) LTD.
Berkeley House Square, London W.1

Library of Congress Catalog Card Number: 62-13104

First Printing, 1962
Second Printing, 1964

PRINTED IN THE UNITED STATES OF AMERICA

Contributors

Numbers in italics indicate the page on which the author's contribution appears.

VERNON AHMADJIAN, Department of Biology, Clark University, Worcester, Massachusetts (*817*)

A. A. BENSON, Department of Agricultural and Biological Chemistry, Pennsylvania State University, University Park, Pennsylvania (*371*)

R. BIEBL, Institute for Plant Physiology, University of Vienna, Austria (*799*)

L. BOGORAD, Department of Botany, University of Chicago, Illinois (*385*)

MARCIA BRODY, Department of Biological Sciences, Hunter College of City University of New York, New York (*3*)

SEYMOUR STEVEN BRODY, IBM Watson Laboratories, Columbia University, New York City, New York (*3*)

C. J. BROKAW, Zoology Department, University of Minnesota, Minneapolis, Minnesota (*595*)

ANNETTE WILBOIS COLEMAN, Biology Department and McCollum-Pratt Institute, Johns Hopkins University, Baltimore, Maryland (*711*)

HERBERT M. CONRAD, Department of Biochemistry, School of Medicine, University of Southern California, Los Angeles, California (*663*)

WILLIAM F. DANFORTH, Department of Biology, Illinois Institute of Technology, Chicago, Illinois (*99*)

MICHAEL R. DROOP, Scottish Marine Biological Association, Millport, Scotland (*141*)

W. T. EBERSOLD, Department of Botany, University of California, Los Angeles, California (*731*)

RICHARD W. EPPLEY, Bioastronautics Laboratory, Northrop Corporation, Hawthorne, California (*255*, *839*)

K. ERBEN, Botanical Institute, University of Cologne, Germany (*701*)

G. E. FOGG,[1] Department of Botany, University College, London, England (*161*, *475*)

L. FOWDEN, Department of Botany, University College, London, England (*189*)

MARTIN GIBBS, Department of Biochemistry, Cornell University, Ithaca, New York (*61*, *91*)

M. B. E. GODWARD, Department of Botany, Queen Mary College, University of London, England (*551*)

PAUL B. GREEN, Division of Biology, University of Pennsylvania, Philadelphia, Pennsylvania (*625*)

ROBERT R. L. GUILLARD, Woods Hole Oceanographic Institution, Woods Hole, Massachusetts (*529*)

PER HALLDAL, Department of Plant Physiology, University of Lund, Sweden (*583*)

EIJI HASE, Institute of Applied Microbiology, University of Tokyo, and Tokugawa Institute for Biological Research, Tokyo, Japan (*617*)

J. W. HASTINGS, University of Illinois, Urbana, Illinois (*687*)

W. HAUPT, Botanical Institute, University of Tübingen, Germany (*567*)

OSMUND HOLM-HANSEN, Botany Department, University of Wisconsin, Madison, Wisconsin (*25*)

[1] Present address: Department of Botany, Westfield College, London, England.

TATSUICHI IWAMURA, Institute of Applied Microbiology, Tokyo University, and Tokugawa Institute for Biological Research, Tokyo, Japan (*231*)

G. JACOBI, Botanical Institute, Academy of Veterinary Medicine, Hannover, Germany (*125*)

R. JAROSCH, Biological Laboratory, Österreichische Stickstoffwerke A. G., Linz/Donau, Austria (*573*)

TERUHISA KATAYAMA, Department of Fisheries, Hiroshima University, Fukuyama, Japan (*467*)

R. W. KRAUSS, Department of Botany, University of Maryland, College Park, Maryland (*673*)

D. R. KREGER, Laboratory of General and Technical Biology, Institute of Technology, Delft, The Netherlands (*315*)[2]

ADOLF KUHL, Institute of Plant Physiology, University of Göttingen, Germany (*211*)

JOYCE C. LEWIN, Scripps Institution of Oceanography, University of California, La Jolla, California (*445, 457*)

RALPH A. LEWIN, Scripps Institution of Oceanography, University of California, La Jolla, California (*841*)

J. W. G. LUND, Freshwater Biological Association, Ambleside, England (*759*)

ERASMO MARRÈ, Institute of Plant Sciences, University of Milan, Italy (*541*)

JOHN J. A. MCLAUGHLIN, Haskins Laboratories, New York City, New York (*823*)

B. J. D. MEEUSE, Botany Department, University of Washington, Seattle, Washington (*289*)

J. D. A. MILLER, Department of Botany, University College, London, England (*357*)

JACK MYERS, Departments of Botany and Zoology, University of Texas, Austin, Texas (*603*)

T. O. M. NAKAYAMA, Department of Food Science and Technology, University of California, Davis, California (*409*)

S. NAKAZAWA, Biology Department, Yamagata University, Yamagata, Japan (*653*)

C. O HEOCHA, Biochemistry Department, University College, Galway, Ireland (*421*)

P. S. O'COLLA, Department of Chemistry, University College, Galway, Ireland (*337*)

CHINKICHI OGINO, Tokyo University of Fisheries, Tokyo, Japan (*437*)

GERHARD RICHTER, Botanical Institute, University of Tübingen, Germany (*633*)

PAUL SALTMAN, Department of Biochemistry, School of Medicine, University of Southern California, Los Angeles, California (*663*)

JEROME A. SCHIFF, Department of Biology, Brandeis University, Waltham, Massachusetts (*239*)

T. I. SHAW, The Laboratory, Marine Biological Association, Plymouth, England (*247*)

ISAO SHIBUYA, Department of Agricultural and Biological Chemistry, Pennsylvania State University, University Park, Pennsylvania (*371*)

PAUL C. SILVA, Department of Botany, University of California, Berkeley, California (*827*)

C. J. P. SPRUIT, Laboratory of Plant Physiological Research, Agricultural University, Wageningen, The Netherlands (*47*)

E. J. STADELMANN, Institute of Botany, University of Freiburg, Switzerland (*493*)

[2] Present address: Laboratory of Biological Ultrastructure Research, University of Groningen, The Netherlands.

B. M. SWEENEY, University of California, La Jolla, California (*687*)

P. J. SYRETT, Botany Department, University College, London, England (*171*)

J. F. TALLING, Freshwater Biological Association, Ambleside, England (*743*)

W. WIESSNER, Institute of Plant Physiology, University of Göttingen, Germany (*267*)

CHARLES S. YENTSCH, Woods Hole Oceanographic Institution, Woods Hole, Massachusetts (*771*)

PAUL A. ZAHL, Haskins Laboratories, New York City, New York (*823*)

Preface

In the last fifty years, phycology has emerged into an experimental phase, leaving behind it those slides of diatom skeletons, so elegantly disposed, which delighted our grandfathers, and those seaweed patterns pressed so prettily by our aunts. This volume summarizes much of the recent progress in the field. It is restricted to the algae, a heterogeneous assemblage of photosynthetic plants and their more or less heterotrophic relatives, whose only common feature is a lack of the terrestrial specializations of bryophytes and tracheophytes. It is further restricted to the biochemistry and physiology of these diverse plants, to those studies which tell us a little about what they are made of, and how they operate in their various activities of growth and propagation. It attempts to review most aspects of what might be called *experimental phycology*, the study of algae as living organisms, and to deal at least briefly with all major topics on which information is available. Its scope is thus wider than that of Fogg's masterly little monograph on algal metabolism (1952). Purely descriptive phases of phycology have been deliberately excluded; for details of anatomy and morphology, of life cycles and of ecology, the reader must seek elsewhere, in more classical texts (e.g., Fritsch, 1935, 1945; Smith, 1951) or in the scattered literature. I have chosen to include only the briefest considerations of applied phycology, such as the extensive use of seaweeds in human and animal diets and of seaweed products in industry, and the possible roles of living algae in the treatment of sewage, in space research, and in the ecology of radioactive wastes or bomb fall-out. Manuals of elysian, cloacal, or industrial phycology remain to be written.

A few aspects of experimental phycology, which should have merited separate chapters in this treatise, could not be discussed because unfortunately we still know so little about them. Thus translocation in the larger Brown Algae,* though strongly indicated on anatomical grounds, has received scant attention from physiologists (cf. Bodenberg, 1927). We urgently need more experimental studies on the physiology of apochlorotic algae, such as the cyanophytes that inhabit the alimentary tracts of vertebrates, the dinoflagellates that parasitize fish or crustaceans, and those pale rhodophytes that grow on their autotrophic allies. We still know nothing about the physiological bases underlying the dependence on their hosts of other obligately epiphytic or endophytic algae. Reviews of algal galls (e.g., Merola, 1956; Tokida, 1958; Kunzenbach and Brucker, 1960) in-

* In this volume, the terms Brown, Green, Red, and Blue-green Algae (capitalized) are used to designate members of the classes Phaeophyta, Chlorophyta, Rhodophyta, and Cyanophyta respectively, without regard to the apparent color of the species in question.

clude a regrettable paucity of experimental data. Likewise in studies of chytrids and other fungal parasites and endophytes of algae the experimental approach has been sadly neglected. It is hoped that these subjects will receive more attention in the near future.

The contributors to the present volume are of a dozen different nationalities, representing the international nature of the science; a voluminous correspondence was entailed as a result. The authors were culled less from the fastnesses of administrative offices, where august authorities wield their worn pens, than from the comparative chaos of laboratory benches where a younger generation of investigators is still involved in experimentation.

In recognition of the need to restrict the size of this compilation to a single volume, each author was initially assigned a maximum length for his contribution. In establishing these limits many border skirmishes arose; I hope my colleagues will in time forgive my obdurate insistence in setting bounds to their understandable enthusiasms. The deadline for the completion of most chapters was nominally May 31, 1960, but some important data published in the following year have also been incorporated.

As a consequence of the wide diversity of subjects and authors, this volume obviously cannot be read from cover to cover like a novel. It is designed as a guide book, primarily for research workers and advanced students, to be useful, not entertaining; and therefore no apology is made for its heterogeneity of style. However, although the articles were written by so many different hands, some coordination was effected by a preliminary circulation and exchange of synopses, and there were a few subsequent transfers of subject matter between the various chapters, in order to eliminate serious overlaps and omissions. The more abstruse syntactical features of some of the contributions were smoothed over as far as possible. For each chapter at least one critical reviewer in the same field provided valued assistance and, in some cases, additional material. A list of the reviewers is appended.

In spite of all our efforts, there are inevitably some shortcomings. Moreover, in some areas information tends to become soon outdated. If you find this treatise unsatisfactory in your own special field, I hope you will keep notes of its deficiencies, so that in a few years, when this introduction to experimental phycology will have become obsolescent, your comments, criticisms, and suggestions may serve in the building of a better Babylon.

September, 1961 RALPH A. LEWIN

REFERENCES

Bodenberg, E. T. (1927). Experiments on conduction in *Nereocystis Luetkeana*. *Publ. Puget Sound Biol. Sta.* **5**, 253–256.

Fogg, G. E. (1953). "The Metabolism of Algae." Methuen, London; Wiley, New York.
Fritsch, F. E. (1935). "The Structure and Reproduction of the Algae," Vol. I. Cambridge Univ. Press, London and New York.
Kunzenbach, R., and Brucker, W. (1960). Zur Bildung von "Tumoren" an Meeresalgen. II. *Ber. deut. botan. Ges.* **73,** 8–18.
Merola, A. (1956). Le galle nelle Alghe. I. Storia della cecidogenesi nelle alghe. *Ann. botan. Rome* **25,** 260.
Smith, G. M. (1951). "Manual of Phycology." Chronica Botanica, Waltham, Massachusetts.
Tokida, J. (1958). A review on galls in seaweeds. *Bull. Japan. Soc. Phycol.* **6,** 93–99.

Preface to Second Printing

The need for a second printing has afforded an opportunity for correcting a few typographical errors and for adding this editorial postscript.

Experimental phycology is clearly blossoming: there have been published of late several significant additions to the literature in this field (Ashida, 1963; Barashkov, 1963; Heim, 1961; Pirson, 1962; Pringsheim, 1963; Takamiya, 1963).

One recent paper, too modestly entitled "The concept of a bacterium" (Stanier and van Niel, 1962), deserves especial comment. When its impact is more widely felt, we may find it convenient to sever taxonomically the procaryotic Cyanophyta from the other (i.e., eucaryotic) algae—as some are indeed doing already—and to consider them with the bacteria. The fine cytological features by which the Procaryota are distinguished correlate closely with biochemical and physiological features, in particular the following:

1. The absence of special intracellular membranes delimiting enzymic apparatus for respiration (mitochondria), photosynthesis (plastids), and gene replication (nuclei).
2. The absence of conspicuous cell vacuoles—with important consequences for osmoregulation.
3. The presence of muramic and diaminopimelic acids in the cell walls.
4. The ability of some Procaryota to fix molecular nitrogen.

Perhaps we have been misled somewhat by the wide scope of the word "algae," and have consequently underplayed the fundamental dichotomy between procaryotic and eucaryotic plants. Perhaps we should now face the problems of the two phycologies.

January 1964 RALPH A. LEWIN

References

Ashida, J., ed. (1963). Studies on microalgae and photosynthetic bacteria. *Plant and Cell Physiol.* (*Special issue*), 636 pp.

Barashkov, G. K. (1963). "Chimiya Vodorosley" (The chemistry of algae), 143 pp. Murmansk Marine Biological Institute, Acad. Sci. USSR, Moscow.

Heim, R., ed. (1961). Chimie et physico-chimie des principes immédiats tirés des algues. *Colloq. Intern. Centre Natl. Recherche Sci.* (*Paris*) **No. 103,** 250 pp.

Pirson, A., ed. (1962). "Beiträge zur Physiologie und Morphologie der Algen." Vorträge aus dem Gesamtgebiet der Botanik (Deutschen Botanischen Gesellschaft) [N.F.] No. 1, 238 pp. Fischer, Stuttgart.

Pringsheim, E. G. (1963). "Farblose Algen," 471 pp. Fischer, Stuttgart.

Stanier, R. Y., and van Niel, C. B. (1962). The concept of a bacterium. *Arch. Mikrobiol.* **42,** 17–35.

Takamiya, A., *et. al.* (1963). A collection of papers dedicated to Professor Hiroshi Tamiya on the occasion of his 60th birthday. *Studies Tokugawa Inst.* (*Tokyo*) **10** (2), approx. 150 pp.

Acknowledgments

Especial thanks are due to Dr. Paul C. Silva, who has acted as taxonomic consultant—virtually as taxonomic co-editor—for all parts of this treatise.

Without the devoted clerical assistance of several colleagues, the editor would have been lost. Warmest thanks are expressed to Miss B. Lorraine Crabtree, Miss Loretta A. Ferguson, Miss M. Joan Smith, Mrs. C. R. Chismore, and Mr. John Emrich.

In addition, the aid of the following scientists, who served as reviewers and mentors for one or more of the chapters in this volume, and of Dr. Ivy Kellerman Reed, who assisted in reading and correcting the proofs, is gratefully acknowledged:

R. E. Alston, *University of Texas, Austin, Texas*
G. O. Arrhenius, *La Jolla, California*
S. Bendix, *Kaiser Foundation, Richmond, California*
J. A. Bentley, *Department of Agriculture and Fisheries, Aberdeen, Scotland*
R. Bloch, *Philadelphia, Pennsylvania*
T. Braarud, *Institute for Marine Biology, Oslo, Norway*
J. Brachet, *University of Brussels, Belgium*
V. Cassie, *Oceanographic Institute, Auckland, New Zealand*
T. J. Chow, *La Jolla, California*
H. Claes, *Florida State University, Tallahassee, Florida*
K. A. Clendenning, *La Jolla, California*
D. L. Correll, *Michigan State University, East Lansing, Michigan*
J. Dainty, *University of Edinburgh, Scotland*
*M. Droop, *Millport, Cumbrae, Scotland*
E. Epstein, *University of California, Davis, California*
J. Foley, *La Jolla, California*
D. L. Fox, *La Jolla, California*
A. W. Frenkel, *University of Minnesota, Minneapolis, Minnesota*
*M. Gibbs, *Cornell University, Ithaca, New York*
P. R. Gorham, *National Research Council, Ottawa, Canada*
C. Gowans, *University of Missouri, Columbia, Missouri*
*P. B. Green, *University of Pennsylvania, Philadelphia, Pennsylvania*
*R. R. L. Guillard, *Woods Hole, Massachusetts*
D. P. Hackett, *University of California, Berkeley, California*
M. Hale, *Smithsonian Institution, Washington, D. C.*
F. T. Haxo, *La Jolla, California*
R. W. Holmes, *La Jolla, California*
S. H. Hutner, *Haskins Laboratories, New York City, New York*
B. S. Jacobson, *University of Texas, Austin, Texas*
L. Jaffe, *Brandeis University, Waltham, Massachusetts*
A. T. Jagendorf, *Johns Hopkins University, Baltimore, Maryland*
N. Kamiya, *Osaka University, Japan*
T. Kaneda, *University of California, Los Angeles, California*
J. Kanwisher, *Woods Hole, Massachusetts*
S. Kelly, *New York State Department of Health, Albany, New York*
B. Lindberg, *Forest Products Research Laboratories, Stockholm, Sweden*
D. J. Manners, *University of Edinburgh, Scotland*
G. C. McLeod, *Carnegie Institution, Stanford, California*
*J. D. A. Miller, *University College, London, England*
*J. Myers, *University of Texas, Austin, Texas*
C. M. Palmer, *U. S. Public Health Service, Cincinnati, Ohio*
A. Pirson, *University of Göttingen, Germany*
R. D. Preston, *The University, Leeds, England*
L. Provasoli, *Haskins Laboratories, New York City, New York*
J. Raper, *Harvard University, Cambridge, Massachusetts*
E. L. Rawitzscher, *University of California, Berkeley, California*
B. E. F. Reimann, *La Jolla, California*
*G. Richter, *University of Tübingen, Germany*
W. Rodhe, *Erken Laboratory, Estuna, Sweden*

G. T. Scott, *Oberlin College, Ohio*
R. Setlow, *Oak Ridge National Laboratory, Tennessee*
A. Shrift, *Kaiser Foundation, Richmond, California*
*C. J. P. Spruit, *Wageningen, Netherlands*
*E. J. Stadelmann, *University of Freiburg, Switzerland*
R. C. Starr, *Indiana University, Bloomington, Ind.*
L. Stoloff, *Marine Colloids, Rockland, Maine*
P. R. Stout, *University of California, Davis, California*
*B. M. Sweeney, *La Jolla, California*
H. Tamiya, *Tokugawa Institute, Tokyo, Japan*
K. V. Thimann, *Harvard University, Cambridge, Massachusetts*
W. H. Thomas, *La Jolla, California*
Y. Tsubo, *Kôbe University, Japan*
B. E. Volcani, *La Jolla, California*
*C. S. Yentsch, *Woods Hole, Massachusetts*
E. G. Young, *National Research Council, Halifax, Canada*
A. Zehnder, *Wettingen, Switzerland*
B. H. Zimm, *La Jolla, California*

* Contributors to this volume.

Contents

—PART I—

Nutrition and Metabolism

1

Light Reactions in Photosynthesis

MARCIA BRODY AND SEYMOUR STEVEN BRODY

2

Assimilation of Carbon Dioxide

OSMUND HOLM-HANSEN

3

Photoreduction and Anaerobiosis

C. J. P. SPRUIT

4
Respiration

Martin Gibbs

5
Fermentation

Martin Gibbs

6
Substrate Assimilation and Heterotrophy

William F. Danforth

7
Enzyme Systems

G. Jacobi

8
Organic Micronutrients

Michael R. Droop

9
Nitrogen Fixation

G. E. Fogg

10
Nitrogen Assimilation

P. J. Syrett

11
Amino Acids and Proteins

L. Fowden

12
Inorganic Phosphorus Uptake and Metabolism

Adolf Kuhl

13
Nucleotides and Nucleic Acids

Tatsuichi Iwamura

14
Sulfur

JEROME A. SCHIFF

15
Halogens

T. I. SHAW

16
Major Cations

RICHARD W. EPPLEY

17
Inorganic Micronutrients

W. WIESSNER

—PART II—
Composition of Cells and Metabolic Products

18
Storage Products

B. J. D. MEEUSE

19
Cell Walls

D. R. Kreger

20
Mucilages

P. S. O'Colla

21
Fats and Steroids

J. D. A. Miller

22
Surfactant Lipids

A. A. Benson and Isao Shibuya

23
Chlorophylls

L. Bogorad

24
Carotenoids

T. O. M. Nakayama

25
Phycobilins

C. Ó hEocha

26
Tannins and Vacuolar Pigments

Chinkichi Ogino

27
Silicification

Joyce C. Lewin

28
Calcification

JOYCE C. LEWIN

29
Volatile Constituents

TERUHISA KATAYAMA

30
Extracellular Products

G. E. FOGG

—PART III—

Physiology of Whole Cells and Plants

31
Permeability

E. J. STADELMANN

32
Salt and Osmotic Balance

ROBERT R. L. GUILLARD

33
Temperature

ERASMO MARRÈ

34
Invisible Radiations

M. B. E. GODWARD

35
Intracellular Movements

W. HAUPT

36
Gliding

R. JAROSCH

37
Taxes

PER HALLDAL

38
Flagella

C. J. BROKAW

39
Laboratory Cultures

JACK MYERS

40
Cell Division

EIJI HASE

41
Cell Expansion

Paul B. Green

42
Nuclear-Cytoplasmic Interactions

Gerhard Richter

43
Polarity

S. Nakazawa

44
Growth Substances

Herbert M. Conrad and Paul Saltman

45
Inhibitors

R. W. Krauss

46
Rhythms

B. M. Sweeney and J. W. Hastings

47
Sporulation

K. Erben

48
Sexuality

Annette Wilbois Coleman

49
Biochemical Genetics

W. T. Ebersold

—PART IV—

Physiological Aspects of Ecology

50
Freshwater Algae

J. F. Talling

51
Soil Algae

J. W. G. Lund

52
Marine Plankton

Charles S. Yentsch

53
Seaweeds

R. Biebl

54
Lichens

VERNON AHMADJIAN

55
Endozoic Algae

JOHN J. A. McLAUGHLIN AND PAUL A. ZAHL

ERRATA

Page	*Corrections and Inserts*
418	between Goodwin and Goodwin and Jaimikorn, insert: Goodwin, T. W., and Gross, J. A. (1958). Carotenoid distribution in bleached substrains of *Euglena gracilis. J Protozool.* **5**(4), 292–295.
843	between Aschner and Aspinall, insert: Ashida, J., *xi*, *xii*
844	between Banse and Barber, insert: Barashkov, G. K., *xi*, *xii*
845	between Bock and Böm-Tüchy, insert: Bodenberg, E. T., *ix*, *x*
846	between Bruce and Bruckmayer, insert: Brucker, W., *ix*, *xi*
850	for Fogg, G. E., read: *ix*, *xi*, 8, *21*, ...
850	for Fritsch, F. E., read: *ix*, *xi*, 154, *157*, ...
853	for Heim, R., read: *xi*, *xii*, 799, *812*
857	between Kuhl and Kupke, insert: Kunzenbach, R., *ix*, *xi*
859	between Menzel and Merrett, insert: Merola, A., *ix*, *xi*
863	for Pirson, A., read: *xi*, *xii*, 14, *21*, ...
863	for Pringsheim, E. G., read: *xi*, *xii*, 100, 106, ...
866	for Smith, G. M., read: *ix*,. *xi*, 448, *455*, 530, ...
867	for Stanier, R. Y., read: *xi*,*xii*, *4*, *18*, ...
868	between Takahashi and Takaoka, insert: Takamiya, A., *xi*, *xii*
868	between Tödt and Tolbert, insert: Tokida, J., *ix*, *xi*
881	line 14 should read: Coccoliths, *frontispiece*, 460–463
882	between Diadinoxanthin and Diatomaceous, insert: Diaminopimelic acid, *xi*
887	insert: Galls, *ix*
895	between Multipolarity and Mutants, insert: Muramic acid, *xi*
908	insert: Translocation, *ix*
918	between *Euastrum didelta* and *Eucheuma*, insert: Eucaryota, *xi*
924	under Phaeophyta, physiology, insert: *ix*
925	between *Prasiola crispa* and Pronoctilucacea, insert: Procaryota, *xi*

—Part I—

Nutrition and Metabolism

—1—

Light Reactions in Photosynthesis

MARCIA BRODY

Department of Biological Sciences, Hunter College of City University of New York, New York

and

SEYMOUR STEVEN BRODY

IBM Watson Laboratories, Columbia University, New York City, New York

I. Introduction[1]

In this article, the authors have tried to summarize some of the more salient aspects of research in the field of photosynthesis. For more detailed information, not confined to studies with algae, the reader is referred to Rabinowitch (1945, 1951, 1956) and to reviews by others (Arnon, 1958, 1959, 1960; Aronoff, 1957; Blinks, 1954; Brown and Frenkel, 1953; Clendenning, 1957, 1960; Duysens, 1956a; Emerson, 1958; Franck, 1951; Lumry *et al.*, 1953; Rosenberg, 1957; Haxo, 1960).

For convenience we have divided the light reactions into two major categories: (1) the *physical* reactions, including the absorption of energy and its transfer between various chloroplast pigments; and (2) the *chemical*

[1] Abbreviations used in this paper: ADP, adenosine diphosphate; ATP, adenosine triphosphate; TPN, triphosphopyridine nucleotide.

reactions, comprising the mechanisms and intermediates of the chlorophyll reactions by which light energy is converted into chemical energy.

II. Photophysical Processes

A. Energy Transfer between Chloroplast Pigments

1. *Role of Accessory Pigments*

a. Energy Transfer as Deduced from Measurements of Oxygen Evolution. Although photosynthesis in algae and higher plants seems to be limited to forms which contain chlorophyll *a*,[2] the contribution of other pigments to this process was proposed as early as the 1880's by Engelmann (1883, 1884), who made ingenious qualitative experiments on the photosynthesis of algae in colored light, using motile aerotactic bacteria as oxygen indicators. He concluded from these experiments that the fucoxanthin of Brown Algae and the phycobilins of Red and Blue-green Algae were as effective as chlorophyll in sensitizing photosynthesis.

From quantitative manometric measurements of the dependence of yield on the wavelength of light, Warburg and Negelein (1923) suggested that some of the light absorbed by the carotenoids in *Chlorella* was probably used for photosynthesis. More direct experiments by Emerson and Lewis (1942, 1943), using the Blue-green Alga *Chroococcus turgidus*, firmly established the validity of Engelmann's proposal by demonstrating that the quantum yield of photosynthesis in the yellow region, where more than half the absorption is attributable to phycocyanin, is as high as that of chlorophyll-sensitized photosynthesis in the Green Alga *Chlorella pyrenoidosa*. The yield of carotenoid-sensitized photosynthesis was found to be low in these algae. In contrast, Dutton and Manning (1941) concluded from measurements of the quantum yield of photosynthesis in "*Nitzschia closterium*" (=*Phaeodactylum tricornutum*) that the main carotenoid of Brown Algae, fucoxanthin, which absorbs light of longer wavelengths than do most carotenoids, was as efficient as chlorophyll. Tanada (1951), in his work with *Navicula minima*, confirmed the activity of fucoxanthin, and also demonstrated that chlorophyll *c* is an effective sensitizer of photosynthesis in this diatom.

[2] Carotene, usually associated with several other carotenoids, is present in all photosynthetic organisms so far examined. However, the fact that photosynthesis can proceed in carotenoid-less mutants of the bacterium *Rhodopseudomonas spheroides* in the absence of oxygen (Sistrom *et al.*, 1956; Cohen-Bazire and Stanier, 1958) indicates that carotenoids are not essential for this process. In support of this, Chance and Sager (1957) showed that a mutant of *Chlamydomonas reinhardtii* can carry out photosynthesis without any evidence for the participation of carotenoids.

b. Concept of Energy Transfer Applied to Photosynthesis Experiments. All of the above-cited experiments indicate that energy absorbed by other pigments, as well as by chlorophyll, can be used for photosynthesis. It was not until 1950 that Arnold and Oppenheimer suggested transfer of energy between dissimilar molecules (from phycobilin to chlorophyll) as a probable explanation for the high yield of photosynthesis of *Chroococcus* in the spectral region where most of the energy is absorbed by phycocyanin. The concept of energy transfer, however, had been applied to photosynthesis by Gaffron and Wohl as early as 1936 to explain the results of experiments by Emerson and Arnold (1932) on the kinetics of photosynthesis in flashing light, although at that time the concept was applied only to the transfer of energy between similar molecules (of chlorophyll).

c. Direct Evidence for Energy Transfer from Fluorescence Experiments. In 1943 Dutton *et al.* showed directly, for the first time, that energy absorbed by one pigment can actually be transferred *in vivo* to another pigment with high efficiency. Their fluorescence experiments with "*Nitzschia closterium*" demonstrated that some of the energy absorbed by the carotenoid fucoxanthin was emitted as fluorescence of chlorophyll to about the same extent as energy directly absorbed by the chlorophyll. The energy emitted from chlorophyll as fluorescence represents only a small fraction, around 3% (Latimer *et al.*, 1956), of the total energy transferred to chlorophyll, the remainder presumably being used in photosynthesis or dissipated by competing processes and by internal conversion into heat. Dutton and Manning's experiments on sensitized fluorescence thus support Engelmann's contention that the accessory pigments act only as light absorbers, and are not themselves photocatalysts.

2. Role of Chlorophyll a as Photocatalyst

a. Low Yields of Photosynthesis and Fluorescence Sensitized by Chlorophyll a. By polarographic measurements of oxygen evolution, Haxo and Blinks (1950) determined action spectra of photosynthesis in a number of different algae, and compared these with the absorption spectra of intact thalli in each case. In Green (*Ulva taeniata*) and Brown Algae (*Coilodesme californica*), the action spectrum closely paralleled the absorption spectrum, except in blue light and in the far red at wavelengths exceeding 700 mμ. However, in Red Algae (*Delesseria decipiens, Porphyra nereocystis, P. naiadum*, and *P. perforata*), action spectra differed markedly from their absorption spectra, exhibiting maxima corresponding to absorption by the phycobilins but little activity in the regions of maximum chlorophyll absorption. Although the low yield in violet light could be partially explained by ineffective absorption by carotenoids, the decline in yield in the red region of the spectrum was evident well within the absorption band of chlorophyll at wavelengths above 650 mμ. Haxo and Blinks concluded that in Red Algae chlorophyll plays only a minor role in photosynthesis, the phycobilins being the pigments of major importance. In contrast to the results of Emerson and Lewis (1942) on *Chroococcus*, they reported that there was only weak chlorophyll activity in the Blue-green Algae, *Anabaena* and *Oscillatoria*.

Duysens (1952), using a similar polarographic technique, measured photosynthesis in the Red Alga *Porphyridium cruentum*, with results in general agreement with those of Haxo and Blinks. Although verifying the low yield of chlorophyll-sensitized photosynthesis, Duysens preferred to retain for chlorophyll the role of primary photocatalyst, since the experiments of Van Norman *et al.* (1948) and French and Young (1952), as well as his own (Duysens, 1951), had shown that energy absorbed by phycoerythrin strongly sensitized chlorophyll fluorescence. Moreover, French and Young had noted that the initial variations in the intensity of fluorescence, which could be correlated with induction phenomena of photosynthesis, occur only in chlorophyll *a* but not in phycobilin.

b. Chlorophyll d Proposed as an "Energy Trap." In an attempt to explain the low yields of directly sensitized chlorophyll fluorescence and of photosynthesis, Duysens (1951) made the *ad hoc* proposal that the chlorophyll *a* in Red Algae was present in two types of units. He postulated that in one type chlorophyll was associated with phycobilin and received its excitation from this accessory pigment, the energy then being used either for photosynthesis or emitted as fluorescence. In the other type, chlorophyll was not associated with phycobilin; energy absorbed in this unit was lost for fluorescence or photosynthesis as a result of transfer to some other "energy sink." Duysens had observed in *Porphyra laciniata* and *Porphyridium cruentum* the fluorescence of an unknown pigment, which was sensitized by light directly absorbed by chlorophyll *a*. He postulated that this unknown pigment might act as an energy sink by virtue of its absorption at wavelengths longer than those of the chlorophyll *a* spectrum, and suggested that it might be chlorophyll *d*, discovered in Red Algae by Manning and Strain (1943).

c. Chlorophyll Efficiency—a Function of Wavelength. M. Brody and Emerson (1959a) measured the dependence of oxygen evolution upon the wavelength of incident light, also using *Porphyridium cruentum*. By employing completely opaque suspensions of cells (a technique first introduced into manometric studies of photosynthesis by Warburg and Negelein (1922–1923), it was possible to make direct efficiency determinations without first comparing action and absorption spectra. Their experiments indicated that, at wavelengths shorter than about 650 mμ, chlorophyll *a* is just as efficient in *P. cruentum* as in other algae, and is at least as efficient as the phycobilins. The results of Haxo and Blinks (1950) and Duysens (1952) gave no indication of the activity of chlorophyll at wavelengths shorter than 650 mμ, because their action spectra and absorption spectra were so similar that small divergencies could be attributed to technical artifacts arising from light scattering, etc. (see Latimer and Rabinowitch, 1956; Shibata *et al.*, 1954; Yocum, 1951; Duysens, 1956*b*). However, at wavelengths longer than 650 mμ, but still within the red absorption band of chlorophyll *a*, a sharp decline in photosynthesis was noted. From considerations of photochemistry one would expect that light absorption by a pigment molecule would lead to equally effective chemical action, independent of the wavelength of exciting light, provided, of course, that the absorption band represents a unique condition of a particular pigment.

d. The Long-Wavelength Decline. The quantum yield measurements of M. Brody (1958) and M. Brody and Emerson (1959a) do not support the hypothesis that chlorophyll *per se* is inactive in Red Algae. Brody considered that wavelengths longer than 650 mμ were beyond the range of maximum efficiency of chlorophyll, and suggested that the decline in this region was comparable to that observed by Emerson and Lewis (1943) for photosynthesis in *Chlorella pyrenoidosa* beyond 680 mμ.

Emerson and co-workers (1956) investigated further the long-wavelength limits for *Chlorella* and other organisms, and compared the results with those obtained by Brody for *Porphyridium*. They found that lower temperatures and supplementary light of shorter wavelengths could extend the range of full photosynthetic efficiency to longer wavelengths.

Emerson *et al.* (1957) and Emerson (1958) suggested that full efficiency of photosynthesis can be realized only when light absorption by chlorophyll *a* is associated with absorption by an accessory pigment having its first excited state at an energy level higher than that of the first excited state of chlorophyll *a*. Emerson visualized two discrete light-requiring steps in normal photosynthesis, one of these requiring quanta energetically larger than those available from light absorption by chlorophyll *a*. He based his conclusions primarily on the observation that the wavelengths most effective in raising the low yield in the red spectral region in *Chlorella* were those where absorption by chlorophyll *b* exceeds that by chlorophyll *a*. (See also Blinks, 1957, 1959, and Myers and French, 1960.) The long-wavelength decline in efficiency has been observed in all classes of algae examined, although in different spectral regions, its onset coinciding with the region where absorption by accessory pigments falls off.

e. Theories Explaining the Long-Wavelength Decline:

(i) *Chlorophyll a and accessory pigments.* According to Emerson's concept, efficient photosynthesis cannot be achieved in organisms lacking accessory pigments. This hypothesis is supported by the observation that algae in which the only chlorophyll is chlorophyll *a*, such as *Ochromonas malhamensis* (Myers and Graham, 1956) and certain Xanthophyta such as *Nannochloris oculata* (Allen, 1958), grow much more rapidly when supplied with an organic substrate than they do by photosynthesis alone, even under optimum conditions of light, etc. (Emerson had just started investigations on these algae when he met with a sad and untimely death.)

Emerson pointed out some of the weaknesses in his explanation of the long-wavelength decline, notably that it failed to account for the corresponding long-wavelength decline in fluorescence yield, and that it conflicted with evidence from fluorescence measurements that energy absorbed at shorter wavelengths may be transferred to chlorophyll *a* with high efficiency. Since it is possible to grow *Monodus* and other algae lacking accessory pigments in media in which they are wholly dependent on the fixation of carbon dioxide for organic synthesis (Miller and Fogg, 1957; Aaronson and Baker, 1960), and since these organisms have no phycobilins and no chlorophyll other than type *a*, one would expect, according to Emerson's explanation, that there would be a decline of efficiency in the blue region of the spectrum, corresponding with diminishing absorption by carotenoids. However, in *Ochromonas danica* and *Polyedriella* sp. the decline in action spectrum for excitation of chlorophyll fluorescence (which, it has been assumed, parallels photosynthesis) occurs at wavelengths similar to those in *Chlorella* (M. Brody, unpublished).

(ii) *Two different electronic transitions.* In seeking a satisfactory interpretation for the long-wavelength decline in yield of both fluorescence and photosynthesis, Franck (1958) suggested that two different but nearly equal electronic transitions (a π–π and an n–π) may be included in the red absorption band. In those molecules in which the n–π is the one of lower energy, and the π–π the one of higher energy, the former transition leads to the formation of a triplet state, while the latter corresponds to a singlet state. Franck postulated that the maximum yield of photosynthesis requires approximately equal numbers of molecules in the excited singlet and triplet states, and attributed the low yield at longer wavelengths to a deficiency of molecules in the singlet state.

(iii) *Monomeric and aggregated forms of chlorophyll a.* The absorption and fluorescence properties of chlorophyll *in vivo* resemble in several respects those in concentrated alcoholic solutions. Under both conditions the chlorophyll exists in two forms, a monomer and a dimer (S. S. Brody, 1958; S. S. Brody and M. Brody, 1961; see also Krasnovsky and Kosobutskaya, 1952, 1953; Lavorel, 1957). Brody and Brody contend that the structure of the chlorophyll aggregate *in vivo* is equivalent to that found *in vitro*. The monomer has a single absorption maximum at the red end of the spectrum, while the dimer is presumed to have two such maxima, one on either side of the monomer peak. This would account for the triple form of the red peak observed *in vivo*, which was resolved by the differential spectrophotometer of French *et al.* (1959), and which had been attributed by them to the presence of three distinct forms of chlorophyll *a*.

At this point it might be well to recall the unknown pigment in *Porphyra laciniata* and *Porphyridium cruentum*, with which Duysens (1951) sought to explain the long-wavelength decline by attributing to it the role of an energy sink. Although at that time it was suggested that the pigment was chlorophyll *d*, M. Brody and Emerson (1959b) subsequently showed that chlorophyll *d* is not present in *Porphyridium*. The fluorescence and absorption properties of the unidentified pigment suggest that it is an aggregated form of chlorophyll similar to that found in "old" cultures of phototrophically grown *Euglena* (French, 1958; M. Brody and Linschitz, 1961). Preponderance of absorption by the dimer may cause the long-wavelength decline. In algae in which the decline begins at shorter wavelengths (e.g., *Porphyridium*), it is postulated that a greater concentration of dimer is present than in forms in which the decline does not occur until longer wavelengths (e.g., *Chlorella*). Indeed, when S. S. Brody and M. Brody (1961) measured the peak ratios of the emission bands of the monomer (685 mμ) and the dimer (720 mμ) at low temperatures, they found that organisms having the largest proportion of monomer to dimer also exhibited full photosynthetic efficiency at the longest wavelengths. (See also Blinks, 1957, 1959, 1960, and Myers and French, 1960.)

Since the decline in photosynthetic yield at long wavelengths was attributed to an excess of excited dimer molecules and a deficiency of excited monomer molecules, at whatever wavelengths this situation obtains, a low efficiency would be anticipated. Perhaps this may account for the anomalous dips in photosynthetic efficiency at 460–500 mμ and 650–670 mμ observed in *Chlorella pyrenoidosa* and *Porphyridium cruentum* (Emerson and Lewis, 1942, 1943; M. Brody and S. S. Brody, 1961). Similarly, the short-wavelength decline (and enhancement effect) reported by Blinks (1960) may be associated with the blue absorption band of the dimer.

At our present state of knowledge we are led to the conclusion that the accessory pigments act solely as energy absorbers, contributing to photosynthesis and chlorophyll fluorescence only in so far as they are able to transfer energy to chlorophyll *a*, whose role remains that of the only photocatalyst in photosynthesis.

B. Energy Transfer; General Considerations

It may be asked, "Of what value are studies of energy transfer in understanding photosynthesis?" Experiments with sensitized fluorescence, by which energy transfer may be measured directly, yield information about the roles of the various pigments as photocatalysts, energy absorbers, or energy traps. Moreover, an elucidation of the mechanism of energy transfer might indicate the strength of coupling of the involved pigments.

Upon the absorption of a photon, an electron is lifted out of the ground state, into an excited state, leaving behind a positive "hole." The electron and the hole may travel together, or one may leave the other behind, or they may move in opposite directions. The first process, described as resonance transfer, may be of either the "fast" (Frenkel, 1931) or "slow" type (see Förster, 1947, 1948, 1951; Förster and Kasper, 1954); the others have been called *electron migration* (see Calvin and Sogo, 1957; Sogo *et al.*, 1957; Commoner *et al.*, 1954; Arnold and Sherwood, 1957).

During the time between its absorption and its photochemical utilization, energy may be transferred between pigment molecules, which may be similar (e.g., between phycoerythrin molecules) or dissimilar (e.g., between phycoerythrin, phycocyanin, and chlorophyll). The former process is called *homogeneous energy transfer*; the latter is called *heterogeneous energy transfer*.

Homogeneous energy transfer *in vivo* was experimentally demonstrated by the depolarization of fluorescence in chloroplast grana of angiosperms (Arnold and Meek, 1956). Examples of heterogeneous energy transfer involving the accessory pigments of algae *in vivo* have already been cited. S. S. Brody and Rabinowitch (1957) measured the time for the energy to be transferred from phycoerythrin to chlorophyll *a* (probably via phycocyanin) in *Porphyridium*; they found the delay in the onset of chlorophyll fluorescence to be about 0.5×10^{-9} sec. The fluorescence lifetime of the excited singlet state of chlorophyll in *Chlorella pyrenoidosa, Porphyridium cruentum* and *Anacystis nidulans* is about 1.5×10^{-9} sec. (Brody and Rabinowitch, 1957).

Intramolecular energy transfer was also shown by means of sensitized fluorescence in purified phycocyanin extracted from *Synechocystis* (Bannister, 1954), and in phycoerythrin from an unspecified source (Vladimirov and Konov, 1959). The time required for this type of transfer of energy from protein to chromophore was found to be of the order of 1.5×10^{-9} sec. (S. S. Brody, 1960).

III. Photochemistry of Photosynthesis

A. General Considerations

Photosynthesis is, of course, a complex of reactions by which light energy is used in the fixation of carbon dioxide. Evidence for a nonphotochemical reaction ("dark reaction") as a component of the photosynthetic process was first detected from the light-saturation effect (Blackman, 1905); further evidence came from experiments with flashing light (Emerson and Arnold, 1932, 1933) and with isotopic carbon-dioxide fixation in the dark (Ruben *et al.*, 1941; Benson and Calvin, 1947). The enzyme-catalyzed dark reactions which lead to carbon-dioxide fixation are discussed by Holm-Hansen (see Chapter 2, this treatise); we present here a digest of present knowledge about the mechanisms whereby the energy of light absorbed by chlorophyll is used to generate the "fuels" needed to drive the dark reactions.

B. Photochemical Studies in Intact Algae

1. *Light-Induced Changes in Absorption*

Little is known about the changes undergone during photosynthesis by chlorophyll and other pigments. A powerful tool for measuring light-sensitized chemical reactions in living systems is difference spectroscopy of irradiated and nonirradiated suspensions of cells (Duysens, 1955; Kok, 1957a). Descriptions of the instruments used for this type of investigation may be found in publications by Kok (1957a), Duysens (1955), Lundegårdh (1954), and Chance (1947, 1953). An instrument designed to measure fast changes in absorption, immediately after flashes of light, was described by Witt (1955a). A summary of the difference spectra obtained with *Nitzschia*, *Chlorella*, *Porphyra*, *Porphyridium*, *Nostoc*, and *Scenedesmus* was given by Kok (1957a).

Absorption changes may arise from modifications in the concentration of pigmented intermediates of the photosynthetic cycle, or from shifts in their absorption maxima. The first mechanism has been used by most workers to interpret their results; the equally feasible second possibility was proposed by Kok (1957a). Some absorption changes have been tentatively correlated with chlorophylls, carotenoids, and cytochromes, while others have been attributed to unknown pigments.

a. Changes Attributed to Chlorophyll. Decreases in absorption at 650 and 475 mμ, 678 and 440 mμ, and 705 and 425 mμ, observed in illuminated cells of *Chlorella* spp., were attributed by Kok (1957a), Coleman *et al.* (1957), and Coleman and Rabinowitch (1959) to chlorophyll *b*, chlorophyll *a*, and an unknown pigment, respectively. Kok (1957a) also suggested, as an alternative interpretation, that these decreases might result from shifts in opposite directions of two unidentified absorption bands, with maxima originally located at 691 and 658 mμ. S. S. Brody and M. Brody (1961) attributed these decreases in absorption either to changes in the structure of the aggregate of chlorophyll, which as a monomer *in vitro* exhibits maxima at 683 and 648 mμ, or to shifts in equilibrium between monomeric and aggregated forms of chlorophyll. It has also been proposed that changes in the red region indicate a transition to a metastable state (Coleman *et al.*, 1957) or a photoreduction of chlorophyll, as in the Krasnovsky reaction (Coleman and Rabinowitch, 1959; Rabinowitch, 1959a, b, c). It should be noted, however, that the changes in the red region are not accompanied by a decrease of the blue (Soret) band or by an increase in the green region, which are characteristic features of both reduced and metastable chlorophyll.

b. Changes Attributed to Carotenoids. Most changes in absorption between 510 and 580 mμ, except those at 565 and 555 mμ (see below), may result from complicated carotenoid transformations (Kok, 1957b; Coleman and Rabinowitch, 1959).

c. Changes Attributed to Cytochromes. Duysens (1955) pointed out that the changes in the absorption spectrum of *Porphyridium* on irradiation are very similar to those involved in the oxidation of cytochrome *f*, namely, decreases in absorption at 420 and 555 mμ (Davenport and Hill, 1952; Lundegårdh, 1954; Hill and Scarisbrick, 1951). However, the increase at 410 mμ, to be expected upon the formation of ferricytochrome, is observed only in Blue-green and Red Algae. Kok (1957a) suggested that the decrease at 555 mμ reflects the disappearance of the *a* band of cytochrome *f*, while the observed increase at 565 mμ may correspond to formation of ferrocytochrome *b* (Hill, 1954).

d. Unspecified Absorption Changes. In many algae other than *Porphyridium*, the spectral modifications induced by illumination are considerably more complicated; for example, in *Chlorella* there is an increase in absorption at 515 mμ and a decrease at 474 mμ, in addition to those changes mentioned above.

These changes were at first attributed to the transformation of a single substance. However, it has been shown that at least three different reactions are responsible for the absorption changes observed in the spectral region between 400 and 550 mμ (Witt, 1955a, b, 1957; Witt *et al.*, 1956, 1958, 1960): (1) a fast, photochemical, temperature-independent transformation in which flashes of duration shorter than about 5×10^{-4} sec. produce an increase in absorption at 515 mμ and a decrease at around 430 mμ; (2) a slow, photochemical, temperature-dependent transition produced when the flashes last longer than 5×10^{-4} sec., resulting in similar changes and an additional decrease in absorption at 475 mμ; (3) a slow, temperature-dependent back-reaction which occurs in the dark and occupies about 10^{-2} sec.

2. Photolysis of Water

Much evidence has substantiated the findings of Ruben *et al.* (1941) that the oxygen evolved during photosynthesis comes from the photolysis of water, although there is no conclusive proof that water is the *sole* source of photosynthetic oxygen (Brown and Frenkel, 1953). Kamen (1956) suggested that the energy used in splitting water may not originate in the primary light reaction, but may be generated from back-oxidations in the dark, and that the evolution of oxygen results from a gradual accumulation of oxidized material. If this is so, then the quantum requirement referred to oxygen evolution is not simply a measure of the efficiency of the primary photochemical reaction, but is rather a measure of the over-all efficiency of reactions occurring both in light and in darkness.

3. Quantum Requirement of Photosynthesis

Since the quantum requirement of photosynthesis is of great theoretical significance, an important problem arose when a number of workers (Petering *et al.*, 1939; Manning *et al.*, 1938, 1939; Magee *et al.*, 1939) were unable to confirm the quantum requirement of 4 obtained by Warburg and Negelein (1922) from manometric experiments with *Chlorella vulgaris*. Using the same biological material and experimental techniques, Emerson and Lewis (1941) repeated the experiments of Warburg and Negelein, although, instead of using time schedules which permitted the inclusion of transient effects, they measured steady-state photosynthesis, and thereby deduced that the quantum requirement was considerably higher, 10 ± 2. After further modification of his methods, Warburg (Warburg and Burk, 1950; Warburg *et al.*, 1950; Burk *et al.*, 1949; Burk and Warburg, 1951) once more repeated his experiments, and then claimed a quantum requirement even lower than his original value of 4.

The discrepancies in the reported values seemed reconcilable in view of the findings of numerous workers, including Kok (1948), van der Veen (1949) and Bassham *et al.* (1955), who reported a non-linear relationship between light intensity and oxygen evolution in the region of the compensation point, where photosynthesis just balances respiration. Franck (1949), reviving an old hypothesis, suggested that the quantum requirement of 4, found *below* the compensation point, represented a reversal of respiration, whereas values of about 8, found *above* the compensation point, reflected the quantum requirement of "true" photosynthesis. However, since attempts by Emerson *et al.* (1957), Warburg *et al.* (1949), and Warburg and Krippahl (1954) failed to confirm the Kok effect, the problem of the quantum requirement is still unresolved.

On the basis of thermodynamic considerations and the preponderance of experimental evidence from manometric (Tanada, 1951; M. Brody and Emerson, 1959a; Yocum and Blinks, 1958; Yuan *et al.*, 1955) and other techniques (Petering *et al.*, 1939; Magee *et al.*, 1939; Tonnelat, 1944; Arnold, 1949), it is becoming increasingly evident that 10 ± 2 quanta are required for the evolution of an oxygen molecule during steady-state photosynthesis at light intensities between the compensation point and a level where some other factor becomes limiting.

C. Photochemical Studies with Chloroplast Preparations

The great advantage of measuring photochemical reactions in chloroplast preparations, rather than in whole cells, lies in the elimination of possible interference by other metabolic processes such as respiration (James and Das, 1957; Burk and Warburg, 1951; Arnon *et al.*, 1956).

1. Hill Reaction

In 1939 Hill discovered that isolated chloroplasts of spinach can liberate oxygen on illumination in the presence of a suitable hydrogen acceptor:

$$A + H_2O \xrightarrow[\text{chloroplast}]{h\nu} AH_2 + \tfrac{1}{2}O_2 \qquad (1)$$

The "Hill reaction" has subsequently been demonstrated with chloroplasts or pigmented fragments isolated from Green Algae (Hill *et al.*, 1953; Punnett and Fabiyi, 1953; Aronoff, 1946; Ehrmantraut and Rabinowitch, 1952; Clendenning and Ehrmantraut, 1950), Blue-green Algae (Wessels and van der Veen, 1956), Brown Algae and Red Algae (McClendon, 1954), as well as with whole algal cells in the presence of quinone (Ehrmantraut and Rabinowitch, 1952; Noack *et al.*, 1939; Warburg and Lüttgens, 1944a, b).

A long-wavelength decline in the action spectrum of the Hill reaction of *Chlorella* cells has been observed (Ehrmantraut and Rabinowitch, 1952); it responds, as does that of photosynthesis, to supplementary light of shorter wavelength (Govindjee *et al.*, 1960).

2. Carbon Dioxide Fixation

Although Boichenko (1943, 1944) reported some reduction of carbon dioxide by preparations of isolated angiosperm chloroplasts, her techniques were questioned by Rabinowitch (1956, see p. 1531). Subsequently, however, when C^{14} became available, Fager (1952a, b) unequivocally demonstrated a light-stimulated fixation of a small amount of CO_2 in a cell-free brei of spinach leaves, and Tolbert and Zill in 1954 produced cell-free preparations of extruded *Chara* or *Nitella* chloroplasts capable of similar assimilatory activity.

It was at first believed that chloroplasts must be intact in order to fix carbon dioxide (Arnon *et al.*, 1954a, 1956), but Thomas *et al.* (1957) demonstrated that, in the presence of all the necessary cofactors, fragments of *Spirogyra* chloroplasts exhibited activity similar to that of whole cells. There is a stoichiometric relationship between oxygen evolution and carbon dioxide uptake by chloroplasts (Arnon *et al.*, 1959), and the same products are formed as in intact cells (Gibbs and Cynkin, 1958).

3. Photophosphorylation and Photoreduction of Pyridine Nucleotide

It is now recognized that the Hill reaction represents only a portion of the photochemical process, whereby water is split, oxygen is evolved, and the liberated hydrogen atoms or electrons can be used for reduction.

Vishniac and Ochoa (1951, 1952), Tolmach (1951), and Arnon (1951) were among the first to demonstrate the photoreduction of pyridine nucleotides in spinach chloroplasts by coupling of the photochemical reaction with an enzyme system. This was confirmed by the work of Vishniac and Rose (1958), among others, who showed that, in illuminated whole cells of *Scenedesmus* and *Chlorella*, hydrogen from water is first transferred to chlorophyll. The subsequent transfer of hydrogen from chlorophyll *a* to TPN was demonstrated by the use of chemically tritiated chlorophyll.

The photoreduction of TPN could be a major reaction in photosynthesis, since its quantum requirement is only about 6 (Duysens and Amesz, 1959). It may be coupled to the formation of ATP by a process of non-cyclic phosphorylation:

$$2\text{ADP} + 2\text{P} + 2\text{TPN} + 2\text{H}_2\text{O} \xrightarrow[\text{chloroplast}]{h\nu} 2\text{ATP} + \tfrac{1}{2}\text{O}_2 + \text{H}_2\text{O} + 2\text{TPNH} \quad (2)$$

Vishniac (1955) and Bassham *et al.* (1954) suggested that ATP might be involved in photosynthesis, since by itself TPNH is not capable of reducing the CO_2 addition product. At about the same time, Arnon demonstrated that photosynthetic phosphorylation could occur in isolated chloroplasts of spinach without concomitant carbon dioxide assimilation

$$\mathrm{ADP} + \mathrm{P} \xrightarrow[\text{chloroplast}]{h\nu} \mathrm{ATP} \tag{3}$$

and that this process could proceed anerobically (Arnon *et al*; 1954b, 1955; Wessels, 1957), which immediately distinguished it from the oxidative phosphorylation occurring in mitochrondria. Chloroplasts isolated from *Spirogyra* were likewise shown to exhibit photosynthetic phosphorylation (Thomas *et al.*, 1957). More recently, Petrack (1959) demonstrated photosynthetic phosphorylation in cell-free particulate preparations of Blue-green Algae, including *Anabaena variabilis,* and found that the particles can be washed practically free of phycocyanin without loss of activity.

REFERENCES

Aaronson, S., and Baker, H. (1960). A comparative biochemical study of two species of *Ochromonas. J. Protozool.* **6,** 282–284.

Allen, M. B. (1958). Studies with green algae that lack chlorophyll *b. In* "The Photochemical Apparatus: Its Structure and Function," pp. 339–343. Brookhaven Natl. Lab., Upton, New York. U. S. Atomic Energy Commission Rept. BNL 512-(C-28).

Arnold, W. (1949). A calorimetric determination of the quantum yield in photosynthesis. *In* "Photosynthesis in Plants" (J. Franck and W. Loomis, eds.), pp. 273–276. Iowa State College Press, Ames, Iowa.

Arnold, W., and Meek, E. (1956). The polarization of fluorescence and energy transfer in grana. *Arch. Biochem. Biophys.* **60,** 82–90.

Arnold, W., and Oppenheimer, J. R. (1950). Internal conversion in the photosynthetic mechanism of blue-green algae. *J. Gen. Physiol.* **33,** 423–435.

Arnold, W., and Sherwood, H. K. (1957). Are chloroplasts semiconductors? *Proc. Natl. Acad. Sci. U.S.* **43,** 105–114.

Arnon, D. I. (1951). Extracellular photosynthetic reactions. *Nature* **167,** 1008–1010.

Arnon, D. I. (1958). Chloroplasts and photosynthesis. *In* "The Photochemical Apparatus: Its Structure and Function," pp. 181–235. Brookhaven Natl. Lab., Upton, New York. U. S. Atomic Energy Commission Rept. BNL 512-(C-28).

Arnon, D. I. (1959). Conversion of light into chemical energy in photosynthesis. *Nature* **184,** 10–21.

Arnon, D. I. (1960). The chloroplast as a functional unit in photosynthesis. *In* "Handbuch der Pflanzenphysiologie" (W. Ruhland, ed.), Vol. V, Part 1, pp. 773–829. Springer, Berlin.

Arnon, D. I., Allen, M. B., and Whatley, F. R. (1954a). Photosynthesis by isolated chloroplasts. *Nature* **174,** 394–396.

Arnon, D. I., Allen, M. B., and Whatley, F. R. (1954b). Photosynthesis by isolated chloroplasts. II. Photosynthetic phosphorylation, the conversion of light into phosphate-bond energy. *J. Am. Chem. Soc.* **76,** 6324–6328.

Arnon, D. I., Whatley, F. R., and Allen, M. B. (1955). Photosynthetic phosphorylation as an anaerobic process. *Biochim. et Biophys. Acta* **16,** 605.

Arnon, D. I., Allen, M. B., and Whatley, F. R. (1956). Photosynthesis by isolated chloroplasts. IV. General concept and comparison of three photochemical reactions. *Biochim. et Biophys. Acta* **20,** 449–461.

Arnon, D. I., Whatley, F. R., Allen, M. B. (1959). Photosynthesis by isolated chloroplasts. VIII. Photosynthetic phosphorylation and the generation of assimilatory power. *Biochim. et Biophys. Acta* **32,** 47–57.

Aronoff, S. (1946). Photochemical reduction of chloroplast grana. *Plant Physiol.* **21,** 293–409.

Aronoff, S. (1957). Photosynthesis. *Botan. Rev.* **23,** 65–107.

Bannister, T. T. (1954). Energy transfer between chromophore and protein in phycocyanin. *Arch. Biochem. Biophys.* **49,** 222–233.

Bassham, J. A., Benson, A. A., Kay, L. D., Harris, A. Z., Wilson, A. T., and Calvin, M. (1954). The path of carbon in photosynthesis. XXI. The cyclic regeneration of carbon dioxide acceptor. *J. Am. Chem. Soc.* **76,** 1760–1770.

Bassham, J. A., Shibata, K., and Calvin, M. (1955). Quantum requirement in photosynthesis related to respiration. *Biochim. et Biophys. Acta* **17,** 332–340.

Benson, A., and Calvin, M. (1947). The dark reductions of photosynthesis. *Science* **105,** 648–649.

Blackman, F. F. (1905). Optima and limiting factors. *Ann. Botany (London)* **19,** 281–295.

Blinks, L. R. (1954). Photosynthetic function of pigments other than chlorophyll. *Ann. Rev. Plant Physiol.* **5,** 93–114.

Blinks, L. R. (1957). Chromatic transients in photosynthesis of red algae. *In* "Research in Photosynthesis" (H. Gaffron, ed.), pp. 444–449. Interscience, New York.

Blinks, L. R. (1959). Chromatic transients in the photosynthesis of a green alga. *Plant Physiol.* **34,** 200–203.

Blinks, L. R. (1960). Action spectra of chromatic transients and Emerson effect in marine algae. *Proc. Natl. Acad. Sci.* **46,** 327–333.

Boichenko, E. A. (1943). Conditions necessary for the activity of chloroplasts outside the cell. *Compt. rend. acad. sci. U.R.S.S.* **38,** 181–184. [Summarized by E. Rabinowitch (1956), pp. 1530–1533.]

Boichenko, E. A. (1944). Catalysts for the action of isolated chloroplasts. *Compt. rend. acad. sci. U.R.S.S.* **42,** 345–347. [Summarized by E. Rabinowitch (1956), pp. 1530–1533.]

Brody, M. (1958). The participation of chlorophyll and phycobilins in the photosynthesis of red algae, and observations on cellular structures of *Porphyridium cruentum.* Doctoral Thesis, University of Illinois, Urbana, Illinois.

Brody, M., and Brody, S. S. (1961). Induced changes in photosynthetic efficiency of pigments in *Porphyridium cruentum.* II. *Arch. Biochem. Biophys.*, in preparation.

Brody, M., and Emerson, R. (1959a). The quantum yield of photosynthesis in *Porphyridium cruentum,* and the role of chlorophyll *a* in the photosynthesis of red algae. *J. Gen. Physiol.* **43,** 251–264.

Brody, M., and Emerson, R. (1959b). The effect of wavelength and intensity of light on the proportion of pigments in *Porphyridium cruentum. Am. J. Botany* **46,** 433–440.

Brody, M., and Linschitz, H. (1961). Fluorescence spectra of green plants and photosynthetic bacteria at room and liquid nitrogen temperature. *Science,* **133,** 705.

Brody, S. S. (1958). A new excited state of chlorophyll. *Science* **128,** 838–839.

Brody, S. S. (1960). Delay in intermolecular and intramolecular energy transfer and lifetimes of photosynthetic pigments. *Z. Elektrochem.* **64,** 187–194.

Brody, S. S., and Brody, M. (1961). Spectral characteristics of aggregated chlorophyll and its possible role in photosynthesis.

Brody, S. S., and Rabinowitch, E. (1957). Excitation lifetime of photosynthetic pigments *in vivo* and *in vitro*. *Science* **125,** 555.

Brown, A. H., and Frenkel, A. W. (1953). Photosynthesis. *Ann. Rev. Plant Physiol.* **4,** 23–58.

Burk, D., and Warburg, O. (1951). Ein-Quanten-Reaktion und Kreisprozess der Energie bei der Photosynthese. *Z. Naturforsch.* **6b,** 12–22.

Burk, D., Hendricks, S., Korzenovsky, M., Schocken, V., and Warburg, O. (1949). The maximum efficiency of photosynthesis: a rediscovery. *Science* **110,** 225–229.

Calvin, M., and Sogo, P. B. (1957). Primary quantum conversion process in photosynthesis: Electron spin resonance. *Science* **125,** 499–500.

Chance, B. (1947). Stable spectrophotometry of small density changes. *Rev. Sci. Instr.* **18,** 601–609.

Chance, B. (1953). The carbon monoxide compounds of the cytochrome oxidases. II. Photodissociation spectra. *J. Biol. Chem.* **202,** 397–416.

Chance, B., and Sager, R. (1957). Oxygen and light induced oxidations of cytochrome, flavoprotein, and pyridine nucleotide in a *Chlamydomonas* mutant. *Plant Physiol.* **32,** 548–560.

Clendenning, K. A. (1957). Biochemistry of chloroplasts in relation to the Hill reaction. *Ann. Rev. Plant Physiol.* **137,** 137–152.

Clendenning, K. A. (1960). Photochemical activity of isolated chloroplasts (Hill reactions). *In* "Handbuch der Pflanzenphysiologie" (W. Ruhland, ed.), Vol. V, Part 1, pp. 736–772. Springer, Berlin.

Clendenning, K. A., and Ehrmantraut, H. C. (1950). Photosynthesis and Hill reactions by whole *Chlorella* cells in continuous and flashing light. *Arch. Biochem.* **29,** 387–403.

Cohen-Bazire, G., and Stanier, R. Y. (1958). Specific inhibition of carotenoid synthesis in a photosynthetic bacterium and its physiological consequences. *Nature* **181,** 250–252.

Coleman, J. W., and Rabinowitch, E. (1959). Evidence of photoreduction of chlorophyll *in vivo*. *J. Phys. Chem.* **63,** 30–34.

Coleman, J. W., Holt, A. S., and Rabinowitch, E. (1957). Reversible bleaching of chlorophyll *in vivo*. *In* "Research in Photosynthesis" (H. Gaffron, ed.), pp. 68–75. Interscience, New York.

Commoner, B., Townsend, J., and Pake, G. (1954). Free radicals in biological materials. *Nature* **174,** 689–691.

Davenport, H., and Hill, R. (1952). The preparation and some properties of cytochrome *f*. *Proc. Roy. Soc.* **B139,** 327–345.

Dutton, H. L., and Manning, W. M. (1941). Evidence for carotenoid-sensitized photosynthesis in the diatom *Nitzschia closterium*. *Am. J. Botany* **28,** 516–526.

Dutton, H. L., Manning, W. M., and Duggar, B. M. (1943). Chlorophyll fluorescence and energy transfer in the diatom *Nitzschia closterium*. *J. Phys. Chem.* **47,** 308–313.

Duysens, L. M. N. (1951). Transfer of light energy within the pigment system present in photosynthesizing cells. *Nature* **168,** 548–550.

Duysens, L. M. N. (1952). The transfer of excitation energy in photosynthesis. Doctoral Thesis, University of Utrecht, Netherlands.

Duysens, L. M. N. (1955). Role of cytochrome and pyridine nucleotide in algal photosynthesis. *Science* **121,** 210–211.

Duysens, L. M. N. (1956a). Energy transformations in photosynthesis. *Ann. Rev. Plant Physiol.* **7,** 25–50.

Duysens, L. M. N. (1956b). The flattening of the absorption spectrum of suspensions, as compared to that of solution. *Biochim. et Biophys. Acta* **19,** 1–12.

Duysens, L. M. N., and Amesz, J. (1959). Quantum requirement for phosphopyridine nucleotide reduction in photosynthesis. *Plant Physiol.* **34**, 210–213.

Ehrmantraut, H., and Rabinowitch, E. (1952). Kinetics of Hill reaction. *Arch. Biochem. Biophys.* **38**, 67–84.

Emerson, R. (1958). The quantum yield of photosynthesis. *Ann. Rev. Plant Physiol.* **9**, 1–24.

Emerson, R., and Arnold, W. (1932). The photochemical reaction in photosynthesis. *J. Gen. Physiol.* **16**, 191–205.

Emerson, R., and Arnold, W. (1933). The separation of the reactions of photosynthesis by means of intermittent light. *J. Gen. Physiol.* **15**, 391–420.

Emerson, R., and Lewis, C. M. (1941). Carbon dioxide exchange and the measurement of the quantum yield of photosynthesis. *Am. J. Botany* **28**, 789–804.

Emerson, R., and Lewis, C. M. (1942). The photosynthetic efficiency of phycocyanin in *Chroöcoccus* and the problem of carotenoid participation in photosynthesis. *J. Gen. Physiol.* **25**, 579–595.

Emerson, R., and Lewis, C. M. (1943). The dependence of the quantum yield of *Chlorella* photosynthesis on wavelength of light. *Am. J. Botany* **30**, 165–178.

Emerson, R., Chalmers, R., Cederstrand, C., and Brody, M. (1956). Effect of temperature on the long-wave limit of photosynthesis. *Science* **123**, 673.

Emerson, R., Chalmers, R., and Cederstrand, C. (1957). Some factors influencing the long-wave limit of photosynthesis. *Proc. Natl. Acad. Sci. U. S.* **43**, 133–143.

Engelmann, T. W. (1883). Farbe und Assimilation. *Botan. Z.* **41**, 1–13.

Engelmann, T. W. (1884). Untersuchungen über die quantitativen Beziehungen zwischen Absorption des Lichtes und Assimilation in Pflanzenzellen. *Botan. Z.* **42**, 81–93, 97–105.

Fager, E. W. (1952a). Photochemical carbon dioxide fixation by cell-free leaf macerates. *Arch. Biochem. Biophys.* **37**, 5–14.

Fager, E. W. (1952b). Photochemical carbon dioxide fixation by a cell-free system. *Arch. Biochem. Biophys.* **41**, 483–495.

Förster, T. (1947). Ein Beitrag zur Theorie der Photosynthese. *Z. Naturforsch.* **2b**, 174–82.

Förster, T. (1948). Zwischenmolekulare Energiewanderung und Fluoreszenz. *Ann. Physik* [6] **2**, 55–75.

Förster, T. (1951). "Fluoreszenz organischer Verbindungen." Vandenhoeck and Ruprecht, Göttingen, Germany.

Förster, T., and Kasper, K. (1954). Ein Konzentrationsumschlag der Fluoreszenz. *Z. physik. Chem. (Frankfurt)* **1**, 275–277.

Franck, J. (1949). An interpretation of the contradictory results in measurements of the photosynthetic quantum yields and related phenomena. *Arch. Biochem.* **23**, 297–314.

Franck, J. (1951). A critical survey of the physical background of photosynthesis. *Ann. Rev. Plant Physiol.* **2**, 53–86.

Franck, J. (1958). Remarks on the long wavelength limits of photosynthesis and chlorophyll fluorescence. *Proc. Natl. Acad. Sci.* **44**, 941–948.

French, C. S. (1958). Various forms of chlorophyll *a* in plants. *In* "The Photochemical Apparatus: Its Structure and Function," pp. 65–74. Brookhaven Natl. Lab., Upton, New York. U. S. Atomic Energy Commission Rept. BNL 512-(C-28).

French, C. S., and Young, V. M. K. (1952). The fluorescence spectra of red algae and the transfer of energy from phycoerythrin to phycocyanin and chlorophyll. *J. Gen. Physiol.* **35**, 873–890.

French, C. S., Brown, J. S., Allen, M. B., and Elliott, R. F. (1959). Types of chlorophyll *a* in plants. *Carnegie Inst Wash. Yr. Book* **58,** 315, 327–331.

Frenkel, L. (1931). On the transformation of light into heat in solids. I and II. *Phys. Rev.* [2] **37,** 17–44, 1276–1294.

Gaffron, H., and Wohl, K. (1936). The theory of assimilation. I and II. *Naturwissenschaften* **24,** 81–90, 103–107.

Gibbs, M., and Cynkin, M. A. (1958). Conversion of carbon-14 dioxide to starch-glucose during photosynthesis by spinach chloroplasts. *Nature* **182,** 1241–1242.

Govindjee, R., Thomas, J. B., and Rabinowitch, E. (1960). "Second Emerson Effect" in the Hill reaction of *Chlorella* cells with quinone as oxidant. *Science* **132,** 421–422.

Haxo, F. T. (1960). Photosynthesis in algae containing special pigments. *In* "Handbuch der Pflanzenphysiologie" (W. Ruhland, ed.), Vol. 5, Part 2, pp. 349–363. Springer, Berlin.

Haxo, F. T., and Blinks, L. R. (1950). Photosynthetic action spectra of marine algae. *J. Gen. Physiol.* **33,** 389–422.

Hill, R. (1939). Oxygen produced by isolated chloroplasts. *Proc. Roy. Soc.* **B127,** 192–210.

Hill, R. (1954). The cytochrome *b* component of chloroplasts. *Nature* **174,** 501–503.

Hill, R., and Scarisbrick, R. (1951). The hematin compounds of leaves. *New Phytologist* **50,** 98–111.

Hill, R., Northcote, D. H., and Davenport, H. E. (1953). Production of oxygen from chloroplast preparations. Active chloroplast preparations from *Chlorella pyrenoidosa*. *Nature* **172,** 948–949.

James, W. O., and Das, V. S. (1957). The organization of respiration in chlorophyllous cells. *New Phytologist* **56,** 325–343.

Kamen, M. D. (1956). Hematin compounds in the metabolism of photosynthetic tissues. *In* "Enzymes: Units of Biological Structure and Function" (O. Gaebler, ed.), pp. 483–498. Academic Press, New York.

Kok, B. (1948). A critical consideration of the quantum yield of *Chlorella* photosynthesis. *Enzymologia* **13,** 1–56.

Kok, B. (1957a). Light induced absorption changes in photosynthetic organisms. *Acta Botan. Neerl.* **6,** 316–336.

Kok, B. (1957b). Changes of absorption spectrum induced by illumination, and their bearing on the nature of the photoreceptor of photosynthesis. *Proc. Intern. Congr. Photobiol., 2nd Congr., Turin, Italy.*

Krasnovsky, A. A., and Kosobutskaya, L. M. (1952). Spectral investigation of chlorophyll during its formation in plant and in colloidal solution of material from etiolated leaves. *Doklady Akad. Nauk S.S.S.R.* **85,** 177–180.

Krasnovsky, A. A., and Kosobutskaya, L. M. (1953). Different conditions of chlorophyll in plant leaves. *Doklady Akad. Nauk S.S.S.R.* **91,** 343–346.

Krasnovsky, A. A., Vojnovskaya, K. K., and Kosobutskaja, L. M. (1952). The natural state of bacteriochlorophyll and the spectral properties of its solutions and solid films. *Doklady Akad. Nauk S.S.S.R.* **85,** 389–392.

Latimer, P., and Rabinowitch, E. (1956). Selective scattering of light by pigment-containing plant cells. *J. Chem. Phys.* **24,** 480.

Latimer, P., Bannister, T. T., and Rabinowitch, E. (1956). Quantum yields of fluorescence of plant pigments. *Science* **124,** 585–586.

Lavorel, J. (1957). Influence of concentration on the absorption spectrum and the action spectrum of fluorescence of dye solutions. *J. Phys. Chem.* **61,** 1600–1605.

Lumry, R., Spikes, J. D., and Eyring, H. (1953). Photosynthesis. *Ann. Rev. Phys. Chem.* **4,** 399–424.

Lundegårdh, H. (1954). On the oxidation of cytochrome *f* by light. *Physiol. Plantarum* **7,** 375–382.

McClendon, J. H. (1954). The physical environment of chloroplasts as related to their morphology and activity *in vitro*. *Plant Physiol.* **29,** 448–458.

Magee, J. L., DeWitt, T. W., Smith, E. C., and Daniels, F. (1939). A photocalorimeter. The quantum efficiency of photosynthesis in algae. *J. Am. Chem. Soc.* **61,** 3529–3533.

Manning, W. M., and Strain, H. H. (1943). Chlorophyll *d*, a green pigment of red algae. *J. Biol. Chem.* **151,** 1–19.

Manning, W. M., Stauffer, J. F., Duggar, B. M., and Daniels, F. (1938). Quantum efficiency of photosynthesis in *Chlorella*. *J. Am. Chem. Soc.* **60,** 266–274.

Manning, W. M., Juday, C., and Wolf, M. (1939). Photosynthesis in *Chlorella*. Quantum efficiency and rate measurements in sunlight. *J. Am. Chem. Soc.* **60,** 274–278.

Miller, J. D., and Fogg, G. E. (1957). Studies on the growth of Xanthophyceae in pure culture. I. The mineral nutrition of *Monodus subterraneus* Petersen. *Arch. Mikrobiol.* **28,** 1–17.

Myers, J., and French, C. S. (1960). Evidences from action spectra for a specific participation of chlorophyll *b* in photosynthesis. *J. Gen. Physiol.* **43,** 723–736.

Myers, J., and Graham, J. R. (1956). The role of photosynthesis in the physiology of *Ochromonas*. *J. Cellular Comp. Physiol.* **47,** 397–414.

Noack, K., Pirson, A., and Michels, H. (1939). Inhibition of assimilation in green algae after depriving of oxygen. *Naturwissenschaften* **27,** 645.

Petering, H. G., Duggar, B. M., and Daniels, F. (1939). Quantum efficiency of photosynthesis in *Chlorella*. II. *J. Am. Chem. Soc.* **61,** 3525–3529.

Petrack, B. (1959). Photophosphorylation in cell-free preparations of blue-green algae. *Federation Proc.* **18,** 302.

Punnett, T., and Fabiyi, O. (1953). Production of oxygen from chloroplast preparations. *Nature* **172,** 947–948.

Rabinowitch, E. (1945). "Photosynthesis and Related Processes," Vol. I: Chemistry of Photosynthesis, Chemosynthesis and Related Processes *in vitro* and *in vivo*. Interscience, New York.

Rabinowitch, E. (1951). "Photosynthesis and Related Processes," Vol. II, Part 1: Spectroscopy and Fluorescence of Photosynthetic Pigments; Kinetics of Photosynthesis, pp. 603–1208. Interscience, New York.

Rabinowitch, E. (1956). "Photosynthesis and Related Processes," Vol. II, Part 2. Kinetics of Photosynthesis (continued). Addenda to Vol. II, Part 1, pp. 1211–2088. Interscience, New York.

Rabinowitch, E. (1959a). General discussion. *In* "Energy Transfer with Special Reference to Biological Systems." *Discussions Faraday Soc. No.* **27,** 252–253.

Rabinowitch, E. (1959b). Primary photochemical and photophysical processes in photosynthesis. *In* "Energy Transfer with Special Reference to Biological Systems." *Discussions Faraday Soc. No.* **27,** 161–172.

Rabinowitch, E. (1959c). Primary photochemical and photophysical processes in photosynthesis. *Plant Physiol.* **34,** 213–218.

Rosenberg, J. L. (1957). Photochemistry of chlorophyll. *Ann. Rev. Plant Physiol.* **8,** 115–136.

Ruben, S., Randall, M., Kamen, M. D., and Hyde, J. L. (1941). Heavy oxygen (O^{18}) as a tracer in the study of photosynthesis. *J. Am. Chem. Soc.* **63,** 877–879.

Shibata, K., Benson, A. A., and Calvin, M. (1954). The absorption spectra of suspensions of living microorganisms. *Biochim. et Biophys. Acta* **15,** 461–470.

Sistrom, W. R., Griffith, M., and Stanier, R. Y. (1956). The biology of a photosynthetic bacterium which lacks colored carotenoids. *J. Cellular Comp. Physiol.* **48,** 473–515.

Sogo, P. B., Pon, N. G., and Calvin, M. (1957). Photo-spin resonance in chlorophyll-containing plant material. *Proc. Natl. Acad. Sci. U.S.* **43,** 387–393.

Tanada, T. (1951). The photosynthetic efficiency of carotenoid pigments in *Navicula minima*. *Am. J. Botany* **38,** 276–283.

Thomas, J. B., Haans, A. J. M., and Van der Leun, A. A. J. (1957). Photosynthetic activity of isolated chloroplast fragments of *Spirogyra*. *Biochim. et Biophys. Acta* **25,** 453–462.

Tolbert, N. E., and Zill, L. P. (1954). Photosynthesis by protoplasm extruded from *Chara* and *Nitella*. *J. Gen. Physiol.* **37,** 575–588.

Tolmach, L. J. (1951). Effects of triphosphopyridine nucleotide upon oxygen evolution and carbon dioxide fixation by illuminated chloroplasts. *Nature* **167,** 946–948.

Tonnelat, J. (1944). Mésure calorimétrique du rendement de la photosynthése. *Compt. rend. acad. sci.* **218,** 430–432.

van der Veen, R. (1949). Induction phenomena in photosynthesis. I. *Physiol. Plantarum* **2,** 217–234.

Van Norman, R., French, C. S., and Macdowall, F. D. H. (1948). The absorption and fluorescence spectra of two red marine algae. *Plant Physiol.* **23,** 455–466.

Vishniac, W. (1955). Biochemical aspects of photosynthesis. *Ann. Rev. Plant Physiol.* **6,** 115–134.

Vishniac, W., and Ochoa, S. (1951). Photochemical reduction of pyridine nucleotides by spinach grana and coupled carbon dioxide fixation. *Nature* **167,** 768–769.

Vishniac, W., and Ochoa, S. (1952). Phosphorylation coupled to photochemical reduction of pyridine nucleotides by chloroplast preparations. *J. Biol. Chem.* **198,** 501–506.

Vishniac, W., and Rose, I. A. (1958). Mechanism of chlorophyll action in photosynthesis. *Nature* **182,** 1089–1090.

Vladimirov, I. U. A., and Konov, S. V. (1959). On the possibility of energy transfer in a protein. *Biophysics* (*U.S.S.R.*) (*Engl. Transl.*) **4,** 26–34.

Warburg, O., and Burk, D. (1950). The maximum efficiency of photosynthesis. *Arch. Biochem. Biophys.* **25,** 410–443.

Warburg, O., and Krippahl, G. (1954). Über Photosynthese-Fermente. *Angew. Chem.* **66,** 493–496.

Warburg, O., and Lüttgens, W. (1944a). The assimilation of carbon dioxide. *Naturwissenschaften* **32,** 161.

Warburg, O., and Lüttgens, W. (1944b). Carbon dioxide assimilation. *Naturwissenschaften* **32,** 301.

Warburg, O., and Negelein, E. (1922). The transformation of energy during the assimilation of carbon dioxide. *Naturwissenschaften* **10,** 647–653. *Z. physik. Chem.* **102,** 235–266.

Warburg, O., and Negelein, E. (1923). Influence of wavelength on the energy change in carbon dioxide assimilation. *Z. physik. Chem.* **106,** 191–218.

Warburg, O., Burk, D., Schocken, V., Korzenovsky, M., and Hendricks, S. (1949). Does light inhibit the respiration of green cells? *Arch. Biochem. Biophys.* **23,** 330–333.

Warburg, O., Burk, D., Schocken, V., and Hendricks, S. (1950). The quantum efficiency of photosynthesis. *Biochim. et Biophys. Acta* **4,** 335–346.

Wessels, J. S. C. (1957). Photosynthetic phosphorylation. I. Photosynthetic phosphorylation under anaerobic conditions. *Biochim. et Biophys. Acta* **25,** 97–100.

Wessels, J. S. C., and van der Veen, R. (1956). Action of some derivatives of phenylurethan and of 3-phenyl-1,1-dimethylurea on the Hill reaction. *Biochim. et Biophys. Acta* **19,** 548–549.

Witt, H. T. (1955a). Rapid absorption changes in the primary process of photosynthesis. *Naturwissenschaften* **42,** 72–73.

Witt, H. T. (1955b). Primary process in photosynthesis. *Z. physik. Chem. (Frankfurt)* **4,** 120–123.

Witt, H. T. (1957). Reaction patterns in the primary process of photosynthesis. *In* "Research in Photosynthesis" (H. Gaffron, ed.), pp. 75–85. Interscience, New York.

Witt, H. T., Moraw, R., and Müller, A. (1956). Zum Primärprozess der Photosynthese an Chlorophyllkörnern ausserhalb der pflanzlichen Zelle. *Z. Elektrochem.* **60,** 1148–1153.

Witt, H. T., Moraw, R., and Müller, A. (1958). Neue Absorptionsänderungen beim Primärprozess der Photosynthese. *Z. physik. Chem. (Frankfurt)* **14** (1/2), 127–129.

Witt, H. T., Moraw, R., Müller, A., Rumberg, B., and Zieger, G. (1960). Kinetische Untersuchungen über die Primärvorgänge der Photosynthese. *Z. Elektrochem.* **64,** 181–187.

Yocum, C. S. (1951). Some experiments on photosynthesis in marine algae. Doctoral Thesis, Stanford University, Palo Alto, California.

Yocum, C. S., and Blinks, L. R. (1958). Light induced efficiency and pigment alterations in red algae. *J. Gen. Physiol.* **41,** 1113–1117.

Yuan, E. L., Evans, R. W., and Daniels, F. (1955). Energy efficiency of photosynthesis in *Chlorella. Biochim. et Biophys. Acta* **17,** 185–193.

—2—

Assimilation of Carbon Dioxide

OSMUND HOLM-HANSEN

Botany Department, University of Wisconsin, Madison, Wisconsin

I. Introduction[1]

A. Autotrophic and Heterotrophic Assimilation

It is well known that the initial photochemical reactions which bring about photolysis of water in photosynthesis can be experimentally separated from the reduction of carbon dioxide. The photochemical reactions, discussed by Brody and Brody in Chapter 1 of this treatise, lead to the generation of high-energy phosphate bonds as in adenosine triphosphate (ATP) and to

[1] Abbreviations used in this chapter: ADP, adenosine diphosphate; ATP, adenosine triphosphate; DHAP, dihydroxyacetone phosphate; DPNH, diphosphopyridine nucleotide (reduced); PEP, phosphoenolpyruvic acid; PGA, 3-phosphoglyceric acid; RuDP, ribulose diphosphate; TPN, triphosphopyridine nucleotide (oxidized); TPNH, triphosphopyridine nucleotide (reduced).

the reduction of pyridine nucleotides (Whatley *et al.*, 1960); these compounds can then serve as the source of energy and "hydrogen" in the reactions undergone by CO_2 during its reduction to the level of carbohydrate (Duysens and Amesz, 1959; Arnon *et al.*, 1959). Since the initial carboxylation reaction and the subsequent reduction of the products of this reaction can be accomplished in the dark, provided that the concentrations of ATP and reduced pyridine nucleotides are high enough, these reactions can be considered as "dark" enzymatic reactions. The photosynthetic incorporation of CO_2 can thus occur either during illumination or in darkness following a period of preillumination.

Assimilation of CO_2 is not restricted to photosynthetic tissue, however; it has been demonstrated in a great variety of plant and animal tissues. This heterotrophic fixation of CO_2 is evidently of fundamental importance in the physiology of many microorganisms, since growth is inhibited in the absence of CO_2 in the environment (Hutner and Provasoli, 1951; Woods and Lascelles, 1954; Davies, 1959). For these organisms, CO_2 is an essential metabolite for normal growth. The non-photosynthetic mechanisms of CO_2 incorporation are not dependent upon preillumination, but occur at a fairly slow but steady rate even after long retention of the cells in the dark. It was thought for many years that the photosynthetic uptake of CO_2 might involve the same reactions as the nonphotosynthetic uptake, but it is now known that the main routes of incorporation of CO_2 in these two processes differ. However, in some organisms there may be one or more reactions common to both its autotrophic and heterotrophic assimilation. Both types of CO_2 incorporation will be discussed in this chapter, which will be concerned not only with the initial carboxylating reactions in algae, but also with some of the subsequent processes which serve to lead the carboxylated intermediates into well-known metabolic pathways of the cell.

The rate of flow of carbon through these metabolic pathways is related to the immediate history of the cells, as well as to the conditions prevailing during the experiment. For studies dealing with variations in metabolic reactions with the age or physiological state of the cell, unicellular algae present excellent experimental material. In Section IV there are discussed some ways in which the assimilation routes of newly incorporated CO_2 are influenced by the experimental conditions or the state of the cells.

B. Choice of Species for Experimental Material

With the exception of some work dealing with the Cyanophyta, most of the CO_2 assimilation studies with algae have employed Green Algae. The species most commonly used have been *Chlorella ellipsoidea*, *C. pyrenoidosa*, *C. vulgaris*, and *Scenedesmus obliquus*. Other Chlorophyta which have been used to some extent include *Ankistrodesmus braunii* (Kessler, 1957; Brown and Weis, 1959) and various species of *Nitella* (Tolbert and Zill, 1954), *Spirogyra* (Thomas *et al.*, 1957), and *Chlamydomonas* (Allen, 1956). Very few marine algae have been investigated in short-term experiments on the incorporation of CO_2, since many, particularly members of the Phaeophyta and the Rhodophyta, present additional technical difficulties. They cannot as readily be grown and maintained in the laboratory, and their high salt content introduces special problems during the chromatographic separation of the extracted compounds. It is nevertheless hoped that these groups will receive more attention in the future. Representatives of other algal phyla have more often been employed to investigate other aspects of photosynthesis, such as the nature and function of accessory photosynthetic pigments, the kinetics of oxygen liberation or CO_2 consumption, the relation of photosynthesis to respiration, etc.

For studies dealing with the assimilation of carbon dioxide, Green Algae such as *Chlorella* and *Scenedesmus* have many important advantages. They are small and unicellular, and can readily be grown in clonal culture, permitting work with large populations which minimizes the effect of individual variation; a suspension of such algae can be pipetted accurately, and can be easily controlled in regard to temperature, light intensity, etc. Equally important is the fact that for these two genera there is an abundance of available information on their growth and physiology. Much of this information has been assembled in the book edited by Burlew (1953). The application of this knowledge to the culturing of the algae for experimentation makes possible the use of uniform and reproducible biological material, which is of the utmost importance in detailed metabolic studies. The information obtained from members of the Chlorophyta regarding the initial reactions, whereby CO_2 is fixed and reduced during photosynthesis, is believed to be valid for algae of all other groups. Justification for such a biochemical extrapolation is furnished by the uniformity of results obtained with photosynthetic bacteria (Stoppani *et al.*, 1955), chemosynthetic bacteria (Trudinger, 1956), Cyanophyta (Norris *et al.*, 1955), Euglenophyta (Lynch and Calvin, 1953), a variety of different Chlorophyta (Norris *et al.*, 1955), and many species of higher plants (Whatley *et al.*, 1960).

The following discussion, then, relies heavily upon data from the unicellular Green Algae; other algal groups are mentioned wherever pertinent information is available. Some of the information concerning the path of carbon in photosynthesis has been obtained from angiosperm chloroplasts since experimental techniques for the isolation of spinach chloroplasts have progressed beyond similar methods for algal material. A few Green Algae such as *Spirogyra* (Thomas *et al.*, 1957), *Chlorella,* and *Chlamydomonas*

(Punnett, 1959) have also been used in experiments with subcellular fractions.

II. Photosynthetic Assimilation of CO_2

The initial carboxylation reaction in photosynthesis and the nature and synthesis of the acceptor molecule for the CO_2 were elucidated chiefly through the investigations of Calvin and his associates. Nearly all their experiments involved carefully controlled suspensions of *Chlorella pyrenoidosa* or *Scenedesmus obliquus,* exposed to $C^{14}O_2$ for varying periods of time under a wide variety of conditions. The algae were then killed in boiling alcohol. For a variety of the photosynthetic products, the total and the specific activities were determined by paper chromatography, and the molecular distribution of C^{14} within each compound was established by degradative methods. For a comprehensive description of the techniques employed, including the continuous-culture devices used for growing algae in reproducible fashion, see Bassham and Calvin (1957), who have summarized the present-day concepts of the reactions believed responsible for the entry of CO_2 into the photosynthetic cycle and the regeneration of the CO_2-acceptor molecule, ribulose diphosphate (RuDP). This is diagramed in Fig. 1 below.

According to this scheme, the first stable intermediate following the initial carboxylation is 3-phosphoglyceric acid (PGA), which, by reversal of the Embden-Meyerhoff pathway, is reduced to triose phosphate and then condensed to yield the six-carbon sugars. Evidence for this comes not only from abundant degradation data, but also from the fact that ruptured spinach chloroplasts in the light can convert PGA to sugar phosphates, but that this reaction can be strongly inhibited by threose-2,4-diphosphate, a specific inhibitor of the triose-phosphate dehydrogenase system (N. G. Pon, personal communication). After the discovery that PGA was the first stable intermediate in CO_2 assimilation, much effort was spent in searching for the postulated C_2 fragment which would serve as the acceptor molecule for CO_2. Later, kinetic data showing the reciprocal relationship between RuDP and PGA under varying conditions of light (Bassham *et al.*, 1956) and CO_2 pressure (Wilson and Calvin, 1955) indicated that the acceptor molecule was actually a five-carbon sugar, ribulose diphosphate.

The regeneration of RuDP presented another difficult problem, the solution of which emerged mainly from analyses of the distribution of radiocarbon within the pentose, hexose, and heptose molecules after the algal cells had been exposed to $C^{14}O_2$ for various periods (Bassham *et al.*, 1954). As shown in Fig. 1, pentose phosphates can apparently be formed in three ways: (*a*) by a trans-ketolase reaction between triose phosphate and

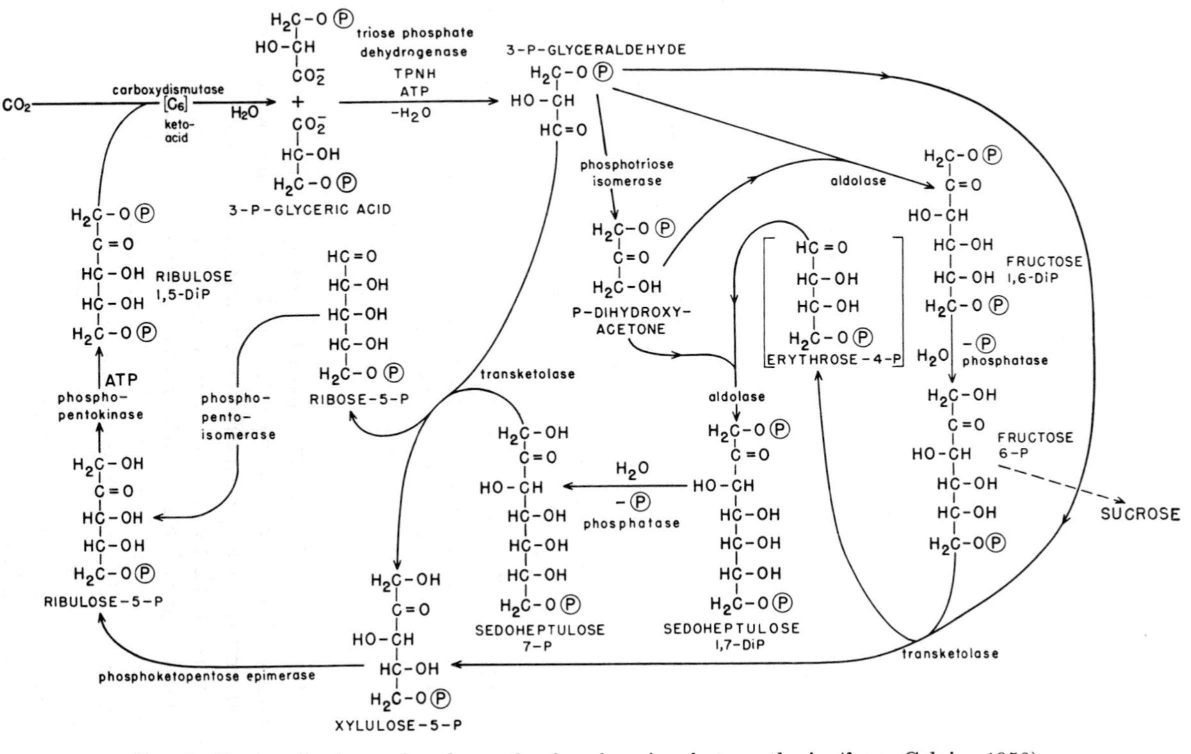

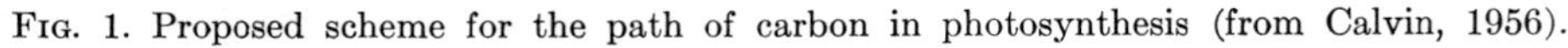
FIG. 1. Proposed scheme for the path of carbon in photosynthesis (from Calvin, 1956).

fructose-6-phosphate ; (*b*) by a trans-ketolase reaction whereby carbon atoms 1 and 2 of sedoheptulose-7-phosphate are condensed with a triose phosphate; (*c*) from the remainder of the sedoheptulose-7-phosphate molecule. The pentose phosphates so formed can then be converted to ribulose-5-phosphate, which with ATP is further phosphorylated to RuDP. The enzymes concerned in the reactions leading from PGA to RuDP are not unique to photosynthetic cells, but are known from a variety of other plant and animal tissues (see Vishniac *et al.*, 1957). Various parts of this cycle about which there is special concern will be discussed separately below.

A. The Initial Carboxylation Reaction

Carboxydismutase, the enzyme catalyzing the reaction of RuDP with CO^2 to yield two molecules of PGA, has been intensively studied in spinach leaves by Weissbach *et al.* (1956), Racker (1957), and Pon (1960). This reaction is also known to occur in non-photosynthetic plant tissue (Fuller and Gibbs, 1959). In spite of extensive efforts, comparatively little is known of the physical and chemical properties of this enzyme, and, in fact, there still exists some doubt concerning the substrates and products of the reaction it catalyzes *in vivo*. Calvin (1956) proposed on theoretical grounds that one of the intermediates in this reaction might be an unstable six-carbon, β-keto acid, which hydrolyzes to yield two molecules of PGA (see Fig. 2). Support for this hypothesis was obtained by Moses and Calvin (1958a), who isolated both the β- and the γ-keto acids (see Fig. 2). The β-keto acid might hydrolyze to yield two molecules of PGA (route II of Fig. 2), or might be reduced first before hydrolysis to yield one molecule of PGA and one molecule of triose phosphate (route I of Fig. 2) (Bassham and Calvin, 1956). The significance of the γ-keto acid is not known.

When the algal cells are illuminated, hydrolysis of the carboxylated intermediate yields only one molecule of PGA together with some unidentified three-carbon molecule at the oxidation level of triose phosphate, whereas in the dark two molecules of PGA are formed (N. G. Pon, personal communication). Studies of carboxydismutase *in vitro* (Racker, 1957) indicated turnover values very much lower than those which must prevail in the intact, functioning cell, suggesting that the natural substrate for this reaction might not be CO_2 but rather a CO_2-complex of some sort. Metzner *et al.* (1957, 1958a, 1958b) presented evidence in favor of an unstable "active CO_2" complex in photosynthesis; but doubt concerning the validity of their data was raised as a consequence of further investigations by Bassham *et al.* (1958), Fuller *et al.* (1958), and Kasprzyk and Calvin (1959). Van Baalen *et al.* (1957) presented evidence for the assimilation of $C^{14}O_2$ into a

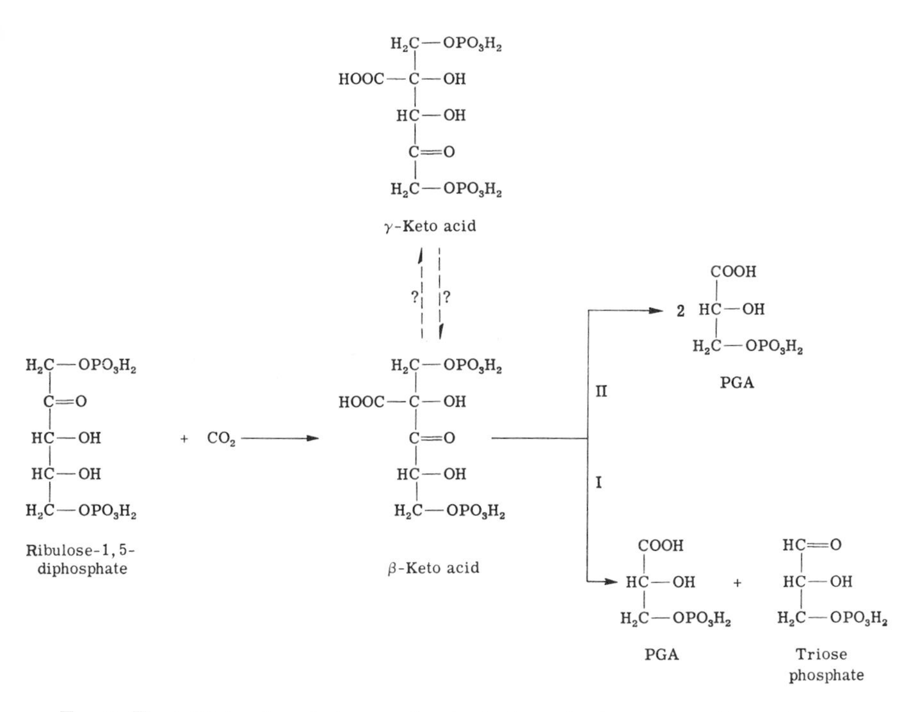

FIG. 2. The initial carboxylation reaction in photosynthesis, showing alternative routes of hydrolysis of the unstable intermediate.

pteridine derivative during short-term incorporation experiments with *Anacystis nidulans*, and suggested that a pteridine–CO_2 complex may act as a donor of C_1 units in the ensuing photosynthetic reactions. Lynen *et al.* (1959) demonstrated with photosynthetic bacteria that in some carboxylation reactions the substrate is not free CO_2, but rather a CO_2–biotin–enzyme complex. Transcarboxylation reactions have also been described whereby C_1 units are transferred from one molecule to another without equilibration with free CO_2 (Swick and Wood, 1960).

B. Other Suggested Carboxylation Reactions in Photosynthesis

Though many of the suggested carboxylation mechanisms for the photosynthetic uptake of CO_2 have been simply reversals of some of the well-known decarboxylation reactions occurring during aerobic respiration, most of the work on this problem, as shown above, has failed to substantiate the validity of these proposals for the Green Algae. However, Fuller and Anderson (1957), using *Chromatium*, reported a light-induced formation of aspartic acid, presumably by the carboxylation of phosphoenolpyruvic acid

(PEP) to oxalacetic acid, which is then transaminated with glutamine to aspartic acid. Gibbs and Kandler (1957), on the basis of experiments with labeled pentoses, hexoses, and heptoses, suggested a carboxylation of a two-carbon fragment derived from pentose, but this hypothesis has not been substantiated by any direct evidence.

It has often been proposed that the first reaction of CO_2 in photosynthesis results in formation of a complex with chlorophyll. Warburg (1957), working with *Chlorella,* again suggested as a first step an initial chlorophyll–CO_2 complex, which is followed by carboxylation of γ-aminobutyric acid to yield glutamic acid; but it is difficult to correlate this hypothesis with the data on photosynthetic intermediates obtained after fixation of $C^{14}O_2$ (Bassham and Calvin, 1957; Gibbs and Kandler, 1957). Although Butler (1960) presented evidence to favor a second CO_2–fixing reaction (i.e., in addition to the carboxydimutase reaction) in leaves of higher plants, he found no evidence for its occurrence with either *Chlorella* or *Scenedesmus.* The reaction proposed to account for this fixation of CO_2 was the reductive carboxylation of pyruvic acid to yield malic acid.

C. Data from Experiments with Isotopic Carbon

Phosphoglyceric acid, when isolated after the algae had been exposed to $C^{14}O_2$ for only a few seconds, was found to contain over 90% of the total carbon fixed. Nearly all the radiocarbon in the PGA molecule was located in the carboxyl group (Calvin and Massini, 1952). When hexose phosphates are isolated after short exposure times, carbon atoms 3 and 4 are the first to become labeled, followed considerably later by carbon atoms 1,2 and 5,6. However, Gibbs and Kandler (1957) demonstrated that after short periods of exposure to $C^{14}O_2$, carbon atom 4 of glucose is of greater specific activity than carbon atom 3, and that carbon atoms 1, 2 incorporate more C^{14} than carbon atoms 5, 6. They proposed three possible explanations for these unexpected results:

(*a*) the addition of CO_2 to the C–4 of RuDP, followed by reduction of the primary addition compound and a rearrangement in which the new carbon atom becomes the C–4 of hexose;

(*b*) a dilution of the dihydroxyacetone phosphate (DHAP) pool, followed by exchange between the "active glycolaldehyde" moieties of fructose-6-phosphate and pentulose phosphate;

(*c*) a cleavage of the RuDP molecule between C–3 and C–4, followed by condensation of the resulting diosephosphate with a reduced form of CO_2 to yield glyceraldehyde-3-phosphate.

In addition, Calvin proposed that this anomaly in the labeling of carbon atoms 3,4 of glucose might be due to a slow isomerization between phosphoglyceraldehyde and DHAP. The phosphoglyceraldehyde (from which originate the carbon atoms 4,5,6 of hexose) would then be understandably of greater specific activity than the DHAP. According to Maruo and Benson (1957), *Scenedesmus* has a high intracellular concentration of diglycerophosphate which, after hydrolysis, could also yield DHAP, thereby decreasing the specific activity of this particular triose, and leading to a lower specific activity of carbon atoms 1,2,3 of hexose. It is possible to explain the observed higher labeling in carbon atoms 1,2 than in 5,6 by postulating a series of back reactions, starting with the pentose molecules, and ultimately re-forming hexose (Wood and Katz, 1958; Bassham and Calvin, 1957). Some of the reactions involved in this proposed recycling sequence involve transketolase enzymes which are believed to transfer C_2 fragments, or "active glycol aldehyde" (Racker *et al.*, 1953).

The synthesis of sedoheptulose-1,7-diphosphate is shown in Fig. 1 to involve erythrose-4-phosphate, which supposedly is derived from fructose-6-phosphate. By isolating and identifying free radioactive erythrose-4-phosphate in short-term photosynthesis experiments with $C^{14}O_2$, Moses and Calvin (1958b) obtained the first experimental evidence for the role of erythrose in the pathway of CO_2 reduction. The reaction between erythrose-4-phosphate and DHAP to yield sedoheptulose-1,7-diphosphate had been previously described in muscle tissue by Horecker *et al.* (1955).

Glycolic acid is one compound which is not represented in Fig. 1, but which is nevertheless very evident on many radioautograms from short-term incorporation experiments with $C^{14}O_2$. Since the metabolic pool of this compound soon becomes highly labeled when algae are exposed to tritiated water (Moses and Calvin, 1959), glycolic acid might in some way be involved in the early CO_2 assimilation reactions. After a few minutes of exposure to $C^{14}O_2$, cultures of *Chlorella pyrenoidosa* were found to contain up to 10% of the fixed radioactivity in glycolic acid, much of which appeared free in the medium (Tolbert and Zill, 1957). High accumulation of glycolic acid in the medium was also noted in cultures of several species of *Chlamydomonas* (Allen, 1956; see also Fogg, Chapter 30, this treatise). There is some evidence that this acid may play a considerable role in the processes of respiration and photosynthesis in leaves of higher plants (Tolbert and Burris, 1950; Tolbert, 1959; Zelitch, 1959). Kornberg and Krebs (1957), working with various heterotrophic bacteria, demonstrated the importance of glycolic acid in general organic acid metabolism and in the synthesis of many cell constituents. Vishniac *et al.* (1957) suggested that, by virtue of its reversible oxidation to glyoxylic acid, glycolic acid may be a substrate for terminal oxidase activity involving reduced diphosphopyridine nucleotide (DPNH), and also may be active in the preservation of the acid-base balance of the cell.

III. Assimilation of CO_2 in the Dark

Although heterotrophic assimilation of CO_2 has been extensively studied in bacterial cells and in mammalian tissue (Wood, 1946; Krebs, 1951; Ochoa, 1951, 1952), comparatively few investigations of nonphotosynthetic CO_2 assimilation have been carried out with algae. In the discussion below, emphasis is of course placed on experiments carried out with algae, but information is also cited from relevant studies of bacteria, higher plants, or animal tissues. No effort has been made to survey the multitude of carboxylation reactions known from these nonalgal groups except where they touch upon observations made in algae. Many of the known carboxylating mechanisms may represent adaptations to a certain condition or environment, such as the hydrogenlyase reaction of *Escherichia coli* and the incorporation of CO_2 into acetate by *Aerobacter*. Other reactions, such as the isocitrate dehydrogenase system (Ochoa, 1945) and the carboxylation of organic acid – coenzyme A complexes (Woessner *et al.*, 1958) may ultimately be found also in algae.

A. Labeling of Krebs-Cycle Intermediates and Related Amino Acids

In addition to the photosynthetic uptake of CO_2, most tissues of plants and animals investigated are able to fix CO_2 *via* other metabolic pathways. Thus, when cells of *Chlorella* or *Scenedesmus* are incubated in the dark with $C^{14}O_2$, there is a slow, steady uptake of the labeled CO_2, amounting usually to less than 1% of the maximum pickup obtainable in the light, and not detectable by the usual manometric techniques. Most metabolic reactions involving the assimilation or liberation of CO_2 are known to be reversible, though this is not true of those mediated by carboxydismutase, pyruvic acid decarboxylase (yielding CO_2 and acetyl-CoA), and possibly oxalacetic acid decarboxylase (yielding CO_2 and pyruvate). Although for most carboxylation reactions which form new carbon-to-carbon bonds, the equilibrium values strongly favor decarboxylation, with appropriate concentrations of reactants and of cofactors such as ATP and reduced pyridine nucleotides, the carboxylation reactions can often be made to occur *in vitro* at an appreciable rate. As these carboxylation reactions are chiefly endergonic (the carboxydismutase reaction is an exception to this), they require expenditure of metabolic energy by the cell (see Calvin and Pon, 1959).

Among the labeled compounds formed from $C^{14}O_2$ by *Chlorella* during dark fixation, most radioactivity resides in carboxy acids and related amino acids. Thus, after one minute of exposure to $C^{14}O_2$ in the dark, the only compounds found to contain C^{14} were malic, citric, fumaric, aspartic, and glutamic acids and alanine (Moses *et al.*, 1959b). Labeled PGA was found after 2 minutes, and was followed by detectable incorporation of C^{14} into sugar phosphates and related compounds. Although the enzymatic reactions responsible for this uptake have not been investigated with algal material *in vitro*, the initial carboxylation reactions involved here are probably similar to those demonstrated in other organisms. Fixed carbon from CO_2 could enter into four-carbon compounds such as malic and aspartic acids by a number of routes: the Wood-Werkman reaction (Suzuki and Werkman, 1958; Ochoa *et al.*, 1948) as in Eq. (1); the malic enzyme reaction (Rutter and Lardy, 1958) as in Eq. (2); the phosphoenolpyruvic carboxylase reaction (Bandurski and Greiner, 1953) as in Eq. (3); and the phosphoenolpyruvic carboxylase reaction (Kurahashi *et al.*, 1957) as in Eq. (4).

$$\text{pyruvic acid} + CO_2 \leftrightarrows \text{oxalacetic acid} \quad (1)$$

$$\text{pyruvic acid} + CO_2 + \text{TPNH} \leftrightarrows \text{malic acid} + \text{TPN} \quad (2)$$

$$\text{phosphoenolpyruvic acid} + CO_2 \leftrightarrows \text{oxalacetic acid} + H_3PO_4 \quad (3)$$

$$\text{phosphoenolpyruvic acid} + CO_2 + \text{inosine diphosphate} \leftrightarrows \text{oxalacetic acid} + \text{inosine triphosphate} \quad (4)$$

There is some question regarding the likelihood of the Wood-Werkman reaction occurring simply as shown in Eq. (1) (Rabinowitch, 1945; Woronick and Johnson, 1960). Once the C^{14} is incorporated into intermediates of the Krebs-cycle, such as malic acid or oxalacetic acid, the labeled amino compounds such as aspartic acid, glutamic acid, glutamine, etc., could be formed by transamination reactions. There are a number of known reactions which could account for the C^{14} in alanine, often one of the predominant compounds resulting from the incorporation of $C^{14}O_2$ in the dark. It could be explained, for instance, by the carboxylation of a C_2 fragment such as that of acetyl phosphate (Lipmann and Tuttle, 1945; Mortlock and Wolfe, 1959) as shown in Eq. (5), or of acetylthioctic acid (Moses *et al.*, 1959b; Reed, 1957), followed by transamination.

$$\underset{\text{Acetyl phosphate}}{CH_3CO\cdot OPO_3^{2-}} + H_2 + CO_2 \leftrightarrows \underset{\text{Pyruvic acid}}{CH_3COCOOH} + HOPO_3^{2-} \quad (5)$$

The label in alanine might also arise in the following way. A carboxylation of pyruvic acid to form malic acid or oxalacetic acid as in Eqs. (1) or (2), followed by equilibration with succinic acid, would tend to distribute the radioactive carbon equally between the two carboxyl groups of oxalacetic acid. On subsequent decarboxylation, this would yield labeled pyruvic acid, which could then be transaminated to form labeled alanine. It is also possible that a carboxydismutase reaction might be occurring in the dark, although at a rate considerably below that in the light. This would of course give rise to labeled PGA, which could be oxidized to pyruvic acid, and then aminated to alanine (Whittingham, 1957). In heterotrophic organisms such as *Astasia* sp. and *Escherichia coli*, RuDP can condense with CO_2 to yield two molecules of PGA (Fuller and Gibbs, 1959). This provides some evidence in support of the second hypothesis.

Studies on the fixation of $C^{14}O_2$ by *Nostoc muscorum* in darkness showed that after 30 seconds radioactivity could be detected only in aspartic acid, but that after 3 minutes C^{14} was also demonstrable in the phosphorylated sugars and in various amino acids. The incorporation pattern was quite similar to that obtained with *Chlorella* (Moses *et al.*, 1959b).

B. Incorporation of $C^{14}O_2$ into Photosynthetic Intermediates

In contrast to the data described above, the distribution of radioactivity in extracts of *Euglena* cells incubated with $C^{14}O_2$ in darkness resembled that obtained during photosynthesis, in that the sugar phosphates and PGA incorporated much of the radioactivity (though only a fraction of that found after photosynthetic $C^{14}O_2$ fixation) (Lynch and Calvin, 1953). This was interpreted as a fixation *via* the photosynthetic pathway, but using energy from endogenous respiration instead of light. An alternative explanation proposed by Moses *et al.* (1959b) is that the label in the sugar phosphates could have originated from the carboxylation of ribulose-5-phosphate to yield 6-phosphogluconic acid, as shown in Eq. (6), which in turn could yield glucose-6-phosphate.

$$\begin{array}{c} H_2C\text{—}OH \\ | \\ C\text{=}O \\ | \\ HC\text{—}OH \\ | \\ HC\text{—}OH \\ | \\ H_2C\text{—}OPO_3H_2 \\ \\ \text{Ribulose-} \\ \text{5-phosphate} \end{array} + CO_2 + TPNH + H^+ \leftrightarrows \begin{array}{c} COOH \\ | \\ HC\text{—}OH \\ | \\ HO\text{—}C\text{—}H \\ | \\ HC\text{—}OH \\ | \\ HC\text{—}OH \\ | \\ H_2C\text{—}OPO_3H_2 \\ \\ \text{6-Phospho-} \\ \text{gluconic acid} \end{array} + TPN^+ \qquad (6)$$

These reactions are known to be reversible (Leloir, 1956). The conversion of glucose-6-phosphate to fructose diphosphate and its ensuing breakdown *via* the glycolytic pathway would give rise to the observed labeled PGA. Similar incorporation of $C^{14}O_2$ in the dark was also found in some experiments with *Nostoc muscorum* (Holm-Hansen, 1956); after 10 minutes exposure to $C^{14}O_2$, the compounds incorporating the most radiocarbon were hexose monophosphate, aspartic acid, PGA, and uridine diphosphoglucose. The Krebs-cycle acids were labeled also, but to a far lesser extent. The sugar diphosphate compounds contained very little C^{14} compared to the amount they contained after exposure of the cells to light.

These results obtained with *Nostoc* differed sharply from those discussed in Section III.A, probably as a result of differences in the culturing of the cells and their treatment both before and during exposure to the $C^{14}O_2$. Some aspects of the variation in results caused by changes either in the age or physiological state of the cell, or in the experimental conditions, are discussed below in Section IV.

C. Synthesis and Function of Carbamyl Phosphate

Norris *et al.* (1955) reported an unknown compound which in *Nostoc muscorum* and *Synechococcus cedrorum* became heavily labeled during photosynthesis with $C^{14}O_2$, but which in 27 other plants of various phyla was either present in very small amounts or absent. This compound was later found to be citrulline (Linko *et al.*, 1957). Radioactive carbon is incorporated into citrulline both in the light and in the dark, but at an increased rate during illumination. The label in citrulline is almost entirely in the ureido carbon atom; on the basis of work with bacteria (Metzenberg *et al.*, 1958; Reichard, 1959), it seems probable that the citrulline might be formed by reactions catalyzed by carbamyl phosphate synthetase, as in Eq. (7), and by ornithine transcarbamylase as in Eq. (8).

$$CO_2 + NH_3 + ATP \leftrightarrows \underset{\text{Carbamyl phosphate}}{H_2N\text{—}CO\text{—}OPO_3H_2} + ADP \qquad (7)$$

$$H_2N\text{—}CO\text{—}OPO_3H_2 + \text{ornithine} \leftrightarrows \text{citrulline} + HPO_4^{2-} \qquad (8)$$

Evidence that citrulline is synthesized in this way has been obtained in cell-free extracts of *Nostoc muscorum* (Holm-Hansen, unpublished results). Ornithine transcarbamylase activity was demonstrated, and labelled CO_2 was incorporated into the ureido carbon atom of citrulline only when ornithine was added as a substrate. Without the added ornithine, the carbamyl phosphate formed apparently did not combine with any substrate and was presumably broken down to CO_2 and NH_3 during the subsequent experimental manipulations. Citrulline probably serves as the precursor of arginine in the algae. Its relation to nitrogen fixation has been discussed in this treatise by Fogg (Chapter 9).

To the author's knowledge, the significance of carbamyl phosphate in algal metabolism has not been investigated further, although in other organisms it is known to be involved in such important biosynthetic pathways as the synthesis of the pyrimidine ring of nucleic acids (Reichard, 1959).

IV. Metabolic Pathways as Influenced by Environmental Conditions

In photosynthetic incorporation experiments, after as little as a minute or two of exposure to the $C^{14}O_2$, the C^{14} can be found widely distributed among the alcohol-soluble compounds as well as in the insoluble residues of protein, fat, and carbohydrate. It thus appears that in photosynthesis the initial products of CO_2 fixation soon enter the many biosynthetic pathways of the cell. The relation of the photosynthetic cycle to the Krebs cycle and to the pentose-phosphate cycle is shown in Fig. 3. Very little is known

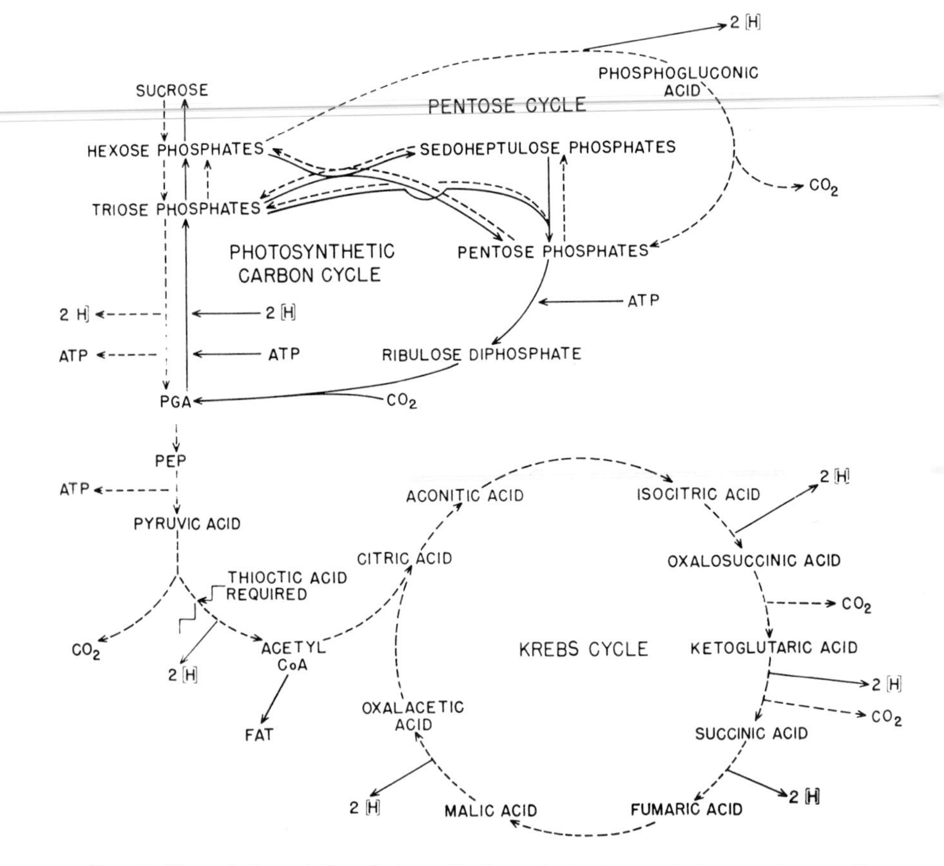

Fig. 3. The relation of the photosynthetic cycle to the respiratory pathways (from Bassham and Calvin, 1956).

regarding the localization of the various metabolic sequences in relation to the distinguishable structural components of the algal cell; most investigations of the relation of subcellular components to metabolic function have employed material from higher plants and animals (e.g., Sissakian, 1958; Arnon *et al.*, 1959; Holm-Hansen *et al.*, 1959b). There is some evidence, however, for a separation of the metabolic pools of certain photosynthetic intermediates within the *Chlorella* cell (Moses *et al.*, 1959a).

The biochemical distribution of the products of CO_2 fixation can be markedly altered by varying one or more of the environmental factors. Thus, when *Chlorella* cells are illuminated in the presence of $C^{14}O_2$, relatively little of the C^{14} is incorporated into Krebs-cycle acids and related amino acids (Calvin and Massini, 1952); but as soon as the light is turned off, appreciable amounts of the radioactive carbon accumulate in these compounds. The suggestion was made that this apparent suppression of the flow of fixed radiocarbon into the Krebs cycle was a consequence of the reduction of thioctic acid in the light, since thioctic acid must be in the oxidized form to be active as a cofactor for the decarboxylation of pyruvic acid to yield acetyl coenzyme A. This suggestion was given experimental support by Gibbs (1953), who studied the labeling of alanine, malic acid, and the hexoses in sunflower leaves. [However, Biswas and Sen (1958) were unable to demonstrate any stimulation of photosynthesis in *Chlorella pyrenoidosa* or *Scenedesmus obliquus* by thioctic acid.] Gaffron (1957), from investigations of the transient exchange of CO_2 in brief light and dark periods, came to the similar conclusion that entry into the Krebs cycle was blocked while the cells were in the light; but he attributed this effect to the photoreduction of the pyridine nucleotides, which in their oxidized form mediate, as coenzymes, decarboxylation steps in the Krebs cycle.

Much of this information was obtained from *Chlorella* cells suspended in distilled water or in a dilute phosphate buffer. It was found, however, that they incorporated much more radioactivity into the amino acids associated with operation of the Krebs cycle when they were suspended in the usual nutrient solution during the experimental period of $C^{14}O_2$ fixation than when the cells were suspended in distilled water (Holm-Hansen *et al.*, 1959a). The most strikingly different incorporation pattern was obtained from cells suspended in dilute ammonium chloride, which not only trebled the total uptake of labeled CO_2, as compared to cells in distilled water, but also increased the percentage of C^{14} found in the alcohol-soluble amino acids from 9.9 to 56.8% of the total fixed radioactivity. This result may have been correlated with a simultaneous incorporation of assimilable nitrogen (see Syrett, Chapter 10, this treatise). Variations in the concentration of the other major-element salts in the usual nutrient solution also influenced the rate of incorporation of CO_2 or the distribution pattern of the newly incorporated carbon, though to a much smaller degree than did ammonium chloride.

In investigations of the normal metabolism of algae it is thus important to study the cells in a suspending medium which neither limits the total rate of photosynthesis nor alters the pattern of incorporation by the deficiency of one or more essential nutrients. Marked depressions of the photosynthetic rate have been shown in *Nostoc* deprived of potassium (Clendenning *et al.*, 1956), and *Chlorella* deprived of potassium, magnesium, or manganese (Hill and Whittingham, 1955). Other environmental conditions such as pH (Ouellet and Benson, 1952), light intensity, temperature (Ouellet, 1951), and the composition of the gas phase (Miyachi *et al.*, 1957), may also have an important influence on the rate of uptake and the distribution of the carbon assimilated.

In addition to the conditions prevailing during the actual course of the experiment, the previous cultural environment of the algal cells may play an important role. Thus, Warburg *et al.* (1956) showed that in *Chlorella* cells grown under simulated "natural" conditions of alternating light and darkness, the quantum yield of photosynthesis was about 4 (i.e., 4 quanta of light were absorbed for each mole of oxygen evolved), whereas when the cells had been grown in light of constant intensity the corresponding value was 10 to 12. Physiological differences between cells of *Chlorella* at different stages of their life cycle have been demonstrated by Tamiya *et al.* (1953a, b), Hase *et al.* (1957), Myers (1957), Sorokin and Krauss (1959), and Sorokin and Myers (1957); see review by Hase (Chapter 40, this treatise). Stange *et al.* (1960), who studied the incorporation of C^{14} both into alcohol-soluble compounds and into alcohol-insoluble residues such as proteins and nucleic acids, found consistent differences according to the age of the cell and its stage of development.

In the above discussion it has been assumed that much of the information obtained from unicellular Green Algae is more or less applicable to other algal groups, although the uniformity which is suggested here applies only to the initial carboxylation reactions and the reduction of the initial products. It is obvious that, beyond the early fixation products, the metabolic routes and incorporation patterns of the newly assimilated CO_2 may vary a great deal from one group of algae to the other, reflecting definitive biochemical differences among the algal phyla in regard to storage products, cell wall constituents, pigmentation, etc. Some of these are discussed in other chapters of this treatise.

References

Allen, M. B. (1956). Excretion of organic compounds by *Chlamydomonas*. *Arch. Mikrobiol.* **24,** 163–168.

Arnon, D. I., Whatley, F. R., and Allen, M. B. (1959). Photosynthesis by isolated chloroplasts. VIII. Photosynthetic phosphorylation and the generation of assimilatory power. *Biochim. et Biophys. Acta* **32**, 47–57.

Bandurski, R. S., and Greiner, C. M. (1953). The enzymatic synthesis of oxalacetate from phosphoryl-enol pyruvate and carbon dioxide. *J. Biol. Chem.* **204**, 781–786.

Bassham, J. A., and Calvin, M. (1956). Photosynthesis. *In* "Currents in Biochemical Research" (D. E. Green, ed.), pp. 29–69. Interscience, New York.

Bassham, J. A., and Calvin, M. (1957). "The Path of Carbon in Photosynthesis." Prentice-Hall, Englewood Cliffs, New Jersey.

Bassham, J. A., Benson, A. A., Kay, L. D., Harris, A. Z., Wilson, A. T., and Calvin, M. (1954). The path of carbon in photosynthesis. XXI. The cyclic regeneration of carbon dioxide acceptor. *J. Am. Chem. Soc.* **76**, 1760–1770.

Bassham, J. A., Shibata, K., Steenberg K., Bourdon, J., and Calvin, M. (1956). Photosynthetic cycle and respiration: light and dark transients. *J. Am. Chem. Soc.* **78**, 4120–4124.

Bassham, J. A., Kirk, M., and Calvin, M. (1958). The effects of hydroxylamine on the $C^{14}O_2$ fixation pattern during photosynthesis. *Proc. Natl. Acad. Sci. U.S.* **44**, 491–493.

Biswas, B. B., and Sen, S. P. (1958). Thioctic acid and photosynthetic fixation of carbon dioxide. *Nature* **181**, 1219–1220.

Brown, A. H., and Weis, D. (1959). Relation between respiration and photosynthesis in the green alga, *Ankistrodesmus braunii*. *Plant Physiol.* **34**, 224–234.

Burlew, J. S , ed. ,(1953). "Algal Culture from Laboratory to Pilot Plant." *Carnegie Inst. Wash. Publ.* **No. 600**.

Butler, W. L. (1960). A secondary photosynthetic carboxylation. *Plant Physiol.* **35**, 233–237.

Calvin, M. (1956). The photosynthetic carbon cycle. *J. Am. Chem. Soc.* **78**, 1895–1915.

Calvin, M., and Massini, P. (1952). The path of carbon in photosynthesis. XX. The steady state. *Experientia* **8**, 445–457.

Calvin, M. C., and Pon, N. G. (1959). Carboxylations and decarboxylations. *J. Cellular Comp. Physiol.* **54**, Suppl. No. 1, 51–74.

Clendenning, K. A., Brown, T. E., and Eyster, H. C. (1956). Comparative studies of photosynthesis in *Nostoc muscorum* and *Chlorella pyrenoidosa*. *Can. J. Botany* **34**, 943–966.

Davies, D. D. (1959). Organic acid metabolism in plants. *Biol. Revs. Cambridge Phil. Soc.* **34**, 407–445.

Duysens, L. N. M , and Amesz, J. (1959). Quantum requirement for phosphopyridine nucleotide reduction in photosynthesis. *Plant Physiol.* **34**, 210–213.

Fuller, R. C., and Anderson, I. C. (1957). CO_2 Assimilation in the photosynthetic purple sulfur bacteria. *Plant Physiol.* **32**, Suppl., p. xvi.

Fuller, R. C., and Gibbs, M. (1959). Intracellular and phylogenetic distribution of ribulose 1,5-diphosphate carboxylase and D-glyceraldehyde-3-phosphate dehydrogenases. *Plant Physiol.* **34**, 324–329.

Fuller, R. C., Anderson, I. C., and Nathan, H. A. (1958). Pteridines in photosynthesis—an artifact of paper chromatography. *Proc. Natl. Acad. Sci. U.S.* **44**, 518–519.

Gaffron, H. (1957). Transients in the carbon dioxide gas exchange of algae. *In* "Research in Photosynthesis" (H. Gaffron, A. H. Brown, C. S. French, R. Livingston, E. I. Rabinowitch, B. L. Strehler, and N. E. Tolbert, eds.), pp. 430–443. Interscience, New York.

Gibbs, M. (1953). Effect of light intensity on the distribution of C^{14} in sunflower leaf metabolites during photosynthesis. *Arch. Biochem. Biophys.* **45,** 156–160.

Gibbs, M., and Kandler, O. (1957). Asymmetric distribution of C^{14} in sugars formed during photosynthesis. *Proc. Natl. Acad. Sci. U.S.* **43,** 446–451.

Hase, E., Morimura, Y., and Tamiya, H. (1957). Some data on the growth physiology of *Chlorella* studied by the technique of synchronous culture. *Arch. Biochem. Biophys.* **69,** 149–165.

Hill, R., and Whittingham, C. P. (1955). "Photosynthesis," pp. 59–60. Wiley. New York.

Holm-Hansen, O. (1956). Dark fixation of $C^{14}O_2$ by *Nostoc muscorum* and *Scenedesmus obliquus. Univ. Calif. (Berkeley) Radiation Lab. Rept.* **No. 3629,** 41–43.

Holm-Hansen, O., Nishida, K., Moses, V., and Calvin, M. (1959a). Effects of mineral salts on short-term incorporation of carbon dioxide in *Chlorella. J. Exptl. Botany* **10,** 109–124.

Holm-Hansen, O., Pon, N. G., Nishida, K., Moses, V., and Calvin, M. (1959b). Uptake and distribution of radioactive carbon from labelled substrates by various cellular components of spinach leaves. *Physiol. Plantarum* **12,** 475–501.

Horecker, B. L., Smyrniotis, P. Z., Hiatt, H. H., and Marks, P. A. (1955), Tetrose phosphate and the formation of sedoheptulose diphosphate. *J. Biol. Chem.* **212,** 827–836.

Hutner, S. H., and Provasoli, L. (1951). The phytoflagellates. *In* "Biochemistry and Physiology of Protozoa" (S. H. Hutner and A. Lwoff, eds.), Vol. 1, pp. 27–128. Academic Press, New York.

Kasprzyk, Z., and Calvin, M. (1959). Search for unstable CO_2 fixation products in algae using low temperature liquid scintillators. *Proc. Natl. Acad. Sci. U.S.* **45,** 952–959.

Kessler, E. (1957). Manganese as a cofactor in photosynthetic oxygen evolution. *In* "Research in Photosynthesis" (H. Gaffron, A. H. Brown, C. S. French, R. Livingston, E. I. Rabinowitch, B. L. Strehler, and N. E. Tolbert, eds.), pp. 243–249. Interscience, New York.

Kornberg, H. L., and Krebs, H. A. (1957). Synthesis of cell constituents from C_2-units by a modified tricarboxylic acid cycle. *Nature* **179,** 988–991.

Krebs, H. A. (1951). Carbon dioxide fixation in animal tissues. *In* "Carbon Dioxide Fixation and Photosynthesis." *Symposia Soc. Exptl. Biol.* **No. 5,** 1–8.

Kurahashi, K., Pennington, R. J., and Utter, M. F. (1957). Nucleotide specificity of oxalacetic carboxylase. *J. Biol. Chem.* **226,** 1059–1075.

Leloir, L. F. (1956). The interconversion of sugars in nature. *In* "Currents in Biochemical Research" (D. E. Green, ed.), pp. 585–608. Interscience, New York.

Linko, P., Holm-Hansen, O., Bassham, J. A., and Calvin, M. (1957). Formation of radioactive citrulline during photosynthetic $C^{14}O_2$-fixation by blue-green algae. *J. Exptl. Botany* **8,** 147–156.

Lipmann, F., and Tuttle, L. C. (1945). On the condensation of acetyl phosphate with formate or carbon dioxide in bacterial extracts. *J. Biol. Chem,* **158,** 505–519.

Lynch, V. H., and Calvin, M. (1953). CO_2 fixation by *Euglena. Ann. N.Y. Acad. Sci.* **56,** 890–900.

Lynen, F., Knappe, J., Lorch, E., Jutting, G., and Ringelmann, E. (1959). Die biochemische Funktion des Biotins. *Angew. Chem.* **71,** 481–486.

Maruo, B., and Benson, A. A. (1957). α-α'-Diglycerophosphate in plants. *J. Am. Chem. Soc.* **79,** 4564–4565.

Metzenberg, R. C., Marshall, M., and Cohen, P. P. (1958). Carbamyl phosphate synthetase: studies on the mechanism of action. *J. Biol. Chem.* **233,** 1560–1564.

Metzner, H., Simon, H., Metzner, B., and Calvin, M. (1957). Evidence for an unstable CO_2 fixation product in algal cells. *Proc. Natl. Acad. Sci. U.S.* **43,** 892–895.

Metzner, H., Metzner, B., and Calvin, M. (1958a). Early unstable CO_2 fixation products in photosynthesis. *Proc. Natl. Acad. Sci. U.S.* **44,** 205–211.

Metzner, H., Metzner, B., and Calvin, M. (1958b). Labile products of early carbon dioxide fixation in photosynthesis. *Arch. Biochem. Biophys.* **74,** 1–6.

Miyachi, S., Hirokawa, T., and Tamiya, H. (1957). The "background" CO_2 fixation occurring in green cells and its possible relation to the mechanism of photosynthesis. *In* "Research in Photosynthesis" (H. Gaffron, A. H. Brown, C. S. French, R. Livingston, E. I. Rabinowitch, B. L. Strehler, and N. E. Tolbert, eds.), pp. 205–212. Interscience, New York.

Mortlock, R. P., and Wolfe, R. S. (1959). Reversal of pyruvate oxidation in *Clostridium butyricum*. *J. Biol. Chem.* **234,** 1657–1658.

Moses, V., and Calvin, M. (1958a). The path of carbon in photosynthesis. XXII. The identification of carboxy-ketopentitol diphosphates as products of photosynthesis. *Proc. Natl. Acad. Sci. U.S.* **44,** 260–277.

Moses, V., and Calvin, M. (1958b). The path of carbon in photosynthesis. XXIII. The tentative identification of erythrose phosphate. *Arch. Biochem. Biophys.* **78,** 598–600.

Moses, V., and Calvin, M. (1959). Photosynthesis studies with tritiated water. *Biochim. et Biophys. Acta* **33,** 297–312.

Moses, V., Holm-Hansen, O., Bassham, J. A., and Calvin, M. (1959a). The relationship between the metabolic pools of photosynthetic and respiratory intermediates. *J. Mol. Biol.* **1,** 21–29.

Moses, V., Holm-Hansen, O., and Calvin, M. (1959b). Nonphotosynthetic fixation of carbon dioxide by three microorganisms. *J. Bacteriol,* **77,** 70–78.

Myers, J. (1957). On uniformity of experimental material. *In* "Research in Photosynthesis" (H. Gaffron, A. H. Brown, C. S. French, R. Livingston, E. I. Rabinowitch, B. L. Strehler, and N. E. Tolbert, eds.), pp. 485–489. Interscience, New York.

Norris, L., Norris, R. E., and Calvin, M. (1955). A survey of the rates and products of short-term photosynthesis in plants of nine phyla. *J. Exptl. Botany* **6,** 64–74.

Ochoa, S. (1945). Isocitric dehydrogenase and carbon dioxide fixation. *J. Biol. Chem.* **159,** 243–244.

Ochoa, S. (1951). Biosynthesis of dicarboxylic and tricarboxylic acids by carbon dioxide fixation. *In* "Carbon Dioxide Fixation and Photosynthesis." *Symposia Soc. Exptl. Biol.* **No. 5,** 29–51.

Ochoa, S. (1952). Enzymatic mechanisms of carbon dioxide fixation. *In* "The Enzymes" (J. B. Sumner and K. Myrbäck, eds.), Vol. II, Part 2, pp. 929–1032. Academic Press, New York.

Ochoa, S., Mehler, A. H., and Kornberg, A. (1948). Biosynthesis of dicarboxylic acids by carbon dioxide fixation. I. Isolation and properties of an enzyme from pigeon liver catalyzing the reversible oxidative decarboxylation of l-malic acid, *J. Biol. Chem.* **174,** 979–1000.

Ouellet, C. (1951). The path of carbon in photosynthesis. XII. Some temperature effects. *J. Exptl. Botany* **11,** 316–320.

Ouellet, C., and Benson, A. A. (1952). The path of carbon in photosynthesis. XIII. pH effects in $C^{14}O_2$ fixation by *Scenedesmus*. *J. Exptl. Botany* **3,** 237–245.

Pon, N. G. (1960). Ph.D. Dissertation. University of California, Berkeley.

Punnett, T. (1959). Stability of isolated chloroplast preparations and its effect on Hill reaction measurements. *Plant Physiol.* **34,** 283–289.

Rabinowitch, E. I. (1945). "Photosynthesis and Related Processes," Vol. I: Chemistry of Photosynthesis, Chemosynthesis and Related Processes *in Vitro* and *in Vivo*, p. 185. Interscience, New York.

Racker, E. (1957). The reductive pentose phosphate cycle. I. Phosphoribulokinase and ribulose diphosphate carboxylase. *Arch. Biochem. Biophys.* **69,** 300–310.

Racker, E., de la Haba, G., and Leder, I. G. (1953). Thiamine pyrophosphate, a coenzyme of transketolase, *J. Am. Chem. Soc.* **75,** 1010–1011.

Reed, L. J. (1957). The chemistry and function of lipoic acid. *Advances in Enzymol.* **18,** 319–348.

Reichard, P. (1959). The enzymatic synthesis of pyrimidines. *Advances in Enzymol.* **21,** 263–294.

Rutter, W. T., and Lardy, H. A. (1958). Purification and properties of pigeon liver malic enzyme. *J. Biol. Chem.* **233,** 374–382.

Sissakian, N. M. (1958). Enzymology of the plastids. *Advances in Enzymol.* **20,** 201–236.

Sorokin, C., and Krauss, R. W. (1959). Maximum growth rates of *Chlorella* in steady-state and in synchronized cultures. *Proc. Natl. Acad. Sci. U.S.* **45,** 1740–1744.

Sorokin, C., and Myers, J. (1957). The course of respiration during the life cycle of *Chlorella* cells. *J. Gen. Physiol.* **40,** 579–592.

Stange, L., Bennett, E. L., and Calvin, M. (1960). Short-time $C^{14}O_2$ incorporation experiments with synchronously growing *Chlorella* cells. *Biochim. et Biophys. Acta* **37,** 92–100.

Stoppani, A. O. M., Fuller, R. C., and Calvin, M. (1955). Carbon dioxide fixation by *Rhodopseudomonas capsulatus*. *J. Bacteriol.* **69,** 491–501.

Suzuki, I., and Werkman, C. H. (1958). Chemoautotrophic carbon dioxide fixation by extracts of *Thiobacillus thiooxidans*. I. Formation of oxalacetic acid. *Arch. Biochem. Biophys.* **76,** 103–111.

Swick, R. W., and Wood, H. G. (1960). The role of transcarboxylation in propionic acid fermentation. *Proc. Natl. Acad. Sci. U.S.* **46,** 28–41.

Tamiya, H., Iwamura, T., Shibata, K., Hase, E., and Nihei, T. (1953a). Correlation between photosynthesis and light-independent metabolism in the growth of *Chlorella*. *Biochim. et Biophys. Acta* **12,** 23–40.

Tamiya, H., Shibata, K., Sasa, T., Iwamura, T., and Morimura, Y. (1953b). Effect of diurnally intermittent illumination on the growth and some cellular characteristics of *Chlorella*. *In* "Algal Culture From Laboratory to Pilot Plant" (J. S. Burlew, ed.). *Carnegie Inst. Wash. Publ.* **No. 600,** 76–84.

Thomas, J. B., Haans, A. J. M., and Van der Leun, A. A. J. (1957), Photosynthetic activity of isolated chloroplast fragments of *Spirogyra*. *Biochim. et Biophys. Acta* **25,** 453–462.

Tolbert, N. E. (1959). Secretion of glycolic acid by chloroplasts. *In* "The Photochemical Apparatus: Its Structure and Function." *Brookhaven Symposia in Biol.* **No. 11,** 271–275.

Tolbert, N. E , and Burris, R. H. (1950). Light activation of the plant enzyme which oxidizes glycolic acid. *J. Biol. Chem.* **186,** 791–804.

Tolbert, N. E , and Zill, L. P. (1954). Photosynthesis by protoplasm extruded from *Chara* and *Nitella*. *J. Gen. Physiol.* **37,** 575–588.

Tolbert, N. E., and Zill, L. P. (1957). Excretion of glycolic acid by *Chlorella* during photosynthesis. *In* "Research in Photosynthesis" (H. Gaffron, A. H. Brown, C. S. French, R. Livingston, E. I. Rabinowitch, B. L, Strehler, and N. E. Tolbert, eds.), pp. 228–231. Interscience, New York.

Trudinger, P. A. (1956). Fixation of carbon dioxide by extracts of the strict autotroph *Thiobacillus dentrificans*. *Biochem. J.* **64,** 274–286.

Van Baalen, C., Forrest, H. S., and Myers, J. (1957). Incorporation of radioactive carbon into a pteridine of a blue-green alga. *Proc. Natl. Acad. Sci. U.S.* **43,** 701–705.

Vishniac, W., Horecker, B. L., and Ochoa, S. (1957). Enzymatic aspects of photosynthesis. *Advances in Enzymol.* **19,** 1–77.

Warburg, O. (1957). Photosynthese. *Angew. Chem.* **69,** 627–658.

Warburg, O., Schröder, W., and Gattung, H. (1956). Züchtung der *Chlorella* mit fluktuierender Lichtintensität. *Z. Naturforsch.* **11b,** 654–657.

Weissbach, A., Horecker, B. L., and Hurwitz, J. (1956). The enzymatic formation of phosphoglyceric acid from ribulose diphosphate and carbon dioxide. *J. Biol. Chem.* **218,** 795–810.

Whatley, F. R., Allen, M. B., Trebst, A. V., and Arnon, D. I. (1960). Photosynthesis by isolated chloroplasts. IX. Photosynthetic phosphorylation and CO_2 assimilation in different species. *Plant Physiol.* **35,** 188–193.

Whittingham, C. P. (1957). Induction phenomena in photosynthetic algae at low partial pressures of oxygen. *In* "Research in Photosynthesis" (H, Gaffron, A, H. Brown, C. S. French, R. Livingston, E. I, Rabinowitch, B. L. Strehler, and N. E. Tolbert, eds.), pp. 409–411. Interscience, New York,

Wilson, A. T., and Calvin, M. (1955). The photosynthetic cycle. CO_2 dependent transients. *J. Am. Chem, Soc.* **77,** 5948–5957.

Woessner, J. T., Jr., Bachhawat, B. K., and Coon, M. J. (1958). Enzymatic activation of carbon dioxide. II. Role of biotin in the carboxylation of β-hydroxyisovaleryl coenzyme A. *J. Biol. Chem.* **233,** 520–523.

Wood, H. G. (1946). The fixation of carbon dioxide and the interrelationships of the tricarboxylic acid cycle. *Physiol. Revs.* **26,** 198–246.

Wood, H. G., and Katz, J. (1958). The distribution of C^{14} in the hexose phosphates and the effect of recycling in the pentose cycle. *J. Biol. Chem.* **233,** 1279–1282,

Woods, D. D., and Lascelles, J. (1954). The no man's land between the autotrophic and heterotrophic ways of life. *In* "Autotrophic Micro-organisms" (B. A. Fry and J. L. Peel, eds.). *Symposium Soc. Gen. Microbiol.* **4,** 1–27.

Woronick, C. L., and Johnson, M. J. (1960). Carbon dioxide fixation by cell-free extracts of *Aspergillus niger*. *J. Biol. Chem.* **235,** 9–15.

Zelitch, I. (1959). The relationship of glycolic acid to respiration and photosynthesis in tobacco leaves. *J. Biol. Chem.* **234,** 3077–3081.

—3—

Photoreduction and Anaerobiosis

C. J. P. SPRUIT

Laboratory of Plant Physiological Research, Agricultural University, Wageningen, The Netherlands

I. Introduction

Since algae produce oxygen during photosynthesis and are therefore not usually associated with environments characterized by low oxygen tensions, it is surprising that cells of a small but increasing number of algal species have been found capable of a highly specialized anaerobic metabolism. First observed in 1939 by Gaffron (1942a) in *Scenedesmus*, similar or related reactions were later encountered in a variety of other species. The early work has been reviewed by Gaffron (1944) and by Rabinowitch (1945); a more recent review is that of Kessler (1960a).

The following reactions, observed in anaerobically adapted algae, will be discussed in this chapter:

a. In the dark

Evolution of hydrogen and carbon dioxide.
Uptake of hydrogen from an atmosphere with a high partial pressure of this gas.
Oxidation of molecular hydrogen by molecular oxygen ("oxy-hydrogen reaction").

Reduction of nitrite, nitrate, or one of a variety of organic acceptors by molecular hydrogen.

b. In the light

Evolution of hydrogen at low partial pressures.

Simultaneous evolution of hydrogen and oxygen in the absence of normal photosynthesis.

Assimilation of carbon dioxide with molecular hydrogen, hydrogen sulfide, or an organic hydrogen donor.

Numerous experiments with algae under low or very low oxygen pressures have been described (e.g., Blinks and Skow, 1938; Damaschke *et al.*, 1953, 1955; Chance and Strehler, 1957a, b; Chance and Sager, 1957; see review by Kessler, 1960b). Since their relevance to the present discussion is not always sufficiently clear, they will not be extensively considered in this chapter.

II. Hydrogenase in Algae

Most plants, when subjected to anaerobic conditions, start to ferment with the production of carbon dioxide and organic acids. Algae behave similarly (Gaffron and Rubin, 1942; see Gibbs, Chapter 5, this treatise), but in some species their behavior changes in character after a certain period, and hydrogen appears among the fermentation products. This hydrogen evolution depends upon a hydrogenase, inactive under aerobic conditions, but transformed to an active enzyme after a suitable period of anaerobiosis. Among the algae in which hydrogenase activity has been observed are representatives of widely differing classes (see Table I).

Usually, hydrogenase activity has been demonstrated by the ability of the algae to carry out "photoreduction" (Section VI). Other tests include the production of hydrogen by anaerobic fermentation, and anaerobic reduction of nitrite in the dark (Kessler, 1956, 1957b). The reduction of methylene blue with hydrogen, a common tool in the study of bacterial hydrogenases, has only occasionally been employed, as has the catalyzed exchange between hydrogen and deuterium oxide (Krasna and Rittenberg, 1954). Manometric methods have been most widely used for the measurement of hydrogen exchange; since other gases are usually also involved, however, special modifications must be introduced. Specific and rapid methods for measuring hydrogen production were described by Damaschke (1957a; see also Damaschke, 1957b; Damaschke and Winkelmann, 1956) and by Spruit (1958). Both methods are based upon electrochemical reactions.

TABLE I
ALGAE IN WHICH HYDROGENASE ACTIVITY HAS BEEN DEMONSTRATED

Alga	Reference
Chlorophyta	
Chlorococcales	
Rhaphidium sp.	Gaffron (1944)
Scenedesmus sp.	Gaffron, (1944)
Scenedesmus obliquus	Gaffron (1944)
Ankistrodesmus sp.	Gaffron (1944)
Ankistrodesmus braunii	Kessler (1956)
Chlorella pyrenoidosa	Damaschke (1957a)
Chlorella vulgaris	Spruit (1954); Kessler and Maifarth (1960)
Coelastrum proboscideum var. *dilatatum*	Kessler and Maifarth (1960)
Selenastrum gracile	Kessler and Maifarth (1960)
Ulvales	
Ulva lactuca	Frenkel and Rieger (1951)
Volvocales	
Chlamydomonas moewusii	Frenkel (1952); Frenkel and Lewin (1954)
Euglenophyta	
Euglena sp.	Krasna and Rittenberg (1954)
Cyanophyta	
Synechococcus elongatus	Frenkel *et al.* (1950)
Synechocystis sp.	Frenkel *et al.* (1950)
Phaeophyta	
Ascophyllum nodosum	Frenkel and Rieger (1951)
Rhodophyta	
Porphyra umbilicalis	Frenkel and Rieger (1951)
Porphyridium cruentum	Frenkel and Rieger (1951)

In contrast to the extensive information now available on the hydrogenases from colorless and purple bacteria, very little is yet known about the properties of algal hydrogenases. Though this may, in part, be due to greater sensitivity of the latter towards oxygen, hydrogenase systems from algae show many points of similarity to those of bacteria.

Krasna and Rittenberg (1954) studied hydrogenase in the bacterium *Escherichia coli* with the aid of the *para→ortho* hydrogen conversion, and by this method also demonstrated the enzyme in species of *Scenedesmus* and *Euglena*. Fisher *et al.* (1954) found that purified hydrogenase from *E. coli* interacts with oxygen, pointing to a two-step oxidation of the enzyme (E):

$$E + O_2 \rightarrow EO_2$$

$$EO_2 \rightarrow E_{ox}$$

The first oxygenation is rapidly reversed by dithionite, but reactivation of the oxidized enzyme is considerably slower. This is strongly reminiscent of the situation in algae, where the initial adaptation, presumably the reduction of E_{ox}, may take a long time. However, the enzyme, if it is temporarily poisoned by the accumulation of a small amount of photosynthetically produced oxygen, rapidly recovers its activity as soon as the oxygen is removed (Rabinowitch, 1945, p. 131). Although the purified bacterial enzyme was found to catalyze the oxy-hydrogen reaction,

$$2H_2 + O_2 \rightarrow 2H_2O$$

it is doubtful whether this reaction alone could account for the oxy-hydrogen reaction of intact cells.

From the viewpoint of hydrogenase activity in photosynthetic organisms, it is important to recall the essential similarity of hydrogenases from the bacteria *E. coli* and *Rhodospirillum rubrum* (Gest, 1952).

Activation rates of the algal hydrogenases (i.e., "adaptation" times) differ considerably. In Gaffron's original experiments with *Scenedesmus* sp., adaptation required several hours at room temperature, but at 36°C. there was full activity as soon as the last traces of oxygen had been removed from the suspension. Damaschke (1957b), working with *Chlorella pyrenoidosa,* mentioned adaptation periods of several days. On the other hand, Kessler and Maifarth (1960) observed adaptation periods ranging from 4 to 40 hours in six different algae, while Spruit (1958), using *Chlorella vulgaris,* and Frenkel and Lewin (1954), using *Chlamydomonas moewusii,* obtained almost instantaneous adaptation at room temperature. Obviously the rate of adaptation must depend upon the activity of some internal system, probably a hydrogen donor, reducing the oxidized enzyme E_{ox} to the active form.

The occurrence of hydrogenases in algae may have some bearing on problems of nitrogen fixation, and *vice versa.* Usually, nitrogen fixation is inhibited competitively by molecular hydrogen, indicating an affinity of the nitrogenase for hydrogen. While symbiotic nitrogen fixation in higher plants is usually favored by a liberal supply of oxygen, this is not true of nitrogen fixation by purple bacteria. Thus Pratt and Frenkel (1959) showed that both hydrogen and oxygen bring about a reversible inhibition of nitrogen fixation in *R. rubrum.* In typically aerobic organisms, such as most algae, it is puzzling to encounter an enzyme such as hydrogenase that becomes activated only after a more or less prolonged exposure to anaerobic conditions. The photochemical reactions in which the algal hydrogenases participate are by no means slow, and the enzyme is presumably present in quantities comparable to those of enzymes of the photosynthetic process. What, then, is the enzyme doing when it is not catalyzing reactions involving molecular hydrogen? There is, as yet, no answer to this question; but in view of the foregoing observations, we may consider for a moment the occurrence in a few algae of enzymes with affinities for all three metabolic gases—hydrogen, nitrogen, and oxygen. Could the partial pressures of these gases, and the relative affinities of the enzymes towards each of them, determine whether they are operating in a hydrogenase, a nitrogenase, or a terminal oxidase ("oxygenase") system? Highly speculative as this may be and notwithstanding some negative reports, it is advisable to look more closely among algae for possible occurrence of the following reactions:

a. nitrogen fixation by algae known to possess a hydrogenase;
b. photochemical hydrogen reactions in all algae for which nitrogen fixation has been demonstrated;
c. evolution of molecular nitrogen under aerobic or anaerobic conditions.

While nothing is yet known of the physiological processes underlying the production of carbon monoxide by certain Brown Algae—7.6% was found in healthy air bladders of *Nereocystis luetkeana* (Langdon, 1917; Rigg and Swain, 1941) and 2% in those of *Pelagophycus porra* (F. T. Haxo, personal communication)—one may speculate that this normally toxic gas is a product of some unusual fermentation reaction.

III. Metabolism of Hydrogen in the Dark

After anaerobic adaptation, some algae evolve hydrogen and carbon dioxide. In the presence of high concentrations of hydrogen, this gas is taken up in considerable quantities for long periods (Gaffron, 1944). In both processes the nature of the hydrogen donors and acceptors is still unknown.

In the oxy-hydrogen reaction, both gases are taken up simultaneously by the algae from a mixture of hydrogen with oxygen below de-adaptation pressure (Gaffron, 1942b). Remarkably, the ratio H_2/O_2 is far below the expected value of 2, and is usually close to 1. In the opinion of the reviewer, this is most easily explained by additional oxygen consumption through respiration. When carbon dioxide is also present in the gas mixture, however, the reaction takes an entirely different course. Manometric measurements (Gaffron, 1942b) indicate the complete combustion of hydrogen, coupled to the reduction of carbon dioxide:

$$4H_2 + 2O_2 \rightarrow 4H_2O$$
$$2H_2 + CO_2 \rightarrow (CH_2O) + H_2O$$
$$\overline{6H_2 + CO_2 + 2O_2 \rightarrow (CH_2O) + 5H_2O}$$

This chemosynthetic reaction is therefore analogous to the one found in the so-called "Knallgas" bacteria. No satisfactory explanation for the influence of carbon dioxide upon the apparent course of hydrogen oxidation has been given.

Kessler (1956, 1957b, 1959) showed that anaerobically adapted *Ankistrodesmus braunii* is able to reduce nitrite to the level of ammonia:

$$HNO_2 + 3H_2 \rightarrow NH_3 + 2H_2O$$

Nitrate and hydroxylamine are also reduced, though much more slowly. Carbon dioxide has a stabilizing effect on the reduction of nitrite with molecular hydrogen. Without carbon dioxide, hydrogen consumption and nitrite reduction slow down long before the reaction is complete, but the addition of 4% carbon dioxide to the gas mixture maintains the initial rate of the reaction for a much longer period. Experiments with labeled carbon dioxide have proved that no appreciable carbon is taken up during the reaction, indicating that the action of carbon dioxide is catalytic in nature. Kessler also observed that methylene blue can be reduced with hydrogen by adapted algae, though more slowly than can nitrite.

IV. Production of Hydrogen and Oxygen in the Light

Illumination of adapted algae at moderate light intensities considerably increases the rate of hydrogen evolution (Gaffron and Rubin, 1942). In contrast to the fermentative production of hydrogen in darkness, this photochemical hydrogen evolution is insensitive to moderate concentrations of dinitrophenol, demonstrating that the two processes are distinct. During photochemical formation of hydrogen, no evolution of oxygen can be demonstrated manometrically, and it was originally assumed that there is no formation of this gas whatsoever. Experiments by Franck *et al.* (1945), with the aid of a very sensitive "phosphorescence quenching" method, demonstrated the production of small amounts of oxygen by illuminated cells of *Scenedesmus* sp. and *Chlorella* sp. under anaerobic conditions similar to those employed by Gaffron. Actually, some of the early experiments by Blinks and Skow (1938) also showed oxygen production by illuminated algae under anaerobic conditions. Spruit (1952, 1958) demonstrated the formation of oxygen during illumination of *Chlorella vulgaris* in anaerobiosis, and Horwitz and Allen (1957a, b) further studied the problem, both by the phosphorescence method of Pringsheim and by mass spectrometry using oxygen isotopes.

The production of oxygen under anaerobic conditions being now well established (Fig. 1), one may ask whether hydrogen and oxygen are formed simultaneously. This was confirmed by Spruit (1958) who, by using electrochemical methods for the measurement of the two gases, also determined

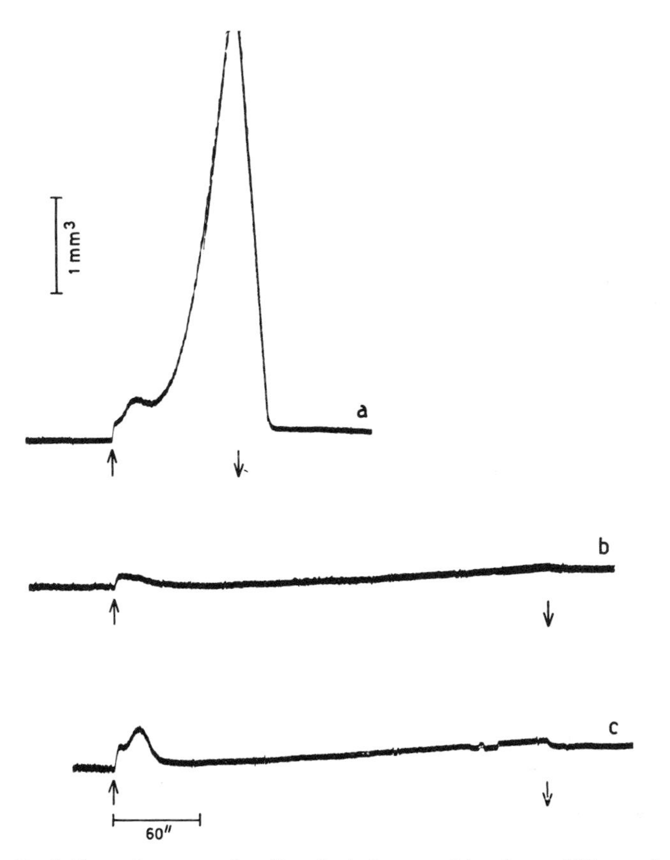

FIG. 1. Evolution of oxygen by illuminated anaerobic algae; 150 mm.³ *Chlorella vulgaris* in 50 ml. buffer, pH 7.0. (*a*) Suspension flushed with nitrogen. (*b*) Aliquot flushed with nitrogen *plus* 8% hydrogen. (*c*) Repeated illumination of *b*. after displacement of hydrogen by nitrogen. Arrows indicate period of illumination. (From Spruit, 1958.)

the ratio in which they are produced. Though considerably influenced by the duration of the illumination, in the absence of secondary reactions the ratio H_2/O_2 is apparently close to 2, suggesting that water is the common source of both gases. As soon as the oxygen pressure rises above a certain low value, however, the hydrogenase is poisoned and hydrogen production stops. On the other hand, higher concentrations of hydrogen also inhibit oxygen evolution (see Fig. 1), perhaps due in part to the oxy-hydrogen reaction. The total quantities of hydrogen and oxygen formed are usually minute, though the initial rates of gas evolution are not much lower than those of normal aerobic photosynthesis.

Although Kessler (1955, 1957c) demonstrated that manganese deficiency in *A. braunii* strongly suppresses photosynthesis but leaves the hydrogen reactions unaffected, Spruit observed no such inhibitory effect of manganese deficiency upon the initial simultaneous production of hydrogen and oxygen by *C. vulgaris* illuminated in anaerobic conditions. It may be that it is only during steady-state photosynthesis that oxygen evolution requires manganese; the problem clearly requires further study (see also Kessler *et al.*, 1957).

Upon illumination, the initial production of oxygen (indicated by the first two peaks in Fig. 1) and hydrogen is independent of the presence of carbon dioxide, which influences neither the duration nor the magnitude of the oxygen peaks, but which decreases the duration of the period before the start of steady oxygen evolution coupled to carbon dioxide reduction (Spruit, 1952, 1958). Similar results were reported by Brackett *et al.* (1957). During this initial phase, hydrogen is evolved even in the presence of carbon dioxide, showing that CO_2 does not initially act as a hydrogen acceptor. Spruit and W. Lindeman (unpublished) have studied the formation of "Calvin-cycle" intermediates during the initial phase.[1] The content of phosphoglyceric acid and ribulose diphosphate is low at the start of the illumination period, which explains the non-participation of CO_2. During illumination, ribulose diphosphate levels rise, but phosphoglyceric acid remains low. In the initial phase inorganic phosphate is rapidly incorporated into organic compounds, and nucleotides, including adenosine triphosphate (ATP) rise sharply. This is in line with the present picture of photosynthesis, and these experiments seem to imply that a little oxygen is required for the carbon dioxide acceptor mechanism (i.e., the Calvin cycle) to reach full capacity.

V. Adaptation and Deadaptation

Some attention has already been devoted to these phenomena in the previous sections. Under conditions which do not prevent reduction of the oxidized hydrogenase E_{ox} to the active form of the enzyme, the activity of the latter is probably determined only by the oxygen concentration in the cells. During illumination, any factor that lowers the oxygen tension should therefore prevent deadaptation, and any factor that prevents reabsorption of oxygen should favor deadaptation. The effects of oxygen, high light intensities, and poisons such as hydroxylamine, *o*-phenanthroline, and cyanide (Gaffron, 1944) could be explained in this way.

Whereas low concentrations of oxygen (0.5%) prevent deadaptation of *Scenedesmus*, after adaptation the cells maintain this condition even in the presence of 1 to 2% oxygen. It was shown by Horwitz (1957) that the

[1] See Holm-Hansen, Chapter 2, this treatise.

discrepancy between these two oxygen levels is due to diffusion limitations, and that the critical oxygen concentrations for adaptation and deadaptation may actually be close together. This is in line with the idea that the activity of the hydrogenase is a function of oxygen pressure, as a consequence of a reversible oxidation of the enzyme (Section II).

Although the initial rates of oxygen and hydrogen production by illuminated cells are high, the total amounts evolved are so low that manometric observation does not normally reveal them. On the other hand, if, during this initial phase, the oxygen tension rises above the critical value for inactivation of the hydrogenase, the cells become deadapted and, in the presence of carbon dioxide, revert to photosynthesis (Section VI). Observed by manometric techniques, such algae would be considered as exhibiting no capacity for anaerobic photoreactions. This may explain why there has been disagreement about the existence of such capacities in some species of algae. It may also be advisable to examine other species with the aid of the modern rapid and specific methods for detection and measurement of oxygen and hydrogen.

VI. The Problem of Photoreduction

The term *photoreduction* is used to describe a number of processes:

a. reduction of dyes and other compounds through the action of light;
b. assimilation of carbon dioxide with organic hydrogen donors;
c. assimilation of carbon dioxide with molecular hydrogen.

Unless stated specifically, the term will be used here only in the last sense.

At suitable light intensities and concentrations of hydrogen and carbon dioxide, adapted algae consume both gases according to the following equation (Gaffron, 1944):

$$CO_2 + 2H_2 \rightarrow (CH_2O) + H_2O$$

This is the same reaction as that observed in purple bacteria. As these have been convincingly shown to produce no oxygen during their photosynthetic activities, and as oxygen was originally believed not to be generated during photoreduction in algae, the latter process was considered to be perfectly analogous with bacterial photosynthesis. In view of the other anaerobic reactions of adapted algae (discussed above), Gaffron also considered the possibility that the over-all reaction of photoreduction was made up of normal photosynthesis *plus* consumption of the photosynthetic oxygen by the oxy-hydrogen reaction:

$$CO_2 + H_2O \rightarrow (CH_2O) + O_2$$

$$2H_2 + O_2 \rightarrow 2H_2O$$

Gaffron later discarded this possibility mainly on the basis of the relative intensitivity of photoreduction to hydroxylamine, a powerful inhibitor of photosynthesis. It is evident that the demonstration of oxygen production under photoreduction conditions necessitates a reconsideration of the problem.

Larsen *et al.* (1952) pointed out that the quantum yield of photosynthesis combined with the oxy-hydrogen reaction might be expected to exceed the quantum yield of aerobic photosynthesis by about 50%, as a consequence of coupling of the oxy-hydrogen reaction with fixation of carbon dioxide in darkness. Although such quantum-yield determinations by Rieke (1949) were not sufficiently conclusive, within the rather wide limits of accuracy, the quantum yields for aerobic photosynthesis and photoreduction were found to be the same.

The problem was reinvestigated by Horwitz and Allen (1957a, b). From experiments with *Scenedesmus*, and using the mass spectrometer and gas atmospheres containing hydrogen, oxygen-32, oxygen-34, carbon dioxide-44, and carbon dioxide-45, these authors calculated the ratio of carbon dioxide consumed to oxygen liberated in the light. This ratio should be close to unity for true photosynthesis, and should approach infinity for photoreduction; a combination of the two processes would give intermediate values. Assuming the "dark reactions" (oxy-hydrogen reaction coupled to carbon dioxide fixation) to continue undisturbed during illumination, values for the ratio CO_2/O_2 were found not to deviate significantly from unity, except in cells poisoned with $5 \times 10^{-4}M$ *o*-phenanthroline. "The data obtained with the mass spectrometer. . . give no reliable evidence for the existence of a true photoreduction in unpoisoned, adapted algae" (Horwitz and Allen, 1957b, p. 57). On the other hand, in experiments with Warburg manometers, the same authors observed rates of carbon dioxide uptake in the light considerably higher than could be accounted for by concurrent photosynthesis and the oxy-hydrogen reaction. In some cases, up to 30% of the carbon dioxide uptake could be ascribed to "true" photoreduction. Horwitz and Allen pointed out that the appreciable oxygen pressures (0.04–0.24%) employed in the mass spectrometer experiments may have been responsible for this discrepancy, but since the evaluation of the data from the mass spectrometer and the Warburg experiments rests upon a number of assumptions, the matter remains unsettled.

Perhaps the most convincing demonstration of what probably must be regarded as true photoreduction was given by Kessler (1957a, c), who found that *A. braunii* cultivated under conditions of extreme manganese deficiency shows no depression of photoreduction, although photosynthesis is strongly inhibited. The latter fact may also explain: (*a*) the protective effect of manganese deficiency against deadaptation of photoreduction, which normally occurs at high light intensities; and (*b*) the stimulation of gas consumption at lower light intensities, since there is no production of oxygen and the hydrogenase therefore remains uninhibited. One possible objection to experiments of this type is that it has not yet been demonstrated that inhibitory treatment (Mn deficiency or poisons) affects photosynthesis similarly at the very low oxygen pressures found during photoreduction. In other words, it is still to be demonstrated that the effect of such inhibitory conditions is independent of oxygen pressures.

Gaffron observed "photoreduction" of carbon dioxide with organic hydrogen donors such as glucose. Such a reaction might be explained as photosynthesis coupled to respiration, *viz.*, oxidation of glucose with photosynthetic oxygen, without evolution of an equivalent amount of carbon dioxide. Under anaerobic conditions, some algae (e.g. *Chlorella*) have in fact been shown to be able to consume considerable amounts of oxygen without concomitant CO_2 production (Spruit and Kok, 1956, cf. their Fig. 6).

Photoreduction of carbon dioxide by the algae *Oscillatoria neglecta* and *Pinnularia* (?) *molaris* with hydrogen sulfide as the hydrogen donor (Nakamura, 1938) can only be mentioned here; it has apparently never been independently confirmed. It should be recalled that *Oscillatoria* is structurally very close to *Beggiatoa*, which may be regarded either as a sulfur bacterium or as an apochlorotic Blue-green Alga, and which normally depends on a chemosynthetic process of this kind.

Badin and Calvin (1950), who studied the fate of labeled carbon dioxide during photosynthesis, photoreduction, and the dark hydrogen-oxygen-carbon dioxide reaction in *Scenedesmus* sp., found no essential difference in the nature of the fixation products from the above reactions, if carried out under comparable conditions (cf. also Brown, 1948). It is therefore of interest to consider whether any alga can grow with photoreduction as its sole source of energy and reduced carbon. Gaffron was unable to demonstrate this in the case of *Scenedesmus* (strain D3); very few other species have been tested. *Ochromonas malhamensis* is capable of slow growth under anaerobic conditions in the light when supplied with certain organic hydrogen donors; however, molecular hydrogen cannot be utilized (Vishniac and Reazin, 1957).

References

Badin, E. J., and Calvin, M. (1950). The path of carbon in photosynthesis. IX. Photosynthesis, photoreduction and the hydrogen-oxygen-carbon dioxide reaction. *J. Am. Chem. Soc.* **72,** 5266–5270.

Blinks, L. R., and Skow, R. K. (1938). The time course of photosynthesis as shown by a rapid electrode method for oxygen. *Proc. Natl. Acad. Sci. U.S.* **24,** 420–427.

Brackett, F. S., Olson, R. A., and Crickard, R. G. (1957). Transients in O_2 evolution by *Chlorella* in light and darkness. I. Phenomena and methods. *In* "Research in Photosynthesis" (H. Gaffron, A. H. Brown, C. S. French, R. Livingston, E. I. Rabinowitch, B. L. Strehler, and N. E. Tolbert, eds.), pp. 412–418. Interscience, New York.

Brown, A. H. (1948). The carbohydrate constituents of *Scenedesmus* in relation to the assimilation of carbon by photoreduction. *Plant Physiol.* **23,** 331–337.

Chance, B., and Sager, R. (1957). Oxygen and light induced oxidations of cytochrome, flavoprotein and pyridine nucleotide in a *Chlamydomonas* mutant. *Plant Physiol.* **32,** 548–561.

Chance, B., and Strehler, B. (1957a). Effects of oxygen upon light absorption by green algae. *Nature* **180,** 749–750.

Chance, B., and Strehler, B. (1957b). Effects of oxygen and red light upon the absorption of visible light in green plants. *Plant Physiol.* **32,** 536–548.

Damaschke, K. (1957a). Verfolgung der Bildung reduzierender Substanzen mit Hilfe der elektrochemischen H_2O_2—Messmethode an inkubierter *Chlorella pyrenoidosa* im Hellen und Dunkeln. *Z. Naturforsch.* **12b,** 150–155.

Damaschke, K. (1957b). Die Wasserstoffgärung von *Chlorella* im Dunkeln nach Anaerobiose unter Stickstoff. *Z. Naturforsch.* **12**b, 441–443.

Damaschke, K., and Lübke, M. (1958). Über die Fähigkeit der *Chlorella pyrenoidosa* zur anaeroben Nitritreduktion. *Z. Naturforsch.* **13b,** 134–135.

Damaschke, K., and Winkelmann, D. (1956). Elektrochemische H_2O_2—Messung und ihre Anwendung beim Zerfall von H_2O_2 durch Katalase. *Z. Naturforsch.* **11b,** 86–89.

Damaschke, K., Tödt, F., Burk, D., and Warburg, O. (1953). An electrochemical demonstration of the energy cycle and maximum quantum yield in photosynthesis. *Biochim. et Biophys. Acta* **12,** 347–355.

Damaschke, K., Rothbühr, L., and Tödt, F. (1955). Photosynthese unter anaeroben Bedingungen. *Z. Naturforsch.* **10b,** 572–578.

Fisher, H. F., Krasna, A. I., and Rittenberg, D. (1954). The interaction of hydrogenase with oxygen. *J. Biol. Chem.* **209,** 569–578.

Franck, J., Pringsheim, P., and Lad, D. T. (1945). Oxygen production by anaerobic photosynthesis of algae measured by a new micromethod. *Arch. Biochem. Biophys.* **7,** 103–142.

Frenkel, A. W. (1952). Hydrogen evolution by the flagellate green alga, *Chlamydomonas moewusii. Arch. Biochem. Biophys.* **38,** 219–230.

Frenkel, A. W., and Lewin, R. A. (1954). Photoreduction by *Chlamydomonas. Am. J. Botany* **41,** 586–589.

Frenkel, A. W., and Rieger, C. (1951). Photoreduction in algae. *Nature* **167,** 1030.

Frenkel, A. W., Gaffron, H., and Battley, E. H. (1950). Photosynthesis and photoreduction by the blue-green alga *Synechococcus elongatus* Näg. *Biol. Bull.* **99,** 157–162.

Gaffron, H. (1942a). The effect of specific poisons upon the photoreduction with hydrogen in green algae. *J. Gen. Physiol.* **26,** 195–217.

Gaffron, H. (1942b). Reduction of carbon dioxide coupled with the oxyhydrogen reaction in algae. *J. Gen. Physiol.* **26,** 241–267.

Gaffron, H. (1944). Photosynthesis, photoreduction and dark reduction of carbon dioxide in certain algae. *Biol. Revs. Cambridge Phil. Soc.* **19,** 1–20.

Gaffron, H., and Rubin, J. (1942). Fermentative and photochemical production of hydrogen in algae. *J. Gen. Physiol.* **26,** 219–240.

Gest, H. (1952). Properties of cell-free hydrogenases of *E. coli* and *Rh. rubrum*. *J. Bacteriol.* **63,** 111–121.

Horwitz, L. (1957). The oxyhydrogen reaction in *Scenedesmus* and its relation to respiration and photosynthesis. *Arch. Biochem. Biophys.* **66,** 23–44.

Horwitz, L., and Allen, F. L. (1957a). Oxygen evolution and photoreduction by adapted *Scenedesmus*. *In* "Research in Photosynthesis" (H. Gaffron, A. H. Brown, C. S. French, R. Livingston, E. I. Rabinowitch, B. L. Strehler, and N. E. Tolbert, eds.), pp. 232–238. Interscience, New York.

Horwitz, L., and Allen, F. L. (1957b). Oxygen evolution and photoreduction in adapted *Scenedesmus*. *Arch. Biochem. Biophys.* **66,** 45–63.

Kessler, E. (1955). The role of manganese in the oxygen-evolving system of photosynthesis. *Arch. Biochem. Biophys.* **59,** 527–529.

Kessler, E. (1956). Reduction of nitrite with molecular hydrogen in algae containing hydrogenase. *Arch. Biochem. Biophys.* **62,** 241–242.

Kessler, E. (1957a). Über die Rolle des Mangans bei Photoreduktion und Photosynthese. *Planta* **49,** 435–454.

Kessler, E. (1957b). Dunkel-Reduktion von Nitrat und Nitrit mit molekularem Wasserstoff. *Arch. Mikrobiol.* **27,** 166–181.

Kessler, E. (1957c). Manganese as a cofactor in photosynthetic oxygen evolution. *In* "Research in Photosynthesis" (H. Gaffron, A. H. Brown, C. S. French, R. Livingston, E. I. Rabinowitch, B. L. Strehler, and N. E. Tolbert, eds.), pp. 243–249. Interscience, New York.

Kessler, E. (1959). Reduction of nitrate by green algae. *Symposia Soc. Exptl. Biol.* **13,** 87–105.

Kessler, E. (1960a). Biochemische Variabilität der Photosynthese: Photoreduktion und verwandte Photosynthesetypen. *In* "Handbuch der Pflanzenphysiologie" (W. Ruhland, ed.), Vol. V: CO_2 Assimilation, pp. 951–965. Springer, Berlin.

Kessler, E. (1960b). Der Einfluss von Sauerstoff auf die Photosynthese. *In* "Handbuch der Pflanzenphysiologie" (W. Ruhland, ed.), Volume V: CO_2 Assimilation, pp. 935–950. Springer, Berlin.

Kessler, E., and Maifarth, H. (1960). Vorkommen und Leistungsfähigkeit von Hydrogenase bei einigen Grünalgen. *Arch. Mikrobiol.* **37,** 215–225.

Kessler, E., Arthur, W., and Brugger, J. E. (1957). Influence of manganese and phosphate on delayed light emission, fluorescence photoreduction and photosynthesis in algae. *Arch. Biochem. Biophys.* **71,** 326–335.

Krasna, A. I., and Rittenberg, D. (1954). The mechanism of action of the enzyme hydrogenase. *J. Am. Chem. Soc.* **76,** 3015–3020.

Langdon, S. C. (1917). Carbon monoxide, occurrence free in kelp. *J. Am. Chem. Soc.* **39,** 149–156.

Larsen, H., Yocum, C. S., and van Niel, C. B. (1952). On the energetics of the photosyntheses in green sulfur bacteria. *J. Gen. Physiol.* **36,** 161–171.

Nakamura, H. (1938). Ueber die Kohlensäureassimilation bei niederen Algen in Anwesenheit des Schwefelwasserstoffs. *Acta Phytochim.* (*Japan*) **10,** 271–281.

Olson, R. A., Brackett, F. S., and Crickard, R. G. (1957). Transients in O_2 evolution by *Chlorella* in light and darkness. II. Influence of O_2 concentration and respiration. *In* "Research in Photosynthesis," (H. Gaffron, A. H. Brown, C. S. French, R. Livingston, E. I. Rabinowitch, B. L. Strehler, and N. E. Tolbert, eds.), pp. 419–429. Interscience, New York.

Pratt, D. C., and Frenkel, A. W. (1959). Studies on nitrogen fixation and photosynthesis of *Rh. rubrum*. *Plant Physiol.* **34,** 333–337.

Rabinowitch, E. I. (1945). "Photosynthesis and Related Processes," Vol. I. Interscience, New York.

Rieke, F. R. (1949). Quantum efficiencies for photosynthesis and photoreduction in green plants. *In* "Photosynthesis in Plants" (J. Franck and W. E. Loomis, eds.), pp. 251–272. Iowa State College Press, Ames, Iowa.

Rigg, G. B., and Swain, L. A. (1941). Pressure-composition relations of the gas in the marine brown alga, *Nereocystis luetkeana*. *Plant Physiol.* **16,** 361–371.

Spruit, C. J. P. (1952). A study of oxidation-reduction potentials in relation to the functions of light and carbon dioxide in *Chlorella* photosynthesis. *Acta Botan. Neerl.* **1,** 551–579.

Spruit, C. J. P. (1954). Photoproduction of hydrogen and oxygen in *Chlorella*. *Proc. 1st Intern. Photobiol. Congr., Amsterdam,* pp. 323–327.

Spruit, C. J. P. (1958). Simultaneous photoproduction of hydrogen and oxygen by *Chlorella*. *Mededel. Landbouwhogeschool Wageningen* **58** (9), 1–17.

Spruit, C. J. P., and Kok, B. (1956). Simultaneous observation of oxygen and carbon dioxide exchange during non-steady state photosynthesis. *Biochim. et Biophys. Acta* **19,** 417–424.

Vishniac, W., and Reazin, G. H. (1957). Photoreduction in *Ochromonas malhamensis*. *In* "Research in Photosynthesis" (H. Gaffron, A. H. Brown, C. S. French, R. Livingston, E. I. Rabinowitch, B. L. Strehler, and N. E. Tolbert, eds.), pp. 238–242. Interscience, New York.

—4—

Respiration[1]

MARTIN GIBBS

Department of Biochemistry, Cornell University, Ithaca, New York

I. Introduction and Definitions[2]

Though a number of reviews of general algal physiology have devoted major sections to the topic of respiration (Blinks, 1951; Fogg, 1953; Krauss,

[1] This review is part of work supported by the National Science Foundation and the United States Air Force through the Air Force Office of Scientific Research.

[2] Abbreviations used throughout this chapter: EMP, Embden-Meyerhof-Parnas pathway; ADP, adenosine diphosphate; ATP, adenosine triphosphate; DPN, diphosphopyridine nucleotide; HMP, hexose monophosphate; TPN, triphosphopyridine nucleotide; CoA, coenzyme A; R. Q., respiratory quotient; TCA, tricarboxylic acid.

1958; Myers, 1951), this chapter is among first efforts to assemble and assess the voluminous and widespread publications on this basic metabolic process.

For the purposes of this paper, *respiration* is considered as the complete oxidation of an organic compound (generally, but not exclusively, a carbohydrate) to carbon dioxide and water, with molecular oxygen serving as the ultimate electron acceptor. *Aerobic fermentation* is defined as the breakdown of a carbohydrate at some low concentration of oxygen, as a result of which a part of the carbon skeleton is converted into compounds normally considered as end products of fermentation, instead of being completely oxidized to carbon dioxide.

II. Rate of Respiration

A. Units

The rate of respiration is usually measured by following manometrically the uptake of oxygen per unit time. Most investigators report their data on the basis of microliters O_2 consumed per milligram dry weight per hour expressing the rate of oxygen uptake defined in this way as the quotient, Q_{O_2}. A minus (−) sign is sometimes used to indicate uptake of oxygen. Respiration is generally determined in darkness by the "direct" method of Warburg (see Umbreit *et al.*, 1957), in which oxygen uptake is determined in the presence of KOH, which absorbs carbon dioxide and thereby keeps its partial pressure close to zero. Since the possibility exists that such measurements might yield atypical values if the absence of carbon dioxide were to affect decarboxylation reactions, the "indirect" method of Warburg (see Umbreit *et al.*, 1957) and a method utilizing a "CO_2 buffer" (Pardee, 1949) have also been employed. In the few instances where the "direct" and "indirect" methods have been compared, there has been no significant difference in the results.

B. Effect of pH

Few workers have given attention to the effect of pH on respiration. This parameter is clearly important; for instance, the Q_{O_2} of *Ochromonas malhamensis* is 27 at pH 5 and only 8 at pH 7 (Reazin, 1954). On the other hand, many Blue-green Algae have a pH optimum for respiration closer to 8, although *Anabaena variabilis*, like *Chlorella pyrenoidosa*, shows little change in Q_{O_2} between pH 4.5 and 9.3. (For a critical discussion see Steemann Nielsen, 1955a).

C. Effect of Depleting Endogenous Reserves

When cells from a growing culture of *Chlorella pyrenoidosa* are aerated in distilled water in darkness, their rate of respiration gradually decreases in the first 24 hours to a value which then tends to remain constant for several days (Cramer and Myers, 1949). In many cases, as in bacteria and fungi, such starved or "resting" cells show a marked response to the addition of an exogenous substrate. Generally the Q_{O_2} of unstarved cells is increased less than 3-fold by the addition of a substrate, whereas cells depleted of endogenous reserves exhibit at least an 8-fold effect. As seen in Table I, the highest relative rates of respiration are generally obtained by the addition of glucose to starved cells.

D. Effect of Temperature

The effect of temperature on the respiration rate seems to divide the algae into two groups. One group, which contains *Chlorella* and *Scenedesmus* spp. has an optimum around 27° (Kandler, 1954), while the second comprises thermophilic Blue-green Algae such as *Oscillatoria subbrevis* (Bünning and Herdtle, 1946; Moyse *et al.*, 1957) and *Anacystis nidulans** (Kratz and Myers, 1955a, b), with an optimum in the range of 40°C.

III. Respiratory Quotient

The respiratory quotient (R.Q.) is defined here as the ratio of CO_2 produced to O_2 consumed.

Some workers have observed that in both glucose respiration and endogenous respiration of autotrophically grown cells the R.Q. is close to unity. All have concluded that the endogenous substrate is some carbohydrate. However, an R.Q. value of 1 is not invariable; for instance, in autotrophically grown *Chlorella pyrenoidosa,* French *et al.* (1934) observed the R.Q. to decrease with time from 1.0 to 0.65, and therefore postulated two endogenous substrates, one a carbohydrate and one a fat. Cramer and Myers (1949) confirmed this observation, but noted that the apparent decrease was actually due to an incidental retention of CO_2 by the suspending fluid. When this source of error was corrected, they noted that the R.Q. remained close to 1, at least during a 2-day starvation period, and thus provided no evidence for respiration of a lipid. Kandler (1954) reported an endogenous R.Q. of 0.8 for *Chlorella pyrenoidosa* cells starved for 48 hours in darkness. Apparently CO_2 retention was not a factor here, since the R.Q. was constant between pH 4.8 and 9.7.

An R.Q. of about 1.5 was exhibited by autotrophically grown *Chlorella pyrenoidosa* when nitrate was substituted for some of the elemental oxygen

TABLE I

RESPIRATION IN MILLILITERS OF O_2 PER GRAM DRY WEIGHT PER HOUR

Organism	Previous conditions	Exogenous substrate	Q_{O_2}	Temperature (° C.)	Reference
I. ALGAE					
Chlorella pyrenoidosa	Grown with photosynthesis, not starved	—	1.0–1.4	25	Genevois (1927)
		Glucose	5.0	25	Genevois (1927)
		Acetaldehyde	3.2	25	Genevois (1927)
		Butyrate	3.4	25	Genevois (1927)
	Grown in darkness with glucose, not starved	Glucose	2.8	20	Genevois (1927)
	Grown with photosynthesis, then starved	—	2.3	27	Kandler (1954)
		Glucose	17.1	27	Kandler (1954)
Scenedesmus sp. (strain D_3)	Grown in darkness with glucose; not starved	—	1.8	20	Gaffron (1939)
		Glucose	3.8	20	Gaffron (1939)
Anabaena variabilis	Growing	—	8.4	25	Webster and Frenkel (1953)
	Starved	—	1.7	25	
Anacystis nidulans	Growing	—	1.6	25	Kratz and Myers (1955a)
	Starved	—	0.3	25	
Ochromonas malhamensis	Growing	—	28	26	Reazin (1954)
	Starved 20 hours	—	19	26	Reazin (1954)

II. OTHER ORGANISMS				
Arum spadix (angiosperm)		31	(?)	James (1953)
Rat liver (mammal)		11	37	Dickens and Simer (1930)
Bakers' yeast (fungus)	—	10	28	Meyerhof (1925)
	Glucose	95	28	Meyerhof (1925)

as the electron acceptor (Warburg and Negelein, 1920). Here, carbon dioxide was produced with limited concomitant oxygen consumption:

$$\begin{array}{l} 2\ CH_2O + HNO_3 \rightarrow 2\ CO_2 + NH_3 + H_2O \\ 4\ CH_2O + 4\ O_2 \rightarrow 4\ CO_2 + 4\ H_2O \\ \hline 6\ CH_2O + HNO_3 + 4\ O_2 \rightarrow 6\ CO_2 + NH_3 + 5\ H_2O \end{array}$$

$$R.Q. = CO_2/O_2 = 6/4 = 1.5$$

(Obviously, since this involves the oxidation of an organic compound, it can correctly be termed respiration, though to distinguish it from oxygen respiration this process has been named "nitrate respiration"). In studies of the same species, Genevois (1927, 1929) observed R.Q. values of 1.26 for the oxidation of acetaldehyde; 1.13 for propionaldehyde; and 1.03 for valeraldehyde.

Gaffron (1939) was the first to observe that cultures of *Chlorella pyrenoidosa*, *Chlorella variegata*, and *Scenedesmus* sp. (strains D_3 and D_1), grown in north-window light with 1.5% glucose, gave R.Q. values of 1.2 to 2.0, as contrasted to 1.0 noted above for autotrophically grown cells. This has been confirmed by others (Kandler, 1954), even in the absence of nitrate, indicating some metabolic transformation of the oxygen-rich foodstuff, carbohydrate, into an oxygen-poor material, probably lipid. This may be illustrated by the following equation for the conversion of a hexose to palmitic acid:

$$4\ C_6H_{12}O_6 + O_2 \rightarrow C_{16}H_{32}O_2 + 8\ CO_2 + 8\ H_2O \qquad R.Q. = 8.0$$

Since a source of nitrogen was not available, there was probably no net protein synthesis, which would have a similar effect (see Syrett, Chapter 10, this treatise).

IV. Respiratory Pathways

A. General Discussion

As indicated elsewhere in this treatise, in the chapter on fermentation (Gibbs, Chapter 5), it has long been felt that the first steps of carbohydrate breakdown are common to respiration and fermentation. This "common pathway" concept was developed originally to account for similarities between the anerobic degradation of glucose to ethanol and carbon dioxide by yeast and the anaerobic dissimilation of glycogen to lactic acid by mammalian muscle. In the present discussion, this chain of reactions also called the Embden-Meyerhof-Parnas (EMP) pathway, is considered to include the steps from glucose to pyruvic acid.

B. The EMP Pathway

Knowledge of the existence and functioning of the EMP pathway in many kinds of living cells is based on: (*1*) the detection and utilization of postulated intermediates; (*2*) the presence of enzymes catalyzing postulated reactions; (*3*) the sensitivity of some of these enzymes to inhibitors such as iodoacetate and fluoride; and (*4*) the degradation of specifically labeled glucose to products labeled as predicted by the EMP pathway (see also Gibbs, Chapter 5, this treatise). The following paragraphs present some of the evidence, on each of these four points, adduced from investigations carried out with algae.

1. Intermediates

Moses *et al.* (1959) observed that *Chlorella pyrenoidosa*, exposed to glucose-C^{14} in the dark, produces the followed labeled intermediates: fructose-1,6-diphosphate; glucose-6-phosphate; fructose-6-phosphate; phosphoglyceric acid; and the amino acid alanine, which metabolically is closely associated with pyruvic acid. *Ochromonas malhamensis*, whose normal respiratory products are carbon dioxide and water, can be induced to accumulate pyruvic acid in the medium by the addition of $2.5 \times 10^{-3} M$ arsenite (Reazin, 1956), a known inhibitor of the pyruvic-acid oxidase system. Kandler (1955) made a similar observation with *Chlorella pyrenoidosa.* Likewise pyruvic acid accumulates in cultures of *Prototheca zopfii*, which normally requires thiamine for the synthesis of diphosphothiamine (cocarboxylase), if this vitamin is omitted from the medium and if glucose or glycerol is supplied as a carbon source (Anderson, 1945).

There are two reports that intact cells of *Chlorella pyrenoidosa* can respire phosphorylated sugars. Genevois (1927) noted oxidation of fructose-6-phosphate, while Emerson *et al.* (1954) observed that both intact and toluene-treated cells could metabolize hexose diphosphate and a hexose monophosphate. However, the purity of the sugar phosphates available in those days might be questioned.

2. Enzymes

Only a few of the enzymes of the EMP pathway have been detected in algae; none has been purified (see Jacobi, Chapter 7, this treatise).

3. Inhibitors

The iodoacetyl radical, either in the form of iodoacetic acid or as iodoacetamide, is an inhibitor of D-glyceraldehyde-3-phosphate dehydrogenase. Kandler (1955) showed that $10^{-4} M$ iodoacetic acid inhibits oxygen uptake in *Chlorella pyrenoidosa* by 55% (see Calo and Gibbs (1960) for a discussion of this inhibitor and its effects on algal respiration) and that 0.1*M* fluoride inhibits respiration by 25%. Holzer (1954) reported that the rate of respiration in this species was reduced to one half by $3 \times 10^{-2}M$ NaF.

4. *Tracer studies*

Sodium arsenite has been used to inhibit respiration of glucose by intact algal cells in order to obtain information about the intermediate reactions (Reazin, 1956). As noted above, pyruvic acid accumulates in such a poisoned system. When either glucose-1-C^{14} or glucose-2-C^{14} was respired by *Ochromonas malhamensis* in the presence of arsenite, the specific activity of isotope in the pyruvic acid was found to be half of that in the glucose originally supplied, which is in accordance with the EMP pathway (see Gibbs, Chapter 5, this treatise). The dissimilation of glucose by this scheme was confirmed by the location of the labeled carbon atom in the product; glucose-1-C^{14} gave rise mainly to methyl-labeled pyruvic acid, whereas glucose-2-C^{14} produced alpha-labeled pyruvic acid. The C_6/C_1 method of Bloom and Stetten (1953) was also used to elucidate glucose respiration by *Ochromonas*. When equal quantities of glucose-1-C^{14} and glucose-6-C^{14} were oxidized, the ratio of labeled C_6 to C_1 in the respired CO_2 was unity, suggesting again that the major, if not the sole, pathway of glucose respiration in this organism is the EMP system. A similar study with *Chlorella pyrenoidosa* gave concordant results (Kandler and Gibbs, 1959).

C. The Pentose-Phosphate Pathway

The initial steps in aerobic and anaerobic carbohydrate respiration in many cells are evidently catalyzed by the EMP pathway. The first unequivocal proof that the glycolytic mechanism may depart from this pattern of reactions came from studies with the heterofermentative bacterium *Leuconostoc mesenteroides* (Gunsalus and Gibbs, 1952). It soon became clear that this departure was not limited to the anaerobic bacteria.

In this review the pentose-phosphate pathway is considered to include not only the enzymes catalyzing the formation of CO_2 and ribulose-5-phosphate from glucose-6-phosphate in what is termed the "hexose-monophosphate pathway" (HMP) or HMP shunt,[3] but also the group of enzymes which catalyze the rearrangement of the atoms in pentose phos-

[3] The term "HMP shunt" had originally a useful purpose in studies with yeast and mammalian tissue, where the major pathway was considered to be the EMP system. Hence the introduction of the word *shunt* to describe the diversion of glucose-6-phosphate into another dissimilation track. However, since in some heterotrophic bacteria the so-called "HMP shunt" is now recognized to be the major pathway, the EMP processes being of lesser importance or altogether negligible, the term *shunt* can be misleading.

phate leading to a resynthesis of glucose-6-phosphate (Gibbs, 1959). This pathway contains two oxidation steps:

glucopyranose-6-phosphate + $TPN^+ \rightleftharpoons$ 6-phosphoglucono-δ-lactone + TPNH + H^+

6-phosphoglucono-δ-lactone + $H_2O \rightleftharpoons$ 6-phosphogluconic acid

6-phosphogluconic acid + $TPN^+ \rightleftharpoons$ ribulose-5-phosphate + CO_2 + TPNH + H^+

This is followed by a series of reactions which can proceed anaerobically, and which represent isomerization and group-transfer reactions of the transaldolase and transketolase type:

ribose-5-phosphate + xylulose-5-phosphate
$\rightleftharpoons$ sedoheptulose-7-phosphate + glyceraldehyde-3-phosphate

sedoheptulose-7-phosphate + glyceraldehyde-3-phosphate
$\rightleftharpoons$ fructose-6-phosphate + erythrose-4-phosphate

The glyceraldehyde-3-phosphate may be converted to pyruvic acid, or it may be condensed to glucose-6-phosphate and re-enter the cycle, which ultimately results in a complete combustion of the hexose chain.

The following statements can be made concerning respiratory metabolism in algae: (*1*) The enzymes of both the HMP pathway and the pentose-phosphate pathway have been detected. (*2*) There is some indirect evidence for the operation of the HMP pathway *in vivo* under aerobic conditions. (*3*) There is *no* evidence for an anaerobic HMP pathway. (*4*) There is *no* evidence for the operation *in vivo* of the complete pentose-phosphate cycle, by which 1 mole of glucose might be completely oxidized to 6 moles of carbon dioxide. Some details on these four points are given below.

1. Enzymes

These are discussed elsewhere in this treatise by Jacobi (see Chapter 7).

2. Evidence for an Aerobic Pathway

As noted above, this evidence is indirect. Richter (1959) detected in extracts of *Anacystis nidulans** both glucose-6-phosphate dehydrogenase and 6-phosphogluconate dehydrogenase, but not fructose-1,6-diphosphate aldolase. Since this aldolase is the enzyme responsible for cleavage of the hexose chain in the EMP pathway, its absence suggests that carbohydrate breakdown must occur by the HMP pathway, presumably by an exclusively aerobic system since this alga is incapable of breaking down glucose anaerobically. *Anacystis* thus occupies a unique position. All other aerobic organisms investigated, unless they contain fructose-1,6-diphosphate aldolase, cleave the hexose derivative 6-phosphogluconate by the Entner-

* Regarding the taxonomy of this organism, see Silva, this treatise, Appendix A, Note 2.

Doudoroff system (Entner and Doudoroff, 1952), whereas the HMP enzymes catalyze its oxidative decarboxylation. Though the Entner-Doudoroff pathway has been detected in heterotrophic and autotrophic bacteria (Szymona and Doudoroff, 1960), Richter did not specify whether this 6-phosphogluconate cleaving system was also present in extracts of *Anacystis*.

The absence of fructose-1,6-diphosphate aldolase is somewhat surprising in view of the presumed importance of this enzyme in the photosynthetic fixation of carbon dioxide. There is no evidence for the fixation of carbon dioxide into hexose by the HMP pathway *in vivo*, although all steps appear to be reversible. In fact, all data so far obtained during short-time photosynthesis experiments indicate that the HMP pathway is exclusively degradative.

3. Quantitative Importance of the HMP Pathway

Many experiments designed to elucidate the pathways of glucose utilization and to determine their relative importance in higher plant tissue have been reviewed by Gibbs (1959). Apparently no papers have dealt specifically with this part of algal respiration.

4. Operation of the Pentose-Phosphate Cycle

Although all of the enzymes of this cycle have been detected in extracts of algae, and although Muntz and Murphy (1957) have provided convincing evidence for the importance of this cycle in animal respiration, we have no direct proof that this cycle for glucose oxidation functions in the living algal cells.

D. The Hexose-Pentose Cycle

In contrast to the pentose-phosphate pathway described above, a "glucuronate-xylulose" cycle, characterized by reactions involving carbon atom 6 of the hexose chain and by non-phosphorylated intermediates, has been identified in certain animal and plant systems (Gibbs, 1959), though not as yet in any alga. However, the report that several Red Algae (*Delesseria sanguinea*, *Phycodrys rubens*, *Polysiphonia violacea*, and *Cystoclonium purpurascens*) apparently lack glucose-6-phosphate dehydrogenase and enzymes of the EMP pathway (Jacobi, 1958) invites speculation on this point.

The oxidation of hexoses prior to phosphorylation to yield the corresponding acids, previously thought to be present only in heterotrophic tissue, has been reported in one of the Red Algae, *Iridophycus flaccidum* (Bean and Hassid, 1956). The enzyme concerned differs from other oxidases of this type in that it catalyzes the oxidation of D-glucose and D-galactose but not D-mannose, indicating a higher specificity with regard to the configuration about C-2 than about C-4. The significance of this direct oxidation process remains to be demonstrated, since the fate of the acids is unknown. One can speculate that D-gluconic acid probably enters the HMP pathway through phosphorylation to form D-6-phosphogluconic acid. It is also possible that D-glucose may be converted to D-glucuronic acid and so enter the glucuronate-xylulose pathway.

The fate of D-galactonic acid or D-galactono-γ-lactone, the presumed product of D-galactose oxidation, will be dealt with in the following section (IV.E).

E. Possible Pathways for the Metabolism of Galactose and Other Hexoses

The interesting reports by Galloway and Krauss (1959a, b) concerning the effects of an antibiotic, polymyxin B, on *Chlorella pyrenoidosa* and *Scenedesmus obliquus* suggest that galactose may be respired by a pathway unlike that of its epimer glucose. This was indicated by the observation that whereas polymyxin B inhibited glucose respiration, the rate of galactose respiration remained unchanged. Although work with animal and higher plant tissues indicates that D-galactose is first converted to D-glucose via D-galactose-1-phosphate and uridine diphosphate derivatives, Galloway and Krauss provided evidence that the initial step in a hitherto unknown pathway is the formation of D-galactose-6-phosphate. The isomerization of the aldose sugar would lead to the formation of D-tagatose-6-phosphate, which can then be further phosphorylated with adenosine triphosphate, uridine triphosphate, or inosine triphosphate, in the presence of rabbit-muscle phosphofructokinase, apparently to yield D-tagatose-1,6-diphosphate (Ling and Byrne, 1954). Tung *et al.* (1954), who reinvestigated the specificity of fructose-1,6-diphosphate aldolase, showed that its activity is not restricted to compounds with a *trans* configuration at C-3 and C-4 (e.g., fructose) but that it will also cleave D-tagatose-1,6-diphosphate between C-3 and C-4, which bear *cis*-hydroxyls. This cleavage to yield D-glyceraldehyde-3-phosphate and dihydroxyacetone phosphate would allow re-entry of the galactose chain into the EMP pathway.

Since in the Red Algae the major reserve products are compounds of glycerol and D-galactose (see Meeuse, Chapter 18, this treatise), and since the enzymes of the EMP pathway are apparently absent from Red Algae (Jacobi, 1957), the postulated galactose respiratory mechanism involving the cleavage of D-tagatose-1,6-diphosphate may not be present here.

Another set of reactions must be postulated for the synthesis of D-glyceric acid from D-galactose:

(*1*) D-galactose→D-tagatose

(*2*) D-tagatose + ATP→D-tagatose-1-phosphate + ADP

(*3*) D-tagatose-1-phosphate→D-glyceraldehyde + dihydroxyacetone phosphate

(*4*) D-glyceraldehyde + DPN^+→D-glyceric acid + DPNH + H^+

No enzyme is known which catalyzes reactions *1* or *2*. Reactions *3* and *4* could respectively be catalyzed by fructose-1-phosphate aldolase and D-glyceraldehyde dehydrogenase (Lamprecht and Heinz, 1958).

Another respiratory mechanism for galactose oxidation might be initiated by the oxidase in *Iridophycus flaccidum* described by Bean and Hassid (1956). The corresponding lactone or uronic acid could be decarboxylated to L-arabinose and then converted directly to α-ketoglutaric acid in the tricarboxylic-acid cycle (Weimberg and Doudoroff, 1955). This postulated mechanism would by-pass both the HMP pathway and the EMP pathway.

F. The Tricarboxylic-Acid Cycle

1. *Occurrence*

Proof of the existence in algae of this cycle, which is responsible for the oxidation of pyruvic acid to carbon dioxide and water, rests largely upon data from growth and respirometric studies. Such experiments present evidence for the classical "Krebs cycle" in algae; they do not indicate whether the cycle plays a major role as a terminal oxidative mechanism in the intact cell, and we still need a critical evaluation of the extent to which algal metabolism proceeds by this pathway. Moreover, it is not yet known whether the individual enzymes are associated and operate as a unit tightly bound to the mitochondria in algae, as is the case in tissues of higher plants and other organisms.

2. *Growth Studies*

These investigations are reviewed by Danforth (see Chapter 6 in this treatise). The only generalization that can be made at present is that, among the acids related to the tricarboxylic-acid cycle, acetic acid is the most widely utilized substrate. It is interesting to note that some algae, notably species of *Chlamydomonas*, though they can oxidize "Krebs-cycle" acids, fail to grow on any of these substrates in the dark (Lewin, 1953). (This situation may be similar to that discussed elsewhere in this treatise under the subject of anerobic growth; see Gibbs, Chapter 5.)

3. *Manometry Studies*

In the course of unsuccessful attempts to obtain cell-free preparations capable of respiratory activity, Millbank (1957) prepared vacuum-dried *Chlorella vulgaris* cells which could oxidize acetate, pyruvate, α-keto-glutarate, succinate, fumarate, and L-malate, at rates of oxygen consumption approaching that obtained with glucose. Malonate, which could presumably penetrate the disorganized cell membranes, did not inhibit respiration even at concentrations of $10^{-2}M$. On the basis of a similar approach, Danforth (1953) concluded that *Euglena gracilis* var. *bacillaris* carried out oxidative metabolism *via* the tricarboxylic acid cycle. In contrast to Millbank's experience, he found that malonate inhibited the oxidation of acetate or of succinate. Merrett and Syrett (1960) likewise reported that in *Chlorella vulgaris* $10^{-2}M$ malonate produced an 87% inhibition of acetate oxidation but only a 26% inhibition of glucose oxidation.

4. *Tracer Evidence*

According to the classical tricarboxylic-acid scheme, the two carbon atoms of acetate are not equivalent in the rate at which they appear as carbon dioxide. Fujita (1959), who investigated the metabolism of acetate-1-C^{14} and acetate-2-C^{14} in *Chlorella ellipsoidea,* confirmed that the carboxyl carbon predominates over the methyl carbon in respiratory CO_2.

Milhaud *et al.* (1956) found that the respiration of pyruvate-2-C^{14} by *Scenedesmus* (Gaffron's strain D-3) in darkness gave rise to labeled acids of the tricarboxylic acid cycle. More direct evidence for the role of this cycle in the oxidative metabolism of algae was obtained from experiments in which labeled substrates were "fed" to the cells. Thus Schou *et al.* (1950) isolated 3-phosphoglyceric acid from *Scenedesmus* (strain D-3) which had been incubated in the dark with glycolic acid-2-C^{14}, and found a distribution of the isotope among the three carbon atoms which is best explained by the entry of glycolic acid into the tricarboxylic-acid cycle prior to its conversion to 3-phosphoglyceric acid. Gibbs (1957) and Schlegel (1959) independently supplied acetate-1-C^{14} and acetate-2-C^{14} to *Chlorella pyrenoidosa,* and on the basis of isotope distribution in the glucose of a polymer (presumably starch) concluded that acetate is first converted by this cycle to oxalacetate and thence to pyruvate and carbohydrate.

The conversion of pyruvate to carbohydrate by the reverse of the EMP pathway may be limited by the energy barrier presented by the conversion of pyruvate to phosphoenolpyruvate. To by-pass this barrier, Krebs and Kornberg (1957) postulated the action of phosphoenolpyruvate carboxykinase, which catalyzes the conversion of oxalacetate to phosphoenolpyruvate with only a small change in free energy; however, isotope labeling patterns do not distinguish between these two alternatives.

5. Other Evidence

Fluoroacetate indirectly "jams" the TCA cycle, since it can substitute for acetate in the "condensing enzyme" reaction to produce a fluorinated tricarboxylic acid, fluorocitrate, which is not subject to the action of aconitase. It is an inhibitor of microbial growth at a concentration of 9.5×10^{-4} M (Jeney and Zsolnai, 1957). Merrett and Syrett (1960) observed that in *Chlorella vulgaris* 0.1 M fluoroacetate reduced the glucose oxidation by 50% and completely inhibited acetate oxidation.

The inhibition of glucose and acetate oxidation by arsenite indicates that pyruvate oxidation and the α-ketoglutarate oxidase system may involve the dithiol lipoic acid, since arsenite is known to combine with lipoic acid, thereby preventing its oxidation and reduction and blocking the oxidative decarboxylation of the α-ketonic acids.

The first quantitative data on acids associated with the tricarboxylic-acid cycle in algae were provided by Jacobi (1958). In young thalli of *Laminaria saccharina* and *L. digitata*, the pyruvic acid concentration was found to be about 10 to 15×10^{-8} moles per gram fresh weight, whereas in old thalli the concentration had fallen to between 2 and 8×10^{-8} moles per gram, respectively. In *Laminaria saccharina* the α-ketoglutaric acid concentration ranged from about 1 to 3×10^{-8} moles per gram fresh weight. These values are only about one hundredth of those reported in baker's yeast, *Saccharomyces*.

G. The "Glyoxylate Cycle"

It has long been known that many flagellates and other unicellular algae can grow with acetate as the sole source of carbon. Since each mole of acetate in the course of one revolution of the tricarboxylic-acid cycle is completely oxidized to 2 moles of carbon dioxide, the plant cannot sustain growth by this pathway alone. However, various investigators have revealed another metabolic pathway, the "glyoxylate cycle," by which acetate can be partially utilized for growth, since by this means 1 mole of a C_4 dicarboxylic acid can be synthesized from 2 moles of acetate, using energy derived from the dissimilation of other acetate molecules.

By enzyme and tracer experiments, Kornberg and Beevers (1957) established that in castor-bean seedlings the glyoxylate cycle can account for the conversion of acetate (in the form of acetyl CoA) to carbohydrate. Although we still have no proof of the existence in acetate-grown algae of the two key enzymes of this cycle, isocitritase and malate synthetase, their presence is to be expected.

H. Electron Transport Pathways

1. Inhibitor Studies

Data from experiments with metabolic inhibitors suggest that in algae the main pathway for the transfer of electrons from reduced pyridine nucleotides to oxygen may differ somewhat from that observed in animal cells.

The effect of cyanide on the respiration of plant cells varies considerably with the organism investigated. $4.5 \times 10^{-4}\ M$ suffices to decrease the rate of respiration of bakers' yeast to 50% of its normal value. The oxygen uptake by many plant tissues is reduced to approximately 20–40% by $10^{-3}M$ cyanide (James, 1953). In contrast, Emerson (1927b) and Genevois (1927) observed that $10^{-4}M$ cyanide *increased* the basal respiratory rate of *Chlorella* cells, though the increased respiration caused by added sugars was inhibited to 30–40% of the control value by the same concentration of poison. From these studies and those of Syrett (1951) it is clear that there exists in *Chlorella* spp. a cyanide-stable respiration which is only inhibited by concentrations in the order of $10^{-2}M$. In fact, Syrett showed that in *Chlorella vulgaris* the rate of oxygen consumption through this cyanide-stable system alone may even reach 30–60% of the total normal rate. Such a highly cyanide-resistant respiration is not common to all algae, however, as is shown by the observation of Webster and Frenkel (1953) that $10^{-4}M$ cyanide reduced both the endogenous and the glucose-supported respiration of *Anabaena variabilis* to 30–40%, while the respiration of another species of Blue-green Algae, *Oscillatoria simplicissima*, was reduced by $10^{-4}M$ hydrogen cyanide to about 20% (Emerson, 1927a).

The inhibition of respiration by cyanide is generally considered to be indicative of the presence of an enzyme system containing a heavy metal. Since this is presumably a cytochrome system, a confirmatory test is for a light-reversible inhibition by CO. In most experiments with this inhibitor, the preparation is exposed to a gas mixture containing CO/O_2 in the ratios of 9/1 to 19/1, and the O_2 uptake is compared with that in mixtures with similar ratios of N_2/O_2. All algae so far tested have demonstrated a high resistance to this poison. The endogenous respiration of a species of *Oscillatoria* was inhibited only 12% by a mixture of 33 CO/1 O_2, while that of *Chlorella pyrenoidosa* was hardly affected at all by a mixture of 19 CO/1 O_2 (Emerson, 1927a). Webster and Frenkel (1953) noted that the basal respiration of *Anabaena variabilis* was likewise relatively insensitive to CO. Glucose-supported respiration was more sensitive than endogenous respiration to CO. *Chlorella pyrenoidosa* respiration was reduced to 83% by a 19/1 mixture of CO/O_2, while in a 14/1 mixture *Anabaena* respiration fell to 58% (Emerson, 1927a).

Such data obtained with inhibitors suggest that part of the respiration of algae is mediated either by a cytochrome system insensitive to carbon monoxide and cyanide or by a flavoprotein with a high affinity for O_2. In the case of *Chlorella pyrenoidosa* the latter possibility may be ruled out by the observation that aerobic respiration continues unimpaired even at a partial pressure of oxygen as low as 5 mm. mercury (Tang and French, 1933).

2. *Spectroscopic Studies*

In cells containing chlorophyll, direct spectroscopic studies of the respiratory pigments, comparable to those of Keilin for nonphotosynthetic cells and tissues, have been limited because the strong absorption bands of the photosynthetic pigments obscure those of the respiratory enzymes. Chance and Sager (1957), applying elegant spectroscopic techniques developed by Chance to a pale-green mutant of *Chlamydomonas reinhardtii*, observed that a pyridine nucleotide, flavoproteins, and cytochromes *b* and *c* are present in amounts typical of other microorganisms. In "difference spectra" the reduced pyridine nucleotide showed a symmetrical absorption band centered at about 345 mμ, while the flavoprotein was observed as an absorption minimum at 470 mμ. Cytochrome *b* was represented by a distinctive α band at 563 mμ and by its corresponding Soret band at 430 to 431 mμ; a cytochrome of the *c* type, characterized by a distinct α band at 551 to 552 mμ, was also detected. In agreement with the CO data for other organisms, discussed previously, these investigators did not detect an absorption band in the Soret region at 445 mμ, which would be characteristic of cytochrome a_3. The α band of cytochrome *a*, usually located at about 605 mμ, was not detected either. Chance and Sager concluded that the concentration of cytochrome a_3 is so low, relative to the amount of cytochrome *c*, that it is doubtful whether this component could function as the terminal oxidase of the pale-green mutant. On the addition of CO to a suspension of these cells, no new major absorption band appeared, indicating the absence of different CO-binding pigments of the type found in many other microorganisms.

A statement dealing with the concentration of pyridine nucleotides is pertinent here. Yeast has a relatively high concentration of DPN, about 1 mg. per gram of fresh weight, whereas leaves of plants such as spinach have a concentration of DPN in the order of 7 to 10 μg. per gram fresh weight (Anderson and Vennesland, 1954). According to Oh-hama and Miyachi (1960), the concentration of DPN in *Chlorella ellipsoidea* is 0.5 to 1 μg. per gram dry weight, equivalent to about 0.05 to 0.1 μg. per gram fresh weight; the TPN concentration is in the same range.

3. *Enzyme Studies*

Several investigators have demonstrated in algal extracts the activity of enzymes apparently associated with the respiratory chain (see Jacobi, Chapter 7, this treatise). The first isolations of pyridine-nucleotide cytochrome-reductase were reported by Katoh (1959a). After triturating the thalli of the Red Algal species *Porphyra tenera,* and centrifuging the homogenate at 24,000 *g* for 30 minutes, he obtained an extract with which he could demonstrate the reduction of *Porphyra* cytochrome 553 (see below) by reduced pyridine nucleotides. The activity of the DPNH-cytochrome reductase was indicated by 146 millimoles of cytochrome 553 reduced in 3 minutes, while the corresponding value for the TPNH-cytochrome reductase was 380. Further purification of such preparations will be needed in order to determine whether the "reductase" is a single enzyme or comprises two separate enzymes. The nature of the prosthetic group of these proteins has not been reported. The *Porphyra* extracts obtained in this way also contained TPNH and DPNH diaphorases.

Katoh (1960b) also demonstrated a TPNH- and DPNH-cytochrome *c* reductase in extracts of *Chlorella ellipsoidea,* using mammalian cytochrome *c* as the oxidant of the reduced pyridine nucleotide. The absorption spectrum of the reductase in its oxidized form shows three maxima at 275, 380, and 465 mμ, suggesting that the enzyme is a flavoprotein, though the prosthetic group was not identified. In addition, this reductase could catalyze the reduction of a copper protein, of unknown function, recently isolated from *C. ellipsoidea* in the same laboratory. The normal potential (E_0') of this copper protein was found to be 0.39 volts at 20°. It is not autoxidizable; when oxidized, it can also be reduced by mammalian cytochrome *c* and *C. ellipsoidea* cytochrome 550. In its oxidized form, the copper protein has a broad absorption band between 500 and 700 mμ, with a maximum at 597 mμ, and is further characterized by a sharp absorption maximum at 278 mμ due to the protein. The role of this protein in the respiratory chain remains to be established.

4. *Isolation of Cytochromes*

Of all the cytochromes so far identified in algae, the one about which most is known is of the cytochrome *c* type. Like its counterpart in mammalian and higher-plant tissue, it is readily extractable with water and is quite stable *in vitro.*

Yakushiji *et al.* (1960) prepared in crystalline form a cytochrome *c* from *Porphyra tenera,* characterized by an α-band at 553 mμ, a β-band at 521 mμ, and an absorption maximum in the Soret region at 416 mμ. The reduced enzyme is not autoxidizable. The purified preparation could be oxidized by cytochrome a_2 from the bacterium *Pseudomonas aeruginosa*, but not by mammalian cytochrome oxidase. Another interesting characteristic is the acidic nature of the protein, as shown by a low isoelectric point of 3.5.

Katoh (1960a) likewise prepared from *Porphyra tenera* a cytochrome with an absorption maximum at 553 mμ and a spectrum in close agreement with that published by Yakushiji *et al.* (1960). He obtained 200 mg. of the crystalline material from 15 kg. of frozen thalli. Whereas the E_0' of mammalian, plant, and yeast cytochrome *c* is about 0.26 volts, the E_0' for cytochrome 553 of *P. tenera* is 0.335 volts, almost identical with that of cytochrome 550 from the bacterium *Rhodospirillum rubrum* ($E_0' = 0.338$ volts). The iron content of cytochrome 553 was found to be 0.51%, from which, assuming one iron atom per mole of protein, one can calculate a molecular weight of 11,000. Other *c*-type cytochromes, detected by Katoh (1959b) in *Grateloupia* sp. and *Gelidium amansii* (Rhodophyta), *Undaria pinnatifida* (Phaeophyta), *Ulva* sp. and *Monostroma nitidum* (Chlorophyta), and *Tolypothrix tenuis* (Cyanophyta), exhibit a close similarity, with practically identical values for their normal potentials ($E_0' = 0.30$ to 0.34 volts) and absorption maxima (α-band, 552 to 553 mμ). Katoh reported that the crude autolyzate of the *Monostroma* contained a *b*-type cytochrome, but he did not characterize this further.

An interesting feature is the apparent inability of particulate fractions from *Porphyra tenera* to catalyze the oxidation of reduced cytochrome 553 in the dark at a sufficiently high rate to account for the full respiratory activity of this alga. Whereas the soluble portions of the extract reduced 380 millimoles of cytochrome 553 in 3 minutes, particulate fractions were only one-hundredth as active, and evidently could not account for more than a few per cent of the oxygen uptake measured manometrically in whole *P. tenera* preparations. Although the oxidation of reduced cytochrome 553 proved sensitive to cyanide, spectrophotometric examination of the particulate fraction provided no evidence for the presence of an *a*-type cytochrome. Though no physiological function can yet be ascribed to cytochrome 553, this heme protein may perhaps participate in certain electron transfers in photosynthesis.

Studies with inhibitors, spectrophotometric observations, and chemical isolation techniques provide some evidence that, with the exception of the terminal oxidase, the components of the respiratory chain in the algae investigated resemble those isolated from mitochondria, which are capable of electron transport coupled to the synthesis of adenosine triphosphate. However, there has been no direct demonstration of oxidative phosphorylation by a particulate fraction of any alga. Evidently the principal respira tory chain is as follows:

substrate→pyridine nucleotide→flavoprotein

→cytochrome *b*→cytochrome *c*→?→oxygen.

Investigations of particulate fractions from the spadix of the skunk cabbage, *Symplocarpus*, revealed an autoxidizable, cyanide-resistant, and CO-insensitive cytochrome of the *b* type, cytochrome b_7 (Chance and Hackett, 1959). It would be of interest to seek a cytochrome of this sort among the algae by the use of similar techniques.

We should now attempt to devise methods for separating those particles capable of carrying out oxidative phosphorylation from those which carry out light-induced reactions. It may be that the cytochromes so far isolated participate only in light-induced reactions, and not in so-called "dark" reactions; but further fractionation studies are needed if we are to resolve some of these possibilities.

5. *Terminal Oxidases Other than Cytochrome Oxidase*

Webster and Frenkel (1953) reported the absence of polyphenol oxidase in extracts of *Anabaena variabilis*; their evidence for an ascorbic acid oxidase is equivocal. A detailed study of an algal terminal oxidase containing copper would be of interest, since polyphenol oxidase from other organisms has been reported to be sensitive to cyanide but fairly resistant to carbon monoxide.

V. Special Aspects of Algal Respiration

A. The Effects of Added Substrates on Respiration

When certain substances are added to cells or tissues, the rate of oxygen uptake rises. In some cases, this may be a chemical effect, due to catalyzed oxidation of the additive, as occurs with ferrous salts or H_2S. In others there is a nonspecific stimulation of respiration, as may occur after addition of small amounts of a metabolic inhibitor (e.g., arsenate, cyanide) or of a surface-active agent. Poisons like dinitrophenol and azide may cause a marked increase in the apparent rate of respiration by uncoupling oxidative phosphorylation. All of these possibilities must be considered if one is to distinguish metabolic substrates by their effects on oxygen uptake. A useful indication that this is the case comes from the stoichiometry of the reaction. Addition of a small amount of a substrate should cause a temporary increase in the respiratory rate, followed by a return to about the endogenous level of respiration. If the effect is only transient, or if the total oxygen uptake exceeds the amount which would be required by the total oxidation of the added substance, then the status of the latter as a respiratory "substrate" must be questioned. This is too rarely done; the data of Watanabe

(1937) on the effects of various acids, alcohols, etc., upon the respiration of *Chlorella ellipsoidea* and of a number of seaweeds provide a dubious case in point.

In manometry studies and in metabolic balance experiments, the question then arises whether, after the addition of a substrate, the endogenous respiratory substrate continues to be converted to carbon dioxide at an unchanged rate, or whether this is somehow suppressed by the exogenous substrate, which then serves as the exclusive source of respiratory energy and carbon dioxide. Two approaches have been used to answer this question. "Indirect" manometric techniques, first described by Barker (1936) with *Prototheca zopfii*, and later used by Taylor (1950) with *Scenedesmus quadricauda* and by Myers (1947) and Syrett (1951) with species of *Chlorella*, in many cases yielded evidence that the endogenous respiration is suppressed by the addition of external substrates. On the other hand, in studies of acetate dissimilation in *Chlamydomonas dysosmos*, stoichiometric considerations indicated that this could not be so (Lewin, 1954).

The so-called "direct" method consists of labeling cells with C^{14} or C^{13} and then following the total amount of isotope in the carbon dioxide produced by respiration in the presence or absence of an unlabeled exogenous substrate. The following investigators have used this method: (*1*) Pirson *et al.* (1955) incubated a uniformly C^{13}-labeled culture of *Chlorella vulgaris* with unlabeled glucose for 6 hours, which resulted in a 30% increase in the evolution of the heavy isotope as respiratory carbon dioxide. (*2*) Kandler (1958a), using C^{14}-labeled *C. pyrenoidosa* cells in a similar technique, found that after the addition of C^{12}-glucose the production of $C^{14}O_2$ in the first few hours increased 4-fold, though after 4 to 5 hours, the observed stimulation of endogenous respiration diminished to approximately 30%. The effects of acetate were somewhat similar. Kandler concluded that glucose stimulated phosphate turnover, which normally limited endogenous respiration. (*3*) Moses and Syrett (1955) likewise labeled *C. vulgaris* with C^{14}, but obtained only a 30% increase of $C^{14}O_2$ output when glucose was supplied to the cells. This result was essentially similar to that obtained by Gibbs and Wood (1952) in experiments with the bacterium *Pseudomonas fluorescens*.

Although most of the "indirect" studies with algae suggested a suppression of endogenous respiration, if one assumes that the "direct" method gives an unequivocal answer, one must conclude, with Kandler and Pirson, that the endogenous respiration is either unaffected or is somewhat stimulated in the presence of a respiratory substrate. However, Moses and Syrett correctly reasoned that interpretations of the isotopic method may be open to doubt, and that though the results obtained so far appear to indicate that endogenous respiration is *not* suppressed by the presence of an exogenous substrate, other interpretations of the data are possible. The effect of an external oxidizable substrate on the endogenous respiration is thus still unresolved. It is not unlikely that the situation may differ from one species to another, or, within one species, according to the physiological state of the cells.

B. Pathway of Endogenous Respiration

There are many published reports that in algae the course of endogenous respiration appears to differ from that of respiration in the presence of an external substrate. In this section, an attempt will be made to assess the basis for this conclusion.

Since the R.Q. of the endogenous respiration in most algae is close to 1, its substrate is most probably a carbohydrate. Although it is not established that the respiratory pathways of an internal and of an external substrate are identical, it is fair to assume that they are similar, and pass through the EMP, the pentose-phosphate and the tricarboxylic-acid cycles.

On the other hand, the strongest evidence for partly separate respiratory pathways is based on inhibitor studies with cyanide, which indicate that the course of electron flow from reduced pyridine nucleotide to oxygen may change according to the substrate. As noted above, in many algae part of the respiration is cyanide-resistant. The endogenous respiration may be stimulated by low concentrations of cyanide, whereas the exogenous respiration is not. From experiments with higher-plant mitochondria, Hackett *et al.* (1960) concluded that when the normal electron-transport pathway is blocked by inhibitors, electrons may be diverted to an alternative non-phosphorylating respiratory pathway. This shunting of the electrons would result in an "uncoupling" of phosphorylation, and might increase the rate of respiration by increasing the availability of ADP. Such a hypothesis could also account for the observation that glucose assimilation is not coupled to the cyanide-resistant respiration (Syrett, 1951). Thus, the phosphorylation of one glucose molecule and its condensation into a polysaccharide takes place at the expense of energy indirectly set free by hydrolysis of 1 molecule of ATP. The combustion of 1 mole of glucose by 6 moles of oxygen potentially enriches the cell by 690 kcal., of which 410 are stored in the form of ATP. (The $\Delta F'$ for the hydrolysis of ATP is taken as 11 kcal.) By the cytochrome-b_7 oxidation shunt, the cell is provided with only about half as much potential energy in the form of ATP, which might all be used in maintaining the metabolic status of the cell, leaving none for polysaccharide formation.

Whereas cyanide may cause a shunting of electrons through the cytochrome-b_7 oxidation route, there is no evidence that this occurs under normal circumstances. Since the rate of respiration of *Chlorella vulgaris* in media containing cyanide plus glucose was never lower than the endogenous rate with cyanide alone (Syrett, 1951), it appears that in the presence of cyanide the respiratory pathways of both glucose and of endogenous substrates may share a common step. However, the data are not yet adequate to establish this. Cyanide may stimulate oxygen consumption at the substrate level, since it produces a considerable increase (89%) in the activity of glucose-6-phosphate dehydrogenase and some increase (up to 31%) of 6-phosphogluconate dehydrogenase activity, probably as a result of cyanohydrin formation (Dickens and Glock, 1951). The interpretation of these observations *in vitro* could be tested by the C_6/C_1 technique (see Section IV.B.4).

C. Relationships to Photosynthesis

According to current biochemical information, photosynthesis takes place in the chloroplasts; the EMP, the HMP, and related enzymes are located in the cytoplasm; and the tricarboxylic-acid cycle together with its associated electron-transport systems are situated in the mitochondria. Since the starting products of photosynthesis and the end products of respiration are similar, the two processes almost certainly interact. Since the end products of photosynthesis provide substrates for respiration, there must be some interchange on this level, too. Other components common to both of these important processes are inorganic phosphate, the adenylates, pyridine nucleotides, and compounds like 3-phosphoglyceric acid and triose phosphate.

Two different approaches have been employed to attack the problem of such interactions. In one, the CO_2 uptake and O_2 output from photosynthesis are distinguished from their respiratory counterparts by ingeniously designed tracer techniques and the use of the mass spectrometer, as employed by Brown and his collaborators. It should be pointed out, however, that in the gas path the mass spectrometer can assay only over-all isotope changes outside the cells, and therefore cannot detect interplay inside the cells where the reactions take place. The second method is to supply labeled compounds to algal cells and then to follow the effect of light on the isotopic composition of released carbon dioxide and of intermediates in the respiratory pathways. This latter technique has the advantage of possibly providing evidence on the traffic between the two vital processes.

In this section, a distinction will be made between oxygen uptake and respiration. The consumption of oxygen by an illuminated alga does not necessarily involve the same carbon pathway enzymes and electron transport system as those which control the release of carbon dioxide. This is an important point, since our definition of respiration is a statement including the status of both gases.

Early experiments by Brown *et al.* (1953) showed that at moderate and low light intensities the oxygen uptake of *Chlorella pyrenoidosa* was unaffected by light, whereas that of other algae, such as *Porphyra umbilicalis* and *Ulva lactuca*, was decreased, especially at high light intensities. Conversely, in *Anabaena variabilis* the oxygen consumption increased at higher light intensities (Brown and Webster, 1953).

In more recent studies by Brown and Weis (1959) on the relation between respiration and photosynthesis in the chlorophyte *Ankistrodesmus braunii* and in the chrysophyte flagellate *Ochromonas malhamensis* (Weis and Brown, 1959), the rates of production and consumption of both carbon dioxide and oxygen were determined simultaneously, in contrast to earlier experiments. With both organisms they made the following general observations: (*1*) at light intensities below the compensation point, illumination had little or no influence on the rate of oxygen consumption; (*2*) higher intensities increased the consumption of oxygen; (*3*) light inhibited the production of carbon dioxide, an effect essentially independent of light intensity between 100 and 1500 lux. The oxygen uptake of unstarved *Ochromonas* cells was unaffected by light, whereas that of starved cells supplied with glucose was nearly doubled by saturating light intensities. Brown and Weis interpreted their data as indicating an interaction between a photosynthetic reductant and respiratory intermediates; they explained the difference in the effect of light on the two gases by postulating that such a reductant could react with two different components of the electron-transport mechanism, with a rate-limiting step between the two sites of reaction. Since the discovery of a pyridine nucleotide transhydrogenase $(TPNH + DPN \rightleftharpoons TPN + DPNH)$ in chlorophyll-containing cells, it now seems possible that illumination may promote the formation of DPNH by way of a photosynthetic pyridine-nucleotide reductase. If this conversion were essentially complete even at low light intensities, it might halt the breakdown of respiratory substrates and thereby result in a decreased output of carbon dioxide. This argument also suggests that the products of the photochemical act (ATP and TPNH) migrate from the chloroplast and are consumed in reactions elsewhere. In agreement with this speculation is the discovery that photophosphorylation induces an increase in glucose uptake by *Chlorella pyrenoidosa* (Kandler, 1950), which perhaps indicates a movement of adenosine triphosphate from the photochemical apparatus outward through the cytoplasm toward the cell membrane.

Experiments of Ryther with the marine flagellate *Dunaliella euchlora* should be mentioned here (Ryther, 1956). Samples of cells uniformly labeled with C^{14} were suspended in a medium containing $C^{12}O_2$ and kept in light or darkness. After 24 hours, the cells in the dark had lost about 20% of their isotope, whereas those in the light had lost none. Correspondingly, experiments with $C^{14}O_2$ showed that at or below the compensation level of illumination the cells took up no exogenous CO_2 at all. The results were interpreted as indicating that respiratory carbon dioxide was used for photosynthesis in preference to external carbon dioxide. Stated in another way, the reassimilation of respired CO_2 was rapid enough during illumination to block its appearance in the medium. However, on the basis of Brown's findings, one could alternatively interpret the lack of carbon dioxide output as evidence for an inhibition of substrate oxidation during photosynthesis.

From similar experiments with *Chlorella pyrenoidosa*, Steemann Nielsen (1955a) concluded that the suppression of CO_2 evolution by light is attributable to an interaction between photosynthesis and respiration. Although at 7000 lux he obtained similar values for the photosynthetic rate by measuring either the uptake of $C^{14}O_2$ or the evolution of oxygen, at 300 lux, which was just below the compensation point, the oxygen method gave a value more than three times that calculated by the C^{14} method.

These data may not be entirely accounted for by a reincorporation of respiratory carbon dioxide by the photosynthetic apparatus. If we agree with Arnon and his colleagues that "assimilatory power" is generated as adenosine triphosphate and reduced pyridine nucleotide, the possibility exists that the reduction of pyridine nucleotide and the evolution of oxygen may be directly correlated with light intensity, whereas the formation of adenosine triphosphate may be brought about only at higher light intensities (cf. Krogmann, 1960). In this way, the $C^{14}O_2$ uptake would be lower than the concomitant oxygen evolution, providing an explanation for the results obtained by Ryther and by Steemann Nielsen.

Calvin and Massini (1952) made the interesting observation that light apparently blocks the entry of photosynthetic intermediates into the tricarboxylic-acid cycle. They attributed this to a maintenance of α-lipoic acid in the reduced state, which thereby suppressed the oxidation of pyruvate. By similar methods Gibbs (1953) confirmed these results. However, it was later shown (Milhaud *et al.*, 1956) that pyruvate-2-C^{14} entered the tricarboxylic-acid cycle as rapidly in the light as in the dark, inducing the Calvin group to modify their hypothesis. They then concluded that there are two sites of pyruvate oxidation in green plant cells. One, situated in the chloroplast, is blocked by light; the other, in the mitochrondria, can proceed equally well in light and in darkness. An alternative interpretation of these data was suggested by Gibbs (1959).

In this connection Moses *et al.* (1959) reported that *Chlorella pyrenoidosa* produced C^{14}-labeled glutamic acid from glucose-C^{14} or acetate-2-C^{14}, in the light as well as in the dark, but not, in short-term experiments, from $C^{14}O_2$, even though photosynthesis was 20 times as rapid as respiration. They suggested that these results indicated a physical separation of the metabolic pools of chemically identical substrates participating both in photosynthesis and in respiration.

Although the effect of illumination on respiration was not the primary purpose of the report of Merrett and Syrett (1960), it seems pertinent to include a brief discussion of it here, since these workers have made an important contribution to the question of the relationship between glucose oxidation and acetate oxidation in *Chlorella vulgaris*. As already mentioned in this chapter, carbohydrates are believed to be respired by a combination of glycolysis and the tricarboxylic-acid cycle. Merrett and Syrett reasoned that, since glucose and acetate are converted to carbon dioxide *via* acetyl CoA, one should expect some metabolic interplay between the oxidations of the two compounds. However, they found that—although there was evidence for interaction in cells from young and exponentially growing cultures—when glucose was supplied to cells from "older" cultures and acetate was added 30 minutes later, the two substrates appeared to be metabolized independently, which suggests that they may be respired by pathways physically separated within the cells.

The interactions between photosynthesis and respiration are not yet sufficiently understood. In studies of the metabolic activities of isolated subcellular particles, considerable progress has been made in recent years; we should now attempt to combine different components in "reconstituted" systems. Possibly, also, the fine structure of mitochondria, of chloroplasts, and of the cytoplasmic ground substance, as revealed by electron microscopy, may provide further clues to enable the biochemist and the cell physiologist to determine more exactly the locations at which respiration and photosynthesis are carried out, and the manner in which they are integrated.

D. Effects of Ionizing Radiation and High Light Intensity on Respiration

Many workers have used ultraviolet irradiation in attempts to inactivate differentially the photosynthetic and respiratory processes. Their data provide some interesting insights into the effect of ionizing radiation, but their primary objective has not been achieved.

Redford and Myers (1951), who studied the effect of 253.7 mμ photons on endogenous respiration, on acetate- and glucose-stimulated respiration, and on photosynthesis in *Chlorella pyrenoidosa*, found the endogenous respiration to be insensitive to the irradiation, whereas the exogenous respiration was highly sensitive. Photosynthesis was also inhibited, but not nearly to the same extent as exogeneous respiration. In a similar study with *Scenedesmus* (strain D_1), Holt *et al.* (1951) found that the endogenous respiration was relatively insensitive but that even the exogenous respiration was much less affected than in *Chlorella*.

Kandler and Sironval (1959) determined the effects of high light intensity (100,000 lux) upon the metabolism of *Chlorella pyrenoidosa*. During the induction phase of bleaching, photosynthesis was strongly inhibited, respiration of endogenous carbohydrate was accelerated by 250%, exogenous respiration was inhibited, and oxidative assimilation decreased. As further evidence for the inhibition of assimilation, they found that in cells which had been brightly illuminated the R.Q. did not decline appreciably, as it did in the controls where presumably glucose was being converted into lipids and proteins. Pre-illumination also decreased the formation of high-energy phosphate compounds in relation to oxygen uptake (P/O ratio), its effect in this respect being similar to that of the "uncoupling" reagent, dinitrophenol.

If ionizing radiation is similar in this respect to high light intensity and likewise acts as an uncoupling agent in algae, this could account for the fact that exogenous respiration is one of the processes most sensitive to ionizing radiation. This hypothesis is based on the assumption that adenosine triphosphate is needed to activate a respiratory substrate, such as glucose or acetate, while the endogenous substrate (probably a polysaccharide), needs only inorganic phosphate for its conversion to units capable of entering the respiratory pathways. With regard to its radiation sensitivity, photosynthesis occupies a position intermediate between the two forms of respiration, in accord with the observation that concentrations of dinitrophenol which considerably inhibit (i.e., uncouple) oxidative phosphorylation, while apparently stimulating endogenous respiration, do not appreciably affect photosynthesis (Kandler, 1958a).

References

Anderson, D. G., and Vennesland, B. (1954). The occurrence of di- and triphosphopyridine nucleotides in green leaves. *J. Biol. Chem.* **207,** 613–620.

Anderson, E. H. (1945). Nature of the growth factor for the colorless alga *Prototheca zopfii. J. Gen. Physiol.* **28,** 287–327.

Barker, H. A. (1936). The oxidative metabolism of the colorless alga *Prototheca zopfii. J. Cellular Comp. Physiol.* **8,** 231–250.

Bean, R. C., and Hassid, W. Z. (1956). Carbohydrate oxidase from a red alga, *Iridophycus flaccidum. J. Biol. Chem.* **218,** 425–436.

Blinks, L. R. (1951). Physiology and biochemistry of algae. *In* "Manual of Physiology" (G. M. Smith, ed.), pp. 263–291. Chronica Botanica, Waltham, Massachusetts.

Bloom, B., and Stetten, DeW., Jr. (1953). Pathways of glucose catabolism. *J. Am. Chem. Soc.* **75,** 5446.

Brown, A. H. (1953). The effects of light on respiration using isotopically enriched oxygen. *Am. J. Botany* **40,** 719–729.

Brown, A. H., and Webster, G. C. (1953). The influence of light on the rate of respiration of the blue-green alga *Anabaena. Am. J. Botany* **40,** 753–758.

Brown, A. H., and Weis, D. (1959). Relation between respiration and photosynthesis in the green alga, *Ankistrodesmus braunii. Plant Physiol.* **34,** 224–234.

Bünning, E., and Herdtle, H. (1946). Über den Stoffwechsel einer thermophilen Blaualge. *Naturwissenschaften* **33,** 159.

Calo, N., and Gibbs, M. (1960). The site of inhibition of iodoacetamide in photosynthesis studied with chloroplasts and cell free preparations of spinach. *Z. Naturforsch.* **15b,** 287–291.

Calvin, M., and Massini, P. (1952). The path of carbon in photosynthesis. XX. The steady state. *Experentia* **8,** 445–457.

Chance, B., and Hackett, D. P. (1959). The electron transfer system of skunk cabbage mitochondria. *Plant Physiol.* **34,** 33–49.

Chance, B., and Sager, R. (1957). Oxygen and light induced oxidations of cytochrome, flavoprotein, and pyridine nucleotide in a *Chlamydomonas* mutant. *Plant Physiol.* **32,** 548–561.

Cramer, M. L., and Myers, J. (1949). Effect of starvation on the metabolism of *Chlorella. Plant Physiol.* **24,** 255–264.

Danforth, W. (1953). Oxidative metabolism of *Euglena. Arch. Biochem. Biophys.* **46,** 164–173.

Dickens, F., and Glock, G. E. (1951). Direct oxidation of glucose-6-phosphate, 6-phosphogluconate, and pentose-5-phosphates by enzymes of animal origin. *Biochem. J.* **50,** 81–95.

Dickens, F., and Simer, F. (1930). The metabolism of normal and tumor tissue. *Biochem. J.* **24,** 1301–1326.

Emerson, R. (1927a). Über die Wirkung von Blausäure, Schwefelwasserstoff und Kohlenoxyd auf die Atmung verschiedener Algen. Doctoral Dissertation, Friedrich-Wilhelms Universität, Berlin.

Emerson, R. (1927b). The effect of certain respiratory inhibitors on the respiration of *Chlorella. J. Gen. Physiol.* **10,** 469–477.

Emerson, R. L., Stauffer, J. F., and Umbreit, W. W. (1954). Relationships between phosphorylation and photosynthesis in *Chlorella. Am. J. Botany* **31,** 107–120.

Entner, N., and Doudoroff, M. (1952). Glucose and gluconic acid oxidation in *Pseudomonas saccharophila. J. Biol. Chem.* **196,** 853–862.

Fogg, G. E. (1953). "The Metabolism of Algae." Methuen, London.

French, C. S., Kohn, H. I., and Tang, P. S. (1934). Temperature characteristics for the metabolism of *Chlorella.* II. The rate of respiration of cultures of *Chlorella pyrenoidosa* as a function of time and of temperature. *J. Gen. Physiol.* **18,** 193–207.

Fujita, K. (1959). Metabolism of acetate in *Chlorella* cells. *J. Biochem.* **46,** 253–268.

Gaffron, H. (1939). Über Anomalien des Atmungsquotienten von Algen aus Zuckerkulturen. *Biol. Zentr.* **59,** 288–302.

Galloway, R. A., and Krauss, R. W. (1959a). Mechanism of action of polymyxin β on *Chlorella* and *Scenedesmus. Plant Physiol.* **34,** 380–389.

Galloway, R. A., and Krauss, R. W. (1959b). Differential action of chemical agents, especially polymyxin β on certain algae, bacteria and fungi. *Am. J. Botany* **46,** 40–49.

Genevois, L. (1927). Über Atmung und Gärung in grünen Pflanzen. *Biochem. Z.* **186,** 461–473.

Genevois, L. (1929). Sur la fermentation et sur la respiration ches les végétaux chlorophylliens. *Rev. gén. botan.* **40,** 654–674, 735–746; **41,** 49–63, 119–128, 154–166; 252–271.

Gibbs, M. (1953). Effect of light intensity on the distribution of C^{14} in sunflower leaf metabolites during photosynthesis. *Arch. Biochem. Biophys.* **45,** 156–160.

Gibbs, M. (1957). The light and dark metabolism of acetate and CO_2 by *Chlorella. Plant Physiol.* **32,** Suppl. XVII.

Gibbs, M. (1959). Metabolism of carbon compounds. *Ann. Rev. Plant Physiol.* **10,** 329–378.

Gibbs, M., and Wood, W. A. (1952). The effect of substrate on the endogenous respiration of *Pseudomonas fluorescens. Bacteriol. Proc.* (*Soc. Am. Bacteriologists*): *Physiol. Bacteria.* No. P11, p. 137.

Gunsalus, I. C., and Gibbs, M. (1952). Position of C^{14} in the products of glucose dissimilation by *Leuconostoc mesenteroides. J. Biol. Chem.* **194,** 871–875.

Hackett, D. P., Rice, B., and Schmid, C. (1960). The partial dissociation of phosphorylation from oxidation in plant mitochondria by respiratory chain inhibitors. *J. Biol. Chem.* **235,** 2140–2144.

Holt, A. S., Brook, I. A., and Arnold, W. A. (1951). Some effects of 2537 Å on green algae and chloroplast preparations. *J. Gen. Physiol.* **34,** 627–645.

Holzer, H. (1954). Chemie und Energetik der pflanzlichen Photosynthese. *Angew. Chem.* **66,** 65–75.

Jacobi, G. (1957). Vergleichend enzymatische Untersuchungen an marinen Grün- und Rotalgen. *Kiel. Meeresforsch.* **13,** 212–219.

Jacobi, G. (1958). Über die Bestimmung stationärer Konzentrationen von Brenztraubensäure und α-Ketoglutarsäure in Laminarien. *Kiel. Meeresforsch.* **14,** 247–250.

James, W. O. (1953). "Plant Respiration." Oxford Univ. Press, London and New York.

Jeney, E., and Zsolnai, T. (1957). Examination of the bacteriostatic and fungistatic effect of fluoroacetic acid. *Zentr. Bakteriol. Parasitenk. Abt. I.* **168,** 453–457.

Kandler, O. (1950). Phosphatspiegelschwankungen bei *Chlorella pyrenoidosa* als Folge des Licht-Dunkel-Wechsels. *Z. Naturforsch.* **5b,** 423–437.

Kandler, O. (1954). Gesteigerter Glucoseeinbau im Licht als Indikator einer lichtabhängigen Phosphorylierung. *Z. Naturforsch.* **9b,** 625–644.

Kandler, O. (1955). Hemmungsanalyse der lichtabhängigen Phosphorylierung. *Z. Naturforsch.* **10b,** 38–46.

Kandler, O. (1958a). The effect of 2,4-dinitrophenol on respiration, oxidative assimilation, and photosynthesis in *Chlorella. Physiol. Plantarum* **11,** 675–684.

Kandler, O. (1958b). Untersuchungen über den Einfluss von Substratzufuhr auf den Abbau zelleigenen Materials bei C^{14}-markierter *Chlorella. Planta* **51,** 167–172.

Kandler, O., and Gibbs, M. (1959). Untersuchungen über den Einfluss der Photosynthese auf die Austauschvorgänge innerhalb des Hexosemoleküls. *Z. Naturforsch.* **14b,** 8–13.

Kandler, O., and Sironval, C. (1959). Photooxidation processes in normal green *Chlorella* cells. II. Effects of metabolism. *Biochim. et Biophys. Acta* **33,** 207–215.

Katoh, S. (1959a). Studies on algal cytochrome. I. Enzymic activities pertaining to *Porphyra tenera* cytochrome 553 in cell-free extracts. *Plant and Cell Physiol.* **1,** 29–38.

Katoh, S. (1959b). Studies on the algal cytochrome of *c*-type. *J. Biochem.* **46,** 629–632.

Katoh, S. (1960a). Crystallization of an algal cytochrome. *Porphyra tenera* cytochrome 553. *Nature* **186,** 138–139.

Katoh, S. (1960b). A new copper protein from *Chlorella ellipsoidea. Nature* **186,** 533–534.

Kornberg, H. L., and Beevers, H. (1957). The glyoxylate cycle as a stage in the conversion of fat to carbohydrate in castor beans. *Biochim. et. Biophys. Acta* **26,** 531–537.

Kratz, W. A., and Myers, J. (1955a). Photosynthesis and respiration of three blue-green algae. *Plant Physiol.* **30,** 275–280.

Kratz, W. A., and Myers, J. (1955b). Nutrition and growth of several blue-green algae. *Am. J. Botany* **42,** 282–287.

Krauss, R. W. (1958). Physiology of the fresh-water algae. *Ann. Rev. Plant Physiol.* **9,** 207–244.

Krebs, H. A., and Kornberg, H. L. (1957). "Energy Transformations in Living Matter." Springer, Berlin.

Krogmann, D. W. (1960). Further studies on oxidative photosynthetic phosphorylation. *J. Biol. Chem.* **235,** 3630–3634.

Lamprecht, W., and Heinz, F. (1958). Isolierung von Glycerinaldehyd Dehydrogenase aus Rattenleber zur Biochemie des Fructosestoffwechsels. *Z. Naturforsch.* **13b,** 464–465.

Lewin, J. C. (1953). Heterotrophy in diatoms. *J. Gen. Physiol.* **9,** 305–313.

Lewin, R. A. (1954). The utilization of acetate by wild-type and mutant *Chlamydomonas dysosmos. J. Gen. Microbiol.* **11,** 459–471.

Ling, K. H., and Byrne, W. L. (1954). Purification of phosphofructokinase from rabbit muscle. *Federation Proc.* **13,** 253.

Merrett, M. J., and Syrett, R. J. (1960). The relationship between glucose oxidation and acetate oxidation in *Chlorella vulgaris. Physiol. Plantarum* **13,** 237–249.

Meyerhof, O. (1925). Über den Einfluss des Sauerstoffs auf die alkoholische Gärung der Hefe. *Biochem. Z.* **162,** 43–86.

Milhaud, G., Benson, A. A., and Calvin, M. (1956). Metabolism of pyruvic-2-C^{14} and hydroxypyruvic acid-2-C^{14} in algae. *J. Biol. Chem.* **218,** 599–606.

Millbank, J. W. (1957). Studies on the preparation of a respiratory cell-free preparation of the alga *Chlorella. J. Exptl. Botany* **8,** 96–104.

Moses, V., and Syrett, P. J. (1955). The endogenous respiration of microorganisms. *J. Bacteriol.* **70,** 201–204.

Moses, V., Holm-Hansen, O., Bassham, J. A., and Calvin, M. (1959). The relationship between the metabolic pools of photosynthetic and respiratory intermediates. *J. Molecular Biol.* **1,** 21–29.

Moyse, A., Couderc, D., and Garnier, J. (1957). L'influence de la température sur la croissance et la photosynthèse d'*Oscillatoria subbrevis. Rev. cytol. et biol. végétales* **18,** 293–304.

Muntz, J. A., and Murphy, J. R. (1957). The metabolism of variously labeled glucose in rat liver *in vivo. J. Biol. Chem.* **224,** 971–985.

Myers, J. (1947). Oxidative assimilation in relation to photosynthesis in *Chlorella. J. Gen. Physiol.* **30,** 217–227.

Myers, J. (1951). Physiology of the algae. *Ann. Rev. Microbiol.* **5,** 157–180.

Oh-hama, T., and Miyachi, S. (1960). Changes in levels of pyridine nucleotides in *Chlorella* cells during the course of photosynthesis. *Plant and Cell Physiol.* **1,** 155–162.

Pardee, A. B. (1949). Measurement of oxygen uptake under controlled pressures of carbon dioxide. *J. Biol. Chem.* **179,** 1085–1091.

Pirson, A., Daniel, A. L., and Becker, E. W. (1955). Zur Beziehung zwischen endogener Atmung und Glucoseatmung bei *Chlorella. Arch. Mikrobiol.* **22,** 214–218.

Reazin, G. H. (1954). On the dark metabolism of a golden-brown alga, *Ochromonas malhamensis. Am. J. Botany* **41,** 771–777.

Reazin, G. H. (1956). The metabolism of glucose by the alga *Ochromonas malhamensis. Plant Physiol.* **31,** 299–303.

Redford, E. L., and Myers, J. (1951). Some effects of ultraviolet radiation on the metabolism of *Chlorella. J. Cellular Comp. Physiol.* **38,** 217–243.

Richter, G. (1959). Comparison of enzymes of sugar metabolism in the photosynthetic algae: *Anacystis nidulans* and *Chlorella pyrenoidosa. Naturwissenschaften* **46,** 604.

Ryther, J. H. (1956). Interrelation between photosynthesis and respiration in the marine flagellate, *Dunaliella euchlora. Nature* **178,** 861–863.

Schlegel, H. G. (1959). Die Verwertung von Essigsäure durch *Chlorella* im Licht. *Z. Naturforsch.* **14b,** 246–253.

Schou, L., Benson, A. A., Bassham, J. A., and Calvin, M. (1950). The path of carbon in photosynthesis. XI. The role of glycolic acid. *Physiol. Plantarum* **3,** 487–495.

Steemann Nielsen, E. (1955a). Influence of pH on the respiration in *Chlorella pyrenoidosa. Physiol. Plantarum* **8,** 106–115.

Steemann Nielsen, E. (1955b). The interaction of photosynthesis and respiration and its importance for the determination of C^{14}-discrimination in photosynthesis. *Physiol. Plantarum* **8,** 945–953.

Syrett, P. J. (1951). The effect of cyanide on the respiration and the oxidative assimilation of glucose by *Chlorella vulgaris. Ann. Botany* **15,** 473–482.

Szymona, M., and Doudoroff, M. (1960). Carbohydrate metabolism in *Rhodopseudomonas spheroides. J. Gen. Microbiol.* **22,** 167–183.

Tang, P. S., and French, C. S. (1933). The rate of oxygen consumption of *Chlorella pyrenoidosa* as a function of temperature and of oxygen tension. *Chinese J. Physiol.* **1,** 353–378.

Taylor, F. J. (1950). Oxidative assimilation of glucose by *Scenedesmus quadricauda. J. Exptl. Botany* **1,** 301–321.

Tung, T. C., Ling, K. H., Byrne, W. L., and Lardy, H. A. (1954). Substrate specificity of muscle aldolase. *Biochim. et Biophys. Acta* **14,** 488–494.

Umbreit, W. W., Burris, R. H., and Stauffer, J. F. (1957). "Manometric Techniques." Burgess, Minneapolis, Minnesota.

Warburg, O., and Negelein, E. (1920). Über die Reduktion der Salpetersäure in grünen Zellen. *Biochem. Z.* **110,** 66–115.

Watanabe, A. (1937). Untersuchungen über die Substrate für Sauerstoffatmung von Süsswasser- und Meeresalgen. Beiträge zur Stoffwechselphysiologie der Algen. *Acta Phytochim. (Japan)* **9,** 235–254, 255–264.

Webster, G. C., and Frenkel, A. W. (1953). Some respiratory characteristics of the blue-green alga, *Anabaena. Plant Physiol.* **28,** 63–69.

Weimberg, R., and Doudoroff, M. (1955). The oxidation of L-arabinose by *Pseudomonas saccharophila. J. Biol. Chem.* **217,** 607–624.

Weis, D., and Brown, A. H. (1959). Kinetic relation between photosynthesis and respiration in the algal flagellate *Ochromonas malhamensis. Plant Physiol.* **34,** 235–239.

Yakushiji, E., Sugimura, Y., Sekuzu, I., Morikawa, I., and Okunuki, K. (1960). Properties of crystalline cytochrome 553 from *Porphyra tenera. Nature* **185,** 105–106.

Fermentation

MARTIN GIBBS

Department of Biochemistry, Cornell University, Ithaca, New York

I. Definition of Terms

Since among algal physiologists there is some disagreement concerning the use of the term *fermentation* (in German, *Gärung*), we will define the term here as the degradation of a carbohydrate molecule into two or more smaller molecules by biological processes not requiring molecular oxygen. In general, the products of fermentation are carbon dioxide, ethanol, lactic acid, and other organic acids.

It has been common to equate the terms *fermentation* and *glycolysis*. The latter process is defined by most reviewers as the degradation of any carbohydrate by fermentative fission into pyruvic acid. The fate of the keto acid is generally not specified.

II. Rate of Fermentation

The rate of fermentation is normally given as milliliters of carbon dioxide formed per gram of dry weight of algae per hour. Following Warburg, this value is generally represented by $Q_{CO_2}^{N_2}$, where the superscript denotes anaerobic conditions.

The manometric technique of Warburg (1926) has usually been employed to measure the rate of algal fermentation. In this method (Gaffron and Rubin, 1942; Genevois, 1927) the cells or tissues are suspended in a bicarbonate buffer (for endogenous rate determinations) or in a similar medium containing carbohydrate (for exogenous rate determinations) and the volume of evolved carbon dioxide is measured. To distinguish between carbon dioxide produced directly by the algae and that due to organic acids produced by fermentation (in German, *fixe Säuren*) reacting with the bicarbonate in the medium, the remaining bicarbonate is titrated. A $Q_{CO_2}^{N_2}$ derived in this way is the number of milliliters of carbon dioxide displaced by acid formation and released directly per gram of dry weight of algae per hour.

Since not all algal fermentations have a common stoichiometry, as contrasted to those of animal tissues which produce only lactic acid, the evaluation of the fermentation rate by $Q_{CO_2}^{N_2}$ determined in a medium containing bicarbonate can be misleading. Thus, an alga producing 3 moles of acetic acid per mole of glucose would have an apparently faster rate of fermentation than one carrying out a yeast-type reaction yielding two moles of carbon dioxide, even though the rate of glucose uptake may be equal. Until specific information is available on the nature of the end products, the symbol $Q_{sugar}^{N_2}$ (i.e., millimoles of sugar consumed per gram dry weight per hour) might prove useful.

Some authors (Gaffron and Rubin, 1942) express the concentrations of

TABLE I

RATES OF FERMENTATION IN MILLILITERS PER GRAM DRY WEIGHT PER HOUR

Organism	Substrate	Temperature (°C.)	$Q_{CO_2}^{NO_2}$	Reference
Algae				
Chlorella pyrenoidosa	(Endogenous)	25	0	Genevois (1927)
	Glucose	25	1.2–1.6	Genevois (1927)
	Fructose	25	1.2	Genevois (1927)
	Mannose	25	1.0	Genevois (1927)
	Galactose	25	0.7	Genevois (1927)
Chlorella pyrenoidosa	Glucose	25	0.85	Gaffron (1939)
Coelastrum proboscideum	Glucose	25	2.4–3.8	Genevois (1927)
Scenedesmus basiliensis	Glucose	25	0.9–2.6	Genevois (1927)
Scenedesmus D_1	Glucose	25	1.25	Gaffron (1939)
Other cell types				
Lactobacillus casei	Glucose	—	316	Stephenson (1949)
Propionibacterium pentosaceum	Glucose	—	20	Stephenson (1949)
Yeast	Glucose	28	200–300	Meyerhof (1925)
Rat, brain cortex	(Endogenous)	37	18	Dickens and Simer (1930)
Rat, liver	(Endogenous)	37	3	Dickens and Simer (1930)

the substance employed and the products formed in terms of volume. If 1 mole of the substance is considered to be a gas occupying 22.4 liters, then one may think of 180 mg. of glucose as equivalent to 22.4 ml. Thus, a $Q^{N_2}_{glucose}$ equal to 5 means that in an atmosphere of nitrogen each gram of dry weight of algae utilizes per hour 5/22.4 millimoles of glucose (equals molecular weight (180) $\times$ (5/22.4)mg. or about 40 mg. glucose).

Some typical examples of $Q^{N_2}_{CO_2}$ values are listed in Table I. It is evident that algal fermentation rates are generally far below those of other unicellular organisms, although they are comparable with those of rat liver.

III. End-Products of Fermentation

Very few data have been reported on this topic. The algal fermentation patterns so far determined fall into three classical types: (i) purely lactic acid; (ii) lactic acid, ethanol, and carbon dioxide; (iii) acetic acid, carbon dioxide, and ethanol. In addition to those products, some produce hydrogen (see Spruit, Chapter 3).

The first investigations concerning algal fermentation were those of Genevois (1927). He noted that *Chlorella pyrenoidosa*, *Coelastrum proboscideum*, and *Scenedesmus basiliensis* produced carbon dioxide and unspecified organic acids from various hexoses under anaerobic conditions. Michels (1940) showed that *C. pyrenoidosa* fermented glucose to lactic acid and other organic acids, whereas *Scenedesmus* yielded mostly carbon dioxide and essentially little organic acid. He noted that the unspecified organic acids of the *Chlorella* fermentation did not include acetic, pyruvic, fumaric, or malic acids. Using *Scenedesmus* (strain D_3) grown with illumination in a medium containing glucose, Gaffron (1939) found that the products of both endogenous fermentation and exogenous glucose fermentation were unspecified organic acids (75%) and carbon dioxide. *Chlorella pyrenoidosa* grown with glucose under the same conditions also yielded essentially unspecified organic acids by fermentation processes. Gaffron and Rubin (1942) later reported that the endogenous fermentation of autotrophically grown *Scenedesmus* (strain D_1) produced hydrogen, carbon dioxide, and traces of lactic acid. On the other hand, lactic acid accounted for approximately 50% of the fermentation acids after glucose was added to a cell suspension. A brief communication by Weis and Mukerjee (1958) indicated the difficulty of predicting algal fermentation types and provided possible reasons for differences in the products reported in earlier studies of similar species. They noted three patterns of acid production among species of *Chlorella* and other algae: (i) *Chlorella pyrenoidosa* (strains Burke and Emerson), *Ankistrodesmus braunii* and *Scenedesmus* (D_3) produced lactic acid as the sole major product; (ii) *Chlorella* sp. (strain Marburg), *C. miniata* and *C. ellipsoidea* produced little or no acid; and (iii) *Chlorella vulgaris* produced acetic acid as the major acid product, and in addition ethanol and carbon dioxide.

Only two investigators have given detailed attention to stoichiometry. The first was Barker (1935) who observed that under anaerobic conditions the colorless alga *Prototheca zopfii* produced 2 moles of D(−)-lactic acid for each mole of glucose consumed. More recent studies with heterotrophically grown *Ochromonas malhamensis* (Reazin, 1956; Reazin and Gibbs, 1956) indicated a reaction which may be written:

$$\text{glucose} \rightarrow 1.60\ CO_2 + 1.75\ \text{ethanol} + 0.18\ \text{lactic acid}$$

IV. Fermentation Inhibitors

Cyanide ($5 \times 10^{-4} M$) affects neither the $Q_{CO_2}^{N_2}$ value of fermentations by carrot root (Marsh and Goddard, 1939) and yeast (Warburg, 1925) nor those of *Chlorella*, *Coelastrum*, *Scenedesmus*, and *Haematococcus* (Genevois, 1927). In air, algal cells poisoned with cyanide show a respiratory quotient slightly higher than 1.

Fluoride inhibits hydrogen evolution as well as lactic acid formation by *Scenedesmus* (strain D_1) (Gaffron and Rubin, 1942).

V. Pathway of Fermentation

James (1953), dealing with the respiration and fermentation of autotrophic tissue, expressed the view that the first steps of carbohydrate degradation are common to both processes. (For a discussion of algal respiration, see Gibbs, Chapter 4). This has given rise to the "common path" theory, stating that in both respiration and fermentation the same enzymes catalyze the conversion of hexose to pyruvic acid, which then has different fates in the presence of oxygen and in its absence. The series of steps by which glucose is converted to pyruvic acid has been termed the Embden-Meyerhof-Parnas (EMP) glycolytic pathway (see Jacobi, Chapter 7). According to this scheme, the carbon atoms of glucose should be distributed among the end products in the following manner:

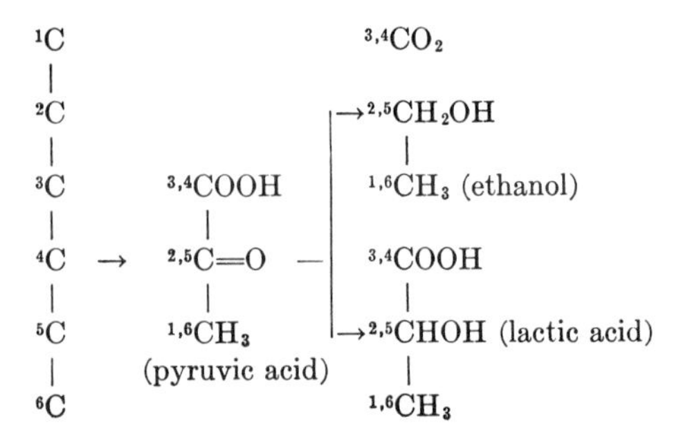

Only with *Ochromonas malhamensis* has this been tested with specifically labeled glucose (Reazin 1956; Reazin and Gibbs, 1956). The results of experiments with glucose-1-C^{14} and glucose-6-C^{14} suggested that *Ochromonas* ferments glucose to carbon dioxide, ethanol, and lactic acid by the EMP pathway. However, since anaerobic pathways other than the EMP have been demonstrated in unicellular, heterotrophic cells (Gunsalus and Gibbs, 1952), it is premature to extrapolate from this one observation to the glycolytic processes of other algae.

VI. Anaerobic Growth

There have apparently been no reports of anaerobic, heterotrophic growth of algae. This is surprising, since both *Prototheca zopfii* and *Ochromonas malhamensis* are capable of carrying on a type of anaerobic metabolism which is known to sustain growth of their biochemical counterparts in the bacterial world. Either the algal cells cannot utilize the energy derived from the fundamental energy-yielding processes under anaerobic conditions, or some other explanation must be sought.

From a consideration of fermentation rates, it would appear that the obligate, aerobic heterotrophy of algae cannot be attributed to a lack of the ability to utilize energy set free during the anaerobic fission of carbohydrates, but rather to a deficiency in the amount of this energy which is available for growth.

Let us assume that a carbohydrate is the substrate for fermentation and respiration in the algal cell. Under these conditions the metabolism of *Chlorella pyrenoidosa*, which has a respiration rate (Q_{O_2}) of about 17 (see Gibbs, Chapter 4 in this treatise, Table I), would correspond to a glucose consumption ($Q_{CO_2}^{N_2}$) of 5 to 6. This is due to the fact that in respiration 6 moles of O_2 are consumed per mole of glucose, whereas in fermentation only 2 moles of lactic acid or 2 moles of CO_2 and 2 moles of alcohol are produced. In respiration 1 mole of glucose oxidized by 6 moles of O_2 sets free approximately 700 kcal, whereas in fermentation anaerobic fission of 1 mole of glucose releases only 60 kcal; only one-twelfth as much. Therefore, to give the same energy yield as in respiration, the rate of fermentation ($Q_{CO_2}^{N_2}$) would have to be $5 \times 12 = 60$. The algal values shown in Table I are only a small fraction of this, whereas *Lactobacillus casei* and *Saccharomyces* spp., which can grow under anaerobic conditions, have $Q_{CO_2}^{N_2}$ values of approximately 300.

Assuming that the factor limiting anaerobic growth of algae is the availability of chemical energy, the next question is, how is this limitation brought about? If we assume that the "common path" theory holds for algae, presumably none of the enzymes catalyzing the conversion of glucose to pyruvic acid is missing. A basic difference between respiration and fermentation lies in the manner whereby reduced pyridine nucleotides (diphosphopyridine nucleotide and triphosphopyridine nucleotide) are reoxidized. Under aerobic conditions, the ultimate oxidizer is molecular oxygen, while in anaerobiosis the oxidant must be a carbon compound, probably pyruvic acid or acetaldehyde, derived from the carbohydrate. Since the synthesis of these compounds is apparently not limiting, it would appear that there is a deficiency of the enzymes catalyzing the reoxidation of the reduced pyridine nucleotides with acetaldehyde or pyruvic acid. In fact, both Jacobi (1957) and Richter (1957) reported either the complete absence or only traces of alcohol dehydrogenase or lactic acid dehydrogenase activity in various marine Green and Red Algae.

References

Barker, H. A. (1935). The metabolism of the colorless alga, *Prototheca zopfii*, Krüger. *J. Cellular Comp. Physiol.* **7,** 73–93.

Dickens, F., and Simer, F. (1930). The metabolism of normal and tumor tissue. *Biochem. J.* **24,** 1301–1326.

Gaffron, H. (1939). Über Anomalien des Atmungsquotienten von Algen aus Zuckerkulturen. *Biol. Zentr.* **59,** 288–302.

Gaffron, H., and Rubin, J. (1942). Fermentative and photochemical production of hydrogen in algae. *J. Gen. Physiol.* **26,** 219–240.

Genevois, L. (1927). Über Atmung und Gärung in grünen Pflanzen. *Biochem. Z.* **186,** 461–473.

Gunsalus, I. C., and Gibbs, M. (1952). The heterolactic fermentation. II. Position of C^{14} in the products of glucose dissimilation by *Leuconostoc mesenteroides*. *J. Biol. Chem.* **194,** 871–875.

Jacobi, G. (1957). Vergleichende enzymatische Untersuchungen an marinen Grün- und Rotalgen. *Kiel. Meeresforsch.* **13,** 212–219.

James, W. O. (1953). "Plant Respiration." Oxford Univ. Press, London and New York.

Marsh, P. B., and Goddard, D. R. (1939). Respiration and fermentation in the carrot, *Daucus carota*. II. Fermentation and the Pasteur effect. *Am. J. Botany* **26,** 767–772.

Meyerhof, O. (1925). Über den Einfluss des Sauerstoffs auf die alkholische Gärung der Hefe. *Biochem. Z.* **162,** 43–86.

Michels, H. (1940). Über die Hemmung der Photosynthese bei Grünalgen nach Sauerstoffentzug. *Z. Botan.* **35,** 241–270.

Reazin, G. H., Jr. (1956). The metabolism of glucose by the alga, *Ochromonas malhamensis*. *Plant Physiol.* **31,** 299–303.

Reazin, G. H., Jr. and Gibbs, M. (1956). Glucose metabolism by the alga *Ochromonas malhamensis*. *Federation Proc.* **15,** 335.

Richter, G. (1957). Nachweiss und quantitative Bestimmung einiger Enzyme des Kohlenhydrat-Stoffwechsels in Grünalgen. *Z. Naturforsch.* **12b,** 662–663.

Stephenson, M. (1949). "Bacterial Metabolism." Longmans, Green, London.

Warburg, O. (1925). Über die Wirkung der Blausäure auf die alkoholische Gärung. *Biochem. Z.* **165,** 196–198.

Warburg, O. (1926). "Stoffwechsel der Tumoren." Springer, Berlin, (English translation by F. Dickens: "The Metabolism of Tumors." Constable, London, 1930.)

Weiss, D. S., and Mukerjee, H. (1958). Algal fermentations. *Plant Physiol. Suppl.* **33,** VIII.

—6—

Substrate Assimilation and Heterotrophy

WILLIAM F. DANFORTH

Department of Biology, Illinois Institute of Technology, Chicago, Illinois

I. General Introduction [1]

There is a great deal of evidence that the "dark metabolism" of photosynthetic plants, including the algae, is essentially similar to that of nonphotosynthetic organisms. From a purely biochemical viewpoint, therefore, we would expect that almost any substrate or intermediate in the major pathways of energy metabolism might substitute for photosynthesis.

This expectation is, in fact, fulfilled only partially, and to varying degrees in various species of algae. Therefore, in dealing with the heterotrophic metabolism of the algae we must be concerned not only with the question of how certain substrates function as carbon and energy sources for growth,

[1] Abbreviations used in this paper: ATP, adenosine triphosphate; DNP, dinitrophenol; DPN, diphosphopyridine nucleotide; TPN, triphosphopyridine nucleotide; RQ, respiratory quotient.

but with the equally difficult questions of why the same compounds support growth of some species and not of others, and why, in a particular species, one substrate will support growth while another, very closely related compound will not.

The subject of heterotrophy in algae is covered in reviews by Myers (1951), Fogg (1953), Krauss (1958), and Pringsheim (1959), and heterotrophy among the algal flagellates by Hutner and Provasoli (1951, 1955).

II. Heterotrophic Growth

A. Introductory Remarks

For the purpose of the discussion which follows, species have been associated, without regard for taxonomy, into groups which permit useful comparison and generalization. This approach was chosen because, in the algal species which have been studied, nutritional behavior shows only slight correlation with taxonomic relationships. It should be borne in mind, however, that with further study the association of some of these groups may prove to be as arbitrary from a biochemical viewpoint as they are from the standpoint of taxonomy.

B. *Chlorella*, *Scenedesmus*, and *Prototheca*

1. *Sugars and Related Compounds*

Table I presents data from several sources concerning the ability of sugars and related compounds to support heterotrophic growth of species of *Chlorella*, of *Prototheca zopfii*, and of a species of *Scenedesmus*. All are members of the Chlorococcales; *Chlorella* and *Scenedesmus* possess chlorophyll and will grow photosynthetically, while *Prototheca* is considered to be the colorless counterpart of *Chlorella* and requires an organic carbon source.

Certain similarities are apparent. All the organisms characteristically utilize glucose and galactose, although a strain of *Chlorella vulgaris* has been reported to be unable to grow on glucose (Finkle *et al.*, 1950). None of the strains tested grew on pentose, sugar alcohols, or phosphorylated sugar derivatives.

There are also conspicuous differences between the species. Only *Scenedesmus* and *Chlorella vulgaris* utilize disaccharides, and each of these uses one or more disaccharides which the other cannot use. There is likewise considerable variation in the ability to utilize specific hexose monosaccharides.

The work of Taylor (1960a, b) on *Scenedesmus quadricauda* throws considerable light on the mechanisms underlying these sugar specificities. He found that the kinetics of uptake of glucose and mannose, sugars which

TABLE I

UTILIZATION OF SUGARS AND RELATED COMPOUNDS BY SEVERAL SPECIES OF CHLOROPHYTA[a]

Substance	*Chlorella vulgaris*[b]	*Chlorella pyrenoidosa*[c]	*Chlorella ellipsoidea*[c]	*Scenedesmus* sp. strain D_3[c]	*Prototheca zopfii*[d]
Glucose	+	+	+	+	+
Fructose	+	?	?	+	+
Galactose	+	+	+	+	+
Mannose	0	0	0	+	+
Sorbose	0				
Rhamnose	0				
Cellobiose	+	0	0	0	
Lactose	+	0	0	+	0
Maltose	0	?	?	+	0
Sucrose	0	0	0	+	0
Arabinose	0	?	?	0	0
Xylose	0	?	?	0	0
Ribose	0	0	0	0	
Adonitol	0				
Dulcitol	0				
Mannitol	0	0	0	0	0
Sorbitol	0				
Glycerol	0	0	0	0	+
Dihydroxyacetone	0				+
Glycerophosphate	0				
Phosphoglycerate		0	0		
Glucose-1-phosphate		0	0		
Glucose-6-phosphate		0	0		
Fructose phosphate		0	0		
Fructose diphosphate		0	0		

[a] KEY: +, supports growth; 0, no growth; ?, doubtful growth.
[b] Neish (1951). [c] Samejima and Myers (1958). [d] Barker (1935a).

stimulated respiration of this species (Taylor, 1950), differed from the kinetics of simple diffusion. Sugars which did not stimulate respiration (fructose, galactose, sorbose, and several pentoses) penetrated much more slowly, following the kinetics of simple diffusion. Glucose uptake was competitively inhibited by mannose and by several structural analogs of glucose. Anaerobiosis, phosphate deficiency, and inhibitors of oxidative phosphorylation also inhibited glucose uptake; in some of these cases the inhibition could be reversed by exogeneous ATP, but not by inorganic phosphate. These findings suggest that the sugars utilized by *Scenedesmus* enter the cell by an active, structurally specific process requiring ATP as an energy source. It has frequently been suggested (for example, cf. Rothstein,

1954b, pp. 40–44) that phosphorylation of the sugars by kinases is directly involved in such uptake mechanisms; Taylor's data are consistent with this possibility.

2. *Fatty Acids and Alcohols*

Barker (1935a) found that *Prototheca* grew readily on all the straight-chain fatty acids from C_2 (acetic) to C_{10}. From C_{10} to C_{16}, only even-numbered acids were tested, and all of these also supported growth, as did isobutyric, isocaproic, α-crotonic, and oleic acids. Of the fatty acids tested, only formic failed entirely to support growth, while growth with isovaleric was poor or questionable. Among the alcohols: ethyl, *n*-propyl, *n*-butyl, and *n*-amyl alcohols supported definite growth; isobutyl and isoamyl alcohols, poor or questionable growth; and methyl and isopropyl alcohols, no growth.

By comparison, the ability of *Chlorella pyrenoidosa* and *Chlorella ellipsoidea* to utilize fatty acids was found to be extremely limited. Both grew on acetate, but not on formate, propionate, or butyrate (Samejima and Myers, 1958). Careful control of pH and substrate concentration was necessary to obtain good growth on acetate, since at pH 4.5, acetate was toxic at concentrations of $0.004M$ or greater, and at pH 6.7, $0.12M$ acetate was toxic. A similar toxicity of weak acids in acidic media is frequently observed. It apparently results from the fact that cell membranes are permeable to undissociated acid molecules, though relatively impermeable to the corresponding anions. Since the concentrations of undissociated molecules inside and outside the cell tend to become equal, and since, when the medium is more acid than the cytoplasm, the acid molecules tend to dissociate on penetrating, high concentrations of the anion may accumulate inside the cell (Jacobs, 1940, Wilson *et al.*, 1959). With *Chlorella*, Erickson *et al.* (1955) found that the toxicity of acetate, as measured by its inhibition of photosynthesis, depended on the concentration of the undissociated acid. In contrast to the usual findings with acid substrates (see below), they reported that stimulation of respiration depended on the concentration of the acetate ion.

3. *Krebs-Cycle Intermediates, Amino Acids, and Related Compounds*

Barker (1935a) found no growth of *Prototheca* in neutral media containing lactic, pyruvic, citric, succinic, fumaric, malic, acetoacetic, glyoxylic, or glycolic acids, glycine, or asparagine; nor did any of these compounds stimulate respiration. Anderson (1945), however, found that pyruvic, lactic, glycolic, and glyoxylic acids did stimulate respiration in more acidic media, and that pyruvic and lactic acids are oxidatively assimilated. It is likely, therefore, that *Prototheca* cells are impermeable to the anions of these acids, some or all of which might support growth in acid media. This possibility should be tested directly, since, as we shall see, the ability of a compound to stimulate respiration does not necessarily prove that the same compound will support growth. Until such tests are made, the ability of *Prototheca* to grow on compounds of this group must remain an open question.

Samejima and Myers (1958) found that glycolate permitted slow growth of *Chlorella ellipsoidea,* but not of *Chlorella pyrenoidosa.* Lactate, pyruvate, α-ketoglutarate, succinate, fumarate, malate, glutamate, and glycine did not support growth of either species. Eny (1950, 1951) found slight or questionable growth and respiratory stimulation of the Wann strain of *Chlorella vulgaris* with most of the Krebs-cycle acids.

The general failure of *Chlorella* to utilize Krebs-cycle intermediates and related compounds can hardly be attributed to enzymatic deficiencies. These compounds have repeatedly been demonstrated to be intermediates in the respiratory pathways and "dark" CO_2 fixation of this species (see, for example, Lynch and Calvin, 1953), and Millbank (1957) has presented strong evidence for the functioning of the Krebs cycle in respiration of *Chlorella vulgaris.* (See also in this treatise Gibbs, Chapter 4; Holm-Hansen, Chapter 2.) It is more likely that the *Chlorella* cell is impermeable to the substrates. However, we shall discuss below similar cases where the unexpected failure of an organism to grow on particular substrates can *not* be attributed to impermeability.

C. *Tribonema* and Some Diatoms

Belcher and Fogg (1958) found that the filamentous xanthophyte *Tribonema aequale* grew well in darkness on glucose at a rate almost identical with that of photoautotrophic growth. Sucrose also supported growth, but only after a three-week adaptive lag period. Acetic and citric acids permitted growth in darkness, though at a lower rate than glucose. Fructose, mannose, galactose, rhamnose, arabinose, xylose, lactose, cellobiose, ethanol, glycerol, fumarate, and succinate did not permit dark growth. However, a number of these stimulated respiration. The addition of succinate, fumarate, and glycerol increased respiration more than did glucose, and ethanol also produced a measurable stimulation.

Tribonema minus has even more limited heterotrophic abilities; of the compounds effective for *T. aequale,* only glucose permitted growth of *T. minus.* Among other Xanthophyta, nutritional abilities range from obligate phototrophy to growth on a wide variety of substrates (Belcher and Miller, 1960).

J. C. Lewin (1953) screened 42 cultures of freshwater and soil diatoms, including at least 7 different species. Of these, she found only 13 capable of heterotrophic growth on the substrates tested. The substrate specificity of *Navicula pelliculosa*, a heterotrophic species, was studied in detail. Of 60 compounds, including the usual sugars, Krebs-cycle and fatty acids, alcohols, amino acids, etc., only glucose supported growth. Lactate, pyruvate, acetate, citrate, and succinate stimulated respiration, while glucose, glycerol, and fructose permitted growth in light in a CO_2-free atmosphere.

In a later study (Lewin and Lewin, 1960), 44 cultures of marine diatoms were similarly tested for their ability to grow in the dark on glucose, lactate, and acetate. Of these, 16 grew only on glucose, 8 on glucose or lactate, 1 on glucose or acetate, 1 on glucose, lactate, or acetate, and 2 only on lactate. The remaining 16 would not grow on any of these substrates. None of the strains tested grew on galactose.

There is no ready explanation for the limited range of substrates which support growth of these organisms. The substrates which stimulate respiration or replace CO_2 obviously can penetrate the cell and are subject to metabolism; moreover, they are substances which *do* support growth in many other algae and which, from their role in carbohydrate metabolism, would be expected to be effective carbon and energy sources.

D. The Acetate Flagellates

The acetate flagellates, or oxytrophs, are a group of plant-like flagellates which grow heterotrophically on acetate and various other fatty acids, Krebs-cycle acids, alcohols, and related compounds, but which are typically incapable of utilizing sugars. The group is taxonomically heterogeneous, including members of the Volvocales (Chlorophyta), Cryptophyta, and Euglenophyta. Since the general biology of the acetate flagellates has been thoroughly reviewed by Hutner and Provasoli (1951, 1955) and by Lwoff (1951), the present discussion will be limited to a few typical, more intensively investigated species.

The nutritional capacities of *Chilomonas paramecium*, a nonphotosynthetic cryptophyte, and *Polytomella coeca* (or *caeca*), a nonphotosynthetic volvocine, have been extensively explored. Table II shows the results of such studies. Both are typical acetate flagellates capable of growth on a rather wide range of substrates. It will be observed that neither grew on any of the sugars or sugar derivatives tested. Possible reasons for this inability to utilize sugars will be discussed later.

Chilomonas grows on most of the intermediates of the Krebs cycle. The exceptions to this rule are the tricarboxylic acids: citric, *cis*-aconitic, and isocitric. Since fluoroacetate and malonate inhibit the oxidation of acetate,

TABLE II

SUBSTRATE SPECIFICITIES OF TWO ACETATE FLAGELLATES[a]

Substrate	*Polytomella coeca*[b]	*Chilomonas paramecium*[c]
Sugars, etc.		
Glucose	0	0
Fructose		0
Galactose		0
Maltose	0	
Sucrose	0	
Trehalose	0	
Sorbitol		0
Glycerol		0
Glyceraldehyde	?	
Dihydroxyacetone		0
Glucose-1-phosphate		0
Hexose diphosphate		0
Krebs-Cycle Acids, etc.		
Lactic		+
Pyruvic	+	+
Citric		0
cis-Aconitic		0
Isocitric		0
α-Ketoglutaric	?	+
Succinic	+	+
Fumaric	0	+
Malic	0	+
Oxalacetic		+
Glycolic		0
Glyoxylic		0
Oxalic		0
Glutaric		0
Malonic		0
Fatty Acids		
Formic		0
Acetic	+	+
Propionic	+	T
Butyric	+	+
Isobutyric	0	0
Valeric	+	T
Caproic	0	+
Isocaproic		0
Heptylic		0
Caprylic	0	+
Alcohols		
Ethyl	+	+
n-Propyl	0	T
Isopropyl	0	0
n-Butyl	+	+
Isobutyl	0	0
sec-Butyl		0
tert-Butyl		0
n-Amyl	+	T
6 Amyl isomers		0
n-Hexyl	0	+
n-Heptyl	0	
n-Octyl	0	
Amino Acids		
20 Amino acids		0

[a] KEY: +, supports growth; 0, no growth; T, toxic; ?, doubtful growth.

[b] Lwoff *et al.* (1949, 1950); Wise (1955, 1959).

[c] Cosgrove and Swanson (1952); Holz (1954).

pyruvate, and γ-ketoglutarate by *Chilomonas* (Holz, 1954), and since citrate accumulates during fluoroacetate poisoning, the Krebs cycle is presumably operative in this species, and the tricarboxylic acids are normal intermediates. Failure of these strongly ionized hydrophilic molecules to penetrate the cell membrane is the most likely explanation for their inability to support growth. Their failure to stimulate respiration (Holz, 1954) is consistent with this interpretation.

The ability of *Polytomella* to grow on Krebs-cycle intermediates is more limited, since fumarate, malate, and probably α-ketoglutrate do not support growth. Here, too, failure to penetrate the cell membrane is probably a controlling factor, since homogenates of *Polytomella* showed dehydrogenase activity for these three substrates (Wise, 1959).

The patterns of fatty-acid utilization in the two organisms differ somewhat. *Chilomonas* utilizes for growth all the even-numbered straight-chain acids up to and including C_8; at similar concentrations the straight-chain odd-numbered acids are toxic while the branched-chain compounds are inert (Cosgrove and Swanson, 1952). *Polytomella*, on the other hand, uses both odd- and even-numbered straight-chain acids up to C_5, but not the longer-chain acids. In this connection, it is worth noting that Cirillo (1955, 1956, 1957) showed that growth of another acetate flagellate, *Polytoma uvella*, on butyrate and caproate is dependent on a two-stage adaptive process. The first step, occurring in a few hours, results in the ability of the cells to oxidize the acids, and seems to involve the synthesis of oxidative enzymes. At this stage the cells are still unable to grow on the acids; their ability to grow develops more slowly, and reaches a maximum only after some ten generations in butyrate media. Adaptation to butyrate is inhibited in the presence of acetate.

For both *Chilomonas* and *Polytomella*, utilization of acidic substrates shows the usual dependence on pH; these substances are used poorly, if at all, in alkaline media, and are toxic in very acid media. In general, the stronger the acid, the lower the optimum pH for growth (Holz, 1954; Wise, 1959).

For *Chilomonas*, there is a qualitative correlation between the ability of substrates to support growth and their ability to stimulate respiration (Holz, 1954). The failure of this organism to grow on any of the 20 amino acids (Holz, 1954) deserves further investigation.

Euglena gracilis forms a bridge between the acetate flagellates and the sugar-utilizing algae. This species includes a number of "varieties" and "strains" which differ widely in their physiology and biochemistry (Baker *et al.*, 1955; Pringsheim, 1955a; Anagnostakos, 1956). Permanently nonphotosynthetic variants of some strains can be obtained by "bleaching" with streptomycin (Provasoli *et al.*, 1948), heat (Pringsheim and Pringsheim,

TABLE III
SUBSTRATE SPECIFICITIES OF TWO STRAINS OF *Euglena gracilis*[a,b]

Substrate	*E. gracilis* var. *bacillaris*	*E. gracilis* "Vischer" strain
Glucose	+	0
Glycerol	0	
Pyruvate	+	
Citrate	0	
Succinate	+	0
Fumarate	+	
Malate	+	0
Glycolate	0	
Acetate	+	+
Butyrate	+	+
Glutamate	+	0
Alanine	+	0
Aspartate	+	0

[a] Cramer and Myers, 1952.
[b] KEY: +, growth; 0, no growth.

1952), or antihistamines (Gross *et al.*, 1955). Flagellates of the genera *Astasia* and *Khawkinea* are probably naturally occurring colorless variants of *Euglena*. The various bleached strains themselves are physiologically different (Gross and Jahn, 1958).

Table III (Cramer and Myers 1952) contrasts the substrates which permit growth in darkness of *Euglena gracilis* of the "Vischer" strain (which is similar or identical to the more commonly studied "Mainx" strain), and of *E. gracilis* var. *bacillaris*. The Vischer strain behaves like an acetate flagellate of very limited versatility, utilizing acetate and butyrate but none of the sugars, Krebs-cycle acids, or amino acids tested. The *bacillaris* strain, on the other hand, grows on glucose, several amino acids, and all of the Krebs-cycle acids tested except citrate. Both the respiration and the growth of *E. gracilis* var. *bacillaris* on acidic substrates show the usual pH dependence (Cramer and Myers, 1952; Danforth, 1953), as does the respiration of *Astasia* (Von Dach, 1942, 1953; F. R. Hunter and J. W. Lee, personal communication). For certain substrates, however, the optimum pH for growth differs greatly from that for respiration (Wilson *et al.*, 1959).

In a colorless variant of the *bacillaris* strain, Danforth and Wilson (1957) found evidence that ethanol-grown cells "adapted" to rapid acetate utilization by an increase in permeability to acetate ions.

The differences in the ability of the several strains of *Euglena gracilis* to utilize exogenous sugar are particularly interesting. The Mainx, Vischer, and related strains, whose heterotrophic abilities are severely restricted in other respects, are apparently entirely incapable of using sugars (Cramer and Myers, 1952; Pringsheim, 1955a; Baker *et al.*, 1955).

Strains of the type represented by *bacillaris* can utilize sugars, but only under very restricted conditions. Cramer and Myers (1952) observed that growth of *bacillaris* on glucose, though negligible in standing cultures, was increased by aeration and still further increased by aeration with additional CO_2. Pringsheim (1955a) found that growth of *bacillaris* and "Z" strains on glucose was very poor unless meat extract was added to the medium. Hutner *et al.* (1956; Baker *et al.*, 1955) obtained similar results using Krebs-cycle acids or related amino acids in place of meat extract. Apparently these organic acids, and similar substances in meat extract, serve the same purpose as CO_2 in the experiments of Cramer and Myers.

Cramer and Myers (1952) also found that growth on glucose was good at pH 4.5, but negligible at pH 6.8. Glucose utilization was strongly dependent on the concentration of sugar, the growth rate increasing with sugar concentration up to 1%. This is in marked contrast to substrates such as acetate and ethanol, where the rates of growth and metabolism are almost independent of concentration over a wide range. Finally, Cramer and Myers found that cells which had been grown in glucose media, when transferred to a fresh medium containing glucose, began to grow immediately, whereas cells which had been grown photosynthetically showed a lag of as much as 10 days in the same medium.

A third group of *E. gracilis* strains grow readily on sugars even in the absence of added CO_2, meat extract, or organic acids (Pringsheim, 1955a; Hutner *et al.*, 1956; Baker *et al.*, 1955). Pringsheim suggested that strains of this type be designated as *E. gracilis* var. *saccharophila*. As far as has been determined, those strains which can utilize sugars at all use both glucose and fructose, but not galactose, mannose, mannitol, or glycerol (Pringsheim 1955a, Hutner *et al.*, 1956). Pringsheim (1955a) found that cells of neither the *bacillaris* nor the *saccharophila* types grow on sucrose, although Hutner *et al.* (1956; Baker *et al.*, 1955) obtained growth of both strains on sucrose.

The causes underlying the inability of acetate flagellates to use exogenous sugars, and the differences between *Euglena* strains in this respect, remain puzzling. All of these organisms possess polysaccharide reserves made up of glucose polymers (see Meeuse, Chapter 18, this treatise). *Polytomella coeca* synthesizes starch by the usual "plant" pathways (Lwoff *et al.*, 1949, 1950; Bebbington *et al.*, 1952a, b). Both *Polytomella* (Wise, 1959) and *Euglena gracilis* var. *bacillaris* (Albaum *et al.*, 1950) contain all the usual phosphorylated intermediates of glycolysis.

Lwoff *et al.* (1949, 1950) have suggested that a deficiency of hexokinase might account for the "glucose block" by preventing the conversion of free glucose to phosphorylated derivatives which are the actual metabolic intermediates. Failure of the hexose phosphates to support growth of *Chilomonas* (Holz, 1954) is not a serious objection to this theory, since permeability to these compounds is a rarity among cells of all types. The writer has found no record of any serious attempt to demonstrate the presence or absence of hexokinase in an acetate flagellate, despite the obvious need for such studies.

Alternatively, failure to utilize sugars might be due to permeability limitations. Cells which readily utilize sugars possess specific, enzyme-like mechanisms for the transfer of sugar molecules across the cell membrane (Cohen and Monod, 1957; LeFevre, 1954; Rothstein, 1954a, b); in the absence of such "permeases" sugars would be expected to penetrate with difficulty, if at all. The demonstration of such a system in *Scenedesmus* (Taylor, 1960a, b) has been described above. If hexokinase and other kinases are actually components of the "permease" systems, as was discussed in connection with sugar uptake by *Scenedesmus*, these two explanations of the "glucose block" would become identical. In the utilization of sugar by *Euglena*, the special role of pH and the requirement for supplementary CO_2, organic acids, or meat extract remain obscure.

Lynch and Calvin (1953) found that in *Euglena gracilis* var. *bacillaris*, unlike other photosynthetic organisms studied, fixation of radioactive CO_2 in darkness produces the same pattern of labeled intermediates as does photosynthetic CO_2 fixation. The metabolic significance of this peculiarity of *Euglena* is still unknown.

E. Obligate Phototrophy

Obligate phototrophy, the total inability to grow in darkness, is one of the most puzzling characteristics found among the algae. Among the Volvocales, a number of obligately phototrophic *Chlamydomonas* species are known (J. C. Lewin, 1950; Wetherell, 1958), although other species are typical "acetate flagellates" (Sager and Granick, 1953) in their dark nutrition. Pringsheim and Pringsheim (1959) found a wide range of nutritional patterns among the colonial Volvocales. Some species appear to be obligate phototrophs, while others will grow heterotrophically on acetate, glucose, or both. (The finding that some strains require acetate even in the light is puzzling and invites further study.) The xanthophytes *Monodus subterraneus* and *Polyedriella helvetica* (Miller and Fogg, 1957, 1958; Belcher and Miller, 1960), several strains of Cyanophyta (Kratz and Myers, 1955), and certain species of *Euglena* (Lwoff, 1951), are obligate phototrophs. Obligate phototrophy appears to be widespread among the diatoms (J. C. Lewin, 1953) and dinoflagellates (Barker, 1935b). In most of these cases, other, closely related species are facultative heterotrophs.

Parker *et al.* (1961) tested the ability of 44 species, in 8 different genera of Chlorococcales, to grow in the dark by utilizing glucose or acetate. Glucose permitted heterotrophic growth of 24 species, some of which could also grow on acetate; the remaining 20 species were incapable of growth in darkness on either substrate. There was some correlation between heterotrophic abilities and taxonomic relationships; none of the 16 species of *Chlorococcum* tested grew heterotrophically under these conditions.

The problem of obligate phototrophy in various species of *Chlamydomonas* has been investigated by J. C. Lewin (1950), R. A. Lewin (1954), and Wetherell (1958). J. C. Lewin tested some 64 carbon compounds, including essentially all of those which have been found to support growth in other algae, and found that none would permit growth of *Chlamydomonas moewusii* in the dark. Acetate, pyruvate, and succinate stimulated respiration, proving that these compounds, at least, penetrated the cell membrane. Miller and Fogg (1958) similarly found that acetate, propionate, succinate, and fumarate stimulate respiration in the obligately phototrophic xanthophyte *Monodus*.

It was thought that light might be required for the synthesis of some metabolite essential to growth, but filtrates, extracts, or hydrolysates of light-grown cells did not permit growth of *Chlamydomonas moewusii* in darkness; nor did similar preparations of dark-grown cells inhibit growth in the light, as would be expected if light were required for the removal of some toxic substance (J. C. Lewin, 1950).

Wetherell (1958) obtained similar results with *Chlamydomonas eugametos*. In darkness cells of this species remained motile and continued to utilize their starch reserves for some days, but growth, protein synthesis, and nitrogen assimilation stopped within the first 9 hours; no substrate was found which permitted growth in the dark or stimulated growth at a limiting light intensity. A mixture of Krebs-cycle acids and related compounds stimulated respiration, though sugars did not. Cell extracts and other potential sources of growth factors were ineffective in permitting growth in darkness.

R. A. Lewin (1954) investigated the facultative heterotroph, *Chlamydomonas dysosmos*, which is normally capable of growth in darkness by the assimilation of acetate. Although acetate would not similarly support growth of a mutant in the dark, it nevertheless permitted growth in the light in a CO_2-free environment. J. C. Lewin (1953) obtained similar results with several diatoms; substances which would not support growth in the dark could replace CO_2 in the light. Such experiments argue strongly against the need for a unique product of light-dependent CO_2 fixation.

The finding of substances which would stimulate respiration but not support growth suggested that the "dark block" might be an inability to couple the oxidative release of energy with assimilation. R. A. Lewin (1954) found evidence supporting this theory from respirometric experiments with an obligately phototrophic mutant of *Chlamydomonas dysosmos*. Acetate stimulated respiration of both the mutant and wild-type cells but, whereas in the mutant the acetate was always completely oxidized, in the wild-type cells a portion of the acetate was apparently assimilated. In the presence of dinitrophenol, which "uncouples" phosphorylation from oxidative metabolism, the wild-type *Chlamydomonas dysosmos* behaved like the mutant. The inability to couple substrate oxidation with assimilation thus appears to provide an entirely satisfactory explanation for obligate phototrophy in this mutant, although it would be desirable to confirm Lewin's conclusions by the use of C^{14}-labeled acetate.

The same explanation proved inadequate for other obligate phototrophs, however. *Chlamydomonas moewusii*, though unable to grow heterotrophically, was found to assimilate acetate in darkness; on the other hand, a mutant of *C. debaryana* could neither oxidize acetate nor grow thereon, although the wild type could do both.

W. H. Thomas (personal communication) found that the dinoflagellate *Gonyaulax polyedra* is incapable of heterotrophic growth on any of the usual substrates which support growth of other algae. None of these compounds stimulated the respiration of intact cells either at pH 7–8 or at pH 3.5. O_2 uptake by cell homogenates was not increased by glucose, hexose diphosphate, acetate, pyruvate, citrate, succinate, or malate. Acetone powders did not take up oxygen or reduce DPN or TPN either in the presence or in the absence of intermediates of the glycolytic pathway or the Krebs cycle.

At present, it appears that obligate phototrophy has different causes in different organisms. In *Chlamydomonas* and *Monodus*, the problem is to explain why substances which can be oxidized, and in some cases assimilated, do not support growth. The results of R. A. Lewin (1954) indicate that several different mechanisms may operate in different species of *Chlamydomonas*. In *Gonyaulax*, on the other hand, it is the total lack of any response to the usual intermediates of carbon and energy metabolism which requires explanation.

F. *Ochromonas*

The nutrition of the chrysophyte *Ochromonas malhamensis* differs considerably from that of the other algae considered here. *Ochromonas* can ingest particulate food, and has for an alga relatively complex nutritional requirements, needing several vitamins and amino acids for optimum growth (Hutner *et al.*, 1953, 1957; Aaronson and Baker, 1959).

O. malhamensis requires sugars for rapid growth in either light or darkness. Glucose and sucrose are both effective. Glycerol permits growth, but at a slower rate than that on sugars. Pringsheim (1955b) presented comparative data on the abilities of sugars and related compounds to support growth of several species and strains of *Ochromonas*. In the absence of these carbon sources, growth in the light is very slow but apparently continuous (Hutner *et al.*, 1953, 1957; Myers and Graham, 1956). Citrate, succinate, glutamate and histidine apparently stimulate growth on sugars (Hutner *et al.*, 1953; but compare Myers and Graham, 1956), but are not utilized as sole carbon sources.

Various sugars, phosphorylated glycolysis intermediates, several Krebs-cycle acids, acetate, triacetin, and higher fatty acids all stimulate respiration (Reazin, 1954). Sugar catabolism follows the Emden-Meyerhof-Parnas glycolysis pathway, to the apparent exclusion of alternative routes. Anaerobically, sugar is fermented to alcohol and carbon dioxide (Reazin, 1956). Acetate appears to be oxidized but not to be assimilated (Reazin, 1954). The photosynthetic rate of *Ochromonas malhamensis* is so low that, in the presence of sugar, photosynthesis does not compensate for respiration, and there is a net *consumption* of oxygen even at saturating light intensity (Myers and Graham, 1956). However, Vishniac and Reazin (1957) found that during growth of *O. malhamensis* in the presence of glucose *and* light, the amount of cellular carbon formed was almost equal to the total carbon of the glucose consumed; in the presence of glycerol and light, the amount of cellular carbon formed exceeded the glycerol carbon consumed by nearly 30%, indicating clearly that considerable light-stimulated CO_2 fixation must have occurred. Since Vishniac and Reazin were also able to demonstrate photoreduction in the presence of organic H-donors, they suggested that the apparent weakness of *Ochromonas* photosynthesis is actually a limitation in the ability to evolve O_2, and that organic substrates stimulate growth in the light by serving not only as sources of carbon but also as H-donors for photoreduction (see Spruit, Chapter 3, this treatise). In another species, *Ochromonas danica*, photosynthesis plays a considerably more important role in nutrition (Pringsheim, 1955b; Aaronson and Baker, 1959).

III. Substrate Assimilation

A. Introductory Remarks

Substrates which support heterotrophic growth serve as both carbon sources and energy sources. A portion of the substrate is oxidized, and the energy released by this oxidation is used to convert the remainder into cellular material. Since Barker's (1935a, 1936) pioneering studies on the nonphotosynthetic alga *Prototheca zopfii*, the mechanisms and the efficiency of this "oxidative assimilation" have attracted increasing interest.

B. Stoichiometry of Oxidative Assimilation.

1. Hexose

Barker (1936) found that during short-term experiments on glucose utilization by *Prototheca* in a nitrogen-free medium only about 36% of the glucose carbon appeared as carbon dioxide, the remaining 64% being assimilated. Under aerobic conditions, there was no indication of the production of any metabolic waste-product other than CO_2. The respiratory quotient (R.Q.) for the oxidative assimilation of glucose was 1.10, as compared to a theoretical R.Q. of 1.00 if carbohydrate were the sole product of assimilation. The theoretical R.Q. for the synthesis of a fat such as tripalmitin is 4.0, while the lack of an endogenous nitrogen source made the synthesis of protein unlikely; hence the main assimilatory product was probably carbohydrate. Under anaerobic conditions, on the other hand, glucose was converted quantitatively (95–99%) to lactic acid, with no assimilation. As might be expected, the cells did not grow anaerobically (Barker, 1935a).

Various other substrates were also studied (Barker, 1936). For all except valeric acid, the carbon balances agreed fairly well with the assumption that carbohydrate is the main assimilatory product, but for valerate the data suggested a product more reduced than carbohydrate.

Barker was unable to show utilization of lactic or pyruvic acid by *Prototheca*, but Anderson (1945) found that these substances are oxidatively assimilated if the medium is sufficiently acidic to permit entry of the undissociated molecules.

Nongrowing cells of several species of *Chlorella* (Myers *et al.*, 1947; Cramer and Myers, 1949; Syrett, 1951; Fujita, 1959) and *Scenedesmus quadricauda* (Taylor, 1950) were found to assimilate 80–90% of the glucose carbon, while the remaining 10–20% appeared as CO_2. In Fujita's experiments, the oxidation-assimilation ratios were determined using radioactive glucose; in the other cases, where respiratory rates were determined manometrically, constant ratios were obtained only if it was assumed that the endogenous metabolism was suppressed during the oxidation of glucose, and oxidation-assimilation ratios were therefore calculated on the basis of this assumption. In contrast, Daniel (1956) found that the endogenous respiration of *Chlorella vulgaris* continued in the presence of glucose, and on this basis concluded that 86–89% of the glucose was assimilated.

The respiratory quotient for the oxidative assimilation of glucose under these conditions is approximately 1.0, as would be expected if carbohydrate were the major assimilatory product.

Taylor (1950) found small but consistent variations in the proportion of the sugar assimilated in media of varying acidity. Syrett (1951) separated the oxidation of glucose into cyanide-sensitive and cyanide-insensitive fractions, and showed that assimilation appeared to be coupled only to the cyanide-sensitive process. Daniel (1956) found that whereas respiration in the presence of glucose was depressed in *Chlorella* cells deficient in potassium, nitrate, or phosphate, the effect of the deficiency on the oxidation-assimilation ratio differed in each case. Potassium deficiency increased the fraction oxidized, nitrate deficiency did not alter the normal ratio, while phosphate deficiency increased the fraction assimilated from about 88% to about 93%. The fact that the percent of assimilation can be increased by such treatments and that a portion of the glucose oxidation is not coupled to respiration suggests that oxidative assimilation of glucose normally occurs at less than maximum efficiency. Syrett (1951) has pointed out that considerably higher efficiencies are theoretically possible on a free-energy basis.

In the above experiments growth was suppressed, either because of the lack of a nitrogen source, or because the cells had previously been incubated without a carbon source in order to deplete endogenous reserves (Myers and Cramer, 1948; Cramer and Myers, 1949). In growing cells, a larger portion of the substrate is oxidized to CO_2 and a smaller proportion assimilated. Thus, Barker (1935a) found that only 44% of the glucose carbon was assimilated during growth of *Prototheca*. In growing *Chlorella pyrenoidosa*, 43–51% is assimilated (Myers and Johnston, 1949; Samejima and Myers, 1958), and in growing *C. ellipsoidea*, 53–59% (Samejima and Myers, 1958).

The lower efficiency of carbon utilization during growth is due in part to the use of energy from substrate oxidation for nitrogen assimilation. Assimilation of nitrate increases the respiratory quotient of glucose metabolism in *Chlorella pyrenoidosa* (Cramer and Myers, 1948), whereas assimilation of ammonia decreases it. Less glucose carbon is assimilated during growth on nitrate than during growth on ammonia or urea (Samejima and Myers, 1958; see also Syrett, Chapter 10, this treatise).

Two other factors probably also contribute to the lower carbon efficiency during growth. First, in the absence of growth, most of the assimilated carbon is stored as carbohydrate reserves. During growth, all the constituents of the cells are synthesized, and the greater number of metabolic steps required would be expected to result in a lower efficiency. Secondly, during long-term growth experiments a greater proportion of the available energy may be used for "maintenance" processes rather than for synthesis. In support of this explanation, Samejima and Myers (1958) found efficiencies of growth to be lower on galactose than on glucose. The energy yield from oxidation of these two substrates should be identical, but growth on galactose is slower than on glucose, so that over any given period the ratio of "maintenance" requirements to galactose utilization is greater than the corresponding ratio for glucose.

2. Acetate

Oxidative assimilation of acetate has been studied in a number of algal species. In the absence of a nitrogen source, colorless cells of *Euglena gracilis* var. *bacillaris* oxidized 42% of acetate carbon and assimilated the remaining 58% (Wilson and Danforth, 1958). The respiratory quotient was 1.02, suggesting that carbohydrate is the main assimilatory product. In contrast to the usual findings with *Prototheca*, *Chlorella*, and *Scenedesmus*, the data for *Euglena* indicated that endogenous respiration continues during the utilization of an exogenous substrate. These conclusions have been directly confirmed by experiments with radioactive carbon (Danforth, 1961). Barker (1936) found that *Prototheca* assimilated 40–50% of acetate carbon under similar conditions. Values in the range of 50–63% assimilation were obtained for nongrowing cells of *Chlorella pyrenoidosa* (Myers *et al.*, 1947), *Chilomonas paramecium* (Blum *et al.*, 1951), and *Polytoma uvella* (Cirillo, 1956).

In *Euglena*, the *proportion* assimilated was constant over a range of conditions which produced considerable changes in the *rate* of acetate metabolism (Wilson and Danforth, 1958). In *Chlamydomonas dysosmos*, on the other hand, very little acetate was assimilated during the first hour of exposure to acetate; appreciable assimilation began on longer exposure. In *C. moewusii* and *C. debaryana* assimilation occurred without such a lag (R. A. Lewin, 1954).

Using radioactive acetate, Fujita (1959) found that in short-term experiments *Chlorella ellipsoidea* assimilated about 80–90% of the acetate carbon, as compared to the 50–60% assimilation usually found in other algae. Since 80% of the assimilated carbon could be recovered in a cell fraction consisting largely of organic acids and free amino acids, what was apparently being measured was the incorporation of acetate into a large "pool" of Krebs-cycle intermediates and related compounds, rather than its complete assimilation into storage carbohydrate. The latter process would require more energy and result in a lower proportion of assimilation.

In growing *C. pyrenoidosa*, acetate is only 26% assimilated (Samejima and Myers, 1958), as compared to about 50% assimilation by nongrowing cells (Myers *et al.*, 1947).

In reproducing cultures of *Chilomonas paramecium*, chemical techniques indicated that about 50% of the acetate carbon was assimilated, while calorimetric measurements indicated 43–47% assimilation (Hutchens *et al.*, 1948; Blum *et al.*, 1951). These efficiencies for growing cells are surprisingly close to values obtained for assimilation in the absence of growth.

The addition of ammonium to the medium slightly lowered acetate assimilation by *Polytoma uvella* from 57% to 51% (Cirillo, 1956).

C. Energy Efficiency of Oxidative Assimilation

Hutchens *et al.* (1948; Blum *et al.*, 1951) made an intensive attempt to obtain accurate values for the free-energy efficiency of acetate metabolism by *Chilomonas*. Combining calorimetric, respirometric, and chemical data, they estimated efficiencies of 27%–31% in the absence of growth, and about 17% during growth. Samejima and Myers (1958) calculated a much higher energy efficiency, 56–58%, for growth of *Chlorella pyrenoidosa* on glucose.

D. Products and Pathways of Assimilation

As has been mentioned, carbohydrate appears to be the major product of assimilation of glucose and acetate by nongrowing algae of a variety of species. The respiratory quotient under such conditions is approximately 1.0. In short-term experiments with *Scenedesmus* and *Chlorella* Taylor (1950) and Syrett (1951) found that about 80% of assimilated glucose carbon can be recovered in cell extracts as nonreducing carbohydrate. The carbohydrate is a mixture of low (alcohol-soluble) and high (alcohol-insoluble) polymers; in *Scenedesmus* (Taylor, 1950) it is hydrolyzable by acid but not by diastase, and therefore is presumably not starch. Fredrick (1951, 1952) found that the extracts of the Blue-green Alga *Oscillatoria princeps* synthesized short, branched-chain polysaccharides similar to animal glycogen. Variant strains, produced by growth at low temperature, formed a less branched polysaccharide; they reverted when grown at ordinary temperatures (Fredrick, 1952, 1953a).

Though as a general rule carbohydrate is the chief assimilatory product in the absence of growth, this does not hold for all substrates. Barker (1936) found that the carbon-oxygen balance of valerate assimilation by *Prototheca* indicated the accumulation of a product more reduced than carbohydrate. For ethanol assimilation by *Euglena*, Wilson and Danforth (1958) found a respiratory quotient of 0.33, as compared to a theoretical value of 0.15 to be expected if the assimilatory product were carbohydrate. These results are most readily explained if it is assumed that both carbohydrate and lipid are synthesized from such substrates, but this explanation requires experimental confirmation.

In nongrowing cells the very high proportion of assimilation directed toward carbohydrate synthesis is probably to be regarded either as an "overflow" mechanism (Myers and Cramer, 1948) which operates when substrate is available in excess of that which can be used for growth, or, in the case of starved cells, as a process for rebuilding depleted reserves. Cells of *Chlorella pyrenoidosa* grown on glucose have an estimated composition of 36% carbohydrate, 49% protein, and 15% lipid (Samejima and Myers, 1958). Bergman (1955) reported a very different composition for heterotrophically grown *C. vulgaris*: 70% carbohydrate, 15% protein, and 15% lipid. Since little or no organic material is excreted, the composition of the cells is an accurate reflection of the organic end-products of glucose metabolism. However, since the relative proportions of these products can be greatly altered by varying the growth conditions (Spoehr and Milner, 1949; Bergmann, 1955), one can thereby account for the discrepant results described above.[2]

From the chemical composition of *Chilomonas* cells harvested from logarithmic growth on acetate, Hutchens and co-workers (Hutchens *et al.*, 1948; Blum *et al.*, 1951) estimated that 39% of the assimilated carbon was converted to starch, 42% to protein, and 19% to fat.

After long-term growth of *Polytomella coeca* on radioactive acetate, Bourne and collaborators (Barker and Bourne, 1955) found about 63% of the cellular radioactivity in starch. Synthesis of starch by *Polytomella* occurs by the phosphorylase mechanism and is entirely similar to that in higher plants. Enzymes for alternative pathways of starch synthesis have not been detected (Lwoff, *et al.*, 1949, 1950; Bebbington *et al.*, 1952a, b; Barker and Bourne, 1955; Bourne *et al.*, 1950; Barker *et al.*, 1953). The same mechanism for polysaccharide synthesis was demonstrated in *Oscillatoria* (Fredrick 1951–1956; Fredrick and Mulligan, 1955), and there is no reason to doubt that this pathway is typical of the algae in general.

Using radioactive acetate, Bevington *et al.* (1950) showed that 78% of the carbon of *Polytomella* starch was derived from the methyl carbon of acetate,

[2] Cells tend to have an extremely high carbohydrate or lipid content and a low protein content when nutritional deficiencies, especially nitrogen deficiency, limit growth (Spoehr and Milner, 1949). Nitrogen had probably been limiting in these experiments of Bergman, since the total nitrogen content of the growth medium, which initially contained 0.81 gm. KNO_3 per liter, was equivalent to only about 3% of the dry weight of the final yield of cells (3.6 gm./liter).

and only 22% from the carboxyl carbon.[3] Carbon atoms 1, 2, 5, and 6 of the glucose units were derived mainly from the methyl carbon of acetate, and atoms 3 and 4 largely from the carboxyl carbon (Bevington *et al.*, 1953). The distribution of methyl and carboxyl carbons was that which would be expected if the acetate were converted to oxalacetate via the Krebs and glyoxylate cycles (see below), the oxalacetate decarboxylated to pyruvate, and the pyruvate converted to carbohydrate by a reversal of glycolysis. Glycolytic intermediates have been demonstrated in *Polytomella* (Wise, 1959), *Euglena* (Albaum *et al.*, 1950), *Chlorella* and *Scenedesmus* (Calvin, 1959).

Evidence has been presented for the operation of the Krebs cycle in *Chlorella* (Millbank, 1957), *Chilomonas* (Holz, 1954), *Euglena* (Danforth, 1953), and *Astasia* (F. R. Hunter and J. W. Lee, personal communication). The assimilation of acetate and related compounds by this pathway poses a problem; oxalacetate is continually being removed for assimilatory purposes, and the cycle cannot be expected to continue without some additional source of oxalacetate. The glyoxylate cycle[4] proposed by Kornberg and Krebs (1957) has been shown to provide the additional oxalacetate in a wide variety of microorganisms (Kornberg, 1959) and in fat-consuming plant tissues (Bradbeer and Stumpf, 1959; Carpenter and Beevers, 1959); however, the only demonstration of the glyoxylate cycle in an alga, *Polytoma uvella*, is the work of Plackett (cited in Kornberg, 1959).

The "tail-to-tail" condensation of two molecules of acetate to form succinate (Seaman and Naschke, 1955) has been suggested as another way of replacing the oxalacetate used up in syntheses. It is doubtful, however, that this reaction is ever a major source of oxalacetate *in vivo* (Kornberg, 1959).

Fats, proteins, and other assimilatory products can be derived from the intermediates of glycolysis and of the Krebs and glyoxylate cycles by routes discussed elsewhere in this treatise.

E. Effect of Light on Substrate Assimilation

In *Chlorella pyrenoidosa* (Myers *et al.*, 1947) and *C. ellipsoidea* (Fujita. 1959) light had no apparent effect on the rate of substrate assimilation. On the other hand, in very brief experiments with *Euglena gracilis* var. *bacillaris*, Lynch and Calvin (1953) found that light increased net acetate assimilation.

[3] Danforth (1961) obtained essentially similar results with *Euglena gracilis*.
[4] See Chapter 4 by Gibbs, in this treatise.

ACKNOWLEDGMENTS

The author is indebted to Dr. William H. Thomas and Dr. Frissell R. Hunter for making available their unpublished results, and to Dr. H. L. Kornberg for his prompt and helpful answer to inquiries. Especial thanks are due to Dr. Joyce C. Lewin for supplying an extensive and valuable bibliography which greatly simplified this task.

REFERENCES

Aaronson, S., and Baker, H. (1959). A comparative biochemical study of two species of *Ochromonas*. *J. Protozool.* **6,** 282–284.

Albaum, H. G., Schatz, A., Hutner, S. H., and Hirshfeld, A. (1950). Phosphorylated compounds in *Euglena*. *Arch. Biochem. Biophys.* **29,** 210–218.

Anagnostakos, N. P. (1956). Carbon sources for green and colorless Euglenida. *J. Protozool.* **3,** Suppl. 12 (abstract).

Anderson, E. H. (1945). Studies on the metabolism of the colorless alga *Prototheca zopfii*. *J. Gen. Physiol.* **28,** 297–327.

Baker, H., Hutner, S. H., and Sobotka, H. (1955). Nutritional factors in thermophily: a comparative study of bacilli and *Euglena*. *Ann. N. Y. Acad. Sci.* **62,** 349–376.

Barker, H. A. (1935a). The metabolism of the colorless alga *Prototheca zopfii* Krüger. *J. Cellular Comp. Physiol.* **7,** 73–93.

Barker, H. A. (1935b). The culture and physiology of the marine dinoflagellates. *Arch. Mikrobiol.* **6,** 157–181.

Barker, H. A. (1936). The oxidative metabolism of the colorless alga, *Prototheca zopfii*. *J. Cellular Comp. Physiol.* **8,** 231–250.

Barker, S. A., and Bourne, E. J. (1955). Composition and synthesis of the starch of *Polytomella coeca*. *In* "Biochemistry and Physiology of Protozoa" (S. H. Hutner and A. Lwoff, eds.), Vol. II, pp. 45–56. Academic Press, New York.

Barker, S. A., Bebbington, A., and Bourne, E. J. (1953). The mode of action of the Q-enzyme of *Polytomella coeca*. *J. Chem. Soc.* pp. 4051–4057.

Bebbington, A., Bourne, E. J., Stacey, M., and Wilkinson, I. A. (1952a). The Q-enzyme of *Polytomella coeca*. *J. Chem. Soc.* pp. 240–245.

Bebbington, A., Bourne, E. J., and Wilkinson, I. A. (1952b). The conversion of amylose into amylopectin by the Q-enzyme of *Polytomella coeca*. *J. Chem. Soc.* pp. 246–253.

Belcher, J. H., and Fogg, G. E. (1958). Studies on the growth of Xanthophyceae in pure culture. *Tribonema aequale* Pascher. *Arch. Mikrobiol.* **30,** 17–22.

Belcher, J. H., and Miller, J. D. A. (1960). Studies on the growth of Xanthophyceae in pure culture. IV. Nutritional types amongst the Xanthophyceae. *Arch. Mikrobiol.* **36,** 219–228.

Bergmann, L. (1955) Stoffwechsel und Mineralsalzernährung einzelliger Grünalgen. II. Vergleichende Untersuchen über den Einfluss mineralischer Faktoren bei heterotropher und mixotropher Ernährung. *Flora (Jena)* **142,** 493–539.

Bevington, J. C., Bourne, E. J., and Wilkinson, I. A. (1950). A microbiological method for the preparation of C^{14}-labelled starch from sodium acetate. *Chem. & Ind. (London)* pp. 691–692.

Bevington, J. C., Bourne, E. J., and Turton, C. N. (1953). Chemical degradation of C^{14}-glucose and its application to C^{14}-starch from *Polytomella coeca*. *Chem. & Ind. (London)* pp. 1390–1391.

Blum, J. J., Podolsky, B., and Hutchens, J. O. (1951). Heat production in *Chilomonas*. *J. Cellular Comp. Physiol.* **37,** 403–426.

Bourne, E. J., Stacey, M., and Wilkinson, I. A. (1950). The composition of the polysaccharide synthesized by *Polytomella coeca*. *J. Chem. Soc.* pp. 2649–2698.

Bradbeer, C., and Stumpf, P. K. (1959). Fat metabolism in higher plants. XI. The conversion of fat into carbohydrate in peanut and sunflower seedlings. *J. Biol. Chem.* **234,** 498–501.

Calvin, M. (1959). Energy reception and transfer in photosynthesis. *In* "Biophysical Science—A Study Program" (J. L. Oncley, ed.), pp. 147–156. Wiley, New York.

Carpenter, W. D., and Beevers, H. (1959). Distribution and properties of isocitricase in plants. *Plant Physiol.* **34,** 403–409.

Cirillo, V. P. (1955). Induction and inhibition of adaptive enzyme formation in a phytoflagellate. *Proc. Soc. Exptl. Biol. Med.* **88,** 352–354.

Cirillo, V. P. (1956). Induced enzyme synthesis in the phytoflagellate, *Polytoma*. *J. Protozool.* **3,** 69–74.

Cirillo, V. P. (1957). Long-term adaptation to fatty acids by the phytoflagellate, *Polytoma uvella*. *J. Protozool.* **4,** 60–62.

Cohen, G. N., and Monod, J. (1957). Bacterial permeases. *Bacteriol. Revs.* **21,** 169–194.

Cosgrove, W. B., and Swanson, B. K. (1952). Growth of *Chilomonas paramecium* in simple organic media. *Physiol. Zoöl.* **25,** 287–292.

Cramer, M., and Myers, J. (1948). Nitrate reduction and assimilation in *Chlorella*. *J. Gen. Physiol.* **32,** 93–102.

Cramer, M., and Myers, J. (1949). Effects of starvation on the metabolism of *Chlorella*. *Plant Physiol.* **24,** 255–264.

Cramer, M., and Myers, J. (1952). Growth and photosynthetic characteristics of *Euglena gracilis*. *Arch. Mikrobiol.* **17,** 384–402.

Danforth, W. F. (1953). Oxidative metabolism of *Euglena*. *Arch. Biochem. Biophys.* **46,** 164–173.

Danforth, W. F. (1961). Oxidative assimilation of acetate by *Euglena*. Carbon balance and effects of ethanol. *J. Protozool.* **8,** 152–158.

Danforth, W. F., and Wilson, B. W. (1957). Adaptive changes in the oxidative metabolism of *Euglena*. *J. Protozool.* **4,** 52–55.

Daniel, A. L. (1956). Stoffwechsel und Mineralsalzernährung einzelliger Grünalgen. III. Atmung und oxydative Assimilation von *Chlorella*. *Flora (Jena)* **143,** 31–66.

Eny, D. M. (1950). Respiration studies on *Chlorella*. I. Growth experiments with acid intermediates. *Plant Physiol.* **25,** 478–495.

Eny, D. M. (1951). Respiration studies on *Chlorella*. II. Influence of various organic acids on gas exchange. *Plant Physiol.* **26,** 268–289.

Erickson, L. C., Wedding, R. T., and Brannaman, B. L. (1955). Influence of pH on 2,4-dichlorophenoxyacetic acid and acetic acid activity in *Chlorella*. *Plant Physiol.* **30,** 69–74.

Finkle, B. J., Appleman, D., and Fleischer, F. K. (1950). Growth of *Chlorella vulgaris* in the dark. *Science* **111,** 309.

Fogg, G. E. (1953). "The Metabolism of Algae." Wiley, New York.

Fredrick, J. F. (1951). Preliminary studies on the synthesis of polysaccharides in the algae. *Physiol. Plantarum* **4,** 621–626.

Fredrick, J. F. (1952). Preliminary studies on the synthesis of polysaccharides in the algae. II. A polysaccharide variant of *Oscillatoria princeps*. *Physiol. Plantarum* **5,** 37–40.

Fredrick, J. F. (1953a). Preliminary studies on the synthesis of polysaccharides in the algae. III. Induction of polysaccharide variants in *Oscillatoria princeps* by low temperatures. *Physiol. Plantarum* **6,** 96–99.

Fredrick, J. F. (1953b). Preliminary studies on the synthesis of polysaccharides in the algae. IV. Branching characteristics of *Oscillatoria* polysaccharides. *Physiol. Plantarum* **6,** 100–105.

Fredrick, J. F. (1954). Preliminary studies on the synthesis of polysaccharides in the algae. V. Kinetics of polysaccharide formation in extracts of *Oscillatoria princeps. Physiol. Plantarum* **7,** 182–189.

Fredrick, J. F. (1956). Physicochemical studies of the phosphorylating enzymes of *Oscillatoria princeps. Physiol. Plantarum* **9,** 446–451.

Fredrick, J. F., and Mulligan, F. J. (1955). Mechanism of action of the branching enzyme from *Oscillatoria* and the structure of branched dextrins. *Physiol. Plantarum* **8,** 74–83.

Fujita, K. (1959). The metabolism of acetate in *Chlorella* cells. *J. Biochem.* (*Tokyo*) **46,** 253–268.

Gross, J. A., and Jahn, T. L. (1958). Some biological characteristics of chlorotic substrains of *Euglena gracilis. J. Protozool.* **5,** 126–133.

Gross, J. A., Jahn, T. L., and Bernstein, E. (1955). The effect of antihistamines on the pigments of green protista. *J. Protozool.* **2,** 71–75.

Holz, G. G. (1954). The oxidative metabolism of a cryptomonad flagellate, *Chilomonas paramecium. J. Protozool.* **1,** 114–120.

Hutchens, J. O. Podolsky, B., and Morales, M. F. (1948). Studies on the kinetics and energetics of carbon and nitrogen metabolism of *Chilomonas paramecium. J. Cellular Comp. Physiol.* **32,** 117–141.

Hutner, S. H., and Provasoli, L. (1951). The phytoflagellates. *In* "Biochemistry and Physiology of Protozoa" (A. Lwoff, ed.), Vol. I, pp. 27–128. Academic Press, New York.

Hutner, S. H., and Provasoli, L. (1955). Comparative biochemistry of flagellates. *In* "Biochemistry and Physiology of Protozoa" (S. H. Hutner and A. Lwoff, eds.), Vol. II, pp. 17–43. Academic Press, New York.

Hutner, S. H., Provasoli, L., and Filfus, J. (1953). Nutrition of some phagotrophic fresh-water chrysomonads. *Ann. N. Y. Acad. Sci.* **56,** 852–862.

Hutner, S. H., Bach, M. K., and Ross, G. I. M. (1956). A sugar-containing basal medium for vitamin B_{12}-assay with *Euglena*; application to body fluids. *J. Protozool.* **3,** 101–112.

Hutner, S. H., Baker, H., Aaronson, S., Nathan, H. A., Rodriguez, E., Lockwood, S., Sanders, M., and Petersen, R. A. (1957). Growing *Ochromonas malhamensis* above 35°C. *J. Protozool.* **4,** 259–269.

Jacobs, M. H. (1940). Some aspects of cell permeability to weak electrolytes. *Cold Spring Harbor Symposia Quant. Biol.* **8,** 30–39.

Kornberg, H. L. (1959). Aspects of terminal respiration in microorganisms. *Ann. Rev. Microbiol.* **13,** 49–78.

Kornberg, H. L., and Krebs, H. A. (1957). Synthesis of cell constituents from C_2-units by a modified tricarboxylic acid cycle. *Nature* **179,** 988–991.

Kratz, W. A., and Myers, J. (1955). Nutrition and growth of several blue-green algae. *Am. J. Botany* **42,** 282–287.

Krauss, R. W. (1958). Physiology of the fresh-water algae. *Ann. Rev. Plant Physiol.* **9,** 207–244.

LeFevre, P. G. (1954). The evidence for active transport of monosaccharides across the red cell membrane. *Symposia Soc. Exptl. Biol.* **No. 8,** 118–135.

Lewin, J. C. (1950). Obligate autotrophy in *Chlamydomonas Moewusii* Gerloff. *Science* **112,** 652–653.

Lewin, J. C. (1953). Heterotrophy in diatoms. *J. Gen. Microbiol.* **9,** 305–313.

Lewin, J. C., and Lewin, R. A. (1960). Auxotrophy and heterotrophy in marine littoral diatoms. *Can. J. Microbiol.* **6,** 127–134.

Lewin, R. A. (1954). The utilization of acetate by wild-type and mutant *Chlamydomonas dysosmos. J. Gen. Microbiol.* **11,** 459–471.

Lwoff, A. (1951). Introduction to biochemistry of protozoa. *In* "Biochemistry and Physiology of Protozoa" (A. Lwoff, ed.), Vol. I, pp. 1–26. Academic Press, New York.

Lwoff, A., Ionesco, H., and Gutmann, A. (1949). Metabolisme de l'amidon chez un flagellé sans chlorophylle incapable d'utiliser le glucose. *Compt. rend. acad. sci.* **228,** 342–344.

Lwoff, A., Ionesco, H., and Gutmann, A. (1950). Synthèse et utilisation de l'amidon chez un flagellé sans chlorophylle incapable d'utiliser le sucres. *Biochim. et Biophys. Acta* **4,** 270–275.

Lynch, V. H., and Calvin, M. (1953). CO_2 fixation by *Euglena. Ann. N. Y. Acad. Sci.* **56,** 890–900.

Millbank, J. W. (1957). Studies on the preparation of a respiratory cell-free extract from the alga *Chlorella. J. Exptl. Botany* **8,** 96–104.

Miller, J. D. A., and Fogg, G. E. (1957). Studies on the growth of Xanthophyceae in pure culture. I. The mineral nutrition of *Monodus subterraneus* Petersen. *Arch. Mikrobiol.* **28,** 1–17.

Miller, J. D. A., and Fogg, G. E. (1958). Studies on the growth of Xanthophyceae in pure culture. II. The relation of *Monodus subterraneus* to organic substances. *Arch. Mikrobiol.* **30,** 1–16.

Myers, J. (1951). Physiology of the algae. *Ann. Rev. Microbiol.* **5,** 157–180.

Myers, J., and Cramer, M. (1948). Metabolic conditions in *Chlorella. J. Gen. Physiol.* **32,** 103–110.

Myers, J., and Graham, J. R. (1956). The role of photosynthesis in the physiology of *Ochromonas. J. Cellular Comp. Physiol.* **47,** 397–414.

Myers, J., and Johnston, J. A. (1949). Carbon and nitrogen balance of *Chlorella* during growth. *Plant Physiol.* **24,** 111–119.

Myers, J., Cramer, M., and Johnston, J. (1947). Oxidative assimilation in relation to photosynthesis in *Chlorella. J. Gen. Physiol.* **30,** 217–227.

Neish, A. C. (1951). Carbohydrate nutrition of *Chlorella vulgaris. Can. J. Botany* **29,** 68–78.

Parker, B. C., Bold, H. C., and Deason, T. R. (1961). Facultative heterotrophy in some chloroccacean algae. *Science* **133,** 761–763.

Pringsheim, E. G. (1955a). Kleine Mitteilungen über Flagellaten und Algen. II. *Euglena gracilis* var. *saccharophila* n. var. und eine vereinfachte Nährlösung zur Vitamin B_{12} Bestimmung. *Arch. Mikrobiol.* **21,** 414–419.

Pringsheim, E. G. (1955b). Kleine Mitteilungen über Flagellaten und Algen. Über *Ochromonas danica* n. sp. und andere Arten der Gattung. *Arch. Mikrobiol.* **23,** 181–192.

Pringsheim, E. G. (1959). Heterotrophie bei Algen und Flagellaten. *In* "Handbuch der Pflanzenphysiologie" (W. Ruhland, ed.), Vol. 11, pp. 303–326. Springer, Berlin.

Pringsheim, E. G., and Pringsheim, O. (1952). Experimental elimination of chromatophores and eye-spot in *Euglena gracilis*. *New Phytologist* **51,** 65–76.

Pringsheim, E. G., and Pringsheim, O. (1959). Die Ernährung koloniebildener Volvocales. *Biol. Zentr.* **78,** 937–971.

Provasoli, L., Hutner, S. H., and Schatz, A. (1948). Streptomycin-induced chlorophyll-less races of *Euglena*. *Proc. Soc. Exptl. Biol. Med.* **69,** 279–282.

Reazin, G. H. (1954). On the dark metabolism of a golden-brown alga, *Ochromonas malhamensis*. *Am. J. Botany* **41,** 771–777.

Reazin, G. H. (1956). The metabolism of glucose by the alga *Ochromonas malhamensis*. *Plant Physiol.* **31,** 299–303.

Rothstein, A. (1954a). Enzyme systems of the cell surface involved in the uptake of sugars by yeast. *Symposia Soc. Exptl. Biol.* **No. 8,** 165–201.

Rothstein, A. (1954b). The enzymology of the cell surface. *In* "Protoplasmatologia" (L. V. Heilbrunn and F. Weber, eds.), Vol. II, Part E5. Springer, Vienna.

Sager, R., and Granick, S. (1953). Nutritional studies with *Chlamydomonas reinhardi*. *Ann. N. Y. Acad. Sci.* **56,** 831–838.

Samejima, H., and Myers, J. (1958). On the heterotrophic growth of *Chlorella pyrenoidosa*. *J. Gen. Microbiol.* **18,** 107–117.

Seaman, G. R., and Naschke, M. D. (1955). Reversible cleavage of succinate by extracts of *Tetrahymena*. *J. Biol. Chem.* **217,** 1–12.

Spoehr, H. A., and Milner, H. W. (1949). The chemical composition of *Chlorella*; effects of environmental conditions. *Plant Physiol.* **24,** 120–149.

Syrett, P. J. (1951). The effect of cyanide on the respiration and oxidative assimilation of glucose by *Chlorella vulgaris*. *Ann. Botany (London)* **15,** 473–492.

Taylor, F. J. (1950). Oxidative assimilation of glucose by *Scenedesmus quadricauda*. *J. Exptl. Botany* **1,** 301–321.

Taylor, F. J. (1960a). The absorption of glucose by *Scenedesmus quadricauda*. I. Some kinetic aspects. *Proc. Roy. Soc.* **B151,** 400–418.

Taylor, F. J. (1960b). The absorption of glucose by *Scenedesmus quadricauda*. II. The nature of the absorptive process. *Proc. Roy. Soc.* **B151,** 483–496.

Vishniac, W., and Reazin, G. H. (1957). Photoreduction in *Ochromonas malhamensis*. *In* "Research in Photosynthesis" (H. Gaffron, A. H. Brown, C. S. French, R. Livingston, E. I. Rabinovitch, B. L. Strehler, and N. E. Tolbert, eds.), pp. 239–242. Interscience, New York.

Von Dach, H. (1942). Respiration of a colorless flagellate, *Astasia klebsii*. *Biol. Bull.* **82,** 356–371.

Von Dach, H. (1953). Effects of some intermediary metabolites on the rate of O_2 consumption of a colorless euglenoid flagellate, *Astasia klebsii*. *Federation Proc.* **12,** 149.

Wetherell, D. F. (1958). Obligate phototrophy in *Chlamydomonas eugametos*. *Physiol. Plantarum* **11,** 260–274.

Wilson, B. W., and Danforth, W. F. (1958). The extent of acetate and ethanol oxidation by *Euglena gracilis*. *J. Gen. Microbiol.* **18,** 535–542.

Wilson, B. W., Buetow, D. E., Jahn, T. L., and Levedahl, B. H. (1959). A differential effect of pH on cell growth and respiration. *Exptl. Cell Research* **18,** 454–465.

Wise, D. L. (1955). Carbon sources for *Polytomella caeca*. *J. Protozool.* **2,** 156–158.

Wise, D. L. (1959). Carbon nutrition and metabolism of *Polytomella caeca*. *J. Protozool.* **6,** 19–23.

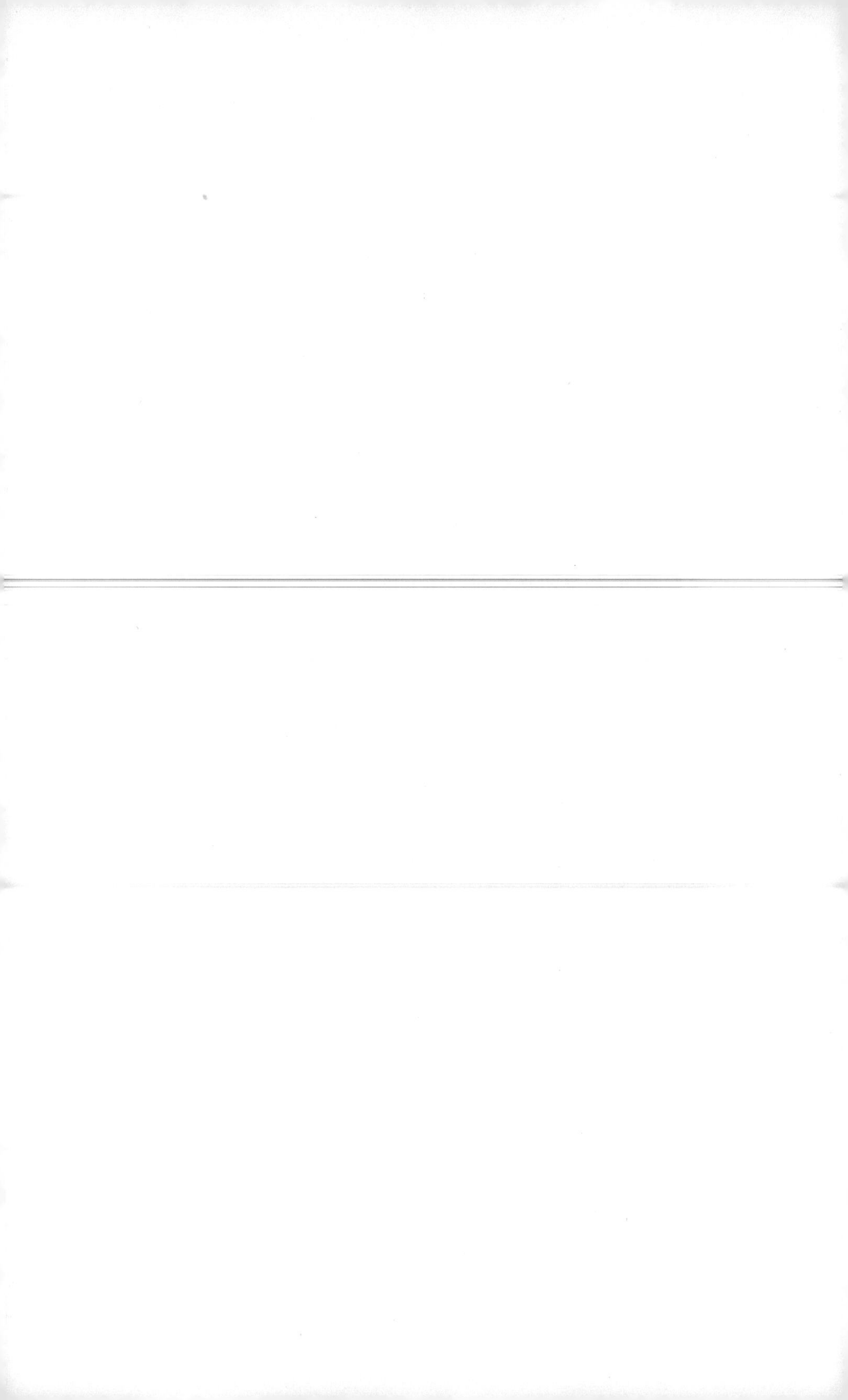

Enzyme Systems

G. JACOBI

Botanical Institute, Academy of Veterinary Medicine, Hannover, Germany

I. Introduction [1]

Although algae have been used extensively for fundamental investigations of plant metabolism, there is still very little information on the enzymes in these organisms. The main reasons for this may be the small amounts of proteins which can be extracted from those unicellular forms, such as *Chlorella* and *Scenedesmus*, which are generally used for physiological and biochemical investigations, and the presence of substances which inhibit enzyme action in many of the larger algae, such as the brown and red seaweeds.

At the present time we have a sound knowledge of the metabolism of a few unicellular forms from manometric and tracer experiments. Many statements on the existence of specific enzyme systems have been based on indirect evidence from such experiments by extrapolation from the knowledge of enzymes found in higher plants and animals. However, a chemical determination or a labeled product does not give a complete picture of the dynamic processes of metabolism in the cell. It is necessary to know how the enzymatic products are formed, especially in cases where several synthetic pathways are possible or where one compound may be attacked by two or

[1] Abbreviations used in this chapter: DPN, diphosphopyridine nucleotide; TPN, triphosphopyridine nucleotide; FAD, flavine adenine dinucleotide.

more enzymes. In such cases it is useful to obtain kinetic and thermodynamic data for the enzymes involved. For a complete picture, it is necessary to combine both approaches: the chemical or tracer experiments and the enzymatic studies. From this point of view, unicellular algae are particularly suitable for metabolic study, since they are not differentiated into specialized tissues and the whole metabolism occurs in each cell.

In the following article, the author refers only to those investigations which offer direct proof for the presence of the enzyme concerned. Other studies, providing only indirect evidence, are excluded, or are referred to only briefly in relation to specific metabolic processes.

II. Preparation of Enzyme Extracts from Algae

The usual method for extracting proteins with high enzymatic activity, by grinding or homogenizing the plant material with a suitable buffer, is often unsuccessful when applied to algae. For unicellular forms with a firm cell wall, an acetone powder often provides the maximum yield of enzymes (Holzer and Holzer, 1952). Aldolase activity, for example, was higher in an acetone powder of *Chlorella pyrenoidosa* than in an extract prepared from cells treated with liquid air. Millbank (1957), in attempts to remove from *C. vulgaris* the permeability barrier to respiratory substrates, tried several methods for disintegration of the cells. With various mechanical methods, with subsonic and ultrasonic disintegration, and with the all-glass Potter homogenizer, he was unable to prepare active extracts. Freezing the cells, extraction with acetone powder, or freezing in solid CO_2 without further treatment resulted in preparations exhibiting endogenous respiration but no stimulation by substrates such as members of the tricarboxylic acid cycle. The only successful method Millbank found for obtaining cell preparations capable of respiration was drying the cells in a dessicator over P_2O_5 at room temperature and then extracting them with water. It should be noted that even after such treatments the respiratory enzymes remained associated with the cell residues. Although these experiments by Millbank demonstrated that some of the enzymes necessary for oxidation of the substrates he tested could not be extracted by any of the various methods used, mechanical disintegration of *C. vulgaris* with the Milner press (Milner *et al.*, 1950) yielded cell-free preparations showing transaminase activity (Millbank, 1953). The difficulties experienced by Millbank may be a feature peculiar to unicellular algae such as *Chlorella*, since cells of *Hydrodictyon*, which also belongs to the Chlorococcales, can be ruptured much more easily. Richter (1957) concluded that treatment with liquid air was as effective as grinding with sand in a mortar in preparing triose isomerase and aldolase from *Hydrodictyon reticulatum*. Jacobi (1957a) pointed out that from thalli of *Ulva lactuca*, a marine Green Alga, more of the enzymes active in glycolysis and in the tricarboxylic acid cycle could be extracted with buffer in the Potter-Elvehjem apparatus than could be extracted by grinding or by disintegration of the cell with liquid air. A phosphate buffer was found to be more effective than diethanolamine. On the other hand, extraction of transaminases is more successful after homogenizing the alga in diethanolamine buffer at pH 8.4 (Jacobi, 1957b). One can usually extract a higher yield of total proteins by the use of 0.25% Na_2CO_3 at pH 10.2, according to Duncan *et al.* (1956), who successfully employed this method for the extraction of carbohydrases. However this solution was found to inactivate transaminases (Jacobi, 1957b).

The extraction of enzymes from Brown Algae presents special difficulties, since they often contain mucilaginous compounds which inhibit enzyme action. No enzyme activity could be demonstrated in an acetone powder of *Laminaria saccharina* (Jacobi, 1958). The inhibitory effect of the slimy extract of *Laminaria* or *Fucus* was demonstrated in purified enzyme preparations with known activity and in the presence of all compounds and substrates necessary for glycolysis. A small residual activity of malic acid dehydrogenase and of the malic enzyme remained after homogenizing the fronds of *Fucus vesiculosus* and *Laminaria saccharina* in 70% saturated ammonium sulfate, although this preparation contained mucilages somewhat inhibitory to most glycolytic enzymes. Another Brown Alga, *Desmarestia viridis*, which possesses a very acid cell sap, viz. pH 1.0 (see Schiff, Chapter 14, this treatise), was also unsatisfactory for the preparation of enzymes. In crude aqueous extracts of Red Algae, the water-soluble pigment phycoerythrin interferes with assays based on ultraviolet absorption by reduced pyridine nucleotide, since the colored extracts have a high absorption in the same range. Clearly, the extraction method of choice varies from one object to another, and there is no generally suitable procedure for preparing enzymes from algae.

III. Enzyme Systems

A. Enzymes of Carbohydrate Metabolism and Related Processes

1. *Carbohydrases*

Several unusual carbohydrates of high or low molecular weight occur as storage products in algae (Lindberg, 1953a, b; see Meeuse, Chapter 18, this treatise). Their presence presumably reflects the existence of special enzyme systems responsible for their synthesis and degradation. Unfortunately we still have no information on such enzymes comparable with that available for starch. The problem of the synthesis of high molecular-weight carbohydrates is not a simple one, as was demonstrated in the seaweed *Iridophycus flaccidum* (Rhodophyta) by Bean and Hassid (1955), using tracer methods. Their results pointed to mechanisms by which either uridine diphosphoglucose or uridine diphosphogalactose was involved in the synthesis of floridean starch. In the same way one might expect trans-glucosylations to operate in the synthesis of such products of Brown Algae as laminarin and alginic acid (Duncan and Manners, 1958; Duncan *et al.*, 1959).

The synthetic carbohydrate pathways of some Green Algae end in starch synthesis, as shown by Richter (1956) and Richter and Pirson (1957), who demonstrated amylase and phosphorylase activity in the freshwater alga *Hydrodictyon reticulatum*, and by Clauss (1959) in similar studies of *Acetabularia mediterranea*. By histochemical methods, Yin (1948) showed that phosphorylase is localized in the chloroplasts of *Spirogyra*, *Zygnema*, and *Cosmarium*. On the other hand, Holzer and Holzer (1952) could not detect phosphorylase in *Chlorella pyrenoidosa* cells, although glucose metabolism in this organism is believed to be similar to that in most others. The lack of phosphorylase activity is in accordance with the demonstration by Taylor (1950) that in *Scenedesmus* most of the glucose taken up by the cells is synthesized into an acid-hydrolyzable polysaccharide not identical with starch.

The problems of carbohydrate degradation and of oligosaccharide synthesis in marine algae have been the subject of studies by Duncan *et al.* (1956). In extracts of *Cladophora rupestris*, *Laminaria digitata*, *Rhodymenia palmata*, and *Ulva lactuca*, they demonstrated enzymes responsible for the cleavage of several α- and β-glucosides, of various polysaccharides, and of xylobiose. However, the activity with certain substrates, as indicated by paper chromatography, was so low that monosaccharides produced by hydrolysis could be demonstrated only after several days. Duncan and Manners (1958) proposed trans-α-glucosidase as the enzyme responsible for oligosaccharide synthesis in *Cladophora rupestris*. They isolated panose,[2] maltotriose, and maltotetraose after incubation of cell extracts with maltose, indicating that α-1:4 and α-1:6 glucosidic linkages both were formed. For the synthesis of β-1:3 linkages of laminaribiose in extracts of *Cladophora rupestris* and *Ulva lactuca*, Duncan *et al.* (1959) assumed that trans-β-glucosylation takes place by the following reactions:

$$\text{cellobiose} + \text{enzyme}\cdot\text{H} \rightarrow \text{glucose} + \text{glucosyl-enzyme}$$

$$\text{glucosyl-enzyme} + \text{H}\cdot\text{OR} \rightarrow \text{enzyme-H} + \text{glucosyl}\cdot\text{OR}$$

where H·OR is an acceptor substrate. This mechanism was investigated

[2] Panose: a triglucose with α-1:6 and α-1:4 linkages.

with mixtures of cellobiose and various sugars as acceptors, and it was shown that D-xylose, though not a substrate for trans-α-glucosylation, can participate in such a trans-β-glucosylation process. The high concentration of cellobiose necessary for this reaction, however, casts some doubt on its occurrence *in vivo*. The low activity and the lack of sufficient specific data to characterize an enzyme point to the need for more precise studies with purified enzyme preparations. Furthermore, it should be noted that several of the substrates used in these studies apparently do not occur as such in the algae, and it would seem profitable in the future to examine the specific action of these enzymes on their natural substrates.

Evidence from pure culture experiments indicated that certain marine diatoms, notably *Nitzschia filiformis* and *N. frustulum*, secrete an extracellular enzyme that digests agar (Lewin and Lewin, 1960).

2. *Enzymes of Glycolysis, the Pentose-Phosphate Cycle, and Carbon Assimilation*

A question arises from the studies of Duncan and Manners (1958), as well as from the amylase activity demonstrated in many seaweeds, concerning the pathway of the metabolic breakdown of glucose or other monosaccharides. If the nonphosphorylated sugars are formed during the hydrolysis of starch or of other poly- or oligosaccharides, two enzymatic reactions may be considered. (*1*) The classical glycolytic degradation, characteristic of most organisms examined, may start with a phosphorylation step mediated by hexokinase. (2) Alternatively, the breakdown reaction may proceed by the direct oxidation of the nonphosphorylated sugar to the corresponding acid via a flavine enzyme, carbohydrate oxidase.

Holzer and Holzer (1952) showed hexokinase to be present in *Chlorella pyrenoidosa*. For many marine algae the situation may be different, since Watanabe (1932) found that their respiration was not stimulated by the addition of glucose, whereas that of *C. pyrenoidosa* showed a significant increase in the presence of this substrate. However, he did not exclude the possibility that permeability barriers might have prevented access of the sugar to the cellular enzyme systems. The lack of a respiratory response to glucose by *Ulva lactuca* agrees with the absence of glucose hexokinase from several marine Green and Red Algae (Jacobi, 1957a, c). Using a purified enzyme preparation from *Iridophycus flaccidum*, Bean and Hassid (1956a, b) confirmed the presence of a carbohydrate oxidase catalyzing the oxidation of glucose or galactose to the corresponding gluconic or galactonic acid. This enzyme contains flavine as a prosthetic group, and H_2O_2 is formed during the reaction. Gluconic acid was the only end-product detected after incubation of the enzyme preparations with glucose (Bean and Hassid, 1956b).

TABLE I

ENZYMES OF CARBOHYDRATE METABOLISM IN MARINE GREEN AND RED ALGAE[a, b]

Alga	Hexo-kinase	Aldolase	Glyceraldehyde-phosphate dehydrogenase		Alcohol dehy-drogenase	Lactic acid dehy-drogenase	Malic acid dehy-drogenase	Malic enzyme	Glucose-6-phosphate dehydro-genase
			DPN	TPN					
Chlorophyta									
Ulva lactuca	−	+	+	+	−	−	+	+	+
Chaetomorpha linum	−	+	+	+	−	−	+		+
Bryopsis plumosa		+	+	+	−	−	+		+
Rhodophyta									
Delesseria sanguinea	−	−	−	−	−	−	+	+	−
Phycodrys rubens		−	−	−	−	−	+	+	−
Polysiphonia violacea	−	−	−	−	−	−	+		−
Cystoclonium purpurascens		−	−	−	−	−	+	+	−

[a] From Jacobi (1957c).

[b] The enzymes in the Red Algae were tested in crude extracts, in extracts from acetone powders, and in several ammonium sulfate fractions; those from Green Algae were tested only in crude extracts.

The course of monosaccharide breakdown and the various steps of glycolysis seem to be similar in all Green Algae and higher plants. Holzer and Holzer (1952) found phosphohexokinase, fructose-diphosphate aldolase, and glyceraldehyde-phosphate dehydrogenase in *Chlorella pyrenoidosa*; Richter (1957) detected aldolase and a high activity of triose isomerase in *Hydrodictyon*; Jacobi (1957a, c) demonstrated aldolase and glyceraldehyde-phosphate dehydrogenase in *Ulva lactuca*, *Chaetomorpha linum*, and *Bryopsis plumosa*; and Fuller and Gibbs (1959) found TPN- and DPN-glyceraldehyde-phosphate dehydrogenase in *Euglena* sp., *Astasia* sp., and in the anomalous Blue-green Alga *Anacytis nidulans*. Several enzymes of the pentose-phosphate cycle were also detected in Green Algae (Cohen, 1950; Jacobi, 1957a, c; Richter, 1959). Jacobi (1957c) found characteristic differences between marine Green and Red Algae (Table I) None of the Red Algae he examined showed activity in tests for aldolase, glyceraldehyde-phosphate dehydrogenase or glucose-6-phosphate dehydrogenase. It is possible that this may have been a result of the central role of galactose, rather than glucose, in the metabolism of these algae, but the possibility was not tested further.

Interesting data on the assimilatory pathway in Red Algae were provided by enzyme studies and tracer experiments on *Iridophycus flaccidum* (Bean and Hassid, 1955). It appears that phosphoglyceric acid is formed as a first product in photosynthesis, but it remains to be established whether this is then reduced to α-glycerophosphate as an intermediate step in the synthesis of galactosyl glyceride and floridoside. The absence of labeled triose phosphates in the experiments of Bean and Hassid, and the absence (according to the arsenate test) of TPN-glyceraldehyde-phosphate dehydrogenase, suggest that there may be different enzymes involved in the reduction of phosphoglyceric acid via phosphoglyceraldehyde and DPN or TPN.

It is evident that the photosynthetic cycle worked out by Calvin and his co-workers is not common to all photosynthetic organisms. The only one of these reactions which might be distributed throughout all plants is the enzymatic carboxylation by ribulose-5-phosphate carboxylase (carboxydismutase), first demonstrated by Quayle *et al.* (1954) in *Chlorella pyrenoidosa*, and later found by Fuller and Gibbs (1959) in *C. variegata*, *Anacystis nidulans*, and species of *Euglena* and *Astasia*. Since Jacobi found the reduction step via the TPN-dependent glyceraldehyde-phosphate dehydrogenase to be absent from various marine Red Algae, and Richter found evidence for the absence of aldolase in one of the Blue-green Algae, *Anacystis nidulans* (see Table II), it is evident that alternative pathways must exist in these classes.

The end-product of glycolysis, pyruvic acid, which is the first compound of the tricarboxylic-acid cycle, can normally be converted to citric acid by

TABLE II

Comparison of the Relative Enzyme Activities in Particle-Free Extracts of *Anacystis nidulans* (Blue-Green Algae) and *Chlorella pyrenoidosa* (Green Algae)[a]

Enzyme	*A. nidulans*		*C. pyrenoidosa*	
	Fraction I[b]	Fraction II[b]	Fraction I[b]	Fraction II[b]
Glucose-6-phosphate dehydrogenase.	335	<0.01	13	<0.01
6-Phosphogluconate dehydrogenase	59	<0.01	1	<0.01
Phosphoribo-isomerase + transketolase	8.4	59	0.3	0.9
Phosphogluco-isomerase	89	<0.01	0.6	<0.01
Triose isomerase	940	9800	2900	23,200
Aldolase	0	0	4340	185
Glyceraldehyde-3-phosphate dehydrogenase (TPN specific)	4.5	3.9	1.6	2.3
Glyceraldehyde-3-phosphate dehydrogenase (DPN specific)		2.5	1.3	5.1
Transaldolase	1.1	<0.01		0.02
Ribulokinase + carboxylating enzyme	7.6[c]		11[c]	

[a] From Richter (1959); units/10 mg protein (not units/mg protein as erroneously reported in the original paper); see this reference for units employed.

[b] Fraction I precipitated by 60% saturated ammonium sulfate; fraction II by 90% saturated ammonium sulfate.

[c] $C^{14}O_2$ fixed per 100 mg. protein per minute (in c.p.m. × 10^3).

oxidative decarboxylation and action of the condensing enzyme. Alternatively, under anaerobic conditions pyruvic acid may be further reduced to lactic acid or, with decarboxylation, to ethyl alcohol. The corresponding enzymes for these reactions are carboxylase, lactic acid dehydrogenase, and alcohol dehydrogenase. Although Watanabe (1949) detected lactic-acid dehydrogenase activity in *Ulva lactuca* by the Thunberg method, in marine Red and Green Algae, including *Ulva*, spectrophotometric evidence indicated the absence of both DPN- and TPN-dependent alcohol dehydrogenases (Jacobi, 1957c). The absence of these enzymes from *Hydrodictyon* was also reported by Richter and Pirson (1957). Evidently, therefore, lactic acid may be converted into other compounds, since the decolorization of methylene blue in Watanabe's tests could not have been directly due to the action of reduced pyridine nucleotides. Furthermore, the equilibrium of the lactic-acid dehydrogenase reaction is very far in the direction of the reduced form, i.e., lactic acid. The use of a purified lactic acid dehydrogenase permits one to determine pyruvic acid enzymatically, as was shown by Jacobi (1958) with *Laminaria saccharina.* Warburg *et al.* (1957a) reported an unusual enzyme system in *Chlorella pyrenoidosa* which converted D(−)-lactic acid to pyruvic acid, whereas the more commonly occurring L-form was not affected.

3. The Tricarboxylic-Acid Cycle and Oxidases

In contrast to the individual variation of enzymes involved in the metabolism of hexoses and pentoses, those of the tricarboxylic-acid cycle seem to be common to all classes of algae. This is especially evident from experiments on respiration using compounds of the tricarboxylic-acid cycle as exogenous substrates. There have also been some quantitative studies of the enzymes concerned. Malic acid dehydrogenase was shown to be present in all groups (Watanabe, 1949; Mehler, 1950; Jacobi, 1957a, c), and isocitric acid dehydrogenase in marine Green and Red Algae (Mehler, 1950). The malic enzyme, catalyzing the oxidative decarboxylation of malic acid with TPN, was demonstrated in several Green and Red Algae (Jacobi, 1957c).

Various studies with respiratory inhibitors have also indicated endogenous respiratory activities in algae, similar to those in other organisms. Watanabe (1937a, b) quantitatively estimated flavoproteins, and found flavines as well as cytochrome *c* to be necessary for the endogenous respiration of *Porphyra tenera* and *Ulva lactuca* (Watanabe, 1949). Jacobi (1957c) noted that in *Polysiphonia violacea* and *Cystoclonium purpurascens* an active DPNH-oxidase interfered with the measurement of pyridine nucleotide enzymes. Katoh (1959, 1960) isolated a cytochrome of the *c*-type from several algae, while from spectrophotometric studies of *Chlorella pyrenoidosa* (Lundegardh, 1954) and *Porphyridium cruentum* (Duysens, 1955) a special cytochrome *f* was shown to be active in photosynthesis. The sequence of redox cofactors such as cytochromes in the electron-transfer pathway of photosynthesis is still under investigation. The mechanism of this light-dependent process is both reductive and oxidative, as shown by Vennesland and his collaborators, who obtained convincing evidence for a light-activated cytochrome-*c*-oxidase in chloroplasts (Nieman and Vennesland, 1959). This "photo-oxidase" has a wide distribution among plants, and has been detected in freshwater and marine Chlorophyta as well as in species of *Fucus* and *Laminaria* (Bishop *et al.*, 1959). In this connection one might also mention the so-called "carotenoid oxygenase" of *Chlorella pyrenoidosa* (Warburg *et al.*, 1958).

Respiratory mechanisms in algae are reviewed by Gibbs (Chapter 4, this treatise).

Preliminary reports of peroxidases in algae were critically examined later from a methodological standpoint (Reed, 1915; Hampton, 1920, cited by Tamiya, 1935). By more refined methods, Rönnerstrand (1943) showed oxidases to be present in a number of marine algae, maximum values being found among the Red Algae. He also estimated laccase and *o*-phenolase by manometric techniques.

The presence of catalase in many algae has been reported by several authors, leading to the belief that catalase activity is a vital feature of photosynthesis. Bergmann (1955) showed that catalase activity in *Chlorella vulgaris* was not influenced by a deficiency of manganese.

B. Enzymes of Nitrogen Metabolism

1. *Nitrate Reductase and Amino-Acid Dehydrogenases*

Inorganic sources of nitrogen for algae include: molecular nitrogen, ammonia, nitrate, and—under certain circumstances—nitrite. The use of molecular nitrogen, by nitrogen fixation, is limited to the Blue-green Algae and bacteria. No enzyme involved directly in nitrogen fixation has been isolated, and the whole process needs further detailed investigation.

From the physiological experiments by Davis (1953), Kessler (1957, 1959), and others, it appeared that there must be an enzymatic nitrate-reducing system in most algae. However, none of these workers demonstrated the enzyme itself, although Evans and Nason (1954) isolated highly purified nitrate reductase from *Neurospora.* Nicholas and Nason (1954a, b) showed that the mechanism of nitrate reductase in this fungus required molybdenum and flavine (FAD) as essential cofactors. All observations and experiments with algae or with algal extracts indicate that the nitrate reductase of algae is similar to the purified mold enzyme.

Omura (1954) extracted nitrate and nitrite reductase from several unicellular freshwater algae, including species of *Chlorella* and *Scenedesmus.* Only the nitrate reductase was liberated from the cells by autolysis. In another extensive study, Takagi and Murata (1954a–e, 1955 a, b, c) reported on the occurrence and properties of nitrate reductase in marine algae. The enzyme was found in various Green, Brown, and Red Algae, but appeared to be absent from species of *Laminaria, Sargassum, Neodilsea,* and *Caulacanthus.* Several compounds were found which could serve as hydrogen donors for crude preparations of this enzyme; either methylene blue or cytochrome *c* could act physiologically as an electron carrier.

The simplest inorganic nitrogen compound which can be incorporated directly into organic substances is ammonia, which is incorporated by amino-acid dehydrogenases in the synthesis of amino acids. The reaction

$$\alpha\text{-keto acid} + NH_4^+ + DPNH \rightleftarrows \text{amino acid} + DPN^+ + H_2O$$

has been tested in algae in two ways. Watanabe (1949), using the Thunberg method, observed decolorization of methylene blue after adding glutamic acid and DPN to extracts of *Ulva lactuca.* With the same alga, Jacobi (1957b) measured spectrophotometrically the oxidation of DPNH with α-ketoglutaric acid and NH_4^+. The enzyme also appears to be active with TPNH, but to a lesser extent. The reductive amination of oxalacetic acid to aspartic acid with DPNH or TPNH was demonstrated in the same preparations, providing evidence for an enzyme activity not previously detected in plants. On the basis of these observations the author concluded that aspartic acid and glutamic acid are both key compounds in the nitrogen metabolism in *Ulva.* The amination of pyruvic acid, however, does not involve pyridine nucleotides.

2. *Enzymes of Intermediate Amino-Acid Metabolism*

If aspartic or glutamic acid is present, the synthesis of other amino acids is possible by transamination. Millbank (1953) first detected transaminases in *Chlorella vulgaris.* Jacobi (1957b), using paper chromatography and spectrophotometry, demonstrated in *Ulva lactuca* preparations active transamination from 14 amino acids, converting α-ketoglutaric acid to glutamic acid. He also found less active but still detectable transamination from 10 amino acids, by which pyruvic acid was converted to alanine. Direct quantitative measurements of glutamic–aspartic transaminase were made by following spectrophotometrically the combined reactions:

$$\alpha\text{-ketoglutaric acid} + \text{aspartic acid} \rightleftarrows \text{glutamic acid} + \text{oxalacetic acid}$$

and

$$\text{oxalacetic acid} + \text{DPNH} + H^+ \rightleftarrows \text{malic acid} + \text{DPN}^+$$

In the same way glutamic-alanine transaminase was coupled with reduction of pyruvic acid. On a protein basis, glutamic-aspartic transaminase was the most active transaminase found in extracts of *Ulva.*

Nutrition experiments have provided some indications of intermediate amino-acid metabolism. Hattori (1958) noted that in the presence of ammonia and urea the arginine content of *Chlorella ellipsoidea* was increased. Since urease was absent, it was concluded that in *Chlorella* the ornithine-citrulline cycle does not operate. On the other hand Reuter (personal communication) observed urease activity in various marine Green, Brown, and Red Algae from the Bay of Naples.

The enzymatic α-decarboxylation of glutamic acid to γ-aminobutyric acid in *Chlorella pyrenoidosa* was described by Warburg *et al.* (1957b). An extracellular glutaminase is produced by *Monodus subterraneus,* according to Miller (1959; see Fogg, Chapter 30, this treatise). Nothing is known about the enzymes of amino-acid activation and peptide synthesis in algae. The enzymatic hydrolysis of gelatin by preparations isolated from *Porphyra* sp. and *Eisenia* sp. was demonstrated by Tazawa and Miwa (1953). These proteinases were active only against proteins, and did not hydrolyze peptides.

C. Variations of Enzyme Activity in Relation to Physiological and Physical Conditions

Most enzymatic studies have been designed to detect and characterize biochemical reactions, and there is very little information on changes of activity in relation to variations of the physiological condition or of the physical environment of the cell. Ultimately, one assumes, all metabolic changes depend on the interactions of enzyme systems with the environment.

The classical example for such an interaction is the activation of hydrogenase in *Scenedesmus* by incubation at low light intensities (Gaffron, 1940). In *Ankistrodesmus braunii* an analogous process was later correlated with nitrite reduction (Kessler, 1957).

Richter and Pirson (1957) demonstrated an adaptation of enzyme activity in *Hydrodictyon* to a rhythmic alternation of periods of light and darkness. Catalase and starch phosphorylase showed a maximum during the period of darkness, whereas the glycolytic enzymes were not influenced. Some observations on the rhythmic activity of luciferase in *Gonyaulax* are discussed elsewhere in this treatise by Sweeney and Hastings (Chapter 46). Sosa-Bourdouil (1946) noted a diminution in the ribonuclease, deoxyribonuclease, and phosphatase activity of male and female conceptacles of *Fucus vesiculosus* subjected to γ-irradiation. However the effects were relatively small, and it was therefore concluded that protective substances might be present in the thallus. Jacobi (1959) observed a decrease in the activity of glyceraldehyde-phosphate dehydrogenase, aldolase, and malic-acid dehydrogenase in *Ulva lactuca* when it was transferred from sea water of 1.5% salinity in the Bay of Kiel to water of 0.4% salinity, in which it survived for several weeks. Clauss (1959) found an increase in the phosphorylase of both nucleated and enucleated cells of *Acetabularia* during growth, although in enucleated cells the increase declined 14 days after enucleation.

In comparative studies of green, bleached, and "dark-grown" strains of *Euglena* sp., Fuller and Gibbs (1959) found the TPN-dependent glyceraldehyde-phosphate dehydrogenase, present in photosynthetic cells, to be absent from both apochlorotic (chloroplast-less) and dark-grown normal cultures.

Variation in the activity of different enzymes during the life cycle has received little attention. Richter and Pirson (1957) found that young cells of *Hydrodictyon* are unable to synthesize starch, in contrast to older ones which do so and which can be shown to contain phosphorylase.

IV. Conclusion

The valuable concepts of comparative biochemistry should not be taken to indicate a complete uniformity in the enzyme systems of all organisms. The biochemical studies of Bean and Hassid (1955, 1956a, b) and the tracer experiments of Bidwell (1957, 1958), who compared the distribution of intermediates in the carbon assimilation of marine Green, Brown, and Red Algae, reveal some of the variety which can be expected in the metabolic patterns of algae. Those forms with an "atypical" metabolism open a wide field for investigations of new biochemical reactions, where one must study the action and isolation of the enzymes involved in order to answer some of the special questions which arise.

Another consideration which should be borne in mind in biochemical work of this sort is the possible influence of the physical circumstances under which the investigations are carried out. It is clear from the above brief review that we still have very little information on metabolic changes which may occur during the life cycle of a living organism. Many such problems remain unexplored. We need a better understanding of the general biology of algae if we are intelligently to interpret their biochemical diversity.

References

Bean, R. C., and Hassid, W. Z. (1955). Assimilation of $C^{14}O_2$ by a photosynthesizing red alga, *Iridophycus flaccidum*. *J. Biol. Chem.* **212,** 411–425.

Bean, R. C., and Hassid, W. Z. (1956a). Carbohydrate oxidase from a red alga, *Iridophycus flaccidum*. *J. Biol. Chem.* **218,** 425–436.

Bean, R. C., and Hassid, W. Z. (1956b). Enzymatic oxidation of glucose to glucosone in a red alga. *Science* **124,** 171–172.

Bergmann, L. (1955). Stoffwechsel und Mineralsalzernährung einzelliger Grünalgen. II. Vergleichende Untersuchungen über den Einfluss mineralischer Faktoren bei heterotropher und mixotropher Ernährung. *Flora (Jena)* **142,** 493–539.

Bidwell, R. G. S. (1957). Photosynthesis and metabolism of marine algae. I. Photosynthesis of two marine flagellates compared with *Chlorella*. *Can. J. Botany* **35,** 945–950.

Bidwell, R. G. S. (1958). Photosynthesis and metabolism of marine algae. II. A survey of rates and products of photosynthesis in $C^{14}O_2$. *Can. J. Botany* **36,** 337–349.

Bishop, N. I., Nakamura, H., Blatt, J., and Vennesland, B. (1959). Kinetics and properties of cytochrome c photooxidase of spinach. *Plant Physiol.* **34,** 551–557.

Clauss, H. (1959). Das Verhalten der Phosphorylase in kernhaltigen und kernlosen Teilen von *Acetabularia mediterranea*. *Planta* **52,** 534–542.

Cohen, S. S. (1950). Studies on the distribution of the oxidative pathway of glucon-6-phosphate utilization. *Biol. Bull.* **99,** 369.

Davis, E. A. (1953). Nitrate reduction by *Chlorella. Plant Physiol.* **28,** 539–544.

Duncan, W. A. M., and Manners, D. J. (1958). Enzyme systems in marine algae. II. Trans-α-glucosylation by extracts of *Cladophora rupestris. Biochem. J.* **69,** 343–348.

Duncan, W. A. M., Manners, D. J., and Ross, A. G. (1956). Enzyme systems in marine algae. I. The carbohydrase activities of unfractionated extracts of *Cladophora rupestris, Laminaria digitata, Rhodymenia palmata* and *Ulva lactuca. Biochem. J.* **63,** 44–51.

Duncan, W. A. M., Manners, D. J., and Thompson, J. L. (1959). Enzyme systems in marine algae. III. Trans-β-glucosylation by extracts of *Cladophora rupestris* and *Ulva lactuca. Biochem. J.* **73,** 295–298.

Duysens, L. N. M. (1955). Role of cytochrome and pyridine nucleotide in algal photosynthesis. *Science* **121,** 210–211.

Evans, H. J., and Nason, A. (1953). The effect of reduced triphosphopyridine nucleotide on nitrate reduction by purified nitrate reductase. *Arch. Biochem. Biophys.* **39,** 234–235.

Fuller, R. C., and Gibbs, M. (1959). Intracellular and phylogenetic distribution of ribulose 1,5-diphosphate carboxylase and D-glyceraldehyde-3-phosphate dehydrogenases. *Plant Physiol.* **34,** 324–329.

Gaffron, H. (1940). Carbon dioxide reduction with molecular hydrogen in green algae. *Am. J. Botany* **27,** 273–283.

Hattori, A. (1958). Studies on the metabolism of urea and other nitrogenous compounds in *Chlorella ellipsoidea.* II. Changes on levels of amino acids and amides during the assimilation of ammonia by nitrogen-starved cells. *J. Biochem.* (*Tokyo*) **45,** 57–64.

Holzer, H., and Holzer, E. (1952). Enzyme des Kohlenhydratstoffwechsels in *Chlorella. Chem. Ber.* **85,** 655–663.

Jacobi, G. (1957a). Enzyme des Kohlenhydratstoffwechsels in Extrakten von *Ulva lactuca. Planta* **49,** 1–10.

Jacobi, G. (1957b). Enzyme des Aminosäure-Stoffwechsels in *Ulva lactuca.* Transaminasen und Aminosäure-Dehydrogenasen. *Planta* **49,** 561–577.

Jacobi, G. (1957c). Vergleichend enzymatische Untersuchungen an marinen Grün- und Rotalgen. *Kiel. Meeresforsch.* **13,** 212–219.

Jacobi, G. (1958). Über die Bestimmung stationärer Konzentrationen von Brenztraubensäure und α-Ketoglutarsäure in *Laminarien. Kiel. Meeresforsch.* **14,** 247–250.

Jacobi, G. (1959). Salinitätswirkung des Seewassers auf die Enzymaktivität von *Ulva lactuca. Kiel. Meeresforsch.* **15,** 161–163.

Katoh, S. (1959). Studies on the algal cytochrome of C-type. *J. Biochem.* (*Tokyo*) **46,** 629–632.

Katoh, S. (1960). Studies on algal cytochrome. I and II. I. Enzymic activities pertaining to *Porphyra tenera* cytochrome 553 in cell-free extract. II. Physico-chemical properties of crystalline *Porphyra tenera* cytochrome 553. *Plant and Cell Physiol.* **1,** 29–38, 91–98.

Kessler, E. (1957). Untersuchungen zum Problem der photochemischen Nitratreduktion in Grünalgen. *Planta* **49,** 505–523.

Kessler, E. (1959). Reduction of nitrate by green algae. *Symposia Soc. Exptl. Biol.* **13,** 87–105.

Lewin, J. C., and Lewin, R. A. (1960). Auxotrophy and heterotrophy in marine littoral diatoms. *Can. J. Microbiol.* **6,** 127–134.
Lindberg, B. (1953a). Low-molecular carbohydrates in algae. I. Investigation of *Fucus vesiculosus. Acta Chem. Scand.* **7,** 1119–1122.
Lindberg, B. (1953b). Low-molecular carbohydrates in algae. II. Synthesis of 1-D-mannitol-monoacetate and 1,6-D-mannitol-diacetate. *Acta Chem. Scand.* **7,** 1123–1124.
Lundegardh, H. (1954). On the oxidation of cytochrome *f* by light. *Physiol. Plantarum* **7,** 375–382.
Mehler, A. H. (1950). Reactions related to photosynthesis and respiration. *Biol. Bull.* **99,** 371.
Millbank, J. W. (1953). Demonstration of transaminase systems in the alga *Chlorella. Nature* **171,** 476–477.
Millbank, J. W. (1957). Studies on the preparation of a respiratory cell-free extract of the alga *Chlorella. J. Exptl. Botany* **8,** 96–104.
Miller, J. D. A. (1959). An extracellular enzyme produced by *Monodus. Brit. Phycol. Bull. No.* **7,** 22.
Milner, H. W., Lawrence, N. S., and French, C. S. (1950). Colloidal dispersion of chloroplast material. *Science* **111,** 633.
Nicholas, D. J. D., and Nason, A. (1954a). Mechanism of action of nitrate reductase of *Neurospora. J. Biol. Chem.* **211,** 183.
Nicholas, D. J. D., and Nason, A. (1954b). Molybdenum as an electron carrier in nitrate reductase. *Arch. Biochem. Biophys.* **51,** 310.
Nieman, R. H., and Vennesland, B. (1959). Photoreduction and photooxidation of cytochrome c by spinach chloroplast preparations. *Plant Physiol.* **34,** 255–262.
Omura, H. (1954). On the nitrate reductase in green algae. *Enzymologia* **17,** 127–132.
Quayle, J. R., Fuller, R. C., Bluson, A. A., and Calvin, M. (1954). Enzymatic carboxylation of ribulose diphosphate. *J. Am. Chem. Soc.* **76,** 3610.
Reed, G. B. (1915). Evidence for the general distribution of oxidases in plants. *Botan. Gaz.* **59,** 407.
Richter, G. (1956). Zur Papierchromatographie von Enzymgemischen aus Grünalgen. *Flora (Jena)* **143,** 161–164.
Richter, G. (1957). Nachweis und quantitative Bestimmung einiger Enzyme des Kohlenhydrat-Stoffwechsels in Grünalgen. *Z. Naturforsch.* **12b,** 662–663.
Richter, G. (1959). Comparison of enzymes of sugar metabolism in two photosynthetic algae: *Anacystis nidulans* and *Chlorella pyrenoidosa. Naturwissenschaften* **46** (21), 604.
Richter, G., and Pirson, A. (1957). Enzyme von *Hydrodictyon* und ihre Beeinflussung durch Beleuchtungsperiodik. *Flora (Jena)* **144,** 562–597.
Rönnerstrand, S. (1943). Untersuchungen über Oxydase, Peroxydase. und Ascorbinsäure in einigen Meeresalgen. Dissertation, Lund University, Sweden.
Sosa-Bourdouil, C. (1946). The enzymatic activity of the antherozoids and the ovules of *Fucus vesiculosus. Bull. muséum natl. hist. nat. (Paris)* **18,** 142.
Takagi, M., and Murata, K. (1954a). Studies on the mechanism of nitrogen assimilation. I. On the measurement of nitrate reductase activity. *Bull. Fac. Fisheries Hokkaido Univ.* **4,** 296–305.
Takagi, M., and Murata, K. (1954b). Studies on the mechanism of nitrogen assimilation in marine algae. II. On the nitrate reductase activity in various species of marine algae. *Bull. Fac. Fisheries Hokkaido Univ.* **4,** 306–309.
Takagi, M., and Murata, K. (1954c). Studies on the mechanism of nitrogen assimi-

lation in marine algae. III. On the optimum pH of the nitrate reductase in some species of marine algae. *Bull. Fac. Fisheries Hokkaido Univ.* **4,** 310–313.

Takati, M., and Murata, K. (1954d). Studies on the mechanism of nitrogen assimilation in marine algae. IV. On the optimum temperature of the nitrate reductase in some species of marine algae. *Bull. Fac. Fisheries Hokkaido Univ.* **5,** 173–175.

Takagi, M., and Murata, K. (1954e). Studies on the nitrogen assimilation in marine algae. V. On the hydrogen donator of nitrate reductase. *Bull. Fac. Fisheries Hokkaido Univ.* **5,** 176–182.

Takagi, M., and Murata, K. (1955a). Studies on the mechanism of nitrogen assimilation in marine algae. VI. The variation of nitrate reductase activity according to the portions and to the frond length of some marine algae. *Bull. Fac. Fisheries Hokkaido Univ.* **6,** 25–28.

Takagi, M., and Murata, K. (1955b). Studies on the mechanism of nitrogen assimilation in marine algae. VII. Effect of various inhibitors on nitrate reductase. *Bull. Fac. Fisheries Hokkaido Univ.* **6,** 29–32.

Takagi, M., and Murata, K. (1955c). Studies on the mechanism of nitrogen assimilation in marine algae. VIII. On the electron carrier of nitrate reductase in *Porphyra yezoensis*. *Bull. Fac. Fisheries Hokkaido Univ.* **6,** 33–36.

Tamiya, H. (1935). Über die Peroxydase in Algen. *Planta* **23,** 284–288.

Taylor, F. J. (1950). Oxidative assimilation of glucose by *Scenedesmus quadricauda*. *J. Exptl. Botany* **1,** 301–321.

Tazawa, Y., and Miwa, A. (1953). Über Algenproteinase. *Botan. Mag. (Tokyo)* **66,** 77–80.

Warburg, O., Gewitz, H. S., and Volker, W. (1957a). D(−)-Milchsäure in *Chlorella*. *Z. Naturforsch.* **12b,** 722–724.

Warburg, O., Klotzsch, H., and Krippahl, G. (1957b). Glutaminsäure-Decarboxylase in *Chlorella*. *Naturwissenschaften* **44,** 235.

Warburg, O., Krippahl, G., Gewitz, H. S., and Volker, W. (1958). Carotinoid-Oxygenase in *Chlorella*. *Z. Naturforsch.* **13b,** 437–439.

Watanabe, A. (1932). Über die Beeinflussung der Atmung von einigen grünen Algen durch Kaliumcyanid und Methylenblau. Beiträge zur Stoffwechselphysiologie der Algen. I. *Acta Phytochim. (Japan)* **6,** 315–335.

Watanabe, A. (1937a). Untersuchungen über die Substrate für Sauerstoffatmung von Süsswasser- und Meeresalgen. Beiträge zur Stoffwechselphysiologie der Algen. II. *Acta Phytochim. (Japan)* **9,** 235–254.

Watanabe, A. (1937b). Über die Verbreitung des Flavins in Meeresalgen. Beiträge zur Stoffwechselphysiologie der Algen. *Acta Phytochim. (Japan)* **9,** 255–264.

Watanabe, A. (1949). Über die Dehydrasen von Meeresalgen. IV. *Acta Phytochim. (Japan)* **15,** 129–141.

Yin, H. C. (1948). Phosphorylase in plastids. *Nature* **162,** 928.

—8—

Organic Micronutrients

MICHAEL R. DROOP

Scottish Marine Biological Association, Millport, Scotland

I. Introduction

A glance at the literature on algal cultivation (e.g., Bold, 1942; Pringsheim, 1946; Provasoli *et al.*, 1957) reveals the reliance that has of necessity been placed on organic ingredients of natural origin and unknown composition, such as soil extract, for cultivating a wide variety of species of algae. Allen (1914) was probably not the first to suggest that algae might need minute amounts of organic growth-promoting substances contained in such materials. This field has advanced largely as a consequence of the pioneering researches of Pringsheim.

II. The B Vitamins

Curiously, of all known vitamins and growth factors, only vitamin B_{12}, thiamine, and biotin have been found to be of any general importance for algae. Instances of other known requirements are comparatively rare. As shown in Table I, some 60% of strains studied have proved to be auxotrophs, i.e., to need an exogenous source of one or more accessory growth

TABLE I

VITAMIN REQUIREMENTS OF ALGAE AS REPORTED IN THE LITERATURE

Division	Total	Auxotrophs	Non-auxotrophs	Strains requiring— only vitamin B_{12}	only thiamine	vitamin B_{12} and thiamine	thiamine and biotin	vitamin B_{12}, thiamine, and biotin
Chlorophyta	47	22	24	10	7	5	0	0
Euglenophyta	10	10	0	0	1	9	0	0
Cryptophyta	11	11	0	2	1	8	0	0
Pyrrophyta	17	17	0	12	0	0	1	4
Chrysophyta	13	12	1	2	1	6	1	2
Bacillariophyta	54	21	33	11	6	4	0	0
Phaeophyta	1	0	1	0	0	0	0	0
Rhodophyta	1	1	0	1	0	0	0	0
Cyanophyta	25	1	24[a]	1	0	0	0	0
Totals	179	95	84	39	16	32	2	6

[a] Cf. Allen (1952), Table 6.

factors; 80% of these require vitamin B_{12}, 53% thiamine, and 10% biotin. However, it would be unwise to place undue weight on these proportions, since they are biased in several ways. For instance, a common method of isolating algae involves purely mineral enrichments which are selective against auxotrophs, whereas in recent years the expanding interest in vitamin requirements had had the opposite effect on the published literature, in which certainly hundreds of algal isolates with no growth-factor requirements have consequently received no mention. Table I is based on species lists compiled by Provasoli (1958a) and Lewin (1959), and on later references summarized in Table II. The appearance of "*Amphora coffaeiformis*" four times in different sections of Table II draws attention to the nutritional variation that may sometimes be found within a single "species" when more than one strain is examined.

Although auxotrophy can be regarded as a limited form of heterotrophy, there is no general correlation between a requirement for vitamins and

TABLE II

ADDITIONAL RECORDS OF VITAMIN REQUIREMENTS IN ALGAE, NOT INCLUDED IN REVIEWS BY PROVASOLI (1958a) OR LEWIN (1959)

Alga	Reference
Species with no vitamin requirement	
Bacillariophyta	
Amphora coffaeiformis	Lewin and Lewin (1960)
Navicula corymbosa	Lewin and Lewin (1960)
N. incerta	Lewin and Lewin (1960)
N. menisculus	Lewin and Lewin (1960)
Nitzschia affinis	Lewin and Lewin (1960)
N. filiformis	Lewin and Lewin (1960)
N. frustulum	Lewin and Lewin (1960)
N. laevis	Lewin and Lewin (1960)
N. lanceolata	Lewin and Lewin (1960)
N. marginata	Lewin and Lewin (1960)
N. obtusa	Lewin and Lewin (1960)
Phaeophyta	
Waerniella lucifuga	Droop (1957a)
Species requiring only thiamine	
Bacillariophyta	
Amphora coffaeiformis	Lewin and Lewin (1960)
A. paludosa var. *duplex*	Lewin and Lewin (1960)
Nitzschia closterium	Lewin and Lewin (1960)

TABLE II—*Continued*

Alga	Reference
Species requiring only vitamin B_{12}	
Chlorophyta	
Balticola buetschlii	Droop (1961)
B. droebakensis	Droop (1961)
Chlamydomonas pulsatilla	Droop (1961)
Platymonas tetrathele	Droop (1961)
Bacillariophyta	
Amphora coffaeiformis	Lewin and Lewin (1960)
A. lineolata	Lewin and Lewin (1960)
Nitzschia frustulum	Lewin and Lewin (1960)
N. ovalis	Lewin and Lewin (1960)
Opephora sp.	Lewin and Lewin (1960)
Cyclotella sp.	Lewin and Lewin (1960)
Rhodophyta	
Goniotrichum elegans	Fries (1959)
Series requiring thiamine and vitamin B_{12}	
Chlorophyta	
Brachiomonas submarina	Droop (1961)
Stephanosphaera pluvialis	Droop (1961)
Volvox globator	Pintner and Provasoli (1959)
Volvox tertius	Pintner and Provasoli (1959)
Bacillariophyta	
Amphipleura rutilans	Lewin and Lewin (1960)
Amphora coffaeiformis	Lewin and Lewin (1960)
Nitzschia closterium	Lewin and Lewin (1960)
Species requiring thiamine, vitamin B_{12}, and biotin	
Pyrrophyta	
Oxyrrhis marina	Droop (1959)

heterotrophy in respect to major carbon or nitrogen sources. The majority of species known to require vitamins are autotrophic in all other respects, though many are facultative heterotrophs, particularly as regards their nitrogen source. On the other hand, there exist examples of obligate or facultative chemotrophs without any vitamin requirements, e.g., *Polytoma uvella, P. obtusum, Haematococcus pluvialis,* and *Chlorella vulgaris.* Indeed, data on some 37 strains of pennate diatoms show a slight, though statistically insignificant, bias towards a reverse correlation (Lewin and Lewin, 1960).

Some facultative chemotrophs are reported to be more exacting in their micronutrient requirements during growth in the dark than during growth in light. However, none of these reports is well substantiated. For instance, Ford (1958) stated that the requirements of *Ochromonas malhamensis* are more complex for maximum growth in darkness, although Hutner *et al.* (1953) reported that the minimum requirements of this strain are the same in light and darkness.

III. Thiamine

Thiamine was the first vitamin to be identified as a growth factor for algae (A. Lwoff and Dusi, 1937a, b, c), but little attention has been given to this vitamin recently, in spite of the fact that almost one third of all the strains of algae studied have proved to require it.

The thiamine molecule, a component of the enzyme cocarboxylase, is itself composed of two portions, a thiazole and a pyrimidine moiety. Some organisms require the intact molecule, others need only the thiazole, others only the pyrimidine, and yet others need both moieties, presumably according to the position of the blocks in the pathways of its biosynthesis. No algae are known to be unable to synthesize the vitamin from its component parts, the requirement being limited to one or both moieties (Table III). The specificity of most thiamine-requiring algae in this respect is still to be ascertained.

The vitamin activity of some thiamine analogs was examined by Lwoff and Dusi (A. Lwoff, 1947) for *Polytoma ocellatum*, *P. caudatum*, *Polytomella caeca*, and *Chilomonas paramecium*. In the thiazole moiety the side chains in the second and fifth positions are necessary for activity, but a certain degree of substitution is tolerated. In the pyrimidine moiety the side chains in positions 2, 4, and 5 are all necessary but, again, some substitution in the aminomethyl chain on position 5 is possible. The pattern is similar to that in the trypanosomid *Strigomonas* (M. Lwoff, 1951).

There are no reports of any algae being able to utilize thiamine pyrophosphate (cocarboxylase) in place of thiamine, as can the bacteria *Streptococcus salivarius* and *Lactobacillus fermenti*, but this is probably a matter of permeability. Cramer and Myers (1950) found the thiamine requirement in *Euglena gracilis* during chemotrophic growth to be spared by glutamic acid, which presumably by-passes some of the acid decarboxylations involved in the photosynthetic assimilation of carbon.

There are no precise data on the physiological level of the thiamine requirement in algae. Some curves for the response to thiamine (Droop,

TABLE III

THIAMINE REQUIREMENT IN ALGAE

Alga	Requirement for—		Reference
	thiazole	pyrimidine	
Chlorophyta			
Polytoma ocellatum	+	−	Lwoff (1947)
Brachiomonas submarina	+	−	Droop (1961)
Stephanosphaera pluvialis	+	−	Droop (1961)
Prototheca zopfii	+	+	Anderson (1945)
Polytomella caeca	+	+	Lwoff (1947)
Chlamydomonas moewusii mutant	−	+	Lewin (1952)
Cryptophyta			
Hemiselmis virescens	+	−	Droop (1958)
Chilomonas paramecium	+	+	Lwoff (1947)
Pyrrophyta			
Oxyrrhis marina	+	−	Droop (1958)
Chrysophyta			
Monochrysis lutheri	−	+	Droop (1958)
Hymenomonas elongata	−	+	Droop (1958)
Prymnesium parvum	−	+	Droop (1958)
Microglena arenicola	−	+	Droop (1958)
Euglenophyta			
Euglena gracilis	−	+	Lwoff (1947)
E. pisciformis	+	+	Lwoff (1947)
E. viridis	+?	−?	Lwoff (1947)
Astasia quartana	+?	−?	Lwoff (1947)
A. chattonii	+?	−?	Lwoff (1947)

1958, 1959) suggest that in pyrimidine-requiring organisms the requirement on a molecular basis is some 200 times as great as that for vitamin B_{12}, and in thiazole-requiring species the requirement is some 4000 times as great.

IV. Vitamin B_{12}

A. General Observations

The isolation of vitamin B_{12} in 1948 was quickly followed by its identification by Hutner and his colleagues (1949, 1953) with hitherto unknown factors needed by the flagellates *Euglena gracilis* and *Ochromonas malhamensis*; and this led to recognition of the prevalence of the requirement for this vitamin among algal flagellates. From the practical standpoint, *Euglena* has proved extremely valuable for the bioassay of the vitamin; *Ochromonas*, though not as sensitive, responds to the same variants of the vitamin as do human beings, so that its use in bioassays eliminates the necessity of using chicks or rats for this purpose (e.g., Coats and Ford, 1955).

In some microorganisms the requirement is spared by methionine; this is so in *Ochromonas* (Hutner *et al.*, 1953), but not, as far as is known, in other algae (which may, of course, be impermeable to this amino acid). In mammals, vitamin B_{12} is known to be involved in resynthesis of the labile methyl groups of methionine from one-carbon precursors (Smith, 1960).

B. Specificity

There are a number of naturally occurring analogs of vitamin B_{12} which can be utilized by some organisms but not by others (Kon, 1955). The molecule of the vitamin consists of a nucleotide attached to a porphyrin-like "planar group" containing a central cobalt atom; the presence and the

TABLE IV

SPECIFICITY OF B_{12}-REQUIRING ALGAE TOWARD VITAMIN B_{12}-VARIANTS[a]

Division	Specificity pattern similar to that of—			Total
	mammals[b]	*Lactobacillus leichmannii*[c]	*Escherichia coli*[d]	
Chlorophyta	9	0	0	9
Euglenophyta	1	3	0	4
Cryptophyta	2	0	1	3
Pyrrophyta	4	3[e]	0	7
Chrysophyta	7	1	0	8
Bacillariophyta	0	0	2	2
Cyanophyta	0	0	1	1
Totals	23	7	4	34

[a] Data from Droop *et al.* (1959).

[b] Responding only when the vitamin contains a benziminazole base in the nucleotide (e.g., vitamin B_{12}, Factor I).

[c] The nucleotide may contain the base adenine or near analog (e.g., pseudovitamin B_{12}, Factors A, G and H).

[d] Responding also to the planar group lacking nucleotide (i.e., Factor B).

[e] These three dinoflagellates respond to Factor A but not to pseudo-vitamin B_{12} (i.e., when the nucleotide contains 6-hydroxy, not 6-amino, purine).

nature of the nucleotide largely determine the vitamin's activity. Only a few B_{12}-requiring organisms (e.g., the *Escherichia coli* mutant now commonly used for assays) respond to vitamin analogs, such as the so-called "factor B," which completely lack the nucleotide moiety. If the base in the nucleotide is adenine or a similar purine (as in pseudo-vitamin B_{12} and factors A, G, and H), some other microorganisms also respond, e.g., *Lactobacillus leichmannii* and *Euglena gracilis*. The true vitamin has a nucleotide containing a benziminazole group, and all B_{12}-requiring organisms respond to it. Algae furnish examples of all three general patterns of specificity (Table IV), but the third, i.e., the mammalian pattern, predominates.

C. Physiological Levels of Requirement

The use of microorganisms for bioassays has led to a special concept of "sensitivity," i.e., the smallest amount of vitamin giving measurably increased growth as compared with a control. Although obviously related to the physiological level of requirement, sensitivity depends to a great extent on technique and has only empirical value. A parameter with slightly more meaning is "relative yield," which we may define as the yield of organisms per dose of vitamin, but which of course is valid only when the yield/dose plot is linear (Droop, 1957b). Another more fundamental relation is that between relative growth rate and nutrient concentration (Monod, 1942),

$$r = r_0 \frac{c}{c_1 + c}$$

c being the concentration and c_1 a constant with the dimensions of a concentration; r, the relative growth rate at concentration c; and r_0, a constant with the dimensions of r. The value of r varies linearly with c, and $c_1 = c$ when $r = \frac{1}{2}r_0$. The value c_1, which we may term the "limiting concentration," and the relative yield are not necessarily interdependent parameters, though each is concerned with an aspect of the physiological level of requirement.

In a study on vitamin B_{12} and the growth of *Ochromonas*, Ford (1958) derived the value for the limiting concentration for growth, in darkness using glucose at 29°C, as 13 mμg./liter, and the relative yield as 0.3 to 1.0 μ^3 of living alga per molecule of vitamin B_{12}. Since 0.3 μ^3 per molecule is also the figure calculated by Droop (1957b) for *Monochrysis lutheri*, *Stichococcus* (?) *cylindricus*, and *Euglena gracilis*, some special significance might be attached to this value. On the other hand, since the requirement of *Ochromonas* for vitamin B_{12} increases 300-fold under conditions of temperature stress (Hutner *et al.*, 1957) we should be cautious in using such figures for further generalizations.

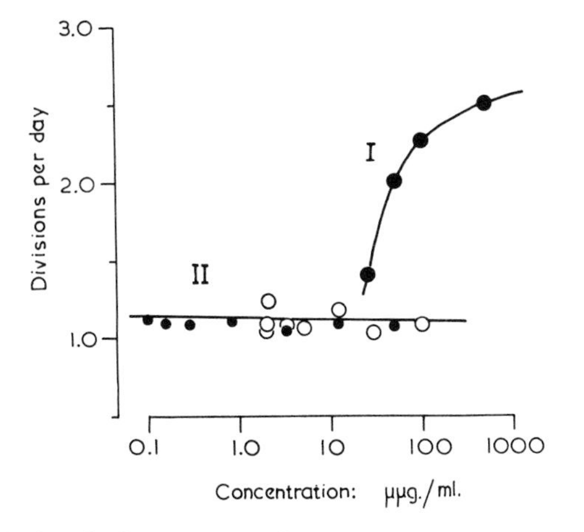

FIG. 1. Effect of vitamin B_{12} concentration on rate of cell division in *Ochromonas malhamensis* (Ford, 1958) and *Monochrysis lutheri* (unpublished). Curve I: *Ochromonas* in dark on glucose at 29°C. Curve II: *Monochrysis* (obligate phototroph) in warm white fluorescent light at 2000 lux. KEY: ●, 20°C.; ○, 15°C.

Although the relative yields for *Ochromonas* and *Monochrysis* in Droop's and Ford's experiments were of the same order, the limiting concentration is at least two orders lower in the case of *Monochrysis*. Rate/concentration curves for *Monochrysis* (Droop, unpublished) and *Ochromonas* (Ford, 1958) are shown in Fig. 1.

A simple, if rough, calculation gives some idea of the scale involved in vitamin B_{12} uptake. Ford's datum of 13 mμg./liter as the limiting concentration for *Ochromonas* can be expressed as 0.006 molecules per μ^3, or about 200 molecules within a radius of 20 μ of the cell. A population of cells of average volume 250 μ^3, multiplying in a medium containing initially, say, 500 mμg./liter, will, by the time the external concentration has been reduced to 13 mμg./liter, contain 99% of the vitamin in the cells, have an internal concentration of 1 to 3 molecules per μ^3 (i.e., 160 to 500 times the external concentration), and yet can multiply at half the maximum rate of nearly three divisions in a day. Each cell is then picking up a molecule every 1.2 to 3.6 minutes against a concentration gradient of 160 to 500. *Monochrysis* would appear to maintain as high a rate, but against a gradient 100 times as steep.

D. Uptake

It is thought that complex formation with a protein, termed the "intrinsic factor," is the mechanism by which vitamin B_{12} is trapped and passed through the cell membrane in vertebrates (Latner, 1955). A stereochemical affinity between the complex and sites on the membrane must be postulated to account for the passage of so large a molecule and for the specificity shown toward complexes formed of various binding proteins and the vitamin. Certain microorganisms likewise bind vitamin B_{12}. Disrupted cells of *Ochromonas* do so, as also does a "binding factor" in cell-free culture fluids of *Ochromonas* and *Euglena* (Ford *et al.*, 1955). Such bound B_{12} is less readily available to *Ochromonas* than is the free vitamin (Ford, 1958), which suggests that, if indeed sequestration is part of the uptake mechanism, it must normally take place at the cell surface, and that liberation of either the binding factor or the complex is a side reaction.

Ford (1958) showed that, although *Ochromonas* cannot use natural vitamin B_{12} analogs (pseudo-vitamin B_{12}, factor A), this is not due simply to a permeability barrier, since low concentrations of the analogs increased the vitamin B_{12} uptake while high concentrations simultaneously decreased uptake and depressed growth, probably by competing for the so-called binding factor either benefically in the solution or harmfully at the cell surface. He considered that the inhibitory action of high concentrations of vitamin B_{12} analogs with substituted benziminazole groups is also due to competition with vitamin B_{12} for binding sites (Ford, 1959). Although several of these analogs have vitamin activity (see also Baker *et al.*, 1959), their utilization by *Ochromonas* is generally less efficient.

5,6-Dimethylbenziminazole, a precursor of vitamin B_{12}, was found to be inhibitory to *Euglena* and *E. coli* (Hendlin, 1953) but not to *Ochromonas* (Ford, 1958). Factor B, another precursor, was also without effect on *Ochromonas*. These precursors evidently do not inhibit vitamin B_{12} uptake by saturating the binding sites as do the natural analogs. The fact that neither moiety of the vitamin molecule is by itself able to saturate the binding protein provides additional support for Smith's contention (1958) that the protein and vitamin molecules are linked at more than one position.

D. W. Woolley reasoned that an analog of a precursor should be inhibitory only to systems able to synthesize the product. Funk and Nathan (1958) and Ford (1959) reported that benziminazole analogs, which presumably do not affect the binding mechanism, have very little inhibitory action on *Ochromonas* and *Euglena*, except at high concentrations known to inhibit other organisms for reasons unconnected with vitamin B_{12} metabolism. However, Funk and Nathan found that in this respect the behavior of *Ochromonas* and *Euglena* does not differ greatly from that of vitamin-B_{12}-synthesizing species of the fungus *Rhizobium*.

The use of algae in the study of vitamin B_{12} has so far been limited to one or two accepted assay organisms, but advantage could be taken of the wider range of algae now available. This includes more than one instance of phylogenetically related organisms (e.g., among the Chrysophyta) with quite different requirements and specificities for vitamin B_{12}, so that one is no longer restricted to a few organisms as different as *Ochromonas*, *Euglena*, and certain bacteria.

V. Other Requirements

Limited though it generally is, the extent of auxotrophy is rather wider in several instances discussed later in connection with temperature and phagotrophy. Apart from these, a requirement for *p*-aminobenzoic acid was recorded for a *Chlamydomonas moewusii* mutant (Lewin, 1952), and stimulation of growth by histidine for *C. chlamydogama* (Provasoli and Pintner, 1953), by proline for *Oxyrrhis marina* (Droop, 1959), and perhaps by uracil for *Amphora coffaeiformis* (referred to as "*A. perpusilla*" by Hutner and Provasoli, 1953). Other amino acid requirements, e.g., valine, alanine, or proline for *Oxyrrhis*, glycine for *Hemiselmis virescens*, and arginine, histidine, or lysine for *C. pulsatilla* (Droop, 1959), are for a major nitrogen source, and probably result from the combination of an inability to reduce nitrate, the unsuitability of ammonium salts in media of a high pH, and an impermeability to other organic nitrogenous compounds.

Plant-growth substances (indolylacetic acid, kinetin, gibberellic acid, etc.) have been reported to be stimulatory to some algae (see Conrad and Saltman, Chapter 44, this treatise), but there are no instances of absolute requirements for this class of compound. The growth-promoting activity of concoctions of natural materials cannot always be accounted for solely in terms of vitamins and other nutrients. In addition to essential metabolites, physically active organic compounds such as chelating agents, pH buffers, etc., may also behave as micronutrients by controlling the availability or toxicity of essential heavy metals (Provasoli *et al.*, 1957). The requirement for a chelating agent may appear to be absolute in the sense that no growth takes place in its absence, but it is seldom specific. Indeed, vitamin-like dose/response curves are sometimes obtained; for instance, *Oxyrrhis* responds quantitatively to minute amounts of citric acid, which can nevertheless be entirely replaced by ethylenediaminetetraacetic acid (Droop, 1959). Delicate pelagic species are particularly sensitive in this respect, and only few pelagic diatoms can be cultivated easily in synthetic media. Thus, growth of *Skeletonema costatum* in such media is promoted by the addition of "Tris"[1] buffer or glycylglycine; although it also responds to various mixtures of amino acids and to reducing agents, this diatom can be shown to have no absolute growth-factor requirement other than for vitamin B_{12} (Droop, 1955, 1957a, 1958). These examples indicate

[1] Tris(hydroxymethyl)aminomethane.

how apparent requirements for organic acids, amino acids, pteridines, etc., may often prove to be artifacts caused by trace-metal unbalance.

VI. Temperature Interactions

The observations that the vitamin B_{12} requirement in rats and chicks is increased by administering thyroid-active materials, which tend to raise the body temperature, led to the hypothesis that the upper temperature tolerance limit may be a manifestation of disorganization of enzyme systems involving vitamin B_{12}, and that, if the end products of the syntheses could be supplied, an organism might be enabled thereby to survive at elevated temperatures. This might provide a useful method of studying the enzyme systems in question (Hutner *et al.*, 1957).

Ochromonas malhamensis was chosen for such studies, since it is an organism of high permeability and pronounced heterotrophic tendencies. Its "natural" temperature limit is 35°C.; below this temperature reproducible growth occurs in defined media containing glucose, with vitamin B_{12}, thiamine, and biotin as the only growth factors. It was found possible to grow this alga at temperatures between 36.0° and 36.8°C. provided that the concentration of vitamin B_{12} was raised to 30 μg./liter and the thiamine to 10 mg./liter. These represent 300-fold increases over the levels ordinarily required. Further increases in temperature led to the emergence of other requirements, e.g., for glycine, valine, and isoleucine, then for phenylalanine, tryptophan, cystine, and lysine, while requirements for certain cations, especially Mg, Fe, Zn, and Mn, rose steeply. Still further increases in temperature (to between 36.7° and 37.0°C.) could be tolerated only in media supplemented by an autoclaved *Ochromonas* cell suspension, which may have included lipid growth factors. Hutner *et al.* noted that, unlike the normal requirement for vitamin B_{12}, the enhanced requirement was spared by folic acid, or by a mixture of purines, pyrimidines, and amino acids, which draws attention to the plurality of roles for vitamin B_{12} in cell metabolism. The particular functions affected by temperature appear to be those concerned with synthesis of folic acid, amino acid, and purine, hence with protein and nucleic acid metabolism.

The behavior of a temperature-resistant strain of *Euglena gracilis* parallels that of *Ochromonas*, but with certain differences (Scher and Aaronson, 1958; Hutner *et al.*, 1958). For instance, the requirement for thiamine, though not that for B_{12}, increased several hundred times between 33.5° and 34.5°C. It is interesting to note that, as temperature and reaction rates rise, the patterns of nutritional requirements in both species become more delicately balanced and specialized and thus resemble more closely the pattern in animals.

In view of the enhanced requirements in *Euglena* and *Ochromonas*, it is surprising that the nutritional pattern in algae at "normal" temperatures should be both so stereotyped and so constant. There are no reports of vanishing auxotrophy toward the lower end of the temperature range. Experiments in the author's laboratory have so far failed to detect a disappearance of the B_{12} requirement in *Monochrysis* at low temperatures. The many algae with an upper temperature limit below 26°C, invite comparison with *Ochromonas* and *Euglena*, for one might suppose that different enzyme systems are involved at lower temperature barriers.

VII. Phagotrophic Algae

Pringsheim's (1959) review on phagotrophic algae provides a broad perspective of the occurrence and general significance of the phenomenon, which, therefore, need only be discussed briefly in connection with organic micronutrients. Phagotrophy—a method of feeding in which particles of food are taken into the cell, digested in food vacuoles, and absorbed through the vacuole membrane—occurs in several algal lines, notably in the Euglenophyta, Chrysophyta, and Pyrrophyta. The nutrition has been studied in one phagotrophic representative of each division: *Peranema trichophorum* (Chen, 1950; Storm and Hutner, 1953); *Ochromonas malhamensis* (Pringsheim, 1952; Hutner *et al.*, 1953; Ford, 1953; Hutner *et al.*, 1957); and *Oxyrrhis marina* (Droop, 1953, 1959).

Phagotrophy does not necessarily entail very complex nutritional requirements, though reliance on this mode of nutrition might be expected to lead to degeneration of such parts of the primitive synthetic apparatus as may become redundant when the prey supplies metabolites essential for the predator. *Oxyrrhis*, though it lacks photosynthetic pigments, has only one organic micronutritional need, a lipid, in addition to three water-soluble vitamins: thiamine, vitamin B_{12}, and biotin. *Peranema*, also lacking pigments, may prove more complex in its requirements, which include riboflavin, several amino acids, and some nucleic acid components and lipids. *Ochromonas*, when grown in the light as discussed previously, requires only the three B vitamins and possibly a Krebs-cycle component, but its enhanced requirements during growth in darkness or at elevated temperatures point to the specialization which may accompany phagotrophy.

The lipids required by *Peranema* appear to be cholesterol and lecithin. *Oxyrrhis* requires an unidentified lipid growth factor, which is present in lemon rind and grass, and which is thermostable, alkali-labile, and colorless, though highly unstable in light when partially purified chromatographically. Vitamins A, D, E, and K, soya sterols and some other plant steroids, cholesterol, soya lecithin, cream, egg yolk, olive oil, and linseed oil, all lack activity (Droop, unpublished data). Though at room temperatures *Ochromonas* has no lipid requirement, Hutner *et al.* (1957) indicated that at higher temperatures, at which this flagellate can be grown if supplied with suspensions of killed cells, a lipid might be among its essential growth factors.

One might expect that lipid requirements would be severely disadvantageous to organisms dependent on dissolved nutrients, and consequently would occur in phagotrophs. Although few instances are known, it appears that lesions in lipid metabolism are not confined to a single locus, since the lipids required by various protozoa include stigmasterol, sitosterol, cholesterol, and linoleic acid. One might even find an organism capable of utilizing ingested chlorophyll for photosynthesis!

VIII. Phylogenetic Implications

A source of continuing speculation is to be found in the obscure affinities of the various algal and protozoan phyla with one another and with the Metazoa and Metaphyta (Pascher, 1917; Fritsch, 1935; Dougherty, 1955; Hutner and Provasoli, 1955).

Phagotrophic nutrition, as previously defined, occurs in certain classes of algae—chiefly flagellates—but not elsewhere in the plant kingdom; among animals it is confined to the less highly evolved phyla. Another nutritional characteristic of algae, as already mentioned, is the tendency for auxotrophy to be limited to one or more of three B vitamins, namely B_{12}, thiamine, and biotin. A third is the tendency for chemotrophic algae to be limited to ethanol and the lower fatty acids as organic substrates for growth (see Danforth, Chapter 6, this treatise). In examples of extended auxotrophy and chemotropy, it is tempting to see phylogenetic links with either Metaphyta or Metazoa, although neither of these regularly share the limitations of the majority of the lower algae. Since many Chlorococcales and pennate diatoms can utilize carbohydrates, they might be considered to exhibit affinities with higher plants, whereas some of the other glucose-utilizing algae, with more complex growth-factor requirements, could be regarded as primitive animals. Vitamin B_{12} specificity also has been used as an argument in favor of the animal affinities of *Ochromonas* and the Chrysophyta in general, especially since in this alga—as in vertebrates—methionine considerably reduces the requirement for the vitamin. However, as Table III shows, mammalian-type specificity is widespread among algae, and it thus loses much of its significance in this respect.

IX. Ecological Implications

Reviews by Hutchinson (1941), Lucas (1947), Saunders (1957), Droop (1957a), Provasoli (1958a, b) and Lewin (1959) discuss the importance of auxotrophy in algal ecology. It has long been the ambition of limnologists and oceanographers to correlate the cycles of phytoplankton abundance with the seasonal turnover of nutrients. Although rough correlations between algal "blooms" and water movements, temperature, light, and inorganic nutrients have often been obtained, the succession of species has usually defied analysis. It appears probable that many of the fluctuations observed in nature can be attributed to changes in the concentrations of various organic micronutrients present in the water.

Thiamine and vitamin B_{12} have been measured in some fresh waters (Hutchinson, 1943; Benoit, 1957) and in the sea (Burkholder and Burkholder, 1956; Cowey, 1956; Daisley and Fisher, 1958; Kashiwada *et al.*, 1957; Starr, 1956; Tomiyama, 1959; Vishniac and Riley, 1959). Sizable variations have been found, but for a growth factor to exert ecological control its fluctuations need to bring it at times below the level of physiological activity, so that it becomes a limiting factor. There is some question as to whether the amounts of vitamin B_{12} measured in the sea, for instance, are low enough to be of general ecological interest (Droop, 1957b; Daisley, 1957; Ford, 1958). The lowest figure recorded is 0.1 mμg./liter, which could be expected ultimately to support a heavy crop of phytoplankton. If, on the other hand, circumstances were such that the relative growth rate assumed overriding importance, concentrations of this order might exert control on the species composition of the plankton in cases where the response of one of its potential components resembled *Ochromonas* rather than *Monochrysis* (see Fig. 1). Furthermore, it is uncertain how much of the vitamin measured by bioassay is available to the algae in question; much of it could be bound and thereby rendered unavailable to nonphagotrophic organisms.

The vitamins of soil and natural waters are mostly of bacterial origin (Ericson and Lewis, 1953; Lochhead, 1957; Burkholder, 1959), though algae themselves cannot be excluded as producers, as Robbins *et al.* (1951), Hashimoto (1954) and Brown *et al.* (1956) have shown. Thus, temperature could influence vitamin production in nature largely by its effect on bacterial growth. It may well be that the sudden outbursts of phytoplankton in springtime follow upon increased bacterial activity resulting from warmer and calmer conditions. Direct effects of temperature on the nutritional requirements of algae also raise ecological and geographical questions. For example, is there generally increased heterotrophy among warm water species, and are organic micronutrients of greater importance in tropical than in polar waters? Is the geographical temperature limit of a circumpolar species merely a nutritional one that can be overstepped under favorable nutrient conditions? Or, more generally, are ecological temperature gradients normally reflected in gradients of increased heterotrophy among the various inhabitants? It is only by the cooperation of physiologists with ecologists that such questions can be answered.

References

Allen, E. J. (1914). On the culture of the plankton diatom *Thalassiosira gravida* Cleve in artificial sea water. *J. Marine Biol. Assoc. United Kingdom* **10,** 417–439.

Allen, M. B. (1952). The cultivation of Myxophyceae. *Arch. Mikrobiol.* **17,** 34–53.

Anderson, E. H. (1945). Studies on the metabolism of the colourless alga *Prototheca zopfii*. *J. Gen. Physiol.* **28,** 297–327.

Baker, H., Frank, O., Pasher, I., Hutner, S. H., Herbert, V., and Sobotka, H. S. (1959). Monosubstituted Vit. B_{12} amides. I. A microbiological study. *Proc. Soc. Exptl. Biol. Med.* **100,** 825–827.

Benoit, R. (1957). Preliminary observations on cobalt and vitamin B_{12} in fresh water. *Limnol. and Oceanog.* **2,** 233–240.

Bold, H. C. (1942). The cultivation of algae. *Botan. Rev.* **8,** 69–138.

Brown, F., Cuthbertson, W. F., and Fogg, G. E. (1956). Vitamin B_{12} activity in *Chlorella vulgaris* Beij. and *Anabaena cylindrica* Lemm. *Nature* **177,** 188.

Burkholder, P. R. (1959). Vitamin producing bacteria in the Sea. *Preprints Intern. Oceanog. Congr., New York, 1959* pp. 912–913.

Burkholder, P. R., and Burkholder, L. M. (1956). Vitamin B_{12} in suspended solids and marsh muds collected along the coast of Georgia. *Limnol. and Oceanog.* **1,** 202–208.

Chen, Y. T. (1950). Investigations on the biology of *Peranema trichophorum* (Eugleninеae). *Quart. J. Microscop. Sci.* **91,** 279–308.

Coats, M. E., and Ford, J. E. (1955). Methods of measurement of vitamin B_{12}. *Biochem. Soc. Symposia (Cambridge, Engl.)* **13,** 36–51.

Cowey, C. B. (1956). A preliminary investigation of the variation of vitamin B_{12} in oceanic and coastal waters. *J. Marine Biol. Assoc. United Kingdom* **35,** 609–620.

Cramer, M., and Myers, J. (1950). Photosynthetic characteristics of *Euglena*. *Am. J. Botany* **37,** 677. (Abstracts of Papers of Botan. Sci. Am. Conf., Columbus, Ohio, September 11–13, 1950.)

Daisley, K. W. (1957). Vitamin B_{12} in marine ecology. *Nature* **180,** 1042.

Daisley, K. W., and Fisher, L. R. (1958). Vertical distribution of vitamin B_{12} in the sea. *J. Marine Biol. Assoc. United Kingdom* **37,** 683–686.

Dougherty, E. C. (1955). Comparative evolution and the origin of sexuality. *Systematic Zool.* **4,** 145–170.

Droop, M. R. (1953). Phagotrophy in *Oxyrrhis marina* Dujardin. *Nature* **172,** 250.

Droop, M. R. (1955). A pelagic marine diatom requiring cobalamin. *J. Marine Biol. Assoc. United Kingdom* **34,** 229–231.

Droop, M. R. (1957a). Auxotrophy and organic compounds in the nutrition of marine phytoplankton. *J. Gen. Microbiol.* **16,** 286–293.

Droop, M. R. (1957b). Vitamin B_{12} in marine ecology. *Nature* **180,** 1041–1042.

Droop, M. R. (1958). Requirement for thiamine among some marine and supra-littoral protista. *J. Marine Biol. Assoc. United Kingdom* **37,** 323–329.

Droop, M. R. (1959). Water-soluble factors in the nutrition of *Oxyrrhis marina*. *J. Marine Biol. Assoc. United Kingdom* **38,** 605–620.

Droop, M. R. (1961). *Haematococcus pluvialis* and its allies. III. Organic nutrition. *Rev. algol.* **4,** 247–259.

Droop, M. R., McLaughlin, J. J. A., Pintner, I. J., and Provasoli, L. (1959). Specificity of some protophytes toward vitamin B_{12}-like compounds. *Preprints Intern. Oceanog. Congr., New York, 1959* pp. 916–918.

Ericson, L. E., and Lewis, L. (1953). On the occurrence of vitamin B_{12} factors in marine algae. *Arkiv Kemi* **6,** 427–442.

Ford, J. E. (1953). The microbiological assay of 'vitamin B_{12}'. The specificity of the requirement of *Ochromonas malhamensis* for cyanocobalamin. *Brit. J. Nutrition* **7,** 299–306.

Ford, J. E. (1958). B_{12}-vitamins and growth of the flagellate *Ochromonas malhamensis. J. Gen. Microbiol.* **19,** 161–172.

Ford, J. E. (1959). The influence of certain derivatives of vitamin B_{12} upon the growth of micro-organisms. *J. Gen. Microbiol.* **21,** 693–701.

Ford, J. E., Gregory, M. E., and Holdsworth, E. S. (1955). Uptake of B_{12}-vitamins in *Ochromonas malhamensis. Biochem. J.* **61,** xxiii.

Fries, L. (1959). *Goniotrichum elegans*. A marine red alga requiring vitamin B_{12}. *Nature* **183,** 558–559.

Fritsch, F. E. (1935). "The Structure and Reproduction of the Algae," Vol. 1. Cambridge Univ. Press, London and New York.

Funk, H. B., and Nathan, H. A. (1958). Inhibition of growth of microorganisms by benzimidazoles. *Proc. Soc. Exptl. Biol. Med.* **99,** 394–397.

Hashimoto, Y. (1954). Vitamin B_{12} in marine and freshwater algae. *J. Vitaminol. (Osaka)* **1,** 49–54.

Hendlin, D. (1953). Discussion of papers on vitamin B_{12}. *Ann. N. Y. Acad. Sci.* **56,** 870–872.

Hutchinson, G. E. (1941). Ecological aspects of succession in natural populations. *Am. Naturalist* **75,** 406–418.

Hutchinson, G. E. (1943). Thiamin in lake waters. *Arch. Biochem.* **2,** 143–150.

Hutner, S. H., and Provasoli, L. (1953). A pigmented marine diatom requiring vitamin B_{12} and uracil. *News Bull. Phycol. Soc. Am.* **6,** 7–8.

Hutner, S. H., and Provasoli, L. (1955). Comparative biochemistry of flagellates. *In* "Biochemistry and Physiology of Protozoa" (S. H. Hutner and A. Lwoff, eds.), Vol. II, pp. 17–44. Academic Press, New York.

Hutner, S. H., Provasoli, L., Stockstad, E. L. R., Hoffmann, C. E., Belt, M., Franklin, A. L., and Jukes, T. H. (1949). Assay of antipernicious anemia factor with *Euglena. Proc. Soc. Exptl. Biol. Med.* **70,** 118–120.

Hutner, S. H., Provasoli, L., and Filfus, J. (1953). Nutrition of some phagotrophic chrysomonads. *Ann. N. Y. Acad. Sci.* **56,** 852–862.

Hutner, S. H., Baker, H., Aaronson, S., Nathan, H. A., Rodriguez, E., Lockwood, S., Sanders, M., and Peterson, R. A. (1957). Growing *Ochromonas malhamensis* above 35°C. *J. Protozool.* **4,** 259–269.

Hutner, S. H., Aaronson, S., Nathan, H. A., Baker, H., Scher, S., and Cury, A. (1958). Trace elements in microorganisms: the temperature factor approach. *In* "Trace Elements" (C. A. Lamb, O. G. Bentley, and J. M. Beattie, eds.), pp. 47–65. Academic Press, New York.

Kashiwada, K., Kakimoto, D., Morita, T., Kanazawa, A., and Kawagoe, K. (1957). Studies on vitamin B_{12} in sea water. II. On the assay method and distribution of this vitamin B_{12} in the ocean. *Bull. Japan. Soc. Sci. Fisheries* **22,** 637–640.

Kon, S. K. (1955). Other factors related to vitamin B_{12}. *Biochem. Soc. Symposia (Cambridge, Engl.)* **13,** 17–35.

Latner, A. L. (1955). Intrinsic factor. *Biochem. Soc. Symposia (Cambridge, Engl.)* **13,** 69–91.

Lewin, J. C., and Lewin, R. A. (1960). Auxotrophy and heterotrophy in marine littoral diatoms. *Can. J. Microbiol.* **6,** 127–134.

Lewin, R. A. (1952). Ultraviolet induced mutations in *Chlamydomonas moewusii* Gerloff. *J. Gen. Microbiol.* **6,** 233–248.

Lewin, R. A. (1959). Phytoflagellates and algae. *In* "Handbuch der Pflanzenphysiologie" (W. Ruhland, ed.), Vol. 14, pp. 401–417. Springer, Berlin.

Lochhead, A. G. (1957). Qualitative studies of soil microorganisms. XV. Capability of the predominant bacterial flora for synthesis of various growth factors. *Soil Sci.* **84,** 395–403.

Lucas, C. E. (1947). The ecological effects of external metabolites. *Biol. Revs. Cambridge Phil. Soc.* **22,** 270–295.

Lwoff, A. (1947). Some aspects of the problem of growth factors for Protozoa. *Ann. Rev. Microbiol.* **1,** 101–114.

Lwoff, A., and Dusi, H. (1937a). La pyrimidine et le thiazole, facteurs de croissance pour le flagellé *Polytoma coeca. Compt. rend. acad. sci.* **205,** 630.

Lwoff, A., and Dusi, H. (1937b). Le thiazole, facteur de croissance pour les flagellés *Polytoma caudatum* et *Chilomonas paramoecium. Compt. rend. acad. sci.* **205,** 756.

Lwoff, A., and Dusi, H. (1937c). Le thiazole, facteur de croissance pour le flagellé *Polytoma ocellatum. Compt. rend. acad. sci.* **205,** 882.

Lwoff, M. (1951). Nutrition of parasitic flagellates (Trypanosomidae, Trichomonadinae). *In* "Biochemistry and Physiology of Protozoa" (A. Lwoff, ed.), Vol. I, pp. 129–176. Academic Press, New York.

Monod, J. (1942). "Recherches sur la croissance bactériennes." Hermann, Paris.

Pascher, A. (1917). Über Flagellaten und Algen. *Ber. deut. botan. Ges.* **32,** 110–114.

Pintner, I. J., and Provasoli, L. (1959). The nutrition of *Volvox globator* and *V. tertius. Congr. intern. botan., 9e Congr., Montreal, Abstr.* pp. 300–301.

Pringsheim, E. G. (1946). "Pure Cultures of Algae." Cambridge Univ. Press, London and New York.

Pringsheim, E. G. (1952). On the nutrition of *Ochromonas. Quart. J. Microscop. Sci.* **93,** 71–96.

Pringsheim, E. G. (1959). Phagotrophie. *In* "Handbuch der Pflanzenphysiologie" (W. Ruhland, ed.), Vol. 11, pp. 179–197. Springer, Berlin.

Provasoli, L. (1958a). Nutrition and ecology of protozoa and algae. *Ann. Rev. Microbiol.* **12,** 279–308.

Provasoli, L. (1958b). Growth factors in unicellular marine algae. *In* "Perspectives in Marine Biology" (A. A. Buzzati-Traverso, ed.), pp. 385–403. Univ. California Press, Berkeley, California.

Provasoli, L., and Pintner, I. J. (1953). Ecological implications of *in vitro* nutritional requirements of algal flagellates. *Ann. N. Y. Acad. Sci.* **56,** 839–851.

Provasoli, L., McLaughlin, J. J. A., and Droop, M. R. (1957). The development of artificial media for marine algae. *Arch. Mikrobiol.* **25,** 392–428.

Robbins, W. J., Hervey, A., and Stebbins, M. E. (1951). Further observations on *Euglena* and vitamin B_{12}. *Bull. Torrey Botan. Club* **78,** 363–375.

Saunders, G. W. (1957). Interrelations of dissolved organic matter and phytoplankton. *Botan. Rev.* **23,** 389–409.

Scher, S., and Aaronson, S. (1958). Nutritional factors in apochlorosis: comparative studies with algae and higher plants. *Brookhaven Symposia in Biol. No.* **11,** 343–347.

Smith, E. L. (1958). Biochemical functioning of vitamin B_{12}. *Nature* **181,** 305–306.

Smith, E. L. (1960). "Vitamin B_{12}." Methuen, London.

Starr, T. J. (1956). Relative amounts of vitamin B_{12} in detritus from oceanic and estuarine environments near Sapelo Island, Georgia. *Ecology* **37,** 658–664.

Storm, J., and Hutner, S. H. (1953). Nutrition of *Peranema. Ann. N. Y. Acad. Sci.* **56,** 901–909.

Tomiyama, T. (1959). Preliminary report of the determination and distribution of vitamin B_{12} in the sea. *Preprints Intern. Oceanog. Congr., New York, 1959,* pp. 941–942.

Vishniac, H. S., and Riley, G. (1959). B_{12} and thiamine in Long Island Sound: patterns of distribution and ecological significance. *Preprints Intern. Oceanog. Congr., New York, 1959,* pp. 942–943.

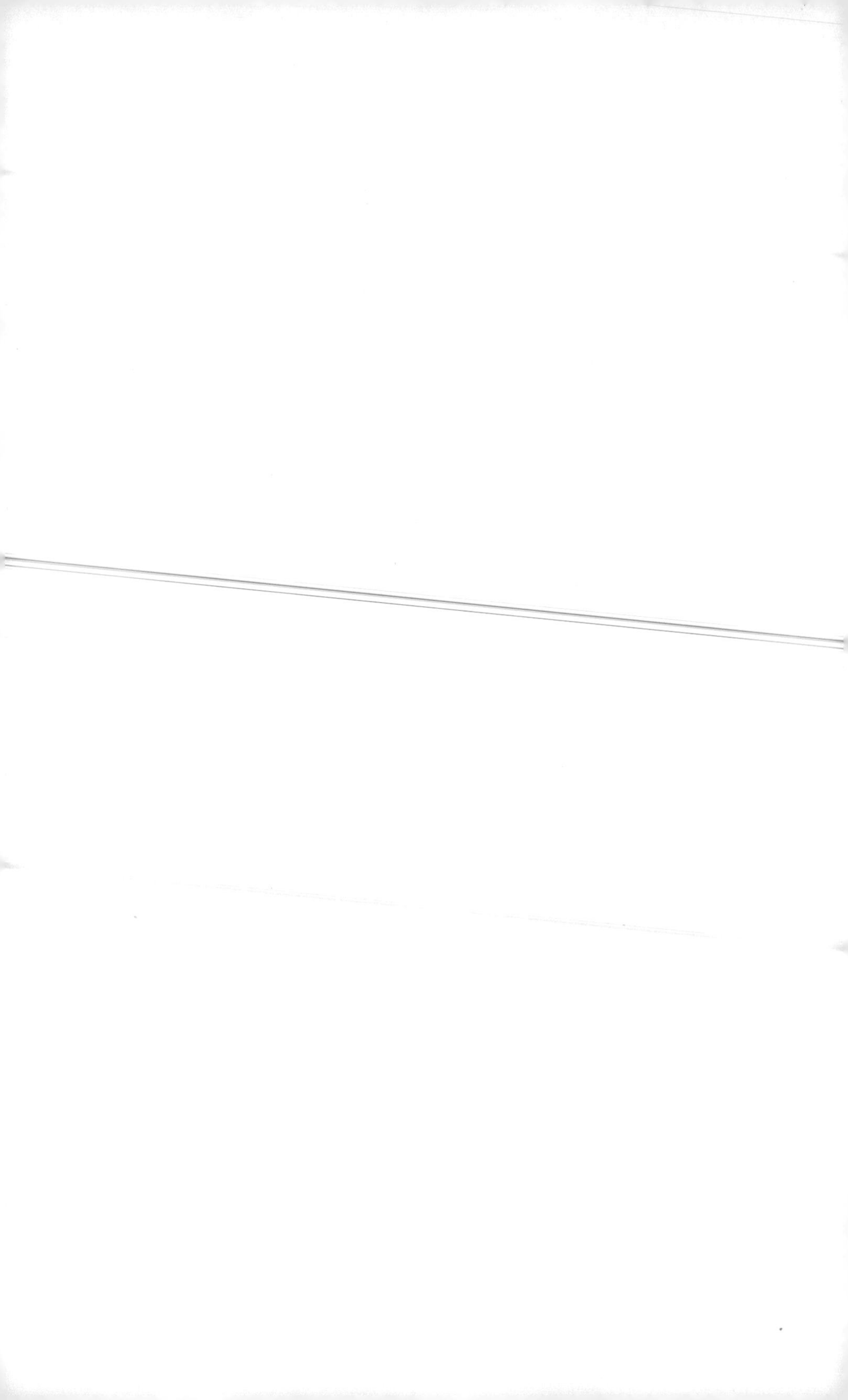

—9—

Nitrogen Fixation

G. E. FOGG[1]

Department of Botany, University College, London, England

I. Introduction

Many Blue-green Algae, like some bacteria, are able to assimilate or "fix" elementary nitrogen, N_2. This process is of obvious interest biochemically, and its interrelations in Blue-green Algae with photosynthesis and growth are particularly worth study both from the purely physiological point of view and because these organisms can be of ecological and economic importance (see Talling, Chapter 50, and Lund, Chapter 51, in this treatise). A general account of biological nitrogen fixation was given by Wilson (1958) and reviews with particular reference to the Blue-green Algae were published by Fogg (1947; 1956a), Fogg and Wolfe (1954), and Allen (1956).

II. Distribution of the Capacity to Fix Nitrogen among the Algae

There is no good evidence for nitrogen fixation by algae belonging to groups other than the Cyanophyta (Fogg, 1956a). The distribution of the property within the Cyanophyta has been considered by Fogg and Wolfe (1954). Definite proof of fixation has been obtained with more than twenty species, and about an equal number have been found in critical tests to lack this capacity. Most of the nitrogen-fixing species so far recognized belong to three genera of the Nostocaceae: *Anabaena*, *Cylindrospermum*, and *Nostoc*. Two other families of the Nostocales, the Rivulariaceae and

[1] *Present address*: Department of Botany, Westfield College, London, England.

the Scytonemataceae, include nitrogen-fixing species, but the property seems to be lacking in another family of the same order, the Oscillatoriaceae. Few species of other orders have been examined, but *Mastigocladus laminosus*, belonging to the morphologically most advanced order, Stigonematales, has been shown to be nitrogen-fixing (Fogg, 1951). Several species of Chroococcales, the unicellular order of Blue-green Algae, have been found not to fix nitrogen, but one supposed member, *Chlorogloea fritschii*, has been shown by Fay (unpublished) to possess the property. Using methods similar to those of Fogg (1951), he found that this alga could fix about 2.4 mg. N per 100 ml. culture in 30 days when grown in the medium of Allen and Arnon (1955) at 25°C. In addition to species studied in pure culture there are many which, from their behavior in impure culture, appear able to fix nitrogen (see, for example, Fogg and Wolfe, 1954; Watanabe, 1959). The general impression is that the Myxophyta include a higher proportion of nitrogen-fixing species than any other major group of microorganisms.

III. The Biochemistry of Nitrogen Fixation

Our scant knowledge of the mechanism of biological nitrogen fixation is mainly derived from studies on the bacteria *Azotobacter*, *Clostridium*, and *Rhizobium* spp. (see Wilson, 1958; Roberts, 1959). Investigations with Blue-green Algae have done little more than confirm that the mechanism in these organisms is generally similar to that in the bacteria.

The kinetics of the process in *Nostoc muscorum* (Allison's strain) were studied by Burris and Wilson (1946), who found that the partial pressure of nitrogen giving half the maximum rate of fixation was of the order of 0.02 atm., and that hydrogen was a specific inhibitor. In these respects the algal system resembled that in *Azotobacter*; however, the effect of carbon monoxide, another specific inhibitor, appeared to depend on the partial pressure of nitrogen, whereas in bacteria the inhibition is noncompetitive. These results were obtained with long-term cultures of *N. muscorum*, and it would be desirable to confirm them using material known to be growing actively throughout the experimental period.

Nostoc muscorum (strain not stated), *Cylindrospermum* sp. (Frenkel and Rieger, 1951), and *Anabaena cylindrica* (Wolfe, 1954b) do not appear to have hydrogenase systems, and thus do not show the correlation between hydrogenase activity and nitrogen fixation which is evident in bacteria (Wilson, 1958).

Like other nitrogen-fixing organisms, Blue-green Algae have a requirement for relatively high concentrations of molybdenum, of the order of 0.1

mg. per liter (Bortels, 1940; Wolfe, 1954a; Allen and Arnon, 1955; Allen, 1956; Arnon, 1958). Molybdenum, which is not replaceable by vanadium (Allen and Arnon, 1955; Arnon, 1958), is required by these algae for nitrate assimilation as well as for nitrogen fixation but not, apparently, for growth with ammonium as the source of nitrogen (Wolfe, 1954a). The finding that molybdenum is an essential component of the nitrate reductase system (Nicholas, 1959) suggests that this element may likewise be a component of an enzyme needed for nitrogen fixation, and that, possibly, the two processes have one or more steps in common. However, molybdenum is known to have several biochemical roles, and the concentrations required for maximum rates of fixation by Blue-green Algae seem higher than would be needed for the formation of an enzyme alone. A theory that the dominant role of molybdenum, in both nitrate reduction and nitrogen fixation in these algae, is that of an inhibitor of phosphatase activity (Fogg and Wolfe, 1954) is supported by observations of Talpasayi (unpublished) that concentrations of molybdenum of the same order as those required for nitrogen fixation do, in fact, have a marked inhibitory action on the phosphatase activity of *Anabaena cylindrica*.

The path of nitrogen in fixation has been investigated in *Nostoc muscorum* (Gerloff's strain) by Magee and Burris (1954). Determination of the distribution of tracer after the organism had been exposed for a short period to elementary nitrogen enriched with N^{15} showed it to be present in the highest proportions in glutamic acid and ammonia, as it was also when the N^{15} was supplied as ammonia. These results are in agreement with the generally accepted idea that ammonia is the key intermediate in fixation and enters into general metabolism by reductive amination of α-ketoglutaric acid (Wilson, 1958). However, Linko *et al.* (1957) found that during short periods of photosynthesis, under conditions which probably allowed nitrogen fixation, up to 20% of the carbon fixed by *Nostoc muscorum* (strain not stated) appeared in citrulline. This did not appear to be functioning in reactions leading to arginine or urea synthesis. From tracer experiments with N^{15}, Leaf *et al.* (1958) concluded that in root nodules of *Alnus* citrulline is formed directly from ammonia produced by fixation of nitrogen; therefore, it may be that in Blue-green Algae glutamic acid and citrulline provide alternative routes of entry of fixed nitrogen into general metabolism.

Biochemical investigation of nitrogen fixation has hitherto been restricted because the enzyme system effecting it could not be studied except in living cells. This restriction may now have been overcome; cell-free preparations capable of fixing nitrogen have been obtained from various organisms, including the Blue-green Alga *Mastigocladus laminosus* (Wilson and Burris, 1960).

IV. The Physiology of Nitrogen Fixation in Blue-Green Algae

A. Nitrogen Fixation and Growth

Generally speaking, nitrogen fixation occurs only in growing cells. Thus, besides the factors mentioned in the previous section as having specific effects on the process, any factor which affects the growth of the alga will have a corresponding effect on nitrogen fixation. [For general accounts of the growth physiology of Blue-green Algae, see Allen (1956); Fogg (1956a, b).] Fixation does not, however, necessarily parallel growth in the presence of an exogenous source of combined nitrogen. Ammonium salts completely suppress the assimilation of elementary nitrogen (Fogg, 1942; Allen, 1956), as would be expected were ammonia the key intermediate in the process. The extent of inhibition by other nitrogen compounds evidently depends on the readiness with which they can give rise to ammonia. Inhibition by urea is complete but that by nitrate is sometimes incomplete (Allen, 1956), there being indications that the previous history of the cells is important and that *Anabaena cylindrica* needs to be adapted to nitrate before it can assimilate it readily (Fogg and Wolfe, 1954).

B. Nitrogen Fixation and Photosynthesis

It may be supposed that the dependence on growth is the result of particularly close intermeshing of nitrogen fixation with other cellular processes. An example of such interdependence is provided by the relation of nitrogen fixation to photosynthesis. These two processes can, of course, be independent of each other, but in Blue-green Algae, in which they may take place in the same cells, it seems that nitrogen fixation depends in rather a direct manner on photosynthesis for the necessary hydrogen donors and carbon skeletons.

Results obtained by Fogg and Than-Tun (1958) suggest that photochemically generated hydrogen donors are freely available for the reduction of nitrogen in *A. cylindrica*. This alga evolves more oxygen when fixing nitrogen in the light than it does under otherwise similar circumstances with nitrogen supplied in an already reduced form. The amount of this "extra" oxygen is equivalent to that of the nitrogen fixed if this is assumed to be reduced to ammonia,

$$N_2 + 3H_2O \rightarrow 2NH_3 + 1\tfrac{1}{2}O_2 \quad (1)$$

Furthermore, oxygen can be evolved by *A. cylindrica* in the absence of carbon dioxide when nitrogen is the only available exogenous acceptor for the hydrogen produced by the photochemical reaction. Humphrey (unpublished) found that the amount of oxygen evolved under these conditions varies greatly according to the physiological condition of the alga and that values such as those reported by Fogg and Than-Tun (1958) are obtained only exceptionally. The evolution of oxygen on illumination of the alga, in a medium which was swept free of nitrogen and carbon dioxide with argon, indicates the photochemical reduction of an endogenous hydrogen acceptor, the presence of which in appreciable amounts in *A. cylindrica* seems plausible on other grounds (Fogg and Than-Tun, 1960). However, substitution of elementary nitrogen for argon may result in oxygen production additional to this, demonstrating that nitrogen can be reduced by photochemically generated hydrogen donors. In these experiments of Humphrey, the ratios of carbon and nitrogen fixed, as deduced from the amounts of oxygen evolved, are similar to the ratio of these elements in the cell material of *A. cylindrica*. Presumably the much higher relative rate of nitrogen fixation that must have occurred in the experiment reported by Fogg and Than-Tun (1958) depended on availability in the cells of unusually large amounts of carbon to accept the nitrogen fixed. Since elementary nitrogen is evidently not a Hill reagent, it is unlikely that its reduction can be purely photochemical, and it must be supposed that, as in the photochemical reduction of nitrate and nitrite in Green Algae (see Syrett, Chapter 10, this treatise), one or more intermediate hydrogen carriers are necessary.

Apart from supplying the hydrogen donor, photosynthesis may also provide directly the carbon skeletons needed to build organic compounds from the nitrogen fixed. It has been amply demonstrated, using C^{14} as a tracer, that both glutamic acid and citrulline, which from the evidence reviewed in Section III appear to be the first organic products of fixation, can be derived rapidly from photosynthetic intermediates in Blue-green Algae (Norris *et al.*, 1955; Linko *et al.*, 1957). A further connection may be by high-energy phosphates. The part played by phosphates in the energy transfers of nitrogen fixation has scarcely been investigated, but it seems probable that compounds of the adenosine triphosphate type are involved and that these might be provided directly by photosynthetic phosphorylation. Indications have been obtained by Talpasayi (unpublished) that photosynthetic phosphorylation in *A. cylindrica* is more rapid when nitrogen fixation is taking place than when it is prevented, but more direct evidence of a connection is still lacking.

It is thus evident that in Blue-green Algae the mechanisms of nitrogen fixation and photosynthesis may interact at various levels, and that correspondingly close kinetic relationships between the two processes may be expected. Fogg and Than-Tun (1960) have shown that there is such a correlation between the rate of photosynthetic carbon assimilation and that of assimilation of elementary nitrogen. They studied the effects of light intensity, carbon dioxide concentration, and concentration of elementary nitrogen, each in relation to temperature, by determination of changes in amounts of cell carbon and cell nitrogen in cultures of *A. cylindrica* grown for 48 hours. Temperature was found to have the most marked differential effect, both low (15°C.) and high (35°C.) temperatures depressing nitrogen assimilation to a greater extent than they depressed carbon assimilation. At any given temperature there was a close correlation between the rates of the two processes (Figs. 1 and 2). The rate of carbon assimilation per unit amount of cell nitrogen was found to be related in the usual way to light intensity, but to be reduced at low nitrogen concentrations. The relative rate of nitrogen assimilation was likewise found not only to be related in the expected way to nitrogen concentration but to increase with light intensity and to be reduced at carbon dioxide concentrations which were limiting for carbon assimilation. These relationships are consonant with the existence of the biochemical connections between the two processes postulated above, but the possibility remains that to some extent they are dependent on differential rates of liberation of nitrogenous and non-nitrogenous extracellular products.

In spite of this close relationship, no obligatory coupling of nitrogen fixation to photosynthesis has been found in the algae which have been examined so far. Some, e.g., Allison's strain of *N. muscorum*, can grow and fix nitrogen in the dark if provided with a suitable carbohydrate (Allison

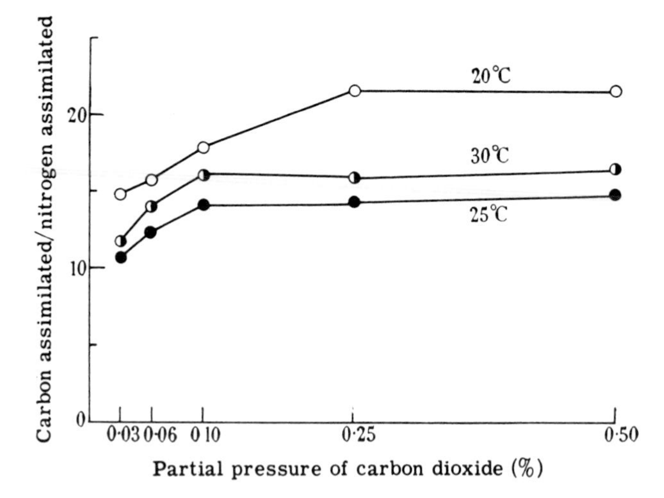

FIG. 1. Effects of partial pressure of carbon dioxide and temperature on the ratio of the amounts of carbon and elementary nitrogen assimilated by *Anabaena cylindrica* in 48-hour cultures (reproduced from Fogg and Than-Tun (1960) by permission of the Council of the Royal Society).

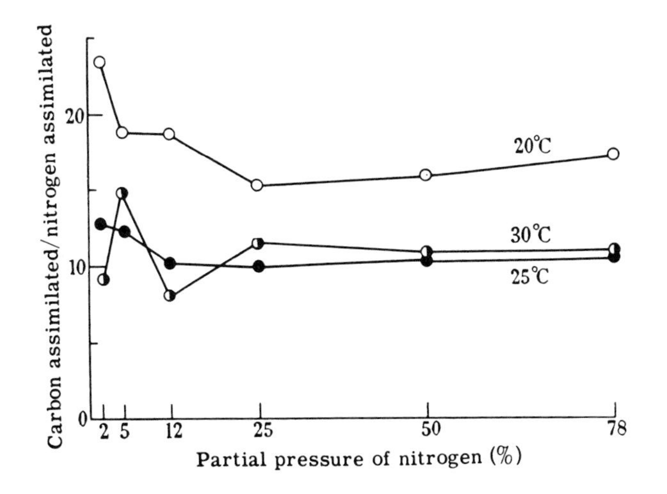

FIG. 2. Effects of partial pressure of nitrogen and temperature on the ratio of the amounts of carbon and elementary nitrogen assimilated by *Anabaena cylindrica* in 48-hour cultures (reproduced from Fogg and Than-Tun (1960) by permission of the Council of the Royal Society).

et al., 1937; Kratz and Myers, 1955). Others, such as *A. cylindrica* (Fogg and Wolfe, 1954) and Gerloff's strain of *N. muscorum* (Kratz and Myers, 1955), are obligate phototrophs, but results obtained by Than-Tun and Harris (quoted by Fogg, 1961), using N^{15} as a tracer, show clearly that *A. cylindrica* is able to fix nitrogen in the dark for a limited period. This contrasts with the situation in the photosynthetic bacterium *Rhodospirillum rubrum*, in which nitrogen fixation stops abruptly when illumination ceases (Pratt and Frenkel, 1959).

C. Extracellular Products of Nitrogen-Fixing Blue-Green Algae

The liberation of substantial amounts of soluble nitrogenous substances from the cells in healthy cultures of nitrogen-fixing Blue-green Algae has been recorded many times (see Fogg, 1952, for references). Most of this extracellular nitrogen is in the form of peptide, free amino acids being present as a rule in small amounts only (Fogg, 1952; Magee and Burris, 1954), though in cultures of *Calothrix brevissima* appreciable amounts of aspartic acid, glutamic acid, and alanine have been found (Watanabe, 1951). There does not appear to be any great proportion of substances specifically associated with the fixation process. In cultures of *A. cylindrica* the relative amount and composition of the extracellular fraction are much the same whether elementary nitrogen, nitrate, or an ammonium salt is the source of nitrogen. Similar extracellular products seem to be liberated by Blue-green Algae which are unable to fix nitrogen (Fogg, 1952). Studies with *A. cylindrica* (Fogg, 1952) and *Nostoc* sp. (Venkataraman, 1961) show that these extracellular products are not the result of autolysis, and that, although their liberation is an invariable concomitant of growth, its relative extent varies with culture conditions, being increased in older cultures, for example, by deficiencies in mineral nutrients such as iron. Henriksson (1957) found that continuous shaking of cultures reduced the amounts of nitrogenous substances liberated into the medium by a strain of *Nostoc* sp. Production of extracellular nitrogenous substances may lead to considerable underestimation of fixation if cell nitrogen only is determined, especially in young cultures in which as much as 60% of the total combined nitrogen may be present in the filtrate. Ecologically, extracellular products may be important as nitrogen sources for other organisms; however this has not been sufficiently investigated. Nitrogen-fixing Blue-green Algae are commonly found in symbiosis with other plants, e.g., with fungi in lichens and endophytically in liverworts, ferns, cycads, and angiosperms (for references see Fogg, 1956), and extracellular products presumably play an important part in these relationships. Henriksson (1957) showed that a *Nostoc* sp. isolated from the lichen *Collema* liberates considerable quantities of nitrogenous substances, and Bond and Scott (1955) put forward reasons for believing that there is transfer of combined nitrogen between the *Nostoc* endophytic in *Blasia* and its hepatic host.

References

Allen, M. B. (1956). Photosynthetic nitrogen fixation by blue-green algae. *Sci. Monthly* **83**, 100–106.

Allen, M. B., and Arnon, D. I. (1955). Studies on nitrogen-fixing blue-green algae. I. Growth and nitrogen fixation by *Anabaena cylindrica* Lemm. *Plant Physiol.* **30**, 366–372.

Allison, F. E., Hoover, S. R., and Morris, H. J. (1937). Physiological studies with the nitrogen-fixing alga *Nostoc muscorum*. *Botan. Gaz.* **98**, 433–463.

Arnon, D. I. (1958). The role of micronutrients in plant nutrition with special reference to photosynthesis and nitrogen assimilation. *In* "Trace Elements" (C. A. Lamb, O. G. Bentley, and J. M. Beattie, eds.), pp. 1–32. Academic Press, New York.

Bond, G., and Scott, G. D. (1955). An examination of some symbiotic systems for fixation of nitrogen. *Ann. Botany (London)* **19**, 67–77.

Bortels, H. (1940). Über die Bedeutung des Molybdäns für stickstoffbindende Nostocaceen. *Arch. Mikrobiol.* **11**, 155–186.

Burris, R. H., and Wilson, P. W. (1946). Characteristics of the nitrogen-fixing enzyme system in *Nostoc muscorum*. *Botan. Gaz.* **108**, 254–262.

Fogg, G. E. (1942). Studies on nitrogen fixation by blue-green algae. I. Nitrogen fixation by *Anabaena cylindrica* Lem. *J. Exptl. Biol.* **19**, 78–87.

Fogg, G. E. (1947). Nitrogen fixation by blue-green algae. *Endeavour* **6,** 172–175.

Fogg, G. E. (1951). Studies on nitrogen fixation by blue-green algae. II. Nitrogen fixation by *Mastigocladus laminosus* Cohn. *J. Exptl. Botany* **2,** 117–120.

Fogg, G. E. (1952). The production of extracellular nitrogenous substances by a blue-green alga. *Proc. Roy. Soc.* **B139,** 372–397.

Fogg, G. E. (1956). Nitrogen fixation by photosynthetic organisms. *Ann. Rev. Plant Physiol.* **7,** 51–70.

Fogg, G. E. (1956b). The comparative physiology and biochemistry of the blue-green algae. *Bacteriol. Revs.* **20,** 148–165.

Fogg, G. E. (1961). Recent advances in our knowledge of nitrogen fixation by blue-green algae. *In* "Symposium on Algology." Indian Council of Agricultural Research, New Delhi.

Fogg, G. E., and Than-Tun. (1958). Photochemical reduction of elementary nitrogen in the blue-green alga *Anabaena cylindrica. Biochim. et Biophys. Acta* **30,** 209–210.

Fogg, G. E., and Than-Tun. (1960). Interrelations of photosynthesis and assimilation of elementary nitrogen in a blue-green alga. *Proc. Roy. Soc.* **B153,** 111–127.

Fogg, G. E., and Wolfe, M. (1954). The nitrogen metabolism of the blue-green algae (Myxophyceae). *In* "Autotrophic Micro-organisms" (B. A. Fry and J. L. Peel, eds.), *Symposia Soc. Gen. Microbiol.* **4,** 99–125.

Frenkel, A. W., and Rieger, C. (1951). Photoreduction in algae. *Nature* **167,** 1030.

Henriksson, E. (1957). Studies in the physiology of the lichen *Collema.* I. The production of extracellular nitrogenous substances by the algal partner under various conditions. *Physiol. Plantarum* **10,** 943–948.

Kratz, W. A., and Myers, J. (1955). Nutrition and growth of several blue-green algae. *Am. J. Botany* **42,** 282–287.

Leaf, G., Gardner, I. C., and Bond, G. (1958). Observations on the composition and metabolism of the nitrogen-fixing root nodules of *Alnus. J. Exptl. Botany* **9,** 320–331.

Linko, P., Holm-Hansen, O., Bassham, J. A., and Calvin, M. (1957). Formation of radioactive citrulline during photosynthetic $C^{14}O_2$ fixation by blue-green algae. *J. Exptl. Botany* **8,** 147–156.

Magee, W. E., and Burris, R. H. (1954). Fixation of N_2 and utilization of combined nitrogen by *Nostoc muscorum. Am. J. Botany* **41,** 777–782.

Nicholas, D. J. D. (1959). Metallo-enzymes in nitrate assimilation of plants, with special reference to micro-organisms. *Symposia Soc. Exptl. Biol.* **13,** 1–23.

Norris, L., Norris, R. E., and Calvin, M. (1955). A survey of the rates and products of short-term photosynthesis in plants of nine phyla. *J. Exptl. Botany* **6,** 64–74.

Pratt, D. C., and Frenkel, A. W. (1959). Studies on nitrogen fixation and photosynthesis of *Rhodospirillum rubrum. Plant Physiol.* **34,** 333–337.

Roberts, E. R. (1959). Some observations on the chemistry of biological nitrogen fixation. *Symposia Soc. Exptl. Biol.* **13,** 24–41.

Venkataraman, G. S. (1961). Nitrogen fixation and production of extracellular nitrogenous substances by an endophytic *Nostoc* strain, isolated from the root nodules of Egyptian clover (*Trifolium alexandrinum*). *In* "Symposium on Algology." Indian Council of Agricultural Research, New Delhi.

Watanabe, A. (1951). Production in cultural solution of some amino acids by the atmospheric nitrogen-fixing blue-green algae. *Arch. Biochem. Biophys.* **34,** 50–55.

Watanabe, A. (1959). Distribution of nitrogen-fixing blue-green algae in various areas of South and East Asia. *J. Gen. Appl. Microbiol.* (*Tokyo*) **5,** 21–29.

Wilson, P. W. (1958). Asymbiotic nitrogen fixation. *In* "Handbuch der Pflanzenphysiologie" (W. Ruhland, ed.), Vol. 8, pp. 9–47. Springer, Berlin.

Wilson, P. W., and Burris, R. H. (1960). Fixation of nitrogen by cell-free extracts from microorganisms. *Science* **131,** 1321.

Wolfe, M. (1954a). The effect of molybdenum upon the nitrogen metabolism of *Anabaena cylindrica*. I. A study of the molybdenum requirement for nitrogen fixation, and for nitrate and ammonia assimilation. *Ann. Botany* (*London*) **18,** 299–308.

Wolfe, M. (1954b). The effect of molybdenum upon the nitrogen metabolism of *Anabaena cylindrica*. II. A more detailed study of the action of molybdenum in nitrate assimilation. *Ann. Botany* (*London*) **18,** 309–325.

—10—

Nitrogen Assimilation

P. J. SYRETT

Botany Department, University College, London, England

I. Introduction[1]

During the past decade, nitrogen assimilation, like other aspects of algal metabolism, has received increased attention, and several reviews have been published (Fogg, 1953, 1959; Fogg and Wolfe, 1954; Kessler, 1959; Krauss, 1958; Myers, 1951; Syrett, 1954). Most work has been done with microscopic algae, including several species of Green Algae and a few Blue-green Algae, diatoms, and members of the Xanthophyta. Few comparative studies have been made, and it is not always clear to what extent results depend on the particular organism used. A few studies have been made with macroscopic genera such as *Ulva* and *Enteromorpha* (Andersson, 1942;

[1] The following abbreviations are used in this chapter: ATP, adenosine triphosphate; DPN and TPN, di- and triphosphopyridine nucleotide; DPNH and TPNH, reduced di- and triphosphopyridine nucleotide.

Kylin, 1945) and *Fucus* (Bauch, 1956), but the difficulties of obtaining such algae in bacteria-free culture hinder their use.

II. Inorganic Nitrogen Sources

All algae with chlorophyll, with one or two exceptions, can apparently utilize either ammonium salts or nitrates when these are supplied at a suitable concentration. *Euglena gracilis* var. *bacillaris* and *Chlamydomonas reinhardtii* cannot use nitrate at all (Cramer and Myers, 1952; R. A. Lewin, unpublished), while *Haematococcus pluvialis* and *Chlamydomonas agloeformis* are reported to use it only in light (Lwoff, 1943). The maximum rate of growth of several commonly investigated algae is the same with either nitrate or ammonium-N (Bongers, 1956; Fogg, 1949; Noack and Pirson, 1939; Samejima and Myers, 1958). Nitrate reduction, however, requires energy, and if the energy supply is limited more growth will occur on ammonium-N. Thus, the efficiency of glucose utilization for cell synthesis in *Chlorella pyrenoidosa* is about 16% greater with ammonium-N than with nitrate (Samejima and Myers, 1958), and the efficiency of light-energy conversion in *Chlorella* spp. is about 30% greater (van Oorschot, 1955). Consequently, at low light intensity, cell synthesis is faster from ammonium-N than from nitrate (Bongers, 1956). Ammonium salts may have undesirable indirect effects, however, since the pH of a poorly buffered medium falls sharply as ammonium-N is assimilated (Pratt and Fong, 1940).

Algae such as *Chlorella* and *Scenedesmus* grow well in 0.01 *M* solutions of nitrates or ammonium salts. Such concentrations are too high for many planktonic algae (Chu, 1942; ZoBell, 1935). Their growth may be inhibited by 0.001 *M* ammonium-N although they tolerate higher nitrate concentrations. The reason for the inhibition by ammonium-N is unknown, but it may be correlated with an increase of internal pH owing to the penetration of undissociated ammonium hydroxide molecules (Blinks, 1951; Shilo and Shilo, 1955).

Ammonium-N is often used preferentially when it is supplied together with nitrate, and in species of *Chlorella* it appears to inhibit nitrate reduction (Ludwig, 1938; Pratt and Fong, 1940; Schuler *et al.*, 1953). This is presumably an indirect effect of ammonium-N, since nitrogen-starved cells of *Chlorella* and *Scenedesmus* assimilate both forms of nitrogen together (Urhan, 1932). Possibly the inhibitor is a product of ammonia assimilation which does not accumulate in nitrogen-deficient cells. It is of interest that nitrite assimilation by a *Mycobacterium* is inhibited by ammonia, methylamine, and dimethylamine, all of which are assimilated, whereas trimethylamine, which is not assimilated, does not inhibit it (De Turk and

Bernheim, 1958). Some algae, e.g., "*Nitzschia closterium*" * (= *Phaeodactylum tricornutum*), like nitrogen-starved *Chlorella*, utilize both ammonium-N and nitrate together (ZoBell, 1935). *Haematococcus pluvialis* is exceptional in preferentially removing nitrate from its medium (Proctor, 1957), but it is not certain whether the nitrate is preferentially used for cell synthesis or whether it accumulates in the cells. Certain marine algae, in particular *Porphyra*, accumulate considerable quantities of nitrate (Suneson, 1933; Syrett, unpublished).

Nitrite, in low concentrations, ca. 0.001 *M*, can also serve as a nitrogen source for many species (Fogg and Wolfe, 1954; Ludwig, 1938; ZoBell, 1935), but higher concentrations are inhibitory.

III. Nitrogen Assimilation by Nitrogen-Deficient Algae

A. Preparation and Properties of Nitrogen-Deficient Cells

Nitrogen-deficient algal cultures, which assimilate ammonium-N and nitrate rapidly, can be readily prepared by subjecting them to nitrogen-starvation. The use of such cultures has greatly aided the study of nitrogen assimilation.

Nitrogen-deficient cells can be prepared in two different ways. (1) Cultures can be grown with a limited amount of available nitrogen, the cells becoming nitrogen-deficient when this is exhausted. (2) Alternatively, nitrogen-sufficient cells can be transferred to a nitrogen-free medium and allowed to photosynthesize. The two methods do not necessarily produce cultures with identical properties (Fogg, 1959).

During the development of nitrogen deficiency, the nitrogen content of the cells falls, e.g., from 8–10% of the dry weight to about 2% in species of *Chlorella* and *Scenedesmus* (Bongers, 1956; Fogg, 1959; Thomas and Krauss, 1955). The chief products of photosynthesis change from protein to carbohydrate, and then to lipid (Bongers, 1956; van Oorschot, 1955). The over-all composition of the cells may continue to change for several weeks. In some nitrogen-deficient algae, e.g., *Chlorella pyrenoidosa*, lipid accumulation begins 4–6 days after the onset of nitrogen starvation, and considerable quantities of fatty material are accumulated (Aach, 1952; Fogg and Collyer, 1953). In others, e.g., *Chlorella vulgaris* and *Chlorella* "strain A," fat accumulation does not begin until much later, and carbohydrate is quantitatively more important (van Oorschot, 1955). As nitrogen deficiency develops, the amount of chlorophyll in the cells decreases faster than the total nitrogen content (Bongers, 1956; Fogg, 1959). The rate of photosynthesis falls, too, though not simply as a consequence of the drop in chloro-

* Regarding the taxonomy of this organism, see Silva, this treatise, Appendix A, Note 22.

phyll content (van Oorschot, 1955). Nothing is known of the changes in enzymatic balance which may occur. In the fungus *Penicillium griseofulvum*, a considerable increase of proteinase activity occurs during nitrogen starvation (Morton *et al.*, 1960). Possibly similar changes occur in algae.

B. Nitrogen Assimilation

Ammonium-N is assimilated rapidly by nitrogen-starved algae, sometimes four to five times more rapidly than by normal cells (Bongers, 1956; Harvey, 1953; Hattori, 1957; Iwamoto and Sugimoto, 1958; Syrett, 1953a). The assimilation takes place at the expense of endogenous carbohydrate reserves, and neither the addition of glucose nor illumination increases the rate. Nitrate and nitrite are assimilated more slowly than ammonium-N, and at about the same rate as that of normal cells (Bongers, 1956; Syrett, 1956a). Although it has been shown with *Chlorella*, *Scenedesmus* and *Chlamydomonas*, for example, that nitrogen-starved cells can assimilate nitrate in darkness, normal cells require light and carbon dioxide (Bongers, 1956; Wetherell, 1958). Evidently nitrogen-starved cells have a carbohydrate reserve which is lacking in normal cells. In experiments of a few hours' duration, no cell division accompanies nitrogen assimilation by these cells, and dry weight does not increase even in the presence of light and carbon dioxide (Bongers, 1956; Syrett, 1956a). In darkness, nitrogen-deficient cells of *Chlorella vulgaris* continue to assimilate ammonium-N until their carbohydrate reserves are exhausted. Nitrate assimilation, however, ceases before complete exhaustion occurs, unless glucose is added. It is not clear why the remaining endogenous carbohydrate is unavailable for the assimilation of nitrate (Syrett, 1956b).

C. The Products of Nitrogen Assimilation

The products of nitrogen assimilation have been studied in a variety of algae, notably species of *Chlorella*, *Scenedesmus*, and *Anabaena* grown in culture (Hattori, 1958; Reisner *et al.*, 1960; Syrett, 1953b, 1956a; Syrett and Fowden, 1952; Thomas and Krauss, 1955; Wolfe, 1954b). As is expected from its greater rate of assimilation, quantitative changes are more pronounced with ammonium-N than with nitrate. There is an immediate increase in cellular, soluble organic-N, followed later by an increase of insoluble N. Eventually, if sufficient carbohydrate is available, most of the added nitrogen is transformed to insoluble N (Syrett and Fowden 1952; Thomas and Krauss, 1955). The formation of amino acids and amides in *Chlorella vulgaris* has been studied in detail by Reisner *et al.* (1960). During the first 15 minutes of ammonium-N assimilation, alanine and glutamine increase markedly, as does arginine to a lesser extent (cf. Syrett and Fowden, 1952; Hattori, 1958). During the later stages of ammonium-N assimilation, glutamic and aspartic acids, lysine, and proline show large increases. The changes when nitrate is assimilated are similar but less striking. Presumably the chief reaction by which ammonia is incorporated into amino acids is the reductive amination of ketoglutaric acid to glutamic acid. Data obtained by feeding $N^{15}H_4^+$ to *Nostoc* are consistent with this view (Wilson, 1952). Other amino acids could then be formed by transamination, as has been demonstrated in *Chlorella vulgaris* (Millbank, 1953). However, alanine might be formed directly from ammonia, as in some other organisms (Fairhurst *et al.*, 1956).

D. Changes in Respiration and Photosynthesis Associated with Nitrogen Assimilation

Ammonium-N assimilation by nitrogen-starved *Chlorella* is accompanied by rapid respiration and by carbohydrate breakdown (Kandler and Ernst, 1955; Hattori, 1957; Iwamoto and Sugimoto, 1958; Syrett, 1953a). The respiration rate rises as soon as ammonium-N is added and may increase four- to fivefold; it often returns to its initial value when assimilation is completed (Syrett, 1953a). Both the ammonium-N assimilation and the increased respiration rate are suppressed by a variety of metabolic inhibitors (Hattori, 1957). The addition of nitrite or nitrate to nitrogen-starved cells results in striking increases in the rate of carbon dioxide production, but oxygen uptake is not stimulated as much as with ammonium-N since nitrite and nitrate can act as electron acceptors instead of oxygen (Hattori, 1957; Syrett, 1955, 1956a; Warburg and Negelein, 1920). The respiration rate of nitrogen-starved cells may be limited by the concentration of a phosphate acceptor which increases when ammonium-N is assimilated (Syrett, 1953b). This view is supported: (1) by the fact that 2,4-dinitrophenol stimulates the respiration of nitrogen-starved cells and yet inhibits ammonium-N assimilation (Hattori, 1957; Syrett, 1954); and (2) by the fall in ATP concentration which occurs immediately after the addition of ammonium-N (Syrett, 1958). However, Holzer (1959) suggested that in yeast ammonium-N may stimulate respiration because the formation of glutamate from α-ketoglutarate and ammonia results in the oxidation of reduced TPN, while Chance (1959) considered that changes in internal pH might be important.

Nitrogen assimilation by nitrogen-starved cells of the diatom *Navicula pelliculosa*, as well as various species of Green Algae, is also accompanied in darkness by much carbon dioxide fixation (Fogg, 1956; Holm-Hansen *et al.*, 1959; Kessler, 1957a; Syrett, 1956c). This is particularly marked when ammonium-N is assimilated, and is presumably a consequence of active amino-acid synthesis.

When ammonium-N is added, the rate of photosynthesis does not increase in such an immediate and marked manner as does respiration. Nevertheless, it is somewhat accelerated, and there is an immediate change in the products of carbon assimilation, amino acids becoming more predominant, while sugars and sugar phosphates become less so (Fogg, 1956; Holm-Hansen *et al.*, 1959). Again, the effects of nitrate are less marked, although in *Chlorella*, and probably in other algae, the photosynthetic quotient may change as a consequence of nitrate reduction (Myers, 1949; Pirson and Wilhelmi, 1950; van Oorschot, 1955). As cells recover from nitrogen deficiency, their chlorophyll content increases (Bongers, 1956; Harvey, 1953; Urhan, 1932) and their rate of photosynthesis and its quantum efficiency recover (Bongers, 1956; van Oorschot, 1955).

IV. Nitrate Reduction and Assimilation

A. Quantitative Relationships

Eight electrons, or hydrogen equivalents, are required to reduce a nitrate ion to the level of ammonium-N. The metabolic consequences of such a reduction are revealed quantitatively in a number of ways. Warburg and Negelein (1920) showed that *Chlorella pyrenoidosa*, suspended in a mixture of 0.01 *M* HNO_3 and 0.1 *M* $NaNO_3$ at pH 2 in darkness, reduced nitrate to ammonium-N. Approximately two "extra" molecules of carbon dioxide were produced for each ammonia molecule formed, in agreement with the equation:

$$2(CH_2O) + NO_3 + 2H^+ \rightarrow NH_4^+ + 2(CO_2) + H_2O \qquad (1)$$

In light, instead of extra carbon dioxide, approximately two extra molecules of oxygen were produced for each ammonia molecule. This has been confirmed by Bongers (1958) with *Scenedesmus*, illuminated, in the absence of carbon dioxide, at the more physiological pH of 7.3. With nitrite, the O_2/NH_3 ratio is 1.5, as expected from the fact that six electrons are required for nitrite reduction.

However, ammonium-N rarely accumulates; usually nitrate-N is converted to organic-N. The reduction of nitrate is then shown by differences in the gas exchange between cells assimilating nitrate and similar cells assimilating ammonium-N. The respiratory quotient (CO_2/O_2) is higher with nitrate and may even exceed 2.0 when nitrate assimilation is rapid (Cramer and Myers, 1948; Hattori, 1957; Kessler, 1952, 1953a; Syrett, 1955; Wolfe, 1954b). When small quantities of nitrogen are assimilated in darkness, 2 extra molecules of carbon dioxide are produced for each nitrate ion assimilated and 1.5 for each nitrite ion (Kessler, 1953a; Syrett, 1955).

Cells growing photosynthetically have a lower photosynthetic quotient (CO_2/O_2) when assimilating nitrate. Cramer and Myers (1948) showed that

Chlorella pyrenoidosa has the composition $C_{5.7}H_{9.8}O_{2.3}N_{1.0}$ whether grown on nitrate or ammonium-N; the photosynthetic quotient was 0.68 with nitrate and 0.94 with ammonium, in close agreement with the values predicted from the equations:

$$1.0(NO_3^-) + 5.7(CO_2) + 5.4(H_2O) \rightarrow C_{5.7}H_{9.8}O_{2.3}N_{1.0} + 8.25O_2 + 1.0(OH^-) \quad (2)$$

$$CO_2/O_2 = 5.7/8.25 = 0.69$$

$$1.0(NH_4^+) + 5.7(CO_2) + 3.4(H_2O) \rightarrow C_{5.7}H_{9.8}O_{2.3}N_{1.0} + 6.25O_2 + 1.0\ H^+ \quad (3)$$

$$CO_2/O_2 = 5.7/6.25 = 0.91$$

It will be seen that for each nitrate molecule assimilated, two extra molecules of oxygen are produced, although the same quantity of carbon dioxide is assimilated as with ammonium-N. Similar equations were given by van Oorschot (1955).

B. Intermediate Stages in Nitrate Reduction

It is generally assumed that, before assimilation, nitrate-N is reduced to ammonium-N in four stages, each of which requires two electrons.

$$HNO_3 \xrightarrow[2(H)]{\text{nitrate reductase}} HNO_2 \xrightarrow[2(H)]{\text{nitrite reductase}} H_2N_2O_2 \xrightarrow[2(H)]{\text{hyponitrite reductase}} NH_2OH \xrightarrow[2(H)]{\text{hydroxylamine reductase}} NH_3$$

Nicholas (1959) reviewed the evidence for this sequence. It is, in brief: (1) some of the postulated intermediates can be detected or made to accumulate; (2) the intermediates can sometimes be assimilated; and (3) enzymes catalyzing the separate steps can be isolated, though in impure form, from some plants. Virtanen and Rautenen (1952), however, suggested that organic-N might be formed directly from hydroxylamine, and McElroy and Spencer (1956) supported this view. Ammonia would not then be an intermediate.

Hyponitrite and hydroxylamine have not been detected in algal cultures growing on nitrate, but nitrite and ammonium-N are frequently present in small amounts. Little ammonium-N accumulates unless assimilation is slow because of an unfavorable pH (Warburg and Negelein, 1920) or the lack of a carbon substrate (Bongers, 1958). In *Ankistrodesmus braunii*, more nitrite accumulates at pH values lower than 4.0 (Kessler, 1953a, b), but Bongers (1956) found most accumulation at pH 7.5 with *Scenedesmus*. The effect of anaerobiosis is also variable; it decreases nitrite accumulation in cultures of *Scenedesmus quadricauda*, *Ankistrodesmus braunii*, and *Chlorella vulgaris* "strain A," but increases accumulation in cultures of other *Chlorella* strains (Kessler, 1953b). The effect of light on nitrite accumulation depends both on its intensity (Bongers, 1956) and on the metabolic state of the cells (Inada, 1958; Mayer, 1952). Since the amount of nitrite which accumulates is small and depends on the difference between its rate of formation and its rate of reduction, such variability in behavior is not surprising.

Nitrite, in low concentration, is assimilated by algae. The evidence for hydroxylamine assimilation is doubtful, because in such low concentrations as $3 \times 10^{-5} M$ it appears to be toxic to species of *Chlorella* and *Scenedesmus* (Bongers, 1956; Inada, 1958; Kessler, 1957a), and therefore probably to most other algae.

Little is known of the enzymes responsible for nitrate reduction in algae. Omura (1954) obtained cell-free extracts which reduced both nitrite and nitrate. In *Chlorella vulgaris,* nitrate reductase is formed in the presence of nitrate, and its formation is partially repressed by ammonia (I. Morris, personal communication). With cell-free extracts containing this enzyme, DPNH (but not TPNH) can act as a hydrogen donor for nitrate reduction. The nitrate reductase of higher plants and fungi is closely associated with molybdenum and is inhibited by cyanide (Nicholas, 1959). It is therefore significant that the reduction of nitrate to nitrite by *Ankistrodesmus* is also very sensitive to cyanide (Kessler, 1953a), and that in *Scenedesmus* and *Anabaena* molybdenum is required for growth on nitrate but not for growth on ammonium-N or urea (Arnon *et al.*, 1955; Ichioka and Arnon, 1955; Wolfe, 1954a). Nitrite reduction by algae is inhibited by 2,4-dinitrophenol (Kessler, 1955a), as is the reductase of the fungus *Neurospora* (Nicholas, 1959). Nitrite reduction by *Ankistrodesmus* is also inhibited by arsenate, indicating that high-energy phosphate may be involved (Kessler and Bücker, 1960). There is evidence for a manganese requirement for hydroxylamine reductase (Nicholas, 1959), and manganese seems likewise to be necessary for the complete reduction of nitrite by *Ankistrodesmus* (Kessler, 1957a). Thus, as far as the limited evidence goes, it suggests that algal enzymes are similar to those of fungi and higher plants.

C. Electron Sources for Nitrate Reduction

The enzymes isolated from *Neurospora* which catalyze the four reduction steps between nitrate and ammonium-N appear to be flavoproteins associated with heavy metal ions. They can all use either DPNH or TPNH as electron donors (Nicholas, 1959). If these reduced coenzymes act as electron donors for nitrate reduction *in vivo*, it is easy to see how reduction could be coupled to respiration.

In addition, some algae with hydrogenase activity, such as *Ankistrodesmus* and *Scenedesmus obliquus* (strain D3), can use molecular hydrogen for nitrate and nitrite reduction (Kessler, 1957a). The reaction is much faster with nitrite than with nitrate and is almost stoichiometric, 2.5–3.0

hydrogen molecules being required to reduce one nitrite ion, in agreement with the equation:

$$NO_2^- + 2H^+ + 3H_2 \rightarrow NH_4^+ + 2(H_2O) \quad (4)$$

The expected value of 3.0 hydrogen molecules is obtained experimentally only when carbon dioxide is present. The reason for this is not clear, since carbon dioxide is not assimilated during the reaction.

It is also possible that a metabolic reductant for nitrate reduction might be formed photochemically (see Rabinowitch, 1945, p. 538; van Niel, 1949).

D. The Effect of Light

In 1920, Warburg and Negelein showed that light stimulates nitrate reduction by *Chlorella*, the reduction being accompanied by oxygen evolution. They considered that, both in light and in darkness, nitrate reduction is coupled with carbohydrate oxidation, but that in light the carbon dioxide which might be expected as a product is assimilated by photosynthesis and replaced by the evolution of oxygen. Although light was thought to stimulate nitrate reduction by increasing the permeability of the cells to nitrate, it could conceivably act in other ways. Photosynthesis might produce organic compounds readily available as electron sources for nitrite or nitrate reduction (Harvey, 1953). Alternatively, nitrate might be reduced by a photochemically produced reductant, and hence coupled more directly with the photosynthetic mechanism (Rabinowitch, 1945; van Niel, 1949). Yet another possibility is that photo-phosphorylation may stimulate *nitrite* reduction (Kessler, 1959). As Kessler pointed out, light may well act in more than one way, since nitrate reduction is a complex process consisting of several stages with differing requirements.

The work of Kok (1951), van Niel *et al.* (1953), and Bongers (1958) strongly suggests that a suitable reductant can be generated photochemically. Van Niel *et al.* found that, at saturating light intensity, the rate of oxygen evolution by *Chlorella pyrenoidosa* was increased by the addition of nitrate while carbon dioxide assimilation was unchanged. Since the carbon dioxide reduction system was already working at full capacity, the evolution of extra oxygen, when nitrate was added, could not be due to the assimilation of carbon dioxide produced during a dark reduction of nitrate. Both Kok and Bongers found that, in the presence of nitrate, illuminated cells produced oxygen even though carbon dioxide was absent. The oxygen evolution showed the same photosynthetic quantum yield whether nitrate, nitrite, or carbon dioxide was being reduced, and the system became saturated at about the same light intensity (Bongers, 1958). In Bongers' experiments, however, nitrate and carbon dioxide apparently competed for the reductant even at saturating light intensities.

Neither nitrite nor nitrate acts as a Hill reagent when supplied alone to chloroplasts isolated from spinach (Kessler, 1955b), but nitrate is reduced to nitrite by illuminated chloroplasts in the presence of TPN and nitrate reductase (Evans and Nason, 1953). Thus, photochemically reduced TPN might be the reductant *in vivo*. A photochemically reduced flavin is another possibility (Stoy, 1956).

Nevertheless, some observations indicate that nitrate reduction in light is more complicated. Davis (1953) found little nitrate reduction by illuminated *Chlorella* in the absence of carbon dioxide, unless glucose was added; he suggested that carbohydrate metabolism was necessary to form the reductant. It is not clear why his results differ from those of Bongers (1958). Bongers used *Scenedesmus* cells from growing cultures and did not include potassium hydroxide in his Warburg vessels, whereas Davis used rather old cultures of *Chlorella* and included a center-well containing alkali in his experimental flasks. Kessler (1959) suggested that perhaps a trace of carbon dioxide is essential for nitrate reduction. Alternatively, since Bongers' cells reduced nitrate to ammonium-N, which accumulated, and since ammonium-N or a product of its metabolism often inhibits nitrate reduction, we may postulate a similar inhibition in the experiments of Davis. Then the addition of glucose, by allowing assimilation, may have prevented ammonium-N accumulation and thereby permitted nitrate reduction.

Kessler (1955b, 1957b, c, 1959) thoroughly investigated the effect of light on nitrate and nitrite reduction by *Ankistrodesmus* in the absence of carbon dioxide. He used anaerobic conditions in order to depress reduction coupled to respiration. Under these conditions, *nitrate* was reduced very slowly but *nitrite* reduction was rapid. Since the rate of nitrite reduction was not much reduced when the temperature was changed from 15° to 4°C, a photochemical reaction is probably involved. Carbon dioxide stimulated the rate of nitrite reduction, which was then much more temperature sensitive. Since the light-induced nitrite reduction became saturated at a rather low light intensity (1000 lux), and since the inhibition of nitrite reduction by 2,4-dinitrophenol may indicate an ATP requirement, Kessler (1955a) suggested that nitrite reduction may be stimulated by photophosphorylation. Harvey (1953) also observed a more pronounced effect of light on the assimilation of nitrite by "*Nitzschia closterium*" (=*Phaeodactylum tricornutum*) than on that of nitrate; and Bongers (1958), too, found that, at high light intensity, the reduction of nitrite by *Chlorella* was faster than that of nitrate. Kessler considered that the slow rate of nitrate reduction in his experiments with *Ankistrodesmus* excludes the possibility of a direct photochemical reduction of nitrate. There is clearly a contradiction between the conclusions of different workers on this subject. It may arise from Kessler's use of anaerobic conditions, although he stated that similar results were obtained in air (Kessler, 1955b). Alternatively, as M. B. Allen suggested (see Kessler, 1957b, p. 255), the interrelationships between nitrate reduction and carbon metabolism may depend on the metabolic condition of the cells.

V. The Assimilation of Organic Nitrogen

A. Urea

Urea can serve as sole nitrogen source for *Chlamydomonas*, *Chlorella*, and other unicellular Green Algae (Arnow *et al.*, 1953; Davis *et al.*, 1953; Droop, 1955; Ellner and Steers, 1955; Ghosh and Burris, 1950; Ludwig, 1938; Moyse, 1956; Ryther, 1954; Sager and Granick, 1953; Samejima and Myers, 1958; Tamiya *et al.*, 1953). It can also be used by members of the Xanthophyta, such as *Tribonema aequale* and *Monodus subterraneus* (Belcher and Fogg, 1958; Miller and Fogg, 1958), by *Monochrysis lutheri* but not by *Prymnesium parvum* among the Chrysophyta (Droop, 1955), and by some, but not all, of the Cyanophyta examined (Allen, 1952; Kratz and Myers, 1955). Urea has been proposed as the best nitrogen source for mass cultures of *Chlorella*, since it does not encourage bacterial contamination, growth of the alga is somewhat faster than on nitrate, the pH remains fairly constant, and a high concentration can be included in the medium without depressing growth (Davis *et al.*, 1953).

The mechanism of urea utilization is obscure. Urease was not detected in *Chlorella pyrenoidosa* or *C. ellipsoidea* (Walker, 1952; Hattori, 1957). It has therefore been suggested that urea is assimilated without a preliminary breakdown to ammonia. One possible mechanism is a combination of urea with ornithine to form arginine, as suggested by Arnow *et al.* (1953), Bollard (1959), Hattori (1958), and Walker (1952). Ornithine might be regenerated from arginine by the reverse action of the urea cycle characteristic of animal tissues (Bollard, 1959; Walker, 1952); one nitrogen atom of the urea molecule would then be incorporated into aspartic acid and the other would appear as ammonia. However, there is little evidence from studies with algae to support this view. Ellner and Steers (1955) reported that *Scenedesmus basiliensis*, growing with 5% carbon dioxide as carbon source, incorporated C^{14} of added C^{14}-urea into guanine without loss of specific activity. If this is so, it must indicate a direct incorporation of the urea carbon. On the other hand, Allison *et al.* (1954) found that, whereas much C^{14} from urea was incorporated by *Nostoc* in the absence of carbon dioxide, little was incorporated in its presence. They considered that urea was first decomposed to carbon dioxide and ammonia before its utilization. Hattori (1960) in his more recent work with *Chlorella ellipsoidea* found no evidence for the incorporation of C^{14}-urea into arginine. He suggested that, during urea assimilation, the carbon of the molecule is released as carbon dioxide while the amide groups combine with an acceptor so that no free ammonia is formed.

B. Amino Acids and Amides

After urea, the organic compounds containing nitrogen most readily available for the growth of algae appear to be the amides, namely acetamide, succinamide, asparagine, and glutamine (Algéus, 1950a, b, 1951; Ludwig, 1938; Miller and Fogg, 1958). Before being utilized by the cells, the amide nitrogen is presumably liberated as ammonia, which was detected in the medium of certain *Scenedesmus* cultures (Algéus, 1950a). *Monodus subterraneus* is particularly interesting in that it excretes an extracellular glutaminase which catalyzes the hydrolysis of the amide group of glutamine. The ammonia thereby liberated serves as a source of assimilable nitrogen, but neither the amino-N nor the carbon of the residual glutamic acid can be used for growth. *Chlorella vulgaris*, on the other hand, does not produce an extracellular glutaminase, but utilizes both the amide and amino-N of glutamine as well as its carbon (Miller, 1959 and unpublished data).

Glycine appears to be the most readily assimilable amino-acid and an excellent nitrogen source for some algae (Algéus, 1948a, b, c; Belcher and Fogg, 1958; Davis *et al.*, 1953). Algéus divided algae into two groups according to the rapidity with which they deaminate glycine in the medium. The first group, typified by *Scenedesmus obliquus*, deaminates glycine so rapidly that ammonium-N accumulates in the medium during growth. In the second group, e.g., *Chlorella vulgaris*, growth appears to be limited by the rate of deamination, and no ammonium-N accumulates.

The other amino acids vary as nitrogen sources. Some can be readily utilized (Algéus, 1949; Arnow *et al.*, 1953; Belcher and Fogg, 1958; Ghosh and Burris, 1950; Miller and Fogg, 1958); others cannot. Hydrolyzates of casein and other proteins, such as peptone and tryptone, contain mixtures of amino acids, many of which can be utilized (Allen, 1952; Ludwig, 1938; Miller and Fogg, 1958).

In experiments on the effects of amino acids on algae, it should be borne in mind that such substances are chelating agents, and that their addition to a medium may influence the growth of algae by sequestering essential cations, an action unrelated to their availability as sources of nitrogen (Hutner *et al.*, 1950; Miller and Fogg, 1958).

C. Other Compounds

Some algae can use the nitrogen in uric acid and xanthine (Droop, 1955; Ludwig, 1938; Miller and Fogg, 1958; Ryther, 1954). Bertha Livingstone (unpublished) showed that *Chlamydomonas moewusii* could quantitatively utilize for growth the nitrogen atoms in the guanine molecule.

References

Aach, H. G. (1952). Über Wachstum und Zusammensetzung von *Chlorella pyrenoidosa* bei unterschiedlichen Lichtstärken und Nitratmengen. *Arch. Mikrobiol.* **17**, 213–246.

Algéus, S. (1948a). Glycocoll as a source of nitrogen for *Scenedesmus obliquus. Physiol. Plantarum* **1**, 66–84.

Algéus, S. (1948b). The utilization of glycocoll by *Chlorella vulgaris. Physiol. Plantarum* **1**, 236–244.

Algéus, S. (1948c). The deamination of glycocoll by green algae. *Physiol. Plantarum* **1**, 382–383.

Algéus, S. (1949). Alanine as a source of nitrogen for green algae. *Physiol. Plantarum* **2**, 266–271.

Algéus, S. (1950a). The utilization of aspartic acid, succinamide and asparagine by *Scenedesmus obliquus. Physiol. Plantarum* **3**, 225–235.

Algéus, S. (1950b). Further studies on the utilization of aspartic acid, succinamide and asparagine by green algae *Physiol. Plantarum* **3**, 370–375.

Algéus, S. (1951). Note on the utilization of glutamine by *Scenedesmus obliquus. Physiol. Plantarum* **4**, 459–460.

Allen, M. B. (1952). The cultivation of Myxophyceae. *Arch. Mikrobiol.* **17**, 34–53.

Allison, R. K., Skipper, H. E., Reid, M. R., Short, W. A., and Hogan, G. L. (1954). Studies on the photosynthetic reaction. II. Sodium formate and urea feeding experiments with *Nostoc muscorum. Plant Physiol.* **29**, 164–168.

Andersson, M. (1942). Einige ernährungsphysiologische Versuche mit *Ulva* und *Enteromorpha. Kgl. Fysiograf. Sällskap. i Lund. Förh.* **12**, 42–52.

Arnon, D. I., Ichioka, P. S., Wessel, G., Fujiwara, A., and Woolley, J. T. (1955). Molybdenum in relation to nitrogen metabolism. I. Assimilation of nitrate nitrogen by *Scenedesmus obliquus. Physiol. Plantarum* **8**, 538–551.

Arnow, P., Oleson, J. J., and Williams, J. H. (1953). The effect of arginine on the nutrition of *Chlorella vulgaris. Am. J. Botany* **40**, 100–104.

Bauch, R. (1956). Biologisch-ökologishe Studien an der Gattung *Fucus.* II. Pigmentsynthese und Pigmentzerstörung beim Blasentang und die Bedeutung tierischer Extracte für die Algenentwicklung im Litoral. *Wiss. Z. Ernst Moritz Arndt-Universität Greifswald* **5**, 317–332.

Belcher, J. H., and Fogg, G. E. (1958). Studies on the growth of Xanthophyceae in pure culture. III. *Tribonema aequale* Pascher. *Arch. Mikrobiol.* **30**, 17–22.

Blinks, L. R. (1951). Physiology and biochemistry of algae. *In* "Manual of Phycology" (G. M. Smith, ed.), pp. 263–292. Chronica Botanica, Waltham, Massachusetts.

Bollard, E. G. (1959). Urease, urea and ureides in plants. *Symposia Soc. Exptl. Biol.* **No. 13**, 304–329.

Bongers, L. H. J. (1956). Aspects of nitrogen assimilation by cultures of green algae. (*Chlorella vulgaris*, strain A and *Scenedesmus.*) *Mededel. Landbouwhogeschool Wageningen* **56**, 1–52.

Bongers, L. H. J. (1958). Kinetic aspects of nitrate reduction. *Neth. J. Agr. Sci.* **6**, 70–88.

Chance, B. (1959). *Comment in* "Regulation of Cell Metabolism" (G. E. W. Wolstenholme and C. M. O'Connor, eds.), pp. 290–292. Churchill, London.

Chu, S. P. (1942). The influence of the mineral composition of the medium on the growth of planktonic algae. Part I. Methods and culture media. *J. Ecol.* **30,** 284–325.

Cramer, M., and Myers, J. (1948). Nitrate reduction and assimilation in *Chlorella pyrenoidosa. J. Gen. Physiol.* **32,** 93–102.

Cramer, M., and Myers, J. (1952). Growth and photosynthetic characteristics of *Euglena gracilis. Arch. Mikrobiol.* **17,** 384–402.

Davis, E. A. (1953). Nitrate reduction by *Chlorella. Plant Physiol.* **28,** 539–544.

Davis, E. A., Dedrick, J., French, C. S., Milner, H. W., Myers, J., Smith, J. H. C., and Spoehr, H. A. (1953). Laboratory experiments on *Chlorella* culture at the Carnegie Institution of Washington department of plant biology. *In* "Algal Culture" (J. S. Burlew, ed.), pp. 105–153. Carnegie Inst., Washington, D. C.

De Turk, W. E., and Bernheim, F. (1958). Effects of ammonia, methylamine and hydroxylamine on the adaptive assimilation of nitrite and nitrate by a *Mycobacterium. J. Bacteriol.* **75,** 691–696.

Droop, M. R. (1955). Some new supra-littoral protista. *J. Marine Biol. Assoc. United Kingdom* **34,** 233–245.

Ellner, P. D., and Steers, E. (1955). Urea as a carbon source for *Chlorella* and *Scenedesmus. Arch. Biochem. Biophys.* **59,** 534–535.

Evans, H. J., and Nason, A. (1953). Pyridine nucleotide-nitrate reductase from extracts of higher plants. *Plant Physiol.* **28,** 233–254.

Fairhurst, A. S., King, H. K., and Sewell, C. E. (1956). Studies in amino-acid biogenesis: The synthesis of alanine from pyruvate and ammonia. *J. Gen. Microbiol.* **15,** 106–120.

Fogg, G. E. (1949). Growth and heterocyst production in *Anabaena cylindrica* Lemm. II. In relation to carbon and nitrogen metabolism. *Ann. Botany (London)* **13,** 241–260.

Fogg, G. E. (1953). "The Metabolism of Algae." Methuen, London.

Fogg, G. E. (1956). Photosynthesis and formation of fats in a diatom. (*Navicula pelliculosa*). *Ann. Botany (London)* **20,** 265–286.

Fogg, G. E. (1959). Nitrogen nutrition and metabolic patterns in algae. *Symposia Soc. Exptl. Biol.* **No. 13,** 106–125.

Fogg, G. E., and Collyer, D. M. (1953). The accumulation of lipides by algae. *In* "Algal Culture" (J. S. Burlew, ed.), pp. 177–181. Carnegie Inst., Washington, D. C.

Fogg, G. E., and Wolfe, M. (1954). The nitrogen metabolism of the blue-green algae (Myxophyceae). *Symposium Soc. Gen. Microbiol.* **No. 4,** 99–125.

Ghosh, P., and Burris, R. H. (1950). Utilization of nitrogenous compounds by plants. *Soil Sci.* **70,** 187–203.

Harvey, H. W. (1953). Synthesis of organic nitrogen and chlorophyll by *Nitzschia closterium. J. Marine Biol. Assoc. United Kingdom* **31,** 477–487.

Hattori, A. (1957). Studies on the metabolism of urea and other nitrogenous compounds in *Chlorella ellipsoidea.* I. Assimilation of urea and other nitrogenous compounds by nitrogen-starved cells. *J. Biochem. (Tokyo)* **44,** 253–273.

Hattori, A. (1958). Studies on the metabolism of urea and other nitrogenous compounds in *Chlorella ellipsoidea.* II. Changes in levels of amino acids and amides during the assimilation of ammonia and urea by nitrogen-starved cells. *J. Biochem. (Tokyo)* **45,** 57–64.

Hattori, A. (1960). Studies on the metabolism of urea and other nitrogenous compounds in *Chlorella ellipsoidea*. III. Assimilation of urea. *Plant and Cell Physiol.* **1,** 107–116.

Holm-Hansen, O., Nishida, K., Moses, V., and Calvin, M. (1959). Effects of mineral salts on short-term incorporation of carbon dioxide in *Chlorella pyrenoidosa*. *J. Exptl. Botany* **10,** 109–124.

Holzer, H. (1959). Enzymic regulation of fermentation in yeast cells. *In* "Regulation of Cell Metabolism" (G. E. W. Wolstenholme and C. M. O'Connor, eds.), pp. 277–287. Churchill, London.

Hutner, S. H., Provasoli, L., Schatz, A., and Haskins, C. P. (1950). Some approaches to the study of the role of metals in the metabolism of microorganisms. *Proc. Am. Phil. Soc.* **94,** 152–170.

Ichioka, P. S., and Arnon, D. I. (1955). Molybdenum in relation to nitrogen metabolism. II. Assimilation of ammonia and urea without molybdenum by *Scenedesmus obliquus*. *Physiol. Plantarum* **8,** 552–560.

Inada, Y. (1958). Stimulatory and inhibitory effect of light on the nitrate assimilation by *Chlorella ellipsoidea*. *J. Gen. Appl. Microbiol.* (*Japan*) **4,** 153–162.

Iwamoto, H., and Sugimoto, H. (1958). Fat synthesis in unicellular algae. III. Absorption of nitrogen by nitrogen-deficient *Chlorella ellipsoidea* cells and its effects on the continuous cultivation of fatty cells. *Bull. Agr. Chem. Soc. Japan* **22,** 410–419.

Kandler, O., and Ernst, H. (1955). Über den Einfluss organischer Säuren auf die Atmung, den Ammoniumeinbau und den Gehalt an freien Aminosäuren von *Chlorella pyrenoidosa*. *Planta* **46,** 46–69.

Kessler, E. (1952). Nitritbildung und Atmung bei der Nitratreduktion von Grünalgen. *Z. Naturforsch.* **76,** 280–284.

Kessler, E. (1953a). Über den Mechanismus der Nitratreduktion von Grünalgen I. Nitritbildung und Nitritreduktion durch *Ankistrodesmus braunii* (Nägeli) Brunnthaler. *Flora* (*Jena*) **140,** 1–38.

Kessler, E. (1953b). Über den Mechanismus der Nitratreduktion von Grünalgen. II. Vergleichend-physiologische Untersuchungen. *Arch. Mikrobiol.* **19,** 438–457.

Kessler, E. (1955a). Über die Wirkung von 2:4-Dinitrophenol auf Nitratreduktion und Atmung von Grünalgen. *Planta* **45,** 94–105.

Kessler, E. (1955b). Role of photochemical processes in the reduction of nitrate by green algae. *Nature* **176,** 1069–1070.

Kessler, E. (1957a). Stoffwechsel physiologische Untersuchungen an Hydrogenase enthaltenden Grünalgen. II. Dunkel-Reduktion von Nitrat und Nitrit mit molekularem Wasserstoff. *Arch. Mikrobiol.* **27,** 166–181.

Kessler, E. (1957b). Contributions to the problem of photochemical nitrate reduction. *In* "Research in Photosynthesis" (H. Gaffron, ed.), pp. 250–256. Interscience, New York.

Kessler, E. (1957c). Untersuchungen zum Problem der photochemischen Nitratreduktion in Grünalgen. *Planta* **49,** 505–523.

Kessler, E. (1959). Reduction of nitrate by green algae. *Symposia Soc. Exptl. Biol.* **No. 13,** 87–105.

Kessler, E., and Bücker, W. (1960). Über die Wirkung von Arsenat auf Nitratreduktion, Atmung und Photosynthese von Grünalgen. *Planta* **55,** 512–524.

Kok, B. (1951). Photo-induced interactions in metabolism of green plant cells. *Symposia Soc. Exptl. Biol.* **No. 5,** 211–221.

Kratz, W. A., and Myers, J. (1955). Nutrition and growth of several blue-green algae. *Am. J. Botany* **42,** 282–287.

Krauss, R. W. (1958). Physiology of the fresh-water algae. *Ann. Rev. Plant Physiol.* **9,** 207–244.

Kylin, A. (1945). The nitrogen sources and the influence of manganese on the nitrogen assimilation of *Ulva lactuca. Kgl. Fysiograf. Sällskap. i Lund Förh.* **15,** 27–35.

Ludwig, C. A. (1938). The availability of different forms of nitrogen to a green alga (*Chlorella*). *Am. J. Botany* **25,** 448–458.

Lwoff, A. (1943). "L'évolution physiologique," pp. 91–93. Hermann, Paris.

McElroy, W. D., and Spencer, D. (1956). Normal pathways of assimilation of nitrate and nitrite. *In* "Inorganic Nitrogen Metabolism" (W. D. McElroy and B. Glass, eds.). Johns Hopkins Press, Baltimore, Maryland.

Mayer, A. M. (1952). Iron, manganese and the reduction of nitrates by *Chlorella vulgaris. Palestine J. Botany, Jerusalem Ser.* **5,** 161–179.

Millbank, J. W. (1953). Demonstration of transaminase systems in the alga *Chlorella. Nature* **171,** 476.

Miller, J. D. A. (1959). An extra-cellular enzyme produced by *Monodus. Brit. Phycol. Bull.* **7,** 22.

Miller, J. D. A., and Fogg, G. E. (1958). Studies on the growth of Xanthophyceae in pure culture. II. The relations of *Monodus subterraneus* to organic substances. *Arch. Mikrobiol.* **30,** 1–16.

Morton, A. G., Dickerson, A. G. F., and England, D. J. F. (1960). Changes in enzyme activity of fungi during nitrogen starvation. *J. Exptl. Botany* **11,** 116–128.

Moyse, A. (1956). Étude de la croissance d'algues monocellulaires (Chlorelles et espèces voisines) en cultures accélérées. II. Vitesse de croissance et milieux de culture. *J. recherches centre natl. recherche sci., Lab. Bellevue* (*Paris*) **No. 35,** 177–192.

Myers, J. (1949). The pattern of photosynthesis in *Chlorella. In* "Photosynthesis in Plants" (J. Franck and W. E. Loomis, eds.), pp. 349–364. Iowa State College Press, Ames, Iowa.

Myers, J. (1951). Physiology of the algae. *Ann. Rev. Microbiol.* **5,** 157–180.

Nicholas, D. J. D. (1959). Metallo-enzymes in nitrate assimilation of plants with special reference to micro-organisms. *Symposia Soc. Exptl. Biol.* **No. 13,** 1–23.

Noack, K., and Pirson, A. (1939). Die Wirkung von Eisen und Mangan auf die Stickstoff assimilation von *Chlorella. Ber. deut. botan. Ges.* **57,** 442–452.

Omura, H. (1954). On the nitrate and nitrite reductase in green algae. *Enzymologia* **17,** 127–132.

Pirson, A., and Wilhelmi, G. (1950). Photosynthese-Gaswechsel und Mineralsalzernährung. *Z. Naturforsch.* **5b,** 211–218.

Pratt, R., and Fong, J. (1940). Studies on *Chlorella vulgaris.* III. Growth of *Chlorella* and changes in the hydrogen-ion and ammonium-ion concentrations in solutions containing nitrate and ammonium nitrogen. *Am. J. Botany* **27,** 735–743.

Proctor, V. W. (1957). Preferential assimilation of nitrate by *Haematococcus pluvialis. Am. J. Botany* **44,** 141–143.

Rabinowitch, E. I. (1945). "Photosynthesis and Related Processes," Vol. I. Interscience, New York.

Reisner, G. S., Gering, R. K., and Thompson, J. F. (1960). The metabolism of nitrate and ammonia by *Chlorella vulgaris. Plant Physiol.* **35,** 48–52.

Ryther, J. H. (1954). The ecology of phytoplankton blooms in Moriches Bay and Great South Bay, Long Island, New York. *Biol. Bull.* **106,** 198–209.

Sager, R., and Granick, S. (1953). Nutritional studies with *Chlamydomomas reinhardi. Ann. N. Y. Acad. Sci.* **56,** 831–838.

Samejima, H., and Myers, J. (1958). On the heterotrophic growth of *Chlorella pyrenoidosa*. *J. Gen. Microbiol.* **18,** 107–117.

Schuler, J. F., Diller, V. M., and Kerslen, H. J. (1953). Preferential assimilation of ammonium ion by *Chlorella vulgaris*. *Plant Physiol.* **28,** 299–303.

Shilo, M., and Shilo, M. (1955). Control of the phytoflagellate *Prymnesium parvum*. *Proc. Intern. Assoc. Theoret. and Appl. Limnol.* **12,** 233–240.

Stoy, V. (1956). Riboflavin-catalysed enzymic photoreduction of nitrate. *Biochim. et Biophys. Acta* **21,** 395–396.

Suneson, S. (1933). Weitere Angaben über die Nitratspeicherung bei den höheren Meeresalgen. *Z. physiol. Chem., Hoppe-Seyler's* **214,** 105–108.

Syrett, P. J. (1953a). The assimilation of ammonia by nitrogen-starved cells of *Chlorella vulgaris*. I. The correlation of assimilation with respiration. *Ann. Botany (London)* **17,** 1–19.

Syrett, P. J. (1953b). The assimilation of ammonia by nitrogen-starved cells of *Chlorella vulgaris*. II. The assimilation of ammonia to other compounds. *Ann. Botany (London)* **17,** 20–36.

Syrett, P. J. (1954). Nitrogen assimilation by green algae (Chlorophyceae). *Symposium Soc. Gen. Microbiol.* **No. 4,** 126–151.

Syrett, P. J. (1955). The assimilation of ammonia and nitrate by nitrogen-starved cells of *Chlorella vulgaris*. I. The assimilation of small quantities of nitrogen. *Physiol. Plantarum* **8,** 924–929.

Syrett, P. J. (1956a). The assimilation of ammonia and nitrate by nitrogen-starved cells of *Chlorella vulgaris,* II. The assimilation of large quantities of nitrogen. *Physiol. Plantarum* **9,** 19–27.

Syrett, P. J. (1956b). The assimilation of ammonia and nitrate by nitrogen-starved cells of *Chlorella vulgaris*. III. Differences of metabolism dependent on the nature of the nitrogen source. *Physiol. Plantarum* **9,** 28–37.

Syrett, P. J. (1956c). The assimilation of ammonia and nitrate by nitrogen-starved cells of *Chlorella vulgaris*. IV. The dark fixation of carbon dioxide. *Physiol. Plantarum* **9,** 165–171.

Syrett, P. J. (1958). Respiration rate and internal adenosine triphosphate concentration in *Chlorella vulgaris*. *Arch. Biochem. Biophys.* **75,** 117–124.

Syrett, P. J., and Fowden, L. (1952). The assimilation of ammonia by nitrogen-starved cells of *Chlorella vulgaris*. III. The effect of glucose on the products of assimilation. *Physiol. Plantarum* **5,** 558–566.

Tamiya, H., Hase, E., Shibata, K., Mituya, A., Iwamura, T., Nihei, T., and Sasa, T. (1953). Kinetics of growth of *Chlorella,* with special reference to its dependence on quantity of available light and on temperature. *In* "Algal Culture" (J. S. Burlew, ed.), pp. 204–232. Carnegie Inst., Washington.

Thomas, W. H., and Krauss, R. W. (1955). Nitrogen metabolism in *Scenedesmus* as affected by environmental changes. *Plant Physiol.* **30,** 113–122.

Urhan, O. (1932). Beiträge zur Kenntis der Stickstoffassimilation von *Chlorella* und *Scenedesmus*. *Jahrb. wiss. Botan.* **75,** 1–44.

van Niel, C. B. (1949). The comparative biochemistry of photosynthesis. *In* "Photosynthesis in Plants" (J. Franck and W. E. Loomis, eds.), pp. 437–496. Iowa State College Press, Ames, Iowa.

van Niel, C. B., Allen, M. B., and Wright, B. E. (1953). On the photochemical reduction of nitrate by algae. *Biochim. et Biophys. Acta* **12,** 67–74.

van Oorschot, J. L. P. (1955). Conversion of light energy in algal culture. *Mededel. Landbouwhogeschool Wageningen* **55,** 225–276.

Virtanen, A. I., and Rautenen, N. (1952). Nitrogen assimilation. *In* "The Enzymes" (J. B. Sumner and K. Myrbäck, eds.), Vol. 2, Part 2, Chapter 76, pp. 1089–1130. Academic Press, New York.

Walker, J. B. (1952). Arginosuccinic acid from *Chlorella pyrenoidosa. Proc. Natl. Acad. Sci. U. S.* **38,** 561–566.

Warburg, O., and Negelein, E. (1920). Über die Reduktion der Salpetersäure in grünen Zellen. *Biochem. Z.* **110,** 66–115.

Wetherell, D. F. (1958). Obligate phototrophy in *Chlamydomonas eugametos. Physiol. Plantarum* **11,** 260–274.

Wilson, P. W. (1952). The comparative biochemistry of nitrogen-fixation. *Advances in Enzymol.* **13,** 345–375.

Wolfe, M. (1954a). The effect of molybdenum upon the nitrogen metabolism of *Anabaena cylindrica.* I. A study of the molybdenum requirement for nitrogen fixation and for nitrate and ammonia assimilation. *Ann. Botany* (*London*) **18,** 299–308.

Wolfe, M. (1954b). The effect of molybdenum upon the nitrogen metabolism of *Anabaena cylindrica.* II. A more detailed study of the action of molybdenum in nitrate assimilation. *Ann. Botany* (*London*) **18,** 309–325.

ZoBell, C. E. (1935). The assimilation of ammonium-nitrogen by *Nitzschia closterium* and other marine phytoplankton. *Proc. Natl. Acad. Sci. U. S.* **21,** 517–522.

—11—

Amino Acids and Proteins

L. FOWDEN

Department of Botany, University College, London, England

I. Introduction

The enormous variation of anatomical structure and morphology encountered among algae is reflected in the gross differences of nitrogenous compounds found among the members of this group. For instance, tenfold differences may occur between the nitrogen content of unicellular algae such as *Chlorella* or *Scenedesmus* and that of some of the large red and brown seaweeds. The total-nitrogen contents of the latter types rarely exceed 2% of their dry weight; values of less than 1% are not uncommon. In contrast, young cultures of *Chlorella* growing under optimum conditions may contain nearly 10% nitrogen, although in old nitrogen-deficient cultures the total-nitrogen content approximates that typical of the seaweeds. Age affects the composition of seaweeds to a smaller degree; moreover, Takagi (1951), Suzuki (1952), Ogino (1955), and Pillai (1957) have shown that there are definite seasonal changes in the nitrogenous constituents of members of the marine Chlorophyta, Rhodophyta, and Phaeophyta.

The cellular nitrogen may be divided, somewhat arbitrarily, into fractions including inorganic-, free α-amino-, free amido-, volatile-, and protein-nitrogen. Volatile nitrogenous compounds, which may include methylamine and trimethylamine (Kapeller-Adler and Csató, 1930; Kapeller-Adler and Vering, 1931), and inorganic nitrogen are usually quantitatively insignificant fractions, although occasional large accumulations of nitrate have been reported in *Laminaria* sp. (Channing and Young, 1953). Protein normally constitutes the main nitrogen fraction. In unicellular algae it represents 80% or more of the total nitrogen, whereas in red and brown seaweeds Channing and Young (1953) found that 70 to 80% of the nitrogen was present as protein. Pillai (1957) obtained a similar result for mature seaweeds, but in young plants where the total nitrogen reached maximum values, there was a sharp decline in the percentage of protein nitrogen.

TABLE I

CONCENTRATIONS OF FREE AMINO ACIDS IN *Chlorella vulgaris* CELLS AT VARIOUS TIMES AFTER SUPPLYING AMMONIA- OR NITRATE-NITROGEN[a,b]

		Time in minutes after nitrogen supplied			
Amino acid	N source	0	15	120	1440
Glutamic acid	NH_3	2.8	2.3	6.4	14.1
	NO_3^-	2.7	2.6	4.6	7.2
Glutamine	NH_3	0.0	3.2	5.6	0.4
	NO_3^-	0.0	0.0	0.6	0.1
Alanine	NH_3	1.5	5.6	7.7	11.8
	NO_3^-	1.2	1.3	2.8	4.9
Aspartic acid	NH_3	0.2	0.2	1.0	1.0
	NO_3^-	0.1	0.2	1.3	0.5
Asparagine	NH_3	0.2	0.4	0.8	1.6
	NO_3^-	0.1	0.1	0.4	1.4
Arginine	NH_3	0.1	0.2	0.7	5.4
	NO_3^-	0.0	0.1	0.1	1.4
Total	NH_3	0.3	0.4	0.8	0.8
	NO_3^-	0.2	0.2	0.4	0.6

[a] Amino acid concentrations are expressed as μmoles/ml. of cell suspension, except in the case of total α-amino-N which is reported as mg./ml.

[b] After Reisner *et al.* (1960).

II. Free Amino Acids and Related Compounds

A. Composition

1. Composition of the Free Amino-Acid Pool

The work of Fowden (1951) established that the free amino-acid pool in algae is composed of compounds similar to those found in flowering plants. Nearly all of the amino acids normally occurring in proteins were present in the free state in cells of *Chlorella pyrenoidosa*, *C. vulgaris*, and *Anabaena cylindrica*, but histidine and tryptophan were consistently not found, presumably because their normal concentrations were too low to be detected by the paper chromatographic methods used. Citrulline and γ-aminobutyric acid were present in both species of *Chlorella*; β-alanine was also detected in *C. vulgaris*.

In the author's opinion, no great value can be attached to comparisons of the *quantitative* amino-acid composition of different algal species (cf. Tsuchiya and Sasaki, 1957), because the exact balance between various free amino acids is so dependent on the nutritional status prevailing immediately before the algae were harvested. The total free α-amino-nitrogen content of *Chlorella vulgaris* cells increased approximately threefold in the 24 hours after either ammonia or nitrate was supplied to a nitrogen-deficient culture, though no cell multiplication occurred (Reisner *et al.*, 1960). The relative increases in certain amino acids (listed in Table I) were considerably greater, especially when ammonium chloride was provided as the nitrogen source. The compositions of the free amino-acid pool at the 15-minute and 24-hour sampling times were markedly different. The presence of arginosuccinic acid (I) was suspected in the earlier samples.

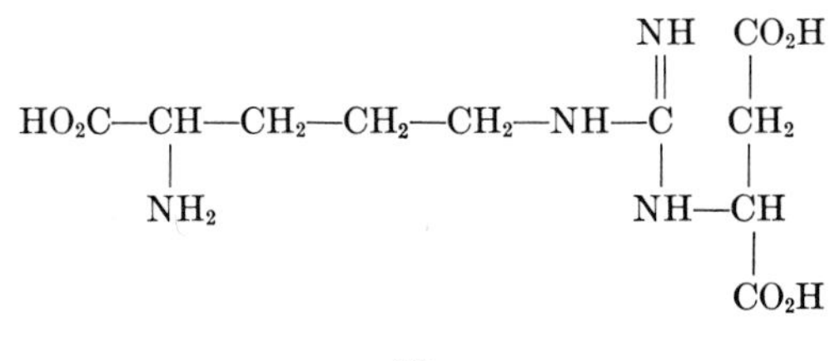

(I)

A comparison of the content of free amino acids in *C. pyrenoidosa* cells grown with either urea or nitrate was made by Champigny (1957). Differences were fairly small except in the cases of glutamic acid, asparagine, and the combined total of serine and glycine. Hattori (1958), who studied the influence of urea as a nitrogen source upon the composition of *C. ellipsoidea*, observed a far greater increase in the nitrogen content of the cells after supplying ammonia than did Reisner *et al.* (1960).

2. *Amino Acids Characteristic of Algae*

In addition to arginosuccinic acid, other less well-known nitrogenous compounds isolated from algae include three imino acids: L-α-kainic and L-α-allokainic acids (II) from *Digenea simplex* (Murakami *et al.*, 1953), and domoic acid (III) from *Chondria armata* (Daigo, 1959). In L-α-kainic acid, the 2- and 3-substituents are in the *trans* position, whereas those on the 3- and 4-carbon atoms are *cis*. In L-α-allokainic acid, both configurations are *trans*. It is claimed that all three imino acids possess useful anthelminthic properties.

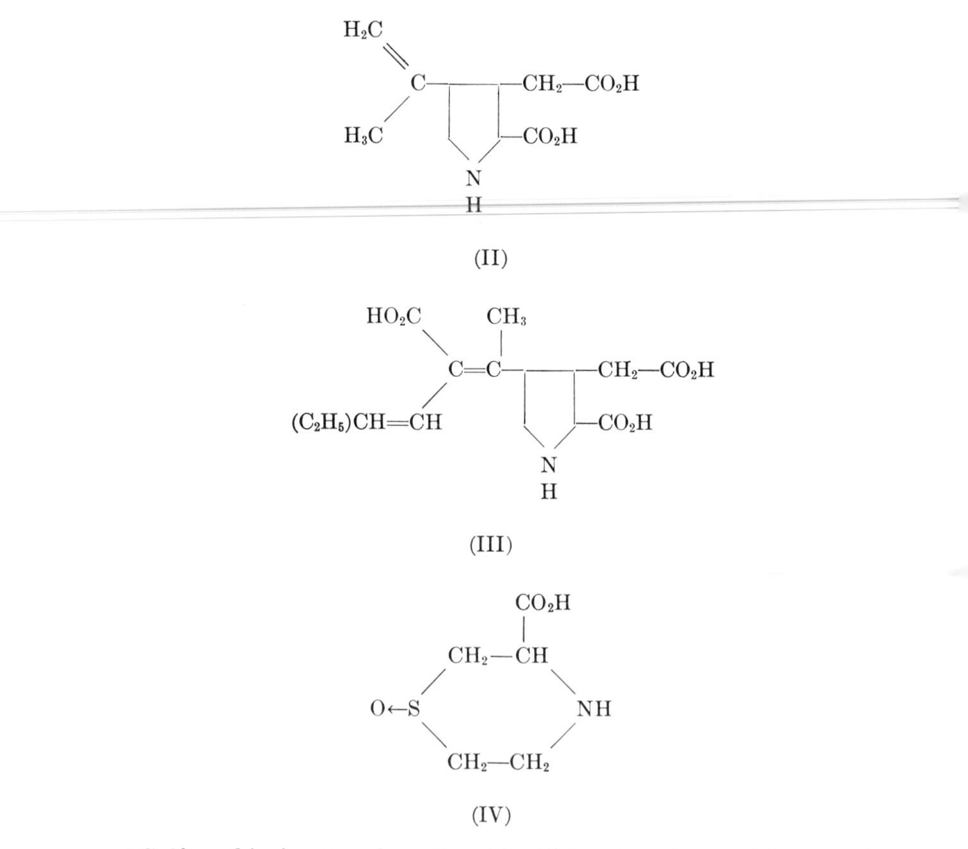

L-1-Sulfoxythiazine-3-carboxylic acid (IV) has been isolated from another of the Red Algae, *Chondria crassicaulis* (Kuriyama *et al.*, 1961). This heterocyclic acid shows a close structural similarity to cycloalliin (5-methyl-1-sulfoxythiazine-3-carboxylic acid), a substance recently isolated from onion (*Allium cepa*) by Virtanen and Matikkala (1959).

Iodoamino acids are characteristic components of marine Brown and Red

Algae, and have been studied particularly in *Laminaria* species. Though diiodotyrosine is the main iodo-organic compound, smaller amounts of monoiodotyrosine and triiodothyronine were indicated by chromatographic and radioautographic methods involving the use of I^{131} (Scott, 1954). The presence of iodoamino acids in one of the Green Algae, *Ulva lactuca*, was also demonstrated by these techniques. Most investigations have been performed in a manner that has made it impossible to conclude with certainty whether or not these compounds exist in the free state as well as in proteins (see sections II.B.2 and IV.D; also Shaw, Chapter 15, this treatise).

3. *Other Nitrogenous Compounds*

Two nitrogenous compounds which cannot be regarded strictly as amino acids, but which are possibly derived from cysteine, have been isolated from Red Algae. L-2-Amino-3-hydroxy-1-propane sulfonic acid (V) (D-cysteinolic acid) was isolated from *Polysiphonia fastigiata* (Wickberg, 1957), in which it represents about 0.1% of the dry weight. *N*-(D2,3-dihydroxy-*n*-propyl)-taurine (VI) (D-glycertyltaurine) is a quantitatively more important constituent of *Gigartina leptorhynchos* (Wickberg, 1956). Both these internal salts are configurationally related to the naturally occurring but rare D-amino acids.

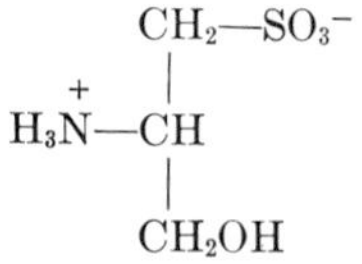

$$\begin{array}{rl} & CH_2{-}SO_3^- \\ & | \\ H_3\overset{+}{N}{-} & CH \\ & | \\ & CH_2OH \end{array}$$

(V)

$$\begin{array}{l} \qquad\quad CH_2{-}SO_3^- \\ \qquad\quad | \\ H_2\overset{+}{N}{-}CH_2 \\ \quad | \\ CH_2 \\ \quad | \\ CHOH \\ \quad | \\ CH_2OH \end{array}$$

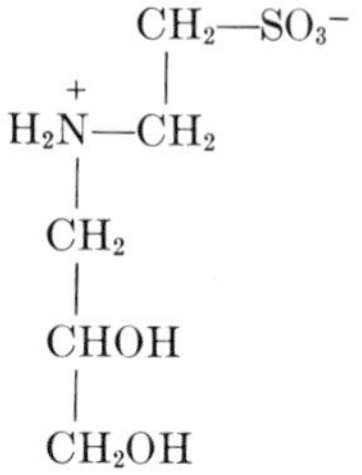

(VI)

In Blue-green Algae (species of *Anabaena, Anacystis, Nostoc, Synechococcus,* and *Synechocystis*) several 6-substituted pteridines (VII) have been found, in concentrations as high as 0.1% of the dry weight (Forrest *et al.*, 1957; Hatfield *et al.*, 1961). Several of the pteridines have been isolated and identified (Forrest *et al.*, 1958, 1959; Van Baalen *et al.*, 1959);

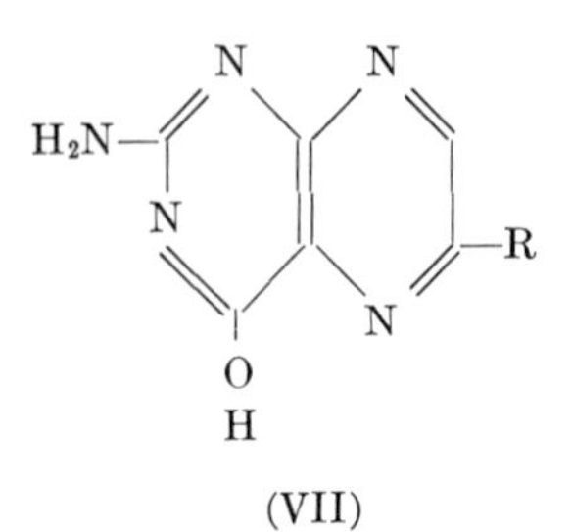

(VII)

some bear various sugars on the side chain. Aside from the incorporation of CO_2 into the glucose moiety of "compound C" during photosynthesis (Van Baalen *et al.*, 1957), there is still no information on their possible function in algae.

B. Metabolism

1. *General Reactions*

It is generally accepted that amino-acid metabolism is closely integrated with the intermediary metabolic systems associated with photosynthesis and respiration. The photosynthetic and respiratory processes are responsible for providing certain of the carbon skeletons of the amino acids, and the problem of amino-acid biogenesis is then largely centered around the processes of amination. Although this connection has now been demonstrated for many plant tissues, it is appropriate to recall that the pioneering work of Calvin and his co-workers was done with unicellular algae, *Chlorella* and *Scenedesmus* (Stepka *et al.*, 1948).

Keto acids form an important class of amino-group acceptors, and it is therefore interesting to note that α-ketoglutarate and pyruvate have high turnover rates in *Chlorella vulgaris* (Millbank, 1957).

Two general methods are available for the conversion of keto acids into amino acids, namely reductive amination and transamination. The formation of glutamic acid from α-ketoglutarate and ammonia by the action of glutamic dehydrogenase is the most familiar example of reductive amination. This is undoubtedly a primary pathway of ammonia assimilation in algae. Jacobi (1957a) showed that *Ulva lactuca* can also assimilate ammonia directly by an analogous reaction to yield aspartic acid. Apparently reduced diphosphopyridine nucleotide proved a better hydrogen donor than reduced triphosphopyridine nucleotide for both dehydrogenase reactions. Pyruvate and α-ketobutyrate could not be aminated in this way.

The formation of glutamine from glutamic acid is catalyzed by glutamine synthetase. This enzyme and the related glutamine transferase have been shown to be present in *Chlorella* sp. and *Scenedesmus* sp. by Loomis (1959).

After the initial formation of glutamate or aspartate in this way, the α-amino group may become distributed among other amino acids by transamination. While there is no doubt that this process plays an important role in algal metabolism, it cannot be accepted as a complete solution to the problem of amino-acid biogenesis until many more of the acceptor keto acids required in transaminations are established with certainty as algal constituents.

Transamination, being a reversible process, is conveniently studied by measuring the rate at which an amino acid transfers its α-amino group to α-ketoglutarate or to some other acceptor keto acid. Millbank (1953) and Jacobi (1957b) determined the relative rates of the transamination reactions between a range of amino acids and α-ketoglutarate or pyruvate, using enzyme preparations from *Chlorella* and *Ulva* respectively. Table II lists some of Jacobi's results; as in many other animal and plant systems investigated, α-ketoglutarate proved to be a better amino-group acceptor than pyruvate; aspartic acid was the best donor amino acid.

Detailed studies of the rapid transamination reaction between γ-amino-

TABLE II

COMPARISON OF THE RATES OF TRANSAMINATION OF VARIOUS AMINO ACIDS WITH PYRUVATE OR α-KETOGLUTARATE, CATALYZED BY EXTRACTS OF *Ulva lactuca*[a,b]

Amino-acid donor	Pyruvate acceptor	α-Ketoglutarate acceptor
Alanine	−	+
Aspartic acid	+	+++
Arginine	(+)	+
α-Aminobutyric acid	+	+
Cystine	−	+
Glycine	+	+
Histidine	−	+
Isoleucine	+	++
Leucine	(+)	++
Methionine	(+)	++
Phenylalanine	(+)	+
Serine	+	+
Threonine	−	++
Tryptophan	−	−
Tyrosine	−	−
Valine	+	++

[a] After Jacobi (1957b).
[b] Rates of transamination: +++, fast; ++, moderate; +, slow; (+), very slow; −, no transamination.

butyric acid and α-ketoglutaric acid catalyzed by extracts of *Chlorella vulgaris* were made by Kating (1958). γ-Aminobutyric acid may possibly be converted into glutamic acid by direct carboxylation, i.e., by reversal of the glutamic decarboxylase action observed in *C. pyrenoidosa* by Warburg and co-workers (see Warburg, 1958). This carboxylation mechanism has featured in recent theories of the Warburg school to explain carbon fixation by *Chlorella* under aerobic conditions. However there are many serious objections to the ready acceptance of the reversibility of the conversion of glutamic acid to γ-aminobutyric acid, and rigid proof of the occurrence of this type of carboxylation is lacking. Certainly work by Kating (1958) using *Chlorella*, and similar studies with bacteria and higher plants, indicate that more probably carbon atoms of γ-aminobutyric acid enter glutamic acid via succinic semialdehyde, succinic acid, and the reaction intermediates of the tricarboxylic-acid cycle leading to α-ketoglutaric acid. This is then converted into glutamic acid by transamination, possibly involving a second γ-aminobutyric acid molecule, or by reductive amination.

2. *Metabolism of Specific Amino Acids*

a. The Ornithine-Citrulline-Arginine Family of Acids. The metabolism of these acids in *Chlorella* has gained a special significance in relation to the role of arginosuccinic acid. The synthesis of this acid by a condensation reaction between arginine and fumarate, in which the guanido-NH_2 group of arginine is attached across the double bond of fumarate, can be catalyzed by extracts of dried cells of *Chlorella pyrenoidosa* (Walker and Myers, 1953). In animal tissues, arginosuccinic acid apparently arises by an alternative mechanism involving the condensation of citrulline with aspartate. Arginine desimidase, an enzyme converting arginine into citrulline and ammonia, was also detected in *Chlorella* extracts (Walker and Myers, 1953).

Citrulline was established as one of the earliest amino acids to become heavily labeled when $C^{14}O_2$ was assimilated photosynthetically by the nitrogen-fixing Blue-green Alga *Nostoc muscorum*, and was proposed as a specific intermediate in nitrogen fixation (Linko *et al.*, 1957). However, both the C^{14} in this experiment and the N^{15} in comparable studies employing root-nodules of *Alnus* (Leaf *et al.*, 1958) were fixed primarily in the carbamino group of citrulline, which is readily explicable in terms of its normal formation from ornithine and carbamyl phosphate.

Walker and Myers (1953) suggested that some of the above reactions may be involved in the metabolism of pyrimidines, according to the accompanying scheme.

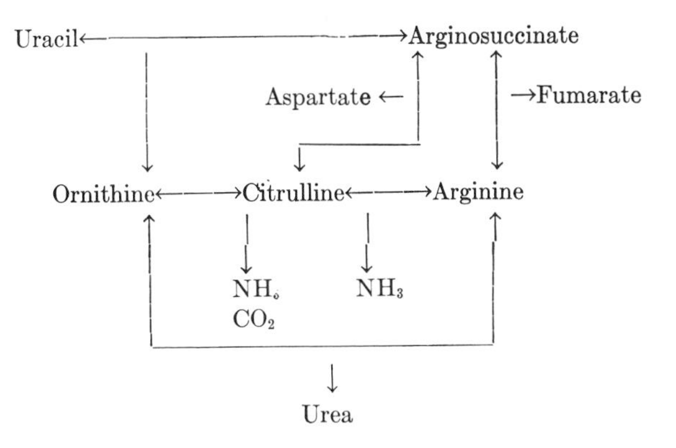

Extracts of *Chlorella pyrenoidosa* catalyze a condensation (analogous to that yielding arginosuccinate) between canavanine and fumaric acid to yield canavanosuccinic acid (Walker, 1953).

A number of algae, including species of *Chlamydomonas*, have been found to contain ornithine δ-transaminase, which catalyzes the following reaction:

$$\text{ornithine} + \alpha\text{-ketoglutaric acid} \leftrightarrows \text{glutamic acid} + \text{glutamic semialdehyde}$$

(Scher and Vogel, 1957). It is perhaps of phylogenetic significance that this enzyme, which is apparently of general occurrence among plants and animals, has not been found in Blue-green Algae or in Gram-negative bacteria.

b. Iodoamino Acids. Many seaweeds have the capacity to concentrate iodide to levels far above that of their environment, and to incorporate a proportion into organic iodocompounds, notably the iodotyrosines. The initial stage of incorporation apparently involves an oxidation of iodide to free iodine, probably dependent on the presence of endogenously produced hydrogen peroxide. Tong and Chaikoff (1955) supplied I^{131}-iodide to excised portions of the lamina of *Nereocystis luetkaena* (Phaeophyta), and observed that much of the iodine became organically bound in 4 hours. When cell-free extracts were used, I^{131} was bound by tyrosine, monoiodotyrosine, and thyronine, the formation of monoiodotyrosine apparently preceding that of diiodotyrosine.

III. Peptides

A. General Considerations

Haas and Hill (1933) suggested that in those marine algae which normally grow in an environment providing low light intensities and frequent periods

TABLE III

QUALITATIVE COMPOSITION OF PEPTIDES FROM MARINE ALGAE[a,b]

	Glycine	Alanine	Arginine	Histidine	Aspartic acid	Glutamic acid
Griffithsia flosculosa	+	+	+	+	−	−
Pelvetia canaliculata	+	+	+	+	−	+
P. canaliculata f. *libera*	+	+	+	+	−	+
Corallina officinalis	+	+	+	+	+	−

[a] After Haas (1950).
[b] +, present; −, absent.

of desiccation, protein metabolism may be extensively modified. The arrested synthesis of protein molecules might be expected to result in an accumulation of amino acids and low molecular-weight peptides. Their studies on the peptides of Brown and Red Algae collected during the winter months developed from this reasoning. In the final paper of a series on this subject, Haas (1950) summarized and extended the earlier work. The amino acids present in hydrolyzates of the isolated peptides, which varied for different species, are listed in Table III.

It must be stressed that there is no proof that other amino acids were absent from any of the peptide preparations, and it seems quite likely that the material from any one species was a mixture of several chemically distinct peptides.

Peptides could not be detected in the summer months, an observation lending support to the suggestion that lack of light may be a factor contributing to peptide accumulation at other times of the year.

Peptides bound to nucleotides or polynucleotides are present in cells of *Chlorella ellipsoidea*, especially at the phase of incipient nuclear division (Hase *et al.*, 1959). The conjugated peptides were separated by zone electrophoresis and paper chromatography into three fractions of different amino-acid composition, although all contained a nucleotide-polynucleotide moiety.

B. Specific Low-Molecular-Weight Peptides

Three chemically distinct peptides, each containing a glutamine residue, have been isolated from marine algae. Ohira (1939, 1940) obtained a peptide (eisenine) from one of the Brown Algae, *Eisenia bicyclis*, and suggested for it the structure L-5-oxopyrrolidine-1-carbonyl-L-glutaminyl-L-alanine. The structure of another peptide, L-5-oxopyrrolidine-1-carbonyl-L-glutaminyl-L-glutamine, isolated from *Pelvetia fastigiata*, has been established with certainty by Dekker *et al.* (1949). Arginylglutamine was isolated from one of the Green Algae, *Cladophora* sp. (Makisumi, 1959), but a related substance, arginylglutamic acid, could not be detected.

IV. Proteins

A. General Considerations

The proteins of algae, like those of other plants, consist of many different molecular species and are perhaps predominantly enzymes possessing specific biological roles. No general method exists for the satisfactory separation of the components of this mixture. In the author's laboratory, an unsuccessful attempt was made to resolve the proteins of *Chlorella* by paper electrophoresis. However, highly purified preparations of certain chromoproteins have been obtained (see Section IV.C; also Ó hEocha, Chapter 25, this treatise).

Most analytical work has been performed on complex mixtures of proteins isolated from whole cells and often designated as the "bulk protein." Data so obtained are "mean" values, and their usefulness is necessarily limited when attempts are made to explain physiological behavior in terms of protein composition. One presumes, *a priori*, that the change in metabolic activity associated with aging should be reflected in an altered balance of the cellular enzymes. Moreover, proteins associated with different intracellular structures, often in the form of larger conjugate molecules, may well be of different composition, and their relative proportions would be expected to vary as the cells grow and divide. However, when the proteins are examined in terms of amino-acid composition (Section IV.B), the averaging effect tends to diminish the magnitude of the predicted changes of composition.

Differences must undoubtedly occur between the protein complements of different algal species, but clear experimental proof of this is still lacking. Although studies of the amino-acid composition of algal proteins are numerous, physicochemical investigations are few. Unpublished results by the author indicate that the protein components of *Chlorella vulgaris*, *Anabaena cylindrica*, and *Tribonema aequale* (Green, Blue-green, and Yellow-green Algae, respectively) show certain similarities during fractional precipitation by increasing concentrations of ammonium sulfate; but the precipitation curve obtained with a protein complex from *Navicula pelliculosa* was markedly different (see Fig. 1).

B. Amino-Acid Composition

With the development of paper chromatographic methods, the study of the amino-acid composition of proteins gained a new impetus, and analyses of the "bulk" proteins from many algae have now been published. However, in the author's opinion, many of these analyses are unreliable and therefore worthless. This condemnation is made on the ground that in many cases

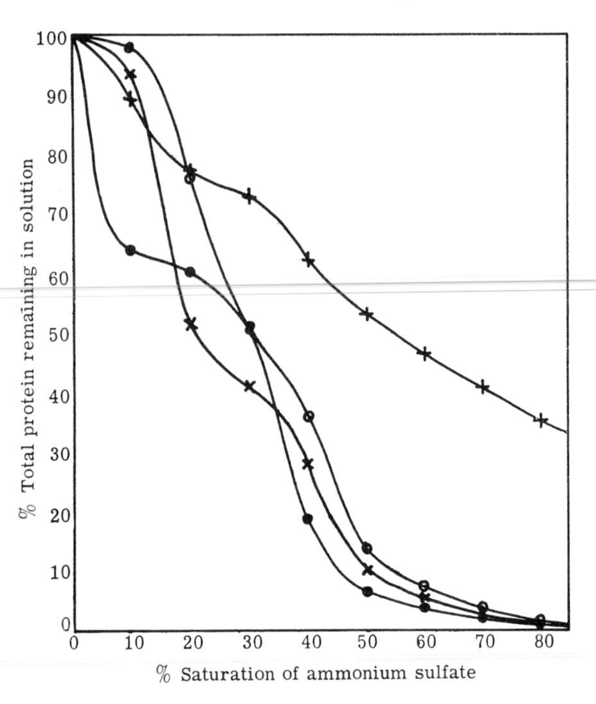

FIG. 1. Fractional precipitation curves obtained when various algal protein solutions are treated with increasing concentrations of ammonium sulfate. KEY: O—O, *Chlorella vulgaris*; ●—●, *Anabaena cylindrica*; ×—×, *Tribonema aequale*; +—+, *Navicula pelliculosa*.

the "total nitrogen" present in the separated constituent amino acids resulting from protein hydrolysis accounts for only a small fraction of the total nitrogen determined in the bulk protein. Losses may arise during the chromatographic assay procedures, but frequently they are incurred during the initial acidic hydrolysis of the protein. If reliable analyses are required, it is absolutely essential first to obtain a somewhat purified protein fraction, largely freed of contaminating carbohydrate, so that it can be hydrolyzed without humin formation. The presence of carbohydrates during acid hydrolysis enhances humin formation with the consequent partial or total destruction of certain amino acids, particularly basic, hydroxy, and sulfur-containing types. Although in practice it may be difficult to obtain a relatively pure protein fraction that nevertheless represents virtually the whole of the cell protein—especially when tough and fibrous marine algae are under investigation—this should be regarded as an indispensable prerequisite for reliable assays.

Protein analyses in which all the component amino acids have been assayed were published by Champigny (1957), Fowden (1952), and Schieler *et al.* (1953) for species of *Chlorella*; by Mazur and Clark (1938), Smith and Young (1953), Coulson (1955), Serenkov *et al.* (1957), Serenkov and Pachomova (1959), and Lewis and Gonzalves (1960) for various marine algae; and by Fowden (1954) for four freshwater species. Fujiwara (1956) and Kimmel and Smith (1958) published complete analyses for the purified chromoproteins, phycocyanin and phycoerythrin. Partial amino-acid compositions of algal proteins are to be found in the publications of Takagi (1953a, b), Channing and Young (1953), and Pillai (1957) for marine species, and of Williams and Burris (1952) for three freshwater Blue-green Algae. Table IV summarizes some of the more complete data obtained.

Comparison of the different analyses is difficult because of uncertainties in the degrees of error involved. Total protein nitrogen is accounted for only in the assays of Fowden (1952, 1954) and Kimmel and Smith (1958). Mazur and Clark (1938), in attempting the first complete amino-acid analyses, claimed to have found sharp differences of composition between the proteins of various algae and suggested that these differences may have phylogenetic importance. However, this implication has not been accepted generally, and the evidence of Channing and Young (1953), Fowden (1954), Williams and Burris (1952), Smith and Young (1953), and Lewis and Gonzalves (1960) suggests that unrelated species may have closely similar protein compositions. Larger differences in the protein composition of various marine algae were reported by Coulson (1955), Serenkov and Pachomova (1959), and Takagi (1953a, b). However the present author is inclined to doubt that there can be such great differences in the tyrosine, tryptophan, threonine, and serine contents of proteins from marine Green, Red, and Brown Algae as those reported by Takagi.

Although the total protein content of *Chlorella* cells is dependent on the age of the culture, the effect of age on the amino-acid composition of the "bulk protein" in the cells is only slight, the largest change noted being in its histidine content (Fowden, 1952). Greater differences in the composition of the protein isolated from young and mature plants of several marine algae were observed by Pillai (1957). Differences in the protein composition of a single species in various seasons were found to be larger than differences between those of different species collected at the same time.

TABLE IV

AMINO-ACID COMPOSITION OF BULK PROTEINS OF SOME ALGAE, AND OF PURIFIED PHYCOBILINS[a]

	(1) *Fucus vesiculosus*	(2) *Chlorella vulgaris*	(2) *Anabaena cylindrica*	(2) *Navicula pelliculosa*	(2) *Tribonema aequale*	(3) Phyco-cyanin	(3) Phyco-erythrin
Alanine	5.4	7.7	6.0	6.5	8.4	10.5	12.8
Arginine	9.4	15.8	11.7	9.2	15.9	15.7	17.9
Aspartic acid	9.0	6.4	6.9	6.4	5.1	7.2	8.9
Cystine	tr	0.2	tr	tr	tr	1.1	2.3
Glycine	5.4	6.2	5.5	6.1	6.2	5.9	5.6
Glutamic acid	11.2	7.8	5.6	4.9	4.6	7.0	5.5
Histidine	1.6	3.3	2.5	2.8	3.7	0.6	1.8
Isoleucine	3.0	3.5	3.9	3.5	4.1	3.7	3.3
Leucine	5.0	6.1	6.2	7.2	6.4	7.1	5.6
Lysine	6.0	10.2	6.6	8.3	9.0	4.2	6.4
Methionine	0.4	1.4	1.2	1.2	1.4	2.1	1.7
Phenylalanine	2.6	2.8	2.9	3.4	2.8	1.5	1.4
Proline	3.3	5.8	5.0	6.2	6.1	2.7	2.1
Serine	3.5	3.3	2.4	4.2	2.4	6.6	6.6
Threonine	3.3	2.9	5.7	4.2	4.0	3.7	3.1
Tryptophan	—	2.1	1.0	1.1	1.8	—	—
Tyrosine	1.2	2.8	1.6	1.9	3.0	3.5	2.7
Valine	3.0	5.5	7.0	7.5	7.5	5.2	6.2
Amide-ammonia	15.3	6.1	8.0	7.1	6.5	7.7	8.7
Total nitrogen	88.6	99.9	89.7	91.7	98.9	102.6	96.0

[a] Data expressed as grams of amino-acid nitrogen/100 gm. protein nitrogen: (*1*) after Smith and Young (1953); (*2*) after Fowden (1954); (*3*) after Kimmel and Smith (1958). ["tr" = trace.]

Many organisms can utilize ammonium salts, nitrates, or organic nitrogen compounds as their sole nitrogen source for growth. Alterations in the nitrogen source available to a cell presumably lead to an altered balance among the enzymes of nitrogen assimilation, and therefore this could possibly be reflected by changes in the amino-acid composition of the "bulk protein." Although general experience seems largely to discount this possibility—since presumably any changes that occur in specific proteins are small in relation to the whole "bulk protein"—Champigny (1957) reported that the composition of the protein of *Chlorella pyrenoidosa* differs markedly according to whether the cells were grown on nitrate or on urea. The very magnitude of the differences reported by the latter author throws doubt on her results.

The amino-acid components of proteinaceous material associated with isolated cell walls of *Chlorella pyrenoidosa* and *Hydrodictyon africanum* were identified chromatographically by Northcote *et al.* (1958, 1960).

C. Specific Proteins

Since they are distinguished by characteristic colors, the biliproteins phycocyanin and phycoerythrin have attracted more attention than any other single molecular species of algal protein. A detailed treatment of these substances is given by Ó hEocha (Chapter 25, this treatise), but for comparative purposes their amino-acid compositions are presented in Table IV.

A purified metalloprotein containing copper has been isolated from *Chlorella ellipsoidea* (Katoh, 1960). This work has particular importance because copper-specific inhibitors suppress the assimilation of carbon dioxide by *Chlorella*, although they have no effect on respiratory activity. The copper-protein readily undergoes oxidation-reduction reactions. The oxidized form can be reduced by such agents as ascorbic acid and hydroquinone, or by reduced cytochrome *c* or a reduced cytochrome 550 (from *Chlorella ellipsoidea*); the reduced form of the protein is not autoxidizable in air. The precise role of this protein in *Chlorella* is still unknown.

A crystalline chlorophyll-lipoprotein complex, similar to that first obtained from clover leaves by Takashima (1952), was successfully isolated from *Chlamydomonas dorsoventralis* (Sherratt and Evans, 1954). Comparable crystals were also obtained from a marine species of *Chlamydomonas*, but a similar extraction technique failed to yield conjugated protein from *Chlorella vulgaris*. The crystalline material from *Chlamydomonas* contained both chlorophyll *a* and chlorophyll *b*. Possibly two independent components may be present in the chlorophyll-lipoprotein complex, containing conjugated chlorophyll *a* and conjugated chlorophyll *b* respectively, as was recently established for the protein complexes from clover (Chiba, 1955).

D. Metabolism

Protein metabolism in algae has not been extensively studied. Relevant information has been gained from investigations of the influence of seasonal or nutritional factors on the protein level (see Section I), or from experiments designed to elucidate the metabolic fate of photosynthetically assimilated carbon atoms. Calvin and his associates established that the proteins of *Chlorella* and *Scenedesmus* are in a state of rapid dynamic equilibrium with simple intermediary metabolites, since carbon atoms from fixed C^{14}-carbon dioxide entered protein molecules in very short periods of time.

The pattern of photosynthetic products in two marine planktonic algae, *Amphidinium carteri* and *Olisthodiscus* sp., was compared with that in *Chlorella pyrenoidosa* (Bidwell, 1957). In all three organisms approximately 50% of the C^{14} atoms fixed from labeled carbon dioxide were found in the protein after 4 hours. The relative specific activities of the constituent amino acids of each protein were closely similar, although the degree of labeling of the free amino acids and of the carbohydrates varied greatly among the three organisms. Bidwell considered that these observations establish the fundamental character of protein synthesis in these algae, and suggested that the process of protein synthesis may utilize not precursors drawn from the free amino-acid pool within the cells, but other intermediates derived immediately from the assimilated carbon atoms. (Compare the alternative hypothesis of protein synthesis propounded by Steward *et al.*, 1958.)

In the presently popular field of study concerning the interdependent roles of subcellular particles such as nuclei, mitochondria, and microsomes in the processes of protein synthesis, the marine algae *Acetabularia* has played a fascinating part. For a summary of this work, which is too extensive for treatment here, the reader is referred to a review by Brachet (1960); see also Richter (Chapter 42 in this treatise).

Little is yet known concerning the distribution of iodoamino acids in algal proteins. Since, when they occur, these amino acids normally constitute only a small percentage of the total weight of protein [the highest value reported seems to be 1.55% (Roche and Lafon, 1949)], it seems probable that they are not present generally in the proteins of the algae in which they occur, but are restricted to a few molecular species of protein possessing special biological properties. Tentatively, the further supposition may be made that the latter proteins arise by a direct *in situ* iodination of certain tyrosine residues already bound in protein molecules. Support for this idea is found in the observations of Tong and Chaikoff (1955), who showed that when I^{131}-iodine was supplied to intact tissues of *Nereocystis luetkeana*, all radioactivity associated with iodotyrosines was present in a protein-bound state.

E. Uses

Although it used to be considered that unicellular algae, particularly *Chlorella*, might help to alleviate an increasing deficit in the world's protein supply, for various reasons, but perhaps primarily due to adverse economic factors, this now seems to have become a pipe-dream. More attention is presently being focused upon the possibility of a large-scale utilization for human consumption of protein preparations extracted from plants now regarded either as weeds or as mere animal fodder. However, the era of "algae as food" did not pass without leaving its mark upon the culinary art, and Morimura and Tamiya (1954) have placed on record a number of tempting recipes utilizing *Chlorella*, while the author recollects reading an account of a *Chlorella* ice-cream being produced in Finland. The high nutrient quality of the proteins of Green Algae in comparison with those of animal or vegetable origin has been established in a variety of animal-feeding trials, one of the most extensive being that of Fink and Herold (1957a, b), who proved that the biological value of protein from *Scenedesmus* was at least as high as that from skim milk or egg white, and superior to that of all other vegetable proteins tested.

Another, more technical use of algal protein is as a source of C^{14}-labelled amino acids (Catch, 1954). When *Chlorella* is grown autotrophically upon C^{14}-carbon dioxide, the amino acids which can be isolated from the hydrolyzed protein have the advantage of being L-isomers in which all the carbon atoms of every amino acid possess the same specific activity as the original carbon dioxide source. A world-wide demand exists for C^{14}-amino acids produced in this way at the Radiochemical Centre, Amersham, Bucks., U.K.

References

Bidwell, R. G. S. (1957). Photosynthesis and metabolism of marine algae. I. Photosynthesis of two marine flagellates compared with *Chlorella*. *Can. J. Botany* **35,** 945–950.

Brachet, J. (1960). Ribonucleic acids and the synthesis of cellular proteins. *Nature* **186,** 194–199.

Catch, J. R. (1954). C^{14}-Amino acids from *Chlorella*. *Proc. Radioisotope Conf., 2nd Conf., Oxford, 1954* (Vol. I: Medical and Physiological Applications) pp. 258–264.

Champigny, M.-L. (1957). Growth studies of unicellular algae in rapid culture. IV. Variations in the composition of the amino acids of *Chlorella pyrenoidosa* in relation to the nitrogen nutrition. *J. recherches centre natl. recherche sci., Lab. Bellevue (Paris)* **No. 8,** 72–76.

Channing, D. M., and Young, G. T. (1953). Amino acids and peptides. X. The nitrogenous constituents of some marine algae. *J. Chem. Soc.* pp. 2481–2491.

Chiba, Y. (1955). Two components in crystalline chlorophyll-lipoprotein. *Arch. Biochem. Biophys.* **54,** 83–92.

Coulson, C. B. (1955). Plant proteins. V. Proteins and amino acids of marine algae. *J. Sci. Food Agr.* **6,** 674–682.

Daigo, K. (1959). Studies on the constituents of *Chondria armata*. III. Constitution of domoic acid. *J. Pharm. Soc. Japan* **79,** 356–360.

Dekker, C. A., Stone, D., and Fruton, J. S. (1949). A peptide from a marine alga. *J. Biol. Chem.* **181,** 719–729.

Fink, H., and Herold, E. (1957a). On the protein quality of a unicellular green alga and its preventative action on liver necrosis. Part I. *Z. physiol. Chem. Hoppe-Seyler's* **305,** 182–191.

Fink, H., and Herold, E. (1957b). On the protein quality of a unicellular green alga and its preventative action on liver necrosis. Part II. *Z. physiol. Chem. Hoppe-Seyler's* **307,** 202–216.

Forrest, H. S., Van Baalen, C., and Myers, J. 1957. Occurrence of pteridines in a blue-green alga. *Science* **125,** 699–700.

Forrest, H. S., Van Baalen, C., and Myers, J. 1958. Isolation and identification of a new pteridine from a blue-green alga. *Arch. Biochem. Biophys.* **78,** 95–99.

Forrest, H. S., Van Baalen, C., and Myers, J. 1959. Isolation and characterization of a yellow pteridine from the blue-green alga, *Anacystis nidulans*. *Arch. Biochem. Biophys.* **83,** 508–520.

Fowden, L. (1951). Amino acids of certain algae. *Nature* **167,** 1030–1031.

Fowden, L. (1952). The effect of age on the bulk proteins of *Chlorella*. *Biochem. J.* **52,** 310–313.

Fowden, L. (1954). A comparison of the compositions of some algal proteins. *Ann. Botany (London)* **18,** 257–266.

Fujiwara, T. (1956). Studies on chromoproteins in Japanese nori (*Porphyra tenera*). II. Amino acid compositions of phycoerythrin and phycocyanin. *J. Biochem. (Tokyo)* **43,** 195–203.

Haas, P. (1950). On certain peptides occurring in marine algae. *Biochem. J.* **46,** 503–505.

Haas, P., and Hill, T. G. (1933). Observations on the metabolism of certain seaweeds. *Ann. Botany (London)* **47,** 55–67.

Hase, E., Mihara, S., Otsuka, H., and Tamiya, H. (1959). New peptide-nucleotide compounds obtained from *Chlorella* and yeasts. *Biochim. et Biophys. Acta* **32,** 298–300.

Hatfield, D. L., Van Baalen, C., and Forrest, H. S. 1961. Pteridines in blue-green algae. *Plant Physiol.* **36,** 240–243.

Hattori, A. (1958). Studies on the metabolism of urea and other nitrogenous compounds in *Chlorella ellipsoidea*. II. Changes in levels of amino acids and amides during the assimilation of ammonia and urea by nitrogen-starved cells. *J. Biochem. (Tokyo)* **45,** 57–64.

Jacobi, G. (1957a). Enzymes in the amino acid metabolism of *Ulva lactuca*. *Naturwissenschaften* **44,** 265–266.

Jacobi, G. (1957b). Enzymes in the amino acid metabolism of *Ulva lactuca*. Transaminases and amino acid dehydrogenases. *Planta* **49**, 561–567.

Kapeller-Adler, R., and Csató, T. (1930). The occurrence of methylated nitrogen compounds in brown algae. *Biochem. Z.* **224**, 378–383.

Kapeller-Adler, R., and Vering, F. (1931). The occurrence of methylated nitrogen compounds in brown algae (II) and in some cold-blooded animals with especial reference to trimethylamine. *Biochem. Z.* **243**, 292–309.

Kating, H. (1958). The reaction between γ-aminobutyric acid and glutamic acid in *Chlorella vulgaris* var. *viridis*. *Planta* **51**, 635–644.

Katoh, S. (1960). A new copper protein from *Chlorella ellipsoidea*. *Nature* **186**, 533–534.

Kimmel, J. R., and Smith, E. L. (1958). The amino acid composition of crystalline pumpkin seed globulin, edestin, C-phycocyanin and R-phycoerythrin. *Bull. soc. chim. biol.* **40**, 2049–2065.

Kuriyama, M., Takagi, M., and Murata, K. (1961). Chemical studies on marine algae. 14. On a new amino acid, "chondrine," isolated from the red alga *Chondria crassicaulis*. *Bull. Soc. Fish. Hokkaido Univ.* **11**, 58–66.

Leaf, G., Gardner, I. C., and Bond, G. (1958). Observations on the composition and metabolism of the nitrogen-fixing root nodules of *Alnus*. *J. Exptl. Botany* **9**, 320–331.

Lewis, E. J., and Gonzalves, E. A. (1960). Amino acid contents of some marine algae from Bombay. *New Phytologist* **59**, 109–115.

Linko, P., Holm-Hansen, O., Bassham, J. A., and Calvin, M. (1957). Formation of radioactive citrulline during photosynthetic $C^{14}O_2$ fixation by blue-green algae. *J. Exptl. Botany* **8**, 147–156.

Loomis, W. D. (1959). Amide metabolism in higher plants. II. Distribution of glutamyl transferase and glutamine synthetase activity. *Plant Physiol.* **34**, 541–546.

Makisumi, S. (1959). Occurrence of arginylglutamine in a green alga, *Cladophora* species. *J. Biochem.* (*Tokyo*) **46**, 63–71.

Mazur, A., and Clark, H. T. (1938). The amino acids of certain marine algae. *J. Biol. Chem.* **123**, 1729–1740.

Millbank, J. W. (1953). Demonstration of transaminase systems in the alga *Chlorella*. *Nature* **171**, 476.

Millbank, J. W. (1957). Keto acids in the alga *Chlorella*. *Ann. Botany* (*London*) **21**, 23–31.

Morimura, Y., and Tamiya, N. (1954). Preliminary experiments in the use of *Chlorella* as human food. *Food Technol.* **8**, 179–182.

Murakami, S., Takemoto, T., and Shimuzu, Z. (1953). Studies on the principles of *Digenea simplex* aq. I. Separation of the effective fraction by liquid chromatography. *J. Pharm. Soc. Japan* **73**, 1026–1029.

Northcote, D. H., Goulding, K. J., and Horne, R. W. (1958). The chemical composition and structure of the cell wall of *Chlorella pyrenoidosa*. *Biochem. J.* **70**, 391–397.

Northcote, D. H., Goulding, K. J., and Horne, R. W. (1960). The chemical composition and structure of the cell wall of *Hydrodictyon africanum*. *Biochem. J.* **77**, 503–508.

Ogino, C. (1955). Biochemical studies on the nitrogen compounds of algae. *J. Tokyo Univ. Fisheries* **41**, 107–152.

Ohira, T. (1939). On a new polypeptide isolated from *Eisenia bicyclis* (Part I). *J. Agri. Chem. Soc. Japan* **15**, 370–376.

Ohira, T. (1940). On a new polypeptide isolated from *Eisenia bicyclis* (Part II). A study of the chemical structure of eisenine. *J. Agr. Chem. Soc. Japan* **16,** 10–11.

Pillai, V. K. (1957). Chemical studies on Indian seaweeds. II. Partition of nitrogen. *Proc Indian Acad. Sci.* **B45,** 43–63.

Reisner, G. S., Gering, R. K., and Thompson, J. F. (1960). The metabolism of nitrate and ammonia by *Chlorella. Plant Physiol.* **35,** 48–52.

Roche, J., and Lafon, M. (1949). The presence of diiodotyrosine in *Laminaria Compt. rend. acad. sci.* **229,** 481–482.

Scher, W. I., and Vogel, H. J. (1957). Occurrence of ornithine δ-transaminase: a dichotomy. *Proc. Natl. Acad. Sci. U. S.* **43,** 796–803.

Schieler, L., McClure, L. E., and Dunn, M. S. (1953). The amino acid composition of *Chlorella. Food Research* **18,** 377–380.

Scott, R. (1954). Observations on the iodo-amino acids of marine algae using iodine-131. *Nature* **173,** 1098–1099.

Serenkov, G. P., and Pachomova, M. V. (1959). [Biochemical investigations of two species of diatoms.) *Vestnik Moskov. Univ.* **1959,** No. 1, 39–45. (In Russian.)

Serenkov, G. P., Pachomova, M. V., and Borisova, I. G. (1957). [A comparative biochemical investigation of two species of Green Algae.] *Vestnik Moskov. Univ.* **1957,** No. 3, 77–85. (In Russian.)

Sherratt, H. S. A., and Evans, W. C. (1954). A crystalline chlorophyll-protein complex from *Chlamydomonas. Nature* **173,** 540.

Smith, D. G., and Young, E. G. (1953). On the nitrogenous constituents of *Fucus vesiculosus. J. Biol. Chem.* **205,** 849–858.

Stepka, W., Benson, A. A., and Calvin, M. (1948). The path of carbon in photosynthesis. II. Amino acids. *Science* **108,** 304.

Steward, F. C., Bidwell, R. G. S., and Yemm, E. W. (1958). Nitrogen metabolism, respiration and growth of cultured plant tissue. *J. Exptl. Botany* **9,** 11–49.

Suzuki, N. (1952). Composition of brown algae. *Bull. Fac. Fisheries, Hokkaido Univ.* **3,** 68–72.

Takagi, M. (1951). Chemical studies on seaweeds. The nitrogen distribution of *Porphyra* by kinds and seasons. *Bull. Fac. Fisheries, Hokkaido Univ.* **2,** 31–42.

Takagi, M. (1953a). Chemical studies on marine algae. VII. Tyrosine and tryptophan contents in various species of marine algae. *Bull. Fac. Fisheries, Hokkaido Univ.* **4,** 86–91.

Takagi, M. (1953b). Chemical studies on marine algae. VIII. Threonine and serine contents. *Bull. Fac. Fisheries, Hokkaido Univ.* **4,** 92–95.

Takashima, S. (1952). Chlorophyll-lipoprotein obtained in crystals. *Nature* **169,** 182.

Tong, W., and Chaikoff, I. L. (1955). Metabolism of I^{131} by the marine alga, *Nereocystis luetkeana. J. Biol. Chem.* **215,** 473–484.

Tsuchiya, Y., and Sasaki, T. (1957). Studies on the savour of marine algae. III. Contents of free amino acids in the laver, *Porphyra tenera. Bull. Japan Soc. Sci. Fisheries* **23,** 230–233.

Van Baalen, C., and Forrest, H. S. 1959. 2,6-Diamino-4-hydroxypteridine, a new, naturally occurring pteridine. *J. Amer. Chem. Soc.* **81,** 1770.

Van Baalen, C., Forrest, H. S., and Myers, J. (1957). Incorporation of radioactive carbon into a pteridine of a blue-green alga. *Proc. Natl. Acad. Sci.* **43,** 701–705.

Virtanen, A. I., and Matikkala, E. J. (1959). The structure and synthesis of cycloalliin isolated from *Allium cepa. Acta Chem. Scand.* **13,** 623.

Walker, J. B. (1953). An enzymic reaction between canavanine and fumarate. *J. Biol. Chem.* **204,** 139–146.

Walker, J. B., and Myers, J. (1953). The formation of arginosuccinic acid from arginine and fumarate. *J. Biol. Chem.* **203,** 143–152.
Warburg, O. (1958). Photosynthesis. *Science* **128,** 68–73.
Wickberg, B. (1956). Isolation of *N*-[D-2,3-dihydroxy-*n*-propyl]-taurine from *Gigartina leptorhynchos. Acta Chem. Scand.* **10,** 1097–1099.
Wickberg, B. (1957). Isolation of 2-L-amino-3-hydroxy-1-propane sulfonic acid from *Polysiphonia fastigiata. Acta Chem. Scand.* **11,** 506–511.
Williams, A. E., and Burris, R. H. (1952). Nitrogen fixation in blue-green algae and their nitrogenous composition. *Am. J. Botany* **39,** 340–342.

—12—

Inorganic Phosphorus Uptake and Metabolism

ADOLF KUHL

Institute of Plant Physiology, University of Göttingen, Germany

I. Introduction[1]

Phosphorus is one of the major nutrient elements required for normal growth of algae (see reviews by Myers, 1951; Ketchum, 1954; Krauss, 1958; Provasoli, 1958). As in all living organisms, compounds containing phosphorus play important roles in nearly all phases of metabolism, particularly in energy transformation reactions. In the physiology of green plants this element gained greater interest when it became evident that phosphorylated compounds participate in the reactions of photosynthesis. Discovery of the latter fact in particular led to a series of investigations which have greatly enlarged our present knowledge of the physiological importance of phosphorus in plants. Many of these investigations have

[1] Abbreviations used throughout this chapter: ADP, adenosine diphosphate; ATP, adenosine triphosphate; DNP, 2,4-dinitrophenol; TCA, trichloroacetic acid.

been carried out with unicellular algae which, for several reasons, are widely used as objects of research in plant physiology. The role of phosphorus in algal metabolism has also attracted the attention of ecologists, since this element is frequently a limiting factor of algal growth in nature (see also Talling, Chapter 50, and Yentsch, Chapter 52, this treatise). Phospholipids are discussed by Benson and Shibuya (Chapter 22).

II. External Factors Influencing the Uptake of Phosphate

A. General Remarks

The absorption of phosphate from the surrounding medium is governed by many conditions, the physiological importance of which has not always been clearly established. For this reason certain aspects of phosphorus metabolism will be discussed here in merely a descriptive manner. General problems of the mechanisms of ion-uptake in algae lie beyond the scope of this article, though we will refer to theoretical considerations whenever possible, especially in cases where they are of assistance in interpreting experimental results. For detailed treatments of ion-uptake mechanisms the reader is referred to Kramer (1956) and Robertson (1958).

B. Light

Since phosphorus plays an important role in photosynthesis, an influence of light on phosphate uptake can be expected. Ketchum (1939b) found experimentally that the uptake of phosphate by P-deficient cells of "*Nitzschia closterium*" (= *Phaeodactylum tricornutum*)* was always significantly greater in the light than in the dark. Phosphorus-deficient *Chlorella pyrenoidosa* cells likewise recovered more rapidly when illuminated than when kept in darkness (Ketchum, 1939b; Scott, 1945b). On the other hand, in normal (i.e., not P-deficient) cells of *C. pyrenoidosa*, Emerson *et al.* (1944) could only demonstrate the promotion of phosphorus uptake by light when carbon dioxide was excluded, while Aronoff and Calvin (1948) failed to detect any influence of light on the rate of radiophosphorus uptake.

Subsequently, however, careful investigations established that even in normal cells the initial entry of phosphate may be accelerated by light. The uptake of phosphate by *Chlorella pyrenoidosa* and *Scenedesmus* (Gaffron's strain D_3) was considerably greater in the light than in the dark (Gest and Kamen, 1948). Although this uptake was partly inhibited by KCN, it was not directly related to the respiratory activity of the cells.

* Regarding the taxonomy of this organism, see Silva, this treatise, Appendix A, Note 22.

Light promoted phosphate uptake by *Chlorella vulgaris* (Wassink *et al.*, 1951a; Wintermans, 1955) and by *C. pyrenoidosa* (Nihei, 1955), especially in the absence of CO_2. In experiments of short duration the P-uptake of *Ankistrodesmus braunii* was slightly enhanced by light (Simonis and Kating; 1956; Jacobi, 1959). Cells of *Scenedesmus quadricauda* behaved similarly (Baslavskaya and Weber, 1959) but a longer period of illumination (10 to 120 minutes) increased the amount of absorbed phosphate from 50 to 320% over that incorporated in darkness. In K-deficient *C. vulgaris* cells Badour (1959) clearly demonstrated an effect of light on the phosphate uptake after the addition of potassium, especially in the presence of CO_2. Kuhl (1957; cf. Pirson and Kuhl, 1958) found that in *Hydrodictyon reticulatum* cultured in intermittent light (12 hours light: 12 hours darkness) the total P-content increased markedly during the light periods. This was due mostly to a reversible accumulation of inorganic phosphate, which in the succeeding dark periods was released in the medium.

Although there are many experimental indications that light influences the uptake of phosphorus, we are still far from being able to explain all the observed effects. A general survey of the literature on the influence of light on the uptake of various ions can be found in an article by Brauner (1956). Since active ion uptake is an energy-requiring process, it seems reasonable to assume that light acts as an energy donor (see Simonis, 1960). However, there is still no direct experimental evidence that the photochemical reactions of photosynthesis per se are involved in phosphate uptake.

C. Phosphate Concentration

In nutrient solutions, as generally in nature, phosphorus is exclusively present as phosphate. It seems quite natural that the concentration of this compound in the medium would influence the rate of its uptake by alga cells. Ketchum (1939a) found that in "*Nitzschia closterium*" (=*Phaeodactylum tricornutum*) the absorption of phosphate per cell in the light was dependent on its concentration in the medium, whereas the uptake by phosphorus-deficient cells in the dark was independent of phosphate concentration (Ketchum, 1939b). Scott (1945b) confirmed this latter observation in his experiments with *Chlorella pyrenoidosa* (except at limiting concentrations of phosphate). The phosphorus content of *C. pyrenoidosa* and *Scenedesmus* (Gaffron's strain D_3) cells was shown to depend strictly on the concentration of phosphate in the medium (Gest and Kamen, 1948); the same is true of the diatoms *Asterionella japonica* (Goldberg *et al.*, 1951) and *A. formosa* (Mackereth, 1953). When phosphate is supplied in excess the phosphorus content per cell is constant, as was shown for *C. pyrenoidosa* by Knauss and Porter (1954).

D. Hydrogen-Ion Concentration

The pH of the medium may alter the rate of phosphate uptake either by a direct effect on the permeability of the cell membrane or by changing the ionic form of the phosphate (Epstein, 1956). The first effect may be closely associated with changes in the activity of enzymes assumed to be situated in the surface layer of the protoplasm and believed to be actively engaged in the absorption process (Rothstein, 1954). The second effect could be explained as a selective absorption of one of the three ionic states of phosphate. This second effect, in particular, is presumed to be considerable, but the following experimental data on the influence of the external pH on phosphate absorption by algae are inconclusive. Scott (1945a) suspended *Chlorella pyrenoidosa* cells for 1 hour in sodium phosphate solutions of various pH-values, and found that the intracellular phosphorus increased significantly in more alkaline solutions. On the other hand, Wintermans (1955) showed that, in the absence of CO_2, illuminated cells of *C. vulgaris* fixed phosphorus only in acid media (around pH 4). In further studies with this species, Badour (1959) found the influence of the external pH on the rate of uptake of phosphate, and on its subsequent distribution into several fractions, to be different in the light and in the dark. Mackereth (1953) found that the greatest uptake of phosphate by *Asterionella formosa* took place at pH values between 6 and 7.

E. Other Factors

A reservoir of available carbohydrates may be of importance in phosphorus uptake because their degradation by oxidative phosphorylation provides energy for the uptake process. Scott (1945b) noted that *Chlorella pyrenoidosa* took up phosphate and potassium in a ratio of 1 atom of phosphorus to 1 atom of potassium, suggesting that both elements combine chemically with some cell constituent. Badour (1959) demonstrated that in *C. vulgaris*, both in the light and in the dark, the absorption of phosphate was promoted by potassium, possibly by its influence on the mechanism of phosphorylation (see Mechsner, 1959). In experiments with *Scenedesmus obliquus*, Krauss and Thomas (1954) noted that during the early stages of growth in a medium deficient of micronutrients the phosphorus content of the cells was appreciably higher than that of normal cells. However, the individual micronutrient responsible for this effect was not determined. For "*Nitzschia closterium*" (=*Phaeodactylum tricornutum*), the nitrate concentration of the medium is another factor influencing the rate of uptake of phosphate (Ketchum, 1939a). A somewhat puzzling mode of behavior of *Asterionella formosa* cells suspended in artificial solutions was reported by

Mackereth (1953): in pure dilute solutions of phosphate dissolved in laboratory-distilled water, very little phosphate was absorbed by the cells compared to that absorbed from natural waters. Since this effect could be partly overcome by adding lake water, it seemed unlikely to be due to toxic substances present in the laboratory water. Further investigations were inconclusive, and Mackereth, like Rodhe (1948), concluded merely that lake waters contain some factor not present in synthetic solutions.

The presence of known organic compounds in the medium may also influence phosphate uptake. Wassink *et al.* (1951b) and Wintermans (1955) observed a suppression of phosphorus uptake by *Chlorella vulgaris* in the light in the presence of glucose (0.2%), which they explained by postulating that the assimilation of glucose and the incorporation of phosphate compete for energy generated by the photochemical reactions of photosynthesis. Simonis and Kating (1956) found a great difference between the rate of uptake of phosphorus by *Ankistrodesmus braunii* cells supplied with glucose (0.15%) *before* being suspended in a phosphate medium and cells incubated with phosphate and glucose *simultaneously*. Whereas in the first case the rate of uptake of phosphate was increased by as much as 20%, in the second case the uptake was only slightly stimulated. Jacobi (1959) observed an accelerated uptake of phosphate by *A. braunii* in the presence of glycolic acid (possibly one of the first products of photosynthesis), suggesting that glycolic acid or one of its derivatives might act as a phosphate acceptor.

Certain algae can also use organic phosphorus compounds to satisfy their requirements for this element (see Provasoli, 1958).

III. Phosphate as an Ecological Factor

The growth of a wide variety of algae in their natural environment as well as in the laboratory has been shown to depend on the amount of available phosphorus. It therefore seems reasonable that the production of phytoplankton may be limited by the supply of phosphate. The ecological role of this anion has been reviewed by Ketchum (1954), Harvey (1957), Hutchinson (1957), Gessner (1958, 1959), Kalle (1958), and Provasoli (1958). To determine whether phosphate is a factor limiting plankton growth, many measurements of its concentration in natural waters have been taken; considerably fewer analyses have been made of the phosphorus content of the plankton alga cells themselves (Gerloff and Skoog, 1954; Al Kholy, 1956). This can vary widely in natural populations as well as in cultures (Lund, 1950), but for certain algae the minimum phosphorus content has been determined. The phosphorus of *Asterionella formosa* ranges between 6×10^{-8} and 4×10^{-6} μg. P per cell (Lund, 1950), which closely corresponds to the minimum requirement (6×10^{-8} μg. P per cell) determined in culture experiments by Mackereth (1953). The minimum amount for *A. japonica*, 5×10^{-8} μg. P per cell, is in the same range (Goldberg *et al.*, 1951). For *Chlorella pyrenoidosa*, Al Kholy (1956) determined both the minimum content, 1×10^{-7} μg. P per cell, and the maximum, 1.5×10^{-6} μg. P per cell. *Scenedesmus quadricauda* in a normal nutrient solution had a content of 8.7×10^{-6} μg. P per cell, which after 3 days in a phosphate-free solution fell to 9.2×10^{-7} μg. P per cell (Franzew, 1932).

A further approach to the problem of phosphate as a limiting factor involves culture experiments. Chu (1942) investigated the requirement of various planktonic algae. Using especially pure chemicals, he found that the nitrogen and phosphorus requirements were approximately the same for different planktonic algae. Good growth occurred in a nutrient solution with 0.1 to 2.0 p.p.m. phosphorus; concentrations below 0.05 p.p.m. limited growth; and an inhibitory effect was observed when the phosphorus exceeded 20 p.p.m. In a later investigation (1943) he found the optimum range of phosphorus concentration for the diatoms *Nitzschia palea* and *Tabellaria flocculosa* to be between 0.018 and 8.9 p.p.m., and those for *Pediastrum boryanum*, *Staurastrum paradoxum*, and *Botryococcus braunii* to fall between 0.09 and 17.8 p.p.m. The concentration required for optimum growth of the planktonic organisms Chu investigated was always higher than the highest concentrations to be found in ordinary natural waters. When nitrate was supplied as the nitrogen source, the upper limit for phosphorus was always higher than when ammonium salts were used.

Rodhe (1948) grouped freshwater algae into three categories according to whether their phosphorus tolerance ranges fall below, around, or above 20 μg. per liter:

1. Low, e.g., *Dinobryon divergens*, *Uroglena americana*
2. Medium, e.g., *Asterionella formosa*
3. High, e.g., *Scenedesmus quadricauda*

Most planktonic algae belong to category *1* or *2*, in which not only a deficiency of phosphate but also a concentration in excess of the upper limit can suppress growth. It seems of great ecological importance that some algae, when provided with a sufficient supply of phosphorus, can take up this element in quantities far in excess of their actual needs (Ketchum, 1939b; Scott, 1945b; Lund, 1950). With these surplus amounts they can then continue to grow even when the external phosphorus supply is depleted. Thus, Franzew (1932) found that *Scenedesmus quadricauda* and *Pandorina morum* grew for a certain time after phosphorus had been completely removed from the culture solution. Using other algae, Al Kholy (1956) and Rodhe (1948) confirmed his results. Gest and Kamen (1948) stated that such "excess phosphate" was not essential and had no influence on the rate of photosynthesis in *Chlorella pyrenoidosa*. Goldberg *et al.* (1951) found that in *Asterionella japonica* a considerable part of the excess phosphorus was loosely bound and readily exchangeable, while a minimum of the cellular content of phosphate was held very tightly. Gerloff and Skoog (1954) determined a 4-fold increase in the phosphorus content of *Microcystis aeruginosa* when the external supply of this element was sufficiently high. Correspondingly, in cultures of cells transferred from a medium with a high concentration to one lacking in phosphorus, growth increased approximately 300%.

The liberation of phosphorus compounds by the breakdown of plankton organisms is also of importance in the phosphate cycle of natural waters (Cooper, 1935; Seiwell and Seiwell, 1938; Hoffmann, 1952; Hoffmann and Reinhardt, 1952; Golterman, 1960; see also Talling, Chapter 50, this treatise), though it has been less extensively studied.

IV. Phosphorus Metabolism in Light and Darkness

Phosphorus compounds are involved in many essential reactions of plant metabolism (see Albaum, 1952). The specific participation of phosphorylated compounds and phosphorylation reactions in photosynthesis is clearly indicated by numerous investigations, several of which have employed algae; see reviews by Strehler (1952), Brown and Frenkel (1953), Arnon (1957, 1960), Kandler (1960), and Simonis (1960). Phosphorylated compounds are involved in both the "light" and "dark" reactions of photosynthesis. Emerson *et al.* (1944), following an earlier suggestion by Ruben (1943), proposed that the sole function of the light absorbed by the chlorophyll system is in the formation of "energy-rich" phosphate bonds, which are then available to drive all succeeding "dark" reactions. In spite of severe criticisms on thermodynamic grounds, much experimental evidence supported these suggestions, e.g., the influence of light and carbon dioxide on the absorption of phosphate and its incorporation into organic compounds (Aronoff and Calvin, 1948; Gest and Kamen, 1948; Wassink *et al.*, 1951a). To overcome interference by secondary reactions, which are to be expected in experiments of longer duration, Kandler (1950) followed only those changes in intracellular inorganic phosphate which occurred immediately after the transfer of *Chlorella pyrenoidosa* from light to darkness or *vice versa*. His results, like those obtained later by Wassink and Rombach (1954), pointed to the formation of energy-rich phosphate bonds by photochemical reactions. In further investigations, Kandler (1954, 1955, 1957) studied the interrelations between phosphate metabolism and photosynthesis under various conditions. Simonis and Kating (1956) and Simonis (1956, 1960) obtained evidence for light-dependent phosphorylation reactions in *Ankistrodesmus braunii*. The mounting body of evidence was further supported by the discovery of "photosynthetic phosphorylation" by Arnon *et al.* (1954) and Arnon (1960), who demonstrated that in isolated chloroplasts from higher plants light energy is used for esterification of inorganic phosphate according to the equation:

$$\text{(inorganic phosphate)} + \text{ADP} \xrightarrow{\text{light}} \text{ATP} \quad (1)$$

In regard to this process, the direct or indirect identification of ATP in algae and its subsequent changes in photosynthetic reactions are of special interest (Albaum *et al.*, 1950; Strehler, 1952, 1953; Goodman *et al.*, 1953; Schwinck, 1956; Bradley, 1957). The participation of phosphorylated compounds in the overall reactions of photosynthesis has also been indicated by the identification of various phosphorylated intermediates (Benson and Calvin, 1950; Gaffron and Fager, 1951; Buchanan *et al.*, 1952).

In view of the close dependence of photosynthesis on phosphate, it could be expected that a deficiency in phosphorus would influence photosynthesis, and, in turn, other metabolic reactions (see Pirson, 1955). In phosphorus-deficient cells of *Ankistrodesmus braunii*, Pirson *et al.* (1952) observed an inhibition of photosynthetic O_2 production which was at least partly reversible by the addition of phosphate. Respiration seemed to be slightly accelerated, especially in the early stages of recovery. In the first stages of P-deficiency, chlorophyll synthesis was apparently enhanced, but in later stages a marked chlorosis appeared. The latter effect was also observed in P-deficient thalli of *Hydrodictyon reticulatum* (Neeb, 1952). Other symptoms of this deficiency were the accumulation of fat, starch, and cell-wall substances, which indicated some interference with nitrogen metabolism. Accumulation of fat was also observed in P-deficient cells of *Chlorella vulgaris* (Bergmann, 1955), which oxidatively assimilated more glucose than did normal cells in spite of a decreased net rate of oxidation (Daniel, 1956).

V. Condensed Inorganic Phosphates

A. General Remarks

With very few exceptions, condensed inorganic phosphates have been found only in bacteria, fungi, and algae, where they have been shown to play important metabolic roles. For information on the chemistry and biology of these polymers the reader is referred to general reviews by Schmidt (1951), Wiame (1958), and Kuhl (1960).

B. Types of Condensed Inorganic Phosphates

Thilo (1959) classified the condensed inorganic phosphates according to their chemical properties, distinguishing between *metaphosphates* with closed-ring molecular structures, of which only the tri- and tetrameta-phosphates have been recognized chemically (Figs. 1 and 2), and *poly-phosphates* with open chains comprising up to 10^6 phosphate units (Fig. 3). Although this distinction has not always been maintained in the published biological literature, we now know that most naturally occurring condensed phosphates are of the latter type. Since the first demonstration of poly-phosphates in an alga, *Chlorella pyrenoidosa*, by Sommer and Booth (1938), these compounds have been found in many other algae, as listed in Table I. Cyclic metaphosphates have been reported only in *Euglena gracilis* (Lohmann, 1958).

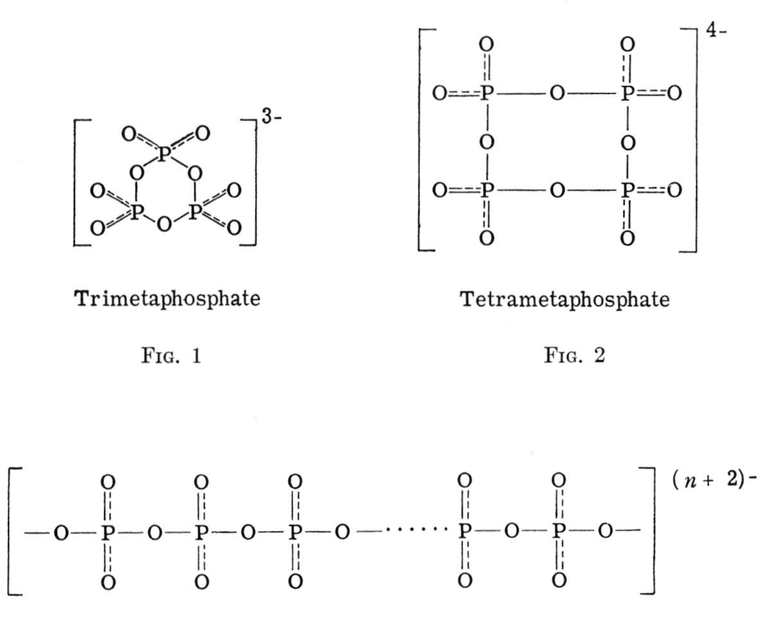

Trimetaphosphate

FIG. 1

Tetrametaphosphate

FIG. 2

Polyphosphate

FIG. 3

Most naturally occurring polyphosphates, which can be detected by both chemical and cytochemical methods, exist in either of two forms: one, easily extracted with cold trichloroacetic acid (TCA); the other, insoluble in cold TCA but extractable with hot TCA, perchloric acid, or dilute

TABLE I

ALGAE IN WHICH POLYPHOSPHATES HAVE BEEN DEMONSTRATED[a]

CYANOPHYTA	
Anabaena variabilis	*Oscillatoria amoena*
Cylindrospermum licheniforme	*O. limosa*
Gloeothece sp.	*Phormidium ambiguum*
Lyngbya aerugineo-coerulea	*P. frigidum*
L. amplivaginata	*P. uncinatum*
Oscillatoria sp.	
CHLOROPHYTA	
Acetabularia mediterranea	*Hydrodictyon reticulatum*
Chlorella sp.	*Mougeotia* sp.
C. ellipsoidea	*Oedogonium* sp.
C. pyrenoidosa	*Rhopalocystis oleifera*
C. vulgaris	*Scenedesmus* sp.
Cladophora sp.	*Spirogyra* sp.
Cosmarium sp.	*Ulothrix* sp.
Enteromorpha sp.	*Zygnema* sp.
BACILLARIOPHYTA	XANTHOPHYTA
Fragilaria sp.	*Vaucheria* sp.
Navicula sp.	
CRYPTOPHYTA	EUGLENOPHYTA
Chilomonas sp.	*Euglena gracilis*
RHODOPHYTA	CHAROPHYTA
Ceramium sp.	*Chara* sp.

[a] See Kuhl (1960).

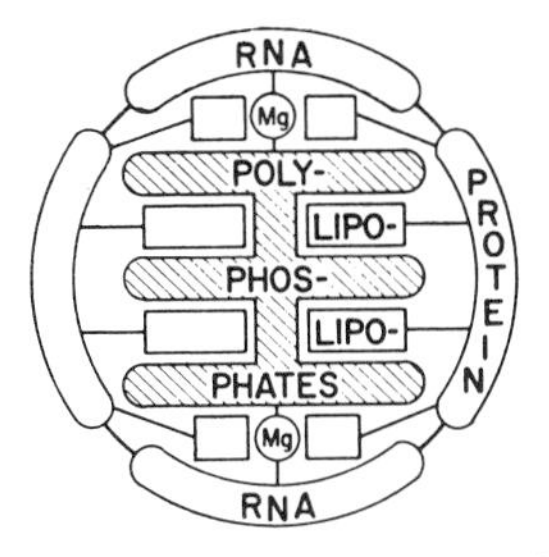

FIG. 4. Diagram indicating possible composition of a bacterial polyphosphate (volutin) granule. After Widra (1959).

alkaline or neutral salt solutions. (For details of extraction procedures see Wintermans, 1955.) The two fractions also differ in their physiological behavior, the TCA-insoluble polyphosphates being apparently more actively involved in metabolism (Wiame, 1949). Both forms are easily hydrolyzed within 7 minutes by 1 *N* HCl at 100° C. and are therefore characterized as "7-minute phosphates."

Polyphosphates show a metachromatic reaction with toluidine blue (Albaum *et al.*, 1950; Goodman *et al.*, 1955; Wintermans, 1955; Nihei, 1955), and can be precipitated from solution with barium salts at low pH values. Using paper chromatography, Thilo *et al.* (1956) demonstrated that the TCA-soluble polyphosphates from *Acetabularia mediterranea* have more than 10 phosphate residues in the molecule. Goodman *et al.* (1955) fractionated the phosphates from *Scenedesmus* sp. by anion-exchange chromatography, and found that the polyphosphates in this alga represent 21% of the total phosphate content.

"Volutin granules" or "metachromatic bodies" have long been recognized as normal constituents of the cells of many microorganisms. Wiame (1946, 1947) and Ebel (1949) showed that they are mainly composed of polyphosphates, and Stich (1953) reported on cytochemical reactions for their identification, including a characteristic metachromatic coloration with toluidine blue (Moreau, 1913; Keck and Stich, 1957). Since the number and size of the "polyphosphate granules" vary widely with the physiological state of the cell (Stich, 1955), they are evidently not mere deposits of excess phosphate, but are actively involved in metabolic processes. Besides the polyphosphates, the metachromatic granules of algae possibly contain other components, such as those indicated by Widra (1959) for bacterial volutin granules (see Fig. 4).

C. The Synthesis of Polyphosphates in Algae

Stich (1953) found that both the number and size of polyphosphate granules in *Acetabularia mediterranea* tended to increase in the light, whereas they decreased in the dark. This led him to suggest a possible connection between photosynthesis and the synthesis of polyphosphate. In support of this he showed that 24 hours after exposing the algae to $P^{32}O_4$ in the light, radiophosphorus accumulated in the granules. In darkness—or in light under the influence of 2,4-dinitrophenol (DNP)—this accumulation was reduced; it was completely suppressed by cyanide. He assumed that polyphosphate synthesis can proceed only if photosynthesis and oxidative phosphorylation are unimpaired (Stich, 1955). The first thorough investigation of the conditions of polyphosphate synthesis in algae was that of Wintermans (1955), who demonstrated that suspensions of *Chlorella vulgaris* cells in the light converted orthophosphate into intracellular polyphosphate, mostly TCA-insoluble. This phosphate fixation could continue for several hours, and was greater in the absence of CO_2 than in its presence. For this reaction light saturation occurred at a much lower intensity than for photosynthesis of similar algal suspensions supplied with 5% CO_2. The maximum accumulation of polyphosphates was observed at a pH of about 4; the rate fell to zero as the pH approached 7–8. The reaction did not require oxygen, and nitrate had no effect, but glucose was somewhat inhibitory. The formation of polyphosphate in the light was much less sensitive to phenylurethane than was photosynthesis, whereas DNP, sodium azide, and sodium fluoride inhibited photosynthesis and polyphosphate formation to about the same degree. Wintermans therefore concluded that energy-rich phosphate groups, formed in the photochemical reaction of photosynthesis, can be accumulated as polyphosphate if the further reactions of photosynthesis are curtailed by lack of CO_2; with this interpretation Wassink (1957) concurred.

Nihei (1955) found that, on illuminating synchronized cultures of *Chlorella ellipsoidea* in the "light cell" stage (L_1—see Hase, Chapter 40, this treatise) in the absence of CO_2, a considerable uptake of phosphate occurred with a simultaneous evolution of oxygen. The photochemically fixed phosphorus was shown to have been incorporated mainly into the fraction insoluble in cold TCA but to be extractable within 5 minutes by a 5% solution of TCA at 100° C. The properties of the major portion of the fixed phosphorus suggest strongly that it was in the form of polyphosphate. Nihei (1957) also found that in synchronous cultures the most active assimilation of phosphorus into polyphosphate took place during the life-cycle stage between "dark-phase cells" and "mature light-phase cells." In the latter cell type the phosphorus in the TCA-insoluble fraction attained its highest level, after which it decreased sharply. He concluded (*1*) that the photochemical incorporation of P in the acid-insoluble fraction probably takes place *via* some substance, presumably ATP, which occurs in the acid-soluble fraction, and (*2*) that the phosphorus incorporated in the acid-insoluble fraction may be further transformed by light-independent reactions, perhaps being transferred in part to the proteins in the residual fraction.

Badour (1959) established that the addition of potassium to K-deficient *Chlorella vulgaris*, in the absence of carbon dioxide and nitrate (the reduction of which requires metabolic energy), accelerated the uptake of phosphorus. Labile "7-minute phosphates" (mainly polyphosphate) accumulated, especially in the light. The importance of potassium in the process of photophosphorylation, which in turn leads to the formation of labile 7-minute phosphates, was demonstrated by Mechsner (1959) in experiments with the same species. However in cultures deficient in magnesium or manganese Badour found a considerable accumulation of phosphorus in the 7-minute fraction. In manganese-deficient cells, large numbers of polyphosphate granules were revealed by cytological stains, but this was not the case in magnesium-deficient cultures.

D. The Role of Polyphosphates in Metabolism

Meyerhof *et al.* (1953) and Yoshida (1955) showed experimentally that the energy of the P—O—P bonds in the polyphosphate molecule is of the same order of magnitude as that of the so-called energy-rich phosphate bond. This provides strong support to a theory first proposed by Hoffmann-Ostenhof and Weigert (1952), who considered the biological functions of the polyphosphates to involve the storing of both phosphate and energy. Polyphosphates should thus act as reservoirs of high-energy phosphate from excess ATP according to the equation:

$$\text{ATP} + (\text{PO}_3^-)_n \rightleftarrows \text{ADP} + (\text{PO}_3^-)_{n+1} \qquad (2)$$

Convincing evidence for this hypothesis is provided by the discovery that some heterotrophic microorganisms, e.g., *Escherichia coli* and *Saccharomyces cerevisiae*, possess enzymes which are able reversibly to transfer phosphate from ATP to polyphosphate (Ebel and Dirheimer, 1957; Kornberg, 1957; Hoffman-Ostenhof and Slechta, 1958). Though this does not prove that polyphosphate is a general storage substance for energy, it is possible that high-energy phosphate stored in this form could be used for some special energy-requiring reactions, such as the synthesis of nucleic acids and proteins, or for processes associated with cell division.

In algae, polyphosphate generally accumulates when the energy from the photochemical processes of photosynthesis cannot be used for normal synthetic or other energy-requiring reactions. However, there is no direct indication that algal cells are able to satisfy even a part of their energy requirement by using polyphosphate as a source of stored energy.

For a detailed treatment of this subject the reader is referred to Kuhl (1960).

ACKNOWLEDGMENTS

For helpful suggestions during the preparation of this manuscript the author is indebted to Dr. Helen M. Habermann (Goucher College, Baltimore, Maryland) visiting investigator at the Institute of Plant Physiology, Göttingen, during the summer of 1960.

REFERENCES

Albaum, H. G. (1952). Metabolism of phosphorylated compounds in plants. *Ann. Rev. Plant Physiol.* **3,** 35–58.

Albaum, H. G., Schatz, A., Hutner, S. H., and Hirshfeld, A. I. (1950). Phosphorylated compounds in *Euglena. Arch. Biochem.* **29,** 210–218.

Al Kholy, A. A. (1956). On the assimilation of phosphorus in *Chlorella pyrenoidosa. Physiol. Plantarum* **9,** 137–143.

Arnon, D. I. (1957). Phosphorus metabolism and photosynthesis. *Ann. Rev. Plant Physiol.* **7,** 325–354.

Arnon, D. I. (1960). The chloroplast as a functional unit in photosynthesis. *In* "Handbuch der Pflanzenphysiologie" (W. Ruhland, ed.), Vol. V, pp. 773–829. Springer, Berlin.

Arnon, D. I., Whatley, F. R., and Allen, M. B. (1954). Photosynthesis by isolated chloroplasts. II. Photosynthetic phosphorylation, the conversion of light into phosphate bond energy. *J. Am. Chem. Soc.* **76,** 6324–6329.

Aronoff, S., and Calvin, M. (1948). Phosphorus turnover and photosynthesis. *Plant Physiol.* **23,** 351–358.

Badour, A. S. S. (1959). Analytisch-chemische Untersuchung des Kaliummangels bei *Chlorella* im Vergleich mit anderen Mangelzuständen. Dissertation, University of Göttingen, Germany.

Baslavskaya, S. S., and Weber, H. (1959). [The effect of light upon transformations of phosphates in plants]. *Doklady Acad. Nauk S.S.S.R.* **124,** 227–230. (In Russian.)

Benson, A. A., and Calvin, M. (1950). Carbon dioxide fixation by green plants. *Ann. Rev. Plant Physiol.* **1,** 25–42.

Bergmann, L. (1955). Stoffwechsel und Mineralsalzernährung einzelliger Grünalgen. II. Vergleichende Untersuchungen über den Einfluss mineralischer Faktoren bei heterotropher und mixotropher Ernährung. *Flora (Jena)* **142,** 493–539.

Bradley, D. F. (1957). Phosphate transients in photosynthesis. *Arch. Biochem. Biophys.* **68,** 172–185.

Brauner, L. (1956). Die Beeinflussung des Stoffaustausches durch das Licht. *In* "Handbuch der Pflanzenphysiologie" (W. Ruhland, ed.), Vol. II, pp. 381–397. Springer, Berlin.

Brown, A. H., and Frenkel, A. W. (1953). Photosynthesis. *Ann. Rev. Plant Physiol.* **4,** 23–58.

Buchanan, J. G., Bassham, J. A., Benson, A. A., Bradley, D. F., Calvin, M., Daus, L. L., Goodman, M., Hayes, P. M., Lynch, V. H., Norris, L. T., and Wilson, A. T. (1952). The path of carbon in photosynthesis. XVII. Phosphorus compounds as intermediates in photosynthesis. *In* "Phosphorus Metabolism" (W. C. McElroy and B. Glass, eds.), Vol. II, pp. 440–466. Johns Hopkins Press, Baltimore, Maryland.

Chu, S. P. (1942). The influence of the mineral composition of the medium on the growth of planktonic algae. I. Methods and culture media. *J. Ecol.* **30,** 284–325.

Chu, S. P. (1943). The influence of the mineral composition of the medium on the growth of planktonic algae. II. The influence of the concentration of inorganic nitrogen and phosphate phosphorus. *J. Ecol.* **31,** 109–148.

Cooper, L. H. N. (1935). The rate of liberation of phosphate in sea water by the breakdown of plankton organisms. *J. Marine Biol. Assoc. United Kingdom* **20,** 197–202.

Daniel, A. L. (1956). Stoffwechsel und Mineralsalzernährung einzelliger Grünalgen. III. Atmung und oxydative Assimilation von *Chlorella*. *Flora (Jena)* **143,** 31–66.

Ebel, J. P. (1949). Participation de l'acide métaphosphorique à la constitution des bactéries et des tissus animaux. *Compt. rend. acad. sci.* **228,** 1312–1313.

Ebel, J. P., and Dirheimer, G. (1957). Relations métaboliques entre polyphosphates inorganiques et adénosine di- et triphosphates. *Compt. rend. soc. biol.* **151,** 979–981.

Eilers, H. (1926). Zur Kenntnis der Ernährungsphysiologie von *Stichococcus bacillaris* Näg. *Rec. trav. botan. néerl.* **23,** 362–395.

Emerson, R. L., Stauffer, J. F., and Umbreit, W. W. (1944). Relationship between phosphorylation and photosynthesis in *Chlorella*. *Am. J. Botany* **31,** 107–120.

Epstein, E. (1956). Uptake and ionic environment (including external pH). *In* "Handbuch der Pflanzenphysiologie" (W. Ruhland, ed.), Vol. II, pp. 398–408. Springer, Berlin.

Franzew, A. W. (1932). Ein Versuch der physiologischen Erforschung der Produktionsfähigkeit des Moskauflusswassers. *Mikrobiologiya* **1,** 122–130. (In Russian. Summarized by Rodhe, 1948.)

Gaffron, H., and Fager, E. W. (1951). The kinetics and chemistry of photosynthesis. *Ann. Rev. Plant Physiol.* **2,** 87–114.

Gerloff, G. C., and Skoog, G. (1954). Cell contents of nitrogen and phosphorus as a measure of their availability for growth of *Microcystis aeruginosa*. *Ecology* **35,** 348–353.

Gessner, F. (1958). Die Binnengewässer. *In* "Handbuch der Pflanzenphysiologie" (W. Ruhland, ed.), Vol. IV, pp. 179–232. Springer, Berlin.

Gessner, F. (1959). "Hydrobotanik," Vol. II: Stoffhaushalt. Deutscher Verlag der Wissenschaften, Berlin.

Gest, H., and Kamen, M. D. (1948). Studies on the phosphorus metabolism of green algae and purple bacteria in relation to photosynthesis. *J. Biol. Chem.* **176,** 299–318.

Goldberg, E. D., Walker, T. J., and Whisenand, A. (1951). Phosphate utilization by diatoms. *Biol. Bull.* **101,** 274–284.

Golterman, H. L. (1960). Studies on the cycle of elements in fresh water. *Acta Botan. Neerl.* **9,** 1–58.

Goodman, M., Bradley, D. F., and Calvin, M. (1953). Phosphorus and photosynthesis. I. Differences in light and dark incorporation of radiophosphate. *J. Am. Chem. Soc.* **75,** 1962–1967.

Goodman, M., Benson, A. A., and Calvin, M. (1955). Fractionation of phosphates from *Scenedesmus* by anion exchange. *J. Am. Chem. Soc.* **77,** 4257–4261.

Harvey, H. W. (1957). "The Chemistry and Fertility of Sea Waters," 2nd ed. Univ. Press, Cambridge, London and New York.

Hoffmann, C. (1952). Weitere Beiträge zur Kenntnis der Remineralisierung des Phosphors bei Meeresalgen. *Planta* **42,** 156–176.

Hoffmann, C., and Reinhardt, M. (1952). Zur Frage der Remineralisation des Phosphors bei Benthosalgen. *Kiel. Meeresforsch.* **8,** 135–144.

Hoffmann-Ostenhof, O., and Slechta, T. (1958). Transferring enzymes in the metabolism of inorganic polyphosphates and pyrophosphate. *Proc. Intern. Symposium Enzyme Chem., Tokyo and Kyoto* **2,** pp. 180–189.

Hoffmann-Ostenhof, O., and Weigert, W. (1952). Über die mögliche Funktion des polymeren Metaphosphats als Speicher energiereichen Phosphats in der Hefe. *Naturwissenschaften* **39,** 303–304.

Hutchinson, G. E. (1957). The phosphorus cycle in lakes. "A Treatise on Limnology," Vol. I, chapter 13, pp. 727–752. Wiley, New York.

Jacobi, G. (1959). Über den Zusammenhang von Glykolsäure und lichtabhängiger Phosphorylierung. *Planta* **53,** 402–411.

Kalle, K. (1958). Das Meerwasser als Mineralstoffquelle der Pflanze. *In* "Handbuch der Pflanzenphysiologie" (W. Ruhland, ed.), Vol. IV, pp. 170–178. Springer, Berlin.

Kandler, O. (1950). Über die Beziehungen zwischen Phosphathaushalt und Photosynthese. I. Phosphatspiegelschwankungen bei *Chlorella pyrenoidosa* als Folge des Licht-Dunkel-Wechsels. *Z. Naturforsch.* **5b,** 423–437.

Kandler, O. (1954). Über die Beziehungen zwischen Phosphathaushalt und Photosynthese. II. Gesteigerter Glukoseeinbau im Licht als Indikator einer lichtabhängigen Phosphorylierung. *Z. Naturforsch.* **9b,** 625–644.

Kandler, O. (1955). Über die Beziehungen zwischen Phosphathaushalt und Photosynthese. III. Hemmungsanalyse der lichtabhängigen Phosphorylierung. *Z. Naturforsch.* **10b,** 38–46.

Kandler, O. (1957). Über die Beziehungen zwischen Phosphathaushalt und Photosynthese. IV. Zur Frage einer stöchiometrischen Beziehung zwischen CO_2-Reduktion und Phosphatumsatz. *Z. Naturforsch.* **12b,** 271–280.

Kandler, O. (1960). Energy transfer through phosphorylation mechanism in photosynthesis. *Ann. Rev. Plant Physiol.* **11,** 37–54.

Keck, K., and Stich, H. (1957). The widespread occurrence of polyphosphate in lower plants. *Ann. Botany (London)* **21,** 611–619.

Ketchum, B. H. (1939a). The absorption of phosphate and nitrate by illuminated cultures of *Nitzschia closterium. Am. J. Botany* **26,** 399–407.

Ketchum, B. H. (1939b). The development and restoration of deficiencies in the phosphorus and nitrogen composition of unicellular plants. *J. Cellular Comp. Physiol.* **13,** 373–381.

Ketchum, B. H. (1954). Mineral nutrition of phytoplankton. *Ann. Rev. Plant Physiol.* **5,** 55–74.

Knauss, H. J., and Porter, J. W. (1954). The absorption of inorganic ions by *Chlorella pyrenoidosa. Plant Physiol.* **29,** 229–234.

Kornberg, S. R. (1957). Adenosine triphosphate synthesis from polyphosphate by an enzyme from *Escherichia coli. Biochim. et Biophys. Acta* **26,** 294–300.

Kramer, P. J. (1956). The uptake of salts by plant cells. *In* "Handbuch der Pflanzenphysiologie" (W. Ruhland, ed.), Vol. II, pp. 290–315. Springer, Berlin.

Krauss, R. W. (1958). Physiology of the freshwater algae. *Ann. Rev. Plant Physiol.* **9,** 207–244.

Krauss, R. W., and Thomas, W. H. (1954). The growth and inorganic nutrition of *Scenedesmus obliquus* in mass culture. *Plant Physiol.* **29,** 205–214.

Kuhl, A. (1957). Über den Phosphathaushalt von *Hydrodictyon reticulatum*. Dissertation, University of Marburg, Germany.

Kuhl, A. (1960). Die Biologie der kondensierten anorganischen Phosphate. *In* "Ergebnisse der Biologie" (H. Autrum, ed.), Vol. 23, pp. 144–186. Springer, Berlin.

Lohmann, K. (1958). Über das Vorkommen kondensierter Phosphate in Lebewesen. *In* "Kondensierte Phosphate in Lebensmitteln" (Symposium April 5–6, 1957, Mainz), pp. 29–44. Springer, Berlin.

Lund, J. W. G. (1950). Studies on *Asterionella formosa* Hass. II. Nutrient depletion and the spring maximum. Part I. Observations on Windermere, Esthwaite Water and Blelham Tarn. Part II. Discussion. *J. Ecol.* **38,** 1–14, 15–35.

Mackereth, F. J. (1953). Phosphorus utilization by *Asterionella formosa* Hass. *J. Exptl. Botany* **4,** 296–313.

Mechsner, K. (1959). Untersuchungen an *Chlorella vulgaris* über den Einfluss der Alkali-Ionen auf die Lichtphosphorylierung. *Biochim. et Biophys. Acta* **33,** 150–158.

Meyerhof, O., Shafas, R., and Kaplan, A. (1953). Heat of hydrolysis of trimetaphosphate. *Biochim. et Biophys. Acta* **12,** 121–127.

Moreau, F. (1913). Les corpuscules métachromatiques chez les algues. *Bull. soc. botan. France* **60,** 123–126.

Myers, J. (1951). Physiology of the algae. *Ann. Rev. Microbiol.* **5,** 157–180.

Neeb, O. (1952). *Hydrodictyon* als Objekt einer vergleichenden Untersuchung physiologischer Grössen. *Flora (Jena)* **139,** 39–95.

Nihei, T. (1955). A phosphorylative process, accompanied by photochemical liberation of oxygen, occurring at the stage of nuclear division in *Chlorella* cells. I. *J. Biochem. (Tokyo)* **42,** 245–256.

Nihei, T. (1957). A phosphorylative process, accompanied by photochemical liberation of oxygen, occurring at the stage of nuclear division in *Chlorella* cells. II. *J. Biochem. (Tokyo)* **44,** 389–396.

Pirson, A. (1955). Functional aspects in mineral nutrition of green plants. *Ann. Rev. Plant Physiol.* **6,** 71–114.

Pirson, A., and Kuhl, A. (1958). Dependence on light of the reversible uptake of inorganic phosphate by *Hydrodictyon*. *Nature* **181,** 921–922.

Pirson, A., Tichy, C., and Wilhelmi, G. (1952). Stoffwechsel und Mineralsalzernährung einzelliger Grünalgen. I. Vergleichende Untersuchungen an Mangelkulturen von *Ankistrodesmus*. *Planta* **40,** 199–253.

Provasoli, L. (1958). Nutrition and ecology of protozoa and algae. *Ann. Rev. Microbiol.* **12,** 279–308.

Robertson, R. N. (1958). The uptake of minerals. *In* "Handbuch der Pflanzenphysiologie" (W. Ruhland, ed.), Vol. IV, pp. 243–279. Springer, Berlin.

Rodhe, W. (1948). Environmental requirements of freshwater plankton algae. *Symbolae Botan. Upsalienses* **10,** 1–149.

Rothstein, A. (1954). The enzymology of the cell surface. *In* "Protoplasmatologia" (L. V. Heilbrunn and F. Weber, eds.), Vol. II: Cytoplasma. E. Cytoplasmaoberfläche, pp. 1–86. Springer, Vienna.

Ruben, S. (1943). Photosynthesis and phosphorylation. *J. Am. Chem. Soc.* **65,** 279–282.

Schmidt, G. (1951). The biochemistry of inorganic pyrophosphates and metaphosphates. *In* "Phosphorus Metabolism" (W. C. McElroy and B. Glass, eds.), Vol. I, pp. 443–475. Johns Hopkins Press, Baltimore, Maryland.

Schwinck, L. (1956). Nachweis von Adenosintriphosphorsäure (ATP) in Grünalgen und *Helodea,* sowie Einbau von radioaktivem Phosphor (P^{32}) bei der Photosynthese. *Planta* **47,** 165–218.

Scott, G. T. (1945a). The influence of H-ion concentration on the mineral composition of *Chlorella pyrenoidosa. J. Cellular Comp. Physiol.* **25,** 37–44.

Scott, G. T. (1945b). The mineral composition of phosphate deficient cells of *Chlorella pyrenoidosa* during the restoration of phosphate. *J. Cellular Comp. Physiol.* **26,** 35–42.

Seiwell, H. R., and Seiwell, G. E. (1938). The sinking of decomposing plankton in sea water and its relationship to oxygen consumption and phosphorus liberation. *Proc. Am. Phil. Soc.* **78,** 465–482.

Simonis, W. (1956). Untersuchungen zur lichtabhängigen Phosphorylierung. II. *Z. Naturforsch.* **11b,** 354–363.

Simonis, W. (1960). Photosynthese und lichtabhängige Phosphorylierung. *In* "Handbuch der Pflanzenphysiologie" (W. Ruhland, ed.), Vol. V, pp. 966–1013. Springer, Berlin.

Simonis, W., and Kating, H. (1956). Untersuchungen zur lichtabhängigen Phosphorylierung. I. Die Beeinflussung der lichtabhängigen Phosphorylierung von Algen durch Glucosegaben. *Z. Naturforsch.* **11b,** 115–172.

Sommer, A. L., and Booth, T. E. (1938). Meta- and pyrophosphate within the algal cell. *Plant Physiol.* **13,** 199–205.

Stich, H. (1953). Der Nachweis und das Verhalten von Metaphosphaten in normalen verdunkelten und trypaflavinbehandelten Acetabularien. *Z. Naturforsch.* **8b,** 36–44.

Stich, H. (1955). Synthese and Abbau der Polyphosphate von *Acetabularia* nach autoradiographischen Untersuchungen des P^{32}-Stoffwechsels. *Z. Naturfosch.* **10b,** 282–284.

Strehler, B. L. (1952). Photosynthesis-energetics and phosphate metabolism. *In* "Phosphorus Metabolism" (W. C. McElroy and B. Glass, eds.), Vol. II, pp. 491–506. Johns Hopkins Press, Baltimore, Maryland.

Strehler, B. L. (1953). Firefly luminescence in the study of energy transfer mechanism. II. Adenosine triphosphate and photosynthesis. *Arch. Biochem. Biophys.* **43,** 67–79.

Tamiya, H., Iwamura, T., Shibata, K., Hase, E., and Nihei, T. (1953). Correlation between photosynthesis and light independent metabolism in the growth of *Chlorella. Biochim. et Biophys. Acta.* **12,** 23–40.

Thilo, E. (1959). Die kondensierten Phosphate. *Naturwissenschaften* **46,** 367–373.

Thilo, E., Grunze, H., Hämmerling, J., and Werz, G. (1956). Über Isolierung und Identifizierung der Polyphosphate aus *Acetabularia mediterranea. Z. Naturforsch.* **11b,** 266–270.

Wassink, E. C. (1957). Phosphate in the photosynthetic cycle in *Chlorella. In* "Research in Photosynthesis" (H. Gaffron, ed.), pp. 333–339. Interscience, New York.

Wassink, E. C., and Rombach, J. (1954). Preliminary report on experiments dealing with phosphate metabolism in the induction phase of photosynthesis in *Chlorella. Koninkl. Ned. Akad. Wetenschap. Proc. Ser. C* **57,** 493–497.

Wassink, E. C., Wintermans, J. F. G. M., and Tjia, J. E. (1951a). Phosphate exchange in *Chlorella* in relation to conditions for photosynthesis. *Koninkl. Ned. Akad. Wetenschap. Proc. Ser. C* **54,** 41–52.

Wassink, E. C., Wintermans, J. F. G. M., and Tjia, J. E. (1951b). The influence of glucose on the changes in TCA-soluble phosphates in *Chlorella* suspensions in relation to conditions of photosynthesis. *Koninkl. Ned. Akad. Wetenschap. Proc. Ser. C* **54,** 496–502.

Wiame, J. M. (1946). Basophilie et métabolisme du phosphore chez la levure. *Bull. soc. chim. biol.* **28,** 552–556.

Wiame, J. M. (1947). Etude d'une substance polyphosphorée, basophile et métachromatique chez les levures. *Biochim. et Biophys. Acta* **1,** 234–255.

Wiame, J. M. (1949). The occurrence and physiological behaviour of two metaphosphate fractions in yeast. *J. Biol. Chem.* **178,** 919–929.

Wiame, J. M. (1958). Accumulation de l'acide phosphorique (Phytine, Polyphosphates). *In* "Handbuch der Pflanzenphysiologie" (W. Ruhland, ed.), Vol. IX, pp 136–148. Springer, Berlin.

Widra, A. (1959). Metachromatic granules of microorganisms. *J. Bacteriol.* **78,** 664–670.

Wintermans, J. F. G. M. (1955). Polyphosphate formation in *Chlorella* in relation to photosynthesis. *Mededel. Landbouwhogeschool Wageningen* **55,** 69–126.

Yoshida, A. (1955). Studies on metaphosphate. II. Heat of hydrolysis of metaphosphate extracted from yeast cells. *J. Biochem.* (*Tokyo*) **42,** 165–168.

—13—

Nucleotides and Nucleic Acids

TATSUICHI IWAMURA

Institute of Applied Microbiology, Tokyo University, and Tokugawa Institute for Biological Research, Tokyo, Japan

I. Introduction [1]

Although the literature contains only a few scattered references on the subject, there is little reason to doubt that RNA and DNA generally have functions in algae similar to those in other plants and animals.

II. Nucleic Acid Metabolism and Algal Growth

A. Steady-State Growth

When grown in limited culture media, algae, like other microorganisms, typically pass through three stages of growth, i.e., the lag, exponential, and stationary phases. In the exponential phase or in the steady state of growth,

[1] The following abbreviations have been used in this chapter: DNA, deoxyribonucleic acid; RNA, ribonucleic acid; A, adenine; C, cytosine; G, guanine; T, thymine; U, uracil.

all the cellular constituents are synthesized in constant proportions, and nucleic acids are no exception. In a steady-state culture of *Polytomella coeca* limited by phosphorus supply, Jeener (1952a, 1953) showed that the syntheses of RNA and of protein proceeded in a fixed ratio, whereas in transitional phases of growth, i.e., at the end of the lag and logarithmic phases, the RNA/protein ratio varied (cf. Section II.C).

B. Synchronous Growth

Changes in the nucleic acid content of *Chlorella ellipsoidea* cells in synchronous culture (see Hase, Chapter 40, this treatise) were studied by Iwamura (1955). RNA was found to be synthesized in proportion to the increase of cellular mass from the beginning of the culture, and almost throughout the life cycle the RNA/protein ratio remained constant. On the other hand, the DNA did not increase appreciably during the growth phase, and began to be synthesized vigorously only at a later stage just prior to cell division. At this stage, the net increase of DNA continued even when the cells were incubated in the dark; meanwhile the RNA content decreased, while the amount of protein showed less change. The *free* nucleotides in *Chlorella* cells were found to consist mainly of mono-, di-, and triphosphates of the four common ribonucleosides (adenosine, guanosine, cytidine, and uridine), among which those of adenosine and uridine were dominant (Iwamura, 1959; Iwamura and Myers, 1959). Despite the conspicuous changes of metabolic state in the course of the life-cycle, the nitrogenous bases of the nucleic acids and of free nucleotides in the cells were found to maintain a constant balance (Iwamura and Myers, 1959).

In *Chlorella pyrenoidosa,* the results by Lorenzen and Ruppel (1960) were similar, except for variations in the RNA/protein ratios. Whether these differences were due to differences in the species examined or in the techniques used for RNA assay or synchronization of growth remains to be seen (cf. Iwamura, 1955; Smillie and Krotkov, 1960a). A similar pattern of DNA synthesis was indicated by cytological observations of nuclear division in a synchronous culture of *Chlamydomonas reinhardtii* (Sueoka, 1960). By using the technique of CsCl density-gradient centrifugation, Sueoka also found that mitotic replication of the DNA molecules occurred in the same manner as in bacteria (Meselson and Stahl, 1958).

C. Effects of Change in Environmental Conditions

The nucleic acids, like other cell constituents, are quantitatively affected by changes of environmental conditions. For instance, in both *Chlorella* and *Acetabularia mediterranea* the ratio of RNA to protein was found to be higher in the light than in the dark (Iwamura, 1955; Richter, 1959b; Lorenzen and Ruppel, 1960). Jeener (1952b) observed that when *Poly-*

tomella cells grown in a phosphorus-deficient medium were transferred to a medium containing phosphorus, the RNA at first decreased and then increased, while protein continued to increase. Newmark and Fujimoto (1959), working with *Chlorella*, reported that nitrogen deficiency caused a decrease of RNA and chlorophyll, whose nitrogen seemed to be utilized for protein synthesis. A marked decrease in the free nucleotides as well as in the RNA and DNA of *Monodus subterraneus* cells occurred under conditions of nitrogen deficiency (Fogg. 1959). Hase *et al.* (1958) found that, when smaller, younger cells of a synchronized *Chlorella* culture were transferred from a normal to a sulfur-free medium, they enlarged to some extent while the RNA and DNA content approximately doubled. Such cells, however, were incapable of cell division. Further synthesis of nucleic acids was evoked when a minute amount of sulfate ($10^{-5}M$) was supplied to the sulfur-starved cells. It was concluded that sulfur plays an essential role in the process of DNA-formation.

D. Effects of Some Chemicals

Brachet (1958) observed that the growth and regeneration of *Acetabularia* were inhibited by various nucleic acid derivatives and their analogs, which he attributed to a disturbance of nucleic acid metabolism. Cells of *Chlorella vulgaris* and *Scenedesmus obliquus* grown in deuterium oxide media showed cytological evidence for disturbances in nucleic acid metabolism and in mitosis (Flauenhaft *et al.*, 1960).

E. Role of the Nucleus

With the unicellular, uninucleate Green Alga *Acetabularia*, extensive cytological and biochemical studies have been performed by Brachet and others (cf. Richter, Chapter 42, this treatise). Investigations of the nuclear control of RNA and protein synthesis in this alga include those of Brachet and Szafarz (1953), Brachet *et al.* (1955), Hämmerling *et al.* (1958), Richter (1959a, b), and Naora *et al.* (1959). Richter (1959a, b) showed that in a growing enucleated portion of the cell the RNA content remained unchanged or rose only slightly, though the soluble protein increased, whereas in a nucleate portion both RNA and protein were actively synthesized. It was therefore concluded that the nucleus plays an important, if not decisive, role in the net synthesis of cytoplasmic RNA.

III. Base Composition of Algal Nucleic Acids

A. General Remarks

Methods for the extraction of nucleic acids from unicellular algae have now been refined by a number of workers (see, for example, Schwinck, 1960;

Smillie and Krotkov, 1960a, b). It has become apparent that biological species differ to some extent in the base composition of their DNA. Although the ratios A/T and G/C tend to be unity, in several Green Algae, as in many bacteria, (G + C) exceeds (A + T), whereas in some diatoms, as in most higher organisms, the reverse is true (Chargaff, 1955; Belozersky and Spirin, 1958; Iwamura and Myers, 1959; Serenkov and Pachomova, 1959; Sueoka, 1960). So far neither 5-methylcytosine nor 5-hydroxymethylcytosine has been reported in the DNA of any alga (see also Biswas, 1956, 1960; Low, 1958).

The base composistion of RNA in the algal cells so far examined has been found to show a predominance of (G + C) or (G + A), due to an appreciable excess of G (Brawerman and Chargaff, 1959b; Iwamura, 1959; Serenkov and Pachomova, 1959; Biswas and Myers, 1960). The RNA of both *Euglena* and *Chlorella* (Brawerman and Chargaff, 1959b; Iwamura and Myers, 1959; Iwamura, 1959) contains two minor components besides the four major nucleotides (adenylic, guanylic, cytidylic, and uridylic acids). One is so-called "pseudo-uridylic acid" (Cohn, 1957) or "the fifth nucleotide" (Davis and Allen, 1957); the other has not yet been identified. In *Chlorella* the relative content of pseudo-uridylic acid was highest (3.0%) in the "supernatant RNA" (Iwamura, 1960). In the RNA of *Anacystis nidulans**, Biswas and Myers (1960) found a new cytidine derivative, which contained 2-*O*-methylribose.

B. Base Composition and Metabolic State

There is now considerable evidence that nucleic acids somehow direct the synthesis of proteins: RNA is actively involved in the condensation of amino acids, while both DNA and RNA may determine the specificity of the proteins synthesized. Thus, by controlling the complement of enzymes, the quantities and qualities of nucleic acids in a cell may exert an appreciable influence on its metabolic state.

Brawerman and Chargaff (1959a, b) compared the base composition of RNA in green, etiolated, and permanently bleached strains of *Euglena gracilis*. They noted that the turnover of both RNA and protein, as evidenced respectively by the incorporation of labeled adenine and leucine, increased during chloroplast formation by etiolated cells transferred to the light, and they concluded that specific RNA molecules were formed during the greening process. Similar results were obtained with this organism by Smillie and Krotkov (1960b). Iwamura and Myers (1959) showed, with synchronously grown *Chlorella*, that base compositions of RNA, DNA, and free nucleotides differed according to the metabolic state of the cells, i.e., during mass increase, during DNA synthesis, or during starvation in dark-

* Regarding the taxonomy of this organism, see Note 2 of the Appendix to this treatise by Silva.

ness. Likewise Newmark and Fujimoto (1959) reported a change in the base composition of *Chlorella pyrenoidosa* RNA after incubation of the exponentially growing cells in a nitrogen-free medium, which shifted the metabolism from protein synthesis to fat synthesis.

IV. Heterogeneity

The heterogeneity of the nucleic acids of an organism can be demonstrated either by fractionation of the isolated nucleic acids or by examining the distribution of nucleic acids among subcellular components. By differential centrifugation Jeener (1952a) separated ribonucleoprotein particles of *Polytomella* into several fractions, which showed differences in the ratio of RNA to protein. Correll *et al.* (1960) similarly isolated and fractionated RNA-polyphosphate complexes from *Anabaena variabilis* cells.

Iwamura (1960) fractionated disrupted *Chlorella ellipsoidea* cells into three subcellular components: broken chloroplasts, ribonucleoprotein particles, and the supernatant. RNA was found to be distributed in all three fractions, whereas DNA occurred chiefly in the broken chloroplasts and the supernatant. The ribonucleoprotein fraction consisted mainly of three kinds of particles, with sedimentation coefficients ($s_{200,w}$) around 35*S*, 50*S*, and 75*S*, respectively (cf. Newmark *et al.*, 1959). The RNA in the supernatant contained more cytidylic and pseudo-uridylic acids and less adenylic and guanylic acids than the RNA in the pellets. The relative amounts of these nucleic acid fractions changed during the life cycle of the alga (Iwamura, unpublished). Newmark *et al.* (1959) noted that, when *Chlorella pyrenoidosa* cells were transferred to a nitrogen-free medium, the amount of 75*S* particles which could be isolated was decreased markedly, but was restored to its original level when the cells were returned to a nitrogenous medium.

The existence of DNA in two subcellular components separable by centrifugation was also demonstrated in *Scenedesmus* (Gaffron's strain D-3) (Iwamura, unpublished). The DNA in the pellet is possibly associated with the chloroplasts, while that in the supernatant may be derived from the nuclei. These two forms of DNA differ in base composition as well as in metabolic activity; in the pellet the DNA from illuminated cells shows much faster turnover in terms of phosphorus than the DNA in the supernatant. Evidence for the existence of DNA in or on the chloroplasts has also been presented by other workers. Thus Brachet (1958) observed incorporation of H^3-thymidine into chloroplasts of *Acetabularia*, and a similar result was reported by Stocking and Gifford (1959) for *Spirogyra*. Flauenhaft *et al.* (1960) cytochemically detected DNA at the periphery of the chloroplasts in *Chlorella vulgaris* cells. According to Biswas (1958), the Blue-green Alga *Nostoc muscorum* also contains two kinds of DNA, differing in base composition and metabolic activity as well as in their solubility in acid and alkali. All these observations, together with those reported for higher plants, indicate that the association of some DNA with chloroplasts (or with their equivalent in the Blue-green Algae) is a general phenomenon, and suggest that it may be actively involved in certain functions of the photosynthetic apparatus.

V. Concluding Remarks

The study of the nucleic acids of algae is a newly developed field, and there still remain many problems for further investigation. For example, the enzymological aspect of nucleic-acid metabolism has so far remained entirely unexplored. Studies with algal species other than those mentioned in this review are also much needed.

References

Belozersky, A. N., and Spirin, A. S. (1958). A correlation of deoxyribonucleic and ribonucleic acids. *Nature* **182,** 111–112.

Biswas, B. B. (1956). Chemical nature of nucleic acids in Cyanophyceae. *Nature* **177,** 95–96.

Biswas, B. B. (1958). Studies on the metabolism of nucleic acid in *Nostoc muscorum*. *In* "Symposium on Recent Advances in the Study of Plant Metabolism," pp. 33–34. University of Allahabad, India.

Biswas, B. B. (1960). Comparative studies on DNA of *Chlorella ellipsoidea* and *Anacystis nidulans*. *Plant Physiol.* **35,** XXX (Supplement).

Biswas, B. B., and Myers, J. (1960). A methyl cytidine from the ribonucleic acid of *Anacystis nidulans*. *Nature* **186,** 238–239.

Brachet, J. (1958). New observations on biochemical interactions between nucleus and cytoplasm in *Amoeba* and *Acetabularia*. *Exptl. Cell Research Suppl.* **6,** 78–96.

Brachet, J., and Szafarz, D. (1953). L'incorporation d'acide orotique radioactif dans des fragments nucléés et anucléés d'*Acetabularia mediterranea*. *Biochim. et Biophys. Acta* **12,** 588–589.

Brachet, J., Chantrenne, H., and Vanderhaeghe, F. (1955). Recherches sur les interactions biochimiques entre le noyau et le cytoplasme chez les organismes unicellulaires. II. *Acetabularia mediterranea*. *Biochim. et Biophys. Acta* **18,** 544–563.

Brawerman, G., and Chargaff, E. (1959a). Changes in protein and ribonucleic acid during the formation of chloroplasts in *Euglena gracilis*. *Biochim. et Biophys. Acta* **31,** 164–171.

Brawerman, G., and Chargaff, E. (1959b). Relation of ribonucleic acid to the photosynthetic apparatus in *Euglena gracilis*. *Biochim. et Biophys. Acta* **31,** 172–177.

Chargaff, E. (1955). Isolation and composition of the deoxypentose nucleic acids and of the corresponding nucleoproteins. *In* "The Nucleic Acids" (E. Chargaff and J. N. Davidson, eds.), Vol. I, pp. 307–371. Academic Press, New York.

Cohn, W. E. (1957). Minor constituents of ribonucleic acids. *Federation Proc.* **16,** 166.

Correll, D. L., Tolbert, N. E., and Ball, R. C. (1960). RNA-polyphosphate complexes from *Anabaena variabilis. Plant Physiol.* **35,** xxx (supplement).

Davis, F. F., and Allen, F. W. (1957). Ribonucleic acid from yeast which contains a fifth nucleotide. *J. Biol. Chem.* **227,** 907–915.

Flauenhaft, E., Conrad, S. M., and Katz, J. J. (1960). Nucleic acid in some deuterated green algae. *Science* **132,** 892–894.

Fogg, G. E. (1959). Nitrogen nutrition and metabolic patterns in algae. *In* "Utilization of Nitrogen and Its Compounds by Plants" (H. K. Porter, ed.), *Symposia Soc. Exptl. Biol. No.* **13,** 106–125.

Hämmerling, J., Clauss, H., Keck, K., Richter, G., and Werz, G. (1958). Growth and protein synthesis in nucleated and enucleated cells. *Exptl. Cell Research Suppl.* **6,** 210–226.

Hase, E., Morimura, Y., Mihara, S., and Tamiya, H. (1958). The role of sulfur in the cell division of *Chlorella. Arch. Mikrobiol.* **31,** 87–95.

Iwamura, T. (1955). Change of nucleic acid content in Chlorella cells during the course of their life-cycle. *J. Biochem.* (*Tokyo*) **42,** 575–589.

Iwamura, T. (1959). Change in content and distribution of purines and pyrimidines in *Chlorella* during its life-cycle. *Federation Proc.* **18,** 252.

Iwamura, T. (1960). Distribution of nucleic acids among subcellular fractions of *Chlorella. Biochim. et Biophys. Acta* **42,** 161–163.

Iwamura, T., and Myers, J. (1959). Change in the content and distribution of the nucleic acid bases in *Chlorella* during the life-cycle. *Arch. Biochem. Biophys.* **84,** 267–277.

Jeener, R. (1952a). Studies on the evolution of nucleoprotein fractions of the cytoplasm during the growth of a culture of *Polytomella coeca.* I. Ribonucleic acid content of cells and growth rate. *Biochim. et Biophys. Acta* **8,** 125–133.

Jeener, R. (1952b). Studies on the evolution of nucleoprotein fractions of the cytoplasm during the growth of a culture of *Polytomella coeca.* II. Rate of synthesis of nucleic acid, studied by means of labelled phosphate. *Biochim. et Biophys. Acta* **8,** 270–282.

Jeener, R. (1953). Ribonucleic acid and protein synthesis in continuous cultures of *Polytomella coeca. Arch. Biochem. Biophys.* **43,** 381–388.

Lorenzen, H., und Ruppel, H. G. (1960). Versuche zur Gliederung des Entwicklungsverlaufs der *Chlorella*-Zelle. *Planta* **54,** 394–403.

Low, E. M. (1958). Composition of the nucleic acids of some algae. *Nature* **182,** 1096.

Meselson, M., and Stahl, F. W. (1958). The replication of DNA in *Escherichia coli. Proc. Natl. Acad. Sci. U. S.* **44,** 671–682.

Naora, H., Richter, G., and Naora, H. (1959). Further studies on the synthesis of RNA in enucleate *Acetabularia mediterranea. Exptl. Cell Research* **16,** 434–436.

Newmark, P., and Fujimoto, Y. (1959). Nucleic acids of nitrogen deficient *Chlorella pyrenoidosa. Federation Proc.* **18,** 293.

Newmark, P., Stephens, J. D., Curry, J. B., and Bower, G. L. (1959). Macromolecular components of *Chlorella pyrenoidosa. Nature* **184,** 963–965.

Richter, G. (1959a). Das Verhalten von Ribonucleinsäure und löslichen Cytoplasmaproteinen in UV-bestrahlten kernhaltigen und kernlosen Zellen von *Acetabularia. Z. Naturforsch.* **14b,** 100–104.

Richter, G. (1959b). Die Auswirkung der Zellkern-Entfernung auf die Synthese von Ribonucleinsäure und Cytoplasmaproteinen bei *Acetabularia mediterranea. Biochim. et Biophys. Acta* **34,** 407–419.

Schwinck, I. (1960). Isolation of DNA from *Chlamydomonas reinhardii*. *J. Protozool.* **7,** 294–297.

Serenkov, G. P., and Pachomova, M. V. (1959). [Physiology and biochemistry of plants. Nucleotide components of DNA and RNA from some algae and higher plants.] *Nauch. Doklady Vyssheĭ Schkoly. Biol. Nauki* **4,** 156–161. (In Russian.)

Smillie, R. M., and Krotkov, G. (1960a). The estimation of nucleic acids in some algae and higher plants. *Can. J. Botany* **38,** 31–49.

Smillie, R. M., and Krotkov, G. (1960b). Phosphorus-containing compounds in *Euglena gracilis* grown under different conditions. *Arch. Biochem. Biophys.* **89,** 83–90.

Stocking, C. R., and Gifford, E. M., Jr. (1959). Incorporation of thymidine into chloroplasts of *Spirogyra*. *Biochem. Biophys. Research Communs.* **1,** 159–164.

Sueoka, N. (1960). Mitotic replication of deoxyribonucleic acid in *Chlamydomonas reinhardii*. *Proc. Natl. Acad. Sci. U. S.* **46,** 83–91.

—14—

Sulfur

JEROME A. SCHIFF

Department of Biology, Brandeis University, Waltham, Massachusetts

I. Introduction

Sulfur exists in nature in several oxidation states, from the most highly reduced form in H_2S and such amino acids as cysteine and methionine to the most common highly oxidized form in sulfate. The most abundant form in natural waters is sulfate, which occurs at a concentration of 0.025 *M* in sea water, ranges from zero to 10^{-4} *M* in fresh water, and reaches 0.4 *M* in some saline lakes (see Hutchinson, 1957). Most algae, like most other microorganisms and higher plants, can utilize and reduce sulfate as their sole source of sulfur.

Gibbs and Schiff (1960) have reviewed the comparative biochemistry of sulfur metabolism and the energy relations of the compounds involved. Although sulfides accumulate in some environments—such as the waters of certain mineral springs, anaerobic marine or freshwater muds, parts of the Black Sea, deep Norwegian fjords, and some lakes and ocean "deeps" (see Wiame, 1958)—there have been no studies on the sulfur nutrition of algae from such environments to determine whether H_2S can be utilized as sole sulfur source for growth. (On the possible utilization of H_2S oxidation as an energy source, see Spruit, Chapter 3, this treatise.) Investigations of the permeability and the intracellular accumulation of sulfate and H_2S are not very informative in this connection, since the degree of sulfur utilization for the synthesis of other compounds remains unknown.

Lewin (1954) presented evidence for the role of reduced sulfur compounds in the uptake of silicon by diatoms, and referred to earlier observations by Harvey (1938–39), Matsudaira (1942), and others on the stimulatory action of sulfides on the growth of certain planktonic algae.

II. Sulfates in Algae

A. Free Sulfates

Experiments on the uptake of the sulfate ion or its exclusion from algal cells were reviewed by Blinks (1951). A later study, using *Chlorella pyrenoidosa*, was carried out by Knauss and Porter (1954). Data on the accumulation of sulfate have been largely confined to certain coenocytic algae, from which it is easy to remove large quantities of vacuolar sap for chemical analyses. Marine species of *Valonia* and *Halicystis* exclude sulfate almost completely. Freshwater species of *Nitella* accumulate appreciable amounts of sulfate, in addition to 1% of the dry weight in the form of nonvacuolar sulfur (Hoagland and Davis, 1923). *Chara ceratophylla*, a plant of brackish waters, is intermediate between marine and freshwater algae in its low accumulation of sulfate (Osterhout, 1931).

Blinks (1951) also reviewed the literature on the exceptionally high acidity of the cells of the marine alga *Desmarestia*. Extracts of some species were found to have pH values of 0.8 to 1.8, which was attributed to the presence of free malic and sulfuric acids. According to Miwa (1953), *D. ligulata*, *D. viridis*, and *D. tabacoides* contain only traces of malic acid but are extremely rich in sulfuric acid. More recent work by Meeuse (1956) showed that the acidity of *D. munda* is due mainly to the dissociation of 0.025 meq. of sulfuric acid, the quantities of organic acids being too low and their pK values too high to account for the observed acidity. Eppley and Bovell (1958) found a concentration of 0.44 N sulfuric acid in *D. munda*, and, by staining with a pH indicator dye, localized the acid in the cell vacuoles. This was consistent with their finding that the reduction of methylene blue by extracts of the plant proceeds at neutrality but not at a low pH, which indicates that the cytoplasm of *D. munda*, like that of most organisms, is probably not far from neutral. Earlier observations on the discoloration of bruised plants support this explanation, since in the undamaged plant the chloroplasts must presumably be insulated from the acid (Blinks, 1951).

B. Sulfate Esters

Polysaccharides esterified with sulfuric acid occur in the Green, Brown, and Red Algae (see O'Colla, Chapter 20, this treatise). The presence of adenosine-3′-phosphate-5′-phosphosulfate in some marine algae (Goldberg and Delbrück, 1959) suggests that this or a similar compound may be active in their synthesis, as it is in incorporation of sulfate into mucopolysaccharides by hen oviduct tissues (Suzuki and Strominger, 1960). Choline sulfate has been isolated from various Red Algae by Lindberg (1955).

III. Other Sulfur Compounds in Algae

In addition to the common amino acids methionine, cystine, and cysteine, some unusual S-containing acids may be found in algae, such as sulfoxythiazine carboxylic acid (see Fowden, Chapter 11, this treatise). Although thioctic (α-lipoic) acid has been invoked as a key metabolite in photosynthesis (Grisebach *et al.*, 1957), it induced no stimulation of CO_2 fixation in short-term experiments with *Chlorella pyrenoidosa* and *Scenedesmus obliquus* (Biswas and Sen, 1958). Some data on the thiamine specificity of algae are presented elsewhere in this treatise by Droop (Chapter 8).

Several interesting "internal salts" which contain sulfur at the oxidation level of sulfite have been isolated from algae. Taurine, *N*-methyl taurine, and *N*-dimethyl taurine have been found in various Red Algae (Lindberg, 1955). *Polysiphonia fastigiata* has yielded 2-L-amino-3-hydroxy-1-propane sulfonic acid (Wickberg, 1957), while *N*-(D-2,3-dihydroxy-*n*-propyl)taurine ("glyceryl taurine") has been isolated from *Gigartina leptorhynchos* (Wickberg, 1956). A sulfolipid isolated from *Chlorella* and other sources has been reported by Benson and is discussed in Chapter 22 of this treatise.

IV. Sulfur Metabolism

A. Sulfate Reduction

There are very few studies on sulfate reduction in algae. Schiff (1959) demonstrated a rapid reduction of sulfate by *Chlorella pyrenoidosa* (Emerson strain). He found evidence for the accumulation of adenosine-3′-phosphate-5′-phosphosulfate ($PAPSO_4$) and *S*-adenosyl methionine when the reduction was inhibited by iodoacetamide, suggesting that the reduction of sulfate might proceed on nucleotide carriers of this type. Though methionine-S^{35} can be incorporated into protein (Schiff, 1956), it does not participate significantly in the synthesis of other sulfur compounds, suggesting that the enzyme which forms *S*-adenosyl methionine from methionine and ATP in rabbit liver (Cantoni, 1953) is probably absent from *C. pyrenoid-*

osa, and that *S*-adenosyl methionine must be formed in some other manner. This is supported by the finding that methionine will not serve as a sulfur source for the growth of this species of *Chlorella*. Inhibition of sulfate reduction by iodoacetamide occurs at several metabolic sites in this organism. If the inhibitor is added before the addition of sulfate, then the uptake of sulfate into the cells is suppressed (Schiff, 1959). Iodoacetamide may either be interfering with sulfate activation, which in other systems has been shown to be sensitive to sulfhydryl inhibitors (Hilz and Lipmann, 1955), or it may merely be suppressing systems which provide energy for the accumulation. If iodoacetamide is added *simultaneously* with sulfate, sulfate is accumulated by the cells but its reduction is suppressed. If the inhibitor is added 10 minutes *after* sulfate, then cysteine and glutathione are converted to *S*-cysteine acetamide and *S*-glutathione acetamide, thereby blocking the incorporation of these substances into protein (Schiff, unpublished).

Wedding and Black (1960), using S^{35}-labeled sulfate, obtained chromatographic evidence for the presence of $PAPSO_4$ in *Chlorella pyrenoidosa*. In support of an active transport mechanism for sulfate uptake, they found that: (1) relatively little of the radiosulfate can be displaced from the cells by desorption with nonradioactive sulfate; (2) the uptake of radiosulfate is dependent on temperature; and (3) within a few minutes at 25° C. the concentration of radiosulfate inside the cells exceeds the external concentration. To assess the importance of oxidative phosphorylation in the process, they compared the effect of dinitrophenol on the sulfate uptake of light-grown cells with its effect on cells first starved for 24 hours in darkness and then suspended in a medium containing sucrose, and concluded that "active" sulfate is not involved in sulfate accumulation. However, it should be borne in mind that light-grown cells are not physiologically comparable with cells starved in darkness (see Hase, Chapter 40, this treatise), and that the depletion of energy stores in the starved cells might retard the uptake of sucrose itself.

Mechanisms of sulfate reduction in other organisms have been investigated by Hilz *et al.* (1959) and Peck (1959).

B. Reduced Forms of Sulfur

1. *Amino Acids and Proteins*

As would be expected, cystine and methionine are normal constituents of the proteins of algae, each constituting 1 to 2% on a *N* basis (Fogg, 1953); they have also been detected as free amino acids in *Chlorella pyrenoidosa* (Wedding and Black, 1960; Schiff, unpublished).

Mandels (1943) observed that sulfur deficiency, in a strain of *Chlorella* designated "Wann #11," resulted in a decrease in the chlorophyll content per cell before division completely ceased. Addition of sulfate to sulfur-deficient cells resulted in an early resumption of chlorophyll synthesis within 5 hours, and of growth within 24 hours. Similar effects followed the addition of suitable quantities of sulfite, thiosulfate, pyrosulfate, or persulfate; H_2S was also effective, but only after a lag, which indicates that it might be necessary that H_2S undergo preliminary oxidation in the medium before being utilized. Addition of cysteine, glutathione, methionine, thioglycolic acid, thiamine, or thiocyanate did not induce recovery. Schiff (1956) found that methionine would not serve as the sole sulfur source for cells of *Chlorella pyrenoidosa*, which became chlorotic in the absence of assimilable sulfur, whereas cystine as sole sulfur source permitted some growth and did not lead to sulfur chlorosis. Contrariwise Shrift (1954a) showed that *Chlorella vulgaris* can utilize D- or L-methionine quite well, but cannot use cystine.

The role of sulfur in the process of cell division in *Chlorella ellipsoidea* was investigated, by different techniques, by Hase (see Chapter 40, this treatise).

2. Sulfonium Compounds

Compounds containing the sulfonium configuration have been isolated from various algae. Earlier observations on the odor of marine algae led Bywood and Challenger (1953; also Challenger *et al.*, 1957) to confirm the evolution of dimethyl sulfide by *Polysiphonia fastigiata* (=*P. lanosa*), *P. nigrescens*, and other organisms. These authors showed that its liberation was probably due to the enzymatic breakdown of some precursor in the plant, since boiling the thalli completely inhibited the formation of dimethylsulfide, whereas grinding or treatment with a dilute base accelerated its evolution. The precursor of dimethyl sulfide in *Polysiphonia fastigiata* was shown to be dimethyl propiothetin (dimethyl-2-carboxyethylsulfonium hydroxide) (Challenger and Simpson, 1948), which was isolated also from *Enteromorpha intestinalis* and *Acrosiphonia centralis* (Challenger *et al.*, 1957). Indirect evidence of its presence in many other marine algae was obtained. Cantoni and Anderson (1956) found that enzymatic cleavage of the thetin to dimethyl sulfide preceeds as follows:

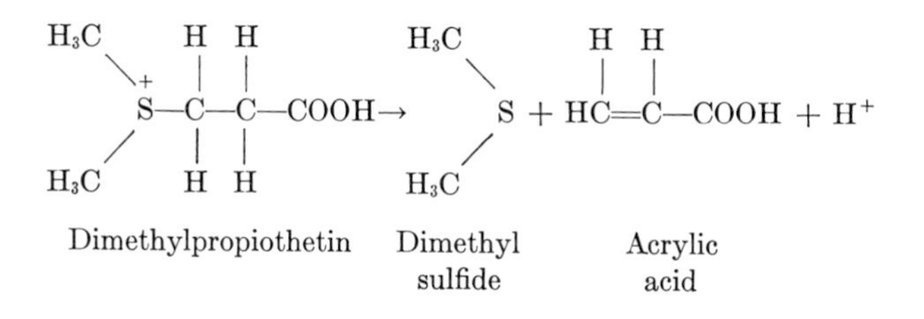

The action of the enzyme on dimethylacetothetin was much slower, while *S*-adenosyl methionine, *S*-dimethylhomocysteine, betaine, and choline were not attacked at all. It may be noted that in liver cells dimethylpropiothetin and dimethylthetin can serve as methyl donors for methionine synthesis from homocysteine (Dubnoff and Borsook, 1948).

S-adenosyl methionine, another sulfonium compound, was found in extracts of *Chlorella pyrenoidosa* by Schiff (1959), who showed that it became labeled when radioactive sulfate was being reduced by the cells.

C. Selenium Inhibition

Shrift (1954a, 1958) showed that selenate is a competitive inhibitor of sulfate utilization for growth in *Chlorella vulgaris*, inhibiting both the absorption of this ion and subsequent sulfur metabolism. Although methionine partially reversed selenate inhibition, it did not do so competitively. Selenomethionine, on the other hand, competed with the absorption and utilization of methionine but did not competitively interfere with sulfate uptake (Shrift, 1954b). When cells were inoculated into a medium containing selenomethionine, cell division was temporarily arrested; but, after division recommenced, cells from such a culture showed no lag on being transferred to a fresh medium containing selenomethionine. Such cells retained the ability to divide without a lag in the presence of selenomethionine even after many generations in the absence of this substance. More recent evidence suggests that the adaptation of cells to selenomethionine involves a transformation of all cells rather than a selection of mutants (Shrift *et al.*, 1961a, b).

During the lag period that followed the transfer of *Chlorella* cells to a medium containing selenomethionine and labeled sulfate, giant cells were formed which continued to synthesize protein and to incorporate sulfate into protein and other fractions (Shrift, 1954b, 1959). Paper chromatography of the protein hydrolyzates, however, showed that methionine was missing, although after the resumption of cell division it was again found in the proteins. Apparently methionine is necessary for normal protein synthesis and cell division in *Chlorella vulgaris*, and the selenium analog temporarily prevents its utilization. How this effect is overcome remains unknown.

References

Biswas, B. B., and Sen, S. P. (1958). Thioctic acid and photosynthetic fixation of carbon dioxide. *Nature* **181**, 1219–1220.

Blinks, L. R. (1951). Physiology and biochemistry of algae. *In* "Manual of Phycology" (G. M. Smith, ed.), pp. 263–284. Chronica Botanica, Waltham, Massachusetts.

Bywood, R., and Challenger, R. (1953). The evolution of dimethyl sulfide by *Enteromorpha intestinalis*. *Biochem. J.* **53,** xxvi.

Cantoni, G. L. (1953). *S*-adenosyl methionine, a new intermediate formed enzymatically from L-methionine and adenosine triphosphate. *J. Biol. Chem.* **204,** 403–416.

Cantoni, G. L., and Anderson, D. G. (1956). Enzymatic cleavage of dimethyl propiothetin by *Polysiphonia lanosa*. *J. Biol. Chem.* **222,** 171–177.

Challenger, F., Bywood, R., Thomas, P., and Hayward, B. J. (1957). The natural occurrence and chemical reactions of some thetins. *Arch. Biochem. Biophys.* **69,** 514–523.

Challenger, R., and Simpson, M. I. (1948). Studies on biological methylation. Part XII. A precursor of dimethyl sulfide evolved by *Polysiphonia fastigata*. Dimethyl-2-carboxyethylsulphonium hydroxide and its salt. *J. Chem. Soc.* **2,** 1591–1597.

Dubnoff, J. W., and Borsook, H. (1948). Dimethylthetin and dimethyl propiothetin in methionine synthesis. *J. Biol. Chem.* **176,** 789–796.

Eppley, R. W., and Bovell, C. R. (1958). Sulfuric acid in *Desmarestia*. *Biol. Bull.* **115,** 101–106.

Fogg, G. E. (1953). "The Metabolism of Algae," p. 101. Wiley, New York.

Gibbs, M., and Schiff, J. A. (1960). The energy relations of chemoautotrophic organisms. *In* "Plant Physiology" (F. C. Steward, ed.), Vol. 1b. Academic Press, New York.

Goldberg, I. H., and Delbrück, A. (1959). Transfer of sulfate from 3′-phosphoadenosine-5′-phosphosulfate to lipids, mucopolysaccharides and aminoalkyl phenols. *Federation Proc.* **18,** 235.

Grisebach, H., Fuller, R. C., and Calvin, M. (1957). Metabolism of thioctic acid in algae. *Biochim. et Biophys. Acta* **23,** 34–42.

Harvey, H. W. (1938–1939). Substances controlling the growth of a diatom. *J. Marine Biol. Assoc. United Kingdom* **23,** 499–520.

Hilz, H., and Lipmann, F. (1955). The enzymatic activation of sulfate. *Proc. Natl. Acad. Sci. U. S.* **41,** 880–890.

Hilz, H., Kittler, M., and Knape, G. (1959). Die Reduktion von Sulfat in der Hefe. *Biochem. Z.* **332,** 151–166.

Hoagland, D. R., and Davis, A. R. (1923). The composition of the cell sap of the plant in relation to the absorption of ions. *J. Gen. Physiol.* **5,** 629–646.

Hutchinson, G. E. (1957). "A Treatise on Limnology," Vol. I: Geography, Physics and Chemistry. Wiley, New York.

Knauss, H. J., and Porter, J. W. (1954). The absorption of inorganic ions by *Chlorella pyrenoidosa*. *Plant Physiol.* **29,** 229–234.

Lewin, J. C. (1954). Silicon metabolism in diatoms. I. Evidence for the role of reduced sulfur compounds in silicon utilization. *J. Gen. Physiol.* **5,** 589–599.

Lindberg, B. (1955). Methylated taurines and choline sulfate in Red Algae. *Acta Chem. Scand.* **9,** 1323–1326.

Mandels, G. R. (1943). A quantitative study of chlorosis in *Chlorella* under conditions of sulfur deficiency. *Plant Physiol.* **18,** 449–462.

Matsudaira, T. (1942). On inorganic sulphides as growth-promoting ingredient for diatoms. *Proc. Imp. Acad.* (*Tokyo*) **17,** 107–116.

Meeuse, B. J. D. (1956). Free sulfuric acid in the Brown Alga, *Desmarestia*. *Biochim. et Biophys. Acta.* **19,** 372–374.

Miwa, T. (1953). Occurrence of abundant free sulphuric acid in *Desmarestia*. *Proc. Pacific Sci. Congr. Pacific Sci. Assoc., 7th Congr.* **5** (*Botany*), 78–80.

Osterhout, W. J. V. (1931). Physiological studies of single plant cells. *Biol. Rev.* **6,** 369–411.

Peck, H. D., Jr. (1959). The ATP-dependent reduction of sulfate with hydrogen in extracts of *Desulfovibrio desulfuricans. Proc. Natl. Acad. Sci. U. S.* **45,** 701–708.

Schiff, J. A. (1956). Preliminary studies on the sulfur metabolism of *Chlorella pyrenoidosa* with sulfur-35. Ph.D. Thesis, University of Pennsylvania, Philadelphia, Pennsylvania. (Dissertation Abstr. No. 17,271.)

Schiff, J. A. (1959). Studies on sulfate utilization by *Chlorella pyrenoidosa* using sulfate-S^{35}. The occurrence of S-adenosyl methionine. *Plant Physiol.* **34,** 73–80.

Shrift, A. (1954a). Sulfur-selenium antagonism. I. Anti-metabolite action of selenate on the growth of *Chlorella vulgaris. Am. J. Botany* **41,** 223–230.

Shrift, A. (1954b). Sulfur-selenium antagonism. II. Antimetabolite action of selenomethionine on the growth of *Chlorella vulgaris. Am. J. Botany* **41,** 345–352.

Shrift, A. (1958). Biological activities of selenium compounds. *Botan. Rev.* **24,** 550–583.

Shrift, A. (1959). Nitrogen and sulfur changes associated with growth uncoupled from cell division in *Chlorella vulgaris. Plant Physiol.* **34,** 505–512.

Shrift, A., Nevyas, J., and Turndorf, S. (1961a). Mass adaptation to selenomethionine in populations of *Chlorella vulgaris. Plant Physiology* **36,** 506–509.

Shrift, A., Nevyas, J., and Turndorf, S. (1961b). Stability and reversibility of adaptation to selenomethionine in *Chlorella vulgaris. Plant Physiol.* **36,** 509–519.

Suzuki, S., and Strominger, J. L. (1960). Enzymatic sulfation of mucopolysaccharides in hen oviduct. *J. Biol. Chem.* **235,** 257–273.

Wedding, R. T., and Black, M. K. (1960). Uptake and metabolism of sulfate by *Chlorella*. I. Sulfate accumulation and active sulfate. *Plant Physiol.* **35,** 72–80.

Wiame, J. M. (1958). Le cycle du soufre dans la nature. *In* "Handbuch der Pflanzenphysiologie" (W. Ruhland, ed.), Vol. IX, pp. 115–116. Springer, Berlin.

Wickberg, B. (1956). Isolation of *N*-(D-2,3-dihydroxy-*n*-propyl)taurine from *Gigartina leptorhynchos. Acta Chem. Scand.* **10,** 1097–1099.

Wickberg, B. (1957). Isolation of 2-L-amino-3-hydroxy-1-propane sulphonic acid from *Polysiphonia fastigiata. Acta Chem. Scand.* **11,** 506–511.

—15—

Halogens

T. I. SHAW
The Laboratory, Marine Biological Association, Plymouth, England

I. Iodine

A. Occurrence and Distribution

In 1811 Courtois discovered the new element iodine in the ash of brown seaweeds, and for 30 years the ash, or kelp, provided the industrial source of the element (Chapman, 1950).

Phaeophyta are particularly rich in iodine; in the Laminariales and Desmarestiales its concentration may occasionally reach 30,000 times that in sea water, and can constitute 1% of the dry weight (Black, 1948). Some Rhodophyta, particularly *Ptilota*, and some Chlorophyta, notably *Codium intricatum*, also contain an appreciable quantity of iodine. However, not all marine algae concentrate the element so intensely; thus *Enteromorpha compressa*, with an iodine content of only 7 mg./kg. dry weight, contains little more than the marine flowering plant *Zostera* (McClendon, 1939). Of freshwater algae less is known, although dried *Vaucheria* is recorded as containing 3.4 mg./kg. iodine, about a hundred times the concentration in many flowering plants (von Fellenberg, 1926).

Algae of related species have widely differing iodine contents, and may vary greatly within a species (Grimm, 1952). Seasonal fluctuations are most marked in sublittoral species, and the time of maximum iodine content differs from one species to another (Black, 1948, 1949; Freundler *et al.*, 1921). Young plants are generally richer in iodine than are old plants, and those from exposed coasts tend to be richer than those from bays or lochs.

B. Localization

In the Brown Algae, iodine is distributed throughout the plant, though the frond is often richer than the stipe (Black, 1948; Rinck and Brouardel, 1949). Cresyl blue, which gives characteristic red crystals in the presence of iodide, forms such crystals within the vacuoles (Mangenot, 1928); autoradiographs of plants which have absorbed I^{131} indicate that the element is distributed in small discrete units throughout the thallus (Roche and Yagi, 1952). In the Florideophyceae specialized cells, *Blasenzellen* or *ioduques*, appear to be particularly rich in iodine (Kylin, 1929).

C. Chemical State

In the past the state of iodine in the algae was the subject of much controversy, but there is now good evidence that it occurs mostly as iodide. It is soluble in water, alcohol, and acetone, precipitated by $PdCl_2$ or $AgNO_3$, and on chromatograms has the same R_F as iodide (Lunde and Closs, 1930; Roche and Yagi, 1952). It is only after the plants have been hydrolyzed for some time with an alkali or a proteolytic enzyme that chromatograms indicate mono- and diiodotyrosine, and occasionally triiodothyronine, in addition to iodide and a fraction which does not migrate. These are the only iodine-containing compounds identified in algal extracts prepared by relatively mild procedures (Tong and Chaikoff, 1955; Scott, 1954). A variety of other compounds, presumably artifacts, occur in extracts treated with oxidizing agents (Masuda, 1933).

D. Iodide Oxidation

A striking feature of many marine algae is their ability to oxidize iodide to I_2 (Kylin, 1929). Fresh plants require oxygen to perform this reaction; they lose their activity if they are heated (Dangeard, 1930, 1933). The reaction has therefore been attributed to an enzyme, "iodide oxidase." In the Brown Algae this enzyme is insoluble; it is associated with the outermost layer of cells, which can be scraped off and dried for a time without losing their iodide-oxidizing activity (Kylin, 1930a). Fluoride and pyrophosphate are described as inhibitors, as is carbon monoxide in the dark. Since attempts to further purify this enzyme have failed, it has been suggested that "iodide oxidase" activity arises from the combined action of a dehydrogenase and a heme compound (Roche *et al.*, 1949). However, since reducing agents mask the presence of "iodide oxidase" by reacting with the I_2 as it is formed, it is possible that apparent inhibitors of "iodide oxidase" act indirectly by causing the plants to release reducing metabolites.

In the Red Algae, by contrast, "iodide oxidase" is a soluble enzyme, which can be concentrated by precipitation with alcohol or ammonium sulfate. Among the inhibitors which interfere with this soluble enzyme is urethane, which is known to inhibit certain dehydrogenases (Gertz, 1926). Sauvageau (1925), finding free iodine in extracts of Red Algae, suggested that I_2 was present within the living cells. However, it is more likely that the molecular iodine he detected had been produced from iodides by the rapid action of the enzyme (Kylin, 1929).

"Iodide oxidase" sometimes reveals its presence in the living thalli of marine algae when slight damage to the cells, such as that caused by drying or warmth, results in the liberation of iodide; this the enzyme oxidizes at the surface of the plant to iodine, which evaporates to produce the phenomenon of "iodovolatilization." Damp starch paper, held in the vicinity of beds of Brown Algae, occasionally turns blue, thereby indicating the presence of free iodine vapor.

Distinct from "iodide oxidase" is an iodine "liberator," which occurs in algal extracts that have been allowed to stand for a time. In the presence of an acid, this liberator converts iodide to I_2; it is probably nitrite, arising from the bacterial reduction of nitrate (Kylin, 1931).

E. Uptake Mechanism

Iodine uptake by the Phaeophyta has been studied with radioactive iodide. I^{131} is taken up very rapidly from sea water; *Laminaria digitata* removes I^{131} from 30 times its own volume hourly (Shaw, 1959). Since "iodide oxidase" occurs at the surface of many algae, it is important to determine the oxidation state in which the iodine enters the plant. Iodate seems to be excluded, since radioactive iodate is not taken up (Baily and Kelly, 1955). Many inhibitors of iodine uptake are compounds which rapidly reduce I_2 or react with it; such compounds include pyruvate, tyrosine, thiosulfate, and thiocyanate. Furthermore, anaerobic conditions, which prevent the action of "iodide oxidase," halt the absorption of I^{131} only when it is added to sea water as iodide, not when it is added as I_2 (Shaw, 1959). There is therefore a body of evidence which suggests that iodine is absorbed not as iodide but as I_2 or in some related oxidation state.

Aqueous solutions of I_2 always contain hypoiodous acid (HIO) and various other forms of iodine, in proportions depending upon the physicochemical conditions. It was found that the rate of iodine uptake by

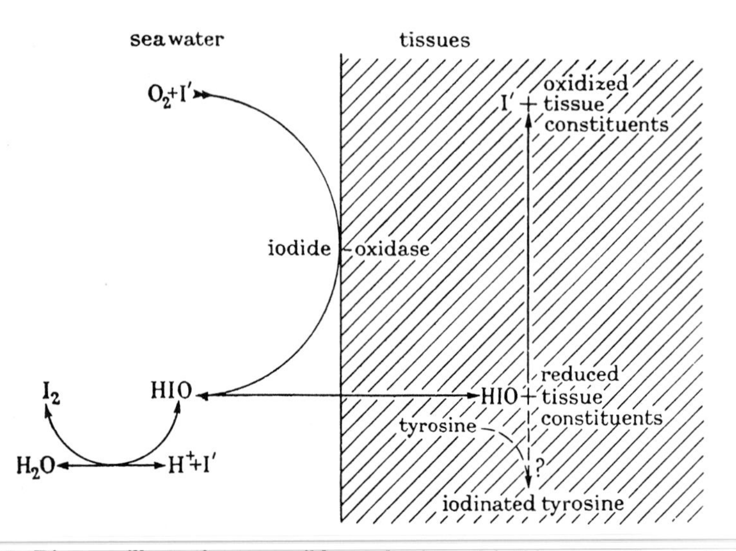

FIG. 1. Diagram illustrating a possible mechanism of iodide absorption by Brown Algae (Shaw, 1959).

Laminaria varies with the concentration of HIO rather than with that of I_2 (Shaw, 1959). It was therefore suggested that, in the course of iodine uptake, iodide is first oxidized to I_2; this hydrolyzes to HIO, which penetrates the cells and is reduced back to iodide. Figure 1 illustrates the hypothesis. The objections to this scheme are that (*a*) "iodide oxidase" is not demonstrable in all algae that concentrate iodine; and (*b*) I^{131} uptake by a species of *Fucus* is unaffected by methyl thiouracil or thiosulfate.

Apart from the inhibitors of iodine uptake mentioned earlier, there are others, including nitrate, perchlorate, and 2,4-dinitrophenol, which neither reduce I_2 nor interfere with "iodide oxidase" (Klemperer, 1957; Shaw, unpublished data).

In certain states of the algae, the uptake of iodine is accompanied by a vigorous burst of respiration. Between 3 and 6 molecules of O_2 are used for each iodide ion taken up. The respiratory quotient is close to unity, which suggests that carbohydrate is used as the substrate (Shaw, 1960).

F. Metabolic Incorporation

Both intact algae and tissue homogenates are able to incorporate radioactive iodide into mono- and diiodotyrosine. In cell-free extracts the monoiodotyrosine predominates. Such preparations, which can iodinate added D- or L-tyrosine, lose their activity on boiling or on treatment with inhibitors which reduce I_2 or react with it. Catalase also blocks the formation of labeled iodotyrosines in the extracts (Tong and Chaikoff, 1955); this strongly suggests that hydrogen peroxide plays some role, probably in the oxidation of iodide. It would indeed be interesting to determine whether the activity of "iodide oxidase," mentioned earlier, is likewise suppressed by catalase.

II. Bromine

Bromine occurs to the extent of 0.03–0.2% by dry weight in several algae (Vinogradov, 1953), and bromides are demonstrable in extracts of a number of Rhodophyta (Kylin, 1929). Sauvageau (1926) claimed to have detected free Br_2 in certain cells of *Antithamnion*, which seemed to convert fluorescein to its brominated derivative, eosin; these cells, however, failed to liberate I_2 from iodides (Kylin, 1930b).

The Rhodomelaceae are peculiarly rich in bromine, though devoid of bromides. Their bromine compounds, forming about 2% of the dry weight of the plants, include brominated phenols (Augier and Mastagli, 1956; Mautner *et al.*, 1953).

III. Chlorine

Chlorine is the major halogen in most, if not all algae (Vinogradov, 1953). Marine forms often contain less chlorine per unit fresh weight than does sea water, although freshwater species are relatively rich in the element when compared with their environment.[1]

The bulk of the chlorine in plants is present as chloride, but an organic compound containing chlorine has been isolated from certain Red Algae; only part of its chlorine seems to be ionic (Ó hEocha *et al.*, 1958, and personal communication).

References

Augier, J., and Mastagli, P. (1956). Sur un composé phénolique bromé extrait de l'algue rouge *Halopitys incurvus*. *Compt. rend. acad. sci.* **242,** 190–192.

Baily, N. A., and Kelly, S. (1955). Iodine exchange in *Ascophyllum*. *Biol. Bull.* **109,** 13–21.

Black, W. A. P. (1948). The seasonal variation in chemical constitution of some of the sub-littoral seaweeds common to Scotland. 1. *Laminaria cloustoni*; 2. *Laminaria*

[1] It is usually taken for granted that the chloride ion is physiologically inert, except as a nonspecific anion. However, some data lead us to question this assumption. Under certain experimental conditions, chloride was reported to promote a fourfold increase in the autotrophic growth (yield) of *Chlorella pyrenoidosa* (Eyster, 1958; Eyster *et al.*, 1958) and a tenfold stimulation in the photochemical reduction of ferricyanide by lyophilized cells of this species (Schwartz, 1956). Ed.

digitata; 3. *Laminaria saccharina* and *Saccorhiza bulbosa*. *J. Soc. Chem. Ind.* (*London*) **67,** 165–176.

Black, W. A. P. (1949). Seasonal variation in chemical composition of some of the littoral seaweeds common to Scotland. 2. *Fucus serratus, Fucus vesiculosus, Fucus spiralis* and *Pelvetia canaliculata*. *J. Soc. Chem. Ind.* (London) **68,** 183–189.

Chapman, V. J. (1950). "Seaweeds and Their Uses." Methuen, London.

Dangeard, P. (1930). Sur l'influence de l'oxygène dans l'iodovolatilisation. *Compt. rend.* **190,** 131–133.

Dangeard, P. (1933). Sur le mécanisme de l'iodovolatilisation et le rôle des cellules iodogènes chez les Laminaires. *Compt. rend. soc. biol.* **113,** 1203–1205.

Eyster, C. (1958). Chloride effect on the growth of *Chlorella pyrenoidosa*. *Nature* **181,** 1141–1142.

Eyster, H. C., Brown, T. E., and Tanner, H. A. (1958). Mineral requirements for *Chlorella pyrenoidosa* under autotrophic and heterotrophic conditions. *In* "Trace Elements" (C. A. Lamb, O. Bentley, and J. M. Beattie, eds.), pp. 157–174. Academic Press, New York.

Freundler, P., Menager, Y., and Laurent, Y. (1921). L'iode chez les Laminaires. *Compt. rend.* **173,** 931–932.

Gertz, O. (1926). Über die Oxydasen der Algen. *Biochem. Z.* **169,** 435–448.

Grimm, M. R. (1952). Iodine content of some marine algae. *Pacific Sci.* **6,** 318–323.

Klemperer, H. G. (1957). The accumulation of iodide by *Fucus ceranoides*. *Biochem. J.* **67,** 381–390.

Kylin, H. (1929). Über das Vorkommen von Jodiden, Bromiden und Jodidoxydasen bei den Meeresalgen. *Z. physiol. Chem., Hoppe-Seyler's* **186,** 50–84.

Kylin, H. (1930a). Über die jodidspaltende Fähigkeit der Phäophyceen. *Z. physiol. Chem., Hoppe-Seyler's* **191,** 200–210.

Kylin, H. (1930b). Über die Blasenzellen bei *Bonnemaisonia, Trailliella* und *Antithamnion*. *Z. Botan.* **23,** 217–226.

Kylin, H. (1931). Über die jodidspaltende Fähigkeit von *Laminaria digitata*. *Z. physiol. Chem., Hoppe-Seyler's* **203,** 58–65.

Lunde, G., and Closs, K. (1930). Über die Bindungsart des Jods bei *Laminaria digitata*. *Biochem. Z.* **219,** 198–217.

McClendon, J. F. (1939). "Iodine and the Incidence of Goiter." Univ. Minnesota Press, Minneapolis, Minnesota.

Mangenot, G. (1928). Sur la signification des cristaux rouges apparaissant, sous l'influence du bleu de crésyl, dans les cellules de certaines algues. *Compt. rend.* **186,** 93–95.

Masuda, E. (1933). Über die Jodverbindungen in Meerestangen. *Proc. Imp. Acad.* (*Tokyo*) **9,** 599–601.

Mautner, H. G., Gardner, G. M., and Pratt, R. (1953). Antibiotic activity of seaweed extracts. II. *Rhodomela larix*. *J. Am. Pharm. Assoc., Sci. Ed.* **42,** 294–296.

Ó hEocha, C., Chonaire, E. N., and O'Hare, P. (1958). Some fluorescent substances of the red algae. *In* "Third International Seaweed Symposium—Abstracts" (C. Ó hEocha, ed.), p. 68. O'Gorman, Galway, Ireland.

Rinck, E., and Brouardel, J. (1949). Gradients de répartition de l'iode, du potassium et de l'eau chez *Laminaria flexicaulis*. *Compt. rend.* **229,** 1167–1168.

Roche, J., and Yagi, Y. (1952). Sur la fixation de l'iode radioactif par les algues et sur les constituants iodés des Laminaires. *Compt. rend. soc. biol.* **146,** 642–645.

Roche, J., Thoai, N-v., and Lafon, M. (1949). Sur le mécanisme enzymatique de la formation d'iode aux depens des iodures par les laminaires. *Compt. rend. soc. biol.* **143,** 1327–1329.

Sauvageau, C. (1925). Sur quelques algues floridées renfermant de l'iode à l'état libre. *Bull. Sta. Biol. Arcachon* **22,** 5–45.

Sauvageau, C. (1926). Sur quelques algues floridées renfermant du brome a l'état libre. *Bull. Sta. Biol. Arcachon* **23,** 5–23.

Schwartz, M. (1956). The photochemical reduction of quinone and ferricyanide by lyophilized *Chlorella* cells. *Biochim. et Biophys. Acta* **22,** 463–470.

Scott, R. (1954). Observations on the iodo-amino-acids of marine algae using iodine-131. *Nature* **173,** 1098–1099.

Shaw, T. I. (1959). The mechanism of iodide accumulation by the brown sea weed *Laminaria digitata.* I. The uptake of I^{131}. *Proc. Roy. Soc.* **B150,** 356–371.

Shaw, T. I. (1960). The mechanism of iodide accumulation by the brown sea weed *Laminaria digitata.* II. Respiration and iodide uptake. *Proc. Roy. Soc.* **B152,** 109–117.

Tong, W., and Chaikoff, I. L. (1955). Metabolism of I^{131} by the marine alga *Nereocystis luetkeana. J. Biol. Chem.* **215,** 473–484.

Vinogradov, A. P. (1953). "The Elementary Chemical Composition of Marine Organisms." Sears Foundation for Marine Research, Memoir II, New Haven, Connecticut.

von Fellenberg, Th. (1926). Das Vorkommen, der Kreislauf und der Stoffwechsel des Jods. *Ergeb. Physiol.* **25,** 176–363.

—16—

Major Cations

RICHARD W. EPPLEY

Bioastronautics Laboratory, Northrop Corporation, Hawthorne, California

I. Introduction

This review will be limited to recent studies of cation distribution and transport at the cellular level. Much of the earlier work on algae has been reviewed by Blinks (1951). The reader is also referred to reviews on mineral nutrition (Krauss, 1953, 1958; Provasoli, 1958), permeability (Collander, 1957) and the electrical properties of cell membranes in relation to cation transport (Blinks, 1949, 1955; Dainty, 1960), and to Chapters 17, 31, and 32 in this treatise, by Wiessner, Stadelmann, and Guillard.

II. Distribution of Cations in Algal Cells

A. Free Space

Briggs and Robertson (1957) have defined *free space* as that part of a cell or tissue into which solute and solvent from the external solution readily penetrate. Differences in actual measurements of free space ("apparent free space") usually depend upon the charge and size of the penetrant molecule. This has given rise to the terms "water free space" and "Donnan free space." In the former, the solute is in equivalent concentration in the free space and the medium. In the latter, the solute distribution is influenced by fixed anions with cation exchange capacity.

Values for apparent free space vary considerably among the different algae. If we think of the free space as the cell walls without the cytoplasm, then this variation probably represents varying amounts of wall material. For *Nitellopsis* a value as low as 1 to 1.5% of the cell volume was obtained (MacRobbie and Dainty, 1958b), this volume being almost certainly associated with the cell walls alone (Dainty, 1960). Larger values of about 20% were found for *Ulva lactuca* (Scott and Hayward, 1954) and *Porphyra perforata* (Eppley and Cyrus, 1960), and 25 to 30% for *Hormosira banskii* (Bergquist, 1958a). However, a value of 21.4% for the free space of *Cladophora gracilis* probably includes part of the cytoplasm, since the cell walls of this alga do not appear to be so voluminous (Follmann and Follmann-Schrag, 1959).

The free space of *Nitella axillaris* (Diamond and Solomon, 1959) and *Chara australis* (Dainty and Hope, 1959) is apparently entirely in the cell wall, and has been shown to consist of an aqueous component and a "Donnan free space." The latter represents that part of the wall which contains a high concentration of fixed anions, mostly in the form of pectic material (Dainty *et al.*, 1960). Sodium or calcium efflux from the wall of *Chara australis* shows at least two phases, indicating different components of the free space, although from both fractions the amounts of sodium or calcium exchanged and the half-times for this exchange vary with the external concentration of the ion. These data imply an entirely physical system involving only diffusion and cation exchange in the *Chara* wall. The system serves as a model for the interpretation of more complex conditions when cytoplasm and vacuoles are also included.

In Brown Algae such as *Ascophyllum nodosum* there is a greater adsorption of cations by acidic polysaccharides of the cell walls, notably alginic acid (Wasserman, 1948). Red Algae such as *Porphyra* also show appreciable cation adsorption, no doubt due to sulfated galactans and polyuronides of the walls (Eppley and Blinks, 1957).

B. Vacuoles

The ionic composition of the vacuole and fluxes across the vacuolar or tonoplast membrane have been evaluated for *Nitellopsis obtusa* (MacRobbie and Dainty, 1958b), for *Nitella* sp. (Diamond and Solomon, 1959) and for *Halicystis ovalis* (Blount and Levedahl, 1960). Potassium efflux from isolated cells of *Nitella* and *Nitellopsis* shows three distinct phases, representing the cell wall, the cytoplasm, and the vacuolar sap. The kinetics of potassium efflux from *Rhodymenia palmata* thalli, which are made up of a variety of cell types, are less easily interpreted in terms of cell morphology (MacRobbie and Dainty, 1958a). Rubidium efflux from *Porphyra* cells likewise shows three components, two of which seem to be cytoplasmic (Eppley, unpublished data). Here the resistance of the membranes to diffusion is not always rate-limiting, as it is in *Nitellopsis* and *Nitella*.

Dainty (1960) emphasized the almost complete absence of cation selectivity in the tonoplast of *Nitellopsis* cells. It seems that neither potassium nor sodium is actively transported across this membrane, although there is good evidence for an inwardly directed "chloride pump" associated with the tonoplast in *Nitellopsis* and in *Halicystis*.

C. Cytoplasm

It is possible experimentally to separate the sap from the cytoplasm of *Nitella* by carefully cutting off the end of a cell, allowing the sap to drain out, and then squeezing out the cytoplasm (Brooks, 1939). However, for accumulated radioisotopes, efflux kinetics may provide equally reliable estimates of the ion concentrations in the various cellular compartments. This kinetic method has the advantage of avoiding contamination between sap and cytoplasm, which tends to occur in the older micrurgical method. With *Nitella axillaris* Diamond and Solomon (1959) achieved a fair agreement between these two methods, obtaining for the potassium ion values between 1.2 and 1.7×10^{-9} moles/cm.^{2} of cytoplasmic area. MacRobbie and Dainty (1958b) also employed both methods in their work with *Nitellopsis* (Fig. 1).

In *Ulva lactuca*, where extracellular cation adsorption is low, total tissue analyses are probably in reasonable agreement with actual cytoplasmic and vacuolar contents, although an acidic polysaccharide, ulvin, between the cell walls (see O'Colla, Chapter 20, this treatise) may introduce an appreciable error in the case of divalent cations. Scott and Hayward (1954) calculated values of about 350 mM for potassium and 190 mM for sodium in the cell fluids. About 90% of the sodium is exchangeable with the medium in a few seconds (Scott *et al.*, 1957). In the cytoplasm of *Porphyra perforata*, the concentrations of cations, corrected for extracellular binding and expressed on a cell-water basis, are about 480 mM for potassium and 80 mM for sodium. All of the measurable calcium appears to be bound to the extracellular materials (Eppley, 1958a).

Rubidium is accumulated in exchange for potassium in *Ulva* and *Porphyra* with no apparent toxicity (Scott and DeVoe, 1957; Eppley, unpublished data). Lithium can penetrate the cytoplasm of *Porphyra* in exchange for sodium (Eppley, 1959), but in this case the cells survive only for a few hours. When thalli of *Ulva* (Scott and Hayward, 1954) or *Porphyra* (Eppley,

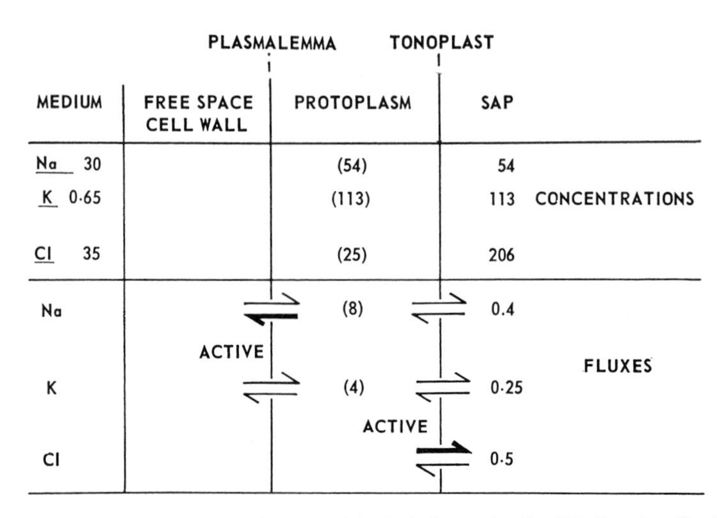

FIG. 1. Scheme for the normal state of ionic balance in the *Nitellopsis* cell, showing ion concentrations (mM) and fluxes (picomoles/sec./cm.^{2}). Figures in parenthesis are provisional. From MacRobbie and Dainty (1958b).

1958b) are soaked in an isotonic sucrose solution, the contents of potassium and sodium fall to very low levels. Long exposure to anoxia or to a K-free artificial sea water reduces the potassium content of *Porphyra* with a concomitant increase in sodium. These effects are reversed when air or potassium salts are added, respectively.

It was pointed out earlier that negative charges in the cell walls of algae may influence the extracellular distribution of cations. Such cation exchange effects might be even greater in the cytoplasm, due to the presence of negatively charged macromolecules such as proteins and nucleic acids, particularly in the vicinity of cytoplasmic organelles. Diamond and Solomon (1959) found that although chloroplasts isolated from *Nitella* have semipermeable membranes, their potassium exchange is very rapid, providing evidence that the chloroplast–cytoplasm system, although structurally heterogeneous, can be regarded as a single compartment in isotopic efflux studies. Thus structural differentiation does not necessarily imply a dual phase diffusion system.

III. Transport of Cations in Algae

A. Direction on Ion Pumps

Since Scott and Hayward (1953c) described a sodium excretion mechanism in *Ulva lactuca*, evidence for "sodium pumps" has been reported for other marine algae such as *Hormosira banksii* (Bergquist, 1958b), *Valonia macrophysa* (Scott and Hayward, 1955), *Rhodymenia palmata* (MacRobbie and Dainty, 1958a), *Porphyra perforata* (Eppley, 1958a), and *Halicystis ovalis* (Blount and Levedahl, 1960). *Nitellopsis obtusa* (MacRobbie and Dainty, 1958b), a brackish-water alga, also displays an active sodium extrusion. The site of such pumps is no doubt the plasmalemma. It would be interesting to examine freshwater algae and higher plants for a counterpart to the sodium pump of marine algae and of animal cells, in which it seems to be ubiquitous. In this regard, Hope and Walker (1960) suggested the action of such a pump for the freshwater *Chara australis*. Lithium is not actively transported in *Porphyra*, and thus cannot substitute for sodium in the sodium pump (Eppley, 1959).

The almost universal accumulation of potassium in the cytoplasm and vacuolar sap of algae has caused considerable speculation as to whether K is actively transported into the cells or whether it is only electrostatically attracted by an electropotential or Donnan system. Where the potential has been measured (Dainty, 1960), K is found to be in electrochemical equilibrium, suggesting that it is not actively transported.

The cytoplasm of those algal cells which have been studied is low in calcium, the plasmalemma apparently acting as a diffusion barrier to this ion (Walker, 1957), although there is no information concerning possible metabolic mechanisms which may exclude it. At low levels the calcium is taken up by cells of *Chlorella pyrenoidosa* in proportion to its concentration in the medium (Knauss and Porter, 1954). There are no data on magnesium transport in algae.

B. Characteristics of Cation Transport

Judicious use of the inhibitors iodoacetate and arsenate has shown that in *Ulva* potassium accumulation and sodium excretion are independent processes (Fig. 2) (Scott and Hayward, 1954). Iodoacetate alone inhibits both processes in the dark, whereas in the presence of both iodoacetate and arsenate only sodium excretion is inhibited. No similar selective protection by arsenate of iodoacetate poisoning was observed in *Valonia* (Scott and Hayward, 1955) or in *Porphyra* (Eppley, 1958a). Cyanide, azide, various urethanes, and *p*-chloromercuribenzoate, as well as anoxia, generally block both K uptake and Na excretion in both *Ulva* (Fig. 2) and *Porphyra*. In *Hormosira*, cyanide, *p*-chloromercuribenzoate, and dinitrophenol influence the distribution of sodium, but only dinitrophenol causes a loss of intracellular potassium (Bergquist, 1958b). Azide, cyanide, dinitrophenol, and exclusion of CO_2 in the light, reduce cesium uptake in *Rhodymenia* (R. Scott, 1954).

Light has a strongly stimulating effect on both cesium uptake (R. Scott,

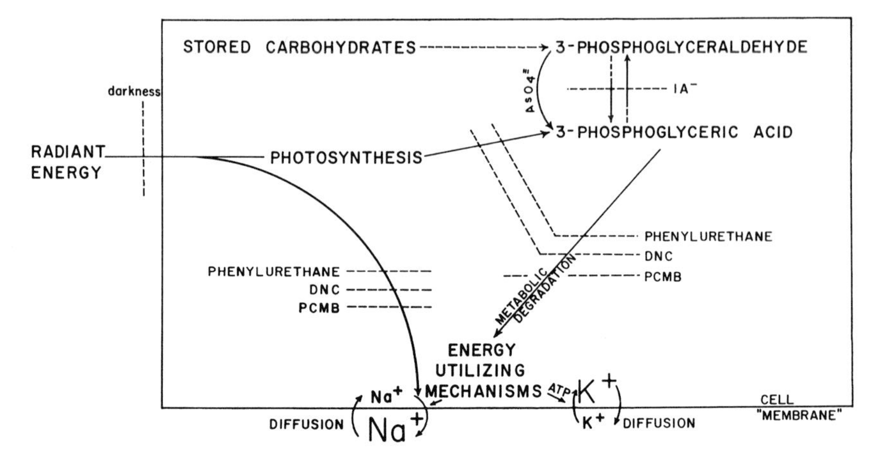

FIG. 2. Diagrammatic representation of the roles of metabolism in sodium exclusion and potassium retention by *Ulva lactuca*. Abbreviations: PCMB = *p*-chloromercuribenzoate; DNC = 4,6-dinitro-*o*-cresol; IA⁻ = iodoacetate; ATP = adenosine triphosphate. From Scott and Hayward (1954).

1954) and K exchange (MacRobbie and Dainty, 1958a) in *Rhodymenia*. Furthermore light prevents the effects of iodoacetate on the Na and K balance in *Ulva* (Fig. 2) (Scott and Hayward, 1953a, c) and stimulates K uptake in *Porphyra* (Eppley, 1958b). The mechanism of light stimulation is not clear, although it may be mediated through photosynthesis (R. Scott, 1954) or through a more direct reducing effect (Scott and Hayward, 1954). Light alters both the influx and efflux of Na in *Chara* (Hope and Walker, 1960). Iaglova (1958) reported light-induced changes in the ion balance of algal cells. Is it possible that some of these effects may be traced to the production of adenosine triphosphate (ATP) in photosynthetic phosphorylation?

Glutamate stimulates the transport of K into and out of cells of *Rhodymenia* in the dark (MacRobbie and Dainty, 1958a). In the dark, pyruvate reverses iodoacetate inhibition of the Na excretion mechanism in *Ulva*, while exogenous phosphoglycerate or ATP offers protection against K loss during iodoacetate poisoning (Scott and Hayward, 1954). Reduced sulfhydryl compounds, including glutathione, thioglycolate, and cysteine, overcome the inhibitory effects of *p*-chloromercuribenzoate in *Porphyra* (Eppley, 1958a).

Potassium exchange in *Ulva* (as measured by the use of isotopes), Na fluxes in *Chara*, and K uptake, Na excretion, and K/Rb exchange in *Porphyra*, all have high temperature coefficients (Scott and Hayward, 1953b; Hope and Walker, 1960; Eppley, 1958b, and unpublished data), suggesting that chemical processes are involved or that diffusion takes place across a high potential-energy barrier.

In media containing rubidium, x-ray doses of 50 to 1000 r. reduced the nonexchangeable Rb level of *Chlorella pyrenoidosa* by as much as 55%, but these doses of irradiation did not reduce the level of rapidly exchangeable Rb (Barber *et al.*, 1957). It is not established whether the reduction of nonexchangeable Rb represents damage to the mechanisms of Rb accumulation and of exchangeable Rb adsorption in the free space. Certainly, at the higher levels of irradiation nonspecific cell damage and membrane leakage would be expected.

C. Mutual Dependence of Transport

The possibility of a linked K–Na pump has been considered for *Porphyra* (Eppley, 1958b, 1959) and *Nitellopsis* (MacRobbie and Dainty, 1958b; Dainty, 1960). In *Porphyra* the presence of an external supply of K or Rb ions is necessary before net Na excretion can be demonstrated, and K accelerates even the passive Na efflux in "Na-free" solutions (Fig. 3). Accumulation of K or Rb occurs as Na is excreted. Thalli of *Porphyra* in which cellular Na has been largely replaced by lithium can still accumulate K, although Li is not excreted (Eppley, 1959). These results suggest either that the postulate of a linked K–Na pump may not be essential, or that the K uptake observed does not represent active transport.

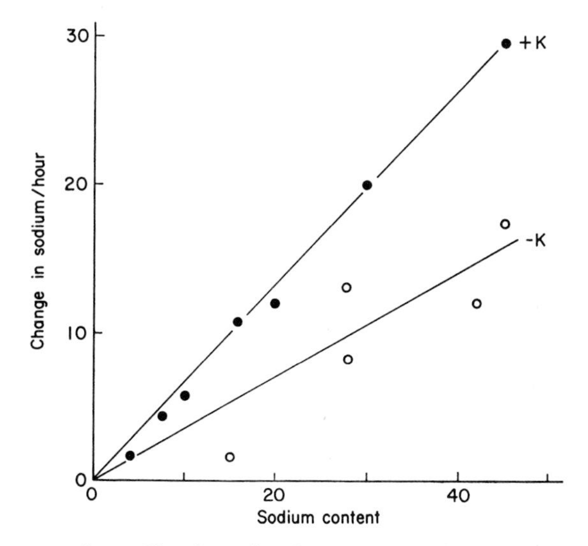

FIG. 3. Passive sodium efflux from *Porphyra perforata* discs into artificial "lithium sea water" in the presence and absence of 25 m*M* KCl. Efflux units are milliequivalents of Na/kg. fresh weight/hour. From Eppley (1959).

D. Mechanisms of Transport

Based largely upon work on various cells and tissues other than algae, several general theories of active transport have been suggested. These include: bacterial permeases—presumably enzymes of the cell surface which catalyze movement across the permeability barrier—(Cohen and Monod, 1957), pinocytosis (Bennet, 1956), contractile proteins (Goldacre, 1952), and redox mechanisms (Conway, 1953; Lundegardh, 1955). Although the redox theories seem to be in greatest favor at the moment, especially in relation to the sodium pump, it should be recalled that ion pumps undoubtedly operate in anaerobic systems, such as yeast, in which the mode of action of a redox system is not readily apparent. Mechanisms utilizing high-energy phosphate, suggested on the basis of indirect evidence, now seem more feasible.

The possibility of ion selectivity and transport dependent upon a K/Na-activated ATP-ase of the membrane, as had been demonstrated in erythrocytes (Post, 1959; Tosteson *et al.*, 1960), awaits verification for algae. It is of interest that K is needed, not only for the activation of this enzyme, but also for amino-acid uptake by certain tumor cells (Riggs *et al.*, 1958; Heinz and Walsh, 1958), and for Na excretion by *Porphyra* (Eppley, 1958b) and by the giant axon of squids (Hodgkin and Keynes, 1955).

Berquist (1959) and Eppley (1960) described transient increases in the respiration of *Hormosira* and *Porphyra* when potassium was added to the medium (sea water). This effect may be due to the activation by potassium of reactions leading to the generation of phosphate acceptors, resulting in an increased rate of oxidative phosphorylation and thus to an increased oxygen consumption, in the manner generally outlined by Laties (1957). Such K-stimulated reactions might result in Na excretion (Eppley, 1960). Bergquist (1959) also considered it likely that the increased oxygen uptake represents an increase in the expenditure of energy needed to accumulate ions.

In his redox theory, Conway (1953) suggested that one or more cytochromes may be involved as Na carriers, while a phosphoinositide is believed to carry Na ions in the salt glands of birds (Hokin and Hokin, 1960). Structural or enzymatic sulfhydryl groups, or both, may play some part in the ion transport mechanisms of yeast, red blood cells, and *Porphyra* (Rothstein, 1959; Passow *et al.*, 1959; Eppley, 1958a). On the basis of their experiments with yeast and with mung bean roots, Passow *et al.* (1959) and Tanada (1955) suggested that the ribonucleic acid in membranes could also be involved in controlling permeability.

There is at present a great need for further studies of submicroscopic membrane structure and of membrane enzymology. An experimental approach might be facilitated if an algal equivalent of the erythrocyte "ghost" were available.

E. Functional Aspects of Transport

Diamond and Solomon (1959) reasserted that the relatively high osmotic pressure of *Nitella* cells, which is maintained by selective ion movements and by inelastic cell walls, confers rigidity on the cells at the cost of very little structural material. The cell walls are rarely more than two microns in thickness.

Scott and Hayward (1954, 1955) and Eppley and Cyrus (1960) emphasized the role of selective ion transport in cellular homeostasis, particularly with respect to the high cytoplasmic ratios of K to Na. *Porphyra* retains a high K content over a wide range of salinities, while Na is excluded. However, the presence of calcium in the medium is necessary for K retention by plants in diluted sea water. Calcium could not be shown to participate directly in K uptake and Na excretion, but may be involved in the maintenance of membrane structure. Biebl (1956) suggested that active transport of cations may allow such algae as *Enteromorpha clathrata* to grow in concentrated brines. The articles by Schwenke (1958) and Stadelmann (Chapter 31, this treatise) should be consulted for other data on the role of calcium in the permeability of algae, and that by Halldal (Chapter 37, this treatise) on the effects of the Ca/Mg ratio on tactic movements by *Platymonas*.

If the substance transported is known to serve as a substrate for metabolism, the functional significance of active transport is clear. Why cells of marine organisms should maintain a low Na and a high K content is less apparent. The "permeases" of algae, and the possible roles of active transport in flotation mechanisms of phytoplankton (see Yentsch, Chapter 52, this treatise) or in calcium carbonate deposition by coralline algae (see Lewin Chapter 28), would provide interesting subjects for study.

References

Barber, D. A., Neary, G. J., and Russell, R. S. (1957). Radiosensitivity of salt uptake in plants. *Nature* **180,** 556–557.

Bennett, H. S. (1956). The concepts of membrane flow and membrane vesiculation as mechanisms for active transport and ion pumping. *J. Biophys. Biochem. Cytol. Suppl.* **2,** 99–103.

Bergquist, P. L. (1958a). Effect of potassium cyanide on apparent free space in a brown alga. *Nature* **181,** 1270.

Bergquist, P. L. (1958b). Evidence for separate mechanisms of sodium and potassium regulation in *Hormosira banksii*. *Physiol. Plantarum* **11,** 760–770.

Bergquist, P. L. (1959). The effects of cations and anions on the respiratory rate of the brown alga, *Hormosira banksii*. *Physiol. Plantarum* **12**, 30–36.

Biebl, R. (1956). Zellphysiologisch-ökologische Untersuchungen an *Enteromorpha clathrata* (Roth) Greville. *Ber. deut. botan. Ges.* **69**, 75–86.

Blinks, L. R. (1949). The source of the bioelectric potentials in large plant cells. *Proc. Natl. Acad. Sci. U.S.* **35**, 566–575.

Blinks, L. R. (1951). Physiology and biochemistry of algae. *In* "Manual of Phycology" (G. M. Smith, ed.), pp. 263–291. Chronica Botanica, Waltham, Massachusetts.

Blinks, L. R. (1955). Some electrical properties of large plant cells. *In* "Electrochemistry in Biology and Medicine" (T. Shedlovsky, ed.), pp. 187–212. Wiley, New York.

Blount, R. W., and Levedahl, B. H. (1960). Active sodium and chloride transport in the single celled marine alga *Halicystis ovalis*. *Acta Physiol. Scand.* **49**, 1–9.

Briggs, G. E., and Robertson, R. N. (1957). Apparent free space. *Ann. Rev. Plant Physiol.* **8**, 11–30.

Brooks, S. C. (1939). Ion exchanges in accumulation and loss of certain ions by the living protoplasm of *Nitella*. *J. Cellular Comp. Physiol.* **14**, 383–401.

Cohen, G. N., and Monod, J. (1957). Bacterial Permeases. *Bacteriol. Rev.* **21**, 169–194.

Collander, R. (1957). Permeability of plant cells. *Ann. Rev. Plant Physiol.* **8**, 335–348.

Conway, E. J. (1953). A redox pump for the biological performance of osmotic work, and its relation to the kinetics of free ion diffusion across a membrane. *Intern. Rev. Cytol.* **2**, 419–445.

Dainty, J. (1960). Ion transport across plant cell membranes. *Proc. Roy. Soc. Edinburgh* **28**, 3–14.

Dainty, J., and Hope, A. B. (1959). Ionic relations of cells of *Chara australis*. I. Ion exchange in the cell wall. *Australian J. Biol. Sci.* **12**, 395–411.

Dainty, J., Hope, A. B., and Denby, C. (1960). Ionic relations of cells of *Chara australis*. II. The indiffusible anions of the cell wall. *Australian J. Biol. Sci.* **13**, 267–276.

Diamond, J. M., and Solomon, A. K. (1959). Intracellular potassium compartments in *Nitella axillaris*. *J. Gen. Physiol.* **42**, 1105–1121.

Eppley, R. W. (1958a). Sodium exclusion and potassium retention by the red marine alga, *Porphyra perforata*. *J. Gen. Physiol.* **41**, 901–911.

Eppley, R. W. (1958b). Potassium-dependent sodium extrusion by cells of *Porphyra perforata*, a red marine alga. *J. Gen. Physiol.* **42**, 281–288.

Eppley, R. W. (1959). Potassium accumulation and sodium efflux by *Porphyra perforata* tissues in lithium and magnesium sea waters. *J. Gen. Physiol.* **43**, 29–38.

Eppley, R. W. (1960). Respiratory responses to cations in the red alga *Porphyra* and their relationships to ion transport. *Plant Physiol.* **35**, 637–644.

Eppley, R. W., and Blinks, L. R. (1957). Cell space and apparent free space in the red alga, *Porphyra perforata*. *Plant Physiol.* **32**, 63–64.

Eppley, R. W., and Cyrus, C. C. (1960). Cation regulation and survival of the red alga, *Porphyra perforata*, in diluted and concentrated sea waters. *Biol. Bull.* **118**, 55–65.

Follmann, G., and Follmann-Schrag, I. A. (1959). Die Wasserführung plasmolysierter Protoplasten von *Cladophora gracilis* (Griff.) Kuetzing unter dem Einfluss von Atmungsgiften. *Z. Naturforsch.* **14b**, 181–187.

Goldacre, R. J. (1952). The folding and unfolding of protein molecules as a basis of osmotic work. *Intern. Rev. Cytol.* **1**, 135–164.

Heinz, E., and Walsh, P. M. (1958). Exchange diffusion, transport and intracellular level of amino acids in Ehrlich's carcinoma cells. *J. Biol. Chem.* **233,** 1488–1493.

Hodgkin, A. L., and Keynes, R. D. (1955). Active transport of cations in giant axons from *Sepia* and *Loligo*. *J. Physiol. (London)* **128,** 28–60.

Hokin, L. E., and Hokin, M. R. (1960). Phosphatidic acid as a carrier for Na ions. *Federation Proc.* **19,** 130.

Hope, A. B., and Walker, N. A. (1960). Ionic relations of cells of *Chara australis*. III. Vacuolar fluxes of sodium. *Australian J. Biol. Sci.* **13,** 277–291.

Iaglova, L. G. (1958). Light-induced changes in the ion balance in algal cells. *Biophysics (Engl. transl.)* **3,** 160–167.

Knauss, H. J., and Porter, J. W. (1954). The absorption of inorganic ions by *Chlorella pyrenoidosa*. *Plant Physiol.* **29,** 229–234.

Krauss, R. W. (1953). Inorganic nutrition of algae. *In* "Algal Culture; from Laboratory to Pilot Plant" (J. S. Burlew, ed.) *Carnegie Inst. Wash. Publ. No.* **600,** pp. 85–102.

Krauss, R. W. (1958). Physiology of the freshwater algae. *Ann. Rev. Plant Physiol.* **9,** 207–244.

Laties, G. G. (1957). Respiration and cellular work and the regulation of the respiratory rate in plants. *Survey Biol. Progr.* **3,** 215–299.

Lundegårdh, H. (1955). Mechanisms of absorption, transport, accumulation, and secretion of ions. *Ann. Rev. Plant Physiol.* **6,** 1–24.

MacRobbie, E. A. C., and Dainty, J. (1958a). Sodium and potassium distribution and transport in the seaweed *Rhodymenia palmata*. *Physiol. Plantarum* **11,** 782–801.

MacRobbie, E. A. C., and Dainty, J. (1958b). Ion transport in *Nitellopsis obtusa*. *J. Gen. Physiol.* **42,** 335–353.

Passow, H., Rothstein, A., and Loewenstein, B. (1959). An all-or-none response in the release of potassium by yeast cells treated with methylene blue and other basic dyes. *J. Gen. Physiol.* **43,** 97–107.

Post, R. L. (1959). Relationship of an ATP-ase in human erythrocyte membrane to the active transport of sodium and potassium. *Federation Proc.* **18,** 121.

Provasoli, L. (1958). Nutrition and ecology of protozoa and algae. *Ann. Rev. Microbiol.* **12,** 279–308.

Riggs, T. R., Walker, L. M., and Christensen, H. N. (1958). Potassium migration and amino acid transport. *J. Biol. Chem.* **233,** 1479–1487.

Rothstein, A. (1959). Role of the cell membrane in the metabolism of inorganic electrolytes by micro-organisms. *Bacteriol. Revs.* **23,** 175–201.

Schwenke, H. (1958). Über einige zellphysiologische Faktoren der Hypotonieresistenz mariner Rotalgen. *Kiel. Meeresforsch.* **14,** 130–150.

Scott, G. T., and DeVoe, R. (1957). The reversible replacement of potassium by rubidium in *Ulva lactuca*. *Biol. Bull.* **112,** 249–253.

Scott, G. T., and Hayward, H. R. (1953a). The influence of iodoacetate on the sodium and potassium content of *Ulva lactuca* and the prevention of its influence by light. *Science* **117,** 719.

Scott, G. T., and Hayward, H. R. (1953b). The influence of temperature and illumination on the exchange of potassium ion in *Ulva lactuca*. *Biochim. et Biophys. Acta* **12,** 401–404.

Scott, G. T., and Hayward, H. R. (1953c). Metabolic factors influencing the sodium and potassium distribution in *Ulva lactuca*. *J. Gen. Physiol.* **36,** 659–671.

Scott, G. T., and Hayward, H. R. (1954). Evidence for the presence of separate mechanisms regulating potassium and sodium distribution in *Ulva lactuca*. *J. Gen. Physiol.* **37,** 601–620.

Scott, G. T., and Hayward, H. R. (1955). Sodium and potassium regulation in *Ulva lactuca* and *Valonia macrophysa*. *In* "Electrolytes in Biological Systems" (A. M. Shanes, ed.), pp. 35–64. American Physiological Society, Washington, D. C.

Scott, G. T., DeVoe, R., Hayward, H. R., and Craven, G. (1957). Exchange of sodium ion in *Ulva lactuca*. *Science* **125,** 160.

Scott, R. (1954). A study of caesium accumulation by marine algae. *Proc. Radioisotope Conf., 2nd Conf., Oxford.* pp. 373–380.

Tanada, T. (1955). Effect of ribonuclease on the absorption of rubidium by excised mung bean roots. *Plant Physiol.* **30,** Suppl. xxxiv.

Tosteson, D. C., Moulton, R. H., and Blaustein, M. (1960). An enzymatic basis for the difference in active cation transport in two genetic types of sheep red cells. *Federation Proc.* **19,** 128.

Walker, N. A. (1957). Ion permeability of the plasmalemma of the plant cell. *Nature* **180,** 94–95.

Wasserman, A. (1948). Cation adsorption by brown algae. The mode of occurrence of alginic acid. *Ann. Botany (London)* **13,** 79–88.

—17—

Inorganic Micronutrients

W. WIESSNER

Institute of Plant Physiology, University of Göttingen, Germany

I. Methodology of Micronutrient Investigations

During the past thirty years the mineral nutrition of algae has become an important subject for botanical and biochemical investigations. Among the essential nutrients are the micronutrients, or trace elements, which must be present in at least very low concentrations if the organisms are to grow and reproduce, and which cannot be replaced by other mineral factors. In this article we will deal chiefly with certain divalent and multivalent metallic ions and with boron. Their participation in many metabolic functions has been demonstrated, and a number of diseases of higher plants and animals have been found to be due to deficiencies in minor elements. Investigations of their role in the metabolism of algae have been stimulated by the need for small and relatively simple organisms suitable for research on fundamental aspects of mineral nutrition, since in this respect algae presumably do not differ appreciably from higher plants (Molisch, 1896).

It is customary to refer to essential elements in the first and second groups of the periodic table as *major cations*, since the amounts generally required are relatively high, and those in the third and fourth groups as *minor elements*, since only relatively small quantities are needed for growth. In this article we will follow this arbitrary distinction, although the relative amounts of nutrients required may vary depending on the organism and on the composition of the nutrient solution. As an example, Molisch (1895) and Pringsheim (1926) demonstrated that different algae require very different quantities of calcium (see also Walker, 1956). Stegmann (1940), who investigated the role of calcium in the metabolism of *Chlorella*, showed that this element is necessary in relatively high concentrations for some metabolic processes but in only small amounts for others. Thus, according to the metabolic function involved, calcium may be considered either a major or a minor element (see Eppley, Chapter 16, this treatise).

Only a brief summary of the more important publications on this subject can be given here. Since most micronutrients are evidently involved in a number of general metabolic reactions, it is difficult to study their influence on special processes. Many problems will remain unsolved until we have more exact, critical, and extensive data on the uptake, storage, and chemical status of the elements within algal cells. In nutrient media the form and relative availability of several micronutrients, particularly heavy metals with several possible valency states, often remain unknown as long as we are restricted to classical methods of chemical analysis. The formation of complexes, now recognized as of great importance in physiology, is only just beginning to be accessible to study by chemical methods.

Successful studies of mineral nutrition depend on appropriate methods for growing organisms and on adequate techniques for distinguishing symptoms of nutrient deficiency. Methods and media for the cultivation of algae have been reviewed extensively (Pringsheim, 1946a, 1954; Brunel *et al.*, 1950; Burlew, 1953; Krauss, 1953; Reisner and Thompson, 1956b; Myers and Clark, 1944; Lorenzen, 1959; see Myers, Chapter 39, this treatise). Although many have been developed for special purposes, there is still no single method of cultivation meeting the following three requirements: (i) sterile mass-culture at constant temperature and with constant aeration; (ii) constant illumination of the cells at every age and density of the culture; and (iii) constant hydrogen-ion concentration. It is therefore sometimes hard to decide whether the effects observed are due to a mineral deficiency or to other changes in culture conditions.

Physiological deficiencies of micronutrient elements in cultures are not always readily achieved. In most experiments the element in question is omitted altogether from the nutrient solution or added in various limiting amounts (Pirson, 1958b). However, accurate figures for the actual concentration present are hard to obtain, because trace elements may also occur as contaminants of other nutrient salts in the medium.

One of the difficulties of working with these elements arises from the present lack of methods for the purification of nutrient salts and the elimination of contaminants introduced thereby. Earlier investigators (e.g., Hopkins and Wann, 1925, 1926, 1927) used the method of base adsorption for removal of elements such as iron from culture solutions by coprecipitation with calcium phosphate. In several investigations citric acid was used as chelating agent, especially for iron (Wann and Hopkins, 1927; Rodhe, 1948; Bitcover and Sieling, 1951). Goldberg (1952) reported that the iron of ferric citrate is not available for growth of the marine diatom *Asterionella japonica*, and that, by the use of citrate, heavy-metal impurities could be neglected in calculating the exact requirement for iron. However, this method cannot be applied to all organisms, as was demonstrated by Wann and Hopkins (1927), who observed that iron in the citrate complex was still available for the growth of *Chlorella* sp.

Since Schatz and Hutner (1949) and Hutner and co-workers (1950) introduced other organic chelating agents, several of these, especially EDTA (ethylenediaminetetraacetate), have been used in culture media for studies of iron requirements. Metal ions bound in such complexes are relatively unavailable to algae, and a marked deficiency can thus be induced. To compensate for the biological unavailability of such bound cations, additional quantities of the elements must be added to the culture medium. If the metals thus added are extremely pure, cationic contaminants are almost completely removed by the chelating agent and may be neglected in comparison with the concentration of the specific ions added to overcome the deficiency. The method proposed by Hutner is based on the assumption that the EDTA cannot be metabolized, and on the fact that the equilibrium between the free metal ion and its complex strongly favors the latter, due to its very weak dissociation, so that ions chelated thereby cannot be absorbed by the algae unless liberated by later changes in the medium. Spencer (1957), too, concluded from experiments with *Phaeodactylum tricornutum* that EDTA is stable and not metabolized. On the other hand Krauss and Specht (1958), using a species of *Scenedesmus* and C^{14}-labeled EDTA, obtained data indicating that this compound can be broken down biologically, and that at least some algae can thereby obtain metals from their chelates.

The precise quantitative determination of the requirement for a micronutrient element in many cases also depends on the composition of the medium and on the concentration of other mineral factors. The antagonism between manganese and potassium, for example, is discussed below.

Research in mineral nutrition is frequently confused by the fact that the lack of any of several elements may produce the same symptoms, usually including retarded growth. For instance, Young (1935) reported experiments in which 33 elements, in addition to major nutrients, increased the growth of *Chlorella* sp. and *Crucigenia* sp., and Emerson and Lewis (1939) found that in trace amounts Mn, Zn, Cu, Mo, B, Ti, Cr, Co, and W are all capable of affecting photosynthesis of *Chlorella pyrenoidosa*. A specific action and an absolute requirement for an element are therefore often difficult to determine.

Likewise the special functions of a chemical element, and secondary effects caused indirectly by the metabolic process affected, are hard to distinguish when the element is involved in a dominant reaction of cell metabolism (Pirson, 1955, 1958a, b). To facilitate the distinction between primary and secondary effects of a deficiency in physiological experiments, Pirson (1937) introduced the "recovery technique." This is based on measurements of the gas exchange in mineral-deficient cells before and after addition of the limiting element. If, immediately or shortly after its addition, the gas exchange returns to normal, it is concluded that the metabolic process of respiration or photosynthesis is directly involved with the trace element. On the other hand, a slow recovery, or no immediate influence of the added element, indicates a more profound cellular disability, for example in protein metabolism, which only indirectly affects the metabolic processes measured.

II. Micronutrient Elements and Algal Ecology

Little has been done in the wide field of the ecological aspects of mineral nutrition in algae. Detailed surveys have been published by Kalle (1958), Gessner (1958), Harvey (1957), and Hutchinson (1957), who also offer many suggestions on possible interrelations between geological, physical, and chemical data and algal distribution. Uspenski (1927) concluded from his experiments with *Spirogyra rivularis*, *Cladophora fracta*, and *Drepanocladus fluitans* that the distribution of many such plants depends on the concentration of "active iron," but Pringsheim (1934) criticized his conclusion mainly on the basis of the method employed for the determination of available iron and its complexes. The possible role of iron as an ecological factor was also discussed by Gran (1933), who suggested that fluctuations in the population density of phytoplankton in the Gulf of Maine could be correlated with periodic rises in the iron concentration of inshore waters and with nutrients and humic compounds washed from the soil near the shore.

Cooper (1937), confirming earlier experiments of Hopkins and Wann (1925, 1927), showed that the solubility of iron hydroxide in alkaline solutions is very low, and that less than $10^{-12}M$ of iron in equilibrium with ferric hydroxide can exist in sea water at pH 8.0 to 8.2. Cooper therefore suggested a direct utilization of particulate ferric hydroxide by algae. Harvey (1937) and Goldberg (1952) concluded from experiments with diatoms that marine phytoplankton can utilize iron in colloidal suspensions. However, the fact that aged nutrient media containing inorganic iron are less favorable for growth than freshly prepared media (Rodhe, 1948; Gerloff *et al.*, 1950) suggests that the slow formation of colloidal complexes, which are only slightly dissociated, might render ferric iron biologically unavailable.

In the Ulcha Reservoir of the Moscow-Volga Canal, Gusseva (1937a, b) found evidence that manganese, at concentrations within its natural range, inhibited *Aphanizomenon flos-aquae,* and concluded that the occasional decline of this alga may be due to concomitant increases of inorganic nitrogen and manganese. Harvey (1939), on the other hand, suggested that in sea water manganese deficiency might be a limiting ecological factor in the development of phytoplankton. The addition of one part of manganese in 10^9 of natural sea water (i.e., $2 \times 10^{-11}M$) proved sufficient to stimulate vigorous growth of *Ditylum brightwellii,* "*Chlamydomonas*" sp. (probably *Dunaliella tertiolecta*), *Chlorella* sp., a cryptomonad, and two chrysomonads.

Gerloff and Skoog (1957a, b) investigated the amounts of iron, manganese, nitrogen, and phosphorus in lake water of Wisconsin and the importance of these elements for the mass appearances or "blooms" of *Microcystis aeruginosa,* and concluded that the limiting factors were apparently always major rather than minor nutrient elements. By an isotope dilution technique Knauss and Porter (1954) determined the quantities of Fe, Mn, Zn, and Cu present in *Chlorella pyrenoidosa* grown in the presence of various concentrations of these elements, and found that the absorption of such cations by the alga is directly proportional to the concentration available in the nutrient solution. However, in several marine algae the contents of Co, Cu, Fe, Mn, Mo, Ni, Pb, Zn, and of some macronutrients showed no obvious correlations when compared with the occurrence of these elements in the water of the Atlantic coast of Canada (Young and Langille, 1958; Wort, 1955). The amounts of iron, boron and zinc were found to be generally higher in Brown Algae than in marine Green Algae (Pillai, 1956); it is probable that in the former the cations may be partly bound by the alginic acid in the walls (cf. Eppley, Chapter 16, this treatise).

III. Specific Micronutrients

A. Iron

1. Growth

Walker (1953, 1954) found about 30 μg. of iron per gram dry weight of *Chlorella pyrenoidosa* cells. Since iron is a constituent of many enzymes, and of cytochromes and certain other porphyrins (Warburg, 1948; Hewitt, 1958), it is not surprising that the most obvious symptom of iron deficiency is a retardation of growth (Eilers, 1926; Hopkins, 1930c; Noack and Pirson, 1939; Pringsheim, 1946b; Myers, 1947; Henkel, 1951; Goldberg, 1952; Pirson *et al.,* 1952; Walker, 1953, 1954). The optimum amount of iron for growth depends on the species investigated and to some extent on the composition of the nutrient solution. Concentrations of 1.8×10^{-7} to

$2.6 \times 10^{-8}M$ were found adequate for the growth of *Chlorella* (Myers, 1947; Hopkins, 1930c), but Goldberg (1952) reported that for satisfactory growth *Asterionella japonica* required only $10^{-10}M$. When EDTA is present in the nutrient solution, larger quantities of iron must be supplied.

2. *Photosynthesis and Pigments*

Reduced rates of growth in media deficient in iron may be due chiefly to a reduction of photosynthesis (Emerson, 1929; Fleischer, 1935; Kennedy, 1940; Pirson *et al.*, 1952), which may in turn result from a decreased chlorophyll content. A direct correlation between photosynthetic activity and chlorophyll content was demonstrated in *Chlorella pyrenoidosa* by Emerson (1929), who concluded that a reduced chlorophyll content is the only factor responsible for the reduced photosynthesis of iron-deficient cells. Pirson *et al.* (1952) found that conditions of iron deficiency in *Ankistrodesmus falcatus* led to a slowly increasing chlorosis which, in its early stages, was paralleled by a decreasing rate of photosynthesis. In applications of the recovery technique to manometric experiments, they found that photosynthetic activity was not restored immediately after the addition of iron to deficient cells. They presumed that a deficiency of iron reduces the synthesis of proteins within the chloroplasts, and that the decreased content of chlorophyll is only an indirect consequence thereof.

Granick (1951, 1955) investigated the metabolism of iron in mutants of *Chlorella vulgaris* and concluded that, in the biosynthesis of chlorophyll, iron is somehow involved in the incorporation of magnesium into the protoporphyrin molecule.

Since the cytochromes are iron porphyrins, it appears that certain enzymes may be involved in the photosynthetic mechanism, as indicated for instance by Duysens (1954) and Chance and Sager (1957), providing a further possible explanation for the adverse effect of iron deficiency on this process.

3. *Respiration*

Except for a slight increase in the respiratory rate of iron-deficient cells of *Ankistrodesmus braunii* on the addition of $3.6 \times 10^{-5}M$ Fe (Pirson *et al.*, 1952; Kessler, 1955), no direct influence of this element on the metabolism of algae in darkness has been observed. However, in *Oscillatoria princeps* Fredrick (1958) reported an activation of phosphorylase by iron.

4. *Nitrogen Metabolism*

The reduction of nitrate in the light, whether an assimilable carbon source is present or not, appears not to be primarily affected by iron deficiency

(Noack and Pirson, 1939). Although in darkness iron-deficient cells become more chlorotic than in light, nitrogen assimilation continues for a longer period.

5. *Relations to Other Elements*

An indication of a relation between iron and manganese, similar to that reported for higher plants, was found in studies of *Chlorella* sp. (Hopkins, 1930a, b), and is probably related to nitrate assimilation as discussed later in this chapter.

B. Manganese

1. *Growth*

Manganese was shown to be necessary for various species of *Chlorella* (Hopkins, 1930a, b; Pirson, 1937; Alberts-Dietert, 1941; Harvey, 1947; Walker, 1953, 1954; Pirson and Bergmann, 1955; Bergmann, 1955; Reisner and Thompson, 1956a, b; Eyster *et al.*, 1956, 1958a, b; Brown *et al.*, 1958), *Ankistrodesmus* (Pirson *et al.*, 1952), several diatoms (Harvey, 1939; von Stosch, 1942; Spencer, 1957), some algal flagellates (Harvey, 1947; Provasoli and Pintner, 1953), *Ulothrix implexa* and *Bangia pumila* (Henkel, 1951) and *Ulva lactuca* (Levring, 1946). However for some organism such as *Scenedesmus quadricauda* a manganese requirement could not so readily be established (Österlind, 1948).

According to Harvey (1947), $1\text{–}4 \times 10^{-8} M$ manganese is necessary for vigorous growth of "*Chlamydomonas*" (=*Dunaliella tertiolecta*) and a marine species of *Cryptomonas*. For *Chlorella pyrenoidosa*, Walker (1953) found 10^{-7} M to be sufficient for growth; Eyster *et al.* (1958a) concluded that as much as $10^{-7}M$ was required for good autotrophic growth of this species, although in heterotrophic cultures grown in darkness a deficiency is manifest only when the concentration of manganese is less than $10^{-10}M$. A similar difference between the manganese requirements under autotrophic and heterotrophic conditions was demonstrated for *C. vulgaris* by Reisner and Thompson (1956a). Pirson (1937), Pirson and Bergmann (1955), and Bergmann (1955) could not demonstrate any requirement for manganese in heterotrophic cultures of this species. In mixotrophic conditions, i.e., cultures grown with glucose in the light, a manganese requirement becomes apparent, as in autotrophic cultures (Bergmann, 1955); however, when the alga is transferred to completely heterotrophic conditions in the dark, manganese is apparently no longer required.

Alberts-Dietert (1941) observed an enlargement of *C. pyrenoidosa* cells in manganese-deficient cultures grown under mixotrophic and heterotrophic conditions. Von Stosch (1942) demonstrated an enlargement of the cells and the absence of normal wall formation in *Achnanthes longipes*, similarly due to manganese deficiency.

2. Photosynthesis

The rate of photosynthesis is always decreased by manganese deficiency (Pirson, 1937; Pirson *et al.*, 1952; Brown, 1954; Arnon, 1954; Kessler, 1955; Kessler *et al.*, 1957; Pirson, 1958a, 1960), both in weak light and at light saturation (Arnon, 1954). In manometric experiments, the addition of manganese to deficient algae is generally followed by an immediate increase in the rate of photosynthesis; after a long exposure of the algae to deficiency conditions, however, recovery is slow and incomplete, this low rate perhaps being attributable to chloroplast damage (Pirson *et al.*, 1952). Brown (1959) showed that under anaerobic conditions $C^{14}O_2$ uptake by *Chlorella pyrenoidosa* is independent of manganese, whereas a dependence can be demonstrated under aerobic conditions.

Chlorosis has never been observed to be induced by the absence of manganese; on the contrary, Bergmann (1955) and Alberts-Dietert (1941) even noted an increase in the chlorophyll content of *C. vulgaris* and *C. pyrenoidosa* per unit dry weight. However, according to Eyster *et al.* (1958b), manganese deficiency raises the sensitivity of the chlorophyll to destruction by light, so that high light intensities may lead to chlorosis as a secondary symptom of the deficiency. Possibly correlated with this is the observation that the fluorescence of chlorophyll in cells of *Ankstrodesmus braunii* is increased by manganese deficiency (Kessler *et al.*, 1957).

The ability to carry out the Hill reaction is completely suppressed by manganese deficiency but can be at once restored by the addition of manganese (Brown, 1954; Eyster *et al.*, 1956, 1958b). Experiments of Kessler (1955) and Kessler *et al.* (1957) with *Ankistrodesmus braunii* likewise suggested that manganese may affect the oxygen-production mechanism in photosynthesis, since photoreduction is not reduced by manganese deficiency. Moreover photoreduction, which in this species is normally demonstrable only at low light intensities, can take place even at 4000 lux in Mn-deficient cultures. So far this effect has not been satisfactorily explained.

3. Respiration

Although a number of enzymes concerned with the breakdown of carbohydrates are activated by manganese (Hewitt, 1958), the effect of manganese deficiency on respiration is not marked, though the rate of endogenous respiration in *Chlorella pyrenoidosa* and *Ankistrodesmus braunii* is reduced thereby (Eyster *et al.*, 1956; Kessler, 1955). The oxidative assimilation of glucose is generally unchanged or may be slightly increased in the early stages of deficiency (Pirson *et al.*, 1952; Brown *et al.*, 1958), but it tends to decrease again with time.

4. *Nitrogen Metabolism*

The rate of autotrophic growth of *C. pyrenoidosa* is reduced in the absence of manganese, whether nitrate or an ammonium salt serves as a nitrogen source (Noack and Pirson, 1939; Noack *et al.*, 1940; Alberts-Dietert, 1941). Under mixotrophic or heterotrophic conditions, however, an influence of manganese deficiency on growth is observed only during nitrate assimilation. Kylin (1943, 1945) likewise noted the importance of manganese for the growth of *Ulva lactuca* in the presence of nitrate. Although, in growth experiments with several species of Chlorophyta, Algéus (1946) did not observe any influence of manganese on the assimilation of nitrate, one role of this cation was demonstrated by Kessler (1955), who showed that in hydrogen-adapted *Ankistrodesmus braunii* the reduction of nitrate with molecular hydrogen is increased in the presence of manganese.

5. *Relations to Other Elements*

Noack and Pirson (1939) and Alberts-Dietert (1941) concluded from their experiments on nitrogen metabolism in *C. pyrenoidosa* that iron may play a role subsidiary to that of manganese in nitrate reduction. Hopkins (1930c) considered that in *Chlorella* sp. manganese may be concerned with the reoxidation of iron after its reduction to the ferrous state. In manganese-deficient cells the concentration of ferrous iron may be too high for good growth, but with an excess of manganese the same concentration may be too low.

Warén (1933) reported an antagonism between calcium and manganese in *Micrasterias rotata*. In this desmid high concentrations of manganese (e.g., $5.7 \times 10^{-4} M$) inhibit growth and cell division at low calcium concentrations (e.g., 10^{-8} to $10^{-5} M$), but promote cell division at high calcium concentrations.

An antagonism between potassium and manganese was observed by Pirson (1937). At low concentrations of potassium, only $10^{-9} M$ of manganese is necessary for healthy growth of *C. pyrenoidosa*, whereas in high potassium concentrations the cells appear to require as much as $10^{-7} M$. High concentrations of sodium and calcium, too, are stated to be antagonistic to manganese. Relations of manganese to phosphorus metabolism were reported by Badour (1959), who found that in *C. vulgaris* the concentration of polyphosphates ("7-minute phosphates"—hydrolyzed within 7 minutes by $1N$ HCl at 100°C.) is raised under conditions of manganese deficiency. A detailed discussion and possible explanations of this effect have not yet been published.

Pringsheim (1946, 1953) showed that the brown color of the cell envelopes—or loricae— of certain algal flagellates (*Anthophysa, Siderodendron, Siphomonas, Trachelomonas*) is due to associations of manganese compounds with hydroxides of iron.

C. Other Micronutrients

1. *Molybdenum*

Molybdenum is known to be closely involved in nitrogen metabolism. General reviews of this function of molybdenum have been presented by Fogg (1956), Wolfe (1954a, b), and Anderson (1956). Bortels (1938, 1940) first demonstrated its requirement for a number of nitrogen-fixing Blue-green Algae (species of *Nostoc, Anabaena,* and *Cylindrospermum*), but its necessity for all groups of algae has not yet been demonstrated. Eyster (1959) showed that both molybdenum and manganese are essential elements for nitrate reduction in *Nostoc muscorum,* and that the minimum amount of molybdenum required for nitrate reduction ($10^{-10}M$) is much less than that required for nitrogen fixation ($10^{-7}M$).

The necessity of molybdenum for the growth of certain Green Algae (notably *Chlorella* spp.) using nitrate, but not ammonium or urea, as a nitrogen source was shown in 1953 by Walker and in 1955 by Arnon and his colleagues. Below $10^{-9}M$, cell division is inhibited (Ichioka and Arnon, 1955), chlorosis appears (Ichioka and Arnon, 1955; Arnon *et al.*, 1955; Loneragan and Arnon, 1954), and much starch tends to be stored (Ichioka and Arnon, 1955). All of these symptoms of deficiency can be alleviated by the addition of molybdate.

The well-known action of molybdenum as a cofactor of nitrate reductase in fungi (Nicholas and Nason, 1955) makes it seem probable that a similar direct action of this element on nitrate reduction in algae may be found. However, molybdenum may play other roles, apart from those in nitrogen metabolism. Wolfe (1954a, b) concluded that the absence of this element affects not only the reduction of nitrate by *Anabaena cylindrica,* but also the synthesis of proteins. He found that in Mo-deficient cells nitrate reduction was promoted by the addition of $10^{-8}M$ molybdenum, but that this effect could be reproduced by the addition of fumarate, succinate, or citrate. Thus, molybdenum appears to control in some way the supply of hydrogen donors or energy carriers necessary for protein synthesis.

Further fields of influence of molybdenum in respiration and photosynthesis should be mentioned, though the importance of the element for these physiological processes is not clear. A deficiency of molybdenum may produce indirect effects by upsetting protein metabolism. Likewise, since the rate of endogenous respiration is higher in deficient than in normal cells (Wolfe, 1954a, b; Loneragan and Arnon, 1954; Arnon *et al.*, 1955), and the chlorophyll content tends to be reduced, net photosynthesis may be suppressed.

2. Vanadium

The role of vanadium in the metabolism of algae has been little investigated, and the necessity for this element remains obscure. According to Arnon and Wessels (1953), in experiments with *Scenedesmus obliquus* grown in media deficient in vanadium the chlorophyll content per cell is little affected, but the dry weight is markedly reduced. In strong light the evolution of oxygen is inhibited, but it can be reactivated after a lag period by the addition of vanadate. From these experiments it appears that vanadium may be involved in photosynthesis, probably in one of the so-called "dark" reactions.

On the other hand, Warburg and his associates (1955a, b) could not demonstrate any effect of vanadium deficiency in *Chlorella pyrenoidosa* grown in strong light, although, surprisingly, they were able to detect deficiency symptoms by manometric experiments at low light intensities. In vanadium-deficient cells the CO_2/O_2 ratio was greater than 1, exceeding that of controls. Only quinquevalent vanadium proved effective for the restoration of normal photosynthesis. A clear explanation of these phenomena remains to be given.

3. Cobalt

Cobalt is known as a constituent of vitamin B_{12}. For details the reader is referred to articles by Hutner and his associates (1950, 1958) and by Droop (Chapter 8, this treatise).

Holm-Hansen and co-workers (1954) showed that cobalt is necessary for the healthy growth of several Blue-green Algae (*Nostoc muscorum*, *Calothrix parietina*, *Coccochloris peniocystis*, *Diplocystis* =*Microcystis aeruginosa*) and demonstrated its replaceability by vitamin B_{12}. Krauss (1955) reported that this element also stimulated the growth of a Green Alga, *Scenedesmus obliquus*. Though Ericson and Lewis (1953) suggested that cobalt should be required by all algae capable of synthesizing vitamin B_{12}, in few cases has this been experimentally established. Scott and Ericson (1955), who used Co^{60} to follow cobalt metabolism in *Rhodymenia palmata*, found no experimental evidence for its incorporation into vitamin B_{12}, although they detected—without identifying—some other stable organic compound containing cobalt. They also found that the uptake of Co^{60} depends on the supply of carbon dioxide, and concluded that certain products of photosynthesis may form complexes with cobalt ions. The role of such products in cell metabolism, specifically in the synthesis of vitamin B_{12}, remains to be elucidated.

4. Zinc

Although the importance of zinc for physiological functions in algae has not been the object of much investigation, it was found necessary for the growth of *Stichococcus bacillaris* (Eilers, 1926). Zinc in concentrations above 10^{-1}–10^{-2} mg./liter (in the range of $10^{-6}M$) was required by several algae in media containing the chelating agent EDTA (Provasoli and Pintner, 1953). In *Chlorella pyrenoidosa*, Walker (1954) obtained symptoms of zinc deficiency, notably a relative decrease in dry weight, at concentrations below $10^{-7}M$, and Stegmann (1940) demonstrated a decrease in chlorophyll formation and in photosynthetic activity in cultures deficient in this element. On the addition of zinc, the recovery of photosynthesis was associated with a steep rise in the chlorophyll concentration. The addition of *o*-phenanthroline, which forms specific complexes with zinc, leads to a decrease in photosynthesis (Warburg and Lüttgens, 1948), reversible by the further addition of zinc. The occurrence of this metal in carbonic anhydrase (Keilin and Mann, 1940) suggests that this element may participate in photosynthesis at the level of CO_2 fixation.

5. Copper

Deficiency symptoms appear in cultures of *Chlorella* sp. when the concentration of copper falls below $10^{-7}M$ (Walker, 1953). Certain algal flagellates also require this element (Provasoli and Pintner, 1953). Copper is needed for the prosthetic group of ascorbic acid oxidase in higher plants (Stotz *et al.*, 1937), and occurs in a number of other oxidizing enzymes. A direct participation of copper in photosynthesis was shown by Green and co-workers (1939), who demonstrated a reversible decrease in the photosynthetic activity of *C. pyrenoidosa* on the addition of organic compounds which form complexes with copper. Respiration is also reduced by copper deficiency.

Toxic effects of copper are evident in *C. vulgaris* at the low concentration of $10^{-7}M$ (Greenfield, 1942). *Ankistrodesmus* cultures can be adapted to tolerate higher concentrations of copper (Kellner, 1955), as indicated by a resumption of growth in sublethal concentrations after an extended lag phase. This apparent adaptation is probably due to mutation, since copper tolerance is retained for several months by cultures grown in the absence of copper; however, mutation has not been confirmed by the use of pure subcultures from single cells.

6. *Sodium*

Anabaena cylindrica and certain other freshwater Blue-green Algae require appreciable concentrations of sodium (Allen, 1952, 1955). Some strains can grow in the presence of either K or Na, but in others K inhibits growth in the absence of Na. The function of Na for these algae is not known (see also Guillard, Chapter 32, this treatise). The apparent absence of a potassium requirement is sufficiently remarkable to warrant a closer examination of the physiology of such organisms.

7. *Boron*

Although there are many known instances of the action of boron in the growth and metabolism of higher plants, a critical examination of requirements for this nonmetallic element in algae is long overdue. Algae only rarely show distinct symptoms of boron deficiency.

Its necessity for the growth of certain diatoms was demonstrated by Hercinger (1940). Suneson (1945) stimulated the development of zygotes and the growth of germlings of *Ulva lactuca* by adding 0.1 to 1.0 mg. of boron to a liter of natural sea water, i.e., 10^{-5} to $10^{-4}M$. According to Eyster (1952, 1959) the minimum amount of boron required for maximum growth of *Nostoc muscorum* is $0.9 \times 10^{-5}M$. Although Stegmann (1940), using quartz vessels, failed to demonstrate any boron deficiency in cultures of *Chlorella pyrenoidosa*, McIlrath and Skok (1957, 1958) found that boron-free cultures of *C. vulgaris* exhibited an increase in cell number and in dry weight per cell when boron was added at an optimum concentration of 0.5 mg./liter. No attempts were made by these authors to demonstrate whether this element could be replaced by others, so these results do not unequivocally establish the essentiality of boron for *C. vulgaris*.

According to Suneson (1945), 10 mg./liter of boron increased the apparent photosynthesis in *Ulva lactuca* and *Fucus vesiculosus*. Eyster (1952) observed a decrease in chlorophyll content and a reduction of growth of *Nostoc muscorum* as a result of boron deficiency.

References

Alberts-Dietert, F. (1941). Die Wirkung von Eisen und Mangan auf die Stickstoffassimilation von *Chlorella*. *Planta* **32,** 88–117.

Algéus, S. (1946). Untersuchungen über die Ernährungsphysiologie der Chlorophyceen. *Botan. Notiser* **1946,** 129.

Allen, M. B. (1952). The cultivation of Myxophyceae. *Arch. Mikrobiol.* **17,** 34–53.

Allen, M. B. (1955). Studies on the nitrogen-fixing blue-green algae. II. The sodium requirement of *Anabaena cylindrica. Physiol. Plantarum* **8,** 653–660.

Allen, M. B. (1956). Photosynthetic nitrogen fixation by blue-green algae. *Sci. Monthly* **83,** 100–106.

Anderson, A. J. (1956). Role of molybdenum in plant nutrition. *In* "Inorganic Nitrogen Metabolism," McCollum-Pratt Inst. Symposium (W. D. McElroy and B. Glass, eds.), pp. 3–58. Johns Hopkins Press, Baltimore.

Arnon, D. I. (1954). Some recent advances in the study of essential micronutrients for green plants. *Congr. intern. botan.* 8ᵉ *Congr. Paris, Sect.* **11/12,** 73–80.

Arnon, D. I., and Wessel, G. (1953). Vanadium as an essential element for green plants. *Nature* **172,** 1039–1040.

Arnon, D. I., Ichioka, P. S., Wessel, G., Fujiwara, A., and Woolley, J. T. (1955). Molybdenum in relation to nitrogen metabolism. I. Assimilation of nitrate nitrogen by *Scenedesmus. Physiol. Plantarum* **8,** 538–551.

Badour, S. S. A. (1959). Analytisch-chemische Untersuchung des Kaliummangels bei *Chlorella* im Vergleich mit anderen Mangelzuständen. Dissertation, University of Göttingen, Germany.

Bergmann, L. (1955). Stoffwechsel und Mineralsalzernährung einzelliger Grünalgen. II. Vergleichende Untersuchungen über den Einfluss mineralischer Faktoren bei heterotropher und mixotropher Ernährung. *Flora (Jena)* **142,** 493–539.

Bitcover, E. H., and Sieling, D. H. (1951). Effect of various factors on the utilization of nitrogen and iron by *Spirodela polyrhiza* (L) Schleid. *Plant Physiol.* **26,** 290–303.

Bortels, H. (1938). Entwicklung und Stickstoffbindung bestimmter Mikroorganismen in Abhängigkeit von Spurenelementen und vom Wetter. *Ber. deut. botan. Ges.* **56,** 153–160.

Bortels, H. (1940). Über die Bedeutung des Molybdäns für stickstoffbindende Nostocaceen. *Arch. Mikrobiol.* **11,** 155–186.

Brown, T. E. (1954). Comparative studies of photosynthesis and the Hill reaction in *Nostoc muscorum* and *Chlorella pyrenoidosa.* Ph.D. Thesis, Ohio State University, Columbus, Ohio.

Brown, T. E. (1959). Anaerobic uptake of $C^{14}O_2$ by manganese-deficient and normal *Chlorella pyrenoidosa. Proc. Intern. Botan. Congr., 9th Congr. Montreal,* Vol. II, Abstracts, p. 49.

Brown, T. E., Eyster, H. C., and Tanner, H. A. (1958). Physiological effects of manganese deficiency. *In* "Trace Elements" (C. A. Lamb, O. G. Bentley, and J. M. Beattie, eds.), pp. 135–155. Academic Press, New York.

Brunel, J., Prescott, G. W., and Tiffany, L. H., eds. (1950). "The Culturing of Algae: A Symposium." The Charles F. Kettering Foundation, Yellow Springs, Ohio.

Burlew, J. S., ed. (1953). "Algal Culture. From Laboratory to Pilot Plant." *Carnegie Inst. Wash. Publ.* **No. 600.**

Chance, B., and Sager, R. (1957). Oxygen and light induced oxidations of cytochrome, flavoprotein, and pyridine nucleotide in *Chlamydomonas* mutant. *Plant Physiol.* **32,** 548–561.

Cooper, L. H. N. (1937). Some conditions governing the solubility of iron. *Proc. Roy. Soc.* **B124,** 299–307.

Duysens, L. M. N. (1954). Reversible chances in the absorption spectrum of *Chlorella* upon irradiation. *Science* **120,** 353–354.

Eilers, H. (1926). Zur Kenntnis der Ernährungsphysiologie von *Stichococcus bacillaris* Näg. *Rec. trav. botan. neerl.* **23,** 362–395.
Emerson, R. (1929). The relation between maximum rate of photosynthesis and concentration of chlorophyll. *J. Gen. Physiol.* **12,** 609–622.
Emerson, R., and Lewis, C. M. (1939). Factors influencing the efficiency of photosynthesis. *Am. J. Botany* **26,** 802–822.
Ericson, L. E., and Lewis, L. (1953). On the occurrence of vitamin B_{12}-factors in marine algae. *Arkiv. Kemi* **6,** 427–442.
Eyster, C. (1952). Necessity of boron for *Nostoc muscorum. Nature* **170,** 755.
Eyster, C. (1959). Mineral requirements of *Nostoc muscorum* for nitrogen fixation. *Proc. Intern. Botan. Congr., 9th Congr., Montreal,* Vol. II, Abstracts, p. 109.
Eyster, C., Brown, T. E., and Tanner, H. A. (1956). Manganese requirement with respect to respiration and the Hill reaction in *Chlorella pyrenoidosa. Arch. Biochem. Biophys.* **64,** 240–241.
Eyster, C., Brown, T. E., and Tanner, H. A. (1958a). Mineral requirements for *Chlorella pyrenoidosa* under autotrophic and heterotrophic conditions. *In* "Trace Elements" (C. A. Lamb, O. G. Bentley, and J. M. Beattie, eds.), pp. 157–174. Academic Press, New York.
Eyster, C., Brown, T. E., Tanner, H., and Hood, S. L. (1958b). Manganese requirement with respect to growth, Hill reaction and photosynthesis. *Plant Physiol.* **33,** 235–241.
Fleischer, W. (1935). The relation between chlorophyll content and rate of photosynthesis. *J. Gen. Physiol.* **18,** 573–597.
Fogg, G. E. (1956). The comparative physiology and biochemistry of the blue-green algae. *Bacterial Revs.* **20,** 148–165.
Frederick, J. F. (1958). Use of chelation phenomenon in studies of the structure and action mechanisms of *Oscillatoria* phosphorylase. *Physiol. Plantarum* **11,** 493–502.
Gerloff, G. C., and Skoog, F. (1957a). Availability of iron and manganese in Southern Wisconsin lakes for the growth of *Microcystis aeruginosa. Ecology* **38,** 551–556.
Gerloff, G. C., and Skoog, F. (1957b). Nitrogen as a limiting factor for the growth of *Microcystis aeruginosa* in Southern Wisconsin lakes. *Ecology* **38,** 556–561.
Gerloff, G. C., Fitzgerald, G. P., and Skoog, F. (1950). The mineral Nutrition of *Coccochloris peniocystis. Am. J. Botany* **37,** 835–840.
Gessner, F. (1958). Die Binnengewässer. *In* "Handbuch der Pflanzenphysiologic" (W. Ruhland, ed.), Vol. IV, pp. 179–232. Springer, Berlin.
Goldberg, E. D. (1952). Iron assimilation by marine diatoms. *Biol. Bull.* **102,** 243–248.
Goldschmidt, V. M., and Peters, C. (1932). Zur Geochemie des Bors. II. *Nachr. Ges Wiss. Göttingen Jahresber. Geschäftsjahr Math. physik. Kl.* **5,** 528–545.
Gran, H. H. (1933). Studies on the biology and chemistry of the Gulf of Maine. II. Distribution of the phytoplankton in August, 1932. *Biol. Bull.* **64,** 159–181.
Granick, S. (1951). Biosynthesis of chlorophyll and related pigments. *Ann. Rev. Plant Physiol.* **2,** 115–144.
Granick, S. (1955). Porphyrin and chlorophyll biosynthesis in *Chlorella. In* "Ciba Foundation Symposium on Porphyrin Biosynthesis and Metabolism" (G. E. W. Wolstenholme and E. C. P. Millar, eds.), pp. 143–152. Churchill, London.
Green, L. F., McCarthy, J. F., and King, C. G. (1939). Inhibition of respiration and photosynthesis in *Chlorella pyrenoidosa* by organic compounds that inhibit copper catalysis. *J. Biol. Chem.* **128,** 447–453.
Greenfield, S. S. (1942). Inhibitory effects of inorganic compounds on photosynthesis in *Chlorella. Am. J. Botany* **29,** 121–131.

Gusseva, K. A. (1937a). [Effect of manganese on the development of algae.] *Mikrobiologiya* **6,** 292–307. (In Russian.)

Gusseva, K. A. (1937b). [Hydro- and microbiological studies of the Ulcha reservoir of the Moscow Volga canal.] *Mikrobiologiya* **6,** 449–464. (In Russian.)

Gusseva, K. A. (1940). [The influence of copper on algae.] *Mikrobiologiya* 0, 480–489. (In Russian.)

Harvey, H. W. (1937). The supply of iron to diatoms. *J. Marine Biol. Assoc. United Kingdom* **22,** 205–219.

Harvey, H. W. (1939). Substances controlling the growth of a diatom. *J. Marine Biol. Assoc. United Kingdom* **23,** 499–520.

Harvey, H. W. (1947). Manganese and the growth of phytoplankton. *J. Marine Biol. Assoc. United Kingdom* **26,** 562–579.

Harvey, H. W. (1957). "The Chemistry and Fertility of Sea Water." Cambridge Univ. Press, London and New York.

Henkel, R. (1951). Ernährungsphysiologische Untersuchungen an Meeresalgen, insbesondere *Bangia pumila. Kiel. Meeresforsch.* **8,** 192–211.

Hercinger, F. (1940). Beiträge zum Wirkungskreislauf des Bors. *Bodenk. u. Pflanzenernähr.* **16,** 141–168.

Hewitt, E. J. (1958). The role of mineral elements in the activity of plant enzyme systems. *In* "Handbuch der Pflanzenphysiologie" (W. Ruhland, ed.), Vol. IV, pp. 427–481. Springer, Berlin.

Holm-Hansen, O., Gerloff, G. C., and Skoog, F. (1954). Cobalt as an essential element for blue-green algae. *Physiol. Plantarum* **7,** 665–675.

Hopkins, E. F. (1930a). The necessity and function of manganese in the growth of *Chlorella* sp. *Science* **72,** 609–610.

Hopkins, E. F. (1930b). Manganese an essential element for a green alga. (Abstract.) *Am. J. Botany* **17,** Suppl., 1047.

Hopkins, E. F. (1930c). Iron-ion concentration in relation to growth and other biological processes. *Botan. Gaz.* **89,** 209–240.

Hopkins, E. F., and Wann, F. B. (1925). The effect of the hydrogen-ion concentration on the availability of iron for *Chlorella* sp. *J. Gen. Physiol.* **9,** 205–210.

Hopkins, E. F., and Wann, F. B. (1926). Relation of the hydrogen-ion concentration to growth of *Chlorella* and the availability of iron. *Botan. Gaz.* **81,** 353–376.

Hopkins, E. F., and Wann, F. B. (1927). Iron requirements for *Chlorella. Botan. Gaz.* **84,** 407–427.

Hutchinson, G. E. (1957). "A Treatise on Limnology," Vol. I. Wiley, New York.

Hutner, S. H. (1948). Essentiality of constituents of sea water for growth of a marine diatom. *Trans. N.Y. Acad. Sci. Ser.* 2, **10,** 136–141.

Hutner, S. H., and Provasoli, L. (1955). Comparative biochemistry of flagellates. *In* "Biochemistry and Physiology of Protozoa" (A. Lwoff, ed.), Vol. II, pp. 17–43. Academic Press, New York.

Hutner, S. H., Provasoli, L., Schatz, A., and Haskins, C. P. (1950). Some approaches to the study of the role of metals in the metabolism of microorganisms. *Proc. Am. Phil. Soc.* **94,** 152–170.

Hutner, S. H., Aaronson, S., Nathan, H. A., Baker, H., Scher, S., and Cury, A. (1958). Trace elements in microorganisms: the temperature factor approach. *In* "Trace Elements" (C. A. Lamb, O. G. Bentley, and J. M. Beattie, eds.), pp. 47–65. Academic Press, New York.

Ichioka, P. S., and Arnon, D. I. (1955). Molybdenum in relation to nitrogen metabolism. II. Assimilation of ammonia and urea without molybdenum by *Scenedesmus. Physiol. Plantarum* **8,** 552–560.

Kalle, K. (1958). Das Meerwasser als Mineralstoffquelle der Pflanze. *In* "Handbuch der Pflanzenphysiologie" (W. Ruhland, ed.), Vol. IV, pp. 170–178.

Keilin, D., and Mann, T. (1940). Carbonic anhydrase, purification and nature of enzyme. *Biochem. J.* **34,** 1163–1174.

Kellner, K. (1955). Die Adaptation von *Ankistrodesmus braunii* an Rubidium und Kupfer. *Biol. Zentr.* **74,** 662–691.

Kennedy, S. R., Jr. (1940). The influence of magnesium deficiency, chlorophyll concentration, and heat treatments on the rate of photosynthesis in *Chlorella. Am. J. Botany* **27,** 68–73.

Kessler, E. (1955). On the role of manganese in the oxygen-evolving system in photosynthesis. *Arch. Biochem. Biophys.* **59,** 527–529.

Kessler, E., Arthur, W., and Brugger, J. E. (1957). The influence of manganese and phosphate on delayed light emission, fluorescence, photoreduction and photosynthesis in algae. *Arch. Biochem. Biophys.* **71,** 326–335.

Knauss, H. J., and Porter, J. W. (1954). The absorption of inorganic ions by *Chlorella pyrenoidosa. Plant Physiol.* **29,** 229–234.

Krauss, R. W. (1953). Inorganic nutrition of algae. *In* "Algal Culture, From Laboratory to Pilot Plant" (J. S. Burlew, ed.). *Carnegie Inst. Wash. Publ.* **No. 600,** 85–102.

Krauss, R. W. (1955). Nutrient supply for large-scale algal cultures. *Sci. Monthly* **80,** 21–28.

Krauss, R. W., and Specht, A. W. (1958). Nutritional requirements and yields of algae in mass culture. *In* "Transactions of the Conference on the Use of Solar Energy. The Scientific Basis" (E. F. Carpenter, ed.), Vol. IV: Photochemical Processes, pp. 12–26. Univ. of Arizona Press, Tucson, Arizona.

Krauss, R. W., and Thomas, W. H. (1954). The growth and inorganic nutrition of *Scenedesmus obliquus* in mass culture. *Plant Physiol.* **29,** 205–214.

Kylin, A. (1943). The influence of trace elements on the growth of *Ulva lactuca. Kgl. Fysiograf. Sällskap. i Lund Förh.* **13,** 202–209.

Kylin, A. (1945). The nitrogen sources and the influence of manganese on the nitrogen assimilation of *Ulva lactuca. Kgl. Fysiograf. Sällskap. i Lund Förh.* **15,** 27–35.

Levring, T. (1946). Some culture experiments with *Ulva* and artificial sea water. *Kgl. Fysiograf. Sällsk. i Lund Förh.* **16,** 45–46.

Lewin, R. A. (1954). A marine *Stichococcus* sp. which requires Vitamin B_{12} (Cobalamin). *J. Gen. Microbiol.* **10,** 93–96.

Lonergan, J. F., and Arnon, D. I. (1954). Molybdenum in the growth and metabolism of *Chlorella. Nature* **174,** 459.

Lorenzen, H. (1959). Die photosynthetische Sauerstoffproduktion wachsender *Chlorella* bei langfristig intermittierender Belichtung. *Flora (Jena)* **147,** 382–404.

McIlrath, W. J., and Skok, J. (1957). Influence of boron on the growth of *Chlorella. Plant Physiol.* **32,** Suppl. xxiii.

McIlrath, W. J., and Skok, J. (1958). Boron requirement of *Chlorella vulgaris. Botan. Gaz.* **119,** 231–233.

Molisch, H. (1895). Die Ernährung der Algen. (Süsswasseralgen, I. Abhandl.) *Sitzber. Akad. Wiss. Wien, Math. naturw. Kl. Abt. I,* **104,** 783–800.

Molisch, H. (1896). Die Ernährung der Algen. (Süsswasseralgen, II. Abhandl.) *Sitzber. Akad. Wiss. Wien, Math. naturw. Kl. Abt. I,* **105,** 633–648.

Myers, J. (1947). Culture conditions and the development of the photosynthetic mechanism. V. Influence of the composition of the nutrient medium. *Plant Physiol.* **22,** 590–597.

Myers, J., and Clark, L. B. (1944). Culture conditions and the development of the photosynthetic mechanism. II. An apparatus for the continuous culture of *Chlorella*. *J. Gen. Physiol.* **28,** 103–112.

Nicholas, D. J. D., and Nason, A. (1955). Molybdenum and nitrate reductase. II. Molybdenum as a constituent of nitrate reduction. *J. Biol. Chem.* **207,** 353–360.

Noack, K., und Pirson, A. (1939). Die Wirkung von Eisen und Mangan auf die Stickstoffassimilation von *Chlorella*. *Ber. deut. botan. Ges.* **57,** 442–452.

Noack, K., Pirson, A., und Stegmann, G. (1940). Der Bedarf an Spurenelementen bei *Chlorella*. *Naturwissenschaften* **28,** 172–173.

Österlind, S. (1948). Growth conditions of the alga *Scenedesmus quadricauda* with special reference to the inorganic carbon-sources. *Symbolae Botan. Upsalienses* **10**(3), 1–141.

Pillai, V. K. (1956). Chemical studies on Indian seaweeds. I. Mineral constituents. *Proc. Indian Acad. Sci.* **B44,** 3–29.

Pirson, A. (1937). Ernährungs- und stoffwechselphysiologische Untersuchungen an *Fontinalis* und *Chlorella*. *Z. Botan.* **31,** 193–267.

Pirson, A. (1955). Functional aspects in mineral nutrition of green plants. *Ann. Rev. Plant. Physiol.* **6,** 71–114.

Pirson, A. (1958a). Manganese and its role in photosynthesis. *In* "Trace Elements" (C. A. Lamb, O. G. Benley, and J. M. Beattie, eds.), pp. 81–98. Academic Press, New York.

Pirson, A. (1958b). Mineralstoffwechsel und Photosynthese. *In* "Handbuch der Pflanzenphysiologie" (W. Ruhland, ed.). Vol. IV, pp. 355–381. Springer, Berlin.

Pirson, A. (1960). Photosynthese und mineralische Faktoren. *In* "Handbuch der Pflanzenphysiologie" (W. Ruhland, ed.), Vol. V, Part 2, pp. 123–151. Springer, Berlin.

Pirson, A., und Bergmann, L. (1955). Manganese requirement and carbon source in *Chlorella*. *Nature* **176,** 209.

Pirson, A., Tichy, C., und Wilhelmi, G. (1952). Stoffwechsel und Mineralsalzernährung einzelliger Grünalgen. I. Vergleichende Untersuchungen an Mangelkulturen von *Ankistrodesmus*. *Planta* **40,** 199–253.

Pollard, A. L., and Smith, P. B. (1951). The adsorption of manganese by polysaccharides. *Science* **114,** 413–414.

Pringsheim, E. G. (1926). Über das Ca-Bedürfnis einiger Algen. *Planta* **2,** 555–568.

Pringsheim, E. G. (1934). Untersuchungen zu Uspenskis Eisenhypothese der Algenverbreitung. *Planta* **22,** 269–312.

Pringsheim, E. G. (1946a). "Pure culture of Algae, Their Preparation and Maintenance." Cambridge Univ. Press, London and New York.

Pringsheim, E. G. (1946b). On iron flagellates. *Phil. Trans. Roy. Soc. London Ser.* **B232,** 311–342.

Pringsheim, E. G. (1953). Observations on some species of *Trachelomonas* grown in culture. *New Phytologist* **52,** 238–266.

Pringsheim, E. G. (1954). "Algenreinkulturen, ihre Herstellung und Erhaltung." Fischer, Jena, Germany.

Provasoli, L., and Pintner, J. J. (1953). Ecological implications of *in vitro* nutritional requirements of algal flagellates. *Ann. N.Y. Acad. Sci.* **56,** 839–851.

Reisner, G. S., and Thompson, J. F. (1956a). Manganese deficiency in *Chlorella* under heterotrophic carbon nutrition. *Nature* **178,** 1473–1474.

Reisner, G. S., and Thompson, J. F. (1956b). The large scale laboratory culture of *Chlorella* under conditions of micronutrient element deficiency. *Plant Physiol.* **31,** 181–185.

Rodhe, W. (1948). Environmental requirements of fresh-water plankton algae. *Symbolae Botan. Upsalienses* **10**(1), 1–149.

Schatz, A., and Hutner, S. H. (1949). An inert metal carrier for culture media. *Bacteriol. Proc. Soc. Am. Bacteriologists*, p. 34.

Scott, R., and Ericson, L. E. (1955). Some aspects of cobalt metabolism by *Rhodymenia palmata* with particular reference to vitamin B_{12} content. *J. Exptl. Botany* **6**, 348–361.

Spencer, C. P. (1957). Utilisation of trace elements by marine unicellular algae. (Symposium) *J. Gen. Microbiol.* **16**, 282–285.

Stegmann, G. (1940). Die Bedeutung der Spurenelemente für *Chlorella*. *Z. Botan.* **35**, 385–422.

Stotz, E., Harrer, C. J., and King, C. G. (1937). A study of ascorbic acid oxidase in relation to copper. *J. Biol. Chem.* **119**, 511–522.

Suneson, S. (1945). Einige Versuche über den Einfluss des Bors auf die Entwicklung und Photosynthese der Meeresalgen. *Kgl. Fysiograf. sällskap. i Lund Förh.* **15**, 185–197.

Url, W. (1955). Resistenz von Desmidiaceen gegen Schwermetallsalze. *Österr. Akad. Wiss. Math. naturw. Kl. Sitzber. Abt. I*, **164**, 207–230.

Uspenski, E. E. (1927). Eisen als Faktor für die Verbreitung niederer Wasserpflanzen. *In* "Pflanzenforschung" (R. Kolkwitz, ed.), Vol. IX, 104 pp. Fischer, Jena, Germany.

von Stosch, H.-A. (1942). Form und Formwechsel der Diatomee *Achnanthes longipes* in Abhängigkeit von der Ernährung. Mit besonderer Berücksichtigung der Spurenstoffe. *Ber. deut. botan. Ges.* **60**, 2–16.

Walker, J. B. (1953). Inorganic micronutrient requirements of *Chlorella*. I. Requirements for calcium (or strontium), copper, and molybdenum. *Arch. Biochem. Biophys.* **46**, 1–11.

Walker, J. B. (1954). Inorganic micronutrient requirement of *Chlorella*. II. Quantitative requirements for iron, manganese, and zinc. *Arch. Biochem. Biophys.* **53**, 1–8.

Walker, J. B. (1956). Strontium inhibition of calcium utilization by green algae. *Arch. Biochem. Biophys.* **60**, 264–265.

Wann, F. B., and Hopkins, E. F. (1927). Further studies on growth of *Chlorella* as affected by hydrogen-ion concentration. *Botan. Gaz.* **83**, 194–201.

Warburg, O. (1948). "Schwermetalle als Wirkungsgruppen von Fermenten." Springer, Berlin.

Warburg, O., und Lüttgens, W. (1948). Photochemische Reduktion von Chinon in grünen Zellen und Granula. *In* "Schwermetalle als Wirkungsgruppe von Fermenten" (O. Warburg, ed.), pp. 170–184. Springer, Berlin.

Warburg, O., Krippahl, G., und Buchholz, W. (1955a). Wirkung von Vanadium auf die Photosynthese. *Z. Naturforsch.* **10b**, 422.

Warburg, O., Krippahl, G., und Schröder, W. (1955b). Wirkungsspektrum eines Photosynthesefermentes. *Z. Naturforsch.* **10b**, 631–639.

Warén, H. (1933). Über die Rolle des Calciums im Leben der Zelle auf Grund von Versuchen an *Micrasterias*. *Planta* **19**, 1–45.

Wiessner, W. (1960). Wachstum und Stoffwechsel von *Rhodopseudomonas spheroides* in Abhängigkeit von der Versorgung mit Mangan und Eisen. *Flora (Jena)* **149**, 1–42.

Wolfe, M. (1954a). The effect of molybdenum upon the nitrogen metabolism of *Anabaena cylindrica*. I. A study of the molybdenum requirement for nitrogen fixation and for nitrate and ammonia assimilation. *Ann. Botany (London)* **18**, 299–308.

Wolfe, M. (1954b). The effect of molybdenum upon the nitrogen metabolism of *Anabaena cylindrica*. II. A more detailed study of the action of molybdenum in nitrate assimilation. *Ann. Botany (London)* **18**, 309–325.

Wort, D. J. (1955). The seasonal variation in chemical composition of *Macrocystis integrifolia* and *Nereocystis luetkeana* in British Columbia coastal waters. *Can. J. Botany* **33**, 323–340.

Young, E. G., and Langille, W. M. (1958). The occurrence of inorganic elements in marine algae of the Atlantic Provinces of Canada. *Can. J. Botany* **36**, 301–310.

Young, R. S. (1935). Certain rarer elements in soils and fertilizers and their role in plant growth. *Cornell Univ. Agr. Expt. Sta. Mem. No. 174.*

—Part II—

Composition of Cells and Metabolic Products

—18—

Storage Products

B. J. D. MEEUSE

Botany Department, University of Washington, Seattle, Washington

I. Introduction

This chapter will be concerned chiefly with β-1:3-linked glucans, such as laminarin, leucosin, paramylon, and callose, which now appear to be quantitatively the most important polysaccharides of this planet. Although it is difficult to delimit cell-wall components from intracellular reserve substances, for practical reasons the algal xylans are not discussed here (see O'Colla, Chapter 20, this treatise). Larger or smaller parts of the field have been reviewed by Oltmanns (1922), Fritsch (1935, 1945), Blinks (1951), Mori (1953), Fogg, (1953, 1956), Augier (1954b), Dillon (1954), Woodward (1954), Whelan (1955), and Lindberg (1956). An excellent discussion of the methods used in the elucidation of carbohydrate structures has been given by Manners (1959).

II. Polyglucans

A. α-1:4-Linked Glucans (Formula I)

1. Floridean Starch

Floridean starch, discovered by Kützing in 1843, forms the typical photosynthetic product of certain Red Algae. It occurs in the form of layered granules which do not seem to be associated with plastids, and which vary considerably in shape and size, from 0.5 to 25 μ, even within a single alga (Meeuse *et al.*, 1960). The frequent occurrence of slightly hollow or bowl-shaped granules may perhaps be attributed to the fact that granules of floridean starch are formed on the outer surface of a plastid, and not within it as in higher plants. Being birefringent, they display between crossed Nicol prisms the familiar dark cross, which suggests a radial arrangement of crystallites similar to that found in starch granules of angiosperms.

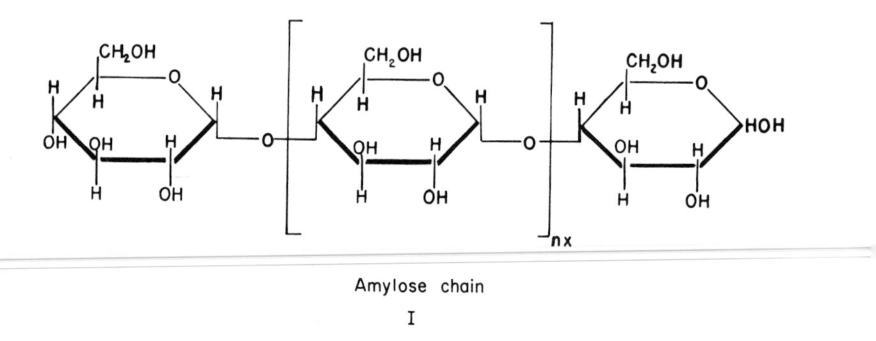

Amylose chain

I

Pure, *native* floridean starch, in the original granular form, has been prepared only by Kylin (1913, 1943a) from *Furcellaria fastigiata*, and by Meeuse and Kreger (1954) and Meeuse *et al.*, (1960) from a number of marine Red Algae, including *Constantinea subulifera* and *Rhodymenia pertusa*. Until the introduction of the "cetavlon" method for the purification of polysaccharides (Barker *et al.*, 1957), all other investigators of floridean starch (Augier, 1934, 1935, 1954a, b; Augier and du Mérac, 1956; Colin, 1934a; Colin and Augier, 1933b; Barry *et al.*, 1949; O'Colla, 1953; Fleming *et al.*, 1956) were obliged to use more or less contaminated preparations from aqueous extracts of the algae, usually *Dilsea edulis*. This circumstance must be directly responsible for the low optical rotation which they found, and indirectly for the claim that β-1:3-glucosidic linkages occur in floridean starch along with the α-1:4-bridges characteristic of higher-plant starches; indeed, some authors seem to have considered floridean starch as a sort of chemical hybrid between true starch (an α-1:4-linked glucan) and the β-1:3-linked glucan, laminarin (Fogg, 1953). The presence of a small number of 1:3-linkages in the molecule of floridean starch from *Dilsea edulis*, purified by the cetavlon method, was demonstrated conclusively by Peat *et al.* (1957, 1959a, b). However, these are α- and not β-bridges, for it was the disaccharide nigerose and not laminaribiose that was demonstrated in a partial acid hydrolyzate of the reserve material. The presence of these few α-1:3-linkages hardly warrants the establishment of a separate category for floridean starch, since they are also found in starch from Golden Bantam corn (Wolfrom and Thompson, 1955).

Floridean starches are essentially identical with the branched or amylopectin fraction of higher-plant starches, on the basis of their reaction with the anthrone reagent (Seifter *et al.*, 1950), their optical rotation, iodine reaction, susceptibility to amylases (Meeuse *et al.*, 1954, 1960) and differential thermal analysis (Morita and Rice, 1955; Morita, 1956), their complete breakdown to glucose by *Cryptochiton* enzymes (Meeuse and Fluegel, 1958a, b), and the x-ray diffraction pattern, which, at least in the starches of *Plocamium*, *Odonthalia* and *Constantinea*, is the same as that of potato starch (Meeuse and Kreger, 1959). Floridean starch is evidently more closely related to the plant amylopectins than to the animal glycogens, since the so-called R-enzyme from broad beans (Hobson *et al.*, 1951), which is without significant action on glycogens, removes branches from the molecule of floridean starch to about the same extent as from that of "waxy" maize starch (Peat *et al.*, 1959b). The basic chain has an average of 15 glucose residues (Fleming *et al.*, 1956; Meeuse *et al.*, 1960; Peat *et al.*, 1959a, b). A major physical difference between floridean starch and higher-plant amylopectin lies in the fact that the former gelatinizes in water only after prolonged boiling (Kylin, 1913; Meeuse *et al.*, 1960); in this respect it is even more extreme than *Maranta* starch, which is the most recalcitrant of the higher-plant starches.

2. Myxophycean Starch (Cyanophyte Starch)

After the accumulation of some rather indirect and incomplete evidence indicating the presence in Blue-green Algae of a starch- or glycogen-like material (Payen, 1938; Kylin, 1943b, c), Hough *et al.* (1952) succeeded in isolating from an unidentified species of *Oscillatoria* a purified polyglucose with all the characteristics of amylopectin. Thus, it had an optical rotation ($[\alpha]_D$) of +188°, was soluble in cold water, gave a reddish-brown color with iodine, and yielded upon methylation and subsequent hydrolysis a mixture of 2,3,4,6-tetra- and 2,3,6-trimethyl-D-glucose, in such proportions as might be expected from a highly branched polysaccharide with predominantly 1:4-linkages and an average chain length of 23 to 26 glucose units. It seems safe to suggest that cyanophyte starch may be generally identified with the amylopectin fraction of higher-plant starches. Within the cells it is not found in the form of clearly discernible granules, but as "undissolved submicroscopic crystals" (Kylin, 1943b, c).

The purely chemical work on the *Oscillatoria* polysaccharide was supplemented by a series of enzymatic investigations (Fredrick, 1951; 1952a, b;

1953a, b; 1954; 1955a, b; 1956; 1957; 1958; 1959a, b; Fredrick and Mulligan, 1955; Fredrick and Mancini, 1955). Undenatured extracts of *Oscillatoria princeps* synthesized a glycogen-like material from α-glucose-1-phosphate, indicating the presence of both amylose phosphorylase (P-enzyme) and "branching factor" (Q-enzyme). A strain described by Fredrick in 1952, however, formed a more insoluble material, with a blue iodine reaction, which he at first interpreted as an unbranched "amylose." Later investigations of this substance (Fredrick, 1953b) revealed that in fact the molecular chains do have branches, but considerably fewer than in the normally synthesized polysaccharide. Variants producing the "amylose" could be obtained by growing single strands of the organism at 5 to 10°C.; when grown at 25 to 32°C., they reverted to a normal pattern of polysaccharide synthesis. Fredrick suggested that these changes could be attributed to some reversible gene alteration. The kinetics of polysaccharide formation in *Oscillatoria* extracts were examined by Fredrick in 1954, and the action of the Q-enzyme by Fredrick and Mulligan in 1955. From extracts of both variants of *O. princeps*, the P- and Q-enzyme were separated by electrophoresis and by chromatography (Fredrick and Mancini, 1955; Fredrick, 1956; 1959b), but no differences were found in the physical characteristics of the enzymes from the two sources. The difference between the polysaccharides in the two strains may be due to a difference in the concentrations of the branching enzyme, or to various modifying factors (Fredrick, 1957, 1958). The evolutionary aspects of such polysaccharide syntheses in algae were discussed by Fredrick in several reviews (1952a, 1955a, 1959a).

3. *Other Algal Starches*

Studies of photosynthesis with radioactive CO_2 (see Holm-Hansen, Chapter 2, this treatise) and of many enzyme systems (see Jacobi, Chapter 7) have shown that the metabolism of Green Algae must be essentially the same as that of land plants. As would therefore be expected, the reserve products of Green Algae are, as a rule, true starches, and glucose occupies a central position in chlorophyte metabolism. Often, as in the Codiaceae and Derbesiaceae, starch is the chief carbohydrate present (Ernst, 1902, 1904a, b; Augier and du Mérac, 1954b); it is possibly formed from glucose produced in photosynthesis (Endo, 1936a). The reserve material in *Caulerpa filiformis* is probably an amylopectin (Mackie and Percival, 1960).

Meeuse and Kreger (1954, 1959) studied the x-ray diffraction pattern of a considerable number of native algal starches. Some (*Hydrodictyon reticulatum*, *Rhizoclonium* sp., *Enteromorpha intestinalis* and *Haematococcus pluvialis*) gave the A-spectrum usually associated with cereal starches; others (*Chara* sp., *Volvox aureus*, *Codium fragile*) gave the B-spectrum of the tuber starches. A number, obviously less "crystalline" than the higher-plant starches, displayed a spectrum with a single broadened diffraction line at 4.51 Å.; this was named the "U-spectrum" (for *Ulva*).

Prototheca zopfii, the colorless counterpart of *Chlorella*, also accumulates a starch-like material (Barker, 1936). Among the flagellates investigated in this connection are *Chilomonas paramecium*, in which Hutchens *et al.* (1948) found evidence for a carbohydrate consisting of about equal parts of amylose and amylopectin; *Polytomella coeca*, in which the reserve material is about 84 to 87% amylopectin (Bourne *et al.*, 1950; Barker *et al.*, 1953); and *Dunaliella bioculata*, which in this respect closely resembles *Polytomella* (Eddy *et al.*, 1958).

B. β-1:3-Linked Glucans

1. General

Until recently it was believed that β-1:3-linked glucans were compounds of minor biochemical importance, found in appreciable quantities only in brown seaweeds (as laminarin) and in yeast cell walls (as yeast glucan). However, various pieces of evidence have now indicated that these glucans are probably the most abundant polysaccharides in nature. Thus Kessler (1958) and Aspinall and Kessler (1957) demonstrated that the callose of grapevine phloem is a β-1:3-linked glucan, a substance which is now known to be widely distributed in the cell walls of other plant tissues and of certain algae such as *Caulerpa* (Eschrich, 1956; Currier, 1957). Paramylon from *Euglena* and leucosin from Chrysophyta and diatoms, the most abundant photosynthetic organisms, are of the same chemical nature.

In higher plants, β-1:3-linked glucan is rapidly produced as a response to physiological and mechanical injury. Feingold *et al.* (1958) and Hassid *et al.* (1959) reported the synthesis *in vitro* of such a glucan from uridine diphosphate glucose, with the aid of a soluble enzyme system obtained from mung beans. No data are yet available on the synthesis of β-1:3-linked glucans in algae.

2. Laminarin

This glucan, in reality a mixture of polysaccharides, was first described by Schmiedeberg in 1885, and has been shown to be ubiquitous in Brown Algae (Quillet, 1958), despite previous claims to the contrary (Kylin, 1944b). Its presence in the Phaeophyta together with the equally ubiquitous mannitol, which shows parallel seasonal variation in quantity, suggests a close biochemical relationship between the two substances. On the basis of experiments with chloroform-treated specimens of *Laminaria flexicaulis*, in which some fructose accumulated, Quillet (1954, 1957) suggested the following pathway:

$$\text{laminarin} \xrightarrow{\text{hydrolysis}} \text{glucose} \rightarrow \rightarrow \rightarrow \text{fructose} \xrightarrow{\text{reduction}} \text{mannitol}$$

In addition, mannitol forms an integral part of the molecule of at least one of the components of natural laminarin.

Initial investigations of laminarin (Kylin, 1913, 1915) were followed by extensive studies of the variation in concentration of the substance in the large European Laminariaceae and Fucaceae with season, depth, current, availability of nutrients, etc. (Lapicque, 1919; Moss, 1948; Black, 1948a, b, c, d, e, 1949, 1950a, b; Black and Dewar, 1949; Baardseth and Haug, 1953; Haug and Jensen, 1954). In *Laminaria digitata*, Black (1954) found laminarin to be invariably absent from the stipe and from the actively growing basal part of the frond, whereas the distal section of the frond exhibits a wide seasonal variation in laminarin content, ranging from zero (January to March) to a peak of 10 to 20% of the dry weight of the frond in late fall. During the period of slow growth in summer, which coincides with a dearth of phosphate and nitrate, the frond of *L. cloustonii* can accumulate as much as 36% laminarin (Black and Dewar, 1949); *Eisenia bicyclis* likewise exhibits a maximum laminarin content in August (Nisizawa, 1938, 1939, 1940a, b). Similar but less spectacular changes are manifest in the European Fucaceae. *Cladostephus spongiosus* (Cladostephaceae), collected near Plymouth, England, at an unspecified season, contained 20% laminarin (Fanshawe and Percival, 1958). Correspondingly, in the Southern Hemisphere, the Brown Algae studied by Sannié (1950, 1951), by von Holdt *et al.* (1955) and by Stewart and Higgins (1960), contained most laminarin in the first months of the year.

The conventional assay procedures for laminarin, such as the method of Cameron *et al.* (1948) discussed extensively by Whelan (1955), are all based on its hydrolysis with hot acid to the easily measurable sugar glucose. They thus lack specificity; moreover, they are founded on the tacit but false assumption (Jensen, 1956a, b) that no free hexose is initially present in Brown Algae. A preferable method is that of Quillet (1958), in which the laminarin is hydrolyzed with snail enzymes to glucose, the specificity and completeness of the reaction being carefully controlled by paper chromatography. However, for species of *Laminaria* and for *Alaria esculenta*, though not for the Fucaceae, the error in figures obtained by the method of Cameron *et al.* is only slight (Jensen, 1956b).

Laminarin occurs in two forms. Soluble laminarin dillsoves as readily in cold water as in hot water, but the so-called "insoluble" laminarin dissolves in water only when heated. The best source of the insoluble form is *Laminaria cloustonii* (Connell *et al.*, 1950; Black *et al.*, 1951a).

A clear extract of the alga in weak hydrochloric acid is neutralized and continuously stirred at room temperatures for 3 days, resulting in the precipitation of laminarin in a remarkably pure state. The yields obtained reached 55 to 65% of the total laminarin in the alga or 16 to 19% of its dry weight. The optical rotation $[\alpha]_D$ is $-14.8°$ in water (Connell *et al.*, 1950). Soluble laminarin can be conveniently obtained by adding alcohol to an extract of *L. cloustonii* after the insoluble laminarin has settled out (Black *et al.*, 1951a). It has also been prepared from *L. digitata*, where it forms the bulk of the reserve material (Black *et al.*, 1951a; Percival and Ross, 1951), from other *Laminaria* species such as *L. flexicaulis* (Colin and Ricard, 1929, 1930a, b), and from *Fucus serratus* (Black *et al.*, 1951a). The optical rotation of soluble laminarin, $[\alpha]_D$ $-12°$, is close to that of the insoluble form (Percival and Ross, 1951). It has not yet been possible to establish clear chemical differences between the two forms; since the difference in solubility cannot be explained merely by polymer size (Friedlaender *et al.*, 1954), it may be due to slight variations in the branching of the molecular chains or to some other minor structural differences.

It is now well established that the main linkages in laminarin are β-1:3-glucosidic bridges (Barry, 1939, 1941). The apparent chain length was reported to be 16 by Barry (1942) and about 20 by Connell *et al.* (1950). However, the reducing power of both the insoluble and the soluble form was found to be unexpectedly low, suggesting that some of the reducing end-groups of the chains were masked (Connell *et al.*, 1950; Percival and Ross, 1951). This was borne out by later research. Mannitol, constituting 2% of the polysaccharide, has been confirmed as an integral part of the molecule (Peat *et al.*, 1955a, b, 1958a, b, 1960; Anderson *et al.*, 1958; Goldstein *et al.*, 1959), forming the terminal non-reducing unit in *some* of the polyglucose chains. Thus laminarin consists of a mixture of two components, distinguished as "laminarose" and "laminaritol" by Smith and co-workers, who were able to separate them by glass-paper electrophoresis (Briggs *et al.*, 1956; Lewis and Smith, 1957). Since it has been established in various ways (Unrau and Smith, 1957; Goldstein *et al.*, 1959) that the degree of polymerization of laminaritol (n = 30) is about twice that of laminarose (n = 15), it is possible that the latter substance is an intermediate in an enzymatic synthesis of laminaritol, or alternatively that laminarose might be an artifact derived from laminaritol, or some similar substance, by acid hydrolysis during isolation. The formula (II) given here (Goldstein *et al.*, 1959) incorporates the features discussed above, as well as the novel idea that one molecule of mannitol may serve as an aglycone residue for two chains of glucose units. Structural formula (II) must now be modified, since there is conclusive evidence that laminarin contains 1:6-glucosidic linkages (Peat *et al.*, 1958b, 1960; Anderson *et al.*, 1958; Smith and Unrau,

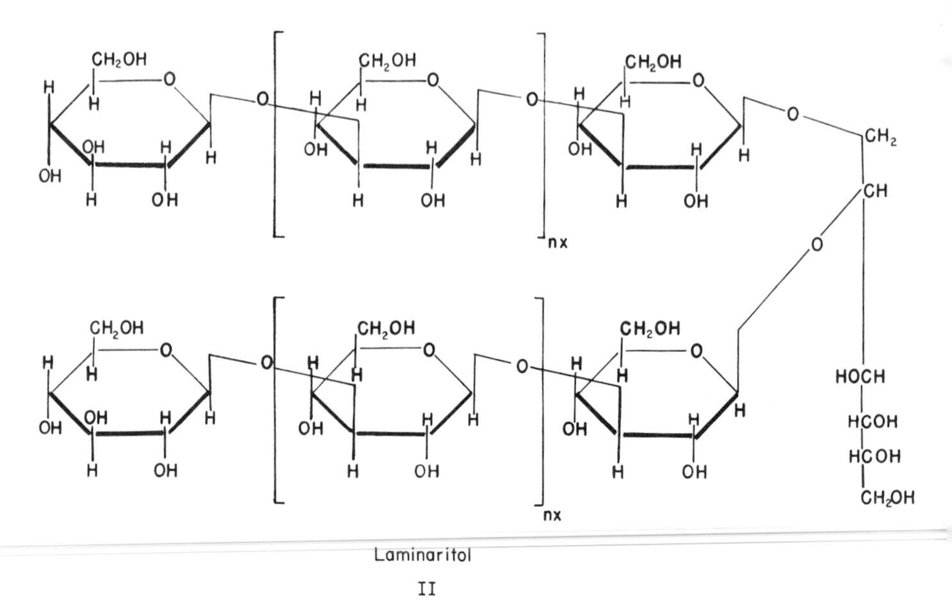

Laminaritol

II

1959b); Beattie *et al.*, 1961) as well as mannose residues (Smith and Unrau, 1959a). Peat *et al.* considered that there was no clear evidence for branching, but Anderson *et al.* were of the opinion that some of the molecules were slightly branched. There is some evidence that laminarin is polydisperse (Broatch and Greenwood, 1956; Anderson *et al.*, 1958).

The breakdown of laminarin can be achieved with a variety of enzymes: preparations from wheat, oats, barley, potato tubers or hyacinth bulbs (Dillon and O'Colla, 1950, 1951); from various seaweeds (Duncan *et al.*, 1956) and *Euglena* (Fellig, 1960 and Meeuse, unpublished data); and from some marine bacteria (Chesters *et al.*, 1956) and the digestive juices of mollusks such as *Tethys* (Nisizawa, 1939), *Helix* (Quillet, 1958), *Cryptochiton* (Meeuse and Fluegel, 1958a, b), and *Tegula* (Galli and Giese, 1959). Though the final breakdown product was reported to be glucose, Dillon and O'Colla found, in addition, laminaribiose and various oligosaccharides, including a tetrasaccharide, as intermediates of hydrolysis or perhaps transglycosylation products (Edelman, 1956). There is thus a possibility that many "laminarinases" may be actually mixtures of several enzymes.

Certain artificially sulfated derivatives of laminarin are of some practical medical importance, since they have been found to inhibit the coagulation of blood (Dewar, 1956).

3. Paramylon (Paramylum)

First described by O. F. Müller in 1786, paramylon—the reserve material typical of the Euglenophyta—was isolated and named by Gottlieb in 1850.

It appears in the cells in the form of highly refractive, optically anisotropic granules or rings, whose constant size and shape within each species is of great taxonomic importance. Whereas Dujardin (1841) tried in vain to break down paramylon with various acids, bases and organic solvents, Gottlieb (1850) and Habermann (1874) succeeded in hydrolyzing it with very strong HCl to glucose, thus proving its relationship to starch. However, in contrast to the latter substance, paramylon does not stain with iodine, nor does it form a paste with water, although under certain conditions the granules swell to some extent.

Elaborate studies of swollen paramylon from species of *Phacus* and *Euglena* (Pochmann, 1956, 1958), in conjunction with studies with polarized light (Kamptner, 1952), indicated that a paramylon granule consists of fibrils wound helically around a long, narrow, axial lumen to form a body displaying rotational symmetry. According to Pochmann, such bodies can grow by a process of "proliferation" in the axial direction to give rise to a whole series of more or less disc-shaped bodies, each possessing the original helical structure. These eventually combine to form second-order granules or chondroliths of ellipsoid or globular shape, which naturally still possess a narrow axial canal. The formation of large paramylon rings is said to result from a lytic process in the center combined with peripheral growth of the granules, and not from hypothetical preformed, ring-shaped cytoplasmic structures. The solid paramylon rods characteristic of certain species are considered to be derived from circular rings through a process of stretching and loss of lumen. Excellent photographs and drawings substantiating these claims were presented by Pochmann (1956, 1958), who attributed the discovery of the helical structure to Carter (1857); see also Heidt (1937) and Kamptner (1952). For more general discussions of paramylon, see Bütschli (1883, 1906), Dangeard (1901), Czurda (1928–1929), Deflandre (1934), Oltmanns (1922), Conrad (1943), Schiller (1952), Küster (1956), Gojdics (1953), and Huber-Pestalozzi (1955).

Paramylon was shown to be a β-1:3-linked glucan by Kreger and Meeuse in 1952, on the basis of x-ray diffraction studies of native paramylon granules from *Euglena viridis*, *E. geniculata*, *E. sanguinea*, and the colorless alga *Astasia ocellata*. The diffraction diagrams obtained from *Euglena* paramylum were practically indistinguishable from those of yeast cell-wall glucan, which is now known to be predominantly β-1:3-linked (Zechmeister and Tóth, 1934, 1936; Hassid *et al.*, 1941; Barry and Dillon, 1943; Houwink *et al.*, 1951; Bell and Northcote, 1950; Northcote and Horne, 1952; Houwink and Kreger, 1953; Peat *et al.*, 1958c). This suggested chemical structure for the paramylon of *Euglena gracilis* was confirmed by Clarke and Stone (1960). Although paramylon from *Astasia ocellata* gave an x-ray pattern somewhat different from that of *Euglena*, a certain correspondence was clear.

Although many euglenids seem to lack the ability to attack free glucose added to the medium—a fact which led to interesting speculations on the presence or absence of crucial glucose-mobilizing enzymes (Lwoff *et al.*, 1949, 1950; Belsky, 1955, 1957; cf. Danforth, Chapter 6, this treatise)—there is evidence that, within the *Euglena* cells, paramylon may be broken down to glucose. From freeze-dried cells of *E. gracilis* var. *bacillaris*, Meeuse (unpublished data) prepared aqueous extracts containing an enzyme which readily hydrolyzed soluble laminarin, a closely related β-1:3-linked glucan. The presence of laminarinase in *Euglena* was reported, independently, by Fellig (1960).

4. *Chrysolaminarin* (*Leucosin, Chrysose*)

In some diatoms and in several members of the Chrysophyta, if not in all, there occur in the cells highly refractive, roundish inclusions which do not stain with lipid reagents and can thus be easily distinguished from the oil droplets which often accompany them. In certain diatoms, where the material may form 15 to 20% of the mass of the cells, it can be induced to crystallize by treatment with a concentrated sodium chloride solution for 5 hours, or with 50% aqueous diacetin at 65°C. for 5 to 30 minutes (von Stosch, 1951). Leucosin formation in the cells of *Ochromonas malhamensis* is enhanced by the addition of sugar to the growth medium. Cells killed by moderate heat undergo autolysis and liberate a reducing sugar, in contrast to cells killed by boiling (Pringsheim, 1952). These facts, and the positive Hotchkiss reaction observed by von Stosch, indicate the carbohydrate nature of the material, for which Pringsheim suggested the name *chrysose*. Meeuse (unpublished) found the autolysis product of leucosin to be glucose, even in cells which have been grown on fructose media. The purified leucosin obtained by repeated hot-water extraction and alcohol precipitation can be quantitatively converted to glucose by digestive enzymes from *Cryptochiton*. The absence of an iodine reaction, the very low optical rotation, and the pronounced resistance to oxidation by periodic acid, all indicate the close relation of this substance to laminarin. Quillet (1955), who isolated 200 mg. of pure leucosin from several kilograms of *Hydrurus foetidus*, suggested the name *chrysolaminarin* on the basis of similar experiments.

The number of glucose units in the molecule of chrysolaminarin, according to Quillet, is at least 8 and almost certainly more. In 1958, Archibald *et al.* showed that the reserve polysaccharide of *Ochromonas malhamensis* has a predominance of β-1:3-glucosidic linkages, in agreement with the conclusions reached independently by Quillet and Meeuse. Archibald *et al.*

postulated a linear structure for the molecule, consisting of 36 to 40 hexose residues. The presence of ~10% non-glucose residues, probably mannose, even in purified preparations, is not surprising in view of the fact that mannose has now been shown to occur also as a minor component of laminarin (Smith and Unrau, 1959a; see above). However, a sample of crystalline chrysolaminarin, extracted from a mixed sample of diatoms, yielded on hydrolysis 99.5% glucose (Beattie *et al.*, 1961). The original material melted at 273°, and in aqueous solution had an optical rotation $[\alpha]_D^{18}$ of −6°. The polymer was found to comprise about 21 glucose units, with mostly 1:3 but also some 1:6 linkages, and with probably one branch per molecule.

The present author recommends that the term *chrysolaminarin* be generally accepted, since the word *leucosin* has sometimes been applied to non-carbohydrate substances inadequately characterized in cells, while Pringsheim's term *chrysose* suggests a free sugar rather than a polysaccharide.

III. Fructosans and Inulin

Though Endo (1936b) and Kylin (1944a) identified free fructose in *Cladophora* species, the storage carbohydrates of most algae are based largely on glucose. However, du Mérac (1953, 1955, 1956) showed that the Dasycladales, a natural order at present comprising about 10 tropical and subtropical genera, are equally well-defined in a biochemical sense, being characterized by the formation of polymers of fructose as reserve materials. Probably the most striking case is presented by *Acetabularia mediterranea*, where spherocrystals of the inulin type, practically indistinguishable from those of *Dahlia variabilis*, appear in alcohol-treated material. The first to notice these crystals was C. Nägeli (1863); the first to recognize their inulin-like nature, Leitgeb (1887). Du Mérac (1953) succeeded in isolating 3.5 gm. of a clearly identified inulin as a white powder from 150 gm. of fresh *Acetabularia mediterranea*, by hot-water extraction and alcohol precipitation. The presence of inulin was qualitatively confirmed for an undetermined species of *Acetabularia* from Curaçao by Meeuse (unpublished). In *Dasycladus vermicularis*, however, du Mérac (1955) found no inulin, although she distinguished at least 5 low-molecular-weight fructosans in addition to free fructose and traces of sucrose. In *Cymopolia barbata* she found sucrose, fructose and at least 8 low-molecular-weight fructosans (du Mérac, 1956).

The predominant linkage types in these algal fructosans remain to be established; also needed are further studies of their synthesis and breakdown in the living plant.

IV. Reserve Substances of Low Molecular Weight

A. Trehalose

Trehalose, 1,1-α-glucosyl-α-glucose, is found not only in many fungi but also in the Blue-green Algae *Calothrix* (Payen, 1938; Kylin, 1943a, b, c) and *Rivularia* (Colin and Payen, 1934; Payen, 1938); in the freshwater Red Algae *Lemanea, Batrachospermum,* and *Tuomeya* (Sauvageau and Denigès, 1930; Colin and Augier, 1933a, b; Augier, 1934, 1935, 1954a); and, in small quantities, in the marine alga *Callithamnion tetricum* (Augier, 1954b). The "trehalose" incorrectly reported by Kylin (1915) in *Rhodymenia palmata* is in reality floridoside (see below; cf. Kylin, 1943a, b, c).

B. Floridoside

Floridoside (III), now recognized as 2-*O*-glycerol-α-D-galactoside (Colin and Guéguen, 1930a, b, c; Guéguen, 1931; Putman and Hassid, 1954), is remarkable among the naturally occurring glycosides in belonging to the

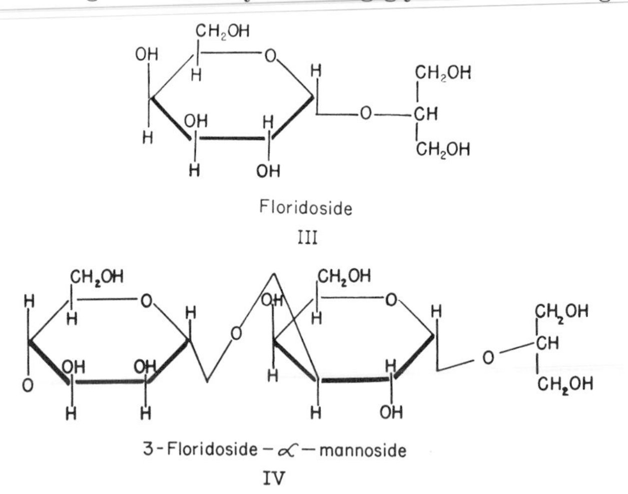

Floridoside

III

3-Floridoside-α-mannoside

IV

α-series. In the Rhodophyta it is widely distributed, and usually abundant (Colin and Guéguen, 1933; Colin, 1934b, 1937; Colin and Augier, 1933a, b; Augier, 1934, 1935, 1946; Augier and Du Mérac, 1954a, 1956; Kylin 1943a, b, c). Its concentration in *Rhodymenia palmata* may reach 15% (Kylin, 1918). The presence of floridoside in the Bangiophyceae indicates biochemical affinities between this group and the Florideophyceae (Augier, 1954a). In addition, certain Bangiophyceae contain isofloridoside, 1-*O*-glycerol-α-D-galactoside (Lindberg, 1955c). Lindberg (1954, 1955a, b) isolated and characterized a derivative, 3-floridoside-α-mannoside (IV), from *Furcellaria fastigiata*.

Bean and Hassid (1955) suggested a mechanism for floridoside biosynthesis in *Iridophycus flaccidum*, by which α-glycerophosphate derived from triose combines with uridine diphosphate galactose to give uridine diphosphate (UDP) and α-D-galactose-2-glyceryl phosphate, which is then hydrolyzed to the free floridoside. Obviously, galacto-waldenase operates in this species, for both UDP-glucose and UDP-galactose appear as labeled compounds within the first 30 seconds of photosynthesis in $C^{14}O_2$. The unusual carbohydrate metabolism of *Iridophycus* is indicated by the presence of an enzyme which can oxidize glucose or galactose to hexonic acids, and maltose, lactose, or cellobiose to the respective aldobionic acid (Bean and Hassid, 1956).

C. Mannoglyceric Acid (Formula V)

Mannoglyceric acid was discovered and identified in a species of Red Alga, *Polysiphonia fastigiata*, by Colin and Augier (1939), and its structure (V) was confirmed by Bouveng *et al.* (1955). It is abundant in the Cera-

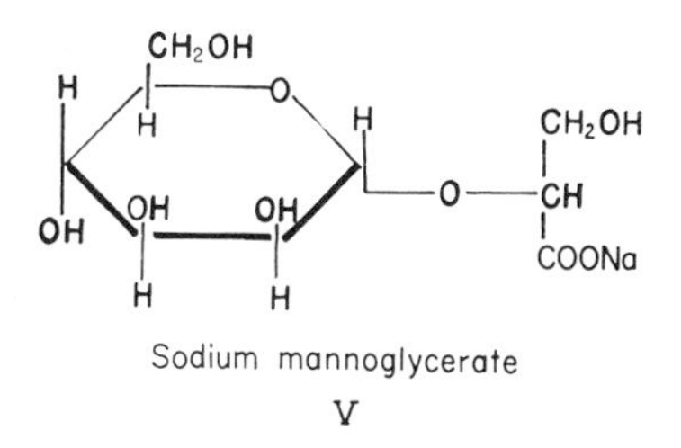

Sodium mannoglycerate

V

miales, especially in the family Rhodomelaceae. In the majority of species examined, this compound occurs in concentrations lower than those of floridoside; however, its concentration exceeds that of the latter in *Ceramium rubrum*, while in *Griffithsia setacea* only mannoglyceric acid occurs. It is also found in the Gigartinales and Cryptonemiales, but not—or only in traces—in the species of Rhodymeniales, Gelidiales, and Nemalionales examined (Augier, 1944, 1946, 1947, 1953, 1954a). Its pronounced seasonal variation has been studied by Hoffmann (1951) and by Augier and Hoffmann (1952).

D. Sucrose

This sugar has been found in appreciable quantities in the Green Algae *Chlorella* (Bassham and Calvin, 1957), *Scenedesmus* (Bassham and Calvin, 1957; Lindberg, 1955a), *Enteromorpha* (Lindberg, 1955a), *Cladophora* sp. (Bidwell, 1958), *Ulva lactuca* (Bidwell, 1958), and in certain of the Dasycladales where it accompanies reserve fructosans (see III above). Bidwell *et al.*

(1952), surveying the free sugars in 10 species of algae, reported the presence of various amounts of sucrose in the Chlorophyta and Rhodophyta, but only traces in the Phaeophyta.

E. Polyhydroxy Alcohols

Sugar alcohols, relatively abundant among algae, are well suited for the role of reserve materials, since they embody more photosynthetic energy, i.e., "reducing power," than do the corresponding sugars. In the Rhodophyta, dulcitol (D-galactitol) has been found in *Iridaea laminarioides* by Hassid (1933), and both dulcitol and sorbitol (D-glucitol) in *Bostrychia scorpioides* by Haas and Hill (1931, 1932). The heptitol volemitol, of interest because of its obvious chemical relation to the 7-carbon sugars implicated in photosynthesis (Bassham and Calvin, 1957; see Holm-Hansen, Chapter 2, this treatise), was demonstrated in a species of Brown Alga, *Pelvetia canaliculata*, by Lindberg and Paju (1954) and Quillet (1957). This is probably the substance isolated by Haas and Hill (1929) and referred to as a "mannitan." Volemitol is absent from other Phaeophyta, but was reported from one of the Red Algae, *Porphyra umbilicalis* (Lindberg, 1955c). The 4-carbon compound erythritol occurs in *Trentepohlia* sp. in concentrations up to 1.4% (Tischer, 1936).

Mannitol (VI), found in *Laminaria saccharina* by Stenhouse in 1844, was long thought to be characteristic of the Phaeophyta, where it occurs generally as a constituent of the cell sap (Haas and Hill, 1929, 1933; Sosa and

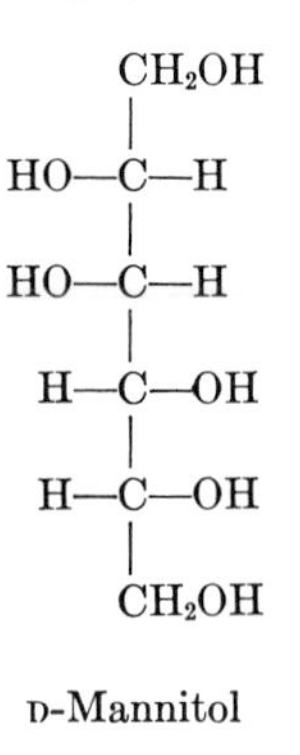

D-Mannitol

VI

Sosa-Bourdouil, 1941; Kylin, 1944b). The mannitol content of the fronds of certain brown seaweeds may vary from a few per cent (dry weight) in early spring to as much as 50% in later summer (Ricard, 1930, 1931; Black, 1948a, b, c, d, e, 1949, 1950a), being also influenced by the physical environment (Black, 1950b; Black and Dewar, 1949; Čmelik and Morovic,

1950; Quillet, 1957). Methods for the large-scale isolation of mannitol from seaweeds have been developed (Black *et al.*, 1951b). Since Lindberg (1955b, c) demonstrated the presence of mannitol in fair concentration in the Red Algae *Furcellaria fastigiata* and *Porphyra umbilicalis*, and Bidwell (1957, 1958) found it in *Olisthodiscus* sp. and *Halosaccion ramentaceum*, it can no longer be regarded as a peculiar feature of the Brown Algae.

1-D-Mannitol monoacetate, 1-D-mannitol monoglucoside, and 1,6-D-mannitol di-β-glucoside were found in *Fucus vesiculosus* and *F. spiralis* by Lindberg (1953a) and Bouveng and Lindberg (1955). The first two compounds also occur in *Desmarestia aculeata* (Bouveng and Lindberg, 1955) and *Laminaria cloustonii* (Lindberg and McPherson, 1954a). The identity of some of the isolated compounds was confirmed by synthesis (Lindberg, 1953b, c). The hypothesis of a close metabolic connection between mannitol and laminarin (see Section II.B.2) received experimental support from Bidwell (1959), who observed that in *Fucus vesiculosus* labeled laminarin is rapidly formed from C^{14}-labeled mannitol.

Laminitol, a *C*-methylinositol, found in several Brown Algae (Lindberg and McPherson, 1954b; Bouveng and Lindberg, 1955) and in *Porphyra umbilicalis* (Lindberg, 1955c), is either 4-*C*-methylmyoinositol or its antipode (Lindberg and Wickberg, 1959).

References

Anderson, F. B., Hirst, E. L., Manners, D. J., and Ross, A. G. (1958). The constitution of laminarin. III. The fine structure of insoluble laminarin. *J. Chem. Soc.*, pp. 3233–3243.

Archibald, A. R., Manners, D. J., and Ryley, J. F. (1958). Structure of a reserve polysaccharide (leucosin) from *Ochromonas malhamensis*. *Chem. & Ind.* (*London*), pp. 1516–1517.

Aspinall, G. O., and Kessler, G. (1957). Structure of callose from the grape vine. *Chem. & Ind.* (*London*), p. 1296.

Augier, J. (1934). Constitution et biologie des Rhodophycées d'eau douce. Thèse Doct. ès Sci., Paris.

Augier, J. (1935). Constitution et biologie des Rhodophycées d'eau douce. *Rev. algol.* **7,** 237–326.

Augier, J. (1944). La répartition du mannoglycérate de sodium chez les Floridées. *Bull. soc. botan. France* **91,** 92–94.

Augier, J. (1946). Nouvelles données sur les glucides des Algues rouges. *Compt. rend. acad. sci.* **222,** 929–931.

Augier, J. (1947). Les glucides et la systématique des Rhodophycées. *Compt. rend. acad. sci.* **224,** 1654–1656.

Augier, J. (1953). La constitution chimique de quelques Floridées Rhodomélacées. *Rev. gén. botan.* **60,** 257–283.

Augier, J. (1954a). Sur la biochimie d'une Lémanéacée nord-américaine, *Tuomeya fluviatilis* Harvey. *Compt. rend. acad. sci.* **239,** 87–89.

Augier, J. (1954b). Biochimie et systématique chez les Rhodophycées. *Congr. intern. botan., 8e Congr., Paris,* **1954**, *Colloq. anal. plantes et problèmes engrais mineraux* Sect. 17, pp. 30–32.

Augier, J., and du Mérac, M. L. R. (1954a). Les sucres solubles des Rhodophycées. *Compt. rend. acad. sci.* **238,** 387–389.

Augier, J., and du Mérac, M. L. R. (1954b). Recherches préliminaires sur le chimisme glucidique de *Codium dichotomum* (Huds.) Setchell. *Bull. lab. maritime Dinard* **No. 40,** 25–27.

Augier, J., and du Mérac, M. L. R. (1956). Sur le chimisme glucidique des Némalionales. *Compt. rend. acad. sci.* **243,** 1785–1787.

Augier, J., and Hoffmann, G. (1952). Sucre et brome dans le *Polysiphonia fastigiata* (Roth) Grev. *Bull. soc. botan. France* **99,** 80–82.

Baardseth, E., and Haug, A. (1953). Individual variation of some constituents in brown algae, and reliability of analytical results. *Norweg. Inst. Seaweed Research, Rept.* **No. 2.**

Barker, H. A. (1936). The oxidative metabolism of the colorless green alga, *Prototheca Zopfii. J. Cellular Comp. Physiol.* **8,** 231–250.

Barker, S. A., Bebbington, A., and Bourne, E. J. (1953). Mode of action of Q-enzyme of *Polytomella coeca. J. Chem. Soc.,* pp. 4051–4057.

Barker, S. A., Stacey, M., and Zweifel, G. (1957). The separation of neutral polysaccharides. *Chem. & Ind. (London),* p. 330.

Barry, V. C. (1939). Constitution of laminarin. Isolation of 2,4,6-trimethylglucopyranose. *Sci. Proc. Roy. Dublin Soc.* **22,** 59–67.

Barry, V. C. (1941). Hydrolysis of laminarin. Isolation of a new glucose disaccharide. *Sci. Proc. Roy. Dublin Soc.* **22,** 423–429.

Barry, V. C. (1942). New method of end-group assay for laminarin and similarly constituted polysaccharides. *J. Chem. Soc.,* pp. 578–581.

Barry, V. C., and Dillon, T. (1943). The glucan of the yeast membrane. *Proc. Roy. Irish Acad.* **B49,** 177–185.

Barry, V. C., Halsall, T. G., Hirst, E. L., and Jones, J. K. N. (1949). The polysaccharides of the Florideae. Floridean starch. *J. Chem. Soc.,* pp. 1468–1470.

Bassham, J. A., and Calvin, M. (1957). "The Path of Carbon in Photosynthesis." Prentice Hall, Englewood Cliffs, New Jersey.

Bean, R. C., and Hassid, W. Z. (1955). Assimilation of $C^{14}O_2$ by a photosynthesizing red alga, *Iridophycus flaccidum. J. Biol. Chem.* **212,** 411–425.

Bean, R. C., and Hassid, W. Z. (1956). Carbohydrate oxidase from a red alga, *Iridophycus flaccidum. J. Biol. Chem.* **218,** 425–436.

Beattie, A., Hirst, E. L., and Percival, E. (1961). Studies on the metabolism of the Chrysophyceae. Comparative structural investigations on leucosin (chrysolaminarin) separated from diatoms and laminarin from the brown algae. *Biochem. J.* **79,** 531–537.

Bell, D. J., and Northcote, D. H. (1950). Structure of a cell-wall polysaccharide of baker's yeast. *J. Chem. Soc.,* pp. 1944–1947.

Belsky, M. M. (1955). Studies on the utilization of glucose by the flagellate *Euglena gracilis. Dissertation Abstr.* **15,** 678.

Belsky, M. M. (1957). The metabolism of glucose and other sugars by the algal flagellate, *Euglena gracilis. Bacteriol. Proc.* **1957,** 123–124.

Bidwell, R. G. S. (1957). Photosynthesis and metabolism of marine algae. I. Photosynthesis of two marine flagellates compared with *Chlorella. Can. J. Bot.* **35,** 945–950.

Bidwell, R. G. S. (1958). Photosynthesis and metabolism of marine algae. II. A survey of rate and products of photosynthesis in $C^{14}O_2$. *Can. J. Botany* **36,** 337–349

Bidwell, R. G. S. (1959). Precursors of polysaccharides in *Fucus vesiculosus. Proc. Intern. Botan. Congr., 9th Congr., Montreal* Vol. 2 (Abstr.), p. 32.

Bidwell, R. G. S., Krotkov, G., and Reed, G. B. (1952). Paper chromatography of sugars in plants. *Can. J. Botany* **30,** 291–305.

Black, W. A. P. (1948a). Seasonal variation in the chemical constitution of some British Laminariales. *Nature* **161,** 174.

Black, W. A. P. (1948b). Seasonal variations in chemical constitution of some of the sublittoral seaweeds common to Scotland. I. *Laminaria cloustoni. J. Soc. Chem. Ind. (London)* **67,** 165–168.

Black, W. A. P. (1948c). Seasonal variation in chemical constitution of some of the sublittoral seaweeds common to Scotland. II. *Laminaria digitata. J. Soc. Chem. Ind. (London)* **67,** 169–172.

Black, W. A. P. (1948d). Seasonal variation in chemical constitution of some of the sublittoral seaweeds common to Scotland. III. *Laminaria saccharina* and *Saccorhiza bulbosa. J. Soc. Chem. Ind. (London)* **67,** 172–176.

Black, W. A. P. (1948e). Seasonal variation in chemical composition of some of the littoral seaweeds common to Scotland. I. *Ascophyllum nodosum. J. Soc. Chem. Ind. (London)* **67,** 355–357.

Black, W. A. P. (1949). Seasonal variation in chemical composition of some of the littoral seaweeds common to Scotland. II. *Fucus serratus, Fucus vesiculosus, Fucus spiralis* and *Pelvetia canaliculata. J. Soc. Chem. Ind. (London)* **68,** 183–189.

Black, W. A. P. (1950a). Seasonal variation in weight and chemical composition of the common British Laminariaceae. *J. Marine Biol. Assoc. United Kingdom* **29,** 45–72.

Black, W. A. P. (1950b). Effect of the depth of immersion on the chemical constitution of some of the sublittoral seaweeds common to Scotland. *J. Soc. Chem. Ind (London)* **69,** 161–165.

Black, W. A. P. (1954). Concentration gradients and their significance in *Laminaria saccharina. J. Marine Biol. Assoc. United Kingdom* **33,** 49–60.

Black, W. A. P., and Dewar, E. T. (1949). Correlation of some of the physical and chemical properties of the sea with the chemical constitution of the algae. *J. Marine Biol. Assoc. United Kingdom* **28,** 673–699.

Black, W. A. P., Cornhill, W. H., Dewar, E. T., and Woodward, F. N. (1951a). Manufacture of algal chemicals. III. Laboratory-scale isolation of laminarin from brown marine algae. *J. Appl. Chem. (London)* **1,** 505–507.

Black, W. A. P., Dewar, E. T., and Woodward, F. N. (1951b). Manufacture of algal chemicals. II. Laboratory-scale isolation of mannitol from brown marine algae. *J. Appl. Chem. (London)* **1,** 414–424.

Blinks, L. R. (1951). Physiology and biochemistry of algae. *In* "Manual of Phycology" (G. M. Smith, ed.), pp. 263–291. Chronica Botanica, Waltham, Massachusetts.

Bourne, E. J., Stacey, M., and Wilkinson, I. A. (1950). Composition of the polysaccharide synthesized by *Polytomella caeca. J. Chem. Soc. (London)*, pp. 2694–2698.

Bouveng, H., and Lindberg, B. (1955). Low-molecular carbohydrates in algae. VII. Investigation of *Fucus spiralis* and *Desmarestia aculeata. Acta Chem. Scand.* **9,** 168–169.

Bouveng, H., Lindberg, B., and Wickberg, B. (1955). Low-molecular carbohydrates in algae. IX. Structure of the glyceric acid mannoside from red algae. *Acta Chem. Scand.* **9,** 807–809.

Briggs, D. R., Garner, E. F., and Smith, F. (1956). Separation of carbohydrates by electrophoresis on glass paper. *Nature* **178,** 154–155.

Broatch, W. N., and Greenwood, C. T. (1956). The molecular weight of laminarin. *Chem. & Ind.* (*London*), p. 1015.

Bütschli, O. (1883). Mastigophora. *In* "Dr. H. G. Bronn's Klassen und Ordnungen des Tier-Reichs. Vol. I. Protozoa." (H. G. Bronn, ed.), pp. 727–730 (Paramylon). Winter, Leipzig und Heidelberg, Germany.

Bütschli, O. (1906). Beiträge zur Kenntnis des Paramylons. *Arch. Protistenk.* **7,** 197–228.

Cameron, M. C., Ross, A. G., and Percival, E. G. V. (1948). Routine estimation of mannitol, alginic acid, and combined fucose in seaweeds. *J. Soc. Chem. Ind.* (*London*) **67,** 161–164.

Carter, H. J. (1857). Additional notes on the freshwater infusoria in the island of Bombay. *Ann. and Mag. of Nat. Hist.* **20,** 34–41.

Chesters, C. G. C., Turner, M., and Apinis, A. (1956). Decomposition of laminarin by micro-organisms. *Proc. 2nd. Internat. Seaweed Sympos.* (*Trondheim, 1955*), pp. 141–144.

Clarke, A. E., and Stone, B. A. (1960). Structure of the paramylon from *Euglena gracilis. Biochim. et Biophys. Acta* **44,** 161–163.

Čmelik, S., and Morovic (1950). Mannitol content in some algae from the Adriatic Sea (with summary in German). *Arhiv Kem.* **22,** 228–235; quoted from *Chem. Abstr.* **45,** 10294*h* (1951).

Colin, H. (1934a). Sur l'amidon des Floridées. *Compt. rend. acad. sci.* **238,** 968–970.

Colin, H. (1934b). Le "sucre" des Floridées. *Bull. muséum natl. hist. nat.* (*Paris*) **6,** 153–155.

Colin, H. (1937). Sur le floridoside, *d*-monogalactoside du glycérol. *Bull. soc. chim. Belges* **4,** (5) 277–281.

Colin, H., and Augier, J. (1933a). Les glucides solubles de *Lemanea nodosa. Compt. rend. acad. sci.* **196,** 1042–1043.

Colin, H., and Augier, J. (1933b). Floridoside, tréhalose et glycogène chez les algues rouges d'eau douce. *Compt. rend. acad. sci.* **197,** 423–425.

Colin, H., and Augier, J. (1939). Un glucide original chez les Floridées du genre *Polysiphonia*, le *d*-mannoside α du 1-glycérate de sodium. *Compt. rend. acad. sci.* **208,** 1450–1453.

Colin, H., and Guéguen, E. (1930a). Le sucre des Floridées. *Compt. rend. acad. sci.* **190,** 653–655.

Colin, H., and Guéguen, E. (1930b). Variations saisonnières de la teneur en sucre chez les Floridées. *Compt. rend. acad. sci.* **190,** 884–886.

Colin, H., and Guéguen, E. (1930c). La constitution du principe sucré de *Rhodymenia palmata. Compt. rend. acad. sci.* **191,** 163–164.

Colin, H., and Guéguen, E. (1933). Le floridoside chez les Floridées. *Compt. rend. acad. sci.* **197,** 1688–1690.

Colin, H., and Payen, J. (1934). Le sucre de *Rivularia bullata. Compt. rend. acad. sci.* **198,** 384–386.

Colin, H., and Ricard, P. (1929). Sur quelques propriétés de la laminarine des Laminaires. *Compt. rend. acad. sci.* **188,** 1449–1451.

Colin, H., and Ricard, P. (1930a). Glucides et dérivés glucidiques des Algues brunes. *Compt. rend. acad. sci.* **190,** 1514–1516.

Colin, H., and Ricard, P. (1930b). Préparation et propriétés de la Laminarine (Laminaroloside) de *Laminaria flexicaulis. Bull. soc. chim. biol.* **12,** 88–96.

Connell, J. J., Hirst, E. L., and Percival, E. G. V. (1950). The constitution of laminarin. I. An investigation on laminarin isolated from *Laminaria cloustoni*. *J. Chem. Soc.*, pp. 3494–3500.

Conrad, W. (1943). Remarques sur le genre *Phacus*. *Mem. Mus. roy. Hist. nat. belg.* **19**, 1–16.

Currier, H. B. (1957). Callose substance in plant cells. *Am. J. Botany* **44**, 478–488.

Czurda, V. (1928–1929). Morphologie und Physiologie des Algenstärkekornes. *Botan. Zentr., Beih.* **45**, Abstr. No. 1, 97–270. (See especially, Part 4, *Eugleninen*, p. 210–215.)

Dangeard, P. A. (1901). Recherches sur les Eugléniens. (Chapter 3, 2°: Le Paramylon.) *Botaniste* **8**, 304–315.

Deflandre, G. (1934). Sur les propriétés optiques du Paramylon (variations de l'anisotropie). *Bull. biol. France et Belg.* **68**, 382–384.

Dewar, E. T. (1956). Sodium laminarin sulphate as a blood anticoagulant. *Proc. 2nd Intern. Seaweed Symposium, Trondheim, 1955*, pp. 55–61.

Dillon, T. (1954). The polysaccharides of the red algae. *Congr. intern. botan., 8th Congr. Paris, Sect.* **17**, 29–30.

Dillon, T., and O'Colla, P. (1950). Hydrolysis of laminarin by wheat β-amylase. *Nature* **166**, 67.

Dillon, T., and O'Colla, P. (1951). The enzymic hydrolysis of 1,3-linked polyglucans. *Chem. & Ind. (London)*, p. 111.

Dujardin, F. (1841). "Histoire naturelle des zoophytes." Paris.

du Mérac, M. L. R. (1953). A propos de l'inuline des Acétabulaires. *Rev. gén. botan.* **60**, 689–705.

du Mérac, M. L. R. (1955). Sur la présence de fructosanes chez *Dasycladus vermicularis* (Scopoli) Krasser. *Compt. rend. acad. sci.* **241**, 88–90.

du Mérac, M. L. R. (1956). Une Néoméridacée fructosanifère: *Cymopolia barbata* L. Harv. *Compt. rend. acad. sci.* **243**, 714–717.

Duncan, W. A. M., Manners, D. J., and Ross, A. G. (1956). Enzyme systems in marine algae. The carbohydrase activities of unfractionated extracts of *Cladophora rupestris, Laminaria digitata, Rhodymenia palmata* and *Ulva lactuca*. *Biochem. J.* **63**, 44–51.

Eddy, B. O., Fleming, I. D., and Manners, D. J. (1958). Alpha-1:4-glucosans. IX. The molecular structure of a starch-like polysaccharide from *Dunaliella bioculata*. *J. Chem. Soc.*, pp. 2827–2830.

Edelman, J. (1956). The formation of oligosaccharides by enzymic transglycosylation. *Advances in Enzymol.* **17**, 189–232.

Endo, S. (1936a). Über das Vorkommen von freiem Zucker bei einigen Grünalgen und seine Beziehung zur Photosynthese. I. *Codium latum* Suringar. *Sci. Repts. Tokyo Kyôiku Bunrika Daigaku. Sect.* **B41**, 223–231.

Endo, S. (1936b). Über das Vorkommen von freiem Zucker bei einigen Grünalgen und seine Beziehung zur Photosynthese. II. *Cladophora Wrightiana* Harvey. *Sci. Repts. Tokyo Kyôiku Bunrika Daigaku. Sect.* **B43**, 291–295.

Ernst, A. (1902). *Siphoneen*studien. I. *Botan. Zentr. Beih.* **13**, 115–148.

Ernst, A. (1904a). *Siphoneen*studien. II. Beiträge zur Kenntnis der *Codiaceen*. *Botan. Zentr. Beih.* **16**, 199–236.

Ernst, A. (1904b). Zur Kenntnis des Zellinhaltes von *Derbesia*. *Flora (Jena)* **93**, 514–582.

Eschrich, W. (1956). Kallose. (Ein kritischer Sammelbericht.) Sammelreferat. *Protoplasma* **47**, 487–530.

Fanshawe, R. S., and Percival, E. (1958). Analysis of the carbohydrates of *Cladostephus* sp. *J. Sci. Food & Agr.* **9**, 241–243.

Feingold, D. S., Neufeld, E. F., and Hassid, W. Z. (1958). Synthesis of a β-1,3-linked glucan by extracts of *Phaseolus aureus* seedlings. *J. Biol. Chem.* **233,** 783–788.

Fellig, J. (1960). Laminarase of *Euglena gracilis. Science* **131,** 832.

Fleming, I. D., Hirst, E. L., and Manners, D. J. (1956). α-1,4 glucosans. IV. A reëxamination of the molecular structure of floridean starch. *J. Chem. Soc.,* pp. 2831–2836.

Fogg, G. E. (1953). "The Metabolism of Algae." Methuen, London.

Fogg, G. E. (1956). The comparative physiology and biochemistry of the blue-green algae. *Bacteriol. Revs.* **20,** 148–165 (see especially p. 158, "Carbohydrates").

Fredrick, J. F. (1951). Preliminary studies on the synthesis of polysaccharides in the algae. *Physiol. Plantarum* **4,** 621–625.

Fredrick, J. F. (1952a). Evolution of polysaccharide synthesis in algae. *Biol. Rev. City Coll. N. Y.* **14,** 26–28.

Fredrick, J. F. (1952b). Synthesis of polysaccharides in the algae. II. A polysaccharide variant of *Oscillatoria princeps. Physiol. Plantarum* **5,** 37–40.

Fredrick, J. F. (1953a). Synthesis of polysaccharides in the algae. III. Induction of polysaccharide variants in *Oscillatoria princeps* by low temperatures. *Physiol. Plantarum* **6,** 96–99.

Fredrick, J. F. (1953b). Branching characteristics of *Oscillatoria* polyglucosides. *Physiol. Plantarum* **6,** 100–105.

Fredrick, J. F. (1954). The synthesis of polysaccharides in algae. V. Kinetics of polysaccharide formation in extracts of *Oscillatoria princeps. Physiol. Plantarum* **7,** 182–189.

Fredrick, J. F. (1955a). Structure of polysaccharides and the biochemical evolution of their synthesis in algae. *Biol. Rev. City Coll. N. Y.* **17,** 33–35.

Fredrick, J. F. (1955b). Proposed structure for straight and branched polymers of glucose. *Physiol. Plantarum* **8,** 288–290.

Fredrick, J. F. (1956). Physicochemical studies of the phosphorylating enzymes of *Oscillatoria princeps. Physiol. Plantarum* **9,** 446–451.

Fredrick, J. F. (1957). Effect of surface activity and chelation phenomena on the activity of polyglucoside-synthesizing enzymes of *Oscillatoria. Physiol. Plantarum* **10,** 844–857.

Fredrick, J. F. (1958). Use of chelation phenomenon in studies of the structure and action mechanism of *Oscillatoria* phosphorylase. *Physiol. Plantarum* **11,** 493–502.

Fredrick, J. F. (1959a). Comparative evolutionary aspects of polyglucoside synthesizing enzymes. *Physiol. Plantarum* **12,** 511–517.

Fredrick, J. F. (1959b). Chromatographic patterns of polysaccharide-synthesizing enzymes of Cyanophyceae. *Phyton* **13,** 15–20.

Fredrick, J. F., and Mancini, A. F. (1955). Paper electrophoresis patterns of enzymes involved in polyglucoside synthesis in *Oscillatoria princeps* and its low temperature strains. *Physiol. Plantarum* **8,** 936–944.

Fredrick, J. F., and Mulligan, F. J. (1955). Mechanism of action of branching enzyme from *Oscillatoria* and structure of branched dextrins. *Physiol. Plantarum* **8,** 74–83.

Friedlaender, M. H. G., Cook, W. H., and Martin, W. G. (1954). Molecular weight and hydrodynamic properties of laminarin. *Biochim. et Biophys. Acta* **14,** 136–144.

Fritsch, F. E. (1935). "The Structure and the Reproduction of the Algae," Vol. I. Cambridge Univ. Press, London and New York.

Fritsch, F. E. (1945). "The Structure and Reproduction of the Algae," Vol. II. Cambridge Univ. Press, London and New York.

Galli, D. R., and Giese, A. C. (1959). Carbohydrate digestion in a herbivorous snail, *Tegula funebralis*. *J. Exptl. Zool.* **140,** 415–440.

Gojdics, M. (1953). The genus *Euglena*. (See especially pp. 20–24, for paramylon.) University Press, Madison, Wisconsin.

Goldstein, I. J., Smith, F., and Unrau, A. M. (1959). Constitution of laminarin. *Chem. & Ind. (London)*, pp. 124–125.

Gottlieb, J. (1850). Über eine neue, mit Stärkemehl isomere Substanz. *Ann. Chem. Pharm.* **75,** 51–61.

Guéguen, E. (1931). Les constituents glucidiques des Algues rouges. Thèse de pharmacie, Sorbonne, Paris.

Haas, P., and Hill, T. G. (1929). An examination of the metabolic products of certain fucoids. II. Mannitol and mannitan. *Biochem. J.* **23,** 1005–1009.

Haas, P., and Hill, T. G. (1931). The occurrence of sugar alcohols in marine algae. Dulcitol. *Biochem. J.* **25,** 1470–1471.

Haas, P., and Hill, T. G. (1932). The occurrence of sugar alcohols in marine algae. II. Sorbitol. *Biochem. J.* **26,** 986–990.

Haas, P., and Hill, T. G. (1933). Observations on the metabolism of certain seaweeds. *Ann. Botany (London)* **47,** 55–67.

Habermann, (no initial) (1874). Über die Oxydationsproducte des Amylums und Paramylums mit Brom, Wasser und Silberoxyd. *Ann. Chem. Pharm.* **172,** 11–15.

Hassid, W. Z. (1933). Occurrence of dulcitol in *Iridaea laminarioides* (Rhodophyceae). *Plant Physiol.* **8,** 480–482.

Hassid, W. Z. (1936). Carbohydrates in *Iridaea laminarioides* (Rhodophyceae). *Plant Physiol.* **11,** 461–463.

Hassid, W. Z., Joslyn, M. A., and McCready, R. M. (1941). The molecular constitution of an insoluble polysaccharide from yeast, *Saccharomyces cerevisiae*. *J. Am. Chem. Soc.* **63,** 295–298.

Hassid, W. Z., Neufeld, E. F., and Feingold, D. S. (1959). Sugar nucleotides in the interconversion of carbohydrates in higher plants. *Proc. Natl. Acad. Sci. U.S.* **45,** 905–915.

Haug, A., and Jensen, A. (1954). Seasonal variation in the chemical composition of *Alaria esculenta*, *Laminaria saccharina*, *Laminaria hyperborea* and *Laminaria digitata* from Northern Norway. *Norweg. Inst. Seaweed Research*, Rept. **No. 4,** pp. 1–13.

Heidt, K. (1937). Form und Struktur der Paramylonkörner von *Euglena sanguinea* (Ehrenberg). *Arch. Protistenk.* **88,** 127–142.

Hobson, P. N., Whelan, W. J., and Peat, S. (1951). Enzymic synthesis and degradation of starch. XIV. R-enzyme. *J. Chem. Soc.*, pp. 1451–1459.

Hoffmann, G. (1951). Variations saisonnières du glucide soluble et du brome total dans *Polysiphonia fastigiata*. *Dipl. ét. sup. Paris.*

Hough, L., Jones, J. K. N., and Wadman, W. H. (1952). The polysaccharide components of certain freshwater algae. *J. Chem. Soc.*, pp. 3393–3399.

Houwink, A. L., and Kreger, D. R. (1953). The cell wall of yeasts—electron microscope and X-ray diffraction study. *Antonie van Leeuwenhoek, J. Microbiol. Serol.* **19,** 1–24.

Houwink, A. L., Kreger, D. R., and Roelofsen, P. A. (1951). Composition and structure of yeast-cell walls. *Nature* **168,** 693–694.

Huber-Pestalozzi, G. (1955). "Das Phytoplankton des Süsswassers," Vol. IV: Eugleninen. Stuttgart, Germany.

Hutchens, J. O., Podolsky, B., and Morales, M. F. (1948). Studies on the kinetics and energetics of carbon and nitrogen metabolism in *Chilomonas paramecium. J. Cellular Comp. Physiol.* **32,** 117–141.

Jensen, A. (1956a). Component sugars of some common brown algae. *Norweg. Inst. Seaweed Research Rept.* **No. 9,** pp. 1–8.

Jensen, A. (1956b). Preliminary investigation of the carbohydrates of *Laminaria digitata* and *Fucus serratus. Norweg. Inst. Seaweed Research Rept.* **No. 10,** pp. 1–11.

Kamptner, E. (1952). Eine polarisationsoptische Untersuchung an Paramylonkörnern von *Euglena* und *Phacus. Österr. Botan. Z.* **99,** 556–588.

Kessler, G. (1958). Zur Charakterisierung der Siebröhrenkallose. *Ber. schweiz. botan. Ges.* **68,** 5–43.

Kreger, D. R., and Meeuse, B. J. D. (1952). X-ray diagrams of *Euglena* paramylon, of the acid-insoluble glucan of yeast cell walls and of laminarin. *Biochim. et Biophys. Acta* **9,** 699–700.

Küster, E. (1956). "Die Pflanzenzelle" (K. Höfler, ed.), Vol. III. Fischer, Jena, Germany.

Kylin, H. (1913). Zur Biochemie der Meeresalgen. *Z. physiol. Chem., Hoppe-Seyler's* **83,** 171–197.

Kylin, H. (1915). Untersuchungen über die Biochemie der Meeresalgen. *Z. physiol. Chem., Hoppe-Seyler's* **94,** 337–425.

Kylin, H. (1918). Weitere Beiträge zur Biochemie der Meeresalgen. *Z.physiol. Chem., Hoppe-Seyler's* **101,** 236–247.

Kylin, H. (1943a). Zur Biochemie der Rhodophyceen. *Kgl. Fysiograf. Sällskap. i Lund Förh.* **13,** 1–13.

Kylin, H. (1943b). Zur Biochemie der Cyanophyceen. *Kgl. Fysiograf. Sällskap. i Lund Förh.* **13,** 1–14.

Kylin, H. (1943c). Verwandtschaftliche Beziehungen zwischen den Cyanophyceen und den Rhodophyceen. *Kgl. Fysiograf. Sällskap. i Lund Förh.* **13,** 1–7.

Kylin, H. (1944a). Zur Biochemie von *Cladophora rupestris. Kgl. Fysiograf. Sällskap. i Lund Förh.* **14,** 1–5.

Kylin, H. (1944b). Über die Biochemie der Phaeophyceen. *Kgl. Fysiograf. Sällskap. i Lund Förh.* **14,** 1–13.

Lapicque, L. (1919). Variations saisonnières dans la composition chimique des algues marines. *Compt. rend. acad. sci.* **169,** 1426–1428.

Leitgeb, H. (1887). Die Inkrustation der Membran von *Acetabularia. Sitzber. math. naturw. Kl. bayer. Akad. Wiss. München, Serie 1,* **96,** 13–37.

Lewis, B. A., and Smith, F. (1957). The heterogeneity of polysaccharides as revealed by electrophoresis on glass fiber paper. *J. Am. Chem. Soc.* **79,** 3929–3931.

Lindberg, B. (1953a). Low-molecular carbohydrates in algae. II. Investigation of *Fucus vesiculosus. Acta Chem. Scand.* **7,** 1119–1122.

Lindberg, B. (1953b). Low-molecular carbohydrates in algae. II. Synthesis of 1-D-mannitol monoacetate and 1,6-D-mannitoldiacetate. *Acta Chem. Scand.* **7,** 1123–1124.

Lindberg, B. (1953c). Low-molecular carbohydrates in algae. III. Synthesis of 1-D-mannitol β-glucoside. *Acta Chem. Scand.* **7,** 1218–1219.

Lindberg, B. (1954). A new glycoside from *Furcellaria fastigiata. Acta Chem. Scand.* **8,** 869.

Lindberg, B. (1955a). Low-molecular carbohydrates in algae. VIII. Investigation of two green algae. *Acta Chem. Scand.* **9,** 169.

Lindberg, B. (1955b). Low-molecular carbohydrates in algae. X. Investigation of *Furcellaria fastigiata. Acta Chem. Scand.* **9,** 1093–1096.

Lindberg, B. (1955c). Low-molecular carbohydrates in algae. XI. Investigation of *Porphyra umbilicalis*. *Acta Chem. Scand.* **9,** 1097–1099.

Lindberg, B. (1956). Low molecular-weight carbohydrates in brown and red algae. *2nd Intern. Seaweed Symposium, Trondheim, 1955,* pp. 33–38.

Lindberg, B., and McPherson, J. (1954a). Low-molecular carbohydrates in algae. V. Investigation of *Laminaria cloustoni*. *Acta Chem. Scand.* **8,** 1547–1550.

Lindberg, B., and McPherson, J. (1954b). Low-molecular carbohydrates in algae. VI. Laminitol, a new C-methyl inositol from *Laminaria cloustoni*. *Acta Chem. Scand.* **8,** 1875–1876.

Lindberg, B., and Paju, J. (1954). Low-molecular carbohydrates in algae. IV. Investigation of *Pelvetia canaliculata*. *Acta Chem. Scand.* **8,** 817–820.

Lindberg, B., and Wickberg, B. (1959). Structure of laminitol. *Arkiv Kemi* **13,** 447–455.

Lwoff, A., Ionesco, H., and Gutmann, A. (1949). Métabolisme de l'amidon chez un flagellé sans chlorophylle incapable d'utiliser le glucose. *Compt. rend. acad. sci.* **228,** 342–344.

Lwoff, A., Ionesco, H., and Gutmann, A. (1950). Synthèse et utilisation de l'amidon chez un flagellé sans chlorophylle incapable d'utiliser les sucres. *Biochim. et Biophys. Acta* **4,** 270–275.

Mackie, I. M., and Percival, E. (1960). Polysaccharides from the green seaweed *Caulerpa filiformis*. II. A glucan of amylopectin type. *J. Chem. Soc.*, pp. 2381–2384.

Manners, D. J. (1959). Structural analysis of polysaccharides. *Roy. Inst. Chem. G. Brit. Ireland, Lectures, Monographs, Repts.* **2,** 1–39. (Meldola Medal Lecture.)

Meeuse, B. J. D., and Fluegel, W. (1958a). Carbohydrases in the sugar-gland juice of *Cryptochiton* (Polyplacophora, Mollusca). *Nature* **181,** 699–700.

Meeuse, B. J. D., and Fluegel, W. (1958b). Carbohydrate-digesting enzymes in the sugar-gland juice of *Cryptochiton stelleri* Middendorff (Polyplacophora, Mollusca). *Arch néerl. zool.* **13,** Suppl. 1, 301–313.

Meeuse, B. J. D., and Kreger, D. R. (1954). On the nature of floridean starch and *Ulva*-starch. *Biochim. et Biophys. Acta* **13,** 593–594.

Meeuse, B. J. D., and Kreger, D. R. (1959). X-ray diffraction of algal starches. *Biochim. et Biophys. Acta* **35,** 26–32.

Meeuse, B. J. D., Andries, M., and Wood, J. A. (1960). Floridean starch. *J. Exptl. Botany* **11,** 129–140.

Mori, T. (1953). Seaweed polysaccharides. *Advances in Carbohydrate Chem.* **8,** 315–350.

Morita, H. (1956). Characterization of starch and related polysaccharides by differential thermal analysis. *Anal. Chem.* **28,** 64–67.

Morita, H., and Rice, H. M. (1955). Characterization of organic substances by differential thermal analysis. *Anal. Chem.* **27,** 336–339.

Moss, B. L. (1948). Studies on the genus *Fucus*. I. On the structure and chemical composition of *Fucus vesiculosus* from three Scottish localities. *Ann. Botany (London)* **12,** 268–279.

Müller, O. F. (1786). "Animalcula Infusoria," p. 135. Copenhagen.

Nägeli, C. (1863). Sphaerocrystalle in *Acetabularia*. *Botan. Mitt. von C. Nägeli* **1,** 206–216.

Nisizawa, K. (1938). Physiological studies on laminarin and mannitol of brown algae. I. Diurnal variation of their content in *Eisenia bicyclis*. *Sci. Repts. Tokyo Kyôiku Bunrika Daigaku Sect.* **B3,** 289–301.

Nisizawa, K. (1939). (Quoted from Mori, 1953.) *J. Chem. Soc. Japan* **60,** 120.

Nisizawa, K. (1940a). Laminarin and mannitol of brown algae. II. Seasonal variation of their content in *Eisenia bicyclis*. *Sci. Repts. Tokyo Kyôiku Bunrika Daigaku Sect.* **B5,** 9–14.

Nisizawa, K. (1940b). Laminarin and mannitol of brown algae. III. Variation of content during growth. *Sci. Repts. Tokyo Bunrika Kyôiku Daigaku Sect.* **B5,** 14–19.

Northcote, D. H., and Horne, R. W. (1952). Chemical composition and structure of the yeast cell wall. *Biochem. J.* **51,** 232–236.

O'Colla, P. (1953). Floridean starch. *Proc. Roy. Irish Acad.* **B55,** 321–329.

Oltmanns, F. (1922). "Morphologie und Biologie der Algen," 2nd ed., Vol. I, pp. 45–50. Euglenaceae. Fischer, Jena, Germany.

Payen, J. (1938). Recherches biochimiques sur quelques Cyanophycées. *Rev. algol.* **11,** 1–99.

Peat, S., Whelan, W. J., and Lawley, H. G. (1955a). Isolation of mannitol from laminarin. *Chem. & Ind.* (*London*) pp. 35–36.

Peat, S., Whelan, W. J., Lawley, H. G., and Evans, J. M. (1955b). The structure of laminarin: isolation of a non-reducing trisaccharide. *Biochem. J.* **61,** x.

Peat, S., Turvey, J. R., and Evans, J. M. (1957). Isolation of nigerose from floridean starch. *Nature* **179,** 261–262.

Peat, S., Whelan, W. J., and Lawley, H. G. (1958a). Structure of laminarin. I. Main polymeric linkage. *J. Chem. Soc.*, pp. 724–728.

Peat, S., Whelan, W. J., and Lawley, H. G. (1958b). Structure of laminarin. II. Minor structural features. *J. Chem. Soc.*, pp. 729–737.

Peat, S., Whelan, W. J., and Edwards, T. E. (1958c). Polysaccharides of baker's yeast. II. Yeast glucan. *J. Chem. Soc.*, pp. 3862–3868.

Peat, S., Turvey, J. R., and Evans, J. M. (1959a). The structure of floridean starch. I. Linkage analysis by partial acid hydrolysis. *J. Chem. Soc.*, pp. 3223–3227.

Peat, S., Turvey, J. R., and Evans, J. M. (1959b). The structure of floridean starch. II. Enzymic hydrolysis and other studies. *J. Chem. Soc.*, pp. 3341–3344.

Peat, S., Whelan, W. J., and Evans, J. M. (1960). The structure of laminarin. III. Synthesis of structural oligosaccharides. *J. Chem. Soc.* pp. 175–178.

Percival, E. G. V., and Ross, A. G. (1951). The constitution of laminarin. Part II. The soluble laminarin of *Laminaria digitata*. *J. Chem. Soc.*, pp. 720–726.

Pochmann, A. (1956). Untersuchungen über Plattenbau und Spiralbau, über Wachstum und Zerteilung der Paramylonkörner. *Österr. Botan. Z.* **103,** (1), 110–141.

Pochmann, A. (1958). Zweiter Beitrag zur Kenntnis der Struktur, Entwicklung und Zerteilung der Paramylonkörner. *Österr. Botan. Z.* **104,** 321–341.

Pringsheim, E. G. (1952). On the nutrition of *Ochromonas*. *Quart. J. Microscop. Sci.* **93,** 71–96.

Putman, E. W., and Hassid, W. Z. (1954). Structure of galactosylglycerol from *Iridaea laminarioides*. *J. Am. Chem. Soc.* **76,** 2221–2223.

Quillet, M. (1954). Sur le métabolisme glucidique des Algues brunes. Présence de fructose chez *Laminaria flexicaulis* en survie dans l'eau de mer chloroformée *Compt. rend. acad. sci.* **238,** 926–928.

Quillet, M. (1955). Sur la nature chimique de la leucosine, polysaccharide de réserve caractéristique des Chrysophycées, extraite d'*Hydrurus foetidus*. *Compt. rend. acad. sci.* **240,** 1001–1003.

Quillet, M. (1957). Volémitol et mannitol chez les Phéophycées. *Bull. lab. maritime Dinard* **No. 43,** 119–124.

Quillet, M. (1958). Sur le métabolisme glucidique des Algues brunes. Présence de petites quantités de laminarine chez de nombreuses nouvelles espèces, réparties dans tout le groupe des Phéophycées. *Compt. rend. acad. sci.* **246,** 812–815.

Ricard, P. (1930). Les constituants glucidiques des algues brunes. *Ann. inst. océanog. (Paris)* **8**, 101–184.

Ricard, P. (1931). Les constituants glucidiques des Laminaires: nature, variations saisonnières. *Bull. soc. chim. biol.* **13**, 417–435.

Richter, G., and Pirson, A. (1957). Enzyme von *Hydrodictyon* und ihre Beeinflussung durch Beleuchtungsperiodik. *Flora (Jena)* **144**, 562–597.

Sannié, C. (1950). Sur la composition d'une algue des îles Kerguelen, *Macrocystis pyrifera* (L.) Ag. *Compt. rend. acad. sci.* **231**, 874–876.

Sannié, C. (1951). Sur la composition des algues des îles Kerguelen, *Macrocystis pyrifera* (L.) Ag. et *Durvillaea antarctica* (Cham) Hariot. *Compt. rend. acad. sci.* **232**, 2040–2041.

Sauvageau, C., and Denigès, G. (1930). Sur le sucre des algues floridées. *Compt. rend. acad. Sci.* **190**, 958–959.

Schiller, J. (1952). Über die Vermehrung des Paramylons und über Alterserscheinungen bei Eugleninen. *Oesterr. Botan. Z.* **99**, 413–420.

Schmiedeberg, J. E. O. (1885). *Tagblatt der 58. Versammlung deutscher Naturforscher und Ärzte in Strassburg,* p. 231. (Quoted from Mori, T.)

Seifter, S., Dayton, S., Novic, B., and Muntwyler, E. (1950). The estimation of glycogen with the anthrone reagent. *Arch. Biochem. Biophys.* **25**, 191–200.

Smith, F., and Unrau, A. M. (1959a). The presence of D-mannose residues in laminarin. *Chem. & Ind. (London)*, p. 636.

Smith, F., and Unrau, A. M. (1959b). On the presence of 1 → 6 linkages in laminarin. *Chem. & Ind. (London)*, p. 881.

Sosa, A., and Sosa-Bourdouil, C. (1941). Sur les *Fucus* et la composition de leurs fructifications. *Bull. lab. maritime Dinard* **23**, 43–47.

Stenhouse, J. (1844). Über das Vorkommen von Mannit in *Laminaria saccharina* und einigen andern Seegräsern. *Ann. Chem. Liebigs* **51**, 349–354.

Stewart, C. M., and Higgins, H. G. (1960). Carbohydrates of *Ecklonia radiata. Nature* **187**, 511.

Tischer, J. (1936). Über die Carotinoide und die Bildung von Jonon in *Trentepohlia* nebst Bemerkungen über den Gehalt dieser Alge and Erythrit. Carotinoide der Süsswasseralgen. II. *Z. physiol. Chem., Hoppe-Seyler's* **243**, 103–118.

Unrau, A. M., and Smith, F. (1957). A chemical method for the determination of the molecular weight of certain polysaccharides. *Chem. & Ind. (London)*, pp. 330–331.

von Holdt, M. M., Ligthelm, S. P., and Nunn, J. R. (1955). South African seaweeds: seasonal variations in the chemical composition of some Phaeophyceae. *J. Sci. Food Agr.* **6**, 193–197.

von Stosch, H. A. (1951). Über das Leukosin, den Reservestoff der Chrysophyceen. *Naturwissenschaften* **38**, 192–193.

Whelan, W. J. (1955). Starch, glycogen, fructosans and similar polysaccharides. *In* "Moderne Methoden der Pflanzenanalyse" (K. Paech and M. V. Tracey, eds.), Vol. II, pp. 145–196. Springer, Berlin.

Wolfrom, M. L., and Thompson, A. (1955). Degradation of amylopectin to nigerose. *J. Am. Chem. Soc.* **77**, 6403.

Woodward, F. N. (1954). Biochemistry of the marine algae. *Congr. intern. botan., 8e Congr. Paris,* 1954, *Rappt. commun. Sect.* **17**, 20–28.

Zechmeister, L., and Tóth, G. (1934). Über die Polyose der Hefemembran. I. *Biochem. Z.* **270**, 309–316.

Zechmeister, L., and Tóth, G. (1936). Über die Polyose der Hefemembran. II. *Biochem. Z.* **284**, 133–138.

Cell Walls

D. R. KREGER

Laboratory of General and Technical Biology, Institute of Technology, Delft, The Netherlands[1]

I. Introduction

The main organic constituents of the cell walls of algae are carbohydrates. However, in the few instances where pure, native walls were investigated in detail, small quantities of lipids and protein were also present (Northcote *et al.*, 1958, 1960).

The cell-wall carbohydrates can be roughly divided into those materials which are soluble in boiling water and those which are not, or only slightly so. The latter usually constitute a more or less firm envelope immediately surrounding each cell, whereas the former are frequently mucilaginous or pectic substances mainly located at greater distances from the cell lumen. The water-insoluble cell-wall carbohydrates discussed in this chapter generally have a less complicated molecular structure than those of the other group (see O'Colla, Chapter 20, this treatise). This feature probably determines to a large extent the properties which render them suitable as skeletal materials; as a consequence, too, they have been more intensively investigated by physical techniques.

Since our knowledge of these compounds is largely dependent on the

[1] Present address: Laboratory of Biological Ultrastructure Research, University of Groningen, The Netherlands.

available methods of analysis and investigation, a brief survey of these methods here precedes the discussion of the carbohydrates themselves.

II. Methods of Investigation

The earlier data on the constitution of the cell walls of algae were based mainly on staining reactions observed under the microscope and on chemical analyses of whole thalli; but the former methods, though still useful, lack specificity (see, e.g., Nicolai and Frey-Wyssling, 1938; Frey, 1950; Nicolai and Preston, 1952; Kinzel, 1953; Roelofsen, 1959, pp. 35–40).

Chemical extraction of cellular components provides material for more reliable chemical analyses. However, exhaustive extraction does not necessarily leave a residue representative of the cell-wall material, since part of the wall may be—and usually is—removed along with the cell contents, while material not originally associated with the wall may remain with the insoluble fraction. Cell walls cleaned by a physical procedure are therefore to be preferred when a more precise picture of cell-wall constitution is desired. Only a few genera, such as *Halicystis* and *Valonia,* have large, easily dissectable "cells"; from most other algae, pure, native cell walls are not easily obtained in quantity, and there have been only a few analyses of such material (Kreger, 1957, 1960; Lewin, 1958; Northcote *et al.*, 1958, 1960).

For details of the chemical procedures of separation, purification, and structural determination of the cell-wall constituents we must refer to books on polysaccharide chemistry; some of the more recent developments are mentioned briefly elsewhere in this treatise by O'Colla (Chapter 20). In the last two decades research on the cell walls of algae has also received considerable impetus from modern techniques such as x-ray diffraction, paper chromatography, and electron microscopy, which justifies a brief discussion of their qualities here.

X-ray diffraction enables one to recognize a substance, such as a carbohydrate, when it is partly or entirely in the crystalline state. A suitably prepared specimen is placed to intercept a narrow beam of x-rays, which are diffracted on its crystal planes at characteristic angles and can then be recorded on a photographic plate. A single crystal gives a pattern of spots, the so-called "Bragg reflections"; though not strictly *reflections*, employment of this term has been sanctioned by usage. A random orientation of small crystallites gives a set of concentric rings. In each case, the relative densities and positions of the images are characteristic of the material under examination. By this means one can identify cell-wall components with a precision otherwise unattainable without time-consuming chemical analyses. Further advantages of this technique are that only very small quantities of material are needed, and that in some cases valuable information can be obtained from merely dried algae. In many algae, however, the native cell wall material does not yield a clear x-ray pattern; but in such cases certain simple extraction and precipitation procedures sometimes suffice to produce material with characteristic reflections (see Figs. 1 and 2, and Sections III.A, B, and C). In addition to indications of a chemical nature, the x-ray method yields data on preferred orientations and sizes of crystallites of the cell-wall components.

Paper chromatography of hydrolyzed cell walls or cell-wall fractions provides information on the kinds of monosaccharide residues which constitute the cell-wall carbohydrates. The technique is of particular importance where the x-ray diagram fails to yield useful data, such as in investigations of amorphous or insufficiently crystalline fractions (see Section IV). Elucidation of the structural relation between the monomers in the original carbohydrates, however, usually entails much more complicated procedures.

Although since 1948 (Preston *et al.*, 1948) electron microscopy has contributed much to our knowledge of the ultrastructure of the cell walls of algae, this technique does not penetrate into molecular structures sufficiently to differentiate the cell-wall components. When, however, the electron microscope can be used also for electron diffraction, it becomes possible to identify a carbohydrate, provided it is sufficiently crystalline and certain other conditions are favorable.

The following discussion will deal mainly with results obtained by these three techniques, since the modern view of algal cell-wall constitution, insofar as the insoluble carbohydrates are concerned, has been largely obtained by these means. References to the earlier literature have been given, for example, by Fritsch (1935, 1945). Among the more recent studies based on staining reactions of cell walls *in situ* we may refer to Kinzel (1950, 1956, 1960).

III. The Water-Insoluble Carbohydrates

The water-insoluble carbohydrates isolated from cell walls of algae include cellulose, "ivory-nut" mannan, xylan, alginic acid, and fucinic acid; the presence of chitin has also been reported. There is generally little information on the forms in which each of these products is chemically or physically associated with other constituents of the native wall.

A. Cellulose

1. *Evidence for Occurrence and Distribution among Algae*

The general properties of cellulose are reviewed by Ott *et al.* (1954) and in other books on natural high polymers. Of the several forms which can be distinguished by their x-ray diffraction patterns, only cellulose I has been definitely identified in native (i.e., untreated) algal walls. Cellulose II and cellulose IV have been revealed after special treatments, as indicated below. In cases where cellulose I is absent, cellulose may nevertheless occur in a different form, such as amorphous cellulose, or as single-chain molecules separated by other carbohydrates, or associated chemically or physically with other constituents.

In a variety of algae, cellulose has been identified by macrochemical investigations of residues obtained by suitable extraction methods. Thus, Naylor and Russell-Wells (1934) demonstrated the presence of cellulose, and in some cases determined its content on a dry-weight basis, in the following Red and Brown Algae: *Corallina officinalis*, 15%; *Bostrychia scorpioides*, 2.6%; *Chondrus crispus*, 2.2%; *Rhodymenia palmata*, 2.1%; *Laminaria saccharina*, 5.7%; *L. digitata*, 3.7%; *Fucus vesiculosus*; *F. serratus*; *Ascophyllum nodosum*; and *Pelvetia canaliculata*. Percival and Ross (1949), in a more detailed study, confirmed its presence in *Fucus vesiculosus*, *Laminaria cloustonii*, and *L. digitata*. Similarly, among the Green Algae, the main cell-well constituent of *Nitella* was identified as cellulose by Hough *et al.* (1952), while Amin (1955) established its presence in *Chara*.

There is still no conclusive evidence for the occurrence of cellulose in the Blue-green Algae (Metzner, 1955). However, in the mucilaginous sheath of some species, after certain treatments, microfibrillar structures can be revealed which, at least in the case of *Nostoc*, appear to be composed of cellulose on the basis of their staining with iodine (Frey-Wyssling and Stecher, 1954).

A few data seem to indicate the absence of cellulose in certain other algae. Among the products of hydrolysis of a *Porphyra* sp., Cronshaw *et al.* (1958) found no glucose; similar results were obtained for cleaned, native cell walls of *Platymonas subcordiformis* (Lewin, 1958) and of *Codium fragile* (Kreger, unpublished). Likewise Frey (1950) was unable to detect cellulose in cuprammonium extracts of pretreated thalli of *Cystococcus* and *Hypnomonas*.

2. *Cellulose I*

a. In the Native Wall. Cellulose I is the modification of cellulose which is found in the cell walls of higher plants; normally their x-ray diagrams exhibit its characteristic spectrum. In the algae it was first observed in *Valonia* (Sponsler, 1930), and further observations indicated its presence in other algae. However, an extensive survey by Nicolai and Preston (1952) showed that, at least in the Green Algae, this is not the form in which it generally occurs, and that cellulose I tends to be limited to certain groups. Table I gives a summary of the results obtained by Nicolai

TABLE I

X-RAY DATA ON THE PRESENCE OR ABSENCE OF CELLLULOSE I IN ALGAE[a, b]

Division and order	Genus	Cellulose I
CHLOROPHYTA		
Chlorococcales	*Hydrodictyon* (2 spp.)	–[b]
Codiales	*Bryopsis*	–[b]
	Codium (2 spp.)	?[b]
Dasycladales	*Acetabularia*	–
Siphonocladales	*Dictyosphaeria*	+
	Siphonocladus	+
Ulotrichales	*Ulothrix* (2 spp.)	–
	Microspora	–
	Stigeoclonium	–
	Trentepohlia	+
	Microthamnion	–
	Chlorochytridion	+
Ulvales	*Enteromorpha* (3 spp.)	–[b]
	Monostroma	–
	Ulva	–[b]
Cladophorales	*Cladophora* (11 spp.)	+
	Aegagropila	+
	Spongomorpha (6 spp.)	–[b]
	Chaetomorpha (3 spp.)	+
	Rhizoclonium (2 spp.)	+
	Urospora	–
Oedogoniales	*Oedogonium*	–
Zygnematales	*Spirogyra* (3 spp.)	?[b]
	Zygnema	–
CHAROPHYTA		
Charales	*Chara*	–[b]
	Nitella	–[b]
XANTHOPHYTA		
Tribonematales	*Tribonema*	+
Vaucheriales	*Botrydium*	+
	Vaucheria (2 spp.)	?

[a] From Nicolai and Preston (1952), except *Chlorochytridion*, which is from Brandenberger and Frey-Wyssling (1947).
[b] Cell-wall constitution discussed in more detail elsewhere in this chapter.

and Preston for washed and dried algae, and includes also observations by other authors.

In many algae where cellulose I is present in the native walls, its x-ray diagram is strikingly sharp, usually revealing a remarkable type of physical organization. The cellulose crystallites are predominantly oriented with the long axes in two directions almost perpendicular to one another and parallel to the wall surface; the crystal lattice plane designated as (101) is also oriented parallel to the wall surface. (See Nicolai and Preston, 1952; and, in particular for *Valonia*, Cronshaw and Preston, 1958.) This so-called "crossed orientation" seems to be associated with the presence of cellulose I, at least in the Cladophorales and Siphonocladales; it is discussed further in Section IV.

Crossed orientation does not occur in *Spirogyra*, for instance, but in this alga the cellulose is remarkable in that it shows a uniplanar orientation with the crystal plane ($10\bar{1}$) or (002)—according to the species—parallel to the wall surface (Kreger, 1957). These different orientations might reflect certain configurational differences between the celluloses in question.

According to the most recent evidence from 7 species of Brown Algae and 15 species of Red Algae, cellulose I does not occur in these two groups (Cronshaw *et al.*, 1958; Myers and Preston, 1959b), in spite of earlier data on Brown Algae discussed below.

b. In the Chemically Purified Wall. In some algae cellulose I reflections have been obtained only after chemical treatment; this suggests that very small crystalline regions of cellulose I may be present in the untreated walls, since cellulose I is never formed *in vitro*. Such regions, too small to produce an x-ray diffraction pattern, might grow to the required size in the course of the chemical treatment, when adjacent cellulose chains, liberated by dissolution of surrounding material, become free to crystallize upon the initial crystallites. By alkaline extraction of *Laminaria cloustonii*, Percival and Ross (1948) prepared a crude fiber giving the complete cellulose I spectrum. Schurz (1953), too, reported a cellulose I spectrum from a Brown Alga, *Fucus serratus*. Among the Green Algae, cell-wall material of *Nitella* sp. yielded the cellulose I spectrum only after acid extraction (personal observation; see also Probine and Preston, 1961).

3. *Cellulose II*

Cellulose II, or regenerated cellulose, is the crystalline modification formed when cellulose is precipitated from a solution; it has been identified in precipitates from neutralized cuprammonium extracts of pretreated *Cladophora*, *Spirogyra*, *Vaucheria*, *Mougeotia*, and *Tribonema* filaments (Frey, 1950). So far, cellulose II has not been definitely demonstrated in any native cell wall, except perhaps by an x-ray diagram of the cell wall of

Halicystis irradiated parallel to the wall surface (Sisson, 1941). Sisson's diagram also revealed a uniplanar orientation of the presumptive cellulose II crystallites, in which plane (101) is parallel to the wall surface. However the presence of an anomalous reflection cast some doubt on the nature of the microcrystalline constituent. After mild chemical treatment a diagram completely matching that of cellulose II was obtained, although the membrane had disintegrated and the material still contained impurities. The latter diagram nevertheless demonstrated the presence of the cellulose type of chain, which in the native wall is probably associated in some way with another constituent (see Section III.C). In fact, the wall of *Halicystis* shows several layers of different microchemical (van Iterson, 1936) and optical (Roelofsen *et al.*, 1953) behavior, indicating that the distribution of its components is not homogeneous.

Cellulose in the form of cellulose II was also obtained from physically cleaned walls of *Hydrodictyon reticulatum* by precipitation from an alkaline solution of cell walls previously solubilized by a brief treatment with boiling dilute hydrochloric acid (Kreger, 1960). The cellulose in the native wall, on the other hand, showed no x-ray evidence for crystallinity; it was associated in some way with an equal quantity of polymannose of the ivory-nut mannan type (see Section III.B) to constitute the bulk of the wall material.

For many algae x-ray diagrams of the cell walls, though reminiscent of those of cellulose II, cannot be regarded as conclusive evidence for the presence of cellulose. This applies to the diagrams obtained from a number of crude preparations of Green Algae (Nicolai and Preston, 1952) and from 15 species of Red Algae (Myers and Preston, 1959b), the walls of which were still contaminated with some or all of the cell contents. In an investigation by Cronshaw *et al.* (1958), who chemically extracted 4 species of Green Algae, 7 of Brown Algae, and 4 of Red Algae in order to separate the "pectin," "hemicellulose," and "α-cellulose" fractions according to methods devised for higher-plant materials, it appeared that the x-ray diagrams (which for the natural material had been very varied) became almost alike after the last extraction, with prominent though somewhat diffuse reflections resembling those of cellulose II. However in no case could complete correspondence be claimed. In most cases the "α-cellulose" fractions, when hydrolyzed, indeed had a high glucose content, but considerable quantities of other monosaccharides were nearly always present (see Table II). In *Porphyra*, as mentioned earlier, no glucose was detected.

4. Cellulose IV

By boiling in glycerol, cellulose II is converted into cellulose IV, a modification with an x-ray diagram much resembling that of cellulose I,

TABLE II

SUGAR COMPONENTS OF VARIOUS FRACTIONS FROM ALGAL CELL WALLS[a, b]

Algae species	Hemicellulose (soluble in alkali)	α-Cellulose residue	
		Before chlorite extraction	After chlorite extraction
GREEN ALGAE			
Chaetomorpha melagonium	ARA	GLU ara	GLU (ara)
Cladophora rupestris		GLU gal ara (xyl)	GLU
Enteromorpha sp.	XYL RHA glu	GLU xyl rha	GLU xyl rha
Ulva lactuca	GAL ARA RHA (xyl)	GLU XYL	GLU XYL
Hydrodictyon africanum	GLU MAN (ara) (xyl)	GLU MAN	
Chlorella pyrenoidosa	GAL man ara xyl rha	{GLU GAL man / ara xyl rha}	
BROWN ALGAE			
Halidrys siliquosa and all other Fucales examined	FUC XYL	GLU fuc xyl	GLU (fuc) (xyl)
Laminaria digitata	FUC XYL	GLU uro (xyl)	GLU uro
L. saccharina	FUC XYL	GLU uro (xyl)	GLU uro
RED ALGAE			
Griffithsia flosculosa	XYL	GLU GAL xyl	GLU (gal) (xyl)
Porphyra sp.	XYL gal man	MAN (xyl)	MAN
Ptilota plumosa	XYL	GLU GAL xyl	GLU (gal) (xyl)
Rhodymenia palmata	XYL	GLU XYL	GLU XYL

[a] Most data from Cronshaw *et al.* (1958); *Chlorella* and *Hydrodictyon* from Northcote *et al.* (1958, 1960).

[b] The component sugars in hydrolyzates were detected by paper chromatography. The following abbreviations are used: ARA = arabinose; FUC = fucose; GAL = galactose; GLU = glucose; MAN = mannose; RHA = rhamnose; XYL = xylose; URO = a uronic acid, not identified further. The density of the spots after spraying with a suitable reagent—a crude indication of the relative quantities present—is typographically indicated by the sequence GLU > glu > (glu).

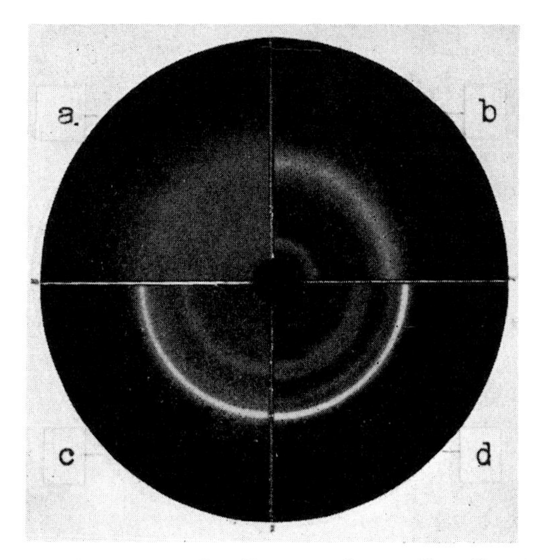

Fig. 1. Quadrants of x-ray powder diagrams from cell walls of: (*a*) *Ulva lactuca* (native walls); (*b*) *Caulerpa* sp. (native walls); (*c*) *U. lactuca* (treated with boiling dilute HCl); (*d*) cellulose I (filter paper). The two inner rings of *d* are replaced by one ring of intermediate position in *c*, as in cellulose IV.

though readily distinguishable therefrom. The appearance of this modification has also sometimes been observed under other conditions (Howsmon and Sisson, 1954). In the cases of *Halicystis* sp. (Sisson, 1941), *Spongomorpha arcta* (Nicolai and Preston, 1952), and *Griffithsia flosculosa* (Myers *et al.*, 1956), the walls of which are not of the cellulose I type, some of the lines of the x-ray spectrum after glycerol treatment are characteristic of cellulose IV. More recently, it has been observed that cell-wall preparations of *Spongomorpha arcta*, *Ulva lactuca*, and *Enteromorpha* sp., cleaned by a physical procedure and treated with dilute acid, yield a spectrum roughly corresponding to that of cellulose IV (Fig. 1*c*) although the native walls are röntgenographically amorphous (Fig. 1*a*). In certain respects the acid-treated material differs from cellulose IV (Kreger, unpublished).

B. Mannan

Mannan was first obtained from *Porphyra umbilicalis*, but has not yet been found in other species of Rhodophyta. It is reported to be alkali-soluble, and to consist of chains of β-1:4-linked D-mannopyranose residues (Jones, 1950).

More recently, mannan was also found in a number of Green Algae. From *Codium fragile*, *Acetabularia calyculus*, and *Halicoryne wrightii*, Iriki and Miwa (1960) isolated it by purification of the walls with hot

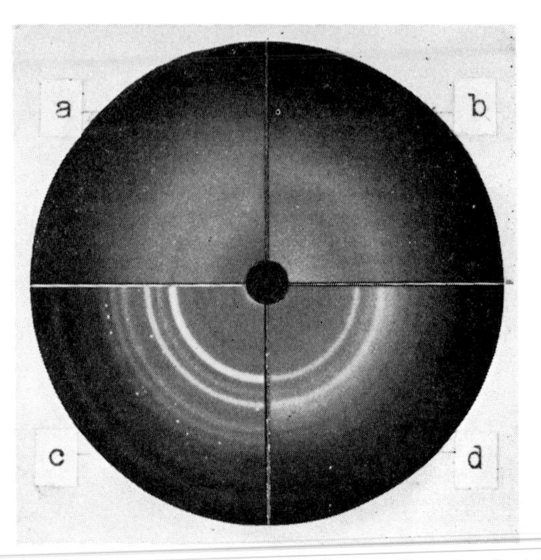

FIG. 2. Quadrants of x-ray powder diagrams from cell walls of: (*a*) *Codium fragile* (native walls); (*b*) *Hydrodictyon reticulatum* (native walls); (*c*) *C. fragile* (treated with boiling dilute HCl); (*d*) *H. reticulatum* (treated with boiling dilute HCl). Sharp lines in (*c*) and (*d*) correspond to those of ivory-nut mannan. Diffuse background in *d* is due to cellulose of low crystallinity.

dilute alkali and acid followed by chlorite bleaching, in the form of a crude fibrous product yielding mainly mannose on hydrolysis. On account of the optical rotation in 50% $ZnCl_2$, it was suggested that the mannans were of the ivory-nut type, i.e., a chain of β-1:4-linked D-mannopyranose residues (Aspinall *et al.*, 1953), similar therefore to the *Porphyra* mannan mentioned above.

Cell walls of *Hydrodictyon reticulatum*, cleaned by a physical procedure and treated with boiling dilute hydrochloric acid, yielded x-ray reflections (Fig. 2*d*) corresponding to the most prominent ones of ivory-nut mannan (Kreger, 1960), and this mannan has been isolated from the acid-treated walls. In the native wall it is associated with cellulose chains (see Section III.A.3) in a predominantly amorphous complex (Fig. 2*b*). Results obtained with other species of *Hydrodictyon* are in essential agreement with these conclusions. Thus after treatment with sulfuric acid the walls of *H. patenaeforme* exhibited an x-ray diffraction pattern indicating the presence of ivory-nut mannan (E. Nicolai; see Kreger, 1960). Hydrolyzates of the cell wall of *H. africanum* contained nearly equal quantities of glucose and mannose, with traces of other monosaccharides (Northcote *et al.*, 1960). Physically cleaned but otherwise untreated *Porphyra umbilicalis* walls yielded the same mannan reflections, as did similar preparations of *Codium fragile* after a brief treatment with acid (Fig. 2*a*,*c*) (Kreger, unpublished).

In hydrolyzates of the *Codium* walls, only mannose was detected (Kreger, unpublished). In *Porphyra* hydrolyzates, however, both mannose and xylose were found, the mannose in this case apparently replacing the glucose which accompanies the xylose in other Red Algae but which is absent from *Porphyra* (Cronshaw *et al.*, 1958). Evidently ivory-nut mannan chains replace cellulose chains in the cell walls of certain Green and Red Algae.

C. Xylan

Since most of the xylans which have been isolated from various Red Algae are water-soluble mucilages, they have been reviewed in Chapter 20 of this treatise by O'Colla. The walls of certain Green Algae have also been shown to contain xylan, and it seems probable that the "callose" which, according to Mirande (1913), replaces cellulose in certain marine Chlorophyta is likewise of this nature.

With cold N NaOH, Mackay and Percival (1959) extracted a xylan from pretreated *Caulerpa filiformis* with a yield approximately 5% of the dry weed. They showed it to consist of straight chains of β-1:3-linked xylose units. By extraction of whole thalli with boiling dilute alkali and acid, followed by chlorite bleaching, Iriki *et al.* (1960) obtained from several Green Algae (*Bryopsis*, *Caulerpa*, *Halimeda*, and *Chlorodesmis* spp.) a crude fiber which, on extraction with cold 10% NaOH, yielded a product containing xylose and glucose residues in the ratio of about 10/1. On account of the low periodate consumption it was concluded that they were 1:3-linked. The x-ray reflections of physically cleaned cell walls of *Caulerpa* spp. (Fig. 1*b*), and the negative birefringence of the walls in tangential view (Kreger, unpublished data), are probably due to the presence of xylan.

A xylan is probably also present in *Halicystis*. Reeves (quoted by Sisson, 1941) found that about 50% of the cell wall consists of cellulose and over 35% of a pentosan (by furfuraldehyde determination). In a hydrolyzate of whole walls Roelofsen *et al.* (1953) found glucose and xylose in proportions estimated at 2/1. Since after preliminary dissolution of 60% of the wall in hot dilute alkali the glucose/xylose ratio in the residue was unchanged, they suggested that the material was basically a xyloglucan. However, on the basis of the above data of Reeves and those of Sisson (see Section III.A.3), it seems more likely that we are dealing with what might be called a "cellulo-xylan," i.e., a complex of cellulose and xylan chains associated in a manner thus far unknown, but comparable to that of the cellulose and mannan chains in *Hydrodictyon* cell walls. This hypothesis is supported by: (*a*) the isolation of a cellulose fraction and a pentosan fraction; and (*b*) the observation that the "anomalous" x-ray reflection of the native wall, which is not produced by the isolated cellulose fraction (Section III.A.3), corresponds with the most intense reflection of the *Caulerpa* wall, indicating that the xylan chains in the *Halicystis* wall are probably of the *Caulerpa* type. It is of interest to point out that the central layer of the *Halicystis* wall, like the *Caulerpa* wall, exhibits negative double refraction (Roelofsen *et al.*, 1953).

D. Alginic Acid

The major constituent of the cell walls of Brown Algae is a polyuronic acid, alginic acid, which has not been found in any other division of algae. It is located mainly in the middle lamella and in the primary wall (Kylin, 1915; Andersen, 1956). It is uncertain whether it occurs naturally as: (*a*) its insoluble Ca salt (Kylin, 1915); (*b*) associated predominantly with various cations such as Ca, Mg, and Na (Wassermann, 1949); or (*c*) mainly in the free state (Bird and Haas, 1931).

The alginic acid content of the brown seaweeds, usually between 10% and 25% on a dry-weight basis, is dependent on the depth at which the algae grow and is subject to considerable seasonal variation (Black, 1948). As pointed out by Jones (1956), such variation is probably just an indirect concomitant of the seasonal rise and fall in the content of food reserves. Magnesium, ammonium, and alkali metal alginates are soluble in water. As originally discovered by Stanford in 1853, on a commercial scale algin is usually solubilized by extraction of the thalli with hot sodium carbonate solutions, whereby sodium alginate is formed by ion exchange. Boiling water alone can also be used (Kylin, 1915). When precipitated from alkaline solutions by acidification with hydrochloric acid, alginic acid forms an insoluble gel, drying irreversibly to a horny substance.

The polymerizing unit was identified as mannuronic acid independently by Nelson and Cretcher (1929), Miwa (1930), and Bird and Haas (1931). Further studies indicated a chain of D-mannuronic acid residues 1:4-linked (Hirst *et al.*, 1939) in the β-configuration, with a degree of polymerization of at least 100 (Chanda *et al.*, 1952). This concept had to be modified when Fischer and Dörfel (1955) found that a considerable proportion of the polymer consists of residues of L-guluronic acid, differing from D-mannuronic acid only by the position of the carboxyl group. As was shown for samples of alginic acid from 22 species of Brown Algae, the ratios of the two types of polymerizing units may vary between about 2/1 and 1/2, with a higher content of mannuronic acid in the Fucales and of guluronic acid in the Dictyotales. The presence of guluronic acid residues in alginic acid was confirmed by Drummond *et al.* (1958) and by Whistler and Kirby (1959).

The distribution of the two different units in the polymer—whether they occur in separate chains or in the same chain—has not yet been elucidated. Use of alginic acid depolymerases, such as those reported from several marine invertebrates (see Eppley and Lasker, 1959; Galli and Giese, 1959) and presumably also secreted by alginolytic bacteria (see Waksman *et al.*, 1934; Chesters *et al.*, 1956), may prove helpful in this connection.

The crystal structure and molecular chain configuration of alginic acid have been discussed by Astbury (1945), Palmer and Hartzog (1945), and Sterling (1957). Sterling suggested that, with reference to the plane of the pyranose rings, the direction of the oxygen bridges in the chain is similar to that in cellulose, i.e., there is an acute angle between the bonds and the ring planes, rather than a right angle as had been earlier assumed. Such a stretched configuration of the units is also claimed for calcium alginate. In the native wall, the molecular chains are arranged in parallel, as is shown by studies of birefringence (Andersen, 1956). Probably their crystallinity is poor because calcium alginate, the form generally supposed to occur in the wall, gives rather diffuse x-ray diagrams (Sterling, 1957).

Alginic acid is reported to exist in two forms, one soluble and the other insoluble in water. The former was obtained by careful dialysis of a sodium alginate solution acidified with acetic acid (Evtushenko, 1954). The latter, the more stable form, is precipitated from alkaline solutions on the addition of mineral acid. According to Evtushenko, it is then in the form of an anhydride, for which he proposes that the term "algin" should be used, retaining the name "alginic acid" for the soluble acid. In current practice, however, "algin" usually refers to the extraction product, sodium alginate (Tseng, 1945).

Haug (1959) reported that sodium alginate is physically inhomogeneous and that two fractions can be separated in 0.6–1.0 N potassium chloride solutions, in which about two-thirds of the alginate is insoluble.

The literature on alginic acid is extensive, since this seaweed product is of considerable industrial application as an emulsifier, as a thickening and stabilizing agent in liquid food preparations, and as a surfacing agent for papers. It has been reviewed by Maass (1959).

E. Fucinic Acid

This acid was reported in appreciable amounts in *Ascophyllum nodosum* and appears to be confined to the larger Brown Algae. It occurs in the form of its calcium salt, together with that of alginic acid from which it cannot be completely freed (Kylin, 1915). It differs from alginic acid in that it stains with iodine in the presence of sulfuric acid, a blue color being obtained at low sulfuric acid concentrations ($\sim$1%), which turns to red at higher concentrations (~50%), in which mixture cellulose stains blue, a distinguishing feature in sections of thalli (Kylin, 1915). Odén (1917) suggested that fucinic acid is a dipentose-dicarboxylic acid, but its chemical nature has not been adequately studied. Since apparently no recent studies have dealt with this substance, a reinvestigation of this material appears desirable.

F. Chitin

On the basis of staining reactions under the microscope, chitin has been reported to occur in some Green Algae (see Fritsch, 1935; Nicolai and Preston, 1952); but only in the outer layer of the cell wall of *Cladophora prolifera* was this corroborated by x-ray diffraction studies (Astbury and Preston, 1940). However, Frey (1950), applying the same method, found no evidence for the presence of chitin either in *C. prolifera* or in *C. glomerata.*

IV. Cell-Wall Constitution and the Microfibrillar Structure

A. The Microfibrillar Structure

Electron microscopy has revealed that at the ultrastructural level cell walls usually exhibit a structure of so-called microfibrils. These are frequently embedded in a substance presenting no conspicuous detail, which looks smooth or slightly grainy and which is usually indicated as the amorphous or continuous matrix. Sometimes the microfibrils are revealed only after chemical extractions. They are usually somewhat flattened, with diameters which generally vary between 100 and 200 Å., though other ranges have also been reported, notably 30–50 Å. for *Chlorella* and *Hydrodictyon* and 100–350 Å. for *Valonia* (Northcote *et al.*, 1958, 1960; Preston and Kuyper, 1951). The microfibrils occur as a network of more or less curved threads, sometimes randomly arranged as in felt, sometimes with definite orientations. Some cell walls exhibit a crossed microfibrillar structure in which layers with bundles of nearly straight microfibrils running in one direction alternate with layers of bundles oriented in another direction, usually almost perpendicular to the former (Fig. 3*a*). (See also Green, Chapter 41 in this treatise.)

Extraction with boiling water or mild chemical agents generally enhances the fibrillar appearance of the cell walls and may lead to the complete removal of nonfibrillar material (Fig. 3*b*); the reverse effect has also been observed. Possibly some of the microfibrils may be artifacts, formed by aggregation of molecular chains during such extraction procedures. Information on the chemical nature of the microfibrils and the continuous matrix has been obtained by considering the electron-microscopic picture

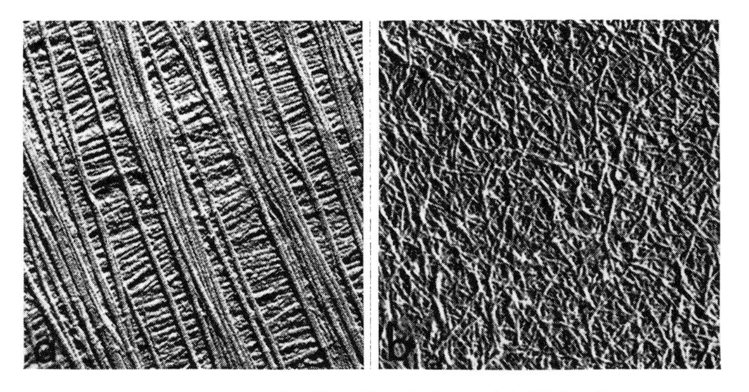

FIG. 3. Electron micrographs of cell walls of algae: (*a*) *Cladophora rupestris,* native wall with lamellae of crossed microfibrils of cellulose I (× 30,000) (Frey and Preston, unpublished); (*b*) *Rhodymenia palmata,* felty network of microfibrils of α-cellulose fraction, containing approximately equal amounts of glucose and xylose residues (× 22,500) (Myers and Preston, 1959a).

after such extractions in conjunction with the x-ray diagram and with the component monosaccharides as revealed by paper chromatography of hydrolyzates. This system was applied by Cronshaw *et al.* (1958) in the extensive investigations already discussed in Section III.A.3, and, with various modification, by Roelofsen *et al.* (1953), Myers and Preston (1959*a,b*), and Northcote *et al.* (1958, 1960). Microfibrillar structure was also revealed in the walls of *Dictyota flabellata* and *Helminthocladia californica* (Dawes *et al.*, 1960).

B. Composition of the Microfibrils

Since the first observations of the microfibrillar structure in an algal cell wall, in studies of *Valonia* by Preston *et al.* (1948), it was evident that the fibrils contain cellulose, because their directions in the crossed fibrillar structure corresponded to the directions of preferred orientation of the cellulose crystallites as revealed by x-ray diffraction. With the electron microscope the conclusion was further substantiated by electron diffraction from selected areas containing the fibrils (Preston and Ripley, 1954; Honjo and Watanabe, 1958). It appears that this structure of crossed, nearly straight microfibrils is characteristic of those algal cell walls that show sharp x-ray spectra of highly crystalline and oriented cellulose I (see Section III.A.2).

When adequately purified, walls with a crossed microfibrillar structure give a high yield of α-cellulose, still in the form of microfibrils. On hydrolysis, this α-cellulose yields more or less pure glucose with no more than a trace of any other sugar, indicating that the fibrils are composed of pure or nearly pure cellulose (Preston and Cronshaw, 1958; Cronshaw *et al.*, 1958).

It should be borne in mind that in the native wall they may have contained some noncellulose components, perhaps on the outside, which were removed by the "purification" treatment. Details of their fine structure have been discussed by Preston (1959).

In nearly all other instances the microfibrillar α-cellulose fractions yield on hydrolysis, in addition to glucose, other monosaccharides in smaller to equal amounts: usually xylose, sometimes also galactose, rhamnose, fucose, and a uronic acid, depending on the taxonomic group of algae (see Table II). These fibrils cannot therefore be of true cellulose. They also differ from the type described above in that they are thinner, their paths are more tortuous and interwoven, and lateral coherence and flattening are less pronounced. The x-ray diagrams do not provide unequivocal evidence for the presence of cellulose.

Myers and Preston (1959*a*) proposed to differentiate between the two kinds of microfibrils by introducing the term "eucellulose" for the highly crystalline material of the former type, retaining the term "cellulose" for the latter, despite its usually high content of non-glucose residues. The advisability of this terminology is questionable. Meanwhile, in those algae where the presence of some form of cellulose has been demonstrated it must be presumed to be present largely in the microfibrillar fraction since glucose is only rarely found in hydrolyzates of the nonfibrillar fraction of the walls (see Table II).

The physical or chemical association of cellulose chains with other constituents prevents or hampers their crystallization, which may explain the inconclusive x-ray diffraction patterns obtained from impure α-cellulose fibrils. But in those algae from which cellulose can be obtained, by whatever means, as cellulose I, the native fibrils must be assumed to contain at least small regions in which the cellulose chains are arranged in the cellulose I pattern and can act as initials for further crystallization during the purification process (see Section III.A.2.*b*). When the walls of certain algae—such as species of *Hydrodictyon*, *Halicystis*, and *Ulva* (Chlorophyta) and *Griffithsia* (Rhodophyta)—are extracted to remove noncellulosic materials, no cellulose I is found in the residue; this indicates that the fibrils, if present in the native walls, could not originally have embodied any crystalline regions of this nature. It should be noted that, in at least one case, i.e., *Porphyra* (Cronshaw *et al.*, 1958), microfibrils have been found which cannot contain any cellulose, because no glucose was obtained in hydrolyzates of the walls, in which ivory-nut mannan may have taken the place of cellulose. It is to be expected that xylan microfibrils may be found in the walls of algae such as *Caulerpa*, which seem to contain straight chains of 1:3-linked xylose residues.

Aspects of microfibril formation in the growing wall have been discussed with particular reference to algae by Preston (1959) and Nicolai and Preston (1959). For a discussion of cell-wall structure in relation to cell enlargement, see Green (Chapter 41, this treatise).

C. Constitution of the Continuous Matrix

The continuous matrix of cell walls dissolves partly as the water-soluble fraction (for constituents, see O'Colla, Chapter 20), and partly in 4 *N* KOH as the hemicellulose fraction. In the four Red Algae investigated by Cronshaw *et al.* (1958), the latter fraction yielded on hydrolysis chiefly xylose; in the Brown Algae, xylose and fucose predominated; while a greater variety of component sugars was found among the Green Algae (see Table II). The percentage composition of the walls of these algae, in fractions of different solubility, is given in Table III. Hemicellulose fractions of the continuous matrix isolated from *Chlorella pyrenoidosa* and *Hydrodictyon africanum*, approximately 30% and 15% of the dry walls, respectively

TABLE III

CELL-WALL COMPOSITION IN SOME ALGAE[a]

Algal species	Water-soluble fraction (%)	Alkali-soluble fraction (%)	Chlorite-soluble fraction (%)	α-Cellulose (%)
GREEN ALGAE				
Chaetomorpha melagonium	41.5	8	9.5	41
Cladophora rupestris	31.5	2	38	28.5
Enteromorpha sp.	30	39	9	21
Ulva lactuca	52	25	4	19
BROWN ALGAE				
Halidrys siliquosa	62	14	10	14
Fucus serratus	44.5	29	13.5	13.5
Himanthalia lorea	67	14	11	8
Ascophyllum nodosum	68.5	16	8.5	7
Pelvetia canaliculata	70	16	12.5	1.5
Laminaria digitata	49	25	6	20
Laminaria saccharina	59	17.5	5.5	18
RED ALGAE				
Griffithsia flosculosa	41.5	14	22.5	22
Porphyra sp.	49	47.5	0	3.5
Ptilota plumosa	36	17.5	23	24
Rhodymenia palmata	50	36.5	6.5	7

[a] From Cronshaw *et al.* (1958).

proved to be electrophoretically homogeneous (Northcote *et al.*, 1958, 1960). The component monosaccharides are included in Table II. Though glucose was absent from the *Chlorella pyrenoidosa* hemicellulose isolated by Northcote *et al.*, Bailey and Neish (1954) obtained from *Chlorella vulgaris* a hemicellulose containing glucose, as well as some rhamnose and xylose residues, which they believed to have been a component of the cell walls.

Hemicellulose fractions may possibly be partly of microfibrillar origin. In *Porphyra* sp., most of the mannose of this fraction certainly originates from microfibrils since the latter dissolve in the alkali. Since the solubility of some carbohydrates in alkali is dependent on their degree of polymerization, essentially similar carbohydrates may be found in both the α-cellulose fraction and in the hemicellulose fraction, the latter containing the shorter molecular chains.

References

Amin, E. S. (1955). The isolation and study of *Chara* cellulose. *J. Chem. Soc.* **1953,** 281–284.

Andersen, G. (1956). On the detection of alginic acid in tissues by means of birefringence. *In* "Second International Seaweed Symposium" (T. Braarud and N. A. Sørensen, eds.), pp. 119–124. Pergamon Press, London.

Aspinall, G. O., Hirst, E. L., Percival, E. G. V., and Williamson, I. R. (1953). The mannans of ivory nut (*Phytelephas macrocarpa*). *J. Chem. Soc.*, pp. 3184–3188.

Astbury, W. T. (1945). Structure of alginic acid. *Nature* **155,** 667.

Astbury, W. T., and Preston, R. D. (1940). The structure of the cell wall in some species of the filamentous green alga *Cladophora. Proc. Roy. Soc.* **B129,** 54–76.

Bailey, J. M., and Neish, A. C. (1954). Starch synthesis in *Chlorella vulgaris. Can. J. Biochem. Physiol.* **32,** 452–464.

Bird, G. M., and Haas, P. (1931). On the nature of the cell wall constituents of *Laminaria* spp. Mannuronic acid. *Biochem. J.* **25,** 403–411.

Black, W. A. P. (1948). The seasonal variation in chemical constitution of the sublittoral seaweeds common to Scotland. *J. Soc. Chem. Ind.* (*London*) **67,** 165–176.

Brandenberger, E., and Frey-Wyssling, A. (1947). Über die Membransubstanzen von *Chlorochytridion tuberculatum* W. Vischer. *Experientia* **3,** 492.

Chanda, S. K., Hirst, E. L., Percival, E. G. V., and Ross, A. G. (1952). The structure of alginic acid II. *J. Chem. Soc.* pp. 1833–1837.

Chesters, C. G. C., Turner, M., and Apinis, A. (1956). Decomposition of laminarin by micro-organisms. *Proc. 2nd. Internat. Seaweed Sympos.* (*Trondheim, 1955*), pp. 141–144.

Cronshaw, J., and Preston, R. D. (1958). A reexamination of the fine structure of the walls of vesicles of the green alga *Valonia. Proc. Roy. Soc.* **B148,** 137–148.

Cronshaw, J., Myers, A., and Preston, R. D. (1958). A chemical and physical investigation of the cell-walls of some marine algae. *Biochim. et Biophys. Acta* **27,** 89–103.

Dawes, C. J., Scott, F. M., and Bowler, E. (1960). Light and electron microscopy study of cell walls of Brown and Red Algae. *Science* **132,** 1663–1664.

Drummond, D. W., Hirst, E. L., and Percival, E. (1958). The presence of L-guluronic acid residues in alginic acid. *Chem. & Ind.* (*London*) p. 1088.

Eppley, R. W., and Lasker, R. (1959). Alginase in the sea urchin, *Strongylocentrotus purpuratus*. *Science* **129,** 214–215.

Evtushenko, V. A. (1954). [The chemical nature of alginic acids. I and II.] *Kolloid Zhur.* **16,** 255–263, 340–344. (In Russian.)

Fischer, F. G., and Dörfel, H. (1955). Die Polyuronsäuren der Braunalgen. *Z. physiol. Chem. Hoppe-Seyler's* **302,** 186–203.

Frey, R. (1950). Chitin und Zellulose in Pilzzellwänden. *Ber. schweiz. botan. Ges.* **60,** 199–230.

Frey-Wyssling, A., und Stecher, H. (1954). Über den Feinbau des *Nostoc*-Schleimes. *Z. Zellforsch. u. mikroskop. Anat.* **39,** 515–519.

Fritsch, F. E. (1935, 1945). "The Structure and Reproduction of the Algae," Vols I. and II. Cambridge Univ. Press, London and New York.

Galli, D. R., and Giese, A. C. (1959). Carbohydrate digestion in a herbivorous snail, *Tegula funebralis*. *J. Exptl. Zool.* **140,** 415–440.

Haug, A. (1959). Fractionation of alginic acid. *Acta Chem. Scand.* **13,** 601–603.

Hirst, E. L., Jones, J. K. N., and Jones, W. O. (1939). The structure of alginic acid. I. *J. Chem. Soc.*, pp. 1880–1885.

Honjo, G., and Watanabe, M. (1958). Examination of cellulose fibre by the low-temperature specimen method of electron diffraction and electron microscopy. *Nature* **181,** 326–328.

Hough, L., Jones, J. K. N., and Wadman, W. H. (1952). An investigation of the polysaccharide components of certain freshwater algae. *J. Chem. Soc.* pp. 3393–3399.

Howsmon, J. A., and Sisson, W. A. (1954). Submicroscopic structure. *In* "Cellulose and Cellulose Derivatives" (E. Ott, H. M. Spurlin, and M. W. Grafflin, eds.), Vol. I, pp. 241–243. Interscience, New York.

Iriki, Y., and Miwa, T. (1960). Chemical nature of the cell-wall of the green algae, *Codium, Acetabularia* and *Halicoryne*. *Nature* **185,** 178–179.

Iriki, Y., Suzuki, T., Nishizawa, K., and Miwa, T. (1960). Xylan of siphonaceous green algae. *Nature* **187,** 82.

Jones, J. K. N. (1950). The structure of the mannan present in *Porphyra umbilicalis*. *J. Chem. Soc.*, pp. 3292–3295.

Jones, R. F. (1956). On the chemical composition of the brown alga *Himanthalia elongata* (L.) S.F. Gray. *Biol. Bull.* **110,** 169–178.

Kinzel, H. (1950). Die Algen *Spirogyra* und *Zygnema* bei Behandlung mit Kupferoxydammoniak. *Mikroskopie* **5,** 89.

Kinzel, H. (1953). Die Bedeutung der Pectin- und Cellulosekomponente für die Lage des Entladungspunktes. *Protoplasma* **42,** 209–226.

Kinzel, H. (1956). Untersuchungen über Bau und Chemismus der Zellwände von *Antithamnion cruciatum* (Ag) Näg. *Protoplasma* **46,** 445–474.

Kinzel, H. (1960). Über den Bau der Zellwände von *Bornetia secundiflora* (J.Ag.) Thur. *Botan. Marina* **1,** 74–86.

Kreger, D. R. (1957). New crystallite orientations of cellulose I in *Spirogyra* cell walls. *Nature* **180,** 914–915.

Kreger, D. R. (1960). An X-ray study of *Hydrodictyon* cell walls. I and II. *Koninkl. Ned. Akad. Wetenschap. Proc. Ser.* **C63,** 613–633.

Kylin, H. (1915). Untersuchungen über die Biochemie der Meeresalgen. *Z. physiol. Chem. Hoppe-Seyler's* **94,** 337–425.

Lewin, R. A. (1958). The cell walls of *Platymonas*. *J. Gen. Microbiol.* **19,** 87–90.

Maass, H. (1959). "Alginsäure und Alginate." Strassenbau, Chemie und Technik Verlagsges., Heidelberg, Germany.

Mackay, I. M., and Percival, E. (1959). The constitution of xylan from the green seaweed *Caulerpa filiformis*. *J. Chem. Soc.* p. 1151.

Metzner, I. (1955). Zur Chemie und zum submikroskopischen Aufbau der Zellwände, Scheiden und Gallerten von Cyanophyceen. *Arch. Mikrobiol.* **22,** 45–77.

Mirande, R. (1913). Recherches sur la composition chimique de la membrane et le morcellement du thalle chez les *Siphonales*. *Ann. sci. nat. Botan. et biol. végétale* **18,** 147–264.

Miwa, T. (1930). Alginic acid. *J. Chem. Soc. Japan* **51,** 738–745.

Myers, A., and Preston, R. D. (1959a). Fine structure in the red algae. II. The structure of the cell wall of *Rhodymenia palmata*. *Proc. Roy. Soc.* **B150,** 447–455.

Myers, A., and Preston, R. D. (1959b). Fine structure in the red algae. III. A general survey of cell-wall structure in the red algae. *Proc. Roy. Soc.* **B150,** 456–459.

Myers, A., Preston, R. D., and Ripley, G. W. (1956). Fine structure in the red algae. I. X-ray and electron microscope investigation of *Griffithsia flosculosa*. *Proc. Roy. Soc.* **B144,** 450–459.

Naylor, G. L., and Russell-Wells, B. (1934). On the presence of cellulose and its distribution in the cell walls of brown and red algae. *Ann. Botany (London)* **48,** 635–641.

Nelson, W. L., and Cretcher, L. H. (1929). The alginic acid from *Macrocystis pyrifera*. *J. Am. Chem. Soc.* **51,** 1914–1922.

Nicolai, E., and Frey-Wyssling, A. (1938). Über den Feinbau der Zellwand von *Chaetomorpha*. *Protoplasma* **30,** 401–413.

Nicolai, E., and Preston, R. D. (1952). Cell-wall studies in the Chlorophyceae. I. A general survey of submicroscopic structure in filamentous species. *Proc. Roy. Soc.* **B140,** 244–274.

Nicolai, E., and Preston, R. D. (1959). Differences in structure and development in the Cladophoraceae. *Proc. Roy. Soc.* **B151,** 244–255.

Northcote, D. H., Goulding, K. J., and Horne, R. W. (1958). The chemical composition and structure of the cell-wall of *Chlorella pyrenoidosa*. *Biochem. J.* **70,** 391–397.

Northcote, D. H., Goulding, K. J., and Horne, R. W. (1960). The chemical composition and structure of the cell-wall of *Hydrodictyon africanum* Yaman. *Biochem. J.* **77,** 503–508.

Odén, S. (1917). Studien über Pektinsubstanzen. II. Zur Kenntniss der Alginsäure und Fucinsäure. *Intern. Z. physik.-chem. Biol.* **3,** 83–93.

Ott, E., Spurlin, H. M., and Grafflin, M. W. (1954). "Cellulose and Cellulose Derivatives." 2nd ed. (Vol. 5 of "High Polymers.") Interscience, New York.

Palmer, K. J., and Hartzog, M. B. (1945). The configuration of the pyranose rings in polysaccharides. *J. Am. Chem. Soc.* **67,** 1865–1866.

Percival, E. G. V., and Ross, A. G. (1948). The cellulose of marine algae. *Nature* **162,** 895.

Percival, E. G. V., and Ross, A. G. (1949). Marine algal cellulose. *J. Chem. Soc.*, pp. 3041–3043.

Preston, R. D. (1959). Wall organization in plant cells. *Intern. Rev. Cytol.* **8,** 33–60.

Preston, R. D., and Cronshaw, J. (1958). Constitution of the fibrillar and non-fibrillar components of the walls of *Valonia ventricosa*. *Nature* **181,** 248–250.

Preston, R. D., and Kuyper, B. (1951). Electron microscopic investigations of the walls of green algae. I. A preliminary account of wall lamellation and deposition in *Valonia ventricosa*. *J. Exptl. Botany* **2,** 247–255.

Preston, R. D., and Ripley, G. W. (1954). Electron diffraction diagrams of cellulose microfibrils in *Valonia*. *Nature* **174,** 76.

Preston, R. D., Nicolai, E., Reed, R., and Millard, A. (1948). An electron microscope study of cellulose in the wall of *Valonia ventricosa*. *Nature* **162,** 665–667.

Probine, M. C., and Preston, R. D. (1961). Cell growth and the structure and mechanical properties of the wall in internodal cells of *Nitella opaca*. I. Wall structure and growth. *J. Exptl. Botany* **12,** 261–282.

Roelofsen, P. A. (1959). "The Plant Cell Wall." Borntraeger, Berlin.

Roelofsen, P. A., Dalitz, V. C., and Wijnman, C. F. (1953). Constitution, submicroscopic structure and degree of crystallinity of the wall of *Halicystis osterhoutii*. *Biochim. et Biophys. Acta* **11,** 344–352.

Schurz, J. (1953). Über die Textur der in *Fucus* vorkommenden Zellulose. *Naturwissenschaften* **40,** 438.

Sisson, W. A. (1941). Some X-ray observations regarding the membrane structure of *Halicystis*. *Contribs. Boyce Thompson Inst.* **12,** 31–44.

Sponsler, O. L. (1930). New data on cellulose space lattice. *Nature* **125,** 633.

Sterling, C. (1957). Structure of oriented gels of calcium polyuronates. *Biochim. et Biophys. Acta* **26,** 186–197.

Tseng, C. K. (1945). The terminology of seaweed colloids. *Science* **101,** 597–602.

van Iterson, G., Jr. (1936). Notes on the structure of the wall of algae of the genus *Halicystis*. *Koninkl. Ned. Akad. Wetenschap. Proc.* **39,** 1066–1074.

Waksman, S. A., Carey, C. L., and Allen, M. C. (1934). Bacteria decomposing alginic acid. *J. Bacteriol.* **28,** 213–220.

Wassermann, A. (1949). Cation adsorption of brown algae: the mode of occurrence of alginic acid. *Ann. Botany (London)* **13,** 79–88.

Whistler, R. L., and Kirby, K. W. (1959). Notiz über die Zusammensetzung der Alginsäure von *Macrocystis pyrifera*. *Z. physiol. Chem. Hoppe-Seyler's* **314,** 46–48.

—20—

Mucilages

P. S. O'COLLA

Department of Chemistry, University College, Galway, Ireland

I. Introduction[1]

This review deals only with water-soluble algal polysaccharides other than the starch-like and laminarin-like storage carbohydrates. The work of Preston (1958) and his associates on cell-wall structure in the algae, as revealed by polarization microscopy, electron microscopy, x-ray diffraction

[1] Abbreviations used in this chapter: A = arabinose; G, glucose; Ga, galactose; GA, glucuronic acid; R, rhamnose; X, xylose.

analysis, and chemical analysis, indicates that these mucilages are mainly constituents of the continuous amorphous phase of cell walls. According to Myers and Preston (1959a, b), the cell walls of the numerous algae examined contain 30 to 70% of water-soluble material. The constituents that have been identified in the hydrolyzates of these extracts are D-glucose, D-mannose, D- and L-galactose, L-rhamnose, D-xylose, L-arabinose, D-glucuronic acid, D-galacturonic acid, L-fucose, D- and L-3,6-anhydrogalactose, and 6-*O*-methyl-D- and L-galactose, as well as sulfuric acid and pyruvic acid. Relatively few mucilages have been subjected to structural examination; those investigated exhibit great diversity in their molecular architecture, corresponding in part to the wide variety of functions which they perform in the plant. On account of their gel-forming properties, some mucilages of the Rhodophyta have been important articles of commerce for a long time and have been studied intensively. It is only in this group that the occurrence of some general structural patterns is detectable at present. An almost inexhaustible field of study remains for the organic chemist and the plant physiologist. Recent developments in methods of investigating polysaccharide structure (Manners, 1959; Bouveng and Lindberg, 1960) and of separating closely related polysaccharides (Neukom *et al.*, 1960) should lead to rapid advances in this field during the next decade. Reviews on algal polysaccharides have been published by Tseng (1946), Percival (1949), Jones and Smith (1949), Humm (1951), Mori (1953), Black (1953), Mautner (1954), Hirst (1958a, b), Araki (1958), Schachat and Glicksman (1959), Schmid (1960), and Smith and Montgomery (1959).

[In this chapter, crude extractives from algae are generally identified by the suffix *-in*, as in *algin*, *carrageenin*, *fucoidin*, *ulvin*. The suffix *-an* is reserved for polymers of a defined monosaccharide, such as *dextran* or *xylan*.]

II. Mucilages of the Rhodophyta

A. General Observations

Analyses of common red seaweeds (Ross, 1953) showed that galactose is the principal component sugar and that many of the polysaccharides have a high sulfate content, confirming earlier investigations (Tseng, 1946) indicating that many water-soluble mucilages were galactan sulfates. In the 1950's considerable advances were made in detecting the minor constituents and their mode of linkage in the molecule. The best known mucilages are agar, obtainable by extraction of species of *Gelidium* and some other genera, and carrageenin, obtained from species of *Chondrus* and *Gigartina*, known collectively as carrageen or Irish moss. Agar and carrageenin have attracted the attention of scientists for at least a hundred years. These terms must now be considered as referring to mixtures of polysaccharides, since agar has been shown to consist of agarose and agaropectin (Araki, 1958), and carrageenin to be a mixture of κ- and λ-fractions (O'Neill *et al.*, 1955).

Structural investigations have shown that some mucilages are closely related to agarose while others resemble κ-carrageenin. A classification of seaweed mucilages, based principally on their physical properties, was proposed by Stoloff and Silva (1957). An important advance in this respect was made by Yaphe (1959), who isolated specific hydrolases for agarose and for κ-carrageenin from marine bacteria. The agarase was used to detect agarose-like material, and the κ-carrageenase was employed to determine quantitively substances akin to the κ-carrageenin of *Chondrus crispus,* in aqueous extracts from thirty species of Red Algae.

With the information now available on their structure, it is convenient to classify the mucilages of the Red Algae empirically as follows:

(1) Mucilages containing agarose-type linkages
(2) Mucilages containing κ-carrageenin-type linkages
(3) Xylans
(4) Other mucilages

Yaphe (1959) suggested that the mucilages of group 2 could be subdivided into those containing: (*a*) a low concentration of κ-carrageenin-like polysaccharides; (*b*) the same content of κ-carrageenin-like polysaccharides as *Chondrus* extract, associated with other unspecified components; and (*c*) mainly κ-carageenin-like polysaccharides.

B. Mucilages Containing Agarose-Type Linkages

1. *Agar*

Agar was first isolated by Payen (1859) from *Gelidium amansii*, which is still the main source of Japanese agar, though commercial samples may also contain lesser quantities of the extracts of *Gelidium subcostatum, G. japonicum, G. pacificum, Pterocladia tenuis, Acanthopeltis japonica, Gracilaria confervoides* (*G. verrucosa*),* *Ceramium rubrum, Ceramium boydenii, Campylaephora hypnaeoides,* etc. (Araki, 1958). In other countries the same weeds are used where they are available, but agar industries have also been based on extracts of *Gelidium cartilagineum* in America, *Gelidium latifolium* in Ireland, and *Ahnfeltia plicata* and *Phyllophora* species in Russia. Further details are given in the review by Schmid (1960).

* Regarding the taxonomy of this organism, see Note 9 of the Appendix by Silva.

Earlier investigations (Tseng, 1946) showed that agar contains D-galactose, small quantities of L-galactose (Pirie, 1936), 3 to 5% sulfate and some 3,6-anhydro-L-galactose (Hands and Peat, 1938; Percival *et al.*, 1938; Araki, 1938). When an anhydrogalactose derivative was first detected in methylated agar, it was considered to be an artifact formed by alkaline desulfatation during the methylation. However, Barry and Dillon (1944) and Percival (1944) showed that the sulfate content of agar was too low to account for the yields of anhydrogalactose actually isolated. Since the development of methods for the preparation and isolation of derivatives of 3,6-anhydro-L-galactose (Fig. 1), this sugar may be obtained

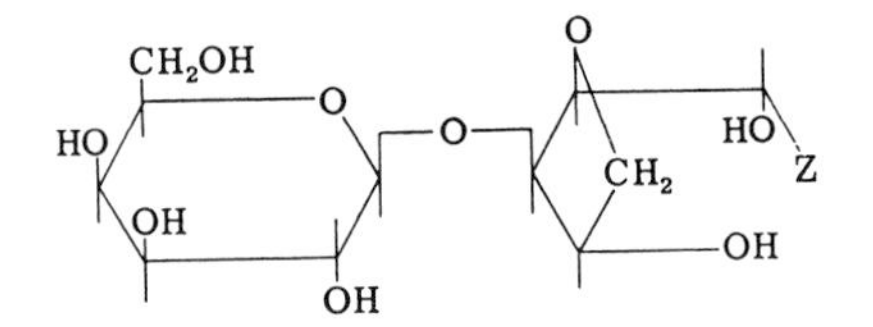

FIG. 1. Derivatives of agarobiose: Z = CHO; $CH(SC_2H_5)_2$; CH_2OH.

readily in high yield from commercial agar (Araki and Hirase, 1956; see Araki, 1958; Clingman *et al.*, 1957).

Important developments in the chemistry of agar have been:

(1) the recognition of two fractions, agarose and agaropectin;
(2) the isolation of agarobiose derivatives from partial hydrolyzates;
(3) the isolation of a disaccharide, neoagarobiose, and a tetrasaccharide, neoagarotetrose, from enzymatic hydrolyzates of agarose;
(4) the detection of a uronic acid (3–7%) and pyruvic acid (1%).

The disaccharide agarobiose, 4-*O*-D-galactopyranosyl-3,6-anhydro-L-galactose, was isolated by partial acid hydrolysis of agar and agarose (Araki and Hirase; see Araki 1958). Cleavage of agar with enzymes, on the other hand, yielded mixtures of oligosaccharides in which galactose occupies the reducing end (Araki and Arai, 1957). All the evidence indicates that agarose is a linear polysaccharide composed of alternating residues of β-D-galactopyranose and 3,6-anhydro-α-L-galactopyranose. The galactose residues are linked through C-1 and C-3, whereas the anhydrogalactose units are linked through C-1 and C-4 (Fig. 2).

The structure of agaropectin has not been established. The major components are D-galactose and 3,6-anhydro-L-galactose, with smaller amounts of sulfate, glucuronic acid, and pyruvic acid. Hirase (1957) showed that the pyruvic acid is connected through ketal linkages with C-4 and C-6 of the D-galactose residue in an agarobiose unit. This indicates that agarobiose is also an important constituent of agaropectin.

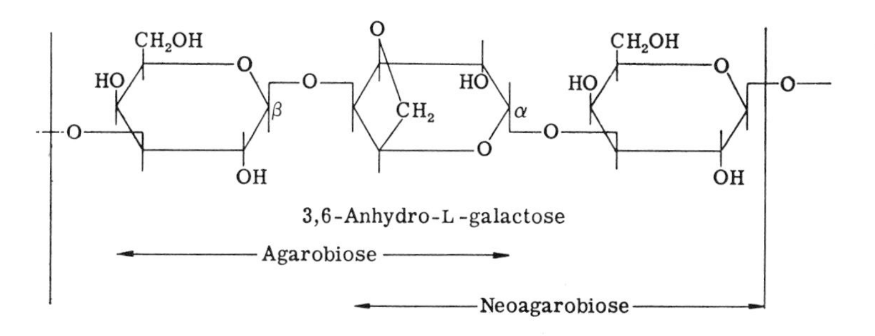

FIG. 2. Agarose.

2. Funorin

Funorin occurs in the aqueous extracts of seaweeds of the genus *Gloiopeltis*, and is extensively used in Japan as a thickener and adhesive. Galactose and sulphuric acid were identified in acid hydrolyzates (Takahashi, 1920). The polysaccharide has a high sulfate content, like carrageenin, but the isolation of agarobiose dimethyl acetal from partially methanolyzed funorin indicates that it is of the agarose type (Hirase *et al.*, 1956; Araki *et al.*, 1958).

C. Mucilages Containing κ-Carrageenin-Type Linkages

1. Carrageenin

Carrageenin is the name given to the polysaccharides extracted by hot water from closely related members of the family Gigartinaceae, including *Chondrus crispus, Gigartina stellata, G. acicularis, G. pistillata,* and *G. radula*. Early studies with extractives from *Chondrus* and *G. stellata* indicated that it was a sulfated galactan, yielding 30–40% galactose and 20 to 30% sulfuric acid on hydrolysis. Dillon and O'Colla (1940, 1951) and Percival and Johnston (1950) showed that the galactose residues were 1:3′-linked, each bearing a sulfate group on C-4, and that an unstable hexose was present. Later Smith and Cook (1953) found that it comprised two components: 40% κ-carrageenin, which can be precipitated from dilute aqueous solutions by potassium chloride, and 60% λ-carrageenin, which remains in solution.

O'Neill *et al.* (1955) showed that κ-carrageenin is composed of D-galactose, 3,6-anhydro-D-galactose, and sulfate groups in the ratios 6:5:7. Later O'Neill (1955), on the basis of mercaptolysis data, proposed a structure for κ-carrageenin consisting of a long chain of alternating β-D-galactopyranose-4-sulfate and 3,6-anhydro-α-D-galactopyranose units (Fig. 3). The galactose residues are linked through C-1 and C-3, whereas the an-

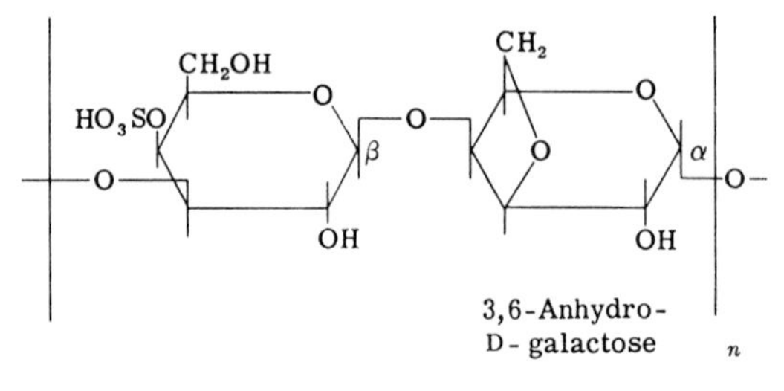

FIG. 3. κ-Carrageenin.

hydrogalactose is linked through C-1 and C-4. Every tenth hexose unit forms a branching point to which is attached, through C-6, a single galactose disulfate. Rees (1961a, b), who indicated that mucilage polysaccharides may be synthesized in Red Algae by enzymatic desulfation of precursors, suggested that carrageenin is formed from a precursor which differs from κ-carrageenin in containing units of galactose-6-sulfate in place of 3,6-anhydrogalactose.

O'Neill *et al.* (1955) showed that λ-carrageenin consists essentially of a sulfated galactan, in which D-galactose residues are linked glycosidically through positions 1 and 3 and bear sulfate groups on position 4 (Fig. 4).

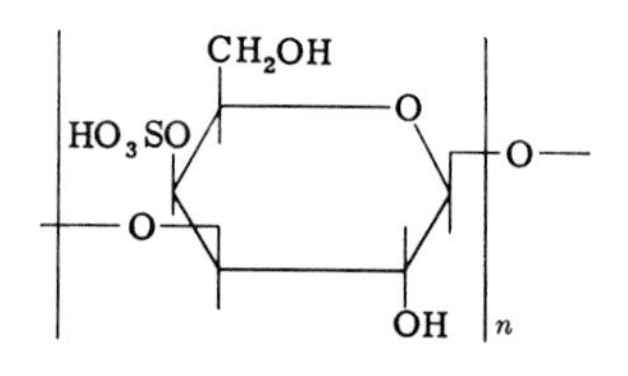

FIG. 4. λ-Carrageenin.

Further degradative studies by Morgan and O'Neill (1959) confirmed this and yielded a homologous series of oligosaccharides composed uniformly of α-1:3′-linked D-galactopyranose residues. They also revealed one residue bearing three contiguous unsubstituted hydroxyl groups, another bearing two such hydroxyl groups, and two 1:6′-linkages for every twenty galactose residues. Further investigation is necessary to determine whether λ-carrageenin is a branched or linear polysaccharide and the exact distribution of all the sulfate groups.

2. Chondrus ocellatus

Extracts of *Chondrus ocellatus* were analyzed by Mori and Tsuchiya (1941). On the basis of data from partial methanolysis, Araki and Hirase (1956) suggested that this polysaccharide is composed of the same 1:3′-linked

α-disaccharide units as κ-carrageenin, but they obtained no evidence concerning the positions of the sulfate groups.

3. Furcellaria fastigiata

Kylin (1943a) found that furcellarin, the mucilage of *Furcellaria fastigiata*, was the potassium magnesium salt of a galactan which contained pentoses. Clancy *et al.* (1958, 1960) showed that it contained galactose, 3,6-anhydrogalactose and sulfate. The de-esterified polysaccharide, obtained in low yield, was considered to be the core of the molecule. By the methylation technique it was shown to be a 1:3′-linked galactan having some branching on C-6. Painter (1960) found the sodium salt to have optical rotation $[\alpha]_D + 75°$ in water, and to contain galactose (43.1%), 3,6-anhydro-D-galactose (30.3%), and SO_3Na (20.1%). It was suggested that this mucilage is like κ-carrageenin, although it has a lower sulfate: hexose ratio of approximately 2:5. This was supported by Yaphe (1959), who deduced from enzymatic hydrolysis experiments that a *Furcellaria fastigiata* extract contained 56% κ-carrageenin-like material.

4. Iridophycus flaccidum

Hassid (1935) concluded that iridophycin, the mucilage of *Iridophycus flaccidum* (= *Iridaea laminarioides*), is a galactan sulfate with each galactose residue esterified on C-6. Mori (1943, 1949) proved conclusively that 1:3′-links were present, and indicated that the sulfate groups are generally on C-6, though some may be also on C-4. The glycosidic linkage probably is of the α-type (Mori and Fumoto, 1949). Yaphe (1959), using the enzymatic technique, found 36% κ-carrageenin-like polysaccharide in the *Iridophycus* mucilage. Iridophycin is also attacked by digestive enzymes of a marine snail, *Tegula* (Galli and Giese, 1959).

5. Hypnea spicifera

By extraction of *Hypnea spicifera* with hot water, Clingman and Nunn (1959) obtained a polysaccharide (16% yield) which could be completely precipitated from aqueous solution by potassium chloride. Analyses showed 47.4% galactose, 31.4% 3,6-anhydro-D-galactose, and 21.2% SO_3Na. The molecular ratios of galactose/3,6-anhydro-D-galactose/SO_3Na are thus 1.4:1.1:1.0. Methylation experiments on de-esterified material indicated that the galactose was joined through positions 1 and 3, while the sulfate was probably on C-4. The polysaccharide strongly resembles κ-carrageenin but it has a different infrared spectrum. It is noteworthy that Yaphe (1959) found 97% κ-carrageenin-like polysaccharide in extracts of a related species, *H. musciformis.*

6. Eucheuma species

Two types of *Eucheuma* mucilages, represented by aqueous extracts of *E. cottonii* from southeast Africa and *E. spinosum* from southeast Asia, are of commercial importance. Proximate analysis indicated that the two mucilages resembled κ-carrageenin, containing galactose, 3,6-anhydrogalactose and sulfate (Hofstede, 1921; de Groot, 1947; Eisses, 1953; Nakamura, 1954, 1958a, b; Zaneveld, 1959). Though both have a sugar/sulfate ratio of 1:1 and an infrared spectrum similar to κ-carrageenin, minor differences in the spectra may be used to distinguish the mucilages from each other and from κ-carrageenin (N. F. Stanley, personal communication).

D. Xylans

1. Rhodymenia palmata ("dulse")

A xylan named dulsin, soluble in dilute acid, was isolated from *Rhodymenia palmata* by Barry and Dillon (1940), and was subsequently examined by the methylation, periodate oxidation, and Barry degradation techniques (Percival and Chanda, 1950; O'Colla *et al.*, 1950; Barry and McCormick, 1957). The molecule evidently consists of approximately forty D-xylose units having two non-reducing end-groups and containing 1:3′- and 1:4′-links in the ratio 1:4. Periodate oxidation studies showed that the 3-linked xylose residues are not linked to one another. Certain bacteria from sheep rumen were found to hydrolyze this xylan, and from the partial hydrolyzate Howard (1957) isolated a series of oligosaccharides. Using the technique of Finan and O'Colla (1955) for the degradation of oligosaccharides from the reducing ends, Howard identified the sequence of linkages in the trisaccharide as β-1:4′, β-1:3′, and in the tetrasaccharide as β-1:4′, β-1:3′, β-1:4′. It is uncertain whether an alkali-soluble xylan isolated from this alga by Myers and Preston (1959a) is identical with that described above.

2. Nemalion multifidum

Hydrolysis of a water-soluble mucilage ($[\alpha]_D$ + 15.5° in water) from *Nemalion multifidum* yielded D-xylose, D-mannose, and sulfuric acid in the molar ratios 7:2:2 (O'Colla and Gardiner, unpublished). The polysaccharide consumed 1 mole of periodate per 312 gm., and the oxidized material still contained xylose and mannose. Methylation studies showed that there are 1:3′- and 1:4′-linked xylose residues, as in *Rhodymenia* xylan. Studies on the partially hydrolyzed xylan of *Nemalion* indicated that the sulfate groups are on the mannose units.

3. Other Red Algae

Ross (1953) showed that polysaccharides containing xylose occur in all of the 26 Red Algae he investigated. In *Rhodochorton floridulum*, xylose is the principal sugar obtained on hydrolysis of the whole weed. However, only a trace of the xylan is water-soluble (O'Colla and Birmingham, unpublished), the bulk of it being alkali-soluble and composed of 1:4′-linked residues like the xylan of the higher plants.

E. Other Mucilages

1. General Observations

Several species of *Porphyra* are used as human foods in various parts of the world. *Porphyra* species have a relatively higher nitrogen content than other seaweeds, but it is doubtful whether our digestive enzymes are capable of breaking down their structural polysaccharides to residues of nutritional value. Earlier investigations of these polysaccharides yielded DL-galactose from *P. laciniata*, *P. tenera* (Oshima and Tollens, 1901), and *P. crispata* (Hayashi, 1941), and also some pentose and methylpentose from *P. tenera* (Hibino, 1942). More recent studies (e.g., Su, 1958) showed that 6-*O*-methyl-D-galactose is a common constituent of the water-soluble mucilages of this group (see below). Since on paper chromatograms 6-*O*-methyl galactose has approximately the same mobility as fucose, early reports of fucose detected by this means alone (e.g., in the hydrolyzate of *P. laciniata* mucilage; O'Colla, 1951) should be accepted with caution.

2. Porphyra Species

The mucilage of *Porphyra capensis* (which in water exhibits an optical rotation of $[\alpha]_D - 60°$ was analyzed by Nunn and von Holdt (1957). The acid hydrolyzate contained D- and L-galactose, 3,6-anhydro-L-galactose, 6-*O*-methyl-D-galactose, and sulfuric acid, in the molar ratios 1:2:1:1. This was the first recorded instance of 6-*O*-methyl-D-galactose in nature.

Porphyra umbilicalis is the edible seaweed known in Wales as "laver" or "laver bread." The hot-water extract contains floridean starch and a sulfated polysaccharide, which can be separated in various ways (Turvey and Rees, 1958). The hydrolyzate of the mucilage contained galactose, 6-*O*-methyl-D-galactose and 3,6-anhydro-D-galactose.

Studies of the structure of *Porphyra* polysaccharides have not been reported.

3. Dilsea edulis

The mucilage of *Dilsea edulis* was examined by Barry and Dillon (1945), Dillon and McKenna (1950a) and Barry and McCormick (1957). It yielded on hydrolysis D-galactose (70%), D-xylose (7%), glucuronic acid (10%), sulfate (9.7%), and a trace of 3,6-anhydrogalactose. Successive degradations revealed a backbone of 1,3′-galactose residues, with the sulfate groups on C-6; the glucuronic anhydride and 3,6-anhydrogalactose were end-groups, and the xylose residues were 1:3′-linked in side-chains (Barry and McCormick, 1957). It is suggested that the mucilage molecule is highly branched, with a core consisting of subunits each with seven galactose and four galactose sulfate residues. To each repeating subunit in the core there must be attached four side-chains whose point of attachment is unknown. The structure suggested for the side-chains is:

—1G4—1G3—1G4—1X3—1G4—1G3—1GA
6
|
SO_3H

4. Dumontia incrassata

The water-soluble mucilage of *Dumontia incrassata*, like that of *Dilsea edulis*, is more readily purified if it is first extracted with 2% hydrochloric acid and then precipitated with ethanol (Dillon and McKenna, 1950b). It contains galactose residues (mainly 1:3′-linked but possibly also some 1:4′-linked), uronic acid groups, and much more sulfate than does *Dilsea* mucilage. Dillon and McKenna suggested the formula

$$[(C_6H_{10}O_5)_9C_6H_6O_5(SO_3)_4]_n$$

5. Laurencia pinnatifida

The mucilage of *Laurencia pinnatifida* may be extracted with water or dilute acids (O'Colla *et al.*, 1958). Galactose, xylose, uronic acid, and sulfuric acid were found in the acid hydrolyzate, in the molar proportions 6:2:1:2; traces of fucose and 3,6-anhydrogalactose were also detected. Degradative analyses indicated a highly branched structure containing mainly 1:3′-linked galactose residues, closely associated in the core of the molecule with xylose and sulfate.

6. Porphyridium cruentum

The extracellular polysaccharide produced in cultures of *Porphyridium cruentum* is electrophoretically homogeneous, with a molecular weight estimated at 100,000. Hydrolysis of this material yielded glucose, galactose, xylose, and some uronic acid, esterified with 10% sulfate (Haxo and Ó hEocha, 1956; R. F. Jones, unpublished).

III. Mucilages of the Chlorophyta

A. General Observations

Polysaccharides containing rhamnose and uronic acid were isolated from *Ulva pertusa* (Miyake *et al.*, 1938) and *Enteromorpha compressa* (Miyake and Hayashi, 1939); Kylin (1946) showed them to be sulfuric acid esters containing a methyl pentose. Norris (1940) also detected a polysaccharide containing rhamnose in *Ulva lactuca.* The mucilage of *Cladophora rupestris* is a galactan sulfate (Kylin, 1944). These mucilages, "ulvin" and "cladophorin," contain acid-resistant polyuronide groups in the molecule and are difficult to hydrolyze. In Table I some of their properties are compared, together with the molar proportions of the sugars present in partial hydrolyzates obtained under similar conditions.

TABLE I

CONSTITUENTS AND PROPERTIES OF THE MUCILAGES OF SOME GREEN ALGAE[a]

Constituent	Molar proportions		
	Ulva lactuca	*Acrosiphonia centralis*	*Cladophora rupestris*
Galactose	—	0.1	2.8
Xylose	1.3	1.6	1.0
Arabinose	—	—	3.7
Rhamnose	4.4	1.4	0.4
Glucose	1.0	1.0	0.2
Mannose	—	0.2	—
Uronic acid (%)	20.8	19.3	3.0
SO_4 (%)	17.5	7.8	16.1
Ash (%)	19.0	10.0	13.7
$[\alpha]_D$ (in H_2O)	−47°	−31°	+69°

[a] From O'Donnell and Percival (1959b).

B. *Ulva lactuca*

Structural studies by Brading *et al.* (1954) on *Ulva lactuca* mucilage, which on hydrolysis yields glucose, xylose, and rhamnose, established the presence of the following residues:

R1—, —1R4—, —1G4—, —1X4—, X1—

It is probable that the sulfate groups are linked to xylose or rhamnose residues.

C. *Cladophora rupestris*

The mucilage of *Cladophora rupestris* was subjected to detailed structural investigation by O'Donnell and Percival (1959a), whose studies revealed a highly branched structure with the following characteristics:

(1) All xylose and two thirds of galactose residues are present as end-groups.
(2) Sulfate groups are evently distributed throughout the molecule, attached to galactose and/or arabinose and rhamnose.
(3) Linkages in the interior of the molecule include the following in the proportions indicated:

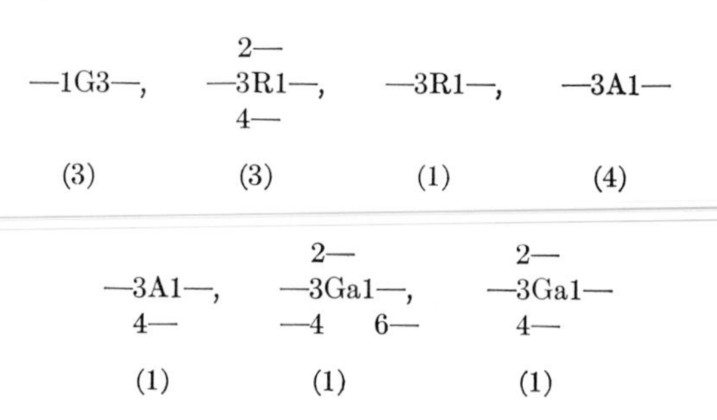

D. *Acrosiphonia centralis**

The polysaccharide of this member of the Cladophoraceae was subjected to detailed structural analysis by O'Donnell and Percival (1959b). Their studies revealed a highly branched molecule, at least part of which consists of 1:3′-linked rhamnose residues bearing glucuronic acid residues as end-groups on C-4. The following residues are present in the proportions indicated:

GA1—,	—4R1—,	—4R1—, 3—	4X1—,	X1—,	—4G1—,	—4G1— 6—
25%	11%	21%	11%	4.2%	1.7%	0.6%

E. *Chlamydomonas* Species

Lewin (1956) examined the extracellular polysaccharides secreted into the medium by 15 species of *Chlamydomonas* in pure culture, and compared them with mucilages extracted from the cells by autolysis, hot water, and hot 1% NaOH. He found that crude polysaccharide represented from 4% to 57% of the total organic matter. Within each strain the constituent

* Regarding the taxonomy of this organism, see Note 48 of the Appendix by Silva.

sugars, liberated by hydrolysis and identified tentatively by paper chromatography, showed no essential differences; however, most strains differed markedly from one another. In *C. ulvaënsis*, glucose and xylose were the chief products of hydrolysis. Arabinose and galactose predominated in all other species, in most cases accompanied by smaller amounts of fucose, mannose, rhamnose, xylose, or uronic acid. Species of the related *Chlorosarcina* and *Gloeocystis* yielded essentially similar results.

F. Other Green Algae

Augier and du Mérac (1954) obtained a mucilage from *Codium dichotomum* that yielded galactose, mannose, and arabinose on hydrolysis. Dzhelileva (1952) estimated seasonal variations in the soluble polysaccharide content of *Enteromorpha intestinalis*, *Chaetomorpha aerea*, *Cladophora utriculosa* and *Codium tomentosum*.

IV. Mucilages of the Phaeophyta

A. General Observations

The principal polysaccharides of the Brown Algae are alginic acid, laminarin and fucoidin. Unlike the other two, alginic acid and its naturally occurring complexes are insoluble in water, but can be extracted with alkali, as is done commercially. As an important cell-wall constituent, it is discussed in this treatise by Kreger (Chapter 19).

Preston (1958) reported that 45 to 70% of the dry weight of seven common brown seaweeds was soluble in hot water. Dewar (1954) detected fucose (9.2%), xylose (1.1%), and galactose (0.4%) in the hydrolyzate of a crude *Fucus vesiculosus* extract. Fanshawe and Percival (1958) found that the relatively small amount of water-soluble carbohydrates in *Cladostephus spongiosus* contains 2% fucoidin associated with a polysaccharide which yields galactose, mannose, and xylose on hydrolysis. The same three sugars were found in hydrolyzates of extracts from twelve other species of Brown Algae (Jensen, 1956a, b). These minor polysaccharides have not been further studied, but fucoidin has been investigated in greater detail.

B. Fucoidin

In several members of the Laminariaceae, 5 to 20% of the total dry matter consists of fucoidin (Black, 1953) which also occurs abundantly in species of *Fucus*, *Chordaria*, etc. It was first isolated by Kylin (1915), who prepared and isolated L-fucose phenylhydrazone from the hydrolyzate. Hoagland and Lieb (1915) found it to contain calcium and sulfate, and further evidence indicated that it is a polysaccharide sulfuric ester (Nelson

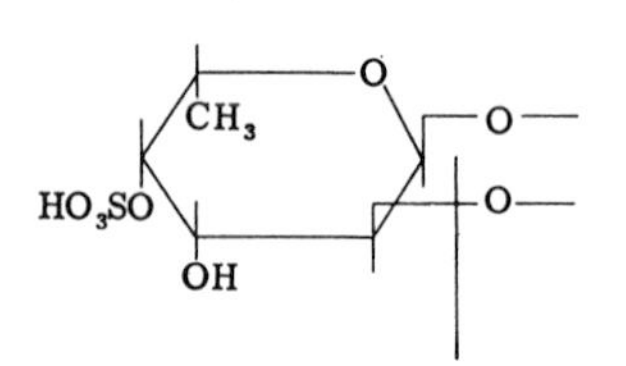

FIG. 5. Fucopyranose sulfate unit.

and Cretcher, 1931; Bird and Haas, 1931). Lunde *et al.* (1937) proposed the formula ($R \cdot O \cdot SO_2OM$), in which R is an L-fucose anhydride residue and M is a metal, mainly calcium; this was later supported by data of Percival and Ross (1950). The ester sulfate groups are stable to alkali. The fucoidin molecule can be represented chiefly as a chain of α-1:2-linked L-fucopyranose units, each bearing a sulfate group on C-4 (Fig.5). However, in chemical degradation studies of purified fucoidin, Percival and Conchie (1950) and O'Neill (1954.) obtained 2-*O*-α-L-fucopyranosyl-L-fucose, and Cote (1959) isolated in addition 3- and 4-*O*-α-L-fucopyranosyl-L-fucose, providing strong evidence for the occurrence of many 1:4'- and 1:3'-linkages. About 10% of the sulfate groups are alkali-labile and hence must be located on C-2 and C-3. Since two components have been revealed by electrophoresis, it is possible that fucoidin—like agar and carrageenin—comprises molecular species sulfated to different levels. A bacterial fucoidinase was isolated by Yaphe and Morgan (1959), and a similar enzyme was demonstrated in the midgut of the snail *Tegula* by Galli and Giese (1959).

C. *Ascophyllum nodosum*

A substance similar to fucoidin was isolated from the reproductive branches of *Ascophyllum nodosum* and *Fucus spiralis* (Dillon *et al.*, 1953). It proved to contain fucose, galactose, and ester sulfate residues in the molar ratios 8:1:9. From a comparative examination of the products of methylation of the polysaccharide and of its desulfated derivative Ó hEocha (1951) concluded that the fucose residues were 1:2'-, 1:3'-, and 1:4'-linked, and that some had two attached sulfate groups.

V. Mucilages of the Cyanophyta

A. General Observations

Payen (1938) found that mucilaginous components were extracted in boiling water from the membranes of some Cyanophyta. Acid hydrolysis of these mucilages yielded sugars which were identified as arabinose and glucose in *Rivularia bullata*, galactose and mannose in *Calothrix pulvinata*, and arabinose in *Nostoc commune*.

B. *Calothrix scopulorum*

Kylin (1943b) found that a hot-water extract of *Calothrix scopulorum* gelled on cooling. A polysaccharide, precipitated from dilute solution by ferric chloride, copper sulfate, or lead acetate, yielded galactose, a pentose, and sulfuric acid on hydrolysis. Kylin concluded that it exists in nature as the calcium or magnesium salt of a polysaccharide sulfuric ester.

C. *Nostoc* Species

A hot-water extract of *Nostoc commune* was examined by Jones *et al.* 1952). The purified mucilage had an equivalent weight of 595; 60% of the polysaccharide was rapidly hydrolyzed by acids, the remainder being much more resistant to hydrolysis. Analyses showed that 30% of the mucilage consists of galacturonic and glucuronic acids, 10% of rhamnose, 25% of D-xylose, and the remainder (35%) largely of galactose, glucose, and an unknown sugar.

The polysaccharide of *N. muscorum* is apparently similar. Biswas (1957) showed by paper chromatography that hydrolyzates contain eight sugars (A, G, Ga, R, X, ribose, and two unidentified components), and studied the incorporation of C^{14} from labeled acetate, carbonate, and formate into its several constituents.

D. *Anabaena cylindrica*

Bishop *et al.* (1954) isolated a polysaccharide from *Anabaena cylindrica* grown in a synthetic culture medium. Prolonged acid hydrolysis yielded yielded glucose, xylose, glucuronic acid, galactose, rhamnose, and arabinose in the molar ratios of 5:4:4:1:1:1. Again, as in the *Nostoc* mucilage, a portion of the polysaccharide showed marked resistance to acid hydrolysis, which may indicate a polyuronide core to which the other sugars are attached.

VI. Mucilages of the Bacillariophyta

Many diatoms produce external gelatinous secretions, a few of which have been partially purified and analyzed by paper chromatography. The capsular material of a freshwater species, *Navicula pelliculosa*, was soluble in dilute alkali, and yielded glucuronic acid on acid hydrolysis (Lewin, 1955); that of a marine diatom, *Phaeodactylum tricornutum*, which could be extracted with hot water, contained xylose, mannose, fucose, and galactose residues (Lewin *et al.*, 1958). The mucilage tubes of *Amphipleura rutilans*, also marine, consisted chiefly of xylose and mannose units, associated with traces of rhamnose and some unidentified components, possibly including protein (Lewin, 1958).

References

Araki, C. (1938). Agar-agar. VII. *l*-Galactose and its derivatives. *J. Chem. Soc. Japan* **59,** 424–432.

Araki, C. (1958). Seaweed polysaccharides. *Proc. Intern. Congr. Biochem. 4th Congr. Symposium* **No. 1,** 1–16.

Araki, C., and Arai, K. (1957). The enzymic hydrolysis of agar. *Bull. Chem. Soc. Japan* **30,** 287–293.

Araki, C., and Hirase, S. (1956). The polysaccharide of *Chondrus ocellata. Bull. Chem. Soc. Japan* **29,** 770–775.

Araki, C., Hirase, S., and Ito, T. (1958). Chemical constituents of the mucilage of *Gloiopeltis furcata. Abstr. 3rd Intern. Seaweed Symposium Galway, Ireland, August, 1958,* pp. 81–82.

Augier, J., and du Mérac, M. L. R. (1954). The carbohydrate composition of *Codium dichotomum. Bull. lab. maritime Dinard* **40,** 25–27.

Barry, V. C., and Dillon, T. (1940). The xylan of *Rhodymenia palmata. Nature* **146,** 620.

Barry, V. C., and Dillon, T. (1944). A formula for agar. *Chem. & Ind. (London)* p. 167.

Barry, V. C., and Dillon, T. (1945). The mucilage of *Dilsea edulis. Proc. Roy. Irish Acad.* **B50,** 349–354.

Barry, V. C., and McCormick, J. (1957). Properties of periodate-oxidized polysaccharides. VI. The mucilage from *Dilsea edulis. J. Chem. Soc.* pp. 2777–2783.

Bird, G. M., and Haas, P. (1931). On the nature of cell wall constituents of *Laminaria* spp. mannuronic acid. *Biochem. J.* **25,** 403–411.

Bishop, C. T., Adams, G. A., and Hughes, E. O. (1954). A polysaccharide from the blue-green alga, *Anabaena cylindrica. Can. J. Chem.* **32,** 999–1004.

Biswas, B. B. (1957). A polysaccharide from *Nostoc muscorum. Sci. and Culture (Calcutta)* **22,** 696–697.

Black, W. A. P. (1953). Constituents of the marine algae. *Ann. Repts. on Progr. Chem. (Chem. Soc. London)* **50,** 322–335.

Bouveng, H. O., and Lindberg, B. (1960). Methods in structural polysaccharide chemistry. *Adv. Carbohydrate Chem.* **15,** 53–89.

Brading, J. W. E., Georg-Plant, M. M. T., and Hardy, D. M. (1954). The polysaccharide from the alga, *Ulva lactuca.* Purification, hydrolysis, and methylation of the polysaccharide. *J. Chem. Soc.* pp. 319–324.

Clancy, M. J., Walsh, K., O'Colla, P. S., and Dillon, T. (1958). The gelatinous polysaccharide of *Furcellaria fastigiata. Abstr. 3rd Intern. Seaweed Symposium, Galway, Ireland, August, 1958,* p. 78.

Clancy, M. J., Walsh, K., O'Colla, P. S., and Dillon, T. (1960). The gelatinous polysaccharide of *Furcellaria fastigata. Sci. Proc. Roy. Dublin Soc.* **A1,** No. 5, 197–204.

Clingman, A. L., and Nunn, J. R. (1959). Red-seaweed polysaccharides. III. Polysaccharide from *Hypnea spicifera. J. Chem. Soc.* pp. 493–498.

Clingman, A. L., Nunn, J. R., and Stephen, A. M. (1957). Agarobiose dimethyl acetal from *Gracilaria confervoides*. *Can. J. Chem.* **35,** 197–301.

Cote, R. H. (1959). Disaccharides from fucoidin. *J. Chem. Soc.* pp. 2248–2252.

de Groot, J. E. (1947). The agar-content of indigenous seaweed. *Chronica Naturae* **103,** 10–12.

Dewar, E. T. (1954). The occurrence of xylose and galactose in brown seaweeds. *Chem. & Ind. (London)* p. 785.

Dillon, T., and McKenna, J. (1950a). The mucilage of *Dilsea edulis*. *Proc. Roy. Irish Acad.* **B53**(6), 45–54.

Dillon, T., and McKenna, J. (1950b). The mucilage of *Dumontia incrassata*. *Nature* **165,** 318.

Dillon, T., and O'Colla, P. (1940). The acetolysis of carrageenin. *Nature* **145,** 749.

Dillon, T., and O'Colla, P. S. (1951). The constitution of carrageenin. *Proc. Roy. Irish Acad.* **B54**(4), 51–65.

Dillon, T., Kristensen, K., and Ó'hEocha, C. (1953). The seed mucilage of *Ascophyllum nodosum*. *Proc. Roy. Irish Acad.* **B55**(9), 189–194.

Dzhelileva, P. D. (1952). Data on the chemical composition of seaweeds on the Black Sea. *Trudy Karadag. Biol. Stantsii Akad. Nauk Ukr. S.S.R.* **No. 12,** 101–110.

Eisses, J. (1953). Seaweeds in the Indonesian trade. *Indonesian J. Natl. Sci.* **109,** 41–56.

Fanshawe, R. S., and Percival, E. (1958). Carbohydrates of *Cladostephus*. *J. Sci. Food Agr.* **9,** 241–243.

Finan, P. A., and O'Colla, P. S. (1955). The stepwise degradation of oligosaccharides. *Chem. & Ind. (London)* pp. 1387–1388.

Galli, D. R., and Giese, A. C. (1959). Carbohydrate digestion in a herbivorous snail, *Tegula funebralis*. *J. Exptl. Zool.* **140,** 415–440.

Hands, S., and Peat, S. (1938). Isolation of anhydro-*l*-galactose derivative from agar. *Nature* **142,** 797.

Hassid, W. Z. (1935). The structure of sodium sulphuric acid ester of galactan from *Iridaea laminarioides*. *J. Am. Chem. Soc.* **57,** 2046–2050.

Haxo, F. T., and Ó'hEocha, C. (1956). Some constituents of *Porphyridium cruentum*. *In* "Proceedings of the Second International Seaweed Symposium, Trondheim, July, 1955" (T. Braarud and N. A. Sorensen, eds.), pp. 23–24. Pergamon Press, London.

Hayashi, J. (1941). Isolation of DL-galactose from *Porphyra crispata*. *J. Soc. Trop. Agr., Taiwan* **13,** 193.

Hibino, J. (1942). The constituents of *Porphyra tenera*. *J. Chem. Soc. Japan* **63,** 1078.

Hirase, S. (1957). Pyruvic acid as a constituent of agar. *J. Chem. Soc. Japan* **30,** 68–78.

Hirase, S., Araki, C., and Ito, T. (1956). Constituents of the mucilage of *Gloiopeltis furcata*. *Bull. Chem. Soc. Japan* **29,** 985–987.

Hirst, E. L. (1958a). Seaweed mucilages. *Abstr. 3rd Intern. Seaweed Symposium, Galway, Ireland, August, 1958* pp. 52–66.

Hirst, E. L. (1958b). Polysaccharides of the marine algae. *Proc. Chem. Soc. (July, 1958)* pp. 177–187.

Hoagland, D. R., and Lieb, L. L. (1915). The complex carbohydrates and forms of sulphur in marine algae of the Pacific coast. *J. Biol. Chem.* **23,** 287–297.

Hofstede, H. (1921). *Algem. Landbouwkundig Weekblad Ned.-Indie* **8,** 319–324.

Howard, B. H. (1957). Hydrolysis of the soluble pentosans of wheat flour and *Rhodymenia palmata* by ruminal micro-organisms. *Biochem. J.* **67,** 643–651.

Humm, H. J. (1951). The red algae of economic importance: agar and related phycocolloids. *In* "Marine Products of Commerce" (D. K. Tressler and J. McW. Lemon, eds.), 2nd ed., rev., pp. 47–93. Reinhold, New York.

Jensen, A. (1956a). Component sugars of some common brown algae. *Norwegian Inst. Seaweed Research Rept.* **9,** 1–8.

Jensen, A. (1956b). Preliminary investigation of the carbohydrates of *Laminaria digitata* and *Fucus serratus. Norwegian Inst. Seaweed Research Rept.* **10,** 1–11.

Jones, J. K. N., and Smith, F. (1949). Plant gums and mucilages. *Advances in Carbohydrate Chem.* **4,** 275–282.

Jones, J. K. N., Hough, L., and Wadman, W. H. (1952). An investigation of the polysaccharide components of certain fresh-water algae. *J. Chem. Soc.* pp. 3393–3399.

Kylin, H. (1915). Zur Biochemie der Meeresalgen. *Z. physiol. Chem. Hoppe-Seyler's* **83,** 171–197.

Kylin, H. (1943a). Zur Biochemie der Rhodophyceen. *Kgl. Fysiograf. Sällskap. Lund, Förh.* **6,** 1–14.

Kylin, H. (1943b). Zur Biochemie der Cyanophyceen. *Kgl. Fysiograf. Sällskap. Lund, Förh.* **13,** 1–13.

Kylin, H. (1944). Zur Biochemie von *Cladophora rupestris. Kgl. Fysiograf. Sällskap. Lund, Förh.* **14,** 1–5.

Kylin, H. (1946). On the nature of the cell wall constituents of the algae. *J. Indian Botan. Soc.* **25** (Iyengar Commemorative Volume), 97–99.

Lewin, J. C. (1955). The capsule of the diatom, *Navicula pelliculosa. J. Gen. Microbiol.* **13,** 162–169.

Lewin, J. C., Lewin, R. A., and Philpott, D. E. (1958). Observations on *Phaeodactylum tricornutum. J. Gen. Microbiol.* **18,** 418–426.

Lewin, R. A. (1956). Extracellular polysaccharides of Green Algae. *Can. J. Microbiol.* **2,** 665–672.

Lewin, R. A. (1958). The mucilage tubes of *Amphipleura rutilans. Limnol. Oceanog.* **3,** 111–113.

Lunde, G., Heen, E., and Oy, E. (1937). Über Fucoidin. *Z. physiol. Chem. Hoppe-Seyler's* **247,** 189–196.

Manners, D. J. (1959). Structural analysis of polysaccharides. *Roy. Inst. Chem. (London) Lectures, Monographs, Repts.* **2,** 1–39.

Mautner, H. G. (1954). The chemistry of brown algae. *Econ. Botany* **8,** 174–192.

Miyake, S., and Hayasi, K. (1939). Polysaccharides of algae. VI. Water-soluble polysaccharide of *Enteromorpha compressa* (L.) Greville. *J. Soc. Trop. Agr., Taihoku Imp. Univ.* **11,** 269–274.

Miyake, S., Hayasi, K., and Takino, Y. (1938). Polysaccharides of algae. II. Water-soluble polysaccharides of *Ulva pertusa* Kjellm. *J. Soc. Trop. Agr. Taihoku Imp. Univ.* **10,** 232–239.

Morgan, K., and O'Neill, A. N. (1959). Degradative studies on λ-carrageenin. *Can. J. Chem.* **37,** 1201–1209.

Mori, T. (1943). The mucilage from Rhodophyceae. V. Chemical constitution of sulfuric acid group-split mucilage of *Iridaea laminarioides. J. Agr. Chem. Soc. Japan* **19,** 297–300.

Mori, T. (1949). The mucilage from Rhodophyceae. VI. The position of the sulfuric acid group in ester combination. *J. Agr. Chem. Soc. Japan* **23,** 81.

Mori, T. (1953). Seaweed polysaccharides. *Advances in Carbohydrate Chem.* **8,** 315–350.

Mori, T., and Fumoto, T. (1949). The mucilage from Rhodophyceae. VII. The kind of glucoside combination and the state of polymerization. *J. Agr. Chem. Soc. Japan* **23,** 81–82.

Mori, T., and Tsuchiya, M. (1941). Mucilage of *Chondrus ocellatus. Econ. Botany* **14,** 585.

Myers, A., and Preston, R. D. (1959a). Fine structure in the red algae. II. The structure of the cell wall of *Rhodymenia palmata. Proc. Roy. Soc.* **B150,** 447–455.

Myers, A., and Preston, R. D. (1959b). Fine structure in the red algae. III. A general survey of cell-wall structure in the red algae. *Proc. Roy. Soc.* **B150,** 456–459.

Nakamura, T. (1954). Mucilage from the red seaweed *Eucheuma spinosum.* Changes in chemical composition due to various treatments. *Bull. Japan. Soc. Sci. Fisheries* **20,** 501–505.

Nakamura, T. (1958a). Mucilage from seaweed *Eucheuma muricatum.* IV. *Bull. Japan. Soc. Sci. Fisheries* **23,** 647–655.

Nakamura, T. (1958b). Mucilage from the seaweed *Eucheuma muricatum.* VI. *Bull. Japan. Soc. Sci. Fisheries* **24,** 285–288.

Nelson, W. L., and Cretcher, L. H. (1931). The carbohydrate acid sulphate of *Macrocystis pyrifera. J. Biol. Chem.* **94,** 2130–2132.

Neukom, H., Deuel, H., Heri, W. J., and Kündig, W. (1960). Chromatographische Fraktionierung von Polysacchariden an Cellulose-Anionenaustauschern. *Helv. Chim. Acta.* **43,** 64–71.

Norris, F. W. (1940). Plant biochemistry. *Ann. Repts. on Progr. Chem.* (*Chem. Soc. London*) **37,** 425.

Nunn, J. R., and von Holdt, M. M. (1957). Red-seaweed polysaccharides. II. *Porphyra capensis* and the separation of D- and L-galactose by crystallization. *J. Chem. Soc.* pp. 1094–1097.

O'Colla, P. S. (1951). Chromatography. *J. Inst. Chem. Ireland.* pp. 27–40.

O'Colla, P. S., Barry, V. C., Dillon, T., and Hawkins, B. (1950). The xylan of *Rhodymenia palmata. Nature* **166,** 787.

O'Colla, P. S., MacCraith, D., and Ní Oláin, R. M. (1958). The mucilage of *Laurencia pinnatifida. Abstr. 3rd Intern. Seaweed Symposium, Galway, Ireland, August, 1958* p. 77.

O'Donnell, J. J., and Percival, E. (1959a). The water-soluble polysaccharides of *Cladophora rupestris.* Barry degradation and methylation of the degraded polysaccharide. *J. Chem. Soc.* pp. 1739–1743.

O'Donnell, J. J., and Percival, E. (1959b). Structural investigations on the water-soluble polysaccharides of the green seaweed *Acrosiphonia centralis. J. Chem. Soc.* pp. 2168–2178.

Ó hEocha, C. (1951). The seed mucilage of *Ascophyllum nodosum.* M.Sc. Thesis, National Univ. of Ireland, Galway.

O'Neill, A. N. (1954). Degradative studies on fucoidin. *J. Am. Chem. Soc.* **76,** 5074–5076.

O'Neill, A. N. (1955). Derivatives of 4-O-β-D-galactopyranosyl-3,6-anhydro-D-galactose from κ-carrageenin. *J. Am. Chem. Soc.* **77,** 6324–6326.

O'Neill, A. N., Smith. D. B., and Perlin, A. S. (1955). Studies on the heterogeneity of carrageenin. *Can. J. Chem.* **33,** 1352–1360.

Oshima, K., and Tollens, B. (1901). Nori from Japan. *Ber. deut. chem. Ges.* **34,** 1422–1424.

Painter, T. J. (1960). The polysaccharides of *Furcellaria fastigiata. Can. J. Chem.* **38,** 112–118.

Payen, J. (1938). Recherches biochimiques sur quelques Cyanophycées. *Rev. Algol.* **11,** 1–99.

Payen, M. (1859). Sur la gélose et les nids de salangane. *Compt. rend. acad. sci.* **49,** 521–530.

Percival, E. G. V. (1944). Ethereal sulphate content of agar specimens. *Nature* **154,** 673–674.

Percival, E. G. V. (1949). Carbohydrate sulphates. *Quart. Revs.* (*London*) **3,** 376–381.

Percival, E. G. V., and Chanda, S. K. (1950). The xylan of *Rhodymenia palmata. Nature* **166,** 787.

Percival, E. G. V., and Conchie, J. (1950). Fucoidin. II. The hydrolysis of a methylated fucoidin prepared from *Fucus vesiculosus. J. Chem. Soc.* pp. 827–832.

Percival, E. G. V., and Johnston, R. (1950). Confirmation of 1,3-linkage in carrageenin and isolation of *l*-galactose derivatives from a resistant fragment. *J. Chem. Soc.* pp. 1994–1998.

Percival, E. G. V., and Ross, A. G. (1950). Fucoidin. I. The isolation and purification of focoidin from brown seaweeds. *J. Chem. Soc.* pp. 717–720.

Percival, E. G. V., Sommerville, J. C., and Forbes, I. A. (1938). Isolation of anhydrosugar derivative from agar. *Nature* **142,** 797–788.

Pirie, N. W. (1936). The preparation of hepta-acetyl-DL-galactose by the acetolysis of agar. *Biochem. J.* **30,** 369–373.

Preston, R. D. (1958). Biophysical and biochemical aspects of some seaweeds. *Abstr. 3rd Intern. Seaweed Symposium, Galway, Ireland, August, 1958* pp. 1–11.

Rees, D. A. (1961a). Biogenesis of 3,6-anhydrogalactose. *Biochem. J.* **78,** 25P.

Rees, D. A. (1961b). The constitution of carrageenin. *Chem. & Ind.* (*London*) p. 793.

Ross, A. G. (1953). Some typical analyses of red seaweeds. *J. Sci. Food Agr.* **7,** 333–335.

Schachat, R., and Glicksman, M. (1959). "Industrial Gums," pp. 149–155. Academic Press, New York.

Schmid, O. J. (1960). Marine red algae. I. Chemical composition. *Botan. Marina* **1,** 54–63.

Smith, D. B., and Cook, W. H. (1953). Fractionation of carrageenin. *Arch. Biochem. Biophys.* **45,** 232–233.

Smith, F., and Montgomery, R. (1959). "The Chemistry of Plant Gums and Mucilages" (Am. Chem. Soc. Monograph Ser. No. 141). Reinhold, New York.

Stoloff, L., and Silva, P. (1957). An attempt to determine possible taxonomic significance of the properties of water-extractable polysaccharides in red algae. *Econ. Botany* **11,** 327–330.

Su, J. C. (1958). Mucilage of *Porphyra crispata. Nature* **182,** 1779–1780.

Takahashi, E. (1920). The mucilaginous substance of Florideae. *J. Coll. Agr., Hokkaido Imp. Univ.* **8,** 183–232.

Tseng, C. K. (1946). Phycocolloids: useful seaweed polysaccharides. *In* "Colloid Chemistry" (J. Alexander, ed.), Vol. VI, pp. 629–734. Reinhold, New York.

Turvey, J. R., and Rees, D. A. (1958). The carbohydrates of red algae: *Porphyra umbilicalis. Abstr. 3rd Intern. Seaweed Symposium, Galway, Ireland, August, 1958* p. 74.

Yaphe, W. (1959). The determination of κ-carrageenin as a factor in the classification of the Rhodophyceae. *Can. J. Botany* **37,** 751–757.

Yaphe, W., and Morgan, M. (1959). Enzymic hydrolysis of fucoidin by *Pseudomonas atlantica* and *Pseudomonas carrageenovora. Nature* **183,** 761.

Zaneveld, J. S. (1959). The utilization of marine algae in tropical South and East Asia. *Econ. Botany* **13,** 89–131.

—21—

Fats and Steroids

J. D. A. MILLER

Department of Botany, University College, London, England

I. Fats

A. Introduction

The following survey will be concerned with the saponifiable algal fats and oils, that is, free fatty acids and their esters, excluding the phospholipids (see Benson, Chapter 22, this treatise).

Fats, owing to their high degree of reduction, constitute a convenient storage material for living organisms. They are widely distributed in algae, especially in spores and other resting stages. It has been suggested that Palaeozoic oil deposits may have been produced by the activity of algae (Traverse, 1955). Until recently algal fats were examined only from general physiological and systematic points of view, but the interest which arose during the war in the possibility of large scale algal cultures for the production of food and industrial materials (see, e.g., Harder and von Witsch, 1942) gave a considerable stimulus to research on these materials. The literature on the nature of algal fats and on the physiology of their formation has been reviewed by von Witsch (1957); the present article will deal mainly with the most recent work and with aspects not dealt with in that review.

B. Chemical Nature of Algal Fats

The acids occurring in algal storage fat range from C_{12} to C_{24}, the even-numbered acids being the only ones of any significance quantitatively. Thus the common saturated straight-chain fatty acids of animals and higher plants, lauric (C_{12}), myristic (C_{14}), palmitic (C_{16}), and stearic (C_{18}), are also of frequent occurrence in algae. In addition Schlenk *et al.* (1960) report traces of arachidic (C_{20}) and behenic (C_{22}) acids in *Chlorella pyrenoidosa*. A considerable degree of unsaturation is almost always found, however, particularly in the higher acids (C_{16} and above): mono-, di-, and trienes are of common occurrence; tetraenes have been encountered; and a C_{20} pentaene is present in the xanthophyte alga *Monodus subterraneus* (Ahrens and Miller, unpublished). There appears to be only one report of a branched-chain fatty acid in an alga (Schlenk *et al.*, 1960). Analyses of algal lipids and the structure of some typical algal triglycerides are given by von Witsch (1957),.while for a more complete account of the chemistry of the naturally occurring fatty acids the reader is referred to Hilditch (1956).

As an example both of the fatty-acid content of algae and of the variation that may occur in a given species under different external conditions, the results of analyses of the fats of *Chlorella pyrenoidosa* by Paschke and Wheeler (1954) and by Schlenk *et al.* (1960) are given in Table I. Paschke and Wheeler employed a high-nitrogen medium; Schlenk *et al.*, a low-nitrogen medium in which the cells were grown for a long period. A notable difference revealed by Table I is the much lower content of trienes and the complete absence of tetraenes in the cells from the low-nitrogen cultures, with a compensating increase in the C_{18} monoene and diene. The percentage of saturated acids (C_{16} plus C_{18}) is similar in each case. Milner (1948), in a less detailed analysis, obtained closely similar values for total C_{16} and C_{18} unsaturated acids in *C. pyrenoidosa* grown in high- and low-nitrogen media. His results were: C_{16} unsaturated, in high-N cells 29.1%, in low-N cells 18.0%; C_{18} unsaturated, in high-N cells 53.9%, in low-N cells 67.1%. It is also of interest that Schlenk *et al.* report traces of odd-numbered acids (C_{15}, C_{17}, and C_{19}) in *Chlorella*. With regard to the major acid components, their findings are in close agreement with those of Mangold and Schlenk (1957), who isolated palmitic, palmitoleic, palmitolinoleic, and palmitolinolenic (C_{16}), and stearic, oleic, linoleic, and linolenic (C_{18}) acids from the lipid of *C. pyrenoidosa* grown in a low-nitrogen medium.

Fatty acids may occur free or, more frequently, as mono-, di-, or triglycerides. Clarke and Mazur (1941) showed that free fatty acids comprised 80% of the total lipid of "*Nitzschia closterium*" (actually *Phaeodactylum tricornutum*)* grown in pure culture, while Ahrens and Miller (unpublished) found that 4.34% of the dry weight of exponentially growing *Monodus subterraneus* is fat, made up as follows: free fatty acids, 9.2%; triglycerides, 60.8%; mono- and diglycerides, 30.0%.

TABLE I

FATTY-ACID COMPOSITION OF *Chlorella pyrenoidosa*

Fatty acid	High-N cells[a] (%)	Low-N cells[b] (%)
Below C_{16}	<2	0.6
C_{16} + C_{18} saturated	17	17.2
C_{16} monoene	4	3.2
C_{16} diene	6	7.0
C_{16} triene	12	5.1
C_{16} tetraene	3	none
Total C_{16} unsaturated	25	15.3
C_{18} monoene	7	34.7
C_{18} diene	11	17.7
C_{18} triene	34	14.6
C_{18} tetraene	1	none
Total C_{18} unsaturated	53	67.0
Above C_{18}	<4	trace

[a] Paschke and Wheeler (1954).
[b] Schlenk *et al.* (1960).

C. Conditions Favoring Fat Formation

1. Physical Conditions

Russell-Wells (1932) determined the fat content of the common British brown seaweeds, and observed that the percentage of crude fats decreased with increasing depth of immersion. Black and Cornhill (1951) found that there is a seasonal variation in the crude fat content of Brown Algae, *Fucus vesiculosus* showing a range of variation from 1.84% of dry weight in February to 3.75% in October. They also found, in agreement with

* Regarding the taxonomy of this organism, see Note 22 of Appendix A by Silva.

Russell-Wells, that the percentage of crude fat varies with depth of immersion, from 8.13% in the exposed alga *Pelvetia canaliculata* to less than 1% in the Laminariaceae. Detailed analyses of *Sargassum ringgoldianum* by Kaneda and Ishii (1950) revealed seasonal variations both in the content and in the degree of unsaturation of the crude oil.

The effect of water deficiency in promoting fat formation has been discussed by Collyer and Fogg (1955), who pointed out that the unusually high fat content found in *Chlorella* by Spoehr and Milner (1949) may have resulted in part from the water deficiency induced in the living cells during the drying process. Evans (1959), using a simple observational technique, also noticed fat accumulation associated with drying in several freshwater diatoms.

Fat production is stimulated by light. Spoehr and Milner (1949) showed that nitrogen-deficient *Chlorella pyrenoidosa* attains a greater R-value (degree of reduction, related to fat content) at high light intensity than at low light intensity. *Chlorella* incorporates exogenous C^{14}-labeled acetate into lipids much more rapidly in the light than in the dark, while during photosynthesis of *Scenedesmus* with $C^{14}O_2$ up to 30% of the radiocarbon incorporated in 5 minutes was found in the lipid fraction (Bassham and Calvin, 1957). Fogg (1956), working with the diatom *Navicula pelliculosa*, studied the effects of light intensity and nitrogen supply on the formation of fat. His results showed that carbon fixed in photosynthesis is rapidly incorporated into fatty substances as well as into sugars and insoluble cell constituents, and supported the idea that photosynthetic carbon assimilation is closely intermeshed with other metabolic processes through common intermediates. The question remained, however, whether the hydrogen donors required for the synthesis of fatty acids are derived directly from the photochemical process or whether they are supplied indirectly *via* the respiratory mechanisms. Coulon (1956) showed that fat formation also proceeds in the absence of photosynthesis, and Iwamoto and Suzuki (1958) concluded that the hydrogen donors for fat synthesis by *Chlorella* in the light are supplied by both photosynthesis and respiration, since a certain amount of carbohydrate is converted to fat even in the dark.

2. *Chemical Conditions*

There is evidence that fat accumulation takes place in many algae as a response to the exhaustion of the nitrogen supply in the medium (see, e.g., Milner, 1948; von Witsch, 1948; Spoehr and Milner, 1949; von Denffer, 1949; Myers, 1951; Fogg, 1959). Aach (1952) analyzed samples from nitrogen-limited *Chlorella* cultures at intervals during growth, and found that the fat content of the cells rose from 22% of dry weight on the second day to 70% on the twenty-fifth day when growth had ceased. Collyer and Fogg (1955) also showed that the proportion of fatty acids in the total algal material of a culture of *Navicula pelliculosa* increases steadily with age, this being associated with decreasing cell-nitrogen content. Growth (i.e., increase in cell number) can continue slowly for some time in a medium entirely depleted of nitrogen, producing cells of high fat content (Iwamoto and Sugimoto, 1958). Intermittent small additions of a nitrogen source accelerate the rate of growth of such cells without seriously affecting their fat content, and this has been made the principle of a method for the semicontinuous cultivation of fatty cells.

A deficiency of phosphorus, magnesium, or potassium does not result in the accumulation of fat (Spoehr and Milner, 1949). The possible mechanism of the induction of fat synthesis by nitrogen deficiency has been discussed by Fogg (1959).

D. Correlation between Fat Formation and Taxonomic Position

It has often been stated that fat accumulation is a characteristic of certain divisions of algae, notably the Xanthophyta and Bacillariophyta (Fritsch, 1935), though Collyer and Fogg (1955) concluded that there are no important differences in the physiological relations of fat accumulation between the representatives of the Chlorophyta, Euglenophyta, Xanthophyta, and Bacillariophyta examined by them. The Rhodophyta and Cyanophyta, which may have a moderately high fat content, appear to differ from the other classes studied in that fat accumulation is not associated with low cell-nitrogen contents. It is clear, however, that the accumulation of high proportions of fat is more dependent on environmental conditions than on an innate tendency of the organism to synthesize fats (Myers, 1951; Collyer and Fogg, 1955).

There is some evidence for general chemical differences between the fats of algae of the various major groups. Lovern (1936), who examined the fats of algae belonging to four divisions, stated that the fatty acid mixture of the Chlorophyta shows a resemblance to that of higher plants, the predominating unsaturated acids being C_{16} and C_{18}. The marine diatom he referred to as "*Nitzschia closterium*" (actually *Phaeodactylum tricornutum*) has a fat similar to that of the Chlorophyta. In the Phaeophyta a lesser degree of unsaturation is found, and this is principally in C_{18} and C_{20} acids. The only species of Red Alga examined by Lovern, *Rhodymenia palmata,* contains a high degree of unsaturation in C_{20} and C_{22} acids, in which respect it resembles marine animals rather than other marine algae. Analyses by Kelly *et al.* (1959) showed that a number of species of marine phytoplankton produce fat approximating in composition to that of marine zooplankton, this being consistent with the fact that the lower animals tend to deposit ingested fat unchanged.

Although the predominant chain length of fatty acids may have some taxonomic significance, it is clear from the results cited in Table I that the degree of unsaturation is not constant but depends, as in the case of higher plants, on external conditions. Nor is it feasible to determine taxonomic affinities by analyzing different organisms grown under a standard set of conditions, owing to the different physical and chemical requirements of different algae (Collyer and Fogg, 1955).

E. Biosynthesis of Fats

The fats of algae consist almost entirely of glycerides of fatty acids with an even number of carbon atoms. The β-oxidative fatty-acid cycle, which can account for the synthesis and degradation of even-numbered acids in two-carbon states, has been demonstrated in higher plants (see Stumpf and Bradbeer, 1959), although not yet in algae. Odd-numbered acids, such as those reported in *Chlorella pyrenoidosa* by Schlenk *et al.* (1960), could arise from even-numbered ones by α-oxidation, involving loss of a single CO_2 per molecule; but here, too, experimental evidence for algae is lacking. The mechanism of formation of unsaturated fatty acids in algae is likewise unknown.

II. Steroids and Other Unsaponifiable Lipids

A. Introduction

The steroids are a group of lipids having in common the cyclopentanoperhydrophenanthrene ring structure shown in Fig. 1. The most widespread of the plant steroids are the sterols, a group of C_{27} to C_{29} secondary alcohols having a hydroxyl group at C–3, an alkyl side-chain at C–17, and one or more double bonds (denoted by Δ in numerical formulas) which may be situated in the rings or the side-chain. They are crystalline solids, characterized on the basis of their optical rotations and melting points, and those of their esters (principally the acetates). All the higher plants appear to contain sterols, the most widely distributed of which are those of the sitosterol group.

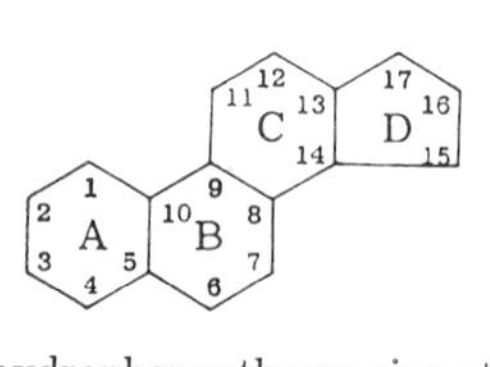

FIG. 1. The cyclopentanoperhydrophenanthrene ring structure common to steroids.

Like the higher plants, the majority of algae investigated hitherto contain sterols in the unsaponifiable lipid fraction, and in addition phytosterolins (sterol glucosides, also common in higher plants) have been found in a few species.

B. Nature and Distribution of Algal Sterols

1. General remarks

Although evidently absent from the Blue-green Algae and from certain bacteria (Bergmann, 1953), sterols appear to be of general occurrence in the remainder of the plant kingdom and among animals. However, the old distinction between phytosterols and zoosterols has been shown to be invalid (Bergmann, 1953; Fieser and Fieser, 1959); for example, chalinasterol from oysters, sea anemones, and the sponge *Chalina arbuscula*, chondrillasterol from the sponge *Chondrilla nucula*, and cholesterol, found in a wide range of animals from the barnacle *Balanus glandula* to vertebrates, have all been subsequently reported as occurring in algae. Hitherto nine different sterols have been identified in algae, and a closer study of their distribution may throw new light on phylogenetic relationships. This may also lead to the discovery of convenient sources of sterols, such as chondrillasterol, which could be employed as starting material for the synthesis of pharmacologically important steroid hormones such as cortisone (Krauss and McAleer, 1953). For these reasons it seems of interest to present a table of algae which have been examined for sterol content (Table II).

The formulas and more important physical constants of the sterols found in algae are given in Table III. Sitosterol from algae, like that of higher plants, is not homogeneous, but comprises a mixture of closely related compounds differing somewhat in melting point and optical rotation (Carter *et al.*, 1939). Further details of the structure and chemistry of the sterols mentioned in Tables II and III are given by Bergmann (1953) and by Fieser and Fieser (1959).

2. Chlorophyta

This division is frequently characterized by the present of sitosterol, the most widely distributed sterol in higher plants, a fact which lends support to the view that the terrestrial plants originated from the Chlorophyta. Curious anomalies are the occurrence of different sterols in two strains of *Scenedesmus obliquus*, and of ergosterol, the major sterol of fungi, in *Chlorella pyrenoidosa*.

3. Chrysophyta, Xanthophyta, and Bacillariophyta

Too few algae belonging to these divisions have been sufficiently examined for sterols to enable any generalization to be made at this time.

TABLE II-1

THE STEROLS OF ALGAE

Organism			
Order	Species	Sterols[a]	Reference[b]
	CHLOROPHYTA		
Chlorococcales	*Chlorella pyrenoidosa*	Ergo	(1)
	Scenedesmus obliquus, strain D-3	Chon	(2)
	Scenedesmus obliquus strain WH-50	?Zymo	(3)
Ulotrichales	*Pleurococcus naegelii*	Sito	(4)
	Trentepohlia aurea	(none)	(4)
Ulvales	*Enteromorpha compressa*	Sito	(4)
	Ulva lactuca	Sito	(4)
Cladophorales	*Cladophora sauteri*	Sito, Fuco	(4)
Oedogoniales	*Oedogonium* sp.	Sito	(4)
Zygnematales	*Zygnema pectinatum*	Sito	(4)
	CHAROPHYTA		
Charales	*Nitella opaca*	Sito, Fuco	(4)
	XANTHOPHYTA		
Vaucheriales	*Botrydium granulatum*	Sito	(4)
	Vaucheria hamata	Sito	(4)
	BACILLARIOPHYTA		
Naviculales	*Navicula pelliculosa*	Chon	(5)
Bacillariales	*Nitzschia closterium*	Fuco, +	(4)
	N. linearis	+	(5)
	CHRYSOPHYTA		
Phaeothamniales	*Apistonema carteri*	Fuco, +	(4)
Ochromonadales	*Ochromonas danica*	Ergo	(6)
Thallochrysidales	*Thallochrysis litoralis*	Fuco, +	(4)
Chromulinales	*Gloeochrysis maritima*	Fuco, +	(4)
	EUGLENOPHYTA		
Euglenales	*Euglena gracilis*	Ergo	(6)

(Continued on pp. 365, 366; for keys to footnotes see pp. 366, 367.)

TABLE II-2

THE STEROLS OF ALGAE (continued)

Organism			
Order	Species	Sterols[a]	Reference[b]
	PHAEOPHYTA		
Ectocarpales	*Pylaiella littoralis*	Fuco	(4)
	Spongonema tomentosum (*Ectocarpus tomentosus*)	Fuco	(4)
Sphacelariales	*Cladostephus spongiosus*	Fuco	(4)
	Sphacelaria pennata (*cirrosa*)	Fuco	(4)
	Stypocaulon scoparium	Fuco	(4)
Dictyotales	*Dictyota dichotoma*	Fuco	(4)
	Padina arborescens	Fuco	(7)
Chordariales	*Heterochordaria abietina*	Fuco	(7)
Dictyosiphonales	*Myelophycus caespitosus*	Fuco	(8)
Laminariales	*Alaria crassifolia*	Fuco	(8)
	Chorda filum	Fuco	(4)
	Costaria costata	Fuco	(7, 8)
	Eisenia bicyclis	Fuco, Sarg	(8)
	Laminaria angustata	Fuco	(7)
	L. digitata	Fuco	(4)
	L. hyperborea (*cloustonii*)	Fuco	(9)
	L. japonica	Fuco	(7)
	L. saccharina	Fuco	(9)
Fucales	*Ascophyllum nodosum*	Fuco	(4, 9)
	Cystophyllum hakodatense	Fuco	(8)
	Fucus ceranoides	Fuco	(4)
	F. evanescens	Fuco	(8)
	F. serratus	Fuco	(9)
	F. spiralis	Fuco	(9)
	F. vesiculosus	Fuco	(4, 9)
	Halidrys siliquosa	Fuco	(4)
	Pelvetia canaliculata	Fuco	(9)
	P. wrightii	Fuco	(8)
	Sargassum ringgoldianum	Fuco, Sarg	(8)
	S. thunbergii	Fuco	(8)

(Continued on p. 366; for keys to footnotes see pp. 366–367.)

4. Phaeophyta

The most regular division is the Phaeophyta in that all its members appear to contain the same sterol, fucosterol, while only two have yet been found which contain in addition the closely related sargasterol. Fucosterol is also, however, of sporadic occurrence in the Chlorophyta, Bacillariophyta, Chrysophyta, and Rhodophyta.

TABLE II-3

THE STEROLS OF ALGAE (continued)

Organism		Sterols[a]	Reference[b]
Order	Species		
	RHODOPHYTA		
Bangiales	*Porphyra umbilicalis*	Fuco	(4)
Nemalionales	*Lemanea mamillosa*	Sito	(4)
Gelidiales	*Acanthopeltis japonica*	Chol, Chal	(12)
	Gelidium amansii	Chol, Chal	(12)
	G. corneum	Sito	(4)
	G. japonicum	Chol	(11)
	G. subcostatum	Chol	(11)
	Pterocladia tenuis	Chol, Chal	(12)
Cryptonemiales	*Corallina officinalis*	+	(4)
	Cyrtymenia sparsa	Chol	(8)
	Gloiopeltis furcata	Chol	(8)
	Grateloupia elliptica	Chol	(8)
	Tichocarpus crinitus	Chol	(8)
Gigartinales	*Ahnfeltia plicata*	+	(4)
	Chondrus crispus	Sito	(4)
	C. giganteus	Chol	(8)
	C. ocellatus	Chol	(8)
	Gigartina stellata	Sito	(4)
	Hypnea japonica	22-DC	(10)
	Iridophycus cornucopiae	Chol	(8)
	Phyllophora membranifolia	Sito	(4)
	Plocamium coccineum	Sito	(4)
	Rhodoglossum pulchrum	Chol	(11)
Rhodymeniales	*Coeloseira pacifica*	Chol	(8)
	Rhodymenia palmata	Fuco	(4)
Ceramiales	*Ceramium rubrum*	Sito	(4)
	Polysiphonia lanosa (*P. fastigiata*)	Sito	(4)
	P. nigrescens	Fuco	(4)
	Rhodomela larix	Chol	(8)
	CYANOPHYTA		
Nostocales	*Oscillatoria rubescens*	(none)	(4)
	Rivularia atra	(none)	(4)
	R. nitida	(none)	(4)

[a] For key to abbreviations of sterols see Table III.

[b] Key to references.

(1) Klosty and Bergmann (1952).
(2) Bergmann and Feeney (1950).
(3) Krauss and McAleer (1953).
(4) Heilbron (1942).
(5) Low (1955).
(6) Stern *et al.* (1960).
(7) Ito *et al.* (1956).
(8) Tsuda *et al.* (1958b).
(9) Black and Cornhill (1951).
(10) Tsuda *et al.* (1959).
(11) Tsuda *et al.* (1957).
(12) Tsuda *et al.* (1958a).

TABLE III

SOME PROPERTIES OF THE STEROLS MENTIONED IN TABLE II

Abbreviation	Sterol	Properties
Ergo	= ergosterol	$C_{28}H_{44}O$, $\Delta^{5,7}$, m.p. 166°C., $[\alpha]_D$ −132°
Chon	= chondrillasterol	$C_{29}H_{48}O$, Δ^7, also unsaturated in side-chain, m.p. 169°C., $[\alpha]_D$ −2°
Zymo	= zymosterol	$C_{27}H_{44}O$, $\Delta^{8(9)}$, also unsaturated in side-chain, m.p. 111°C., $[\alpha]_D$ +49°
Sito	= sitosterol	$C_{29}H_{50}O$, Δ^5, m.p. 127–138°C., $[\alpha]_D$ −37°
Fuco	= fucosterol	$C_{29}H_{48}O$, Δ^5, also unsaturated in side-chain, m.p. 124°C., $[\alpha]_D$ −38°
Sarg	= sargasterol	$C_{29}H_{48}O$, Δ^5, also unsaturated in side-chain. (The 20-epimer of fucosterol, giving more levorotatory derivatives.)
Chal	= chalinasterol	$C_{28}H_{46}O$, Δ^5, also unsaturated in side-chain, m.p. 142°C., $[\alpha]_D$ −42°
Chol	= cholesterol	$C_{27}H_{46}O$, Δ^5, m.p. 148°C., $[\alpha]_D$ −40°
22-DC	= 22-dehydrocholesterol	$C_{27}H_{44}O$, Δ^5, also unsaturated in side-chain, m.p. 135°C., $[\alpha]_D$ −57°
+	= unidentified sterols	

5. *Rhodophyta*

A much more complex picture is presented by this, the most intensively studied class from the point of view of sterol content. The most striking feature is that all the Japanese Red Algae hitherto examined are reported to contain cholesterol, with the exception of *Hypnea japonica* which contains 22-dehydrocholesterol, whereas the British species are reported to contain either sitosterol or fucosterol, but never cholesterol. There seems to be no correlation between taxonomic position and sterol content in the Rhodophyta; for instance, British members of the genera *Gelidium* and *Chondrus* contain sitosterol, whereas Japanese algae assigned to these same genera contain cholesterol. The structure and physical constants of the sitosterol group and of cholesterol, and those of their respective derivatives, seem to differ sufficiently to preclude the possibility of confusion having arisen. The significance of this difference in the sterol content of British and Japanese Rhodophyta remains to be explained.

6. *Cyanophyta*

The absence of sterols from the three Blue-green Algae examined by Carter *et al.* (1939) was tentatively correlated with the lack of sexuality in this division, though it was pointed out that the apparent absence of sterol from *Trentepohlia aurea*, a species of Green Alga which reproduces sexually, is anomalous.

C. Conditions Affecting Formation of Sterols

Krauss and McAleer (1953) found that *Scenedesmus obliquus* when grown at a high light intensity had a higher sterol content (0.35% of dry weight) than when grown in low light (0.11%). There appeared to be no significant difference in the sterol content of cells from high-nitrogen and low-nitrogen cultures at low light intensity. According to the quantitative studies of Black and Cornhill (1951), who developed a colorimetric method for estimating fucosterol, the sterol content of Brown Algae, like their content of saponifiable fat, is correlated with the mean depth of immersion. Thus the exposed alga *Pelvetia canaliculata* contains about three times as much fucosterol, on a percentage dry weight basis, as does the sublittoral species *Laminaria hyperborea*. A seasonal variation in the total sterol content of *Sargassum ringgoldianum*, with a minimum in July after the shedding of gametes, was reported by Kaneda and Ishii (1950).

The sterol content of certain algae is considerably higher than that quoted above for *Scenedesmus*. Exponentially growing *Monodus* contains 0.88% free sterols on a dry weight basis (Ahrens and Miller, unpublished), while *Ochromonas danica* was reported to contain 1.16% ergosterol (Stern *et al.*, 1960). These latter authors also showed that a functional chloroplast is not essential for the synthesis of ergosterol in *Euglena gracilis*, since aplastidic mutants and wild-type cells grown chemo-organotrophically in the dark contain about as much sterol as photosynthetically grown cells.

D. Other Unsaponifiable Lipids in Algae

A phytosterolin (sterol glucoside) was found by Heilbron *et al.* (1935) in the unsaponifiable fraction of the lipid of *Nitella opaca* and *Oedogonium* sp. In addition a number of algae examined by them contained the paraffin hydrocarbon hentriacontane, $C_{31}H_{64}$. Other unsaponifiable substances found in algal lipid have been reviewed by Black (1953); they include hydrocarbons such as $C_{18}H_{36}$ and $C_{20}H_{34}$, sesquiterpenes and sesquiterpene alcohols. A remarkably high content of unsaponifiable oil (22% of the dry weight), probably containing sterols, was found in old cultures of the green alga *Botryococcus braunii* (Belcher, 1957; cf. Traverse, 1955).

E. Biosynthesis of Sterols

The biosynthesis of cholesterol is described by Fieser and Fieser (1959, see Chapter 13). Ottke *et al.* (1950) have shown that acetate is the sole carbon source of ergosterol in a mutant strain of the mould *Neurospora*. There appear to have been no comparable studies on sterol biosynthesis in algae.

REFERENCES

Aach, H. G. (1952) Über Wachstum und Zusammensetzung von *Chlorella pyrenoidosa* bei unterschiedlichen Lichtstärken und Nitratmengen. *Arch. Mikrobiol.* **17,** 213–246.

Bassham, J. A., and Calvin, M. (1957). "The Path of Carbon in Photosynthesis." Prentice-Hall, Englewood Cliffs, New Jersey.

Belcher, J. H. (1957). Ph.D. thesis, University of London, England.

Bergmann, W. (1953). The plant sterols. *Ann. Rev. Plant Physiol.* **4,** 383–426.

Bergmann, W., and Feeney, R. J. (1950). Sterols of algae. I. The occurrence of chondrillasterol in *Scenedesmus obliquus*. *J. Org. Chem.* **15,** 812–814.

Black, W. A. P. (1953). Constituents of the marine algae. *Ann. Repts. on Progr. Chem.* (*Chem. Soc. London*) **50,** 322–335.

Black, W. A. P., and Cornhill, W. J. (1951). A method for the estimation of fucosterol in seaweeds. *J. Sci. Food Agr.* **2,** 387–390.

Carter, P. W., Heilbron, I. M., and Lythgoe, B. (1939). The lipochromes and sterols of the algal classes. *Proc. Roy. Soc.* **B128,** 82–108.

Clarke, H. T., and Mazur, A. (1941). The lipids of diatoms. *J. Biol. Chem.* **141,** 283–289.

Collyer, D. M., and Fogg, G. E. (1955). Studies on fat accumulation by algae. *J. Exptl. Botany* **6,** 256–275.

Coulon, F. (1956). Über die Fettbildung und den Plastidenformwechsel bei *Nitzschia palea*. *Arch. Protistenk.* **101,** 443–476.

Evans, J. H. (1959). Value of a simple observational technique for determining the fat content of algal cells. *Nature* **184,** 1737–1738.

Fieser, L. F., and Fieser, M. (1959). "Steroids." Reinhold, New York; Chapman and Hall, London.

Fogg, G. E. (1956). Photosynthesis and formation of fats in a diatom. *Ann. Botany* (*London*) **20,** 265–285.

Fogg, G. E. (1959). Nitrogen nutrition and metabolic patterns in algae. *Symposia Soc. Exptl. Biol.* **No. 13,** 106–125.

Fritsch, F. E. (1935). "The Structure and Reproduction of the Algae," Vol. I. Cambridge Univ. Press, London and New York.

Harder, R., and von Witsch, H. (1942). Bericht über Versuche zur Fettsynthese mittels autotropher Mikroorganismen. *Forschungsdienst* **16,** 270–275.

Heilbron, I. M. (1942). Some aspects of algal chemistry. *J. Chem. Soc.* **1942,** 79–89.

Heilbron, I. M., Parry, E. G., and Phipers, R. F. (1935). The algae. II. The relationship between certain algal constituents. *Biochem. J.* **29,** 1376–1381.

Hilditch, T. P. (1956). "The Chemical Constitution of Natural Fats." Chapman and Hall, London.

Ito, S., Tamura, T., and Matsumoto, T. (1956). Fucosterol of some brown algae. *Nippon Daigaku Kôgaku Kenkyûsho Ihô* **13,** 99–103; *Chem. Abstr.* **53,** 13276d, 1959.

Iwamoto, H., and Sugimoto, H. (1958). Fat synthesis in unicellular algae. III. Absorption of nitrogen by nitrogen-deficient *Chlorella* cells and its effects on the continuous cultivation of fatty cells. *Bull. Agr. Chem. Soc. Japan* **22,** 410–419.

Iwamoto, H., and Suzuki, A. (1958). Fat synthesis in unicellular algae. IV. Nocturnal effect on fat accumulation in *Chlorella*. *Bull. Agr. Chem. Soc. Japan* **22,** 420–425.

Kaneda, T., and Ishii, S. (1950). Studies on the oil and sterol of algae. I. Seasonal variation of *Sargassum*-lipids. *Bull. Japan. Soc. Sci. Fisheries* **15,** 608–610.

Kelly, P. B., Reiser, R., and Hood, D. W. (1959). The origin of the marine polyunsaturated fatty acids. Composition of some marine plankton. *J. Am. Oil Chemists' Soc.* **36,** 104–106.

Klosty, M., and Bergmann, W. (1952). Sterols of algae. III. The occurrence of ergosterol in *Chlorella pyrenoidosa. J. Am. Chem. Soc.* **74,** 1601.

Krauss, R. W., and McAleer, W. J. (1953). Growth and evaluation of species of algae with regard to sterol content. *In* "Algal Culture From Laboratory To Pilot Plant" (J. S. Burlew, ed.) *Carnegie Inst. Wash. Publ.* **No. 600,** 316–325.

Lovern, J. A. (1936). Fat metabolism in fishes. IX. The fats of some aquatic plants. *Biochem. J.* **30,** 387–390.

Low, E. M. (1955). Studies on some chemical constituents of diatoms. *J. Marine Research (Sears Foundation)* **14,** 199–204.

Mangold, H. K., and Schlenk, H. (1957). Preparation and isolation of fatty acids randomly labeled with C^{14}. *J. Biol. Chem.* **229,** 731–741.

Milner, H. W. (1948). The fatty acids of *Chlorella. J. Biol. Chem.* **176,** 813–817.

Myers, J. (1951). Physiology of the algae. *Ann. Rev. Microbiol.* **5,** 157–180.

Ottke, R. C., Tatum, E. L., Zabin, I., and Bloch, K. (1950). Ergosterol synthesis in *Neurospora. Federation Proc.* **9,** 212.

Pascher, A. (1921). Über die Übereinstimmungen zwischen den Diatomeen, Heterokonten und Chrysomonaden. *Ber. deut. botan. Ges.* **39,** 236–248.

Paschke, R. F., and Wheeler, D. H. (1954). The unsaturated fatty acids of the alga *Chlorella. J. Am. Oil Chemists' Soc.* **31,** 81–85.

Russell-Wells, B. (1932). Fats of brown seaweeds. *Nature* **129,** 654–655.

Schlenk, H., Mangold, H. K., Gellerman, J. L., Link, W. E., Morrissette, R. A., Holman, R. T., and Hayes, H. (1960). Comparative analytical studies of fatty acids of the alga *Chlorella pyrenoidosa. J. Am. Oil Chemists' Soc.* **37,** 547–552.

Spoehr, H. A., and Milner, H. W. (1949). The chemical composition of *Chlorella*; effect of environmental conditions. *Plant Physiol.* **24,** 120–149.

Stern, A. I., Schiff, J. A., and Klein, H. P. (1960). Isolation of ergosterol from *Euglena gracilis*; distribution among mutant strains. *J. Protozool.* **7,** 52–55.

Stumpf, P. K., and Bradbeer, C. (1959). Fat metabolism in higher plants. *Ann. Rev. Plant Physiol.* **10,** 197–222.

Traverse, A. (1955). Occurrence of the oil-forming alga *Botryococcus* in lignites and other Tertiary sediments. *Micropaleontologist* **1,** 343–349.

Tsuda, K., Akagi, S., and Kishida, Y. (1957). Discovery of cholesterol in some red algae. *Science* **126,** 927–928.

Tsuda, K., Akagi, S., and Kishida, Y. (1958a). Steroid studies. VIII. Cholesterol in some red algae. *Pharm. Bull. (Tokyo)* **6,** 101–104.

Tsuda, K., Akagi, S., Kishida, Y., Hayatsu, R., and Sakai, K. (1958b). Untersuchungen über Steroide. IX. Die Sterine aus Meeresalgen. *Pharm. Bull. (Tokyo)* **6,** 724–727.

Tsuda, K., Sakai, K., Tanabe, K., and Kishida, Y. (1959). Isolation of 22-dehydrocholesterol from *Hypnea japonica. Pharm. Bull. (Tokyo)* **7,** 747–749.

von Denffer, D. (1949). Die planktische Massenkultur pennater Grunddiatomeen. *Arch. Mikrobiol.* **14,** 159–202.

von Witsch, H. (1948). Physiologischer Zustand und Wachstumintensität bei *Chlorella. Arch. Mikrobiol.* **14,** 128–141.

von Witsch, H. (1957). Die Fette der Algen, Stoffwechselphysiologie der Fette und Fettähnlicher Stoffe. *In* "Handbuch der Pflanzenphysiologie" (W. Ruhland, ed.), Vol. 7, pp. 50–58. Springer, Berlin.

—22—

Surfactant Lipids[1]

A. A. BENSON AND ISAO SHIBUYA

Department of Agricultural and Biological Chemistry, Pennsylvania State University, University Park, Pennsylvania

I. The Role of Surfactant[2] Lipids in Chloroplast and Mitochondrial Structures

A. Structural Considerations

The primary activity of most algal cells is photosynthesis. In young cells, at least, there is little accumulation of stored fats, and the chloroplasts possess most of the cell lipid. The lipids of algae, like those of chloroplasts of higher plants, are largely surfactant[2] molecules which function both as structural elements and as metabolites in the photosynthetic organelles. Algal chloroplasts contain lamellae in which the molecular arrangements of the pigments, proteins, and lipids possess a high degree of orientation, and which are characterized by interfaces of large total area (Goedheer, 1957; Sager, 1958; Gibbs, 1960). It is reasonable to consider that the surfactant lipids, which include chlorophyll, compete for adsorption and therefore orientation at these interfacial boundaries. It may be this physical orientation which stabilizes the lamellar structure and mediates its unique

[1] Preparation of this manuscript and the previously unpublished research it reports were supported by the United States Atomic Energy Commission, the National Science Foundation, and the Pennsylvania Agricultural Experiment Station.

[2] A *surfactant* is a substance with *surface-active* properties.

photochemical function (Benson, 1961). Trurnit *et al.* (1961) studied the surfactant properties of chloroplast lipids and noted their strong adsorption at the water-air interface. Mitochondrial structures of algae have not been examined extensively, but they appear to resemble those of higher plants and mammals. Their large membrane areas likewise appear to be stabilized by adsorption of surfactant lipids.

The electrical charge possessed by several of these surfactants may lead to ionic interfaces which mediate physical processes of the chloroplast as well as inorganic ion transport across the cell membrane. Analogous activities of lipid membranes in animal cells result in the conduction of nerve impulses and in salt transport.

Active transport of carbohydrates through cell membranes may be mediated and specifically controlled by the reversible formation of galactolipids in the membrane. Only glucose and galactose support growth of such algae as *Chlorella* (see Danforth, Chapter 6). The cells possess high concentrations of galactosyl diglycerides, which suggests that carbohydrate transport may involve the extracellular formation of galactolipids from uridine diphosphate glucose and uridine diphosphate galactose. These could penetrate the thin lipid membrane to allow release of the free sugar by phosphorolysis inside the cell, thereby regenerating the diglyceride transport vehicle. Such a mechanism is consistent with the current concepts of the relationships between galactoside permease and β-galactosidase (cf. Maio and Rickenberg, 1960; Horecker *et al.*, 1960).

B. Metabolism of Algal Lipids

Surfactant lipids of *Chlorella* and *Scenedesmus* were shown to be actively engaged in the photosynthetic metabolism (Miller, 1962; Ferrari, 1959;

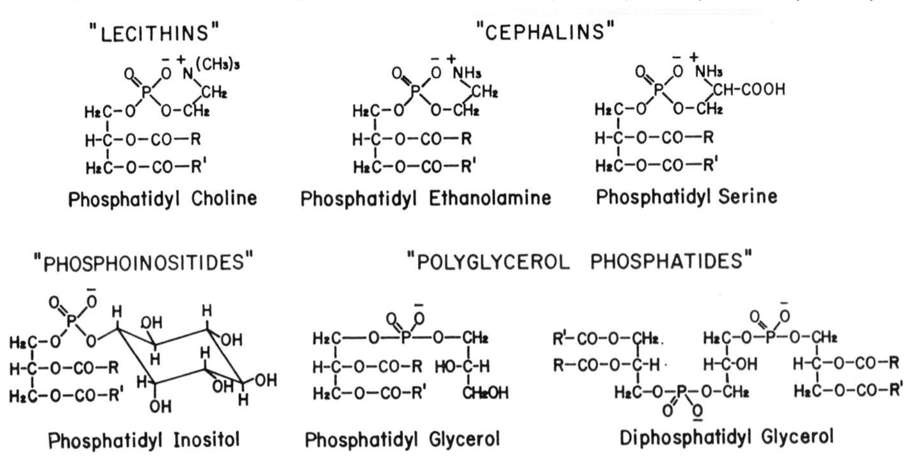

FIG. 1. The glycerol phosphatides found in algae.

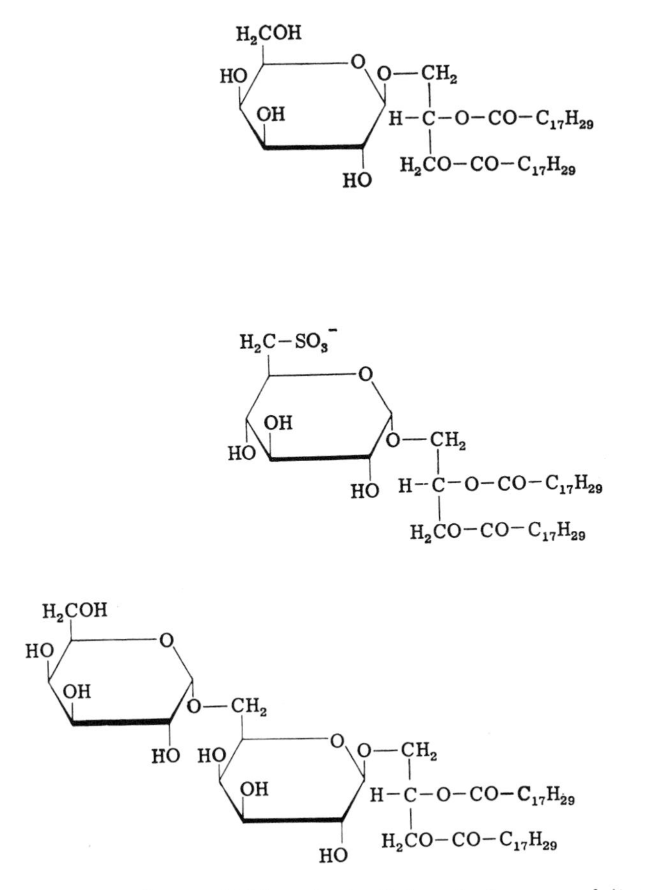

FIG. 2. Glycolipids of algae and higher plants: *O*-β-D-galactopyranosyl-(1→1′)-2′,3′-di-*O*-acyl-D-glycerol; *O*-[6-sulfo-6-deoxy-α-D-glucopyranosyl]-(1→1′)-2′,3′-di-*O*-acyl-D-glycerol; *O*-[α-D-galactopyranosyl-(1→6)-β-D-galactopyranosyl-(1→1′)]-2′,3′-di-*O*-acyl-D-glycerol.

Ferrari and Benson, 1961). Since the most actively involved were phosphatidyl glycerol (Fig. 1) and the galactose moiety of the galactosyl diglycerides (Fig. 2), it was concluded that these groups were oriented in the lamellar membranes on the side facing the enzymes of carbohydrate synthesis (Benson, 1961). The rapid labeling of galactolipids during photosynthesis in $C^{14}O_2$ is largely a result of hexose-C^{14} exchange in the galactolipids (Ferrari and Benson, 1961). Labeling of the fatty acids is relatively slow. The formation of digalactosyl diglyceride and its higher homologs was considered to be the result of successive galactosylations.

A striking example of lipid metabolism in algae or higher-plant chloroplasts is that of the major glycerol phosphatide, phosphatidyl glycerol

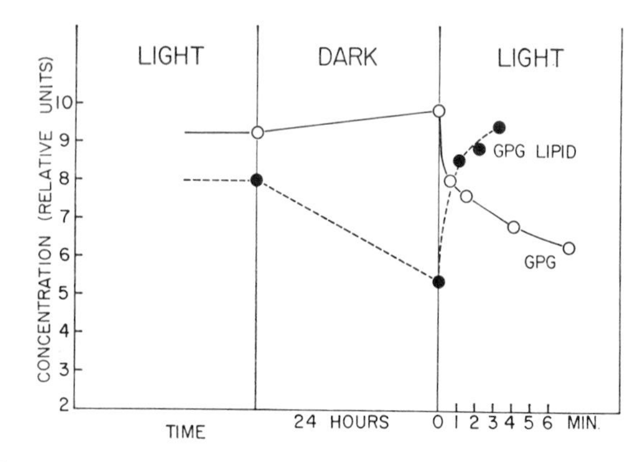

FIG. 3. Effects of illumination upon the concentrations of phosphatidyl glycerol (GPG lipid) and diglycerophosphate (GPG) in *Chlorella pyrenoidosa*. Algae were cultured 6 days in the light in P^{32} medium prior to paper chromatographic analysis of the labeled components (Miller, 1962).

(Fig. 3). Its deacylated derivative, diglycerophosphate, is the most abundant organic phosphorus compound in *Scenedesmus obliquus* and *Chlorella pyrenoidosa* cells grown at a moderate light intensity (Benson and Maruo, 1957, 1958). The concentration of this compound exceeds that of phosphoglycerate and other phosphorylated metabolites by a factor as high as 10. Illumination of the cells leads to the rapid synthesis of phosphatidyl glycerol and the disappearance of diglycerophosphate. This change is probably associated with glycerophosphate dehydrogenase activity in the chloroplasts:

$$\text{dihydroxyacetone phosphate} + \text{DPNH} \rightarrow \text{glycerophosphate} + \text{DPN}^{+}$$ [3]

Deacylated phosphatides such as the glycerophosphoryl esters of choline and ethanolamine, normally present in only small amounts, assume high concentrations in older cultures of *Chlorella pyrenoidosa* (Miller, 1962); but, like diglycerophosphate, they diminish on illumination and consequent lipid synthesis. Their metabolic turnover is not as rapid as that of diglycerophosphate.

II. Lipid Composition of Algae

A. Phospholipids

The main phosphatides of *Chlorella pyrenoidosa* (Table I) are phosphatidyl glycerol (Benson and Maruo, 1958), phosphatidyl choline (lecithin)

[3] *Abbreviations:* DPN^+, diphosphopyridine nucleotide; DPNH, reduced diphosphopyridine nucleotide.

TABLE I

LIPID CONCENTRATIONS IN *Chlorella pyrenoidosa*

Deacylated derivative	Molarity × 10^3 [a]
β-D-Galactosyl glycerol, G-Gal	18[b]
α-D-Galactosyl-(1→6)-β-D-galactosyl glycerol, G-Gal-Gal	13
6-Sulfo-α-D-quinovopyranosyl glycerol, G-Quin-SO_3H	6.5
Glycerophosphoryl glycerol, GPG	6[c]
Glycerophosphoryl choline, GPC	5
1-*O*-Glycerophosphoryl myoinositol, GPI	2
Glycerophosphoryl ethanolamine, GPE	0.1
Glycerophosphoryl serine, GPS	0.05
Glycerol	7

[a] Averages of several determinations from growing cells.

[b] Determined from C^{14} radioactivity in extracts of *Chlorella* prepared after 5 days of culture in $C^{14}O_2$. Activities in phosphorus compounds were standardized against known concentrations of phosphatides.

[c] Determined by neutron activation of phosphorus compounds on paper chromatograms.

and phosphatidyl inositol (Fig. 1). It appears that active transphosphatidylation occurs among these compounds during metabolism (Kiyasu *et al.*, 1960).

1. *Phosphatidyl Glycerol*

As in chloroplasts of higher plants (Wintermans, 1960), phosphatidyl glycerol is the most concentrated phosphatide in young *Chlorella* cells. It is readily estimated by deacylation of extracted lipids and two-dimensional paper chromatography of the resultant glycerophosphoryl esters (Fig. 4).

2. *Diphosphatidyl Glycerol (Cardiolipin)—(Fig. 1)*

The tetraacyl diphosphatide, cardiolipin (Macfarlane, 1958; Macfarlane and Wheeldon, 1959), long considered a lipid characteristic of cardiac tissue, has now been found also in various mammalian mitochondria. Miller (1962) found especially high concentrations of diphosphatidyl glycerol in *Chlorella pyrenoidosa* which had been cultured under phosphate deficiency and restored to a sufficient medium (Fig. 4). One can surmise that the excess "reducing power" (e.g., cytidine triphosphate, DPNH) available to illuminated phosphorus-deficient cultures may lead to synthesis of such a lipid. The association of cardiolipin with mitochondrial structures and its high concentration in *Chlorella* cells and in the chromatophores of

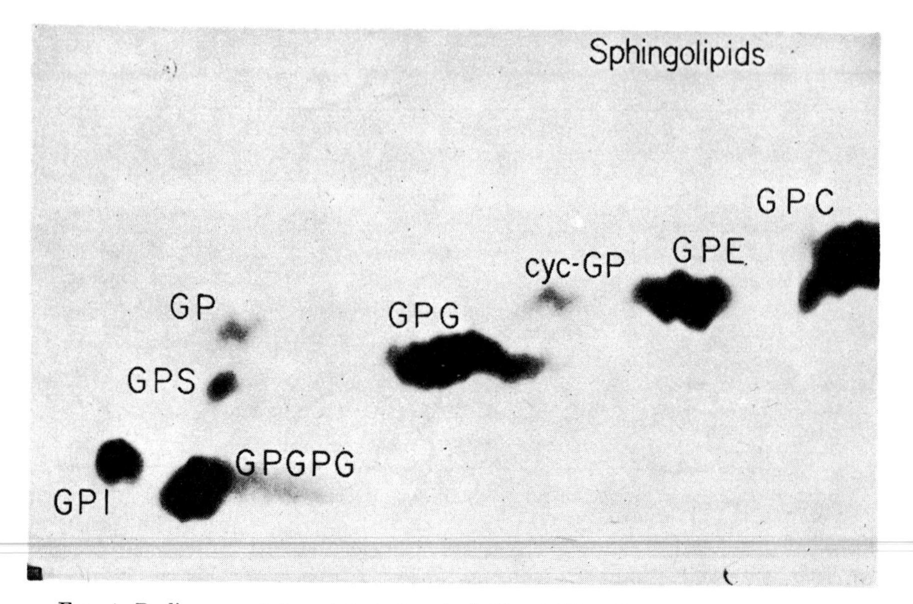

FIG. 4. Radiogram of deacylated phospholipids of *Chlorella pyrenoidosa* cultured in radiophosphate. Chromatogram was developed in x-direction in phenol:water (100:40) and in y-direction in butanol:propionic acid:water (140:71:100) (Miller, 1962).

photosynthetic bacteria (Benson and Strickland, 1960) suggest that this dipolar lipid may function in stabilizing lipid-aqueous interfaces in the chloroplast (Strickland and Benson, 1960).

3. Phosphatidyl Inositol (Fig. 1)

The inositide in all algae which have been examined comprises 15 to 20% of the total phosphatides. Although it is involved in rapid transphosphatidylation, its concentration does not vary markedly (Lepage *et al.*, 1960). There is no evidence for a di- or triphosphoinositide in algae.

4. Phosphatidyl Choline and Phosphatidyl Ethanolamine (Fig. 1)

The concentration of phosphatidyl ethanolamine in algae is generally low (6–10%) and that of phosphatidyl serine is even lower (0.5–2%). Phosphatidyl choline, however, comprises 30 to 40% of the phosphatides of *Chlorella pyrenoidosa* and *Scenedesmus obliquus*. It should be borne in mind that these lipids are zwitterionic and can present only a cationic surface charge if intermolecular forces, such as those of protein-bound argininium groups, attract the phosphate ion to release the cationic group, $—N^{+}(CH_3)_3$. The deacylated derivatives, such as glycerophosphoryl choline, are not ionically absorbed by ion-exchange resins, but behave as nonpolar molecules (Maizel *et al.*, 1956).

B. Galactolipids of Algae

1. *The Galactosyl Diglycerides* (*Fig. 2*)

a. Structure. All algae so far investigated contain two β-linked galactosyl diglycerides (Carter *et al.*, 1956). The monogalactolipid is generally predominant. The digalactolipid, 1′-*O*-[α-D-galactosyl-(1→6)-β-D-galactosyl]-2′,3′-di-*O*-acylglycerol, and its higher homologs contain α-linked hexose moieties. Specific enzymatic cleavage by α- and β-galactosidases is then possible. A lipase, active in homogenates of *Chlorella ellipsoidea*, removes one of the two fatty acid groups to yield a "lysogalactolipid."

b. Analysis. The deacylated galactolipids are readily detected on paper chromatograms by spraying with the periodate-Schiff reagent (Buchanan *et al.*, 1950). Their glycosidic linkages can be determined by the action of specific α- and β-galactosidases followed by co-chromato-

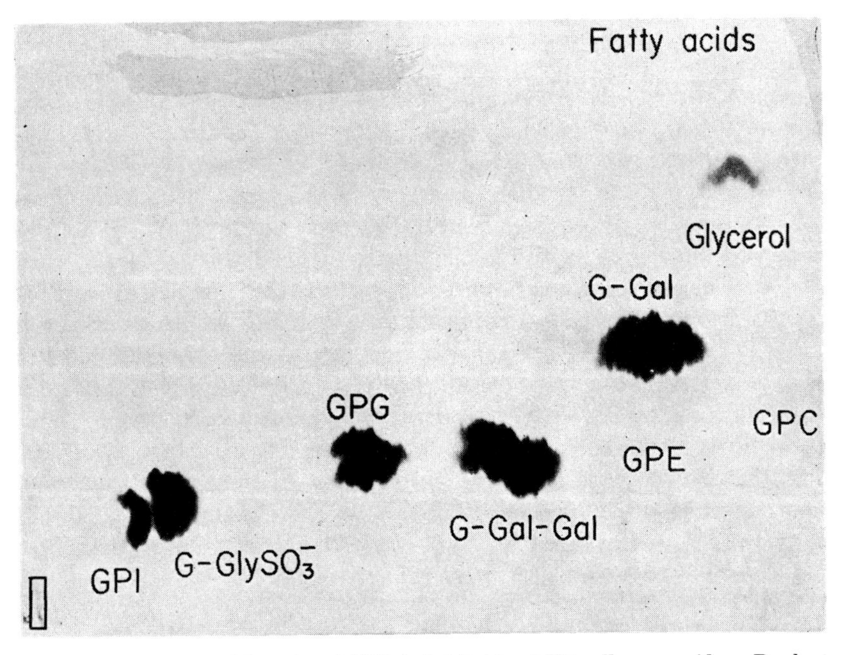

Fig. 5. Radiogram of deacylated C^{14}-labeled lipids of *Chlorella pyrenoidosa*. Products from lecithin and cephalin, faintly visible on the original film, are not reproduced (Ferrari and Benson, 1961).

graphy with authentic compounds. After extended periods of $C^{14}O_2$ fixation the lipid components of algal cells may be estimated from their radioactivity on paper chromatograms (Fig. 5). Wintermans (1960) applied colorimetric methods for the same purpose.

2. *Relation of the Galactolipids to the Galactosyl Glycerols (Floridoside and Isofloridoside) of Red Algae*

For a general review of glycolipids, see Lederer (1958). The occurrence of floridoside and isofloridoside in algae is reviewed by Meeuse (see Chapter 18, this treatise).

Since the deacylated derivatives of the glycerol phosphatides and the sulfolipid occur in the free form in *Scenedesmus obliquus*, it was suspected that floridoside (Fig. 6), the major soluble carbohydrate of the Red Algae,

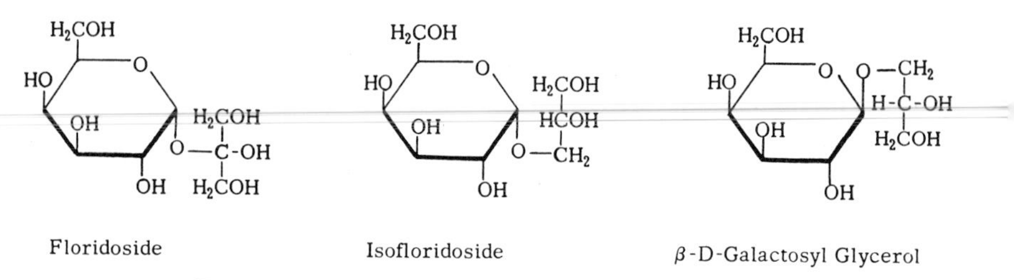

FIG. 6. Floridoside and analogous galactosyl glycerols.

might be related to a floridoside lipid (Putman and Hassid, 1954; Su, 1956). This, however, is not the case. Floridoside has not been found among the deacylation products of lipids from any of the algae listed in Table II. Isofloridoside (Fig. 7), 1-*O*-(α-D-galactosyl)glycerol (Lindberg, 1955), does not occur among the hydrolysis products of the lipids of *Porphyra umbilicalis*. The mechanism of biosynthesis of the α- and β-galactosyl glycerols must therefore be unrelated. Bean and Hassid (1955), in studies of floridoside synthesis by *Iridophycus flaccidum*, identified phosphofloridoside, the formation of which they attributed to a reaction between glycerophosphate and uridine diphosphate galactose. In *Porphyra umbilicalis*, the only alga so far found which produces isofloridoside, the specific galactosylation of the C-2 of glycerol is lacking. The 1-glyceryl-α-D-galactoside is also formed. One may surmise that the α-configuration of these compounds specifically limits their diffusion through the β-linked lipids of the cell membranes, and thus tends to keep them within the cell.

C. Sulfoquinovosyl [4] Diglyceride

When cells of *Chlorella pyrenoidosa* or other algae are cultured in nutrient media containing sulfate labeled with S^{35}, the major alcohol-

[4] Quinovose is 6-deoxyglucose.

TABLE II

PERCENTAGE COMPOSITION OF ALGAL LIPIDS

Deacylated derivative	Rhodophyta					Chlorophyta	
	Porphyra umbilicalis	*Porphyridium cruentum*[a]	*Iridophycus flaccidum*[b]	*Polysiphonia harveyi*	*Rhodymenia palmata*[c]	*Ulva lactuca*	*Scenedesmus obliquus*
G-Gal[d]	63[e]	15[f]	53[e]	46[e]	48[e]	36[e]	32[e]
G-Gal-Gal	10	40	11	16	16	43	37
G-QuinSO_3H	3	25	14	10	10	9	11
Glycerol	5	5	5	10	—	1	5
GPG	6	5	5	4	4	6	7
GPC	9	5	4	—	—	—	5
GPI	2	—	2	3	—	—	2
GPE	1	—	1	2	—	6	1
GP	—	—	5	8	—	—	—
GPGPG	—	—	—	3	—	—	1

[a] Kindly provided by Dr. Mary Belle Allen.

[b] Kindly provided by Professor L. R. Blinks.

[c] Dr. Gérard Milhaud, Institut Pasteur, Paris; unpublished results.

[d] KEY: G-Gal, galactosyl diglyceride; G-Gal-Gal, digalactosyl diglyceride; G-QuinSO_2H, sulfoquinovosyl diglyceride; Glycerol, glycerides; GPG, phosphatidyl glycerol; GPC, phosphatidyl choline; GPI, phosphatidyl inositol; GPE, phosphatidyl ethanolamine; GP, phosphatidic acid; and GPGPG, diphosphatidyl glycerol, cardiolipin.

[e] Determined by chromatographic analysis of labeled products of photosynthesis in $C^{14}O_2$.

[f] Estimated from relative intensities of chromatographic spots.

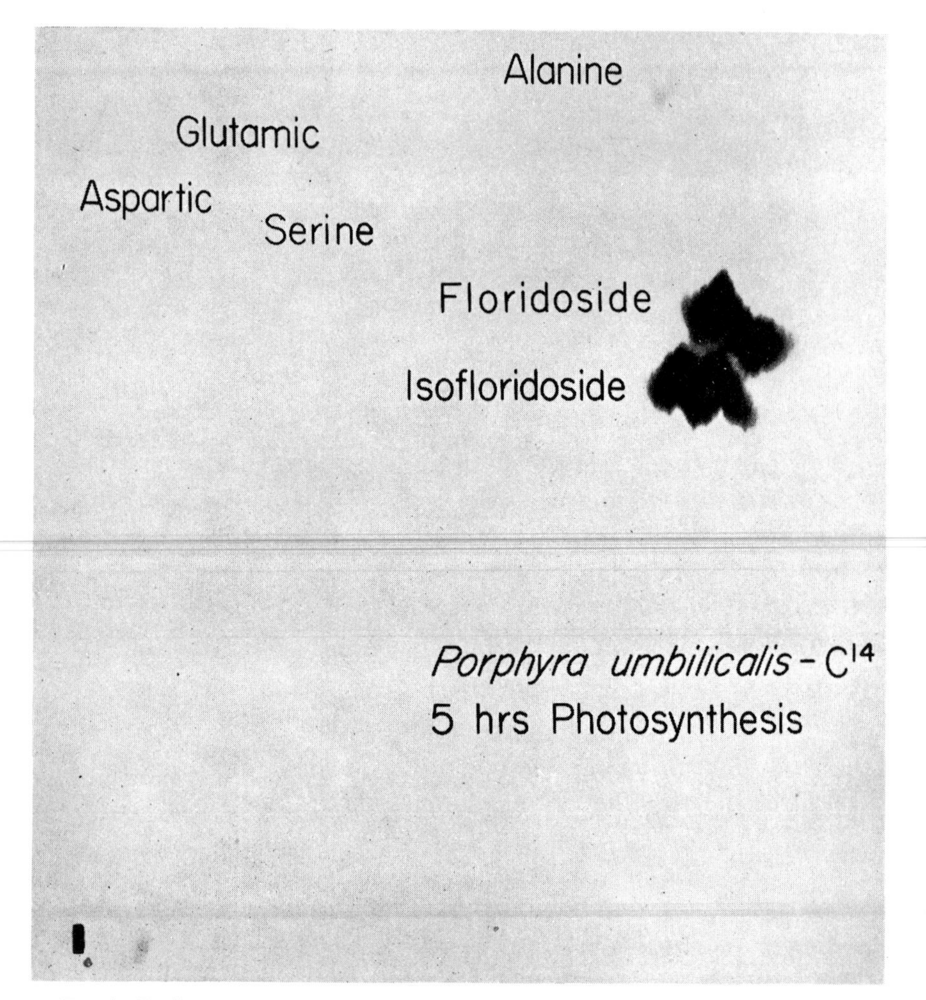

FIG. 7. Radiogram of floridoside and isofloridoside produced by *Porphyra umbilicalis* during 5 hours photosynthesis in $C^{14}O_2$ (cf. Lindberg, 1955).

soluble labeled product is an anionic lipid structurally related to the galactosyl diglycerides (Benson *et al.*, 1959). It differs from the latter in being an α-glycoside, although it can be cleaved by the enzyme β-galactosidase. Daniel *et al.* (1961) have shown that this diglyceride is a derivative of 6-sulfo-D-quinovose, i.e., of 6-sulfo-6-deoxy-D-glucose). This compound possesses a $—CH_2—SO_3H$ group with a direct C—S bond and is therefore a sulfolipid rather than a sulfatide ($—CH_2—O—SO_3H$), which would be anticipated from the general occurrence of sulfate esters in algal polysaccharides. Its synthesis from $S^{35}O_4^{=}$ or from $C^{14}O_2$ is rapid; one or two minutes suffice to accumulate detectable radioactivity. Although the path of its biosynthesis is not yet understood, this compound could act as a reservoir for photosynthetically reduced sulfate. Upon depletion of sulfur in the medium, the lipid sulfur diminishes, being utilized in the synthesis of cellular protein. The dependence of sulfolipid concentration upon sulfate nutrition in *Chlorella ellipsoidea* was investigated by Hase *et al.* (1959).

D. The Fatty Acid Components

The two fatty acids bound in the phosphatidyl esters and the glycosyl diglycerides have not been investigated specifically. The similarity of algal lipids to those of angiosperm chloroplasts suggests that the fatty acids involved may be highly unsaturated. Zill and Harmon (1959) reported that the sulfolipid of spinach chloroplasts contained 90% linolenic acid, and that a 100% linolenyl monoglyceride could be isolated from barley. Schlenk *et al.* (1960) found only 14% linolenic acid esters in the total lipids of 12-week-old cultures of *Chlorella pyrenoidosa.* Younger cultures of *Chlorella ellipsoidea* (4 days old) contained 90% of unsaturated fatty acids in their lipids.

Plasmalogens, acetal glycerolphosphatides, have not been found in algae or other plants. Low concentrations of sphingomyelins, however, have been found in *Scenedesmus obliquus* (Miller, 1962).

References

Bean, R. C., and Hassid, W. Z. (1955). Assimilation of $C^{14}O_2$ by a photosynthesizing Red Alga, *Iridophycus flaccidum*. *J. Biol. Chem.* **212,** 411–425.

Benson, A. A. (1961). Lipid function in the photosynthetic structure. *In* "Light and Life" (W. D. McElroy, ed.), McCollum-Pratt Symposia. Johns Hopkins Press, Baltimore, Maryland. pp. 392–396.

Benson, A. A., and Maruo, B. (1957). Radiochemical identification of diglycerophosphate and its probable role in lipid synthesis by plants. *In* "Proceedings of the International Conference on Radioisotopes in Scientific Research," Vol. IV: Research with Isotopes in Plant Biology (R. C. Extermann, ed.), pp. 510–519. Pergamon Press, London.

Benson, A. A., and Maruo, B. (1958). Plant phospholipids. I. Identification of the phosphatidyl glycerols. *Biochim. et Biophys. Acta* **27,** 189–195.

Benson, A. A., and Strickland, E. H. (1960). Plant phospholipids. III. Identification of diphosphatidyl glycerol. *Biochim. et Biophys. Acta* **41,** 328–333.

Benson, A. A., Daniel, H., and Wiser, R. (1959). A sulfolipid in plants. *Proc. Natl. Acad. Sci. U. S.* **45,** 1582–1587.

Buchanan, J. G., Dekker, C. A., and Long, A. G. (1950). The detection of glycosides and non-reducing carbohydrate derivatives in paper partition chromatography. *J. Chem. Soc.* **1950,** 3162–3167; cf. *J. Chem. Soc.* **1956,** 2818–2823.

Carter, H. E., McCluer, R. H., and Slifer, E. D. (1956). Lipids of wheat flour. I. Characterization of galactosylglycerol components. *J. Am. Chem. Soc.* **78,** 3735–3738.

Daniel, H., Miyano, M., Mumma, R. O., Yagi, T., Lepage, M., Shibuya, I., and Benson, A. A. (1961). The plant sulfolipid. Identification of 6-sulfo-quinovose. *J. Amer. Chem. Soc.* **83,** 1765.

Ferrari, R. A. (1959). Phytosynthesis of the lipids. Ph.D. thesis, Pennsylvania State University, University Park, Pennsylvania.

Ferrari, R. A., and Benson, A. A. (1961). The path of carbon in photosynthesis of the lipids. *Arch. Biochem. Biophys.* **93,** 185–192.

Gibbs, S. P. (1960). The fine structure of *Euglena gracilis* with special reference to the chloroplasts and pyrenoids. *J. Ultrastructure Research* **4,** 127–148.

Goedheer, J. C. (1957). Optical properties and *in vivo* orientation of photosynthetic pigments. Dissertation, Utrecht, Netherlands.

Hase, E., Otsuka, H., Mihara, S., and Tamiya, H. (1959). Role of sulfur in cell division of *Chlorella,* studied by the technique of synchronous culture. *Biochim. et Biophys. Acta* **35,** 180–189.

Horecker, B. C., Thomas, J., and Monod, J. (1960). Galactose transport in *Escherichia coli.* II. Characteristics of the exit process. *J. Biol. Chem.* **235,** 1586–1590.

Kiyasu, J., Paulus, H., and Kennedy, E. P. (1960). An enzymatic reaction of CDP-diglyceride with L-alpha-glycerophosphate. *Federation Proc.* **19,** 233.

Lederer, E. (1958). Glycolipides des bactéries, plantes et animaux inférieurs. *Colloq. Ges. physiol. Chem.* **8,** *Mosbach/Baden,* 1957 119–146.

Lepage, M., Mumma, R. O., and Benson, A. A. (1960). Plant phospholipids. II. Isolation and structure of glycerophosphoryl inositol. *J. Am. Chem. Soc.* **82,** 3713–3715.

Lindberg, B. (1955). Low-molecular carbohydrates in algae. XI. Investigation of *Porphyra umbilicalis. Acta Chem. Scand.* **9,** 1097–1099.

Macfarlane, M. G. (1958). The structure of cardiolipin. *Nature* **182,** 946.

Macfarlane, M. G., and Wheeldon, L. W. (1959). Position of the fatty acids in cardiolipin. *Nature* **183,** 1808.

Maio, J. J., and Rickenberg, H. V. (1960). The β-galactosidase of mouse strain L-cells and mouse organs. *Biochim. et Biophys. Acta* **37,** 101–106.

Maizel, J. V., Benson, A. A., and Tolbert, N. E. (1956). Identification of phosphoryl choline as an important constituent of plant saps. *Plant Physiol.* **31,** 407–408.

Miller, J. A. (1962). Effects of phosphate deficiency on the metabolism of *Chlorella.* Ph.D. thesis, Pennsylvania State University, University Park, Pennsylvania.

Putman, E. W., and Hassid, W. Z. (1954). Structure of galactosylglycerol from *Iridaea laminarioides. J. Am. Chem. Soc.* **76,** 2221–2223.

Sager, R. (1958). The architecture of the chloroplast in relation to its photosynthetic activities. *In* "The Photochemical Apparatus, Its Structure and Function" (R. C. Fuller, ed.), *Brookhaven Symposia in Biol. No.* **11,** 101–117.

Schlenk, H., Mangold, H. K., Gellerman, J. L., Link, W. E., Morrissette, R. A., Holman, R. T., and Hayes, H. (1960). Comparative analytical studies of fatty acids of the alga *Chlorella pyrenoidosa. J. Am. Oil Chem. Soc.* **37,** 547–552.

Strickland, E. H., and Benson, A. A. (1960). Neutron activation paper chromatographic analysis of phosphatides in mammalian cell fractions. *Arch. Biochem. Biophys.* **88,** 344–348.

Su, J.-C. (1956). Studies on saccharides of marine algae. I. Identification of floridoside and a DL-galactan in a red alga, *Porphyra crispata* Kjellm. *J. Chinese Chem. Soc. (Taiwan)* [II] **3,** 65–71.

Trurnit, H., Colmano, G., and Zill, L. P. Surface adsorption properties of chloroplast lipids. To be published.

Wintermans, J. F. G. M. (1960). Concentrations of phosphatides and glycolipids in leaves and chloroplasts. *Biochim. et Biophys. Acta* **44**, 49–54.

Zill, L. P., and Harmon, E. A. (1959). Major lipid components of the chloroplast. Fatty acids of freshwater and marine algae. *Federation Proc.* **18**, 359.

Chlorophylls

L. BOGORAD

Department of Botany, University of Chicago, Illinois

I. Introduction

Since the review by Strain (1951) on the properties of the five main chlorophylls and their distribution among algal groups, a great deal has been learned about some aspects of chlorophyll metabolism. The use of algae in investigations of this subject has been motivated sometimes by an interest in these organisms *per se*, but at least as frequently by the realization that experimentation with microscopic algae affords great operational advantages. Research has been conducted primarily on two genera, *Chlorella* and *Euglena*, though studies on forms of chlorophylls *in vivo* have been extended to several others.

Trends of current research activities are reflected in the organization of this chapter. Spectral properties of chlorophylls and the distribution of

TABLE I

Distribution of Chlorophylls, Biliproteins, and Carotenoids among Algae[a, b]

Pigments	Charophyta	Chlorophyta: "Siphonales"[c]	Chlorophyta: Others	Euglenophyta	Xanthophyta	Chrysophyta	Bacillariophyta	Phaeophyta	Pyrrophyta	Cryptophyta	Cyanophyta	Rhodophyta
Chlorophylls[1]												
a	+	+	+	+	+	+	+	+	+	+	+	+
b	+	+	+	+	−	−	−	−	−	−	−	−
c	−	−	−	−	−	?	+	+	+	+	−	−
d	−	−	−	−	−	?	−	−	−	−	−	±
e	−	−	−	−	[2]	?	−	−	−	−	−	−
Biliproteins[3]												
Phycocyanin				−	−	−	−	−	−	+[4]	+	+
Phycoerythrin			?[5]	−	−	−	−	−	−	+	+	+
Carotenes[6]												
α-Carotene		+	±				±	±		+[8]		±
β-Carotene	+	+	+	+	+	+[7]	+	+	+		+	+
γ-Carotene	+											
Lycopene	+											
ε-Carotene							+[7]			+[8]		
Unknown				+								

Xanthophylls[6]		L, N, S Sx V, Z	As[10] L[11] N V Z	As[10] L,[12] N,[12] Un		F L[7] Un	Dd Dt F	Flx,[7] F, L,[13] V	Dd Dn P	?Z[8]	Apn,[9] Apl[9] Flc,[9] ±L Mn,[14] Ml, O,[15] Z[15]	T,[11] ±Z, L[11]

[a] KEY

+ = Present
— = Absent
? = Insufficient information
Apn = Aphanicin
Apl = Aphanizophyll
As = Astaxanthin (euglenarhodone)
Dd = Diadinoxanthin
Dt = Diatoxanthin
Dn = Dinoxanthin
Flc = Flavacin
Flx = Flavoxanthin
F = Fucoxanthin
L = Lutein
Mn = Myxoxanthin (aphanin, echinenone)
Ml = Myxoxanthophyll
N = Neoxanthin
O = Oscilloxanthin
P = Peridinin (sulcatoxanthin)
S = Siphonein
Sx = Siphonoxanthin
T = Taraxanthin
V = Violaxanthin
Z = Zeaxanthin
Un = Unknown

[b] REFERENCES

[1] See Strain (1958) and Haxo and Fork (1959).
[2] Only observed in *Tribonema bombycinum*.
[3] For further details see Ó hEocha, Chapter 25 in this treatise, Table I.
[4] Reported also in *Cyanidium caldarium*; see Silva, Appendix A, note 11.
[5] Reported in *Palmellococcus miniatus*.
[6] See Nakayama (Chapter 24 in this treatise) and Strain (1958).
[7] Strain (1951).
[8] Haxo and Fork (1959).
[9] Tischer (1938, 1939).
[10] Tischer (1936, 1941, 1944).
[11] Carter *et al.* (1939).
[12] Goodwin and Jamikorn (1954a, b).
[13] Strain *et al.* (1944).
[14] Goodwin and Taha (1950).
[15] Karrer and Rutschmann (1944).

[c] Codiales and Derbesiales; see Silva, Appendix A, Note 40.

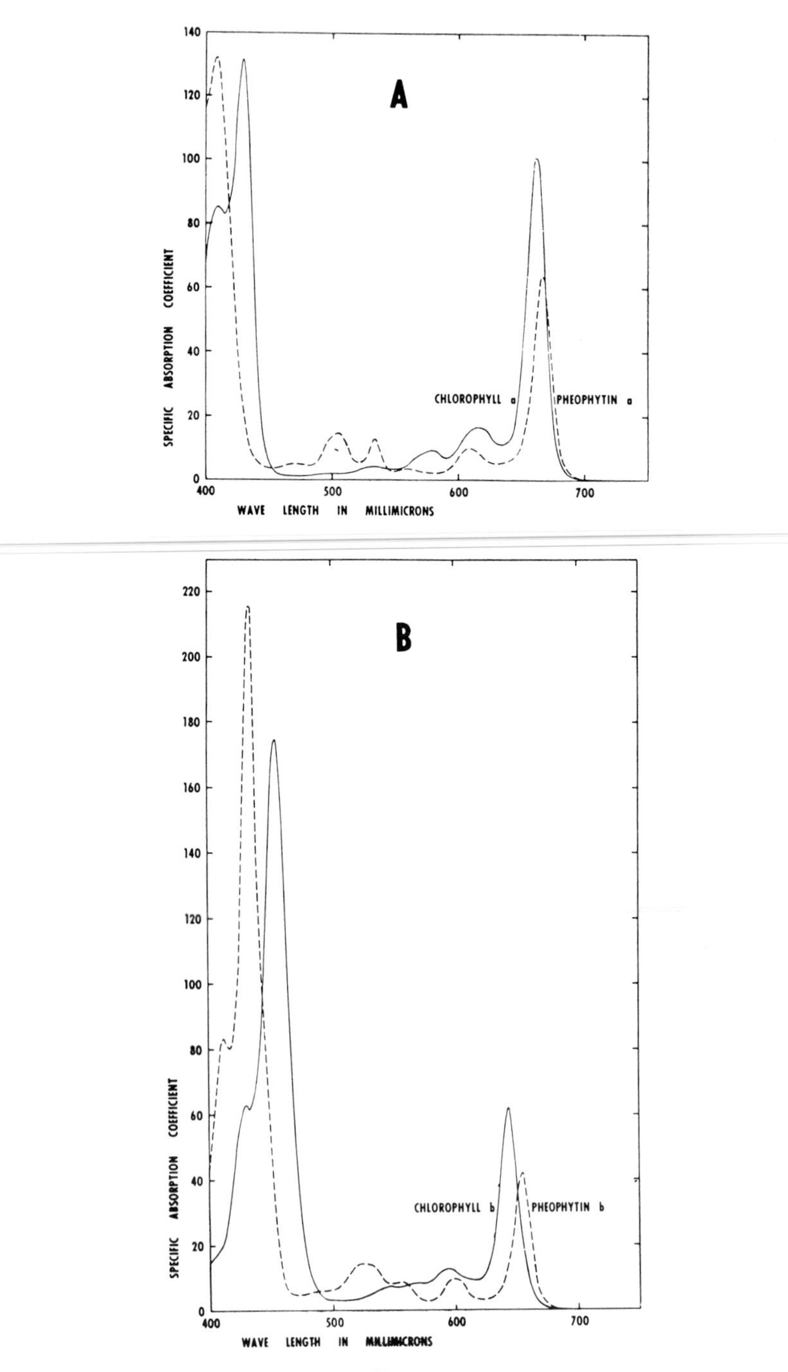
A
SPECIFIC ABSORPTION COEFFICIENT
WAVE LENGTH IN MILLIMICRONS
CHLOROPHYLL a
PHEOPHYTIN a
B
SPECIFIC ABSORPTION COEFFICIENT
WAVE LENGTH IN MILLIMICRONS
CHLOROPHYLL b
PHEOPHYTIN b

FIG. 1.

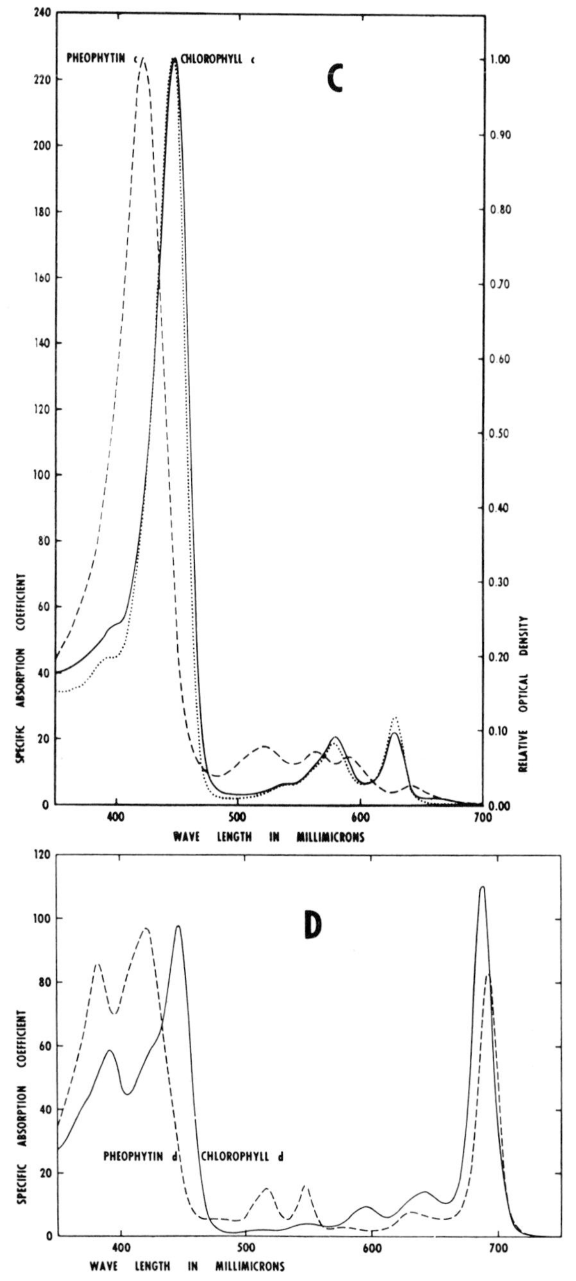

FIG. 1. Absorption spectra of various chlorophylls and their corresponding pheophytins dissolved in ether. KEY: A, chlorophyll *a*; B, chlorophyll *b*; C, chlorophyll *c*; D, chlorophyll *d*. (From Smith, J. H. C. (1954). *Carnegie Inst. Wash. Yr. Book* **53,** 170.)

these pigments among algal groups are presented in tabular form. More extensive treatment is given to such rapidly expanding areas of investigation as the forms of chlorophylls *in vivo*, their biosynthesis, and certain other physiological aspects of their metabolism.

II. The Distribution and Properties of Chlorophylls

A. Extraction

The distribution of chlorophylls *a*, *b*, *c*, *d*, and *e* among the various classes of algae is presented in Table I.

The absorption spectra in ether of chlorophylls *a*, *b*, *c*, and *d* and of their corresponding pheophytins are shown in Fig. 1. The absorption maxima for these chlorophylls in ether are shown in Table II; those for chlorophyll *e* in methanol are reported to be at 415 and 654 mμ (Strain, 1951).

Procedures for the isolation and estimation of chlorophylls have been critically evaluated by Smith and Benitez (1955), who have reviewed the spectrophotometric data on individual pigments and have presented formulas for the quantitative estimation of chlorophylls from absorption data on crude extracts which may contain more than one of these pigments.

It is generally more difficult to extract chlorophylls from algae than from leaves of higher plants. Eighty per cent acetone, a favorite extractant for the latter, is ineffective with many algae; extraction with hot or cold

TABLE II

ABSORPTION MAXIMA (mμ) AND SPECIFIC ABSORPTION COEFFICIENTS[a] OF CHLOROPHYLLS DISSOLVED IN ETHER[b]

Chlorophyll *a*	410 (85.2)	430 (131.5)	533.5 (4.22)	578 (9.27)		615 (16.3)	662 (100.9)
Chlorophyll *b*	430 (62.7)	455 (174.8)	549 (7.07)	595 (12.7)			644 (62)
Chlorophyll *c*[c]		447 (227)		579 (20.6)		628 (22.0)	
Chlorophyll *d*[c]		447 (97.8)	512 (1.98)	548 (403)	595 (9.47)	643 (14.3)	688 (110.4)
Protochlorophyll *a*		432 (325.5)	535 (7.2)	571 (14.9)		623 (39.9)	

[a] *Specific absorption coefficient* (numbers in parentheses) = D/lC, where D = optical density; C = concentration of the pigment in grams per liter; and l = inside length of absorption cell in centimeters.

[b] Data from Smith and Benitez (1955).

[c] Provisional values, based on magnesium determinations and an assumed molecular weight of 893.48 for chlorophylls *c* and *d*.

methanol is usually more satisfactory. Strain (1958) recommended a mixture of two volumes of methanol to one volume of petroleum ether (boiling point about 50°) as a general extracting solvent when petroleum ether solutions are desired for chromatography of chlorophylls.

B. Chlorophylls *in vivo*

The first indications that chlorophylls *in vivo* might differ from presumably identical pigments dissolved in organic solvents came from the observation that the absorption maxima of chlorophylls *in vivo* were at wavelengths about 10 to 20 mμ longer than those in organic solvents. [See discussion by Rabinowitch (1951), Vol. II (1), Chapter 22.]

Until recently accurate absorption spectra of living algal cells were obtained only with difficulty by using integrating spheres or by placing the sample very close to the photocell of the spectrophotometer (see French and Young, 1956). In 1954, Shibata *et al.* devised a set of simple procedures for obtaining excellent absorption spectra of translucent materials, such as algal suspensions, with commercial spectrophotometers. The most convenient consists of interposing a sheet of some light-diffusing material such as opal glass between the sample and the light-sensing element of the spectrophotometer, as close to the sample as possible (Shibata, 1958, 1959). A similar diffusing plate is placed in a comparable position with respect to the blank control.

By such methods, absorption spectra in the range of 200 to 800 mμ were obtained for living cells of *Chlorella pyrenoidosa*, *Euglena gracilis*, *Scenedesmus* sp., *Porphyridium cruentum*, *Synechococcus cedrorum*, and *Ectocarpus siliculosus* (Shibata *et al.*, 1954).

Several forms of chlorophyll *a* have been detected *in vivo* by French and co-workers using the opal-glass technique in conjunction with a derivative spectrophotometer (French, 1957), which automatically plots the first derivative of the absorption spectrum of the material being examined, and thereby facilitates the precise location of absorption maxima. (At the wavelength of maximum or minimum absorption on the usual absorption curve, the derivative crosses the zero line; slight deviations in the position of an absorption maximum are thus made conspicuous. Furthermore, slight deviations in the slope of the usual absorption curve become peaks on the derivative curve, revealing trace amounts of substances whose absorption bands overlap those of the predominant material.)

Derivative spectra of intact algae or broken algal cells in suspension have been reported (French, 1958; French and Elliott, 1958) for the following groups:

(1) Green Algae and euglenids, which both contain chlorophylls *a* and *b*:

Carteria eugametos, Chlorella vulgaris, Coccomyxa rayssiae, Haematococcus lacustris, Ulva sp., and *Euglena gracilis.*

(2) Algae which contain chlorophylls *a* and *c*: *Navicula* sp., *Cryptomonas* sp., *Gonyaulax* sp.; or *a* and *d*: *Gigartina papillata, Iridophycus flaccidum.*

(3) Algae which contain chlorophyll *a* and phycobilins: *Porphyra naiadum* (now preferably named *Smithora naiadum**), *Porphyridium aerugineum*, and *Nostoc* sp.

(4) Algae which contain chlorophyll *a* but no other chlorophylls or phycobilins: *Callithamnion* sp., *Botrydiopis alpina, Ochromonas danica, Vischeria stellata, Nannochloris* sp., and dark-grown cells of a *Chlorella vulgaris* mutant (which, however, also produces chlorophyll *b* when illuminated).

From an analysis of these data and of derivative spectra of chlorophylls *a* and *b* in organic solvents, French *et al.* (1959) concluded that there are *in vivo* at least three types of chlorophyll *a*, and that two or more of these occur together in all plants which they studied. These chlorophylls have absorption maxima *in vivo* at 673, 683, and 694 (±2) mμ (they are therefore designated chl. a_{673}, chl. a_{683}, and chl. a_{694}) and fluorescence maxima at 677, 692, and 707 mμ, respectively. Krasnovsky and Kosobutskaja (1953) have suggested that the 673 mμ peak is that of "monomeric" chlorophyll *a* and the 682 mμ peak that of a "polymeric" form. (See Brody and Brody, Chapter 1 in this treatise.)

Brown (1959) reported unsuccessful attempts to isolate different forms of chlorophyll *a* from cells of *Euglena, Chlorella*, and leaves of Swiss chard (*Beta vulgaris*). Wassink *et al.* (1939) were likewise unable to extract spectrophotometrically distinguishable bacteriochlorophylls from purple bacteria.

Chlorophyll *b in vivo* absorbs light of wavelengths between 620 and 670 mμ. In all plants which have been studied except *Ulva* sp., chlorophyll *b* contributes to the absorption spectrum a single maximum at 650 ± 2 mμ; the derivative spectrum of *Ulva* shows a double peak in this region, which suggests that two forms of chlorophyll *b* may occur in the living thallus (French, 1958).

Wolken and co-workers reported the isolation of "chloroplastin," a pigment-protein complex obtained from frozen and thawed cells of *Euglena gracilis* by grinding and then extracting them with 1 to 2% digitonin. This material, of a molecular weight estimated to be about 40,000, appeared homogeneous in the analytical ultracentrifuge and upon electrophoresis. The argument that "chloroplastin" is a native complex, and not an artifact produced by coprecipitation or adsorption of the pigments upon the

* See Silva, Appendix A, note 4.

protein, is based principally on the following observations. (1) Some preparations of chloroplastin in light, but not in darkness, catalyze the reduction of 2,6-dichlorobenzenone indophenol. (2) Complexes of chlorophyll with peptone or with a protein such as gelatin, though they absorb light at about the same wavelength as freshly prepared chloroplastin, fail to catalyze such a reduction (Wolken, 1958).

It seems clear from spectral data that functional chlorophylls *in vivo* are not in the same molecular state as chlorophylls dissolved in organic solvents, but the nature of these differences remains unresolved. Do the spectral variants differ merely as a consequence of the state of association of chlorophyll molecules, as Krasnovsky suggested? Can chlorophylls *in vivo* occur in complexes with more than one kind of protein to produce the observed spectral varieties of chlorophyll? Or are there chemically different chlorophylls *a*, for example, which are labile or interconvertible *in vitro* but relatively stable *in vivo*?

III. The Biosynthesis of Chlorophylls

A. General Remarks

Historically the earliest landmarks to be established on the biosynthetic pathway of such porphyrins as chlorophyll were: (1) the demonstration that upon illumination of etiolated leaves of angiosperms the appearance of chlorophyll *a* paralleled the disappearance of protochlorophyll *a* (Smith, 1948); (2) the discovery by Granick (1948a) of a mutant of *Chlorella vulgaris* which accumulated protoporphyrin IX but no chlorophyll; and (3) the observations, especially those of Shemin, Neuberger, and their collaborators, which showed that avian erythrocytes *in vitro* could utilize acetate and the α-carbon and nitrogen of glycine as the sole sources of the carbon and nitrogen atoms of heme (iron protoporphyrin IX). (See, for example, Wittenberg and Shemin, 1949; Muir and Neuberger, 1950; Grinstein *et al.*, 1949; Shemin and Wittenberg, 1951). Figure 2 shows the probable relative positions of these and other intermediates in chlorophyll biosynthesis. This scheme is presented here as a framework for discussing research in which algae have been used to explore the synthetic pathway, as well as for reviewing current knowledge of chlorophyll biosynthesis in algae and other plants. Although the literature on porphyrin biosynthesis is extensive (see Bogorad, 1960), the following discussion is largely confined to investigations of algal material.

B. The Utilization of Glycine and Acetate

Della Rosa *et al.* (1953) supplied *Chlorella vulgaris*, growing for 5 to 6 days under fluorescent light, with glycine-1-C^{14}, glycine 2-C^{14}, acetate-

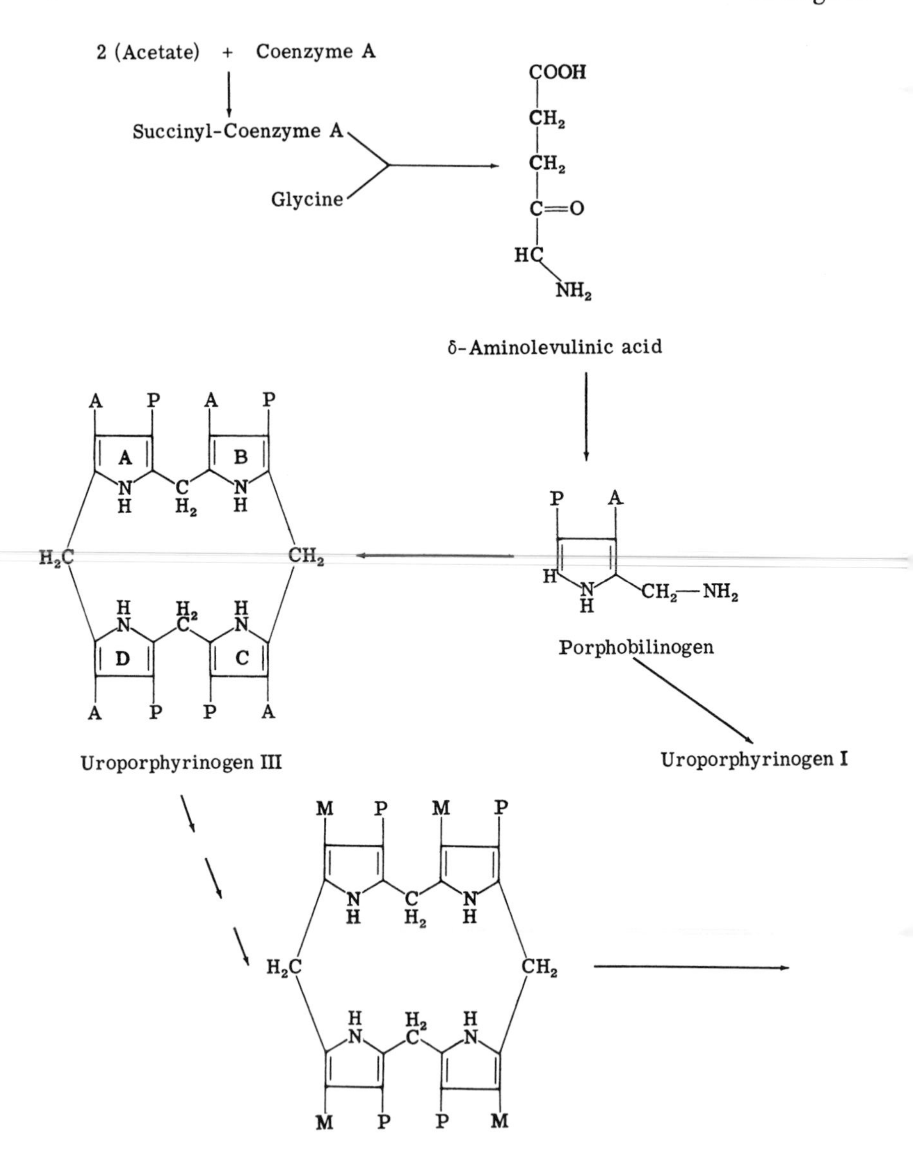

1-C^{14}, or acetate-2-C^{14}. They isolated the chlorophylls produced, prepared their methyl pheophorbides, and determined their radioactivity. The results of these experiments were in general agreement with those on avian erythrocytes mentioned above, though the Della Rosa group found that a little of the carboxyl carbon of glycine, as well as its α carbon atom,

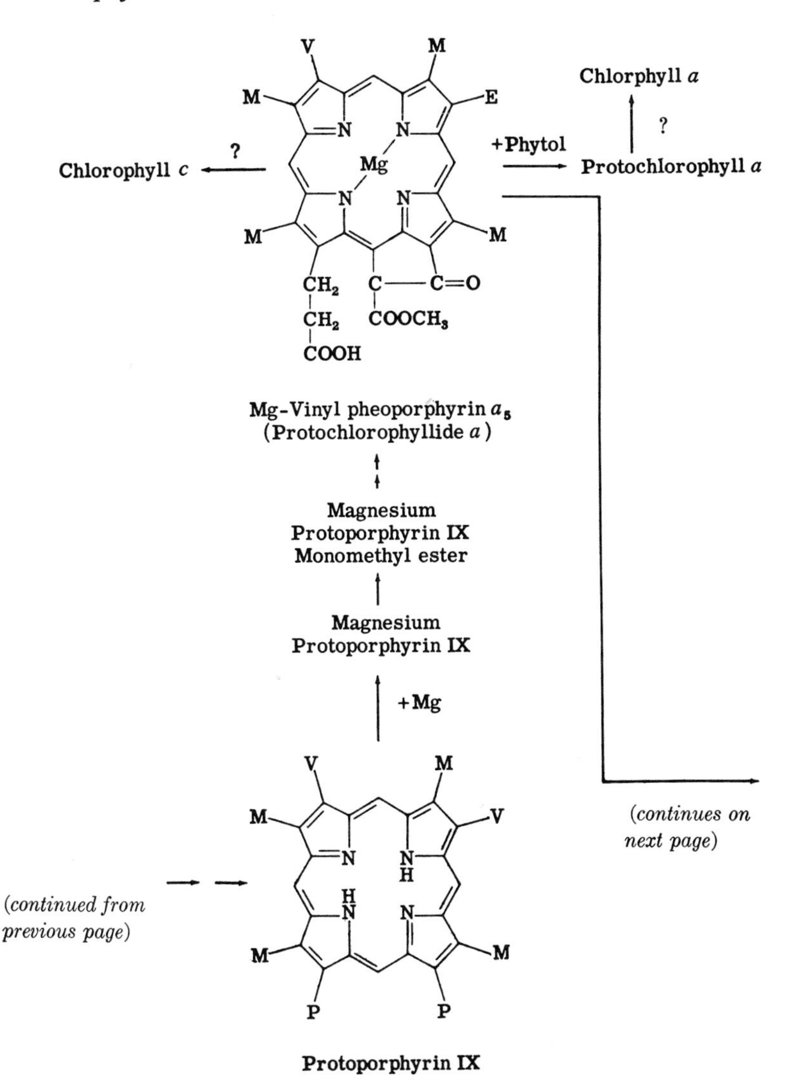

Protoporphyrin IX

could be incorporated into chlorophylls. They recognized the possibility that the carboxyl carbon of glycine might be released as $C^{14}O_2$ in the course of metabolism, and then only indirectly incorporated *via* photosynthesis.

C. δ-Aminolevulinic Acid to Porphobilinogen

The work of Shemin and his co-workers culminated in the discovery that δ-aminolevulinic acid is a specific precursor of heme, and that its

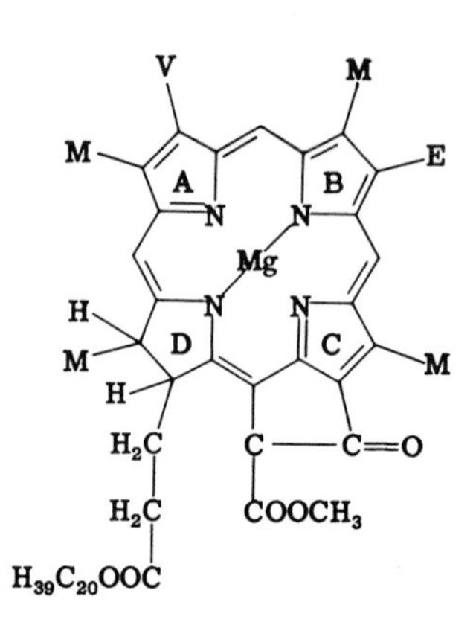

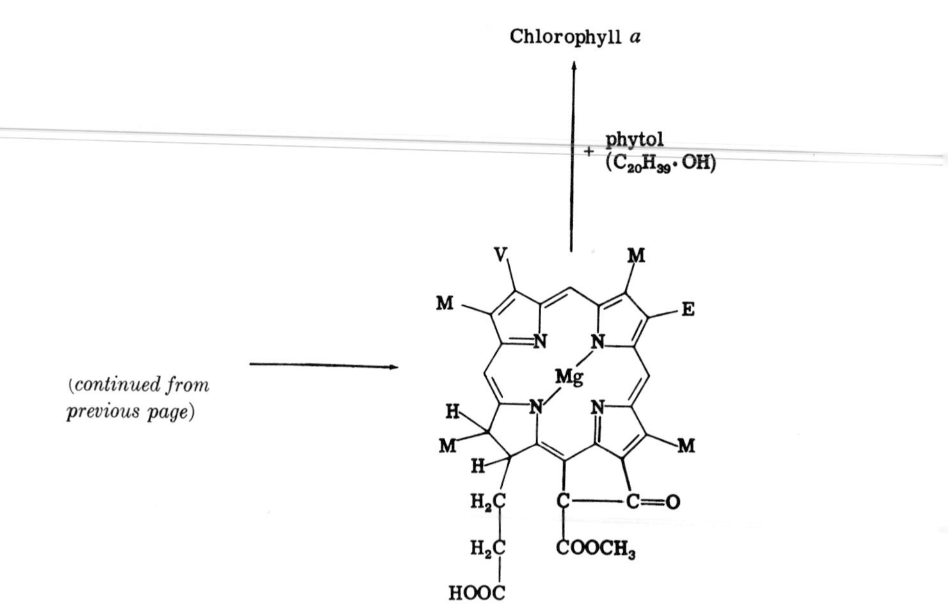

Chlorophyllide *a*

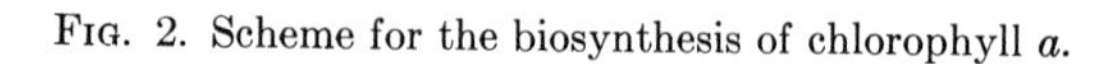

FIG. 2. Scheme for the biosynthesis of chlorophyll *a*.

KEY: A, $—CH_2—COOH$
P, $—CH_2—CH_2—COOH$
M, $—CH_3$
V, $—CH{=}CH_2$
E, $—CH_2—CH_3$

synthesis from succinyl-coenzyme A and glycine appears to require pyridoxal phosphate (Shemin and Russel, 1953; Kikuchi *et al.*, 1958). Granick (1954) showed that extracts of *Chlorella* cells can catalyze the condensation of two molecules of δ-aminolevulinic acid to form one molecule of a pyrrole, porphobilinogen, in much the same way as can preparations of vertebrate liver and red blood cells (Schmid and Shemin, 1955; Gibson *et al.*, 1955; Granick, 1954, 1958).

D. Porphobilinogen to Uroporphyrinogen III

The next step in chlorophyll biosynthesis, the condensation of four molecules of porphobilinogen to form a cyclic tetrapyrrole, can be catalyzed by frozen and thawed preparations of *Chlorella vulgaris* (Bogorad and Granick, 1953b). The intermediates between porphobilinogen and protoporphyrin IX (see Fig. 2) are also formed.

Uroporphyrins have one acetic and one propionic acid substituent on each of the pyrrole residues; the isomers are distinguished by the arrangement of the side chains. Of the four possible isomers of uroporphyrin, only the I and III isomers and their derivatives have been recovered from material of biological origin. Uroporphyrin I is found in the urine of mammals with congenital porphyria; uroporphyrin III, because of the disposition of the propionic and acetic acid side chains on ring D, is most closely related to the chlorophylls and to iron protoporphyrin IX (heme). In the initial studies on the condensation of porphobilinogen by *Chlorella* preparations, both isomers I and III of uroporphyrin were formed. However, it was found that heating the preparation at 55°C. for 30 minutes, though it did not appreciably alter the enzymatic activity with respect to the rate of consumption of porphobilinogen, resulted in the production of only isomer I.

These studies with *Chlorella* preparations, which were confirmed with vertebrate erythrocytes (Booij and Rimington, 1957), led to the suggestion that the formation of uroporphyrinogen III from porphobilinogen may involve at least two interacting enzymes, uroporphyrinogen-I synthetase and uroporphyrinogen-III co-synthetase (Bogorad, 1960).

E. Uroporphyrinogen III to Protoporphyrin IX

Since preparations of *Chlorella* can catalyze the synthesis of coproporphyrin III and protoporphyrin IX from uroporphyrin*ogen* III[1] but not

[1] Porphyrins are oxidized porphyrin*ogens*. The pyrrole rings of a porphyrin*ogen* are linked by saturated, methylene bridges. Upon oxidation, which may occur spontaneously in air, the bridges become unsaturated and a colored porphyrin is formed. Porphyrin*ogens*, unlike porphyrins, cannot chelate metals.

from uroporphyrin III (Bogorad, 1958), the former is presumably an intermediate in chlorophyll biosynthesis. The decarboxylation of the four acetic acid side chains of uroporphyrinogen III to methyl groups results in the formation of coproporphyrinogen III. All four decarboxylations appear to be catalyzed by a single enzyme, since intermediate porphyrins with 7, 6, and 5 carboxyl groups per molecule have been found in reaction mixtures of uroporphyrinogen III and uroporphyrinogen decarboxylase from chicken erythrocytes (Mauzerall and Granick, 1958).

As already indicated, crude *Chlorella* preparations can catalyze the formation of protoporphyrin IX from porphobilinogen or from uroporphyrinogen III, presumably via coproporphyrinogen III. Furthermore, cell-free extracts of *Euglena gracilis* capable of catalyzing the conversion of coproporphyrinogen III to protoporphyrin IX were prepared by Granick and Mauzerall (1958). The isolation of a *Chlorella vulgaris* mutant which accumulates hematoporphyrin IX and monovinyl monohydroxyethyl deuteroporphyrin IX (as well as uro- and coproporphyrins) suggested that these compounds, or the corresponding porphyrinogens, might be intermediates in this part of the biosynthetic path (Bogorad and Granick, 1953a; Granick *et al.*, 1953). However, since hematoporphyrin, hematoporphyrinogen, and diacetyl deuteroporphyrinogen did not serve as substrates for the synthesis of protoporphyrin IX in a frozen and thawed *C. vulgaris* system (Marks and Bogorad, unpublished), the compounds accumulated by the mutant cultures probably arise by spontaneous degradation of the true intermediate(s).

The enzyme coproporphyrinogen oxidase, which catalyzes the oxidative decarboxylation of the propionic-acid residues on rings A and B of coproporphyrinogen III to vinyl groups, thereby producing protoporphyrinogen IX, has been demonstrated in preparations of vertebrate liver mitochondria (Granick and Sano, 1961), but not yet in any alga. However, a heterotrophic mutant of *Chlorella*, which accumulated protoporphyrin IX but no chlorophylls, provided the first evidence for the position of protoporphyrin in the synthetic pathway shown in Fig. 2 (Granick, 1948a).

F. Mg-Protoporphyrin IX to Protochlorophyllide *a*

Magnesium protoporphyrin IX is considered to follow protoporphyrin IX in the biosynthetic chain because it accumulates in cultures of another *Chlorella* mutant (Granick, 1948b); the mechanism by which the magnesium is incorporated is unknown.

Yet another mutant when grown in darkness accumulates magnesium vinyl pheoporphyrin a_5, i.e., protochlorophyllide *a* or protochlorophyll lacking the phytol side chain (Granick, 1950). Protochlorophyllide *a* is formed by the following modifications of Mg protoporphyrin: (1) esterification of the carboxyl group of the propionic acid residue on ring C; (2) cyclization of the propionic acid group on ring C to form a cyclopentanone ring; and (3) reduction of one vinyl group to an ethyl group. Apparently (1) occurs first, since a magnesium protoporphyrin IX monomethyl ester can be isolated from etiolated barley seedlings supplied with δ-aminolevulinic acid in darkness (Granick, 1959). No other intermediates are known in the span between magnesium protoporphyrin IX and protochlorophyllide *a*; consequently, the sequence in which changes (2) and (3) occur cannot yet be given.

G. Protochlorophyllide *a* to Chlorophyll *a*

Most algae which have been grown heterotrophically form chlorophylls equally well in darkness and in light. Exceptional in this respect are some species of *Euglena* and *Chlorella*, and certain induced mutants of *Chlorella vulgaris* and *C. pyrenoidosa* (e.g., Granick, 1950; Allen, 1958) and of *Chlamydomonas reinhardtii* (Sager, 1958) which become green only in light. Wild-type cells of *Cyanidium caldarium** form neither phycobilins nor chlorophyll *a* in darkness, but a commonly occurring mutant type can form these pigments even in darkness (Allen, 1959; Nichols and Bogorad, 1960). Such a relatively high frequency with which the cells may develop the ability to become green in darkness is not known among angiosperms, all of which require light for chlorophyll synthesis.

The action spectra for the initial production of chlorophyll *a* in etiolated leaves of higher plants (Frank, 1946; Smith, 1948; Smith and Young, 1956) and in dark-grown cells of *Euglena gracilis* var. *bacillaris* (Wolken and Mellon, 1956; Nishimura and Huzisige, 1959) correspond closely to the absorption spectrum of protochlorophyll *a* (or of protochlorophyllide *a*, which has an identical spectrum). As chlorophyll is formed, protochlorophyll (or protochlorophyllide) *a* correspondingly decreases.

Protochlorophyllide *a* has been found in leaves of etiolated angiosperms (Loeffler, 1955) as well as in the *Chlorella* mutant mentioned above (Granick, 1950). Protochlorophyll *a* occurs in etiolated leaves and in *Euglena gracilis* grown for many cell generations in darkness. From the data on *Euglena* published by Nishimura and Huzisige (1959), one cannot conclude whether the pigment they isolated was protochlorophyll *a* or protochlorophyllide *a*. Though both are probably present in etiolated angiosperm tissues, only the protochlorophyllide is converted upon exposure to light, first to chlorophyllide *a*, and then, presumably by esterification with phytol, to chlorophyll *a* (Wolff and Price, 1957; Virgin, 1960). The fate of the protochlorophyll *a* is unknown. The *Chlorella* mutant described by

* Regarding the taxonomy of this organism, see Note 11 of Appendix A in this treatise by Silva.

Granick (1950) which accumulates protochlorophyllide *a* (magnesium vinyl pheoporphyrin a_5) in darkness, resembles higher plants in that it forms normal chlorophyll in the light, thereby providing evidence that, at least in *Chlorella*, chlorophyll *a* is normally produced via protochlorophyllide *a*. It is uncertain whether any plant forms chlorophyll *a* directly from protochlorophyll *a*.

Soluble preparations of protochlorophyllide holochrome (i.e., protochlorophyllide *plus* protein) from etiolated bean leaf tissue are converted by the action of light into the corresponding chlorophyllide holochrome (Smith *et al.* 1957; Smith, 1958). Similar investigations might be undertaken with extracts of algal material, such as Granick's protochlorophyllide-accumulating mutant or dark-grown cells of *Euglena*.

In *Euglena gracilis* (Nishimura and Huzisige, 1959), as in higher plants (Withrow *et al.*, 1956; Wolff *et al.*, 1957; Virgin, 1958), there are indications that light influences not only the conversion of protochlorophyll(ide) to chlorophyll(ide) but also the steady-state production of chlorophyll, perhaps by affecting chloroplast development.

H. Other Chlorophylls

There are few experimental data on the origin of chlorophylls other than type *a*. Blass *et al.* (1959) studied the labeling with C^{14} of chlorophylls *a* and *b* in *Chlorella pyrenoidosa* and *Scenedesmus obliquus* illuminated in the presence of $C^{14}O_2$. After one hour, the chlorophyll *a* extracted from *Scenedesmus* contained three times as much radioactivity per molecule as did chlorophyll *b*, which led Blass *et al.* to conclude that chlorophylls *a* and *b* are not rapidly interconverted, but that chlorophyll *a* may be a precursor of chlorophyll *b*. Similar conclusions can be drawn from data obtained by Becker and Sheline (1955), who incubated *Chlorella pyrenoidosa* with $Mg^{28}(NO_3)_2$ and C^{14}-glucose for 18 hours in light, extracted and separated the chlorophylls, and found approximately 2.5 times as much Mg^{28} and C^{14} in type *a* as in type *b*. The existing information on the biosynthesis of these two chlorophylls does not permit a firm decision on their metabolic relationship.

Chlorophyll *a* appears first upon illumination of dark-grown *Euglena* cells, as in etiolated leaves of angiosperms, but the separation of the production of these two pigments by a photochemical step is known only in certain mutants of *Chlorella pyrenoidosa* which produce chlorophyll *a* in darkness but both chlorophylls *a* and *b* in light (Allen, 1958).

Chlorophyll *c* appears to resemble protochlorophyllide in that: (1) it is not esterified with phytol; (2) the pyrrole ring D is not reduced; and (3) a cyclopentanone ring is apparently present. However, the magnesium is more tightly held in chlorophyll *c* than in chlorophylls *a* or *b* or in protochlorophyll. On the basis of these observations, Granick (1949) suggested that chlorophyll *c* might be a modified magnesium pheoporphyrin derived from protochlorophyllide *a*.

There is no information on the biochemical origin of chlorophylls *d* and *e*. According to Holt and Morley (1960), chlorophyll *d* differs from chlorophyll *a* only in the presence of a formyl group instead of a vinyl group on ring A (see Fig. 3).

IV. The Chlorophyll Content of Algae

A. General Remarks

Rabinowitch (1945, Vol. I, pp. 410–411) has tabulated values obtained by various workers for the concentration of chlorophylls in some species of Green, Brown, Red, and Blue-green Algae. Many of the reports cited are based on analyses of plant material collected in nature. Studies on algae cultured under controlled conditions in the laboratory reveal that mineral nutrition, light intensity, and cell age can all be major factors in influencing the concentration of chlorophyll in an individual organism.

B. Mineral Nutrition

Hase *et al.* (1957) reported that cells of *Chlorella ellipsoidea* are "etiolated" when grown in N- or Mg-free medium but are normal in color if P or K is omitted. Other examples of the effects of mineral deficiencies upon chlorophyll content are reviewed by Wiessner (Chapter 17 in this treatise).

C. Light Intensity, Temperature, and Chlorophyll Accumulation

Sargent (1940) reported that chlorophyll constituted 6.6% of the dry weight of "shade-grown" cells of *Chlorella pyrenoidosa* but only 3.3% of the dry weight of "sun-grown" cells. Myers' (1946) more detailed studies on this species revealed some of the problems involved in establishing such values. Cell number, cell size, and chlorophyll content were measured and plotted as functions of the intensity of light under which the algae were cultured. Although over a considerable range the chlorophyll per milliliter of packed cells was inversely related to light intensity, there was a direct relation between the intensity of light and cell size. The chlorophyll content *per cell* was essentially constant over a range of intensities from 60 to 3600 lux (meter-candles).

It is clear from Myers' work that the chlorophyll content of *Chlorella* cells is independent of their size, at least when this is controlled by light intensity. Is this a direct manifestation of the semi-independence of the chloroplasts, or is the comparatively low chlorophyll content of the large cells in these experiments an indirect consequence of the bleaching of chlorophyll at the relatively high light intensities required for the induction of large cells? The observations of Tamiya *et al.* (1953) on synchronously grown *Chlorella ellipsoidea* (see below) argue against the latter explanation.

From studies of a number of *Chlorella pyrenoidosa* mutants, Allen (1958) concluded that bleaching of chlorophyll by bright light is most rapid in those mutants which lack chlorophyll *b*, suggesting that in some way chlorophyll *b* may act to stabilize and protect the chlorophyll *a* in the algal cell. She also noted (Allen, 1959) that bright blue-green cultures of *Cyanidium caldarium* develop at about 3000 lux, but that at high light intensities—around 9000 to 15,000 lux—cultures lose not only much of their chlorophyll but also almost all their phycocyanin.

Euglena gracilis apparently requires light for chlorophyll maintenance, since green cultures transferred to darkness gradually become almost yellow. Pheophytin (chlorophyll *minus* magnesium) is the only chlorophyll degradation product found in such cultures (Wolken *et al.*, 1955). Nothing more is known about the biochemistry of chlorophyll breakdown in any plant, to the reviewer's knowledge.

Interactions between the effects of temperature and light intensity on the accumulation of chlorophyll have been frequently noted. An intensive study of *Anacystis nidulans** was made by Halldal (1958), using cultures of cells dispersed on an agar plate across which he imposed crossed gradients of temperature (15–44°C.) and light intensity (750–9000 lux). Absorption spectra of samples, determined at intervals by the opal-glass technique of Shibata (1959), revealed that cells exposed to about 3000 lux at 30 to 40°C. produced the most chlorophyll during the first 24 hours. The position at which maximum chlorophyll accumulation occurred changed as the culture aged.

Halldal also observed that although the absorption spectra of cells from most regions of the agar plate showed a maximum at about 670 mμ, cells grown at about 25°C. showed a maximum at 675 mμ, due to a decrease in the ratio of chlorophyll a_{673} to a_{683}. Within the range of culture conditions examined, this ratio varied from about 1.2 to 3.7, being highest in cells grown for 48 hours at 4000 lux and at 30 to 38°C. The ratio was particularly low in cells grown in dim light at the extremes of the temperature range, indicating either that chlorophyll a_{673} is less susceptible to bleaching, or that the formation of this form of pigment requires more light than

* Regarding the taxonomy of this organism, see Note 2 of Appendix A in this treatise by Silva.

does that of chlorophyll a_{683}. The biosynthetic relation between these two pigment types is not clear.

Brown and French (1959) studied the effects of intense illumination by red light (>620 mμ) on the bleaching of various chlorophylls in a colloidal suspension prepared from *Chlorella pyrenoidosa* cells broken by passage through a needle-valve device. From complex changes observed in the absorption spectrum they concluded: (1) that chlorophyll a_{683} bleached more rapidly than a_{673} (in agreement with one interpretation of the observations of Halldal, and with data of Vorobeva and Krasnovsky); (2) that a_{695}, if present, was particularly sensitive to bleaching; and (3) that b_{653}, the only form of chlorophyll *b* detected in their preparations, was more stable to light than any of the varieties of chlorophyll *a*.

D. Cell Age and Chlorophyll Metabolism

Experiments with the crossed-gradient plate also demonstrated that the chlorophyll content of various algae can change rapidly as the cells age and the culture medium becomes depleted of mineral nutrients; the zone of greenness generally moves from lower to higher temperatures within the same light-intensity range (Halldal and French, 1958). In studies on the pigmentation of cells in laboratory cultures, it is difficult to separate the effects of changes in the nutrient medium from those of aging of the cells themselves, unless one works with synchronized cultures.

"Young" cells of *C. ellipsoidea*, especially when illuminated, rapidly become rich in chlorophyll; as their photosynthetic rate increases, they produce much cell material but relatively little additional chlorophyll (see Hase, Chapter 40, this treatise). The final composition of a culture in a steady state of growth thus depends on light intensity and temperature, and integrated values for chlorophyll per cell may be quite different under different conditions of illumination. The possible pertinence of these observations to those of Sargent (1940) and Myers (1946) is obvious.

In synchronized cultures of *Chlorella pyrenoidosa* grown at 30°C. in alternating periods of 16 hours of light and 12 hours of darkness, Pirson *et al.* (1959) observed that sudden exposures to a low temperature (4°C.) 7 to 8 hours after the initiation of illumination resulted in a transient inhibition of cell division and growth and a reversible loss of as much as 50% of the chlorophyll. The disappearance of chlorophyll could not be attributed entirely to photo-oxidation, because it continued even after the cells had been returned to darkness.

During the second light period following the cold treatment, chlorophyll synthesis proceeded at nearly the normal rate although the cell number remained practically constant. Complete recovery from the cold shock was attained within 75 hours after treatment.

E. Treatments Which Specifically Reduce Chlorophyll Content

Streptomycin (Provasoli *et al.*, 1948), heat (Pringsheim and Pringsheim, 1952), chlortetracycline (Robbins *et al.*, 1953), acriflavine (Lwoff, 1950), and pyribenzamine (Gross *et al.*, 1955) have all been reported to induce in *Euglena* cultures the development of strains devoid of chlorophyll.

Provasoli *et al.* (1948) observed that permanently colorless strains of *E. gracilis* var. *bacillaris* were produced after white (dark-grown) or green (light-grown) cultures had been exposed to streptomycin at concentrations of about 1 mg./ml. Similar effects were observed in proliferating cultures exposed to the antibiotic in light or darkness, and in nonproliferating cultures incubated with streptomycin in darkness. Since the chloroplasts become disorganized under these conditions (Provasoli *et al.*, 1948; Wolken, 1956), the effect on chlorophyll synthesis is probably secondary, the primary effect being on chloroplast multiplication, maturation, or maintenance. An elegant confirmation of this was provided by De Deken-Grenson and Messin (1958).

Wolken (1956) reported that when green cells of *E. gracilis* were exposed to streptomycin concentrations of 2.5 mg./ml., their content of chlorophylls and carotenoids declined, and pheophytin accumulated, in about the same way as when green *Euglena* cells are transferred to darkness. The chloroplasts were observed to swell, and 65% of the pigment was lost within 70 hours of the initial exposure to streptomycin. Is the chlorophyll lost because the plastids are disrupted, or do the plastids break down as a consequence of failure to replace the chlorophyll which is turning over? In the latter case, are whole plastids, or grana, or individual lamellae discarded, or are individual chlorophyll or holochrome molecules the units of turnover?

References

Allen, M. B. (1958). Possible functions of chlorophyll *b*. Studies with green algae that lack chlorophyll *b*. *Brookhaven Symposia in Biol. No.* **11,** 339–342.

Allen, M. B. (1959). Studies with *Cyanidium caldarium*, an anomalously pigmented chlorophyte. *Arch. Mikrobiol.* **32,** 270–277.

Becker, R. S., and Sheline, R. K. (1955). Biosynthesis and exchange experiments on some plant pigments using Mg^{28} and C^{14}. *Arch. Biochem. Biophys.* **54,** 259–265.

Blass, U., Anderson, J. M., and Calvin, M. (1959). Biosynthesis and possible functional relationships among the carotenoids; and between chlorophyll *a* and chlorophyll *b*. *Plant Physiol.* **34,** 329–333.

Bogorad, L. (1958). The enzymatic synthesis of porphyrins from porphobilinogen III. Uroporphyrinogens as intermediates. *J. Biol. Chem.* **233,** 516–519.

Bogorad, L. (1960). The biosynthesis of protochlorophyll. *In* "Comparative Biochemistry of Photoreactive Systems" (M. B. Allen, ed.), pp. 227–256. Academic Press, New York.

Bogorad, L., and Granick, S. (1953a). Protoporphyrin precursors produced by a *Chlorella* mutant. *J. Biol. Chem.* **202,** 793–800.

Bogorad, L., and Granick, S. (1953b). The enzymatic synthesis of porphyrins from porphobilinogen. *Proc. Natl. Acad. Sci. U.S.* **39,** 1176–1188.

Booij, H. L., and Rimington, C. (1957). Effect of preheating on porphyrin synthesis by red cells. *Biochem. J.* **65,** (Proc. of 356th meeting, communications) 4 pp.

Brown, J. S. (1959). Studies on the components of chlorophyll *a*. *Carnegie Inst. Wash. Yr. Book* **58,** 330–331.

Brown, J. S., and French, C. S. (1959). Absorption spectra and relative photostability of different forms of chlorophyll in *Chlorella*. *Plant Physiol.* **34,** 305–309.

Carter, P. W., Heilbrun, I. M., and Lythgoe, B. (1939). The lipochromes and sterols of the algal classes. *Proc. Roy. Soc.* **B128,** 82–109.

De Deken-Grenson, M., and Messin, S. (1958). La continuité génétique des chloroplastes chez les Euglènes. I. Mécanisme de l'apparition des lignées blanches dans les cultures tractées par la streptomycine. *Biochim. et Biophys. Acta* **27,** 145–155.

Della Rosa, R. J., Rocco, J., Altman, K. I., and Saloman, K. (1953). The biosynthesis of chlorophyll as studied with labelled glycine and acetic acid. *J. Biol. Chem.* **202,** 771–779.

Frank, S. R. (1946). Effectiveness of the spectrum in chlorophyll formation. *J. Gen. Physiol.* **29,** 157–179.

French, C. S. (1957). Derivative spectrophotometry. *Carnegie Inst. Wash. Yr. Book* **56,** 281–282.

French, C. S. (1958). Various forms of chlorophyll *a* in plants. *Brookhaven Symposia in Biol. No.* **11,** 305–309.

French, C. S., and Elliott, R. F. (1958). The absorption spectra of chlorophylls in various algae. *Carnegie Inst. Wash. Yr. Book* **57,** 278–286.

French, C. S., and Young, V. M. K. (1956). The absorption, action, and fluorescence spectra of photosynthetic pigments in living cells and in solutions. *In* "Radiation Biology" (A. Hollaender, ed.), Vol. III, p. 343–392. McGraw-Hill, New York.

French, C. S., Brown, J. S., Allen, M. B., and Elliott, R. F. (1959). The types of chlorophyll *a* in plants. *Carnegie Inst. Wash. Yr. Book* **58,** 327–330.

Gibson, K. D., Neuberger, A., and Scott, J. J. (1955). The purification and properties of δ-aminolevulinic acid dehydrase. *Biochem. J.* **61,** 618–629.

Goodwin, T. W., and Jamikorn, M. (1954a). Studies in carotenogenesis. II. Carotenoid synthesis in the alga *Haematococcus pluvalis*. *Biochem. J.* **57,** 376–380.

Goodwin, T. W., and Jamikorn, M. (1954b). Studies in carotenogenesis. Some observations on carotenoid synthesis in two varieties of *Euglena gracilis*. *J. Protozool.* **1,** 216–219.

Goodwin, T. W., and Taha, M. M. (1950). A study of the carotenoid echinenone and myxoxanthin with special reference to their probable identity. *Biochem. J.* **48,** 513–514.

Granick, S. (1948a). Protoporphyrin-9 as a precursor of chlorophyll. *J. Biol. Chem.* **172,** 717–727.

Granick, S. (1948b). Magnesium protoporphyrin as a precursor of chlorophyll in *Chlorella*. *J. Biol. Chem.* **175,** 333–342.

Granick, S. (1949). The pheoporphyrin nature of chlorophyll *c*. *J. Biol. Chem.* **179,** 505.

Granick, S. (1950). Magnesium vinyl pheoporphyrin a_5, another intermediate in the biological synthesis of chlorophyll. *J. Biol. Chem.* **183,** 713–730.

Granick, S. (1954). Enzymatic conversion of δ-aminolevulinic acid to porphyrin. *Science* **120,** 1105.

Granick, S. (1958). Porphyrin biosynthesis in erythrocytes. Formation of δ-aminolevulinic acid in erythrocytes. *J. Biol. Chem.* **232,** 1101–1117.

Granick, S. (1959). Magnesium porphyrins formed by barley seedlings treated with δ-aminolevulinic acid. *Plant Physiol.* **34,** xviii.

Granick, S., and Mauzerall, D. (1958). Enzymatic formation of protoporphyrin from coproporphyrinogen III. *Federation Proc.* **17,** 233 (Paper No. 919).

Granick, S., and Sano, S. (1961). Mitochondrial coproporphyrinogen oxidase and the formation of protoporphyrin. *Federation Proc.* **20,** 376.

Granick, S., Bogorad, L., and Jaffe, H. (1953). Hematoporphyrin IX, a probable precursor of protoporphyrin in the biosynthetic chain of heme and chlorophyll. *J. Biol. Chem.* **202,** 801–813.

Grinstein, M., Kamen, M. D., and Moore, C. V. (1949). The utilization of glycine in the biosynthesis of hemoglobin. *J. Biol. Chem.* **179,** 359.

Gross, J. A., Jahn, T. L., and Bernstein, E. (1955). The effect of antihistamines on the pigments of green protista. *J. Protozool.* **2,** 71–75.

Halldal, P. (1958). Pigment formation and growth of blue-green algae in crossed-gradients of light intensity and temperature. *Physiol. Plantarum* **11,** 401–420.

Halldal, P., and French, C. S. (1958). Algal growth in crossed gradients of light intensity and temperature. *Plant Physiol.* **33,** 249–252.

Hase, E., Morimura, Y., and Tamiya, H. (1957). Some data on the growth physiology of *Chlorella* studied by the technique of synchronous culture. *Arch. Biochem. Biophys.* **69,** 149–165.

Haxo, F. T., and Fork, D. C. (1959). Photosynthetically active accessory pigments of cryptomonads. *Nature* **184,** 1056.

Holt, A. S., and Morley, H. V. (1960). Recent studies of chlorophyll chemistry. *In* "Comparative Biochemistry of Photoreactive Systems" (M. B. Allen, ed.), pp. 169–179. Academic Press, New York.

Karrer, P., and Rutschmann, J. (1944). Beitrag zur Kenntnis der Carotenoids aus *Oscillatoria rubescens. Helv. Chim. Acta* **27,** 1691–1695.

Kikuchi, G., Shemin, D., and Bachmann, B. (1958). The enzymic synthesis of δ-aminolevulinic acid. *Biochim. et Biophys. Acta* **28,** 219–220.

Krasnovsky, A. A., and Kosobutskaya, L. M. (1953). [Different conditions of chlorophyll in plant leaves.] *Doklady Akad. Nauk SSSR.* **91,** 343–346. (In Russian.)

Loeffler, J. E. (1955). Precursors of protochlorophyll in etiolated barley seedlings. *Carnegie Inst. Wash. Yr. Book* **54,** 159–160.

Lwoff, A. (1950). "Problems of Morphogenesis in Ciliates." Wiley, New York.

Mauzerall, D., and Granick, S. (1958). Porphyrin biosynthesis in erythrocytes. III. Uroporphyrinogen and its decarboxylase. *J. Biol. Chem.* **232,** 1141–1162.

Muir, H. M., and Neuberger, A. (1950). The biogenesis of porphyrins. 2. The origin of the methyne carbons. *Biochem. J.* **47,** 97–104.

Myers, J. (1946). Culture conditions and the development of the photosynthetic mechanism. III. Influence of light intensity on cellular characteristics of *Chlorella. J. Gen. Physiol.* **29,** 419–427.

Nichols, K. E., and Bogorad, L. (1960). Studies on phycobilin formation with mutants of *Cyanidium caldarium. Nature* **188,** 870–872.

Nishimura, M., and Huzisige, H. (1959). Studies on the chlorophyll formation in *Euglena gracilis* with special reference to the action spectrum of the process. *J. Biochem.* (*Tokyo*) **46,** 225–234.

Pirson, A., Lorenzen, H., and Koepper, A. (1959). A sensitive stage in synchronized cultures of *Chlorella. Plant Physiol.* **34,** 353–355.

Pringsheim, E. G., and Pringsheim, O. (1952). Experimental elimination of chromatophores and eye-spot in *Euglena gracilis. New Phytologist* **51,** 65–76.

Provasoli, L., Hutner, S. H., and Schatz, A. (1948). Streptomycin-induced chlorophyll-less races of *Euglena. Proc. Soc. Exptl. Biol. Med.* **69,** 279–282.

Rabinowitch, E. I. (1945, 1951, 1956). "Photosynthesis and Related Processes": Vol. I, 1945; Vol. II (1), 1951; Vol. II (2), 1956. Interscience, New York.

Robbins, W. J., Hervey, A., and Stebbins, M. E. (1953). *Euglena* and Vitamin B_{12}. *Ann. N. Y. Acad. Sci.* **56,** 818–830.

Sager, R. (1958). The architecture of the chloroplast in relation to its photosynthetic activity. *Brookhaven Symposia in Biol. No.* **11,** 101–117.

Sargent, M. C. (1940). Effect of light intensity on the development of the photosynthetic mechanism. *Plant Physiol.* **15,** 275–290.

Schmid, R., and Shemin, D. (1955). The enzymatic formation porphobilinogen from δ-aminolevulinic acid and its conversion to protoporphyrin. *J. Am. Chem. Soc.* **77,** 506–507.

Shemin, D., and Russell, C. S. (1953). δ-Aminolevulinic acid, its role in the biosynthesis of porphyrins and purines. *J. Am. Chem. Soc.* **75,** 4873–4874.

Shemin, D., and Wittenberg, J. (1951). The mechanism of porphyrin formation. The role of the tricarboxylic acid cycle. *J. Biol. Chem.* **192,** 315–334.

Shibata, K. (1958). Spectrophotometry of intact biological materials. Absolute and relative measurements of their transmission, reflection, and absorption spectra. *J. Biochem. (Tokyo)* **45,** 599–623.

Shibata, K. (1959). Spectrophotometry of translucent biological materials. *In* "Methods of Biochemical Analysis" (D. Glick, ed.), Vol. VII, p. 77–109. Interscience, New York.

Shibata, K., Benson, A. A., and Calvin, M. (1954). The absorption spectra of suspensions of living microorganisms. *Biochim. et Biophys. Acta* **15,** 461–470.

Smith, J. H. C. (1948). Protochlorophyll, precursor of chlorophyll. *Arch. Biochem. Biophys.* **19,** 449–454.

Smith, J. H. C. (1958). Quantum yield of the protochlorophyll-chlorophyll transformation. *Brookhaven Symposia in Biol. No.* **11,** 296–302.

Smith, J. H. C., and Benitez, A. (1955). Chlorophylls: Analysis in plant materials. *In* "Modern Methods of Plant Analysis" (K. Paech and M. V. Tracey, eds.), Vol. IV, pp. 142–196. Springer, Berlin.

Smith, J. H. C., and Young, V. M. K. (1956). Chlorophyll formation and accumulation in plants. *In* "Radiation Biology" (A. Hollaender, ed.), Vol. III, p. 393–442. McGraw-Hill, New York.

Smith, J. H. C., Kupke, D. W., Loeffler, J. E., Benitez, A., Ahrne, I., and Giese, A. T. (1957). The natural state of protochlorophyll. *In* "Research in Photosynthesis" (H. Gaffron, ed.), pp. 464–474. Interscience, New York.

Strain, H. H. (1951). The pigments of algae. *In* "Manual of Phycology" (G. M. Smith, ed.), pp. 243–262. Chronica Botanica, Waltham, Massachusetts.

Strain, H. H. (1958). "Chloroplast Pigments and Chromatographic Analysis." 32nd Annual Priestley Lectures, Penn. State Univ., University Park, Pennsylvania.

Strain, H. H., Manning, W. M., and Hardin, G. J. (1944). Xanthophylls and carotenes of diatoms, brown algae, dinoflagellates, and sea-anemones. *Biol. Bull.* **86,** 169–191.

Tamiya, H., Iwamura, T., Shibata, K., Hase, E., and Nihei, T. (1953). Correlation between photosynthesis and light-independent metabolism in the growth of *Chlorella. Biochim. et Biophys. Acta* **12,** 23–40.

Tischer, J. (1936). Über das Euglenarhodon und andere Carotenoide einer roten Euglene. *Z. physiol. Chem. Hoppe-Seyler's* **239,** 257–269.

Tischer, J. (1938). Über die Polyenpigmente der Blaualge *Aphanizomenon flos-aquae*. *Z. physiol. Chem. Hoppe-Seyler's* **251,** 109–128.

Tischer, J. (1939). Über die Polyenpigmente der Blaualge *Aphanizomenon flos-aquae*. *Z. physiol. Chem. Hoppe-Seyler's* **260,** 257–271.

Tischer, J. (1941). Über die Identität von Euglenarhoden mit Astacin. *Z. physiol. Chem. Hoppe-Seyler's* **267,** 281–284.

Tischer, J. (1944). Über die Carotenoide von *Haematococcus pluvalis*. *Z. physiol. Chem. Hoppe-Seyler's* **281,** 143–155.

Virgin, H. I. (1958). Studies on the formation of protochlorophyll and chlorophyll *a* under varying light treatments. *Physiol. Plantarum* **11,** 347–362.

Virgin, H. I. (1960). Pigment transformations in leaves of wheat after irradiation. *Physiol. Plantarum* **13,** 155–164.

Wassink, E. C., Katz, E., and Dorrestein, D. (1939). Infra-red absorption spectra of purple bacteria. *Enzymologia* **7,** 113–129.

Withrow, R. B., Wolff, J. B., and Price, L. (1956). Elimination of the lag phase of chlorophyll synthesis in dark-grown bean leaves by a pretreatment with low irradiances of monochromatic energy. *Plant Physiol.* **31,** xiii.

Wittenberg, J., and Shemin, D. (1949). The utilization of glycine for the biosynthesis of both types of pyrroles in protoporphyrin. *J. Biol. Chem.* **178,** 47–51.

Wolff, J. B., and Price, L. (1957). Terminal steps of chlorophyll biosynthesis in higher plants. *Arch. Biochem. Biophys.* **72,** 293–301.

Wolff, J. B., Price, L., and Withrow, R. B. (1957). Stimulation of protochlorophyll synthesis in dark-grown bean leaves by irradiation with low energy. *Plant Physiol.* **32,** ix.

Wolken, J. J. (1956). Molecular morphology of *Euglena gracilis* var. *bacillaris*. *J. Protozool.* **3,** 211–221.

Wolken, J. J. (1958). The chloroplast structure, pigments, and pigment-protein complex. *Brookhaven Symposia in Biol. No.* **11,** 87–100.

Wolken, J. J., and Mellon, A. D. (1956). The relationships between chlorophyll and the carotenoids in the algal flagellate, *Euglena*. *J. Gen. Physiol.* **39,** 675–685.

Wolken, J. J., Mellon, A. D., and Greenblatt, C. L. (1955). Environmental factors affecting growth and chlorophyll synthesis in *Euglena*. I. Physical and chemical. II. The effectiveness spectrum for chlorophyll synthesis. *J. Protozool.* **2,** 89–96.

—24—

Carotenoids

T. O. M. NAKAYAMA

Department of Food Science and Technology, University of California, Davis, California

I. Introduction

Carotenoids are yellow, orange, or red pigments of aliphatic or alicyclic structure composed of isoprene units, usually eight, linked so that the methyl groups nearest the center of the molecule are in the 1,6-position while all other lateral methyl groups are in the 1,5-position. The series of conjugated double bonds constitutes the chromophoric system of the carotenoid (Karrer and Jucker, 1950). In general, natural carotenoids may be divided into two classes: (i) oxygen-free hydrocarbons, the carotenes; and (ii) their oxygenated derivatives, the xanthophylls (Berzelius, 1837). The numbering system for the carbon atoms of carotenoids is illustrated for α-carotene in Fig. 1.

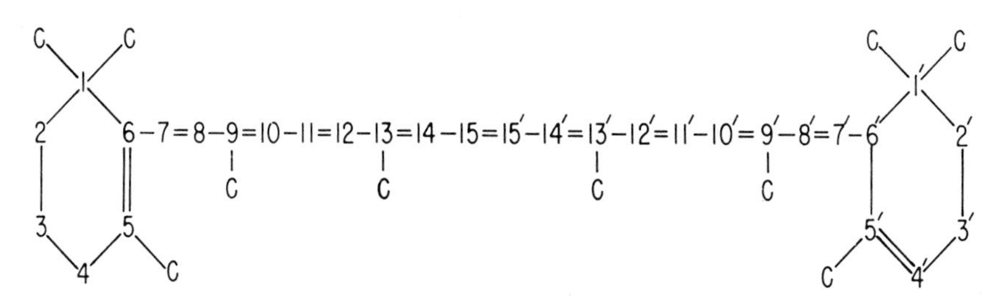

FIG. 1. The numbering system of carotenoids illustrated for α-carotene according to the American Chemical Society Committee on Nomenclature (Anonymous, 1946).

II. Methodology

The examination of carotenoids *in vitro* generally involves an extraction of the source material by acetone, methanol, or some other organic solvent. Mixtures of pigments have been traditionally separated by adsorption chromatography (Tswett, 1906, 1911). Later refinements employ reverse-phase column chromatography using polyethylene and methanol (Blass *et al.*, 1959), partition chromatography on silica gel (Purcell, 1958), liquid-liquid partition (Curl, 1953), and paper chromatography (Jensen and Liaaen Jensen, 1959). The choice of method depends on the pigments involved. The characterization of carotenoids which occur in small amounts generally involves spectroscopy, determination of partition coefficients, chromatographic behavior, etc. Pigments isolated and crystallized in sufficient amounts should also be analyzed for unsaturation, functional groups, degradation products, etc. (See Goodwin, 1955; Strain, 1945; Karrer and Jucker, 1950; Zechmeister, 1950; Zechmeister and Cholnoky, 1943.)

III. Algal Carotenoids

The distribution of carotenoids in the various classes of algae has been reviewed by Heilbron (1942, 1946), Cook (1945), Strain (1951, 1958), Strain *et al.* (1944), and Goodwin (1960a, b); it is summarized in Table I.

In general, the principal carotene is β-carotene, although exceptions occur in the Siphonales (Strain, 1951), in *Cryptomonas ovata* (Haxo and Fork, 1959), and in *Phycodrys sinuosa* (Larsen and Haug, 1956), where α-carotene predominates. ε-Carotene occurs only as a minor component. It is in the xanthophylls that the greatest differences are found. The Green Algae closely resemble the higher plants in this respect, notably in the predominance of lutein, with lesser amounts of violaxanthin, neoxanthin, and zeaxanthin. Astaxanthin esters, found in certain encysted algae, are extraplastidic (Goodwin and Jamikorn, 1954a).

The carotenoids of the Blue-green Algae are found, together with other photosynthetic pigments, in the peripheral region of the protoplast (Fogg, 1950) and occur in a structurally undefined photosynthetic system termed the chromatoplasm (Thomas, 1960). The occurrence of echinenone in plants was believed to be restricted to the Cyanophyta until reported in strains of *Euglena gracilis* by Goodwin and Gross (1958) and Krinsky and Goldsmith (1960).

Fucoxanthin is the principal carotenoid in the Phaeophyta, Bacillariophyta, and Chrysophyta. The structure of this carotenoid is as yet uncertain although several possibilities have been proposed. Liaaen and Sorensen (1956) showed that fucoxanthin contains an isolated carbonyl group and a second carbonyl in conjugation with an allene group, which would account

TABLE I

PRINCIPAL CAROTENOIDS OF ALGAE

Division	Carotenes	Xanthophylls	Reference
CHLOROPHYTA			
	α-Carotene		Strain (1951)
	β-Carotene		Carter *et al.* (1939)
	γ-Carotene		Karrer *et al.* (1943)
		Lutein	Carter *et al.* (1939)
		Violaxanthin	Strain (1951)
		Neoxanthin	Strain (1951)
		Astaxanthin (euglenarhodone)	Tischer (1941, 1944)
EUGLENOPHYTA			
	β-Carotene		Strain (1951)
		Lutein[a]	Goodwin and Jamikorn (1954b)
		Neoxanthin	Goodwin and Jamikorn (1954b)
		Astaxanthin (euglenarhodone)	Tischer (1936)
BACILLARIOPHYTA			
	β-Carotene	Fucoxanthin	Strain *et al.* (1944)
	ε-Carotene	Diatoxanthin	Strain *et al.* (1944)
		Diadinoxanthin	Strain *et al.* (1944)
CHRYSOPHYTA			
	β-Carotene	Lutein?	Carter *et al.* (1939)
		Fucoxanthin	Dales *et al.* (1960)
		Diadinoxanthin	Dales *et al.* (1960)
XANTHOPHYTA			
	β-Carotene	?	Carter *et al.* (1939) Strain *et al.* (1944)
CRYPTOPHYTA			
	α-Carotene		Haxo and Fork (1959)
	ε-Carotene		
		Zeaxanthin?	
PYRROPHYTA			
	β-Carotene		Strain (1951)
		Peridinin (sulcatoxanthin)	Strain (1951)
		Diadinoxanthin	Strain (1951)
		Dinoxanthin	Strain (1951)
PHAEOPHYTA			
	β-Carotene		Strain *et al.* (1944)
		Lutein	Strain *et al.* (1944)
		Fucoxanthin	Strain *et al.* (1944)
		Diatoxanthin	Strain *et al.* (1944)
		Violaxanthin	Strain (1951)

[a] This was later shown to be a misidentification; the predominant xanthophyll in *E. gracilis* is antheraxanthin (Krinsky and Goldsmith, 1960).

TABLE I—Continued

PRINCIPAL CAROTENOIDS OF ALGAE

Division	Carotenes	Xanthophylls	Reference
CYANOPHYTA			
	β-Carotene		Heilbron (1946)
		Lutein	Heilbron (1946)
		Myxoxanthin	Cook (1945)
		(aphanin, echinenone)	Goodwin and Taha (1950)
		Myxoxanthophyll	Karrer and Rutschmann (1944)
		Aphanizophyll	Tischer (1938, 1939)
		Oscilloxanthin	Karrer and Rutschmann (1944)
		Zeaxanthin	Karrer and Rutschmann (1944)
		Aphan˙cin	Tischer (1938)
		Flavacin	Tischer (1938, 1939)
RHODOPHYTA			
	α-Carotene		Strain (1951)
	β-Carotene		Carter *et al.* (1939)
		Lutein	Carter *et al.* (1939)
		Taraxanthin	Carter *et al.* (1939)

for its extreme sensitivity to alkali. This was demonstrated by isolating five main products after treatment of purified fucoxanthin with 0.08% K_2CO_3 in methanol at room temperature for periods of from 12 minutes to 190 hours. Infrared spectra showed that the allenic group was still present, but that the conjugated carbonyl absorption had disappeared. The discovery of the allene group (Torto and Weedon, 1955) represents a major contribution in the determination of the structure of this interesting carotenoid, one of the last major carotenoids still undefined. However, the structures of many of the other xanthophylls, such as peridinin, have yet to be worked out. The oxygens are present in hydroxyl, epoxy, and carbonyl groups; in algal carotenoids methoxyl and carboxyl groups have not yet been found, though the latter have been implicated (Karrer and Rutschmann, 1944).

In some algae the gametes differ from each other in their carotenoid content. Male gametes of *Ulva lactuca*, for example, contain twice as much carotene as do female gametes (Haxo and Clendenning, 1953). Even more remarkable is the distinction in the case of fucoids—*Fucus serratus*, *F.*

spiralis, *F. vesiculosus*, and *Ascophyllum nodosum*—in which the predominant pigment of the antherozoids is β-carotene, while that of the egg cells is fucoxanthin, associated in the latter case with chlorophyll (Carter *et al.*, 1948). The functional significance of such differences, which also occur in certain fungi (e.g., *Allomyces*) is unknown.

IV. Biosynthesis

Though many important natural carotenoids can now be synthesized in the laboratory (see reviews by Isler and Zeller, 1957; Isler and Montavon, 1958; Isler *et al.*, 1958), progress in the elucidation of the biogenesis of carotenoids has been slower. This problem has centered around two general questions. (1) How is the molecule assembled? (2) How are the different carotenoids elaborated? The synthetic pathway of the skeleton is presumably identical in all organisms which make carotenoids *de novo*. A singularly definitive work was reported by Grob (1959), who found that all 40 carbon atoms of β-carotene originated from either the methyl or carboxyl carbons of acetate, the sole carbon source supplied in this case for *Mucor hiemalis*. Grob's conclusion is not inconsistent with current concepts involving hydroxymethyl glutaric, mevalonic, and dimethyl acrylic acids as intermediates (Anderson *et al.*, 1960; Yokoyama *et al.*, 1960; Goodwin, 1959a,b).

A formal scheme of biosynthesis showing the interrelationships of the various tomato carotenoids and their more saturated isologs was proposed by Porter and Lincoln (1950). This hypothesis assumes that two C_{20} compounds condense to a C_{40} isoprenoid hydrocarbon, tetrahydrophytoene ($C_{40}H_{76}$), which by a series of quadri-dehydrogenations in turn gives rise successively to phytoene, phytofluene, ζ-carotene, tetrahydrolycopene (neurosporene), and lycopene. γ-Carotene is formed by cyclization of one terminal end of lycopene, and β-carotene by further cyclization of the other end. Xanthophylls arise by oxidation of carotenes.

Rabourn (1957) proposed a general scheme of carotene biosynthesis which is more in accord with the structures of phytoene ($C_{40}H_{64}$) (Rabourn and Quackenbush, 1956) and phytofluene ($C_{40}H_{62}$) (Zechmeister, 1958). In Rabourn's scheme, ζ-carotene ($C_{40}H_{60}$) is the pivotal carotene from which lycopene, γ-carotene and β-carotene can arise by independent routes, the first two not being obligatory precursors of β-carotene.

Claes (1954, 1956, 1957, 1958, 1959) presented evidence, from work with x-ray mutants of *Chlorella vulgaris*, that the immediate precursors of the carotenes are their more saturated isologs. Strain 5/871 synthesizes only phytoene, while strain 5/515 produces the hydrocarbons phytoene, phytofluene, and ζ-carotene. In both strains, there is a complete block of xanthophyll synthesis. In strain 9a, xanthophylls are present, but α- and β-carotenes are almost or completely absent. Another strain, 5/520, is especially interesting because, when grown heterotrophically with glucose in darkness, it forms almost no chlorophyll and only very small amounts of xanthophylls, together with phytoene, phytofluene, ζ-carotene, protetrahydrolycopene (poly-*cis*-neurosporene), and prolycopene (poly-*cis*-lycopene). However, under anaerobic illumination of dark-grown cultures, α-carotene, β-carotene, and "carotene X" are formed, at the expense of the aforementioned compounds.

Thus light appears to replace a biosynthetic step occurring in the wild type in darkness. A subsequent period of aerobic treatment in the dark promotes the synthesis of xanthophylls with a concomitant, almost quantitative, reduction in the amount of carotenes. Aside from the appearance of carotene X where one would expect γ-carotene, this series of transformations resembles the general features of the Porter-Lincoln hypothesis. Elucidation of the structure of carotene X may help to define its role. Similarly it remains to be determined whether the various neurosporenes isolated from different sources are identical, and, if so, whether they correspond to 5,6,5′,6′-tetrahydrolycopene, as indicated by Eugster *et al.* (1956), or to 7,8-dihydrolycopene.

The interrelationships of three typical xanthophylls of Green Algae—lutein, violaxanthin, and neoxanthin—are not yet clear. Work on *Chlorella pyrenoidosa* and *Scenedesmus obliquus* by Blass *et al.* (1959), using C^{14} as a tracer, indicated that the carotenes were labeled earlier than xanthophylls. The observation that lutein and violaxanthin became equally labeled in such experiments is in agreement with a postulate of Sapozhnikov *et al.* (1959) that lutein can be formed from violaxanthin by the action of light, and that the reverse reaction can be induced by oxygen in the dark.

The occurrence of zeaxanthin in *Fucus vesiculosus* was shown to be a *post mortem* change. Liaaen and Sorensen (1956) showed that fresh tissue contains virtually none, whereas portions of thallus kept under anaerobic conditions in sterile sea water showed a production of zeaxanthin and a corresponding reduction of violaxanthin. The amounts of β-carotene and fucoxanthin remained unchanged.

The origin of astaxanthin in *Haematococcus pluvialis* and in certain *Euglena* species (e.g., *E. sanguinea*) has not yet been established. This pigment, which is characteristic of Crustacea, occurs esterified in *Haematococcus* (Goodwin and Jamikorn, 1954a). Gilchrist and Green (1960) showed that in the brine shrimp, *Artemia salina*, astaxanthin esters arose from the carotenoids (which did not include astaxanthin) of the algae ingested, *Dunaliella tertiolecta* and *Phaeodactylum tricornutum*, probably via a keto carotenoid. It would be interesting to compare the course of formation of astaxanthin in this crustacean with that in algae.

V. Functions

The occurrence of carotenoids as integral parts of the photosynthetic apparatus has led to speculations on their role in photosynthesis. (For other possible functions, see Goodwin, 1959a). It has been suggested that they are involved in the transfer of energy to chlorophyll (Duysens, 1956; Blinks, 1954), in energy mediation and electron transport (Calvin, 1958; Platt, 1959), in coenzyme functions (Warburg and Krippahl, 1954), in O_2 transport (Dorough and Calvin, 1951; Cholnoky *et al.*, 1956), and in protection from photo-oxidation (Noack, 1925; Griffiths *et al.*, 1955).

The utilization of light absorbed by carotenoids in photosynthesis has been demonstrated in Green Algae (Haxo and Blinks, 1950), in diatoms (Dutton and Manning, 1941; Wassink and Kersten, 1946; Tanada 1951), in Brown Algae (Haxo and Blinks, 1950), and in dinoflagellates (Haxo, 1960). A system consisting of micelles of chlorophyll and fucoxanthin *in vitro* transferred energy absorbed by fucoxanthin almost quantitatively to chlorophyll (Teale, 1958).

A protective function was clearly defined by Stanier and his associates in mutants of *Rhodopseudomonas spheroides* (Griffiths *et al.*, 1955) and in diphenylamine-treated cells of *Rhodospirillum rubrum* (Cohen-Bazire and Stanier, 1958). In these bacteria, the absorption of light by chlorophyll under aerobic conditions was lethal in the absence of carotenoids. The *Chlorella* mutants of Claes reacted similarly; only strain 5/520, which was able to convert precursors to normal carotenoids, could survive in the light. Work by Sager and Zalokar (1958) on a mutant of *Chlamydomonas reinhardii* with a greatly reduced carotenoid content showed that, although the organism could photosynthesize, it could not survive prolonged illumination. These experiments would seem to preclude an obligatory role for carotenoids in photosynthesis. The problem of formulating a mechanism for the "normal" role of carotenoids, the absence of which would result in photo-oxidative death of mutants, has been discussed by Calvin (1955).

Experiments *in vitro* by earlier workers (Noack, 1925; Aronoff and Mackinney, 1943; Pepkowitz, 1943) showed a protective action of carotenoids upon chlorophyll in illuminated solutions. Claes and Nakayama (1959b) found that aerobic destruction of illuminated chlorophyll is alleviated only when the conjugated chain length of the carotenoid molecule is at least equivalent to that of neurosporene, with nine double bonds.

The photosensitized oxidation of unsaturated compounds by molecular oxygen in the presence of dyes, including chlorophyll, was studied by Schenck and his collaborators. Schenck (1953) formulated a mechanism whereby an excited chlorophyll molecule combines reversibly with an oxygen molecule to form a reactive peroxide, which can then transfer the oxygen to another molecule. The protective function of a carotenoid *in vivo* might result from its acceptance of the otherwise destructive oxygen molecule, assuming of course that necessary mechanisms exist for the reduction of the carotenoid back to its original state. However, in studies by Sironval and Kandler (1958), Kandler and Sironval (1959), and Kandler and Schötz (1956) on the kinetics of photo-oxidation in *Chlorella pyrenoidosa* and *C. vulgaris*, the bleaching of the pigments was secondary to inhibitory effects on photosynthesis and on oxidative phosphorylation.

The interaction of excited chlorophyll molecules with poly-*cis*-carotenes in the *Chlorella* mutant strain 5/520 of Claes was shown to result in isomerization of the latter to the *trans* forms (Claes and Nakayama, 1959a). Fujimori and Livingston (1957) showed carotenoids to be efficient quenchers of the triplet state of chlorophyll, while experiments by Arnold and Maclay (1958) demonstrated that dried films of chlorophyll and carotene may exhibit characteristics of semiconductors.

The sensitive spectrophotometric methods of Chance and associates enabled direct studies to be made on carotenoids *in situ*. Examination of strains of *Chlamydomonas* and *Chlorella* revealed transitory changes at 518 mμ in response to light and/or oxygen (Chance, 1958; Sager, 1958). Although a discussion of the possible roles of carotenoids in energy absorption and transfer is beyond the intent of this review, it is expected that an elucidation of the processes related to energy conversions and oxygen evolution will lead to an understanding of the role of carotenoids in tissues containing chlorophyll.

The present evidence does not enable us to decide whether the protective effect of the carotenoids results solely from their being preferred acceptors of oxygen in the presence of chlorophyll and light, or whether they remove energy from the excited chlorophyll by a physiologically advantageous pathway. A third possibility is that carotenoids function in a less direct manner, involving some other cell constituent as yet unrecognized.

The carotenoids of algae include a variety of both common and unusual types. Further work will undoubtedly provide new information on the role of carotenoids in plants, the interrelationships of the xanthophylls, their functions as provitamin A, and their distribution in nature.

References

Anderson, D. G., Nogard, D. W., and Porter, J. W. (1960). The incorporation of mevalonic acid -2-C^{14} and dimethyl acrylic acid -3-C^{14} into carotenes. *Arch. Biochem. Biophys.* **88,** 68–77.

Anonymous. (1946). The nomenclature of the carotenoid pigments. *Chem. Eng. News* **24,** 1235–1236.

Arnold, W., and Maclay, H. K. (1958). Chloroplasts and chloroplast pigments as semiconductors. *Brookhaven Symposia in Biol. No.* **11,** 1–9.

Aronoff, S., and Mackinney, G. (1943). The photo-oxidation of chlorophyll. *J. Am. Chem. Soc.* **65,** 956–958.

Berzilius, J. J. (1837). Über die gelbe Farbe der Blätter im Herbst. *Ann. Chem. (Liebigs)* **21,** 257–262.

Blass, U., Anderson, J. M., and Calvin, M. (1959). Biosynthesis and possible functional relationships among the carotenoids; and between chlorophyll A and chlorophyll B. *Plant Physiol.* **34,** 329–333.

Blinks, L. R. (1954). The photosynthetic function of pigments other than chlorophyll. *Ann. Rev. Plant Physiol.* **5,** 93–114.

Calvin, M. (1955). Function of carotenoids in photosynthesis. *Nature* **176,** 1215.

Calvin, M. (1958). From microstructure to macrostructure and function in the photochemical apparatus. *Brookhaven Symposia in Biol. No.* **11,** 160–180.

Carter, P. W., Heilbron, I. M., and Lythgoe, B. (1939). The lipochromes and sterols of the algal classes. *Proc. Roy. Soc.* *B***128,** 82–109.

Carter, P. W., Cross, L. C., Heilbron, I. M., and Jones, E. R. (1948). The lipochromes of the male and female gametes of some species of the Fucaceae. *Biochem. J.* **43,** 349–352.

Chance, B. (1958). Oxygen-linked absorbency changes in photosynthetic cells. *Brookhaven Symposia in Biol. No.* **11,** 74–86.

Cholnoky, L., Gyorgyfy, C., Nagy, E., and Panczel, M. (1956). Function of carotenoids in chlorophyll-containing organs. *Nature* **178,** 410–411.

Claes, H. (1954). Analyse der biochemichen Synthesekette für Carotenoide mit Hilfe von *Chlorella*-Mutanten. *Z. Naturforsch.* **9b,** 461–470.

Claes, H. (1956). Biosynthese von Carotenoiden bei *Chlorella.* II. Tetrahydrolycopin und Lycopin. *Z. Naturforsch.* **11b,** 260–266.

Claes, H. (1957). Biosynthese von Carotenoiden bei *Chlorella.* III. Untersuchungen bei die lichtabhängige Synthese von α und β Carotin und Xanthophyllen bei der *Chlorella*-Mutante 5/520. *Z. Naturforsch.* **12b,** 401–407.

Claes, H. (1958). Biosynthese von Carotenoiden bei *Chlorella.* IV. Die Carotinsynthese einer *Chlorella*-Mutante bei anaerober Belichtung. *Z. Naturforsch.* **13b,** 222–224.

Claes, H. (1959). Biosynthese von Carotenoiden bei *Chlorella.* V. Die Trennung von Licht- und Dunkelreaktionen bei der lichtabhängigen Xanthophyllsynthese von *Chlorella. Z. Naturforsch.* **14b,** 4–7.

Claes, H., and Nakayama, T. O. M. (1959a). Isomerization of poly *cis* carotenes by chlorophyll *in vivo* and *in vitro*. *Nature* **183,** 1053.

Claes, H., and Nakayama, T. O. M. (1959b). Das photoxydative Ausbleichen von Chlorophyll *in vitro* in Gegenwart von Carotinen mit verschiedenen chromophoren Gruppen. *Z. Naturforsch.* **14b,** 746–747.

Cohen-Bazire, G., and Stanier, R. Y. (1958). Inhibition of carotenoid synthesis in photosynthetic bacteria. *Nature* **181,** 250–252.

Cook, A. H. (1945). Algal pigments and their significance. *Biol. Rev. Cambridge Phil. Soc.* **20,** 115–132.

Curl, A. L. (1953). Application of countercurrent distribution to Valencia orange juice carotenoids. *J. Agr. Food Chem.* **1,** 456–460.

Dales, R. P. (1960). On the pigments of the Chrysophyceae. *J. Marine Biol. Assoc. United Kingdom* **39,** 693–699.

Dorough, G. D., and Calvin, M. (1951). The path of oxygen in photosynthesis. *J. Am. Chem. Soc.* **73,** 2362–2365.

Dutton, H. J., and Manning, W. M. (1941). Evidence for carotenoid-sensitized photosynthesis in the diatom *Nitzschia closterium*. *Am. J. Botany* **28,** 516–526.

Duysens, L. N. M. (1956). Energy transformations in photosynthesis. *Ann. Rev. Plant Physiol.* **7,** 25–50.

Eugster, C. H., Linner, E., Trivedi, A. H., and Karrer, P. (1956). Synthese eines 6,7,6′,7′ Tetrahydrolycopins und dessen Beziehung zum Neurosporin. *Helv. Chim. Acta* **39,** 690–698.

Fogg, G. E. (1950). The comparative physiology and biochemistry of Blue-Green algae. *Bacteriol. Revs.* **20,** 148–165.

Fujimori, E., and Livingston, R. (1957). Interaction of chlorophyll in its triplet state with oxygen, carotene, etc. *Nature* **180,** 1036–1038.

Gilchrist, B. M., and Green, J. (1960). The pigments of *Artemia*. *Proc. Roy. Soc.* *B***152,** 118–136.

Goodwin, T. W. (1955). Carotenoids. *In* "Modern Methods of Plant Analysis" (K. Paech and M. V. Tracey, eds.), Vol. III, pp. 272–311. Springer, Berlin.

Goodwin, T. W. (1957). Carotenoids as photoreceptors in plants. Atti 2nd Congr. intern. Fotobiologia. pp. 361–369. Turin, Italy, June, 1957.

Goodwin, T. W. (1959a). The biosynthesis and function of the carotenoid pigments. *Advances in Enzymol.* **21,** 295–368.

Goodwin, T. W. (1959b). A comparison of the incorporation of labelled CO_2, acetate and mevalonate into carotenoids in a number of carotenogenic systems. *In* CIBA Foundation Symposium on "Biosynthesis of Terpenes and Sterols" (G. E. W. Wolstenholme and C. M. O'Connor, eds.), pp. 279–294. Little, Brown, Boston.

Goodwin, T. W. (1960a). Chemistry, biogenesis and physiology of the carotenoids. *In* "Handbuch der Pflanzenphysiologie" (W. Ruhland, ed.), Vol. V, Part 1, pp. 394–436. Springer, Berlin.

Goodwin, T. W. (1960b). Algal carotenoids. *In* "Comparative Biochemistry of Photoreactive Systems" (M. B. Allen, ed.), pp. 1–10. Academic Press, New York.

Goodwin, T. W., and Jamikorn, M. (1954a). Studies in carotenogenesis. II. Carotenoid synthesis in the alga *Haematococcus pluvialis*. *Biochem. J.* **57,** 376–380.

Goodwin, T. W., and Jamikorn, M. (1954b). Studies in carotenogenesis. Some observations on carotenoid synthesis in two varieties of *Euglena gracilis*. *J. Protozool.* **1,** 216–219.

Goodwin, T. W., and Taha, M. M. (1950). A study of the carotenoids echinenone and myxoxanthin with special reference to their probable identity. *Biochem. J.* **48,** 513–514.

Griffiths, M., Sistrom, W. R., Cohen-Bazire, G., and Stanier, R. Y. (1955). Function of carotenoids in photosynthesis. *Nature* **176,** 1211–1215.

Grob, E. C. (1959). Biosynthesis of carotenoids by micro-organisms. *In* "CIBA Foundation Symposium on Biosynthesis of Terpenes and Sterols" (G. E. W. Wolstenholme and C. M. O'Connor, eds.), pp. 267–278. Little, Brown, Boston.

Haxo, F. T. (1960). Wavelength dependence of photosynthesis. *In* "Comparative Biochemistry of Photoreactive Systems" (M. B. Allen, ed.), pp. 329–360. Academic Press, New York.

Haxo, F. T., and Blinks, L. R. (1950). Photosynthetic action spectra of marine algae. *J. Gen. Physiol.* **33,** 389–422.

Haxo, F. T., and Clendenning, K. A. (1953). Photosynthesis and phototaxis in *Ulva lactuca* gametes. *Biol. Bull.* **105,** 103–114.

Haxo, F. T., and Fork, D. C. (1959). Photosynthetically active accessory pigments of cryptomonads. *Nature* **184,** 1051–1052.

Heilbron, I. M. (1942). Some aspects of algal chemistry. *J. Chem. Soc.* 79–89; *Nature* **149,** 398–400.

Heilbron, I. M. (1946). Twenty years and onward. *Chem. Eng. News* **24,** 1035–1039.

Isler, O., and Montavon, M. (1958). Synthèses dans le domaine des caroténoides. *Chimia (Switz.)* **12,** 1–42.

Isler, O., and Zeller, P. (1957). Total synthesis of carotenoids. *Vitamins and Hormones* **15,** 31–71.

Isler, O., Ofner, A., and Siemers, G. F. (1958). Industrial syntheses of carotenoids as useful food colors. *Food Technol.* **12,** 1–9.

Jensen, A., and Liaaen Jensen, S. (1959). Quantitative paper chromatography of carotenoids. *Acta Chem. Scand.* **13,** 1863–1868.

Kandler, O., and Schötz, F. (1956). Untersuchungen über die photooxydative Farbstoffzerstörung und Stoffwechselhemmung bei Chlorellamutanten und panaschierten Oenotheren. *Z. Naturforsch.* **11b,** 708–718.

Kandler, O., and Sironval, C. (1959). Photoxidation processes in normal green *Chlorella* cells. II. Effects on metabolism. *Biochim. et Biophys. Acta* **33,** 207–215.

Karrer, P., and Jucker, E. (1950). "Carotenoids." Elsevier, Amsterdam.

Karrer, P., and Rutschmann, J. (1944). Beitrag zur Kenntnis der Carotenoids aus *Oscillatoria rubescens. Helv. Chim. Acta.* **27,** 1691–1695.

Karrer, P., Fatzer, W., Favarger, M., and Jucker, E. (1943). Die Antheridienfarbstoffe von Chara-Arten (Armleuchtergewäsche). *Helv. Chim. Acta* **26,** 2121–2122.

Krinsky, N. I., and Goldsmith, T. H. (1960). Carotenoids of *Euglena gracilis* (Z strain). *Arch. Biochem. Biophys.* **91,** 271–279.

Larsen, B. and Haug, A. (1956). Carotene isomers in some red algae. *Acta Chem. Scand.,* **10,** 470–472.

Liaaen, S., and Sorensen, N. A. (1956). Postmortal changes in the carotenoids of *Fucus vesiculosus. In* "Second International Seaweed Symposium" (T. Braarud and N. A. Sorensen, eds.), pp. 25–32. Pergamon Press, London.

Noack, K. (1925). Photochemische Wirkungen des Chlorophylls und ihre Bedeutung für die Kohlensäureassimilation. *Z. Botan.* **17,** 481–548.

Pepkowitz, L. (1943). The stability of carotene in acetone and petroleum ether extracts of green vegetables. *J. Biol. Chem.* **149,** 465–471.

Platt, J. R. (1959). Carotene-donor-acceptor complexes in photosynthesis. *Science* **129,** 372–374.

Porter, J. W., and Lincoln, R. E. (1950). I. *Lycopersicon*: Selections containing a high content of carotenes and colorless polyenes. II. The mechanism of carotene biosynthesis. *Arch. Biochem.* **27,** 390–403.

Purcell, A. E. (1958). Partition separation of carotenoids by silica-methanol columns. *Anal. Chem.* **30,** 1049–1051.

Rabourn, W. J. (1957). "A Unified Concept of Carotene Phytosynthesis." Abstract of papers presented at the 132nd meeting of the American Chemical Society, New York, p. 88c.

Rabourn, W. J., and Quackenbush, F. W. (1956). The structure of phytoene. *Arch. Biochem.* **61,** 111–118.

Sager, R. (1958). The architecture of the chloroplast in relation to its photosynthetic activities. *Brookhaven Symposia in Biol.* **11,** 101–117.

Sager, R., and Zalokar, M. (1958). Pigments and photosynthesis in a carotenoid-deficient mutant of *Chlamydomonas. Nature* **182,** 98–100.

Sapozhnikov, D. I., Maevskaya, A. N., Krasovskaya-Antropova, T. A., Prialgauskaĭte, L. L., and Turchina, V. S. (1959). The effect of anaerobic conditions on change in the relations of the main carotenoids in green leaves. *Biokhēmiya* (English translation) **24,** 39–41.

Schenk, G. O. (1953). Zur Reaktion photochemisch angeregter Molekeln mit O_2. *Z. Naturw.* **41,** 452–453.

Sironval, C., and Kandler, O. (1958). Photoxidation processes in normal green *Chlorella* cells. I. The bleaching process. *Biochim. et Biophys. Acta* **29,** 359–368.

Strain, H. H. (1945). "Chromatographic Adsorption Analysis." Interscience, New York.

Strain, H. H. (1949). Functions and properties of the chloroplast pigments. *In* "Photosynthesis in Plants" (J. Franck and W. E. Loomis, eds.), Iowa State College Press, Ames, Iowa, pp. 133–178.

Strain, H. H. (1951). Pigments. *In* "Manual of Phycology" (G. M. Smith, ed.) Chronica Botanica, Waltham, Massachusetts, pp. 243–262.

Strain, H. H. (1958). "Chloroplast Pigments and Chromatographic Analysis." Pennsylvania State Univ. Press, University Park, Pennsylvania.

Strain, H. H., Manning, W. M., and Hardin, G. J. (1944). Xanthophylls and carotenes of diatoms, brown algae, dinoflagellates, and sea-anemones. *Biol. Bull.* **86,** 169–191.

Tanada, T. (1951). The photosynthetic efficiency of carotenoid pigments in *Navicula minima. Am. J. Botany* **38,** 276–283.

Teale, F. W. J. (1958). Carotenoid-sensitized fluorescence of chlorophyll *in vitro. Nature* **181,** 415–416.

Thomas, J. B. (1960). Chloroplast structure. *In* "Handbuch der Pflanzenphysiologie" (W. Ruhland, ed.), Vol. V, Part 1, pp. 511–565. Springer, Berlin.

Tischer, J. (1936). Über das Euglenarhoden und andere Carotenoide einer roten Euglene. *Z. physiol. Chem., Hoppe-Seyler's* **239,** 257–269.

Tischer, J. (1938). Über die Polyenpigmente der Blaualge *Aphanizomenon flos-aquae. Z. physiol. Chem., Hoppe-Seyler's* **251,** 109–128.

Tischer, J. (1939). Über die Polyenpigmente der Blaualge *Aphanizomenon flos-aquae. Z. physiol. Chem., Hoppe-Seyler's* **260,** 257–271.

Tischer, J. (1941). Über die Identität von Euglenarhodon mit Astacin. *Z. physiol. Chem., Hoppe-Seyler's* **267,** 281–284.

Tischer, J. (1944). Über die Carotenoide von *Haematococcus pluvialis. Z. physiol. Chem., Hoppe-Seyler's* **281,** 143–155.

Torto, F. G., and Weedon, B. C. L. (1955). Spectral properties and oxidation of fucoxanthin. *Chem. and Ind.* 1219–1220.

Tswett, M. (1906). Adsorptionsanalyse und chromatographische methods. *Ber. deutsch. botan. Ges.* **24,** 384–393.

Tswett, M. (1911). Üeber den makro- und microchemischen Nachweis des Carotins. *Ber. deutsch. botan. Ges.* **29,** 630–636.

Warburg, O., and Krippahl, G. (1954). Über Photosynthese-Fermente. *Angew. Chem.* **66,** 493–496.

Wassink, E. C., and Kersten, J. A. H. (1946). Observations sur le spectre d'absorption et sur le rôle des caroténoides dans la photosynthèse des diatomées. *Enzymologia* **12,** 3–32.

Yokoyama, H., Chichester, C. O., and Mackinney, G. (1960). Formation of carotene *in vitro. Nature* **185,** 687–688.

Zechmeister, L. (1950). "Progress in Chromatography 1938–1947." Chapman and Hall, London.

Zechmeister, L. (1958). Some *in vitro* conversions of naturally occurring carotenoids. *Fortschr. Chem. org. Naturstoffe* **15,** 31–82.

Zechmeister, L., and Cholnoky, L. (1943). "Principles and Practice of Chromatography." Wiley, New York.

—25—

Phycobilins

C. Ó hEOCHA

Biochemistry Department, University College, Galway, Ireland

I. Introduction

Photosynthetically active red and blue biliproteins, called phycoerythrins and phycocyanins, respectively, have been isolated only from algae. Their prosthetic groups or chromophores are tetrapyrroles known as phycobilins. Unlike the chlorophylls, phycobilins are not readily released from associated proteins; consequently it is the biliproteins rather than the free phycobilins which have been most generally studied. In this review, emphasis will be placed on recent work; extensive references to the earlier literature are given by Boresch (1932), Rabinowitch (1945; 1951; 1956) and Haxo and Ó hEocha (1960). The role of phycobilins as accessory pigments in photosynthesis has been discussed by Brody and Brody in Chapter 1 of this treatise.

II. Distribution and Formation of the Phycobilin-Proteins (Biliproteins)

Algal biliproteins, which are of general occurrence in the Rhodophyta, Cyanophyta, and Cryptophyta, have also been reported in one or two members of the Chlorophyta. Among the Rhodophyta *Rytiphlaea tinctoria*, is exceptional, in that it owes its red coloration to the nonproteinaceous floridorubin (Feldmann and Tixier, 1947). Table I lists only a small frac-

TABLE I

DISTRIBUTION OF THE ALGAL BILIPROTEINS

Algal group and species	Biliproteins[a]	Reference
RHODOPHYTA		
Class: Bangiophyceae		
Order: Porphyridiales		
Porphyridium cruentum	B-PE; R-PC; AlloPC	Haxo *et al.* (1955); Ó hEocha (1955)[b]
Order: Bangiales		
Porphyra tenera	R-PE; C-PC; AlloPC	Lemberg (1930); Lemberg and Bader (1933); Hattori and Fujita (1959b)[b]
Porphyra perforata	R-PE; R-PC; AlloPC	Haxo *et al.* (1955)[b]; Jones and Blinks (1957)[b]
Smithora naiadum	B-PE; C-PC[c]; AlloPC	Airth and Blinks (1956); French *et al.* (1956)[b]
Class: Florideophyceae		
Order: Nemalionales		
Rhodochorton rothii	R-PE; PC	Ó hEocha (1955,[b] 1958)
Rhodochorton floridulum	B-PE; PC	Ó hEocha and Ó Carra (1961)
Order: Gigartinales		
Plocamium pacificum	R-PE	Ó hEocha (1958)
Order: Cryptonemiales		
Grateloupia sp.	R-PE; C-PC; AlloPC	Hattori and Fujita (1959b)[b]
Order: Rhodymeniales		
Rhodymenia palmata	R-PE; R-PC; AlloPC	Raftery and Ó hEocha (unpublished); Ó hEocha (1960)[b]
Order: Ceramiales		
Ceramium rubrum	R-PE; R-PC; AlloPC	Svedberg and Katsurai (1929)[b]; Ó Carra and Ó hEocha (unpublished)
Polysiphonia urceolata	R-PE; C-PC; AlloPC	Hattori and Fujita (1959b)[b]
CYANOPHYTA		
Tolypothrix tenuis	C-PE; C-PC; Allo PC	Hattori and Fujita (1959a)[b]
Arthrospira maxima	C-PC; AlloPC	Ó hEocha (1958)[b]
Phormidium ectocarpi	C-PE	Kylin (1937); Ó hEocha (1955; 1960)[b]
CRYPTOPHYTA		
Hemiselmis virescens	PC[d]	Allen *et al.* (1959)[b]
Cryptomonas ovata	PE	Haxo and Fork (1959); Allen *et al.* (1959)[b]
Sennia sp.	PE; PC	Ó hEocha and Raftery (1959)
Cyanidium caldarium	C-PC	Allen (1959)[b]

tion of the algae whose biliproteins have been studied. A selection was made on the basis of (*a*) a study having been made of the isolated biliproteins of the alga, and (*b*) an identified biliprotein being of interest in relation to the classification of the alga.

Biliproteins are located in the lamellae of chloroplasts of Red Algae (Brody and Vatter, 1959). Electron micrograph studies indicate that Blue-green Algae also contain chloroplast-like structures (Elbers *et al.*, 1957) and it seems likely that the biliproteins are located in their lamellae rather than in free chromatophores or in the cytoplasm (Thomas and de Rover, 1955).

The biliprotein content of marine littoral Red Algae varies with season. It accounted for 1.9% (dry weight) of *Ceramium rubrum* in December to February, but dropped to about half this value in March (Lemberg, 1928). On the other hand, this species contained relatively more phycocyanin in March than in December. The phycocyanin content is less in deep-growing algae than in related species from the intertidal or upper sublittoral zones (Kylin, 1937; Jones and Blinks, 1957).

High biliprotein concentrations occur in algae cultured in white light of low intensity (Myers and Kratz, 1955; Brody and Emerson, 1959). A 24% yield (dry weight) of C-phycocyanin was obtained from *Anacystis nidulans** grown in such light (Myers and Kratz, 1955). Light intensity also affects the relative proportions of the different biliproteins in some algae. Thus, Halldal (1958) reported that *Anabaena* sp., which synthesizes only phycocyanin at a high light intensity, forms phycoerythrin in addition at lower intensities. On the other hand, phycoerythrin is not formed by *Anacystis nidulans* grown under any light condition.

The quality of the light, too, influences the relative proportions of the biliproteins in algae. Hattori and Fujita (1959a) reported that the ratio of phycoerythrin to phycocyanin in *Tolypothrix tenuis* depended on whether the alga was grown in fluorescent or incandescent light. The dry-weight yield of phycoerythrin varied from 9.0% (fluorescent light) to 0.5% (incandescent light). In a chlorophyll-less mutant of *Cyanidium caldarium*,* phycocyanin was formed in light of either 450 or 600 mμ, but not in light of 550, 660, or 700 mμ (Nichols and Bogorad, 1960). Brody and Emerson (1959) found that at low light intensities the formation of phycoerythrin by

---

[a] PE = phycoerythrin; PC = phycocyanin. The prefixes are explained in Section IV.A. The biliprotein composition of some algae depends on light conditions during growth.

[b] Publication which includes absorption spectra of some or all of the pigments.

[c] The phycocyanin of *S. naiadum* var. *australis* differs spectrally from C-phycocyanin (Ó hEocha and Haxo, 1960).

[d] A spectrally different phycocyanin was obtained from *H. virescens*(?) from another location (Ó hEocha and Raftery, 1959). See page 427.

* Regarding the taxonomy of these organisms, see notes 2 and 11 of Appendix A in this treatise by Silva.

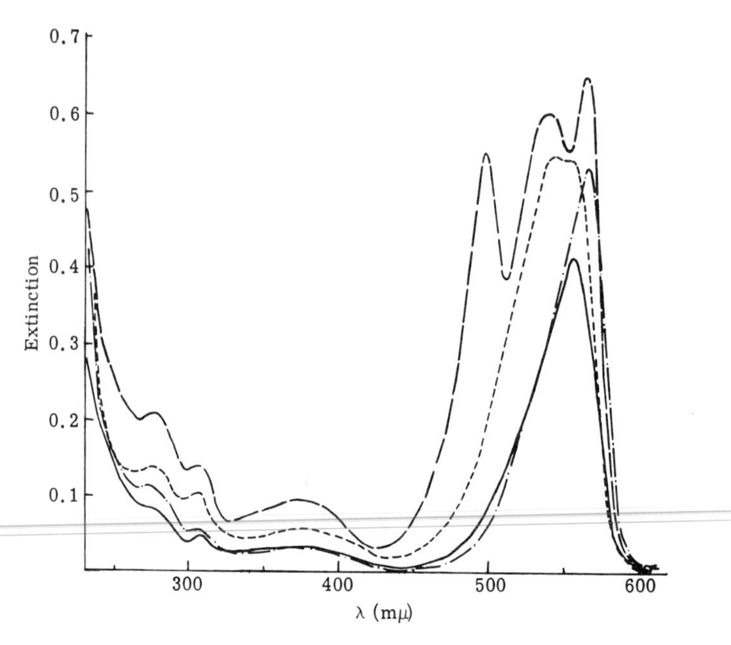

FIG. 1. Absorption spectra of aqueous solutions of phycoerythrins (pH 6–7): — — — —, R-phycoerythrin (*Ceramium rubrum*); - - - - - -, B-phycoerythrin (*Porphyridium cruentum*); –·–·–, C-phycoerythrin (*Phormidium persicinum*); ———, cryptomonad phycoerythrin (*Hemiselmis rufescens*).

Porphyridium cruentum was stimulated more by green light (546 mμ) than by blue (436 mμ), whereas at high intensities the reverse was true. Phycoerythrin from this alga displays a maximum light absorption at 540 to 560 mμ and a minimum at 430 to 440 mμ (Fig. 1). Complementary chromatic adaptation (that is, the enhanced formation of the pigment that most strongly absorbs the incident light) is effective only at low light intensity (Brody and Emerson, 1959). These results emphasize the importance of varying both the intensity and the color of the illumination used in studying the reactions leading to biliprotein synthesis.

As would be expected from the nitrogenous nature of biliproteins, culture media deficient in nitrogen limit their formation (Fogg, 1952; Hattori and Fujita, 1959c). Pre-illuminated N-deficient cultures of *Tolypothrix* formed biliproteins in the dark when nitrate was added to the medium. The ratio of the pigments was affected by the character of the light used during the pre-illumination period: green light favored the synthesis of phycoerythrin; red light, that of phycocyanin. In cultures grown heterotrophically in darkness, phycocyanin was still formed, but not phycoerythrin (Hattori and Fujita, 1959c; Fujita and Hattori, 1960a, b). Deficiencies of molybdenum and sodium reduce the phycocyanin but not the chlorophyll content of Blue-green Algae (Fogg, 1952; Allen and Arnon, 1955), whereas an iron deficiency reduces the concentration of both pigments (Boresch, 1921).

III. Isolation and Purification of the Biliproteins

When their cellular structures are broken, algae containing biliproteins release these brilliantly colored proteins into distilled water or buffer solutions. Among the mechanical methods which have been used to effect the breaking of the cells are maceration in a blender, grinding, sonic oscillation, and repeated freezing and thawing. For the purification of pigments, the aqueous extract is centrifuged to remove cell fragments. These operations should be performed in the cold and dark.

Fractionation of the biliproteins and the removal of other water-soluble algal constituents were originally achieved by fractional precipitation, followed by crystallization from ammonium sulfate solution; but this is a tedious process, particularly if the protein to be isolated is present in low concentration. Swingle and Tiselius (1951) showed that phycoerythrins and phycocyanins could be separated by adsorption chromatography on columns of tricalcium phosphate gel, and biliproteins from many sources have since been purified by this method (Krasnovsky *et al.*, 1952; Haxo *et al.*, 1955; Jones and Blinks, 1957; Ó hEocha and Haxo, 1960). Phosphate buffers of increasing concentration elute the proteins, phycoerythrin usually preceding phycocyanins. Direct adsorption on tricalcium phosphate may be used in a batch process for large-scale preparations of biliproteins (Tiselius, 1954). Hattori and Fujita (1959a, b) combined the direct adsorption method with fractional precipitation to obtain crystalline preparations of a number of biliproteins.

Photomicrographs of biliprotein crystals have been published by several workers (Lemberg, 1930; Fujiwara, 1955; Hattori and Fujita, 1959a). The hydrogen-ion concentration at which precipitation occurs affects the shape of biliprotein crystals (Bouillene-Walrand and Delarge, 1937).

IV. Physical Properties of the Biliproteins

A. Absorption Spectra

The absorption spectra of aqueous biliprotein solutions correspond very closely to their spectra in intact algae (cf. Haxo and Blinks, 1950), although extraction may lead to a slight shift of the absorption maxima to shorter wavelengths (Emerson and Lewis, 1942; Halldal, 1958).

The visible spectra of phycoerythrins may have one, two, or three peaks.

The prefixes C-, B-, and R- are used to distinguish these three types of phycoerythrin (Fig. 1)[1]. Specific extinction coefficients for R-phycoerythrin (as well as for some phycocyanins) have been published by Lemberg (1928, 1930) and Svedberg and Katsurai (1929). The wavelength and relative extinctions at the absorption maxima depend on the species and the method of preparation of the phycoerythrins. For example, some R-phycoerythrin variants from different Red Algae lack a distinct absorption maximum at about 540 mμ which is present in others (Svedberg and Eriksson, 1932; Haxo *et al.*, 1955; Ó hEocha, 1960). The shoulder at 560–565 mμ, which characterizes chromatographically purified as well as once-crystallized B-phycoerythrin in neutral solutions, disappears when the protein is twice recrystallized (Airth and Blinks, 1956). Prolonged extraction converts the single-peaked C-phycoerythrin of *Phormidium* spp. to an artifact, a chromoprotein having two visible absorption maxima (Ó hEocha and Haxo, 1960). This spectral perturbation is caused by the action on C-phycoerythrin of proteolytic enzymes, found to be most active in extracts of cells from 4-month-old cultures (Ó hEocha and Curley, 1961).

The absorption maximum of cryptomonad phycoerythrin from different sources varies from 545 to 568 mμ (see, for example, Fig. 1). Since the pigmented proteins of cryptomonads differ from phycoerythrins of Red and Blue-green Algae in some other properties, their position in the nomenclature system of phycobilin pigments must await further study (Allen *et al.*, 1959; Ó hEocha and Raftery, 1959; Haxo and Fork, 1959).

Absorption spectra of representative phycocyanins are given in Fig. 2. Although some workers (e.g., Svedberg and Katsurai, 1929) reported that *Porphyra tenera* contains R-phycocyanin, others obtained C-phycocyanin from this alga (Lemberg, 1930; Fujiwara, 1955; Hattori and Fujita, 1959b). Hattori and Fujita claimed that R-phycocyanin is not a pure pigment, but a mixture of phycoerythrin and C-phycocyanin (see also Blinks, 1954). They and others (e.g., French *et al.*, 1956) obtained C-phycocyanin from different species of Red Algae (Table I), but it remains to be seen whether the carefully purified R-phycocyanin of *Ceramium rubrum*, for example, is a mixture. Although samples differing slightly in the ratio of the extinction maxima can be prepared (Haxo and Ó hEocha, 1960), the absorption spectrum of R-phycocyanin from this and other algae (see Fig. 2) lacks the slight phycoerythrin peak at about 510 mμ which is exhibited by the contaminated C-phycocyanin of *Porphyra tenera* (Hattori and Fujita, 1959b).

Allophycocyanin, which is widely distributed among Rhodophyta and Cyanophyta (Table I), accounts for a relatively high proportion of the

[1] These letters were originally selected to designate phycoerythrins occurring in Cyanophyta, Bangiales, and other Rhodophyta, respectively.

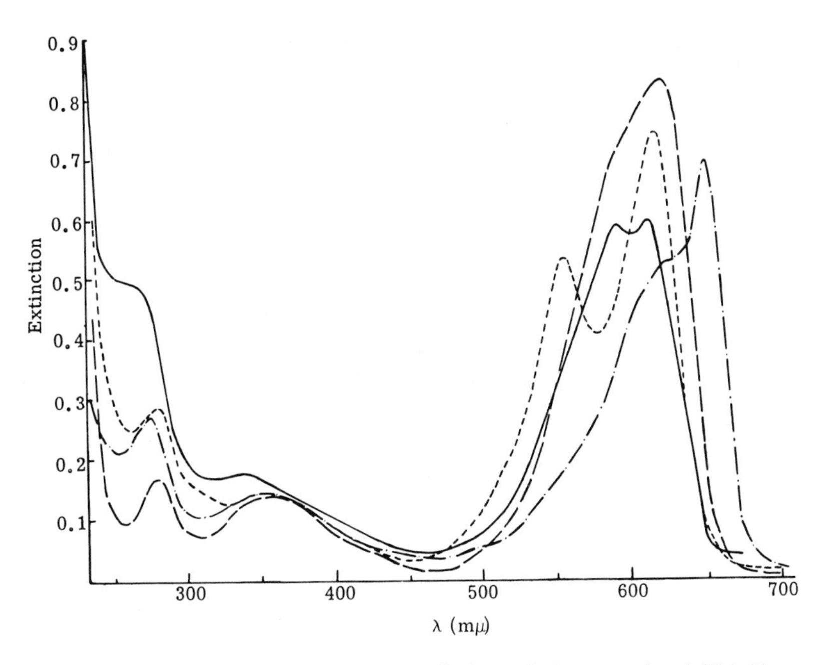

FIG. 2. Absorption spectra of aqueous solutions of phycocyanins (pH 6–7):
- - - - - -, R-phycocyanin (*Rhodymenia palmata*);
— — — —, C-phycocyanin (*Lyngbya lagerheimii*);
–·–·–, Allophycocyanin (*Ceramium rubrum*);
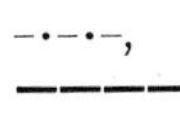, Cryptomonad phycocyanin (*Hemiselmis virescens*), Plymouth strain No. 157).

biliproteins of some species of Blue-green Algae (Hattori and Fujita, 1959b). The spectral properties of cryptomonad phycocyanins differ from those of all other algae examined. The spectrum given in Fig. 2 is that of one type of cryptomonad phycocyanin. Another, having an additional absorption maximum at 645 mμ, was encountered by Allen *et al.* (1959).

The ultraviolet absorption spectra of phycoerythrins and phycocyanins have maxima in common at about 275 and 365 mμ (Figs. 1 and 2); phycoerythrins possess an additional peak at 305–310 mμ (Haxo *et al.*, 1955; Hattori and Fujita, 1959b; Ó hEocha, 1960).

It was originally thought that there was some relation between absorption spectra and the taxonomic positions of the algal sources of the biliproteins. However, as can be seen from Table I, many types of biliproteins seem to be randomly distributed among the various families and species. In fact, there are cases on record of algae which had been classified in the same species, but obtained from different locations, differing in the spectral characteristics of their phycocyanins, e.g., *Hemiselmis virescens* (Allen *et al.*, 1959; Ó hEocha and Raftery, 1959) and *Smithora naiadum* (Ó hEocha and Haxo, 1960).

B. Fluorescence Spectra

Biliproteins do not fluoresce appreciably while within the chloroplasts, but immediately after they are released into the vacuole or into aqueous solution their brilliant fluorescence becomes apparent (McClendon and Blinks, 1952). When suitably irradiated, phycoerythrins emit orange light, and phycocyanins emit red light. The fluorescence spectra of R- and B-phycoerythrins display maxima at 578 mμ, while the allophycocyanin peak lies at 663 mμ (French and Young, 1956; French *et al.*, 1956). The latter authors reported that the maximum for C-phycocyanin isolated from a species of Red Alga, *Porphyra naiadum* (now renamed *Smithora naiadum*), lies at 637 mμ, whereas the peak of the C-phycocyanin curve which Duysens (1952) obtained from *Oscillatoria* sp. is apparently at 650–655 mμ.

C. Molecular Weights

The molecular size of the biliproteins is pH-dependent. The molecular weights of R- and B-phycoerythrins are highest (about 290,000) near their isoelectric point (pH 4.3–4.5) (Eriksson-Quensel, 1938; Krasnovsky *et al.*, 1952; Airth and Blinks, 1956). The only recorded value for C-phycoerythrin is 226,000 at pH 7.2 (Hattori and Fujita, 1959a). R-Phycoerythrin is stable in a pH range of 3 to 10, but the molecule breaks down into smaller units in more alkaline media. R-Phycocyanin (molecular weight 273,000) has a more limited stability range (pH 2.5–6.0), splitting reversibly into half-molecules at pH 7.0–8.5. C-Phycocyanin (pI 4.7[2]) is also unstable in alkali, breaking down to units of about one-sixth the original molecular weight at pH 12.0 (Svedberg and Katsurai, 1929). Allophycocyanin has a molecular weight of 134,000 at pH 7.2 (Hattori and Fujita, 1959a).

V. Amino-Acid Composition of the Proteins

The phycobilin chromophore accounts for about 4% of the mass of C-phycocyanin (Clendenning, 1954), the remainder of the molecule being composed of amino-acid residues (Kimmel and Smith, 1958). There is evidence that carbohydrate may be a component of some biliproteins (Fujiwara, 1955), and this may explain the low yield (88.8%) of amino-acid residues obtained on hydrolysis of R-phycoerythrin (Kimmel and Smith, 1958). If the carbohydrate were a sulfate ester, characteristic of marine algae, this could explain the fact that many workers have failed to

[2] pI = isoelectric point.

account for the sulfur content of their biliprotein preparations in terms of S-containing amino acids.

The analyses of Jones and Blinks (1957), Kimmel and Smith (1958) and others, show that biliproteins contain the seventeen amino acids usually encountered in the proteins of higher plants. A feature of them all is the excess of the acidic amino acids (glutamic and aspartic acids) over the basic acids (lysine, histidine, and arginine). This is true even of the biliproteins of *Rhodymenia palmata*, the bulk proteins of which are known to contain more basic than dicarboxylic acid residues (cf. Ó hEocha, 1960). Another feature of most biliproteins is their high content of acids such as alanine and leucine, which contain hydrophobic groups. No correlation is apparent between the amino-acid composition and the spectral properties or sources of the biliproteins. Methionine has been identified as the N-terminal amino-acid residue of R-, B-, and C-phycoerythrins (Ó Carra and Ó hEocha, unpublished), and alanine as the C-terminal residue of R-phycoerythrin (Raftery and Ó hEocha, unpublished).

VI. Preparation and Properties of the Phycobilins

It was shown by Lemberg (1930) and Lemberg and Bader (1933) that the phycobilins are bile pigments, the chemistry of which is reviewed by Lemberg and Legge (1949), Gray (1953), and Stevens (1959).

Hot acid hydrolysis of C-phycocyanin yields a mesobiliviolin (I)—see Fig. 3 (Curve B).

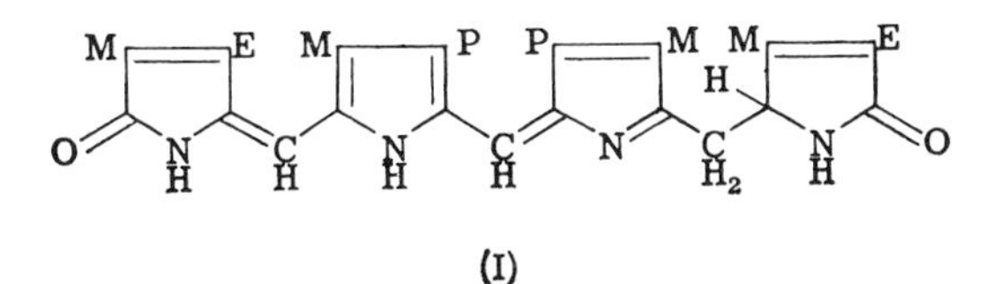

(I)

Mesobiliviolin IX α

(M, CH_3; E, C_2H_5; P, CH_2CH_2COOH).

Clendenning (1954) obtained a 4 percent yield of the tetrapyrrole from C-phycocyanin, indicating about sixteen chromophoric groups per molecular weight unit (see Lemberg and Legge, 1949). Acid hydrolysis of C-phycocyanin at room temperature yields a blue-green bile pigment—not mesobiliviolin (Fig. 3). The pigment derived from *Arthrospira maxima* phycocyanin is unstable in acid, being converted after some hours to a mesobiliviolin (Ó hEocha, 1958); that from other algae (e.g., *Nostoc muscorum*, *Anabaena cylindrica*) is stable, and resembles mesobiliverdin (II) in many

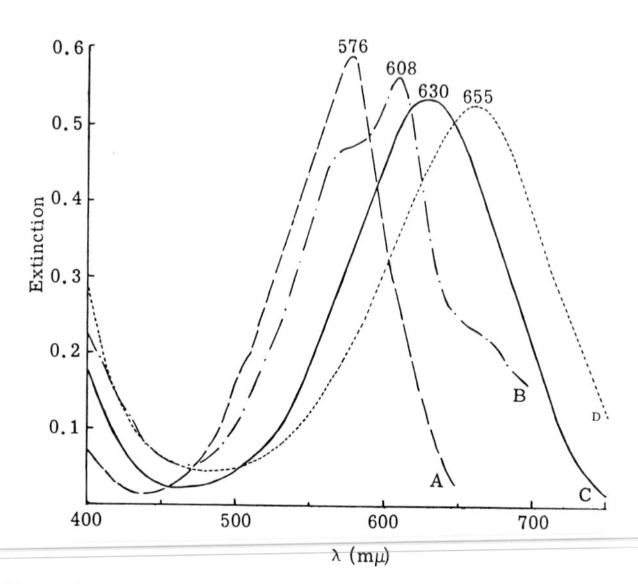

FIG. 3. Absorption spectra, in acid chloroform, of phycobilins obtained by concentrated hydrochloric acid hydrolysis of:

C-phycoerythrin (from *Phormidium fragile*) for 30 minutes at room temperature. (Curve *A*.)

C-phycocyanin (from *Arthrospira maxima* or *Nostoc muscorum*) for 20 minutes at 80°C. Mainly a mesobiliviolin hydrochloride. (Curve *B*.)

C-phycocyanin (from *Arthrospira maxima*) for 30 minutes at room temperature. This pigment is converted to a mesobiliviolin (cf. curve *B*) after hydrolysis for 7 hours. (Curve *C*.)

C-phycocyanin (from *Nostoc muscorum*) for 12 hours at room temperature. A similar curve is obtained after 30 minutes' hydrolysis, but the yield is low. (Curve *D*.)

Note added in proof: C-phycocyanins from *Arthrospira* and *Nostoc* yield the same phycobilin (Curve C) when treated with 12N HCl; the reported differences in their prosthetic groups are now attributed to differences in the concentrations of acid employed during hydrolysis.

of its properties (Ó hEocha and Lambe, 1961). The absorption spectrum of the native chromophore of C-phycocyanin (as hydrochloride) is considered to be curve C (Fig. 3; see footnote to legend). However, its chemical structure is still uncertain. Alkaline hydrolysis of C-phycocyanin yields mesobiliverdin (II) (Lemberg and Bader, 1933).

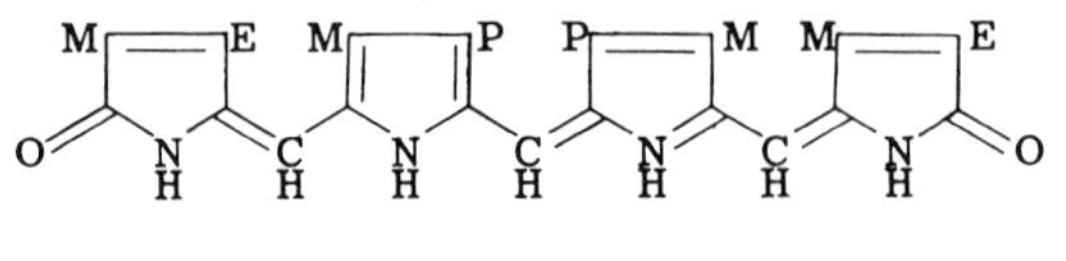

(II)

Mesobiliverdin

Cold acid hydrolysis of R-phycocyanin gives phycoerythrobilin, the phycoerythrin chromophore, in addition to a pigment of the mesobiliverdin type (Ó hEocha, 1960). Hot acid hydrolysis of R-phycoerythrin yields a chromopeptide, the prosthetic group of which was identified as mesobilirhodin (mesobilierythrin) (Lemberg, 1930; Lemberg and Bader, 1933). Two structures, IIIa (Siedel) and IIIb (Lemberg), have been proposed for this pigment (see Lemberg and Legge, 1949; Rabinowitch, 1956).

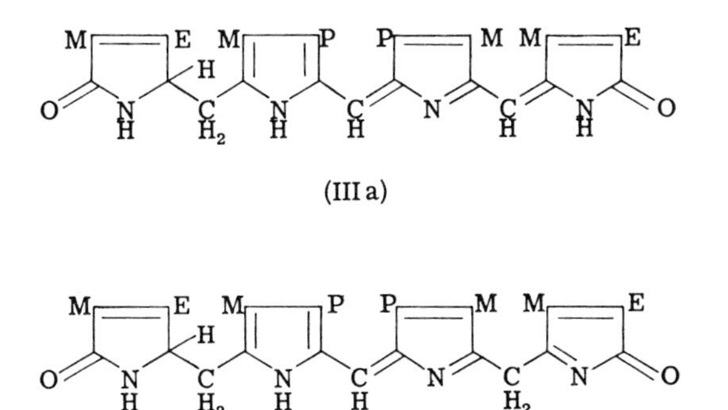

Mesobilirhodin (proposed structures)

When hydrolyzed with acid for short periods in the cold, phycoerythrins yield a free tetrapyrrolic phycoerythrobilin (Fig. 3) (Ó hEocha, 1958; Ó hEocha and Raftery, 1959). This pigment is unstable in concentrated hydrochloric acid, and its principal absorption maximum is then shifted from 576 to 500 mμ (determined in acidified chloroform), which may indicate isomerization from a structure of type (IIIa) to (IIIb) (Ó hEocha *et al.*, 1961). Such an isomerization would decrease the number of conjugated double bonds in the molecule from seven to five, leading to a spectral shift of the order observed (see Lemberg and Legge, 1949, p. 107). R-phycoerythrin, which is distinguished from the others by an absorption maximum at about 495 mμ (Fig. 1), contains two quite dissimilar chromophoric groups. Phycoerythrobilin is one, while the other possesses a urobilinoid conjugated system such as (IIIb) (Ó hEocha, 1960; Ó hEocha and Ó Carra, 1961), and is not readily dissociated from the protein.

Phycobilin solutions are not fluorescent, but most of them combine with zinc ions to form brilliantly fluorescent salt complexes with characteristic absorption spectra (Ó hEocha, 1958). The low ash content of purified biliproteins (Lemberg, 1928; Hattori and Fujita, 1959a) indicates that the phycobilins are not complexed with metals in the native state.

The biosynthesis of phycobilins has not been subjected to detailed study, but in view of the known formation of animal bile pigments by oxidation of the α-methine group of porphyrins (see Gray, 1953; Pullman and Perault, 1959), it is likely that phycobilin formation involves a similar oxidation of a porphyrin. Experiments using radioactive carbon (Garnier, 1959) and algal mutants (Nichols and Bogorad, 1960) indicated that the biosyntheses of chlorophyll *a* and of phycocyanins proceed by different pathways. Nichols and Bogorad speculated that heme proteins may serve as photoreceptors, and possibly also as precursors, in the biosynthesis of phycocyanins.

VII. Phycobilin-Protein Linkages

Since concentrated acid is required to liberate the phycobilins from the biliproteins, Lemberg (1930) postulated that they are attached through peptide linkages involving their propionic acid side-chains and amino groups in the protein. Ó Carra and Ó hEocha (unpublished; see Ó hEocha, 1960) found that the ε-amino groups of lysine in R-phycoerythrin do not participate in such linkages. Jones and Fujimori (1961) reported that sulfhydryl and disulfide groups are involved in producing the 540 and 565 mμ absorption peaks of this biliprotein.

The fluorescence of biliprotein solutions is quenched by dilute acid (Lemberg, 1930) and by concentrated urea solutions (Ó hEocha and Ó Carra, 1961). The changes in absorption spectra accompanying these treatments suggest that the labile chromophore–protein linkages of phycoerythrins (Lemberg, 1930) may be hydrogen bonds joining pyrrole nitrogen atoms of phycoerythrobilins to the proteins. These linkages appear to be essential for the fluorescence of these phycoerythrobilins which are responsible for the long wavelength absorption maximum (556–566 mμ) of phycoerythrins (Fig. 1). However, the phycobilin responsible for the 495 mμ peak of R-phycoerythrin is not involved in such labile bonding (Ó hEocha and Ó Carra, 1961).

Acknowledgment

Preparation of this chapter was supported by the U. S. Air Force through its European Office (Contract No. AF 61(052)-409).

References

Airth, R. L., and Blinks, L. R. (1956). A new phycoerythrin from *Porphyra naiadum*. *Biol. Bull.* **111**, 321–327.

Allen, M. B. (1959). Studies with *Cyanidium caldarium*, an anomalously pigmented chlorophyte. *Arch. Mickrobiol.* **32**, 270–277.

Allen, M. B., and Arnon, D. I. (1955). Studies on nitrogen-fixing blue-green algae. II. The sodium requirement of *Anabaena cylindrica*. *Physiol. Plantarum* **8**, 653–660.

Allen, M. B., Dougherty, E. C., and McLaughlin, J. J. A. (1959). Chromoprotein pigments of some cryptomonad flagellates. *Nature* **184,** 1047–1049.

Blinks, L. R. (1954). The role of accessory pigments in photosynthesis. *In* "Autotrophic Microorganisms" (B. A. Fry and J. L. Peel, eds.), pp. 224–246. Cambridge Univ. Press, London and New York.

Boresch, K. (1921). Ein Fall von Eisenchlorose bei Cyanophyceen. *Z. Botan.* **13,** 64–78.

Boresch, K. (1932). Algenfarbstoffe. *In* "Handbuch der Pflanzenanalyse" (G. Klein, ed.), Vol. III, pp. 1382–1410. Springer, Vienna.

Bouillene-Walrand, M., and Delarge, L. (1937). Contribution a l'étude des pigments végétaux. I. Extraction et crystallization de la phycocyanine de *Phormidium uncinatum* Gom. *Rev. gén. botan.* **49,** 537–557.

Brody, M., and Emerson, R. (1959). The effect of wavelength and intensity of light on the proportion of pigments in *Porphyridium cruentum. Am. J. Botany* **46,** 433–440.

Brody, M., and Vatter, A. E. (1959). Observations on cellular structures of *Porphyridium cruentum. J. Biophys. Biochem. Cytol.* **5,** 289–294.

Clendenning, K. A. (1954). Recent investigations of photosynthesis and the Hill Reaction. *Congr. intern. botan., 8e Congr., Paris, 1954* Section 11: Physiologie végétale, pp. 21–35.

Duysens, L. N. M. (1952). "Transfer of Excitation Energy in Photosynthesis." Kemink, Utrecht, Netherlands.

Elbers, P. F., Minnaert, K., and Thomas, J. B. (1957). Submicroscopic structure of some chloroplasts. *Acta Botan. Neerl.* **6,** 345–350.

Emerson, R., and Lewis, C. M. (1942). The photosynthetic efficiency of phycocyanin in *Chroococcus* and the problem of carotenoid participation in photosynthesis. *J. Gen. Physiol.* **25,** 579–595.

Eriksson-Quensel, I.-B. (1938). The molecular weights of phycoerythrin and phycocyan. I. *Biochem. J.* **32,** 585–589.

Feldmann, J., and Tixier, R. (1947). Sur la floridorubine, pigment rouge des plastes d'une Rhodophycée., (*Rytiphloea tinctoria* (Clem.) C. Ag.) *Rev. gén. botan.* **54,** 341–353.

Fogg, G. E. (1952). The production of extracellular nitrogenous substances by a blue-green alga. *Proc. Roy. Soc.* **B139,** 372–397.

French, C. S., and Young, V. K. M. (1956). The absorption, action and fluorescence spectra of photosynthetic pigments in living cells and in solutions. *Radiation Biol.* **3,** 343–392.

French, C. S., Smith, J. H. C., Virgin, H. I., and Airth, R. L. (1956). Fluorescence-spectrum curves of chlorophylls, phaeophytins, phycoerythrins, phycocyanins and hypericin. *Plant Physiol.* **31,** 369–374.

Fujita, Y., and Hattori, A. (1960a). Formation of phycoerythrin in preilluminated cells of *Tolypothrix tenuis* with special reference to nitrogen metabolism. *Plant and Cell Physiol.* **1,** 281–292.

Fujita, Y., and Hattori, A. (1960b). Effect of chromatic lights on phycobilin formation in a Blue-green Alga, *Tolypothrix tenuis. Plant and Cell Physiol.* **1,** 293–303.

Fujiwara, T. (1955). Studies on chromoproteins in Japanese Nori (*Porphyra tenera*). I. A new method for the crystallization of phycoerythrin and phycocyanin. *J. Biochem. (Tokyo)* **42,** 411–417.

Garnier, J. (1959). Sur l'influence de la lumière sur la vitesse de formation des différents pigments d'*Oscillatoria subbrevis* Schmidle (Cyanophycées). *Compt. rend. acad. sci.* **248,** 724–727.

Gray, C. H. (1953). "The Bile Pigments." Methuen, London.

Halldal, P. (1958). Pigment formation and growth in blue-green algae in crossed gradients of light intensity and temperature. *Physiol. Plantarum* **11,** 401–420.

Hattori, A., and Fujita, Y. (1959a). Crystalline phycobilin chromoproteins obtained from a blue-green alga, *Tolypothrix tenuis. J. Biochem.* (*Tokyo*) **46,** 633–644.

Hattori, A., and Fujita, Y. (1959b). Spectroscopic studies on the phycobilin pigments obtained from blue-green and red algae. *J. Biochem.* (*Tokyo*) **46,** 903–909.

Hattori, A., and Fujita, Y. (1959c). Effect of pre-illumination on the formation of phycobilin pigments in a blue-green alga, *Tolypothrix tenuis. J. Biochem.* (*Tokyo*) **46,** 1259–1261.

Haxo, F. T., and Blinks, L. R. (1950). Photosynthetic action spectra of marine algae. *J. Gen. Physiol.* **33,** 389–422.

Haxo, F. T., and Fork, D. C. (1959). Photosynthetically active accessory pigments of cryptomonads. *Nature* **184,** 1051–1052.

Haxo, F. T., and Ó hEocha, C. (1960). Chromoproteins of algae. *In* "Handbuch der Pflanzenphysiologie" (W. Ruhland, ed.), Vol. V, pp. 497–510. Springer, Berlin.

Haxo, F. T., Ó hEocha, C., and Norris, P. (1955). Comparative studies of chromatographically separated phycoerythrins and phycocyanins. *Arch. Biochem. Biophys.* **54,** 162–173.

Jones, R. F., and Blinks, L. R. (1957). The amino acid constituents of the phycobilin chromoproteins of the red alga *Porphyra. Biol. Bull.* **112,** 363–370.

Jones, R. F., and Fujimori, E. (1961). Interaction between chromophore and protein in phycoerythrin from the red alga, *Ceramium rubrum. Physiol. Plantarum* **14,** 253–259.

Kimmel, J. R., and Smith, E. L. (1958). The amino acid composition of crystalline pumpkin seed globulin, edestin, C-phycocyanin and R-phycoerythrin. *Bull. soc. chim. biol.* **40,** 2049–2065.

Krasnovsky, A. A., Evstigneev, V. B., Brin, G. P., and Gavrilova, V. A. (1952). The isolation of phycoerythrin from red seaweed and its spectral and photochemical properties. (In Russian.) *Doklady Akad. Nauk S.S.S.R.* **82,** 947–950.

Kylin, H. (1937). Über die Farbstoffe und die Farbe der Cyanophyceen. *Kgl. Fysiograf. Sällskap. Lund, Förh.* **7,** 131–158.

Lemberg, R. (1928). Die Chromoproteide der Rotalgen. I. *Ann. Chem. Liebigs* **461,** 46–89.

Lemberg, R. (1930). Chromoproteide der Rotalgen. II. Spaltung mit Pepsin und Säuren. Isolierung eines Pyrrolfarbstoffs. *Ann. Chem. Liebigs* **477,** 195–245.

Lemberg, R., and Bader, G. (1933). Die Phycobiline der Rot-Algen. Überführung in Mesobilirubin und Dehydromesobilirubin. *Ann. Chem. Liebigs* **505,** 151–177.

Lemberg, R., and Legge, J. W. (1949). "Hematin Compounds and Bile Pigments." Interscience, New York.

McClendon, J. H., and Blinks, L. R. (1952). Use of high molecular weight solutes in the study of isolated intracellular structures. *Nature* **170,** 577–578.

Myers, J., and Kratz, W. A. (1955). Relations between pigment content and photosynthetic characteristics in a blue-green alga. *J. Gen. Physiol.* **39,** 11–22.

Nichols, K. E., and Bogorad, L. (1960). Studies of phycobilin formation with mutants of *Cyanidium caldarium. Nature* **188,** 870–872.

Ó hEocha, C. (1955). The comparative biochemistry of phycoerythrins and phycocyanins. Doctoral thesis. University of California, Los Angeles, California.

Ó hEocha, C. (1958). Comparative biochemical studies of the phycobilins. *Arch. Biochem. Biophys.* **73,** 207–219; **74,** 493.
Ó hEocha, C. (1960). Chemical studies of phycoerythrins and phycocyanins. *In* "Comparative Biochemistry of Photoreactive Systems" (M. B. Allen, ed.), pp. 181–203. Academic Press, New York.
Ó hEocha, C., and Curley, D. (1961). Enzymically-induced spectral perturbations in C-phycoerythrin. *Abstr. Intern. Congr. Biochem., Moscow, 5th Congr.* Section 22.20, p. 447.
Ó hEocha, C., and Haxo, F. T. (1960). Some atypical algal chromoproteins. *Biochim. et Biophys. Acta* **41,** 515–519.
Ó hEocha, C., and Lambe, R. F. (1961). The prosthetic group of C-phycocyanin. *Arch. Biochem. Biophys.* **93,** 458–460.
Ó hEocha, C., and Ó Carra, P. (1961). Spectral studies of denatured phycoerythrins. *J. Am. Chem. Soc.* **83,** 1091–1093
Ó hEocha, C., and Raftery, M. (1959). Phycoerythrins and phycocyanins of cryptomonads. *Nature* **184,** 1049–1051.
Ó hEocha, C., Ó Carra, P., and Carroll, D. (1961). Phycoerythrobilin, a prosthetic group of algal chromoproteins. *Biochem. J.* **80,** 25 p.
Pullman, B., and Perault, A-M. (1959). On the metabolic breakdown of hemoglobin and the electronic structure of the bile pigments. *Proc. Natl. Acad. Sci. U.S.* **45,** 1476–1480.
Rabinowitch, E. I. (1945). "Photosynthesis," Vol. I: Chemistry of Photosynthesis, Chemosynthesis and Related Processes *in vitro* and *in vivo.* Interscience, New York.
Rabinowitch, E. I. (1951). "Photosynthesis," Vol. II, Part 2: Spectroscopy and Fluorescence of Photosynthetic Pigments; Kinetics of Photosynthesis. Interscience, New York.
Rabinowitch, E. I. (1956). "Photosynthesis," Vol. II, Part 2: Kinetics of Photosynthesis (continued). Addenda to Volume I and Volume II, Part 1. Interscience, New York.
Stevens, T. S. (1959). The pyrrole pigments. *In* "Chemistry of Carbon Compounds" (E. H. Rodd, ed.), Vol. IV, B, pp. 1104–1162. Elsevier, Amsterdam.
Svedberg, T., and Eriksson, I-B. (1932). The molecular weights of phycocyan and of phycoerythrin. III. *J. Am. Chem. Soc.* **54,** 3998–4010.
Svedberg, T., and Katsurai, T. (1929). The molecular weights of phycocyan and of phycoerythrin from *Porphyra tenera* and of phycocyan from *Aphanizomenon flos-aquae. J. Am. Chem. Soc.* **51,** 3573–3583.
Swingle, S. M., and Tiselius, A. (1951). Tricalcium phosphate as an adsorbent in the chromatography of proteins. *Biochem. J.* **48,** 171–174.
Thomas, J. B., and de Rover, W. (1955). On phycocyanin participation in the Hill reaction of the blue-green alga *Synechococcus cedrorum. Biochim. et Biophys. Acta* **16,** 391–395.
Tiselius, A. (1954). Chromatography of proteins on calcium phosphate columns. *Arkiv Kemi* **7,** 443–449.

Tannins and Vacuolar Pigments

CHINKICHI OGINO

Tokyo University of Fisheries, Tokyo, Japan

I. Tannins of Phaeophyta

The aqueous extracts of fresh thalli of some Phaeophyta are colored blue-violet or red-violet by the addition of iron salts, are astringent to the taste, and precipitate gelatin (Suehiro and Searashi, 1948; Ogino and Taki, 1957). These properties taken as a whole are characteristic of tannin. In general, species of *Sargassum* show a rather strong color reaction with iron salts. Indeed, aqueous extracts of *Sargassum ringgoldianum* have been used in the tanning industry (Noguchi, 1943). On the other hand, extracts of other species such as *Undaria pinnatifida, Nemacystus decipiens*, etc., completely fail to give color reactions characteristic of tannins (Suehiro and Searashi, 1948).

The tannic substances of Phaeophyta have been considered to be stored in so-called "physodes" or "fucosan vesicles." These are vesicles which contain a colorless, highly refractile, acidic fluid which stains red with vanillin and hydrochloric acid (Crato, 1892, 1893; Kylin, 1912, 1913, 1918, 1938). Crato named these vesicles "physodes" and stated that they contain phloroglucinol or its derivatives, whereas Kylin employed the terms "fucosan vesicles" and "fucosan" for the vesicles and their contents respectively. According to Kylin, fucosan is a nonglycosidic, tannin-like substance which is stained red with vanillin and hydrochloric acid, possesses strong reducing action, produces precipitates with lead acetate, and in aqueous solution is astringent to the taste. Fucosan is readily oxidized in air, resulting in the formation of a brown or black pigment, "phycophaein." By cytochemical observation of preparations stained with cresyl blue, neutral red, and dichlorphenol-indophenol, one may distinguish three or four types of physode, each associated with a particular fucosan (Chadefaud, 1932, 1934; Ando, 1951, 1958a, b). Ando distinguished fucosans *a*, *b*, and *c*, of which fucosan *b* appeared chemically identical with Kylin's fucosan, while fucosan *a* was considered to be a phenolic glucoside and fucosan *c* an essential oil consisting of terpenes. Mangenot (1921, 1922) observed precipitates of phenolic substances in the cell vacuoles of Phaeophyta, and referred to them as "fucosan granules."

Ogino and Taki (1957) isolated a tannin from methanol extracts of fresh *Sargassum ringgoldianum*. The method of isolation and the properties of the purified material are summarized in Tables I and II respectively. The tannin so obtained is a lustrous, hygroscopic powder, pale yellowish-brown when fresh but blackening in air. In solution it is acidic, astringent to the taste; it is adsorbed on hide powder. It is almost completely precipitated when boiled with 10% sulfuric acid. The molecule contains a phloroglucinol nucleus; it has no sugar residue. From these properties Ogino and Taki concluded that their preparation was a "condensed" tannin or catechol, according to the classification of Freudenberg (1932). Paper chromatog-

TABLE I

ISOLATION OF TANNIN FROM *Sargassum ringgoldianum*

1 KG. FRESH SAMPLE

↓ Extracted with methanol (2 liters) for 30 hours at room temperature in N_2

FILTRATE

↓ Concentrated *in vacuo* under a stream of N_2; residues dissolved in water

AQUEOUS SOLUTION

↓ Added NaCl, filtered; filtrate shaken with ether

AQUEOUS LAYER

↓ Shaken with ethyl acetate

ETHYL ACETATE LAYER

↓ Dehydrated with anhydrous Na_2SO_4; concentrated *in vacuo* under a stream of N_2; residue dissolved in acetone

SOLUBLE FRACTION

↓ Dried in a vacuum desiccator

TANNIN POWDER (2.1 GM.)

TABLE II

CHEMICAL PROPERTIES OF THE TANNIN FROM *Sargassum*

Reagent	Solubility, color, or precipitation reaction
H_2O	Soluble
Acetone	Soluble
Ethyl acetate	Soluble
Pyridine	Soluble
Methyl-, ethyl-, and amyl alcohols	Soluble
Chloroform	Insoluble
Carbon tetrachloride	Insoluble
Ether	Insoluble
Mohr's salt (ferrous ammonium sulfate)	Bluish-violet
$FeCl_3$	Reddish-violet
NaCN	Yellow
NaOH	Dark green
Lead acetate	Yellow precipitate
Lead acetate and acetic acid	Pale-green precipitate
Gelatin	White precipitate
$CuSO_4$	Greenish-yellow precipitate
Bromine water	Slight precipitate
HCl and formalin (heating)	Completely precipitated
Fehling's solution	Reduced

raphy of their preparation in the solvent mixture of *n*-butanol:acetic acid: 1% aqueous $NaHSO_3$(4:1:5 v/v, upper layer) gave 4 spots in addition to the bulk of the tannin, which did not move from the origin. However, it remained uncertain whether more than one kind of tannin existed as such in the living alga or whether they were formed secondarily during the preparation of the samples. Since this material has now been shown conclusively not to be a polysaccharide, it is felt that the misleading term "fucosan" should be avoided, and that the materials should hence forth be referred to simply as "phaeophyte tannins."

Larsen and Haug (1958) and Haug and Larsen (1958) extracted with dilute acid from *Ascophyllum nodosum* reducing substances which could be adsorbed on hide powder and which gave a red coloration with vanillin and hydrochloric acid. They reported that the salinity of the sea water in which the algae grow may influence the content of such substances.

The black pigment produced when the thalli of various Phaeophyta are washed ashore and die on exposure to air was studied by Takahashi (1931). He isolated this pigment from air-dried samples of *Ecklonia cava* and purified it to some extent. It is insoluble in water and forms phloroglucinol upon potash fusion. It is probably an oxidation product of tannin, and may be identical with "phycophaein." Its insolubility suggests the reason why tannins cannot be extracted from air-dried samples (Ogino and Taki, 1957).

II. Tannins of *Spirogyra* (Chlorophyta; Zygnematales)

The presence of tannin in the vacuoles of *Spirogyra* (Green Algae) was mentioned by De Vries in 1885 (Oltmanns, 1923). Van Wisselingh (1914) studied the *Spirogyra* tannin from a physiological viewpoint, and reported that the tannin content of the cells varied in the course of conjugation.

Nakabayashi (1954, 1955a, b) and Nakabayashi and Hada (1954a, b) isolated the tannin from *Spirogyra arcta* by the following method. A fresh sample of alga, washed with distilled water, is extracted twice with acetone under reflux for a total of 5 hours. The acetone solutions are combined, the acetone is distilled off, and the residues are dissolved in water. The aqueous solution is decolorized with activated carbon, and filtered through a layer of asbestos. The filtrate is shaken with chloroform to remove other impurities, and the crude aqueous solution of tannin is extracted with ether in a continuous extraction apparatus for 20 hours. After distillation of most of the ether, the residue is dried in a vacuum desiccator.

The tannin thus obtained was a white amorphous powder, soluble in hot water, acetone, ether, alcohol, and ethyl acetate, only slightly in cold water, and insoluble in benzene and chloroform. The aqueous solution was colored deep blue after addition of ferric chloride and showed other reactions characteristic of tannins. Paper chromatography in *n*-butanol:acetic acid:1% aqueous $NaHSO_3$(4:1:5 v/v, upper layer) gave only one spot. Its absorption spectrum in alcoholic solution had a maximum at 280 mμ. This tannin proved to be glycosidic: when hydrolyzed by a tannase from *Aspergillus niger*, or by boiling with 5% sulfuric acid, the tannin yielded glucose and gallic acid in a molecular ratio of 1:5. Hydrolysis of the methylated derivative yielded 3,4,5-trimethylgallic acid together with a small amount of 4,5-dimethylgallic acid and monomethyl glucose. From these findings Nakabayashi concluded that *Spirogyra* tannin was chiefly pentagalloyl-D-glucose, together with a small amount of a galloylglucose possessing two or more *m*-digalloyl groups.

The biological role of tannins in algae is unknown, but it is perhaps significant that the tannin content declines when *Spirogyra* filaments are kept for a few days in the laboratory (Clendenning *et al.*, 1957).

III. Vacuolar Pigments of Zygnematales

Blue-purple and red-purple pigments dissolved in the cell vacuolar sap have been observed in several species of Zygnemataceae, e.g., *Pleurodiscus purpureus, Temnogametum uleanum, T. thaxteri, Mougeotia capucina,* and *Zygogonium ericetorum* (Transeau, 1951), and from the following genera in other families of the Zygnematales: *Mesotaenium, Ancylonema,* and *Penium* (Lagerheim, 1895). Blue zygospore cell walls have been reported in over forty species of the Zygnemataceae.

Lagerheim (1895) made detailed but superficial studies on the vacuolar pigment of *Zygnema purpureum,* which he called a "phycoporphyrin." Mainx (1923) reported that the phycoporphyrin of *Zygnema purpureum* might be regarded as a kind of anthocyanin. Alston (1958) extracted the purplish vacuolar pigment of *Zygogonium ericetorum* with boiling water, obtaining a deep purple, almost ink-like solution; this strongly absorbed light in the ultraviolet as well as in visible regions of the spectrum, but there were no sharp absorption peaks, only a broad maximum at about 500–550 mμ. The solution became reddish-brown on the addition of ammonium hydroxide and was instantly decolorized by hydrochloric acid, indicating that it was not an anthocyanin. Under ultraviolet light the pigment showed a weak yellow-green fluorescence. These properties of the pigment agreed closely with those of Lagerheim's "phycoporphyrin" and the "anthocyanin-like" pigment of Mainx. However, the pigment also showed positive reactions for tannin, leading Alston to conclude that the pigment was probably an iron-tannin complex. His conclusion was reinforced by the striking resemblance between the vacuolar pigment and an iron–gallic acid complex prepared in the laboratory. The presence of ferric ion in cells producing tannin would result in the formation of a colored product. In fact, Allen and Alston (1959) found that filaments of *Spirogyra pratensis,* when cultured in a medium containing iron sequestrene (Versenate), develop a purple pigment which, in its reactions to ammonia and hydrochloric acid, resembles an iron–tannin complex.

Alston pointed out that in the surface layers of an acid bog, which represents the characteristic habitat of many of the purple algae, ferric iron might be expected to be readily available. Indeed, this is further suggested by the observations that cells of *Zygogonium ericetorum* frequently have yellow or brown encrustations—presumably hydrated iron oxides—on the walls (Transeau, 1951), and that when gallic acid is added to bog water it produces a purple coloration similar to that in the vacuoles of this alga. Water colored purple by iron–tannin complexes has even been found in natural pools (Dowgiallo and Fischer, 1960).

Thus, Mainx's description of the purple vacuolar pigment as "anthocyanin-like" is probably erroneous, and the term "phycoporphyrin" of Lagerheim is also misleading. Alston therefore has proposed that, until the precise molecular structure of this algal pigment is determined, it would be best to refer to it as an "iron–tannin" complex.

IV. Other Pigments

Two possibly related but little-known algal pigments which would well merit further investigation are those responsible for the blue color of the marine littoral diatom *Navicula ostrearia* and for the lilac-tinted holdfasts of the giant kelp, *Macrocystis pyrifera*, along the shores of Southern California and Baja California.

According to early studies by Bocat (1907), the blue pigment of *N. ostrearia*, which he called "marennine," is soluble in water but not in benzene, petroleum ether, or amyl alcohol. In acidic solutions marennine is violet, with three spectroscopic absorption bands; from such a solution Bocat was able to obtain it in crystalline form. In alkaline solutions it turns green, with two absorption bands, and is ultimately precipitated. Its general behavior led Bocat to conclude that it is a chromoprotein. There are also blue pigments of unknown composition associated with the plastids of certain dinoflagellates (Geitler, 1924); these, however, may prove on further examination to be phycobilins.

Preliminary investigations of the *Macrocystis* pigment indicate that it occurs predominantly in the cytoplasm of the superficial cells of the haptera; that it can be extracted with alcohol, but not water; and that it is evidently not an anthocyanin, a phycobilin, or a carotenoid (S. Krugman and F. T. Haxo, personal communication). It has an absorption maximum at about 550 mμ.

References

Allen, M. A., and Alston, R. E. (1959). Formation of purple pigment in *Spirogyra pratensis* cultures. *Nature* **183,** 1064–1065.

Alston, R. E. (1958). An investigation of the purple vacuolar pigment of *Zygogonium ericetorum* and the status of "algal anthocyanins" and "phycoporphyrins." *Am. J. Botany* **45,** 688–692.

Ando, Y. (1951). On the so-called "fucosan" in marine Phaeophyceae. *Botan. Mag. (Tokyo)* **64,** 192–195.

Ando, Y. (1958a). Studies on physodes. I. *Bull. Japan. Soc. Phycol.* **6,** 28–34.

Ando, Y. (1958b). Studies on physodes. II. *Bull. Japan. Soc. Phycol.* **6,** 45–50.

Bocat, L. (1907). Sur la marennine de la diatomée bleue; comparaison avec la phycocyanine. *Compt. rend. soc. biol.* **62,** 1073–1075.

Chadefaud, M. (1932). Sur les physodes des Phéophycées. *Compt. rend. acad. sci.* **194,** 1675–1677.

Chadefaud, M. (1934). Signification morphologique des physodes des Phéophycées. *Compt. rend. acad. sci.* **198,** 2114–2116.

Clendenning, K. A., Brown, T. E., and Walldov, E. E. (1957). Natural inhibitors of the Hill reaction. *In* "Research in Photosynthesis," (H. Gaffron *et al.*, eds.). Interscience, New York, pp. 274–284.

Crato, E. (1892). Die Physode, ein Organ des Zellenleibs. *Ber. deutsch. botan. Ges.* **10,** 295–302.

Crato, E. (1893). Über die Hansteen'schen Fucosankörner. *Ber. deutsch. botan. Ges.* **11,** 235–241.

Dowgiallo, A., and Fischer, E. (1960). Chemical and microbiological identification of the violet water-coloring agent in a pool of Puszcza Kampinoska. *Polskie Arch. Hydrobiol.* **7,** 159–170.

Freudenberg, K. (1932). Die natürlichen Gerbstoffe. *In* "Handbuch der Pflanzenanalyse" (G. Klein, ed.), Vol. III, pp. 344–412. Springer, Vienna.

Geitler, L. (1924). Über einige wenig bekannte Süsswasserorganismen mit roten oder blau-grünen Chromatophoren. *Rev. Algol.* **1,** 357–375.

Haug, A., and Larsen, B. (1958). Influence of habitat on the chemical composition of *Ascophyllum nodosum. Nature* **181,** 1224.

Kylin, H. (1912). Über die Inhaltskörper der Fucoideen. *Arkiv Botan.* **11,** 1–26.

Kylin, H. (1913). Zur Biochemie der Meeresalgen. *Z. physiol. Chem. Hoppe-Seyler's* **83,** 171–197.

Kylin, H. (1918). Über die Fucosanblasen der Phaeophyceen. *Ber. deutsch. botan. Ges.* **36,** 10–19.

Kylin, H. (1938). Bemerkungen über die Fucosanblasen der Phaeophyceen. *Kgl. Fysiograf. Sällskap. Lund Förh.* **8,** 1–10.

Lagerheim, G. (1895). Über das Phycoporphyrin, einen Conjugatenfarbstoff. *Videnskapsselskapets-Skrifter. I. Mat.-naturv. Kl., Kristiania* **5,** 3–25.

Larsen, B., and Haug, A. (1958). Chemical composition of brown alga *Ascophyllum nodosum* (L.) Le Jol. Presence of reducing compounds in *Ascophyllum nodosum. Nature* **181,** 1224–1225.

Mainx, F. (1923). Über eine Zygnematacee mit roten Zellsaftfarbstoff. *Lotos* **71,** 183–186.

Mangenot, G. (1921). Sur les "grains de fucosane" des Phéophycées. *Compt. rend. acad. sci. Roumanie* **172,** 126–129.

Mangenot, G. (1922). Recherches sur les constituants morphologiques du cytoplasma des algues. *Arch. Morphol. gén. exptl.* **9,** 1–330.

Nakabayashi, T. (1954). Studies on tannins of *Spirogyra arcta.* III. Estimation of tannin in *Spirogyra arcta. J. Agr. Chem. Soc. Japan* **28,** 958–961.

Nakabayashi, T. (1955a). Studies on tannins of *Spirogyra arcta.* IV. Formation of alkylgallate from tannic acid by fungous tannase. *J. Agr. Chem. Soc. Japan* **29,** 161–165.

Nakabayashi, T. (1955b). Studies on tannins of *Spirogyra arcta.* V. On the structure of *Spirogyra* tannin. *J. Agr. Chem. Soc. Japan* **29,** 897–899.

Nakabayashi, T., and Hada, N. (1954a). Studies on tannins of *Spirogyra arcta.* I. Isolation of tannins. *J. Agr. Chem. Soc. Japan.* **28,** 387–391.

Nakabayashi, T., and Hada, N. (1954b). Studies on tannins of *Spirogyra arcta.* II. Investigation of the becoming of methylgallate and the sugar in *Spirogyra* tannin. *J. Agr. Chem. Soc. Japan* **28,** 788–791.

Noguchi, E. (1943). Utilization of *Sargassum Ringgoldianum.* (Preliminary report). *Bull. Japan. Soc. Sci. Fisheries* **12,** 52–53.

Ogino, C., and Taki, Y. (1957). Studies on the tannin of brown alga, *Sargassum ringgoldianum* Harv. *J. Tokyo Univ. Fisheries* **43,** 1–5.

Oltmanns, F. (1923). "Morphologie und Biologie der Algen," Vol. III, pp. 45–46. Fischer, Jena, Germany.

Suehiro, Y., and Searashi, T. (1948). The utilization of the inedible sea-weeds. *Bull. Physiog. Sci. Research Inst., Tokyo Univ.* **1,** 44–50.

Takahashi, T. (1931). Studies on *Ecklonia cava.* III. On black pigment. *Tokyo Kôgyô Shikensho Hôkoku* **26,** 1–9.

Transeau, E. M. (1951). "The Zygnemataceae." Ohio State Univ. Press, Columbus, Ohio.

van Wisselingh, C. (1914). Über den Nachweis des Gerbstoffes in der Pflanze und über seine physiologische Bedeutung. *Botan. Centr. Beih.* **A32,** 155–217.

—27—

Silicification[*]

JOYCE C. LEWIN

Scripps Institution of Oceanography, University of California, La Jolla, California

I. Algae That Deposit Silica

Silicified structures occur in several algal classes. All cells of diatoms (Bacillariophyta) are characteristically enclosed in a silica wall or frustule, described in more detail below. Among the Chrysophyta, the vegetative cells in some genera bear numerous discrete siliceous scales; in addition, many species form cysts with silicified walls. The silicoflagellates, which may be related to the Chrysophyta (Fritsch, 1935), have internal skeletons of silica. Lastly, certain of the Xanthophyta have been reported to form silicified cell walls.

Studies of the fine structure of diatom frustules and of the silica scales of Chrysophyta have been made by a number of workers. Bibliographic indexes to the diatom species examined with the electron microscope, now numbering some 500 species, have been published by Desikachary (1957), by Okuno and Kurosawa (1957), and by Hendey (1959). Electron micrographs of silica scales and spines of *Mallomonas* and *Synura* have been published (Fott and Ludvík, 1957; Harris and Bradley, 1957, 1960; Takahashi, 1959). Such investigations have shown that the silicified valves of Bacillariophyta (see Figs. 1 and 2) and the scales of Chrysophyta (see Figs. 3 and 4) are generally perforated plates of extreme complexity, which may

* Contribution from Scripps Institution of Oceanography, New Series.

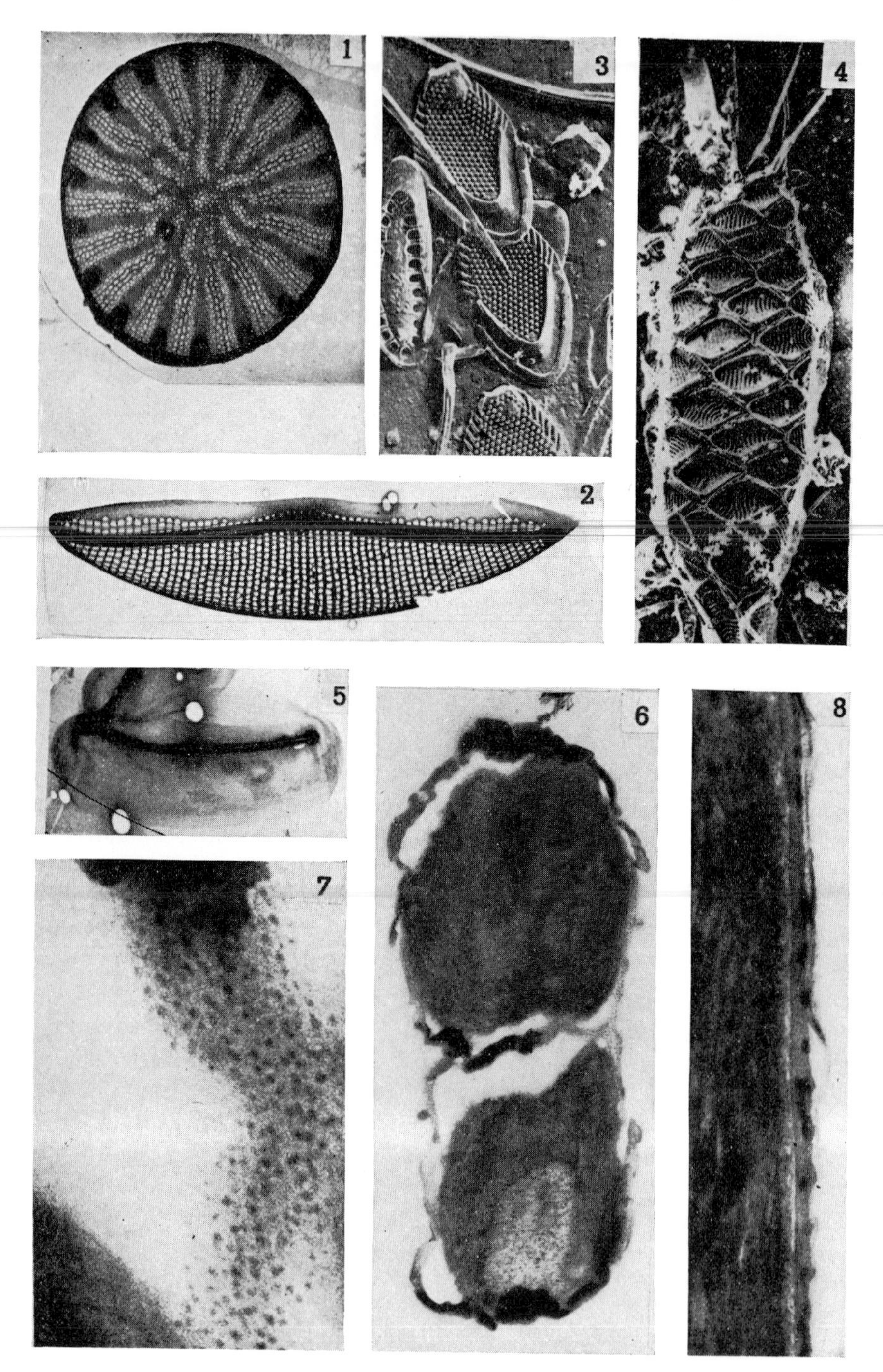
1
3
4
2
5
6
8
7

bear thickened ribs or struts, spines, bristles, etc. In the case of the more complex diatoms, each valve may be composed of two or more parallel layers held together by a system of perpendicular septa (see Desikachary, 1957; Hendey, 1959). The stereoscopic electron micrographs of Helmcke and Krieger (1953, 1954) reveal especially well the changes of structural complexity with depth.

II. Silicification in the Life Cycle of Diatoms

During the past 75 years there have been many descriptive studies on the structure of the silica frustule and on the mode of cell division in diatoms. These early observations have been well reviewed by Hustedt (1930) and by Fritsch (1935). The diatom wall consists essentially of two parts, which are not firmly united but fit into each other like the halves of a box. In the process of cell division, the two newly formed daughter protoplasmic bodies withdraw slightly from each other in the center of the mother cell, and new siliceous structures are deposited on the freshly exposed surfaces. Ultimately the halves of the parent cell wall become completely separated. The new cytoplasmic surfaces of the daughter cells are thus the main active sites of the silicification process, although further deposition of silica can occur in the future development of the cell, probably as a thickening of the existing shells.

Diatoms have not as yet been intentionally grown devoid of their silica shells. Nonsilicified protoplasmic masses have been reported in diatom cultures (Bachrach and Lefèvre, 1929; Wiedling, 1941; von Stosch, 1942; Hendey, 1946), but in all cases except that described by von Stosch (see Section IV) the environmental factors eliciting this condition were unknown. Such naked cells were enclosed within a tenuous membrane, probably the organic part of the cell wall, but they bore no markings char-

---

PLATE I

FIG. 1. Silica valve of a centric diatom, *Cyclotella* sp. (×5610).

FIG. 2. Silica valve of a pennate diatom, *Amphora* sp. ×2400.

FIG. 3. Silica scales and bristles of *Mallomonas papillosa*, also an odd scale of *Synura* sp. (×7578). [From Harris and Bradley (1957); Plate IV, Fig. 9.]

FIG. 4. *Mallomonas coronata*, replica of a complete organism showing silica scales (×3800). [From Harris and Bradley (1957); Plate II, Fig. 1.]

FIG. 5. Organic cell wall of *Phaeodactylum tricornutum* (oval cell form) bearing silica valve (×5610).

FIG. 6. Transverse section through recently divided cell of *Navicula pelliculosa* (×18,360).

FIG. 7. Section of incompletely silicified cell wall of *Navicula pelliculosa*. (Enlarged portion of Fig. 6, ×165,240.)

FIG. 8. Section through the wall of a cell of *Nitzschia angularis* (×49,000). [From Sarah P. Gibbs, Biological Laboratories, Harvard University (unpublished).]

acteristic of the species (Hendey, 1945). Naked chrysophyte flagellates devoid of their customary sheath of silica scales have also been reported to occur, e.g., in *Mallomonas* and *Synura* (see Smith, 1950; Harris, 1953).

In most rigidly silicified diatom species, cell division leads to the formation of smaller and smaller cells as the population multiplies. The formation of auxospores is a process, generally associated with sexual reproduction, making possible the return of the cells to maximum size. During the process of auxospore formation, the cells are released from their enclosing silica frustules and expand within the zygote membranes. The surface of the mature auxospore (i.e., the zygote membrane) may then become weakly silicified, but with a pattern differing from that of the parental valve (see Reimann, 1960). After a lapse of time the cytoplasm contracts within the zygote membrane and becomes enclosed in new silica shells. The process of auxospore formation has been described in detail for various diatoms by Geitler (1932), Iyengar and Subrahmanyan (1944), and von Stosch (1958). Thus silicification also occurs as a consequence of auxospore formation.

III. Composition of the Silica in the Walls

A. Chemical Composition

The siliceous structures of the Bacillariophyta, like those of radiolarians, sponges, and higher plants, are composed of hydrated amorphous silica (i.e., polymerized silicic acid) in a high degree of purity. (An anomalous report by Brandenberger and Frey-Wyssling (1947) of quartz associated with cellulose in the cell walls of *Chlorochytridion tuberculatum*, a species of Green Alga, would be worth further investigation.) A chemical analysis of diatom walls from marine plankton indicated 96.5% SiO_2 and only approximately 1.5% of Al_2O_3 or Fe_2O_3; no titanium or alkaline earth metals could be detected (Rogall, 1939). Desikachary (1957), however, found that acid-cleaned walls of *Desmogonium guyanense* contained aluminum, as indicated by spectroscopic and microchemical tests.

The cleaned and dried frustules of the freshwater diatom *Navicula pelliculosa* contain about 9.6% water (Lewin, 1957), while those of the marine species *Biddulphia sinensis* and *Coscinodiscus concinnus* contain only about 1.9% [Rogall (1939); cf. 5 to 13% water in the siliceous spicules of various sponges (Vinogradov, 1953)]. The rate of solution of a sample of diatom silica in water decreases with time (Jørgensen, 1955a, 1957; Lewin, 1961), suggesting that the amorphous silica within a given species exists in different states of hydration.

B. Physical Composition

Einsele and Grim (1938) stated that the specific gravity of diatom silica is 2.07; Hull *et al.* (1953) gave a value of 2.00 for fossil walls. Published values for the refractive index, 1.434 (see Hustedt, 1930, p. 31) and 1.40 to 1.42 (Hart, 1957), indicate higher water contents than would be indicated by the specific-gravity values quoted above. It would be valuable to have a set of comparative determinations for the specific gravity, refractive index, and degree of hydration of various freshwater and marine species, both recent and fossil; such data are apparently not yet available.

High-resolution electron micrographs of thin parts of diatom shells reveal a foam-like or sponge-like substructure which suggests that the walls are composed of silicic acid gel (Helmcke, 1954). Geissler (1958) showed that the silica walls of diatoms adsorb positively charged colloids to the same degree as do dried silica gels. A heterogeneous structure of small particles, about 10 $m\mu$ in diameter, has been revealed in the highly magnified siliceous cysts of a chrysomonad (Hovasse and Joyon, 1957). Similar particles, approximately 6 $m\mu$ in diameter, are visible in sections of partially silicified walls of recently divided *N. pelliculosa* cells (see Figs. 6 and 7). One may obtain an independent indication of the size of the colloidal particles comprising a silica gel by measuring the specific surface area of the gel, e.g., by the nitrogen adsorption method (Iler, 1955). The value obtained for silica valves from *N. pelliculosa* cells grown in culture (see Lewin, 1961) was found to be about 123 $m.^2/gm.$, which is about five times the "visible" surface and thus indicates a high degree of porosity. If the silica consisted entirely of spherical particles, it can be calculated that their average diameter would be about 22 $m\mu$.

IV. Physiological Studies on Silica Deposition

Some physiological studies on the uptake of silicic acid from solution and on the deposition of silica in walls have been made with diatoms, but nothing is known of the process in other algae.

Diatoms apparently have an absolute requirement for silicon if cell division is to take place. With adequate amounts of all other nutrients in the culture medium, the final yield of cells is proportional to the amount of silicon added (Jørgensen, 1952, 1957; Lewin, 1955a).

Silicon is usually added to culture media in the form of sodium metasilicate, $Na_2SiO_3 \cdot 9H_2O$. In solution this forms orthosilicic acid, $Si(OH)_4$, which around neutrality is only slightly ionized (Greenberg and Sinclair, 1955). Orthosilicic acid, which reacts readily with ammonium molybdate to give the yellow silico-molybdate employed in the colorimetric test for silicon, is probably the form of silicon used by diatoms for growth. There is evidence that more highly polymerized forms cannot be used. A substituted orthosilicic acid, in which one hydroxyl group has been replaced by a methyl group, is likewise inactive (Lewin, 1955a). Furthermore, at least for the growth of *Navicula pelliculosa*, silicon cannot be replaced by any of the elements similar to it in physical and chemical properties or in atomic radius, such as Ge, C, Sn, Pb, As, P, B, Al, Mg, or Fe (Lewin, unpublished).

Since the silica shells of some species are characteristically thicker than those of others (Einsele and Grim, 1938), the amount of silica, when expressed as a percentage of the dry weight, varies over a wide range. In some of the more silicified species, 50% of the dry weight is silica (Vinogradov, 1953). In *N. pelliculosa* the silica content may be as low as 4% of the dry weight (Lewin, 1957); in *Phaeodactylum tricornutum* it is less than 1% (Lewin *et al.*, 1958). Within a given diatom species, too, the amount of silica per cell can vary, dependent upon the concentration of Si in the growth medium and the rate of cell division. As a rule, rapidly dividing cells deposit thinner shells than do more slowly dividing cells (Lund, 1950; Jørgensen, 1955b; Lewin, 1957).

When some nutrient other than Si is limiting, so that the cells cease to divide, *Navicula pelliculosa* can nevertheless continue to take up additional Si. The amount of frustule silica per cell can double under such conditions, indicating thickening of the existing shells (Lewin, 1957). The silicon taken up is equivalent to the amount needed for the production of the two new shells that would be formed if the cells were able to divide. Von Stosch (1942) found that in a culture medium lacking manganese the cells of *Achnanthes longipes* lost the ability to form new shells. The cytoplasm appeared to become detached from the wall and ultimately emerged from it altogether. In this way nearly all of the smaller cells become naked protoplasts, while the larger cells gave rise to swollen, spherical plasmodia containing hundreds of nuclei. When Mn was supplied to such cells they formed silica shells, though of an abnormal structure. Normal silica shells typical of the species were produced only after auxospore formation.

Studies with synchronously dividing cells of *N. pelliculosa* have shown that the cells take up Si from solution at a constant rate throughout the cell division cycle (Lewin, unpublished). There is not, as might have been expected, a sudden uptake of Si just prior to the formation of the two new valves. In this connection it may be mentioned that new silica valves are deposited on freshly exposed cytoplasmic surfaces not only during the process of cell division and after auxospore formation, but also when cells are plasmolyzed. Küster-Winkelmann (1938) observed that when cells of *Achnanthes longipes* were placed in hypertonic media, the cytoplasm contracted, shrinking away from the cell wall, and new walls were deposited within the old frustule. Conversely, Hendey *et al.* (1954) observed that the mean apical length of *Skeletonema costatum* cells was increased when the cells were allowed to undergo a cell division in sea water diluted to one fifth salinity. The use of hypertonic and hypotonic media may prove helpful in experimental studies of the physiology of silica deposition.

Diatoms may remain viable for many weeks in Si-depleted cultures, especially in dim light or in darkness. In bright light they continue to photosynthesize and accumulate intracellular reserves or, in the case of *N. pelliculosa,* quantities of a mucilaginous capsular material (Lewin, 1955b). After a prolonged period in bright light they eventually bleach and die.

"Si-deficient" diatoms, harvested from Si-depleted cultures, have been used for studying the process of silicic acid uptake (Lewin, 1954, 1955c) When transferred to a medium containing Si—but lacking an assimilable nitrogen source— and kept in the dark so that no cell division can take place, such Si-deficient cells continue to remove silicic acid from solution. It has been shown by such techniques that energy-yielding processes are necessary for Si-uptake (Lewin, 1955c). Such agents as cyanide, fluoride, iodoacetate, arsenite, azide, and fluoroacetate inhibit Si-uptake at concentrations which are also inhibitory to respiration. 2,4-Dinitrophenol, which uncouples oxidative phosphorylation from respiration, inhibits Si-uptake by *N. pelliculosa* at concentrations which considerably stimulate O_2-uptake. The respiration of Si-deficient cells of *N. pelliculosa,* depleted of endogenous reserves by starvation in the dark in mineral medium for 2 to 4 days, is stimulated by such substrates as glucose and lactate; at the same time, the potential rate of silicon uptake is also enhanced (Lewin, 1955c).

Silicon uptake is completely suppressed in cells washed repeatedly with distilled water, although respiration is not affected (Lewin, 1954). Cadmium chloride ($10^{-3}M$) has the same effect. Reduced sulfur compounds, such as sodium thiosulfate, glutathione, L-cysteine, DL-methionine, or ascorbic acid plus sulfate, restore the ability of washed cells to remove Si from solution. Evidently either sulfhydryl groups in the cell membrane or enzymes containing sulfhydryl groups are somehow involved in Si uptake.

Although amorphous silica is slowly soluble at the pH of natural waters, the silica shells of living diatoms are somehow protected against solution. After the death of the cells, the silica usually dissolves (see Lewin, 1961). In certain environments, however, the walls of planktonic diatoms may settle and accumulate on the bottom faster than they dissolve, thereby forming a diatomaceous ooze.

V. Integration of Organic Matter with Silica in the Cell Walls

The cytoplasm of a diatom cell is limited by a membrane, as are other plant cells, and is enclosed by a cell wall. The latter is composed of hydrated amorphous silica together with an organic material which has been called "pectin" mainly because it can be stained with ruthenium red. However, since the true chemical nature of the organic component of the wall is at present unknown, it is unfortunate that the term "pectin" has been so widely used in reference to this material. The spatial relation of the organic component to the silica parts of the wall is also unclear. Liebisch (1928, 1929) considered that the two constituents were distinct, and that the wall comprised an inner "pectic" layer and an outer layer of silica. On the other hand, Mangin (1908) believed that the "pectin" of the cell wall was impregnated throughout with silica. Since silica valves of some diatoms are made up of distinguishable layers, the process of deposition probably occurs at different levels within the matrix (Desikachary, 1957; Hendey, 1959; Reimann, 1960). One may conceive of other mechanisms, such as deposition on an organic surface which has a pre-established topography and which serves as a template for the final pattern.

From the few electron micrographs of sections of diatom cells so far examined (J. C. Lewin; S. P. Gibbs; B. Reimann: all unpublished), it has not yet been possible to determine the spatial relationship between the silica and the organic material in the walls (see, for example, Fig. 8). Electron micrographs of sections of the chrysophyte *Synura caroliniana* revealed a distinct protoplasmic membrane presumably responsible for the deposition of the overlying silica scales (Manton, 1955).

The oval cells of *Phaeodactylum tricornutum*, an atypical diatom, are incompletely silicified, having only one silica valve on each cell (Lewin *et al.*, 1958). When such cells are subjected to osmotic shock, the cell contents lyse, leaving the organic parts of the wall still bearing the silica valve (see Fig. 5). This indicates a close relationship between the two wall components. In *Navicula pelliculosa* both valves are silicified, but at the edges in the region of the girdle bands silicification tapers off. In sections, this weakly silicified margin appears to consist of a membrane with embedded electron-dense particles (see Figs. 6 and 7), again indicating a close association between the two components of the wall. On the other hand, as is well-known, the organic constituents may be destroyed by incineration or by treatment with corrosive acids, leaving the mature silica valves intact. According to reports in the literature, the siliceous component may be removed by the action of hydrofluoric acid or by prolonged treatment with steam, leaving the organic membrane undissolved, still bearing in replica (more correctly, *in intaglio*) the typical markings of the silica valve (see Fritsch, 1935).

Although the silica and the organic material of the wall are closely integrated, the whole process of deposition must ultimately be controlled by the living cytoplasm, and thus the cytoplasmic membrane itself must also be involved. New cell walls may be deposited and completely silicified within 10 to 20 minutes after the division of the protoplast (Reimann, 1960). The polymerization process thus takes place very rapidly, with silicification beginning and spreading from certain defined centers. Physicochemical studies by Holt and Bowcott (1954) showed that silicic acid polymerizes to form a rigid network after the molecules have been adsorbed on a protein monolayer, presumably through the formation of hydrogen bonds between hydroxyl groups of the silicic acid and keto-imino groups of the proteins (Clark *et al.*, 1957). Similar reactions might also occur in the diatom cell, but we have still no indication as to the nature of the binding material.

VI. Conclusion

For a satisfactory analysis of the process of silica deposition in the walls of algae, information is needed on mechanisms controlling:

(*a*) the accumulation of silicic acid from the medium;
(*b*) its dehydration and polymerization to form silica gel particles;
(*c*) the deposition of submicroscopic siliceous particles in specific patterns;
(*d*) the stabilization of such shells, while they constitute an integral part of the living cell, against the forces of re-solution.

Unfortunately the information at present available on any of these points is only fragmentary.

References

Bachrach, E., and Lefèvre, M. (1929). Contribution à l'étude du rôle de la silice chez les êtres vivants. Observations sur les Diatomée. *J. physiol. et pathol. gén.* **27,** 241–249.

Brandenberger, E., and Frey-Wyssling, A. (1947). Über die Membransubstanzen von *Chlorochytridion tuberculatum* W. Vischer. *Experentia* **3,** 492–493.

Clark, S. G., Holt, P. F., and Went, C. W. (1957). The interaction of silicic acid with insulin, albumin, and nylon monolayers. *Trans. Faraday Soc.* **53,** 1500–1508.

Desikachary, T. V. (1957). Electron microscope studies on diatoms. *J. Roy. Microscop. Soc.* **76,** 9–36.

Einsele, W., and Grim, J. (1938). Über den Kieselsäuregehalt planktischer Diatomeen und dessen Bedeutung für einige Fragen ihrer Ökologie. *Z. Botan.* **32,** 545–590.

Fott, B., and Ludvík, J. (1957). Die submikroskopische Struktur der Kieselschuppen bei *Synura* und ihre Bedeutung für die Taxonomie der Gattung. *Preslia* **29,** 5–16.

Fritsch, F. E. (1935). "The Structure and Reproduction of the Algae," Vol. 1. Cambridge Univ. Press, London and New York.

Geissler, U. (1958). Das Membranpotential einiger Diatomeen und seine Bedeutung für die lebende Kieselalgenzelle. *Mikroskopie* **13,** 145–172.

Geitler, L. (1932). Der Formwechsel der pennaten Diatomeen (Kieselalgen). *Arch. Protistenk.* **78,** 1–226.

Greenberg, S. A., and Sinclair, D. (1955). The polymerization of silicic acid. *J. Phys. Chem.* **59,** 435–440.

Harris, K. (1953). A contribution to our knowledge of *Mallomonas*. *J. Linnean Soc. London Botany* **55,** 88–102.

Harris, K., and Bradley, D. E. (1957). An examination of the scales and bristles of *Mallomonas* in the electron microscope using carbon replicas. *J. Roy. Microscop. Soc.* **76,** 37–46.

Harris, K., and Bradley, D. E. (1960). A taxonomic study of *Mallomonas*. *J. Gen. Microbiol.* **22,** 750–776.

Hart, T. J. (1957). Notes on practical methods for the study of marine diatoms. *J. Marine Biol. Assoc. United Kingdom* **36,** 593–597.

Helmcke, J.-G. (1954). Die Feinstruktur der Kieselsäure und ihre physiologische Bedeutung in Diatomeenschalen. *Naturwissenschaften* **11,** 254–255.

Helmcke, J.-G., and Krieger, W. (1953, 1954). "Diatomeenschalen im Electronenmikroscopischen Bild." Parts I and II. Transmare-Photo, Berlin. (2nd ed., Engelmann-Cramer, Weinheim, Germany.)

Hendey, N. I. (1945). Extra-frustular diatoms. *J. Roy. Microscop. Soc.* **65,** 34–39.

Hendey, N. I. (1946). Diatoms without siliceous frustules. *Nature* **158,** 588.

Hendey, N. I. (1959). The structure of the diatom cell wall as revealed by the electron microscope. *J. Quekett Microscop. Club* **5,** 147–175.

Hendey, N. I., Cushing, D. H., and Ripley, G. W. (1954). Electron microscope studies of diatoms. *J. Roy. Microscop. Soc.* **74,** 22–34.

Holt, P. F., and Bowcott, J. E. L. (1954). The interaction of proteins with silicic acid. *Biochem. J.* **57,** 471–475.

Hovasse, R., and Joyon, L. (1957). Les kystes siliceux de la Chrysomonadine *Hydrurus foetidus* Kirchner. *Compt. rend. acad. sci.* **244,** 1675–1678.

Hull, W. Q., Keel, H., Kenny, J., and Gamson, B. W. (1953). Diatomaceous earth. *Ind. Eng. Chem.* **45,** 256–269.

Hustedt, F. (1930). Die Kieselalgen Deutschlands, Österreichs und der Schweiz. *In* "Dr. L. Rabenhorst's Kryptogamen-Flora von Deutschland, Österreich und der Schweiz" (L. Rabenhorst, ed.), Vol. VII, Part 1, pp. 12–216. Akademische Verlagsgesellschaft, Leipzig, Germany.

Iler, R. K. (1955). "The Colloid Chemistry of Silica and Silicates." Cornell Univ. Press, Ithaca, New York.

Iyengar, M. O. P., and Subrahmanyan, R. (1944). On the reduction division and auxospore formation in *Cyclotella meneghiniana* Kütz. *J. Indian Botan. Soc.* **23,** 125–153.

Jørgensen, E. G. (1952). Effects of different silicon concentrations on the growth of diatoms. *Physiol. Plantarum* **5,** 161–170.

Jørgensen, E. G. (1955a). Solubility of the silica in diatoms. *Physiol. Plantarum* **8,** 846–851.

Jørgensen, E. G. (1955b). Variations in the silica content of diatoms. *Physiol. Plantarum* **8,** 840–845.

Jørgensen, E. G. (1957). Diatom periodicity and silicon assimilation. *Dansk Botan. Arkiv* **18,** 6–54.

Küster-Winkelmann, G. (1938). Über die Doppelschalen der Diatomeen. *Arch. Protistenk.* **91,** 237–266.

Lewin, J. C. (1954). Silicon metabolism in diatoms. I. Evidence for the role of reduced sulfur compounds in Si utilization. *J. Gen. Physiol.* **37,** 589–599.

Lewin, J. C. (1955a). Silicon metabolism in diatoms. II. Sources of silicon for growth of *Navicula pelliculosa*. *Plant Physiol.* **30,** 129–134.

Lewin, J. C. (1955b). The capsule of the diatom *Navicula pelliculosa*. *J. Gen. Microbiol.* **13,** 162–169.

Lewin, J. C. (1955c). Silicon metabolism in diatoms. III. Respiration and silicon uptake in *Navicula pelliculosa*. *J. Gen. Physiol.* **39,** 1–10.

Lewin, J. C. (1957). Silicon metabolism in diatoms. IV. Growth and frustule formation in *Navicula pelliculosa*. *Can. J. Microbiol.* **3,** 427–433.

Lewin, J. C. (1961). The dissolution of silica from diatom walls. *Geochim. et Cosmochim. Acta.* **21,** 182–198.

Lewin, J. C., Lewin, R. A., and Philpott, D. E. (1958). Observations on *Phaeodactylum tricornutum*. *J. Gen. Microbiol.* **18,** 418–426.

Liebisch, W. (1928). *Amphitetras antediluviana* Ehrbg., soweit einige Beiträge zum Bau und zur Entwicklung der Diatomeenzelle. *Z. Botan.* **20,** 225–271.

Liebisch, W. (1929). Experimentelle und kritische Untersuchungen über die Pektinmembran der Diatomeen unter besonderer Berücksichtigung der Auxosporenbildung und der Kratikularzustände. *Z. Botan.* **22,** 1–65.

Lund, J. W. G. (1950). Studies on *Asterionella formosa* Hass. II. Nutrient depletion and the spring maximum. *J. Ecol.* **38,** 15–35.

Mangin, M. L. (1908). Observations sur les diatomées. *Ann. sci. nat.* (*botan.*) **8,** 177–219.

Manton, I. (1955). Observations with the electron microscope on *Synura caroliniana* Whitford. *Proc. Leeds Phil. Soc.* (*Sci. Sect.*) **6,** 306–316.

Okuno, H., and Kurosawa, K. (1957). Index of diatoms researched with the electron microscope. *Bull. Fac. Textile Fibers, Kyoto Univ.* **2,** 43–60.

Reimann, B. (1960). Bildung, Bau und Zusammenhang der Bacillariophyceenschalen. *Nova Hedwigia* **2,** 349–373.

Rogall, E. (1939). Über den Feinbau der Kieselmembran der Diatomeen. *Planta* **29,** 279–291.

Smith, G. M. (1950). "The Fresh-Water Algae of the United States." McGraw-Hill, New York.

Takahashi, E. (1959). Studies on genera *Mallomonas, Synura,* and other plankton in fresh-water by electron microscope. *Bull. Yamagata Univ. Agr. Sci.* **3,** 117–151.

Vinogradov, A. P. (1953). "The Elementary Chemical Composition of Marine Organisms" Memoir II. Sears Foundation for Marine Research. Yale Univ. Press, New Haven, Connecticut.

von Stosch, H. A. (1942). Form und Formwechsel der Diatomee *Achnanthes longipes* in Abhängigkeit von der Ernährung. Mit besonderer Berücksichtigung der Spurenstoffe. *Ber. deut. botan. Ges.* **60,** 2–16.

von Stosch, H. A. (1958). Kann die oogame Araphidee *Rhabdonema adriaticum* als Bindeglied zwischen den beiden grossen Diatomeengruppen angesehen werden? *Ber. deut. botan. Ges.* **71,** 241–249.

Wiedling, S. (1941). A skeleton-free diatom. *Botan. Notiser* pp. 33–36.

Calcification *

JOYCE C. LEWIN

Scripps Institution of Oceanography, University of California, La Jolla, California

I. Introduction

The deposition of $CaCO_3$ by living algae is an important reaction, especially in tropical seas. Regrettably, there have been few critical physiological or biochemical studies of this process, and there is no experimental evidence that a calcareous skeleton is essential to the growth and development of these forms. The small amount of information now available is summarized in this chapter in the hope of stimulating more intensive investigations of the problem with some of the newer techniques now available, such as have been employed, for example, in studies of calcification in mollusks (Hammen and Wilbur, 1959; Watabe and Wilbur, 1960).

II. Algae That Deposit Calcium Carbonate

Algae that deposit calcium carbonate as part of the structure of their cell walls are summarized in Table I, based on data from Pia, 1934; Fritsch, 1935, 1945; Vinogradov, 1953; and Lowenstam, 1954.[1] According to Horn

* Contribution from Scripps Institution of Oceanography, New Series.

[1] An anomalous report by Walter-Lévy and Strauss (1954) of calcite in *Liagora viscida* and of aragonite in *Tenarea* (= *Lithophyllum*) *tortuosa* suggested a re-examination of these species. Samples collected at Cap de Troc, France (Pyr.-Or.), by Dr. P. C. Silva were therefore subjected to x-ray diffraction analysis by Dr. T. J. Chow. Only aragonite was found in *L. viscida*, and only calcite in *T. tortuosa*, both algae thus conforming with other members of their respective orders (see Table I).

TABLE I
ALGAE CHARACTERIZED BY CALCAREOUS SKELETONS

Algae	Habitat of calcified species		Nature of deposit	Examples
	Marine or freshwater habitat	Geographical location		
Division: Chrysophyta				
Family: Coccolithophoraceae	Planktonic; mostly marine	Mostly tropical to subtropical	Calcite	*Syracosphaera; Coccolithus*
Division: Chlorophyta				
Order: Dasycladales				
Family: Dasycladaceae	Marine	Chiefly tropical and subtropical	Aragonite	*Neomeris; Cymopolia; Acetabularia*
Order: Caulerpales				
Family: Udoteaceae	Marine	Chiefly tropical and subtropical	Aragonite	*Udotea; Halimeda; Penicillus*
Division: Charophyta				
Order: Charales	Freshwater	Tropical, temperate	Calcite	*Chara*
Division: Rhodophyta				
Class: Florideophyceae				
Order: Nemalionales	Marine	Chiefly tropical and subtropical	Aragonite	*Liagora*
Order: Bonnemaisoniales	Marine	Chiefly tropical and subtropical	Aragonite	*Galaxaura*
Order: Cryptonemiales	Marine	Tropical, temperate and polar	Calcite	All Corallinaceae, including: *Corallina; Melobesia; Lithophyllum; Amphiroa*
Division: Phaeophyta				
Order: Dictyotales	Marine	Chiefly tropical and subtropical	Aragonite	Some species of *Padina*

af Rantzien (1959), calcification of the oosporangia is a highly specific feature of many of the Charophyta, whereas calcification of the vegetative parts of the thallus is nonspecific and appears to be correlated with high concentrations of $Ca(HCO_3)_2$ in the habitat. In addition to those groups mentioned in Table I, calcification may occur in other algae, such as several Cyanophyta (species of *Rivularia, Schizothrix*, etc.; Fritsch, 1945, p. 868), Chlorophyta (such as *Chaetophora tuberculosa* and the desmids *Cosmarium quadratum* and *Oocardium stratum*; Wallner, 1933, 1934) and the rare freshwater species of Brown Alga, *Pleurocladia lacustris* (Fritsch, 1945, p. 55). Prát (1925) cultured some of the calcareous Blue-green Algae and found that the incrustations, although a natural and characteristic part of these algae, were not necessary for their growth. In all such Green and Blue-green Algae examined, the material proved to be calcite. In certain cases the form of the crystals was found to be characteristic of the species, even within a uniform habitat (Wallner, 1934).

III. Nature of the Deposition

A. Calcium Carbonate Crystal Organization

Under ordinary conditions anhydrous $CaCO_3$ exists in nature in two crystalline forms, calcite (rhombohedral) and aragonite (orthorhombic), which differ in such properties as structure, hardness, specific gravity, solubility, etc. Calcareous algae precipitate carbonate as either calcite or aragonite. No algae depositing a mixture of these two minerals have been recorded to date (Chave, 1954), although such mixtures are common among the shells of mollusks and other invertebrates. Dispersed crystalline hydrates, such as $CaCO_3 \cdot 6H_2O$, are found in some algae, but they do not form coherent skeletons (Vinogradov, 1953). The presence of associated sulfates or phosphates has apparently not been reported in algal skeletons.

It is among the chrysophyte flagellates classified as Coccolithophoraceae that calcification has developed the greatest structural complexity, with the appearance of crystal arrangements in a variety of form and pattern. The number of coccoliths borne on a cell depends on its size and age; for example, each mother cell of *Coccolithus pelagicus* bears from 25 to 30, while in the daughter cells the number varies from 8 to 17 (Parke and Adams, 1960). The sizes of individual coccoliths range from 1 to 35μ according to species (Kamptner, 1954).

Coccoliths have been classified into two basic types: (*a*) *heterococcoliths*, usually discs or ovals made up of crystallite fibers ("lamellae" according to Halldal and Markali, 1955) arranged radially, often associated with rim elements, center pieces, etc.; and (*b*) *holococcoliths*, made up of recognizable rhombohedral or hexagonal microcrystals of calcite.

In nearly all species, each coccolith comprises only one crystal type. However, in *Calyptrosphaera papillifera* crystal dimorphism occurs within each coccolith, the major part being an assemblage of individual hexagonal prisms arranged in a perfect network of crystals with the *c*-axes normal to the surface of the cell, while the rim consists of calcite rhombohedra (Halldal and Markali, 1954). Coccolith dimorphism may occur on the same cell, as in *Zygosphaera divergens*, where both disciform and zygoform holococcoliths are produced (Halldal and Markali, 1955). The discovery by Parke and Adams (1960) of two distinct types of cells (a motile phase bearing holococcoliths and a nonmotile phase bearing heterococcoliths) in the life history of one coccolithophorid has already upset the system of classification proposed by Halldal and Markali (1955).

All members of the Corallinaceae deposit $CaCO_3$ in the form of calcite, but here the skeletons are much less complex, being composed of individual crystals a few tenths of a micron long with the *c*-axis perpendicular to the longitudinal axis of the cell-wall fibers (Baas-Becking and Galliher, 1931).

Only the rhombohedral form of calcite was deposited by the several (unspecified) species of the Blue-green Algae maintained in culture by Ulrich (1928). Calcite first appears in the cell sheaths as separate crystals, which then gradually fuse until all of the mucilage may be replaced by calcium carbonate (see Fritsch, 1945, p. 869).

On the other hand, the calcified members of the Dasycladales and Caulerpales (Green Algae) and of the Nemalionales and Bonnemaisoniales (Red Algae) deposit acicular crystals of aragonite (see Lowenstam, 1955).

B. Associated Mineral Components

Chave (1952) showed that magnesium carbonate forms a solid solution with calcium carbonate in high magnesian calcitic skeletons, although in natural minerals such a solid solution is formed only under high temperature conditions (Goldsmith *et al.*, 1955). Aragonitic skeletons are lower in magnesium than are those composed of calcite. Thus, aragonitic skeletons from *Halimeda*, *Acetabularia*, *Galaxaura*, *Penicillus*, and *Liagora* contain about 1% $MgCO_3$, whereas calcitic skeletons from Corallinaceae (*Lithothamnium*, *Lithophyllum*, *Corallina*, *Amphiroa*, *Bossea*, and *Goniolithon*) contain considerably more, the recorded values ranging from 7% to almost 30% $MgCO_3$ (Chave, 1954). In contrast, however, the calcareous parts of the Coccolithophoraceae, Characeae, and other algae depositing calcite contain little Mg or only traces (Vinogradov, 1953, p. 62). The calcite coccoliths from *Syracosphaera carterae* contained only 0.1 to 0.15% of Mg (Lewin and Chow, 1961).

In all groups of calcitic organisms a linear relationship exists between the magnesium content of the skeleton and the temperature of the water in which the organism lived. This correlation, although still clear, is less marked in the marine algae than in any other group of organisms (Chave, 1954). Haas *et al.* (1935), studying calcium and magnesium metabolism in calcareous algae such as *Corallina squamata*, found seasonal changes in the amount of these elements on a dry-weight basis. Chave (1954) showed that the ratio of Mg to Ca in the calcite laid down by *Lithothamnium* during the summer is 2 to 4% higher than in the calcite deposited in the winter. According to Baas-Becking and Galliher (1931), the Mg content may fluctuate appreciably within the same organism, and in many cases may be essentially zero.

Strontium carbonate is incorporated to a greater extent in aragonite than in calcite (Thompson and Chow, 1955): e.g., the aragonitic skeleton of a Green Alga *Halimeda* (species not specified) contains about 1.3% $SrCO_3$ (Emery *et al.*, 1954). Strontium analyses by Lowenstam indicated $SrCO_3$ values up to 2.3% for aragonitic Green Algae (see Revelle and Fairbridge, 1957), whereas calcitic Corallinaceae (species of *Bossea*, *Calliarthron*, *Corallina*, *Lithophyllum*, and *Lithothamnium*) analyzed by Thompson and Chow (1955) contained only 0.25 to 0.37% $SrCO_3$. Corresponding values based on analyses of skeletal material were published by Odum (1957), who found Sr/Ca ratios of about 11–13 $\times$ 10^{-3} in aragonitic Green Algae, 3–4 $\times$ 10^{-3} in the calcite of Corallinaceae, and about 0.5 $\times$ 10^{-3} in the calcite of freshwater Charophyta. Somewhat higher ratios (equivalent to 3.9 to 7.6 $\times$ 10^{-3}) were obtained for corallines by Goldsmith *et al.* (1955). The incorporation of Sr into the calcite coccoliths of *Syracosphaera carterae* grown in laboratory cultures containing various amounts of strontium was briefly investigated by Lewin and Chow (1961). Within the limits of tolerance the cells showed a marked ability to discriminate against Sr, taking up only 2% of the amount which would be expected if there had been no discrimination.

IV. Organic Matrix of the Calcified Walls

A feature common to all calcareous algae is the presence of gelatinous or mucilaginous substances associated with the cell walls. This organic material may be involved in the deposition and organization of the crystalline mineral, although of course most algae with gelatinous surfaces are not calcareous.

The organic cell walls of corallines (Red Algae) are not composed of cellulose, being insoluble in ammoniacal copper hydroxide and soluble in alkali. They contain appreciable amounts of so-called "pectic materials" which can be stained with ruthenium oxychloride or aniline blue, and which are believed to contain methoxy groups (Baas-Becking and Galliher, 1931). "Pectic substances," associated with "callose," were reported in the cell walls of calcified species of Green Algae belonging to the Udoteaceae (Mirande, 1913, p. 190). The main cell-wall constituents of *Udotea, Chlorodesmis,* and *Halimeda* sp.—all belonging to this family—were later identified as β-1:3-linked xylans; whereas fibers isolated from decalcified walls of *Acetabularia calyculus* and *Halicoryne wrightii* (Dasycladaceae) proved to consist chiefly of β-1:4-linked mannose units (Iriki and Miwa, 1960; see Kreger, Chapter 19, this treatise).

Organic scales having the same outlines as the coccoliths remain after decalcification of coccolithophore cells with acid (Braarud *et al.*, 1952) and can also be found in young cells before the scales have become calcified (Parke and Adams, 1960). Neither this matrix material of the scales nor other organic cell-wall constituents of the coccolithophorids has been investigated chemically. The cell walls of *Pleurochrysis scherffelii* gave no cellulose reaction with zinc chloriodide, but they stained with methylene blue, gentian violet, and safranin (Pringsheim, 1955). The outer slime layer stained similarly but more weakly. In the "*Crystallolithus*" phase of *Coccolithus pelagicus*, the inner layer of the cell wall gave a reaction with Schultze's solution suggesting cellulose, but the outer striated hyaline layer in which the coccoliths are embedded gave no reaction for "pectic" material with ruthenium red (Parke and Adams, 1960).

V. Discussion

The amount of $CaCO_3$ dissolved in sea water has a considerable effect on the distribution of calcareous algae and the composition of their skeletons. Tropical waters with a high temperature and a low partial pressure of CO_2 are usually saturated or supersaturated with $CaCO_3$, the concentrations being especially high in the upper layers, whereas in the seas of polar regions the saturation does not even reach 90% (see Vinogradov, 1953, p. 572). The distribution of algae with calcified skeletons shows a close correlation with the degree of saturation of sea water by $CaCO_3$. Corallines and other calcareous algae are important contributors to the enormous formation of reef limestone, which occur chiefly between the 30° S. and 30° N. parallels (Vinogradov, 1953, p. 572). Likewise, the planktonic Coccolithophoraceae are much more abundant in tropical and temperate regions of the sea (see Braarud *et al.*, 1952).

It has commonly been assumed that $CaCO_3$ deposition is a consequence of the extraction of carbon dioxide from the water during photosynthesis, shifting the equilibrium of Eq. (1) towards the right:

$$Ca^{++} + 2HCO_3^- \rightleftharpoons CaCO_3 + H_2O + CO_2 \qquad (1)$$

One might therefore expect all photosynthetic marine algae to be coated with a layer of $CaCO_3$. Looked at from this viewpoint, the problem is not so much why only certain algae deposit $CaCO_3$, but why most seaweeds *do not* do so. Many algae with apparently high rates of production of organic matter live immediately adjacent to $CaCO_3$ depositors, yet show no calcification. Even among those algae whose surface is mineralized, only certain parts of the surface become coated, other areas of the same organism remaining free of $CaCO_3$; this distribution is not necessarily correlated with the distribution of photosynthetic tissue. If sea water were initially just saturated with $CaCO_3$, then approximately 2 g. of $CaCO_3$ could be deposited for each gram of carbon dioxide reduced in photosynthesis and incorporated in the plant tissue, and the ratio of $CaCO_3$ to organic carbon by weight should be about 8:1, as pointed out by Revelle and Fairbridge (1957). However, these authors reported that in *Halimeda* fragments analyzed by Emery *et al.* (1954) the ratio of $CaCO_3$ to organic carbon was nearly 30:1, and they therefore considered it likely that the aragonite initially precipitated during photosynthesis in heavily supersaturated tropical waters might then act as a nucleus for further precipitation.

Craig (1953, 1954) studied the isotopic fractionation of C^{12} and C^{13} in the organic tissues and mineralized parts of calcified algae. In aragonite-producing Green Algae, namely species of *Halimeda*, *Acetabularia*, and *Penicillus*, the organic matter was much enriched in C^{12} compared to oceanic bicarbonate, while the $CaCO_3$ skeletons were enriched in C^{13}, indicating that the dissolved CO_2 in the sea water is partitioned into heavy and light components during photosynthesis (see Revelle and Fairbridge, 1957, p. 262; Parke and Epstein, 1960). However, in the one calcitic Red Alga (a species of *Corallina*) studied by Craig, both the $CaCO_3$ and the organic carbon were enriched in C^{12} compared to oceanic bicarbonate, suggesting that metabolic processes other than photosynthesis were involved in the precipitation. This is in agreement with the observation that large amounts of $MgCO_3$ are somehow metabolically incorporated to form a solid solution in algal calcite (see Revelle and Fairbridge, 1957, p. 262). The data on the isotopic composition of oxygen in the carbonate skeletons of species of *Halimeda*, *Rhipocephalus*, *Penicillus*, and *Udotea* (Lowenstam and Epstein, 1957) also suggest that such algal carbonate is precipitated under non-equilibrium conditions (Parke and Epstein, 1960).

The secretion of coccoliths *inside* the cells of coccolithophorids, as reported by a number of workers, presents special physiological problems. Parke and Adams (1960) presented evidence that these structures may originate in vesicles deep within the cell, though how they are formed and how they are later arranged on the cell surface is not known. If these observations are correct, then the accumulation and deposition of $CaCO_3$ cannot be merely a reaction occurring at the cell surface as a result of changes in CO_2 tension due to photosynthesis, since the $CaCO_3$ has somehow first to be accumulated and incorporated within the cell. The observations that cells in old cultures often lose their coccoliths (Mjaaland, 1956), and that these are not immediately replaced, also indicate that more than surface reactions are involved.

Horn af Rantzien (1959) suggested that calcification of the fertilized oosporangia among the Charophyta is an active physiological process correlated with the high concentration of calcium salts, possibly including succinate, in the cell sap.

Note added in proof: Some chemical and physiological data on calcified algae are reviewed in a recent book on the subject by Johnson (1961).

References

Baas-Becking, L. G. M., and Galliher, E. W. (1931). Wall structure and mineralization in coralline algae. *J. Phys. Chem.* **35,** 467–479.

Braarud, T., Gaarder, K. R., Markali, J., and Nordli, E. (1952). Coccolithophorids studied in the electron microscope. Observations on *Coccolithus huxleyi* and *Syracosphaera carterae*. *Nytt Mag. Botan.* **1,** 129–133.

Chave, K. E. (1952). A solid solution between calcite and dolomite. *J. Geol.* **60,** 190–192.

Chave, K. E. (1954). Aspects of the biogeochemistry of magnesium. I. Calcareous marine organisms. *J. Geol.* **62,** 266–283.

Craig, H. (1953). The geochemistry of the stable carbon isotopes. *Geochim. et Cosmochim. Acta* **3,** 53–92.

Craig, H. (1954). Carbon 13 in plants and the relationships between carbon 13 and carbon 14 variations in nature. *J. Geol.* **62,** 115–149.

Emery, K. O., Tracey, J. I., and Ladd, H. S. (1954). Geology of Bikini and nearby atolls. *U.S. Geol. Survey. Profess. Papers.* **No. 260–A.**

Fritsch, F. E. (1935). "The Structure and Reproduction of the Algae," Vol. I. Cambridge Univ. Press, London and New York.

Fritsch, F. E. (1945). "The Structure and Reproduction of the Algae," Vol. II. Cambridge Univ. Press, London and New York.

Goldsmith, J. R., Graf, D. L., and Joensuu, O. I. (1955). The occurrence of magnesian calcites in nature. *Geochim et Cosmochim. Acta* **7,** 212–230.

Haas, P., Hill, T. G., and Karstens, W. K. H. (1935). The metabolism of calcareous algae. II. The seasonal variation in certain metabolic products of *Corallina squamata* Ellis. *Ann. Botany (London)* **49,** 609–619.

Halldal, P., and Markali, J. (1954). Morphology and microstructure of coccoliths studied in the electron microscope. *Nytt Mag. Botan.* **2,** 117–118.

Halldal, P., and Markali, J. (1955). Electron microscope studies on coccolithophorids from the Norwegian Sea, the Gulf Stream and the Mediterranean. *Avhandl. Norske Videnskaps Akad. Oslo I. Mat.-Naturv. Kl.* **No. 1,** 5–30.

Hammen, C. S., and Wilbur, K. M. (1959). Carbon dioxide fixation in marine invertebrates. I. The main pathway in the oyster. *J. Biol. Chem.* **234,** 1268–1271.

Horn af Rantzien, H. (1959). Recent charophyte fructifications and their relations to fossil charophyte gyrogonites. *Arkiv. Botan. Ser. 2.* **4,** 165–332.

Iriki, Y., and Miwa, T. (1960). Chemical nature of the cell wall of the green algae, *Codium, Acetabularia,* and *Halicoryne. Nature* **185,** 178–179.

Johnson, J. H. (1961). Limestone-Building Algae and Algal Limestones. 297 pp. Johnson Publishing Company, Boulder, Colo.

Kamptner, E. (1954). Untersuchungen über den Feinbau der Coccolithen. *Arch. Protistenk.* **100,** 1–90.

Lewin, R. A., and Chow, T. J. (1961). La enpreno de strontio en kokolitoforoj. *Plant and Cell Physiol.* **2,** 203–208.

Lowenstam, H. A. (1954). Factors affecting the aragonite:calcite ratios in carbonate-secreting marine organisms. *J. Geol.* **62,** 284–322.

Lowenstam, H. A. (1955). Aragonite needles secreted by algae and some sedimentary implications. *J. Sediment. Petrol.* **25,** 270–272.

Lowenstam, H. A., and Epstein, S. (1957). On the origin of sedimentary aragonite needles of the Great Bahama Bank. *J. Geol.* **65,** 364–375.

Mirande, R. (1913). Recherches sur la composition chimique de la membrane et la morcellement du thalle chez les Siphonales. *Ann. sci. nat.* (*9e sér., botan.*) **18,** 147–264.

Mjaaland, G. (1956). Some laboratory experiments on the coccolithophorid *Coccolithus huxleyi. Oikos* **7,** 251–255.

Odum, H. T. (1957). Biogeochemical deposition of strontium. *Inst. Marine Sci.* **4,** 38–114.

Parke, M., and Adams, I. (1960). The motile (*Crystallolithus hyalinus* Gaarder and Markali) and non-motile phases in the life history of *Coccolithus pelagicus* (Wallich) Schiller. *J. Marine Biol. Assoc. United Kingdom* **39,** 263–274.

Parke, R., and Epstein, S. (1960). Carbon isotope fractionation during photosynthesis. *Geochim. et Cosmochim. Acta* **21,** 110–126.

Pia, J. (1934). Die Kalkbildung durch Pflanzen. *Botan. Centr. Beih.* **A52,** 1–72.

Prát, S. (1925). The culture of calcareous Cyanophyceae. *Studies Plant Physiol. Lab. Charles Univ. Prague* **3,** 86–88.

Pringsheim, E. G. (1955). Kleine Mitteilungen über Flagellaten und Algen. I. Algenartige Chrysophyceen in Reinkultur. *Arch. Mikrobiol.* **21,** 401–410.

Revelle, R., and Fairbridge, R. (1957). Carbonates and carbon dioxide. *In* "Treatise on Marine Ecology and Paleoecology," (J. W. Hedgpeth, ed.), Vol. I. *Geol. Soc. Am. Mem.* **No. 67,** Chapter 10, pp. 239–296.

Thompson, T. G., and Chow, T. J. (1955). The strontium-calcium ratio in carbonate-secreting marine organisms. *Papers in Marine Biol. and Oceanog. Deep-Sea Besearch* **3** (Suppl.) 20–39.

Ulrich, F. (1928). Über die Wachstumsform des organogen abgeschiedenen Kalkspates und ihre Beeinflussung durch das Kristallisationsmedium. *Z. Krist.* **66** 513–515.

Vinogradov, A. P. (1953). "The Elementary Chemical Composition of Marine Organisms," Mem. II. Sears Foundation for Marine Research, Yale University, New Haven, Connecticut.

Wallner, J. (1933). *Oocardium stratum* Naeg., eine wichtige tuffbildende Alge Südbayerns. *Planta* **20,** 287–293.

Wallner, J. (1934). Zur Kenntnis des unter pflanzlichem Einfluss gebildeten Kalkspates. *Planta* **23,** 51–55.

Walter-Lévy, L., and Strauss, R. (1954). Contribution à l'étude des concrétions minérales chez les végétaux. *Compt. rend. acad. sci.* **239,** 897–899.

Watabe, N., and Wilbur, K. M. (1960). Influence of the organic matrix on crystal type in molluscs. *Nature* **188,** 334.

—29—

Volatile Constituents

TERUHISA KATAYAMA

Department of Fisheries, Hiroshima University, Fukuyama, Japan

I. Introduction

Studies on the volatile constituents of seaweed are comparatively few. Heilbron and Phipers (1935) detected the presence of terpenes in the non-saponifiable matter of *Fucus vesiculosus*. Haas (1935) found that *Polysiphonia fastigiata* produced dimethyl sulfide, and showed that treatment of this seaweed with hot water inhibited the evolution of this compound. Obata *et al.* (1951) also isolated dimethyl sulfide from *Ulva pertusa*, and a similar substance was detected in cultures of the diatom *Phaeodactylum tricornutum* (Armstrong and Boalch, 1960). Its biochemistry is reviewed by Schiff (Chapter 14, this treatise). Systematic studies of various other volatile constituents of algae have been carried out by Katayama and Tomiyama (1951) and Katayama (1953–1961).

II. Distillation of Volatile Constituents

Since a considerable quantity of material has to be used in order to collect volatile constituents in amounts sufficient for chemical identification, sun-dried or air-dried thalli, which are easily handled and stored, were generally employed although some volatile components may thereby have been lost. Occasionally fresh fronds were used.

Certain volatile constituents may vaporize when algal fronds are heated slightly. Haas (1935) demonstrated the evolution of dimethyl sulfide from *Polysiphonia fastigiata* by passing a flow of air over fresh plants heated to

TABLE I

VOLATILE COMPONENTS DETECTED IN VARIOUS SEAWEEDS

Compound	Chlorophyta			Phaeophyta		Rhodophyta	
	Ulva pertusa	*Enteromorpha* sp.	*Codium fragile*	*Sargassum* sp.	*Laminaria* sp.	*Porphyra tenera*	*Digenea simplex*
Sulfur compounds							
Dimethyl sulfide	+	+	+	−	−	−	−
Methanethiol	−	−	−	+	+	+	+
Acids							
Formic	+	+	+	+	+	+	+
Acrylic	+	+	+	−	−	−	−
Acetic	+	+	+	+	+	+	+
Propionic	+	+	+	+	+	+	+
Butyric	+	+	+	+	+	+	+
Isovaleric	+	+	+	+	+	+	+
n-Caproic	+	+	+	+	+	+	+
Caprylic	+	+	+	+	+	+	+
Myristic	+	+	+	+	+	+	+
Palmitic	+	+	+	+	+	+	+
Linoleic	+	+	+	+	+	+	+
Aldehydes							
Furfural	+	+	+	+	+	+	+
Methyl furfural	+	+	+	+	+	+	+
n-Valeraldehyde	+	+	+	+	+	+	+
Benzaldehyde	+	+	+	+	+	+	+
Alcohols							
Furfuryl	+	+	+	+	+	−	−
Unidentified R_f 0.21[a]	+	+	+	−	−	−	−
Unidentified R_f 0.26[a]	−	−	−	−	−	+	+
Unidentified R_f 0.35[a]	−	−	−	+	+	−	−
Terpenes[b]							
1,8-Cineol	+	+	+	+	+	+	+
p-Cymene	+	+	+	+	+	+	+
Linalool	+	+	+	+	+	+	+
Geraniol	+	+	+	+	+	+	+
Phenols							
p-Cresol	+	+	+	+	+	+	+
Hydrocarbons							
Heneikosane	+	+	+	+	+	+	+

[a] Chromatostrip; solvent 15% ethyl acetate in *n*-hexane.

[b] The presence of α-pinene and *d*-limonene in *Ulva pertusa* and *Laminaria* sp. has been confirmed by gas chromatography.

TABLE II

Yields of Volatile Constituents in Steam Distillates of Marine Algae

Division	Species	Volatile constituents (% dry weight)
Chlorophyta	*Ulva pertusa*	0.019
	Enteromorpha sp.	0.021
	Codium fragile	0.034
Phaeophyta	*Laminaria* sp.	0.051
	Hizikia fusiforme	0.053
	Undaria pinnatifida	0.054
	Sargassum sp.	0.062
	Scytosiphon sp.	0.053
Rhodophyta	*Digenea simplex*	0.098
	Gracilaria sp.	0.094
	Gelidium sp.	0.130
	Porphyra tenera	0.053

30°C., and then through a saturated solution of mercuric chloride in ethanol, in which dimethyl sulfide was trapped and precipitated as a complex salt. Obata *et al.* (1951) employed a similar method.

A greater variety of volatile constituents, however, can be obtained by subjecting air- or sun-dried fronds to steam distillation. Takaoka and Ando (1951) separated the steam distillate of *Dictyopteris divaricata* into fatty acids and neutral compounds, and identified cadinene, a sesquiterpene, in the latter fraction. The results of some analyses of marine algae by Katayama and Tomiyama (1951) and Katayama (1953–1961) are summarized in Table I.

The amounts of volatile constituents differ according to habitat (Katayama, 1956a), season of collection (Ando, 1953a, b; Katayama, 1956a), and the portion of the thallus examined (Katayama, 1960b). As a rule, larger yields were obtained from the five species of Phaeophyta and the four Rhodophyta than from the three Chlorophyta examined (Table II).

III. Methods of Identification

The general method of analysis employed is summarized in Table III.

IV. Correlation of Volatile Constituents with Characteristic Odors of Seaweeds

The quantities of volatile constituents obtained from dried *Laminaria* are shown in Table IV. In Table I are summarized the volatile constituents

TABLE III

GENERAL SCHEME FOR SEPARATION OF VOLATILE CONSTITUENTS

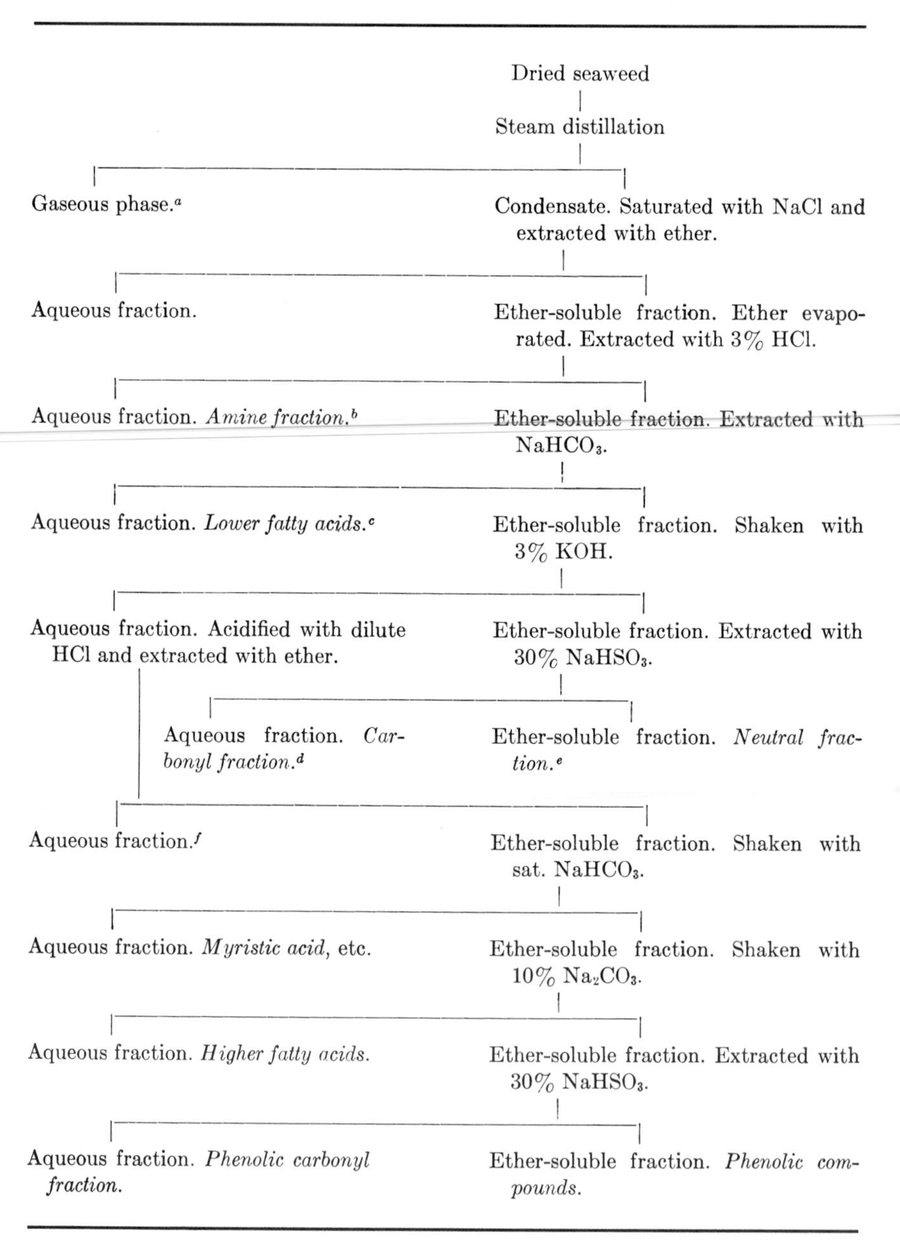

separated by the foregoing method from three species of Chlorophyta: *Ulva pertusa* (Katayama and Tomiyama, 1951; Katayama, 1953, 1955a, 1955b), *Enteromorpha* sp. (Katayama, 1955c), and *Codium fragile* (Katayama, 1958b); two species of Phaeophyta: *Sargassum* sp. (Katayama, 1955d) and *Laminaria* sp. (Katayama, 1958c); and two species of Rhodophyta: *Porphyra tenera* (Katayama, 1956a) and *Digenea simplex* (Katayama, 1958a).

As shown in Table I, fatty acids, phenolic compounds, and carbonyl compounds, together with neutral compounds with higher boiling points, were isolated from all the seaweeds examined. Differences, however, were observed among the more volatile of the neutral compounds, and in those containing sulfur. Dimethyl sulfide was obtained only in the distillates from Chlorophyta, although it was first discovered by Haas in *Polysiphonia* (Rhodophyta). Its distribution correlated with that of acrylic acid, these two compounds being products of the enzymatic cleavage of dimethyl propiothetin (see Schiff, Chapter 14, this treatise). Methanethiol was obtained from Phaeophyta and Rhodophyta, but not from the Chlorophyta examined. Furfuryl alcohol was present in both the Chlorophyta and Phaeophyta but not in the Rhodophyta examined. In general, the great-

---

[a] The vapor phase that did not condense during steam distillation was passed through a saturated solution of mercuric chloride. The mercuric chloride double salts of dimethyl sulfide and methanethiol were isolated in this way.

[b] The amine fraction was separated into primary and tertiary amines, which were then crystallized and identified as their picrates.

[c] The fatty-acid fractions were analyzed by paper chromatography and by silica-gel column chromatography with methanol as the stationary phase and iso-octane as the mobile phase. The acids present in each fraction were characterized as their *p*-bromophenacyl esters.

[d] The carbonyl fraction was further fractionated on a "chromatoplate" (Kirchner *et al.*, 1951; Miller and Kirchner, 1952, 1953), followed by treatment with 2,4-dinitrophenylhydrazine, and the benzene extract of each hydrazone band was column-chromatographed with benzene on activated alumina and identified.

[e] The neutral components, chiefly terpenes and alcohols, were further fractionated in a modified Widmer micro-still (Katayama, 1955a), and purified by the use of 15% ethyl acetate in *n*-hexane on a mixture of silica gel and gypsum in a "chromatostrip" (Kirchner *et al.*, 1951; Miller and Kirchner, 1952, 1953). They were identified by R_f values, pressure *versus* boiling-point curves, and the colors developed on one-dimensional chromatograms sprayed with a 3:1 mixture of concentrated sulfuric and nitric acids. No long-chain alcohols were detected.

[f] The higher fatty acids were separated into saturated and unsaturated acids by the *Pb*-salt alcohol method (Kita, 1957). In the saturated fatty-acid fraction, palmitic acid (C_{16}) was identified as its *p*-bromophenacyl ester. The unsaturated acid fraction was saturated by hydrogenation, and the presence of linoleic acid (C_{18}) was thereby determined, together with a C_{16}-unsaturated fatty acid.

TABLE IV

YIELDS OF DIFFERENT VOLATILE CONSTITUENTS FROM 80 KG. OF DRIED *Laminaria* SP.[a]

Fraction	Odor	Yield (gm.)
Methanethiol (as $HgCl_2$ salt)	Unpleasant	0.1
Acetone and propionaldehyde	Unpleasant	0.6
Trimethylamine (as hydrochloride)	Unpleasant, fishy	0.05
Amine fraction (as hydrochloride)	Unpleasant	1.1
Lower fatty acid fraction	Unpleasant	5.9
Higher fatty acid fraction	Unpleasant	6.1
Phenol fraction	Unpleasant	3.0
Carbonyl fraction	Aromatic	1.9
Neutral fraction (terpenes, etc.)	Aromatic	7.3

[a] From Katayama (1958c).

est differences were observed between the volatile fractions of Chlorophyta and Rhodophyta. It may be that these volatile constituents are linked with biochemical processes characteristic of the particular divisions. It is hoped that these observations will soon be extended both qualitatively and quantitatively by the use of gas chromatography.

V. Antibiotic Activity

The antibacterial action of the lower fatty acid, higher fatty acid, carbonyl, and terpene fractions was determined on the basis of turbidity and plate-culture observations. All fractions at dilutions of 2000–3000 inhibited the growth of *Micrococcus pyogenes* var. *aureus* (*Staphylococcus aureus*) strain 209p. and of *Escherichia coli* (Katayama, 1956b, 1960a). The carbonyl and terpene fractions were found to be paralytic or toxic to certain annelids (Katayama, 1956c, 1956d); while in dilutions of certain acid or neutral fractions the movement of the nematode *Ascaris* was depressed or contorted (Bando and Katayama, 1955).

REFERENCES

Ando, Y. (1953a). Studies on essential oil of sea weeds-II. On the essential oil contents of various kinds of sea weeds. *Bull. Japan. Soc. Sci. Fisheries* **19,** 713–716.

Ando, Y. (1953b). Studies on essential oil of sea weeds-III. The constituents of the oil of *Dictyota dichotoma. Bull. Japan. Soc. Sci. Fisheries* **19,** 717–721.

Armstrong, F. A. J., and Boalch, G. T. (1960). Volatile organic matter in algal culture media and sea water. *Nature* **185,** 761–762.

Bando, T., and Katayama, T. (1955). Pharmacological action of volatile constituents of seaweed against *Ascaris suilla.* I. *Nippon Yakurigaku Zasshi* **51,** 40–41.

Haas, P. (1935). The liberation of methylsulfide by seaweed. *Biochem. J.* **29,** 1297–1299.

Heilbron, I. M., and Phipers, R. F. (1935). The lipochromes of *Fucus vesiculosus. Biochem. J.* **29,** 1369–1375.

Katayama, T. (1953–1961). Chemical studies on volatile constituents of seaweed. (1953): *Bull. Japan. Soc. Sci. Fisheries* **19,** 793–797; (1955a): *Ibid.* **21,** 413–415; (1955b): *Ibid.* **21,** 416–419; (1955c): *Ibid.* **21,** 420–424; (1955d): *Ibid.* **21,** 425–428; (1956a): *Ibid.* **22,** 244–247; (1956b): *Ibid.* **22,** 248–249; (1956c): *Ibid.* **22,** 251–252; (1956d): *Ibid.* **22,** 253–256; (1958a): *Ibid.* **24,** 205–208; (1958c): *Ibid.* **24,** 346–354; (1959): *Ibid.* **24,** 925–932; (1961): *Ibid.* **27,** 75–84.

Katayama, T. (1958b). Chemical studies on volatile constituents of seaweed. *J. Fac. Fisheries Animal Husbandry, Hiroshima Univ.* **2,** 67–77.

Katayama, T. (1960a). Structure and antibacterial activity of terpenes. *Bull. Japan. Soc. Sci. Fisheries* **26,** 29–32.

Katayama, T. (1960b). Classification of seaweed through volatile constituents and biochemical significance of their existence. *Bull. Japan. Soc. Phycol.* **8,** 79–84.

Katayama, T., and Tomiyama, T. (1951). Chemical studies on volatile constituents of seaweed. I. Fractionation of volatile constituents of *Ulva pertusa* K. *Bull. Japan. Soc. Sci. Fisheries* **17,** 122–127.

Kirchner, J. G., Miller, J. M., and Keller, G. J. (1951). Separation and identification of some terpenes by a new chromatographic technique. *Anal. Chem.* **23,** 420–425.

Kita, G. (1957). "Yushi oyobi Sono Shikenho" ("Fats and Oils, and Methods of Their Examination") (in Japanese) p. 707. Shibundo, Tokyo.

Miller, J. M., and Kirchner, J. G. (1952). Some improvements in chromatographic techniques for terpenes. *Anal. Chem.* **24,** 1480–1482.

Miller, J. M., and Kirchner, J. G. (1953). Chromatostrips for identifying constituents of essential oils. *Anal. Chem.* **25,** 1107–1109.

Obata, Y., Igarashi, H., and Matano, K. (1951). Studies on the flavor of seaweed. I. On the laver-like flavor constituents of green algae. *Bull. Japan. Soc. Sci. Fisheries* **17,** 60–62.

Takaoka, M., and Ando, Y. (1951). Studies on essential oil of seaweed. I. *J. Chem. Soc. Japan* **72,** 999–1004.

—30—

Extracellular Products

G. E. FOGG[1]

Department of Botany, University College, London, England

I. Introduction

It is generally recognized that healthy cells of bacteria and fungi may liberate organic substances into the medium in which they grow, but the idea that algae may produce extracellular substances in a similar fashion is not often considered. This may be because Pütter (1908) greatly overemphasized the quantitative importance of such excretion by algae, and several subsequent investigators (e.g., Krogh *et al.*, 1930) were concerned to show that it takes place to a much lesser extent than he supposed. Certainly, from the information available it was reasonable for Myers (1951) to conclude that "conservation of carbon and a minimal level of excretion is probably generally characteristic of the green algae." However, more recent investigation has increasingly tended to show that extracellular products of algae belonging to various groups are, at least under

[1] Present address: Department of Botany, Westfield College, London, England.

certain circumstances, considerable in amount and probably ecologically important.

II. General Considerations

Extracellular products are here defined as soluble substances liberated from healthy cells, as distinct from substances set free by injured cells or by autolysis or decomposition of dead ones. Mucilages will not be considered in any detail, although in current bacteriological usage such materials are referred to as extracellular products (Wilkinson, 1958). It is not always clear whether organic matter reported to be present in culture filtrates is extracellular in this restricted sense or not. Where analysis shows that a particular substance or class of substances constitutes the greater part of the organic matter in a culture filtrate, it is likely that it has not been derived by cell breakdown; indeed, some investigators have shown that the production of certain extracellular substances is relatively high during active growth when few, if any, cells are moribund or dead. Except where otherwise mentioned, the results discussed below were obtained with pure bacteria-free cultures; results obtained with mixed or contaminated cultures must be regarded with reserve.

Various circumstances can result in the escape of substances from a living cell. In the simplest situation the cell membrane may be permeable to a metabolic intermediate so that diffusion takes place until an equilibrium concentration is reached in the external medium. At equilibrium, the amount of substance liberated would then be proportional to the volume of the medium, and independent of the number of algal cells, at least in small cultures but not, of course, in lakes or the sea. Such a relationship has apparently not been reported; in two instances the amounts of extracellular products were found to be proportional to the amount of alga, and not related to the volume of the medium (Fogg, 1952, 1958). However, in a dynamic system in which metabolites are simultaneously produced and consumed perhaps this simple relationship is not to be expected. Another possibility is that a by-product of metabolism may accumulate extracellularly, like alcohol in yeast fermentation or organic acids in cultures of molds supplied with an excess of sugars. This does not appear to occur to any great extent when algae are cultured chemotrophically in air, but it may be appreciable under anaerobic conditions. For example, *Ochromonas malhamensis* can ferment glucose almost quantitatively to carbon dioxide and alcohol (Reazin, 1956). Some of the nitrogenous substances liberated by Blue-green Algae seem resistant to metabolic breakdown; they are evidently by-products of protein synthesis since they are produced most abundantly during rapid growth (Fogg, 1952). Perhaps many of the antibiotic substances produced by algae may be similarly regarded as by-products of metabolism. Other substances found in culture filtrates may be derived by solution or hydrolysis of capsular materials. This seems to be the origin of the extracellular polysaccharides of *Chlamydomonas* spp. studied by Lewin (1956), for their amount is proportional to that of the cell material and there are only slight chemical differences between them and the capsular polysaccharides of the same cells.

When the effect of different conditions on the production of extracellular substances is considered, its general dependence on growth must be taken into account. It is not valid to make direct comparison of either the absolute or relative amounts of these substances present at a given time after inoculation in cultures grown under different conditions, if, as usually happens, the different conditions result in different rates of growth. Quantitative comparisons can be made only at equivalent stages of growth, considering the amount of extracellular products as a function of growth in cell material. An allometric method of doing this has been suggested by Fogg (1952).

Measurements of the amounts of extracellular material can have no precise significance unless the physiological history of the system and its environmental conditions are defined. Nevertheless, it may be worth giving a few values. Braarud and Føyn (1930) found extracellular organic matter amounting to 30% of the total organic matter synthesized in cultures of a marine species of *Chlamydomonas*, and more recently Allen (1956) reported values of 10 to 45% for healthy cultures of freshwater species of this genus. Fogg (1952) found that from 5 to 60% of the nitrogen fixed by *Anabaena cylindrica* may appear extracellularly in combined form. In short-term experiments Tolbert and Zill (1957) discovered that extracellular glycolic acid accounted for 3 to 10% of the total carbon fixed by photosynthesizing cells of *Chlorella pyrenoidosa*. Clearly the production of extracellular substances by algae can sometimes be quantitatively considerable.

III. The Nature of the Extracellular Products Formed by Algae

A. Acids

The colorless alga *Prototheca zopfii* was found by Barker (1935) to produce considerable quantities of L-lactic acid, together with lesser amounts of succinic acid, during fermentation of glucose under semi-anaerobic conditions. Thiamine-deficient *Prototheca* cells liberate pyruvic acid (Anderson, 1945). Under anaerobic conditions, species of *Chlorella* and other Green Algae ferment sugars with the accumulation of small amounts of organic acids, perhaps mainly acetic and lactic acids (Gaffron, 1939;

Syrett, 1958). *Ochromonas malhamensis* also produces traces of lactic acid during fermentation of glucose (Reazin, 1956).

Algae growing autotrophically may also liberate organic acids. Goryunova (1950) reported the presence of oxalic, tartaric, succinic, and other acids in filtrates from cultures of *Oscillatoria splendida*. Various *Chlamydomonas* species liberate appreciable quantities of glycolic, oxalic, and, probably, pyruvic acids (Allen, 1956). Lewin (1957) confirmed the formation of glycolic acid as an extracellular product of two species of *Chlamydomonas*. Using radiocarbon as a tracer, Tolbert and Zill (1956, 1957) found that during short-term photosynthesis experiments *Chlorella pyrenoidosa* liberates glycolic acid as the principal extracellular organic product, its concentration in actively growing cultures reaching 3 to 8 mg. per liter. The rapid excretion of glycolate was found to be dependent on factors influencing the uptake of bicarbonate ion, i.e., presence of bicarbonate, aerobic conditions, light, and pH, thus suggesting a Donnan equilibrium between the two ions across the cell membrane. No other organic acid was found to behave in this way. Rapid excretion and absorption of glycolic acid by freshwater phytoplankton has also been demonstrated by Fogg and Nalewajko (unpublished).

Indications that long-chain fatty acids may be commonly liberated by algae are discussed in Section III.E.

B. Amino Acids and Peptides

Many investigators have found that certain Blue-green Algae liberate substantial amounts of soluble substances containing peptide-, amide-, and traces of free amino-nitrogen (see Fogg, Chapter 9, this treatise). Algae belonging to other divisions, e.g., *Chlorella pyrenoidosa, Chlamydomonas* spp., *Tribonema aequale, Navicula pelliculosa,* and *Ectocarpus confervoides,* liberate similar substances in smaller amounts (Fogg and Westlake, 1955; Allen, 1956; Fogg and Boalch, 1958). *Anabaena cylindrica* produces several extracellular polypeptides which do not give a ninhydrin reaction until hydrolyzed and which contain non-amino as well as amino acids (Pietruszko and Dixon, unpublished).

C. Carbohydrates

Goryunova (1950) reported that *Oscillatoria splendida* excretes polysaccharides into the medium, a process which she believed to be related to trichome movement (see Jarosch, Chapter 36, this treatise). Fogg (1952) found as much as 7 mg/litre of pentose or pentosan in filtrates from *Anabaena cylindrica* cultures. Bishop *et al.* (1954) later identified this material as a homogeneous complex polysaccharide consisting of glucose, xylose, glucuronic acid, galactose, rhamnose, and arabinose, in molar ratios of 5:4:4:1:1:1. Allen (1956) demonstrated the production of extracellular polysaccharides by *Chlamydomonas* spp. Lewin (1956), who examined 18 species of Green Algae, most of which belonged to this genus, found that they all produced some soluble polysaccharide in the medium. In all species but one, galactose and arabinose were the main components of this material; in the exceptional species, *C. ulvaënsis*, glucose and xylose predominated. Minor components of the polysaccharides included fucose, rhamnose, mannose, uronic acids, and several unidentified substances, presumably also sugars. Tolbert and Zill (1956) showed that in acidic media, especially at pH values below 3.5, most of the labeled organic matter excreted by *Chlorella pyrenoidosa* during short-term photosynthesis with C^{14}-bicarbonate was in the form of sucrose.

Guillard and Wangersky (1958) detected carbohydrate with *N*-ethyl carbazole in filtrates from cultures of marine or brackish-water flagellates, including *Dunaliella* spp., *Pyramimonas* sp., *Chlamydomonas* sp., *Monochrysis lutheri*, *Isochrysis galbana*, *Prymnesium parvum*, and *Rhodomonas* (?) sp. In all cases, accumulation of extracellular carbohydrate was considerable only after the cessation of exponential growth; it may have originated from moribund cells damaged during filtration. The greatest amounts were liberated by the chrysophyte species (for *P. parvum* the highest concentration found was 123 mg. per liter), perhaps as a result of the release of leucosin, which has been reported to occur in some of these forms. Guillard and Wangersky did not determine the composition of the carbohydrates liberated by the species which they studied.

D. Vitamins and Growth Substances

Lewin (1958) detected thiamine in the culture medium of *Coccomyxa* sp., but otherwise there appears to be no published evidence for the extracellular liberation of vitamins by algae. Substances responsible for the growth-promoting effects of one algal species on another, as described by Lefèvre and Jakob (1949), may be of this nature, but none has been chemically characterized. However, plant hormones of the auxin type, which have been demonstrated in extracts of *Chlorella* and various plankton algae, have also been found in filtrates from cultures of *Anabaena cylindrica* and from lake water containing a nearly unialgal growth of *Oscillatoria* sp. (Bentley, 1958, 1960; see Conrad and Saltman, Chapter 44, this treatise).

E. Auto-inhibitors and Antibiotics

There is much evidence suggesting that certain species of algae liberate extracellular products which inhibit their own growth or that of other organisms. Auto-inhibition of growth, apparently because of specific metabolic products accumulating in the medium, has been recorded for *Nostoc punctiforme* (Harder, 1917), a strain of *Chlorella vulgaris* (Pratt and Fong, 1940) and *Nitzschia palea* in impure culture (von Denffer, 1948; Jørgensen, 1956), but the substances concerned have not been fully characterized. Many workers have reported that certain algal species inhibit the growth of others in mixed cultures (Flint and Moreland, 1946; Lefèvre and Jakob, 1949; Lefèvre *et al.*, 1952; Rice, 1954; Jørgensen, 1956; Proctor, 1957; Zavarzina, 1959). Perhaps some such effects are due to specific antibiotics; but since, in general, these have been neither isolated nor chemically characterized, and since the conditions of culture have not always been strictly controlled, the existence of these putative antibiotics may be questioned. Proctor (1957) showed clearly that inhibition of *Haematococcus pluvialis* by *Chlamydomonas reinhardtii* was due to a fatty acid; however, since this is evidently liberated on the death of the *Chlamydomonas* cells, it is therefore not an extracellular product *sensu stricto*. On the other hand, in a careful study of the growth of two planktonic diatoms, *Asterionella formosa* and *Fragilaria crotonensis*, in mixed culture and in media containing culture filtrates, Talling (1957) found no evidence for the production by either species of any extracellular substance which appreciably modified the growth of the other. It should be mentioned, however, that these experiments were conducted in the presence of bacteria.

When vigorously growing cultures of algae are exposed to contamination, it is often observed that relatively few bacteria develop, an effect which could conceivably be due to their suppression by antibacterial agents. Substances having pronounced antibacterial activity, obtained from filtrates of *Chlorella pyrenoidosa* cultures, have been shown to be oxidation products of unsaturated fatty acids (Spoehr *et al.*, 1949). Bacteriostatic substances, perhaps of a similar nature, are produced by *Oscillatoria splendida* (Goryunova, 1950) and by *Stichococcus bacillaris* and *Protosiphon botryoides* (Harder and Opperman, 1953). Steemann Nielsen (1955) noted indications suggesting that in the light *Chlorella* sp. and the marine diatom *Thalassiosira nana* may produce substances inhibiting bacterial respiration. Sieburth (1959, 1960) presented evidence that the thermolabile antibiotic agent produced by *Phaeocystis pouchetii*, a common unicellular alga of marine plankton, is acrylic acid. Henriksson (1960) reported the presence of substances inhibiting the growth of the mycobiont from the lichen *Collema* in filtrates from cultures of various nonsymbiotic Blue-green Algae, as well as from cultures of the phycobiont, *Nostoc* sp.

F. Toxins

Certain planktonic algae, occasionally abundant in fresh water and in the sea, have attracted attention because of their lethal effects on fish and other higher animals. Often these effects are attributable to secondary conditions, such as oxygen deficiency in the water resulting from bacterial decomposition of a dense "bloom," but sometimes it is evident that specific toxins are produced. It is not always clear whether these are liberated from living cells, but mention may be made here of a few of the more important items in the extensive literature on the subject. Ballantine and Abbott (1957) reviewed the occurrence and physiological effects on animals of toxic marine flagellates, among which dinoflagellates of the genera *Gymnodinium* and *Gonyaulax* are most prominent. The toxin of *Gymnodinium veneficum* is water-soluble and ether-insoluble, acid-labile, and evidently of high molecular weight; it can kill various mollusks, fish, and mammals by acting specifically on the nervous system (Abbott and Ballantine, 1957). The brackish-water chrysophyte *Prymnesium parvum*, which is occasionally responsible for mass mortality of fish, produces a definitely extracellular toxin, which is nondialyzable, acid-labile, and thermolabile (Shilo *et al.*, 1953; Shilo and Aschner, 1953; McLaughlin, 1958). Reich and Rotberg (1958) showed that it is formed in greater amounts in diluted seawater or in media with a high concentration of calcium ion. Occasionally, domestic birds and mammals die as a consequence of drinking water containing dense blooms of Blue-green Algae, among which species of *Microcystis* and *Anabaena* seem to be particularly poisonous (Olson, 1951, 1960; Rose, 1953; Ingram and Prescott, 1954; Gorham, 1960). The toxic substance from these algae was found to be readily soluble in water but not in organic solvents, and to be thermostable in neutral solutions but destroyed in alkaline media (Shelubsky, 1951). At least four toxic factors from algal blooms have now been recognized on the basis of the speed at which they act and the symptoms which they produce in mice (Gorham, 1960). One produces fast deaths and is algal in origin. The other three produce slow deaths and are bacterial in origin. The "fast-death" factor has so far been found to be produced only by certain strains or blooms of *Microcystis*. It is an acidic, probably cyclic, peptide (Bishop *et al.*, 1959). (A rapidly-acting toxin from certain blooms of *Anabaena* (Olson, 1960) may or may not be similar.) Although it can be detected in the filtrates of young, actively-growing cultures of *Microcystis*, it is mostly liberated by destruction or decomposition of the cells (Hughes *et al.*, 1958), and therefore may not be truly extracellular in the sense adopted in this chapter.

G. Enzymes

The ability of certain algae such as *Nostoc punctiforme* (Harder, 1917) to grow chemotrophically using starch or other substances of high molecular weight as carbon sources suggests that extracellular enzymes may be produced. Pringsheim (1951) observed that *Nitzschia putrida* is able to liquify gelatine, and thus may excrete a proteolytic enzyme. In investigating the utilization of glutamine by *Monodus subterraneus*, Miller (1959) found that culture filtrates contain an active glutaminase. This enzyme is specific, attacking neither isoglutamine nor asparagine. It is liberated from healthy cells and is produced irrespective of the presence of glutamine in the medium. Several other members of the Xanthophyta, and *Chlorella vulgaris*, produce no extracellular glutaminase (Miller, unpublished results).

H. Miscellaneous

Fogg and Boalch (1958) attributed the yellow color of filtrates from pure cultures of *Ectocarpus confervoides* to three or more substances, with absorption maxima in the region of 400 mμ and properties suggesting that they might be carotenoid–protein complexes. The dinoflagellate *Amphidinium carteri* was found by Wangersky and Guillard (1960) to liberate a substance which on spontaneous hydrolysis yields a low molecular-weight organic base, perhaps an analog of acetylcholine, apparently responsible for the fishy odor of old cultures. Armstrong and Boalch (1960) also detected organic sulfides and other volatile substances, absorbing ultraviolet light in the region 200 to 250 mμ, in filtrates from pure cultures of *Ectocarpus confervoides* and from unialgal cultures of *Enteromorpha* sp., *Phaeocystis pouchetii*, and *Phaeodactylum tricornutum*. Since many algae have characteristic odors, the production of more or less specific volatile extracellular products is evidently a common phenomenon (see Katayama, Chapter 29, this treatise).

IV. Ecological Effects of Extracellular Products of Algae

A. Occurrence in Natural Waters

Substances found in algal culture media may, if also liberated in natural environments, produce a variety of ecological effects, some of which could conceivably be of considerable importance. A general review of this subject has been published by Saunders (1957); reference should also be made to a discussion by Wilkinson (1958) of the functions of extracellular polysaccharides of bacteria. Although conditions in laboratory cultures differ from those under which algae grow in nature, the available evidence suggests that similar extracellular products are liberated by algae to an appreciable extent under natural conditions. Natural waters normally contain detectable amounts of dissolved organic substances, some of which are of kinds known to be excreted by algae in culture. Vallentyne (1957) has given an excellent critical review of our knowledge of the dissolved organic substances of lake and sea water. In analyses of sea water, Koyama and Thompson (1959) found several organic acids, among which glycolic acid is of particular interest in view of its excretion by certain algae in culture; the presence of glycolic acid in concentrations of the order of 1 mg. per liter in both sea water and fresh waters has been confirmed by Fogg and Nalewajko (unpublished). Wangersky (1952) reported the presence of a "rhamnoside", in concentrations of up to 100 mg per liter, associated with phytoplankton growth in inshore water of the Gulf of Mexico. In offshore sea water Armstrong and Boalch (1960) found volatile organic substances similar to those obtained from algal cultures, while Bentley (1958, 1960) detected auxin-like substances, similar to those obtained from algal cells, in filtrates of both inshore and offshore sea water. Although growth-promoting substances are undoubtedly present in most natural waters, there is still no conclusive proof that they were liberated by algae.

Direct evidence that phytoplankton liberates extracellular products under natural conditions has been obtained by using radiocarbon as a tracer. In experiments carried out by the *in situ* method of Steeman Nielsen (1952), the membrane filtrate from samples of lake water to which C^{14}-bicarbonate had been added was found to contain C^{14}-labeled organic matter. Since this labeling was dependent on light and on the amount of phytoplankton, it evidently denoted the liberation of products of photosynthesis from algal cells. In preliminary experiments carried out in Lake Erken, Sweden, when *Gloeotrichia* and *Aphanizomenon* predominated in the plankton, the extracellular products amounted to about 1.5% of the total carbon fixed (Fogg, 1958). In water from Windermere, English Lake District, at times when diatoms were the most abundant plankton algae, further experiments, with an improved method for the recovery of organic carbon from filtrates, demonstrated apparent liberation of extracellular products to the extent of 3 to 90% of the total carbon fixed. The proportion of extracellular products tended to increase with depth (Fig. 1), and it appears likely that the dissolved organic material liberated under these conditions is largely in the form of glycolate (Fogg and Nalewajko, unpublished).

It is clear that extracellular products liberated from phytoplankton cells form only a minor proportion of the total dissolved organic material in natural waters. The variations in the total dissolved organic matter both in lakes (Domogalla *et al.*, 1925) and in the sea (Duursma, 1960) suggest that the major part originates by decay of phytoplankton and other material. However, variations in the concentration of particular substances may be related to phytoplankton periodicity. Thus Wangersky (1959), who analyzed sea water from Long Island Sound for dissolved carbohydrate through a complete annual plankton cycle, found none during the spring diatom maximum and significant amounts only towards the end of a dinoflagellate maximum in July. The concentrations he de-

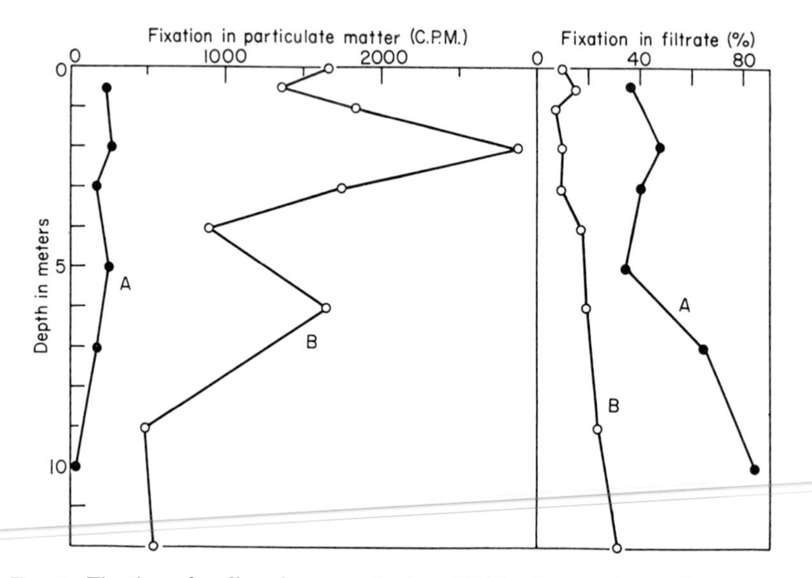

FIG. 1. Fixation of radiocarbon, supplied as C^{14}-bicarbonate, in particulate matter and in organic substances in filtrates from water samples from Windermere, English Lake District. The experiments were carried out by the *in situ* method of Steemann Nielsen (1952) over 24-hour periods: (*A*) March 16–17, 1960; (*B*) June 6–7, 1959. (C.P.M. = counts per minute.) Fixation in dissolved organic matter is expressed as percentage of total fixation. (Fogg and Nalewajko, unpublished results.)

tected, 0.5 to 1.5 mg. per liter, were of an order of magnitude less than those found in filtrates from bacteria-free cultures of the same dinoflagellate.

B. Utilization as Nutrients

Substances liberated from healthy algal cells may be used directly as carbon sources by other organisms. Although the actual concentrations of dissolved organic acids and carbohydrates in natural waters may be low, their rate of turn-over is probably sufficiently high to be ecologically important. *Chlorella* rapidly absorbs C^{14}-glycolate from solutions of concentrations similar to those in which glycolate is found in natural waters (Fogg and Nalewajko, unpublished), and presumably bacteria, fungi, and protozoa could absorb it at comparable rates. Lewin (personal communication) has isolated several strains of marine and freshwater bacteria able to use glycolate as a carbon source. Many bacteria presumably derive much of their carbon from dissolved organic substances, of which algal products may form an appreciable proportion. However, the theory of Pütter (1908) that aquatic animals might subsist largely on extracellular products of algae seems untenable (Krogh *et al.*, 1930); among other reasons for scepticism, the low surface/volume ratios of even the smaller zooplankton do not favor the absorption of organic substances from dilute solution. If algae liberate vitamins, these might also play a significant part in the nutrition of zooplankton. Extracellular products are probably of importance as nutrients in the symbiotic associations in which algae participate. Doubtless the chemotrophic partner obtains some of its organic carbon from them (e.g., Muscatine and Hand, 1958); similarly various organisms have been shown to derive at least part of their nitrogen from extracellular products liberated by nitrogen-fixing Blue-green Algae with which they are associated (see Fogg, Chapter 9, and Ahmadjian, Chapter 54, this treatise).

C. Antibiosis and Growth Promotion

Lucas (1947, 1958), in particular, emphasized the importance of traces of dissolved organic substances in determining the qualitative characteristics of aquatic communities. Extracellular products of algae may be important in this way. It is tempting to explain the fluctuations of phytoplankton, for example, in terms of growth substances or antibiotics which, released by one dominant species, condition the water so as to favor another species which consequently tends to succeed it (see Saunders, 1957). As we have seen, there are plenty of indications that algae liberate such growth-promoting and inhibiting substances; but since unequivocal information on their existence and nature is scanty, their importance in aquatic ecology remains to be established. The effect of auxins on the growth of algae has been reviewed elsewhere in this treatise by Conrad and Saltman (Chapter 44).

Pathogenic and coliform bacteria die out rather quickly in sewage-oxidation ponds in which there is abundant algal growth; this could perhaps be a consequence of the production of antibacterial agents such as those discussed above. There is more definite evidence for adverse effects of some of the extracellular products of algae on animals: the toxic effect of certain algae on fish and other animals under natural conditions is well documented, and it seems probable that the "rhamnoside" in sea water, which has been reported to influence the filter-feeding rate of oysters, is an extracellular product of phytoplankton. It has even been suggested that Blue-green Algae may help to control mosquitoes in rice fields (Griffin and Rees, 1956), but it is not known whether extracellular toxins are involved.

D. The Formation of Complexes

The value of added chelating agents in algal culture media is now well known. Besides maintaining micro-nutrients in available and nontoxic forms, chelators may also affect algal growth by altering the effective balance between the major nutrient ions (Fogg and Miller, 1958; Miller and Fogg, 1958). Such extracellular products of algae as organic acids and polypeptides can be expected to chelate inorganic ions; the extracellular polypeptide of *Anabaena cylindrica*, for instance, was shown to form complexes with various ions (Fogg and Westlake, 1955). Electrometric titration and other chemical evidence show that these substances are able to chelate cupric ions, for example, thereby enabling an alga to grow in the presence of otherwise toxic concentrations. Such complex formation may explain why treatment of reservoirs with copper sulfate is not always successful in suppressing algal growth. Complex formation by extracellular products may also favor algae by maintaining nutrient substances in solution, and thus available for growth, in waters from which they would otherwise be precipitated. Various other possibilities are discussed by Saunders (1957). Though proof that such effects are appreciable in natural environments has yet to be obtained, it seems possible that modification of the ionic environment may be an important biological role of extracellular metabolites.

References

Abbott, B. C., and Ballantine, D. (1957). The toxin from *Gymnodinium veneficum* Ballantine. *J. Marine Biol. Assoc. United Kingdom* **36,** 169–189.

Allen, M. B. (1956). Excretion of organic compounds by *Chlamydomonas*. *Arch. Mikrobiol.* **24,** 163–168.

Anderson, E. H. (1945). Studies on the metabolism of the colorless alga *Prototheca zopfii*. *J. Gen. Physiol.* **28,** 297–327.

Armstrong, F. A. J., and Boalch, G. T. (1960). Volatile organic matter in algal culture media and sea water. *Nature* **185,** 761–762.

Ballantine, D., and Abbott, B. C. (1957). Toxic marine flagellates; their occurrence and physiological effects on animals. *J. Gen. Microbiol.* **16,** 274–281.

Barker, H. A. (1935). The metabolism of the colourless alga, *Prototheca zopfii* Krüger. *J. Cellular Comp. Physiol.* **7,** 73–93.

Bentley, J. A. (1958). Role of plant hormones in algal metabolism and ecology. *Nature* **181,** 1499–1502.

Bentley, J. A. (1960). Plant hormones in marine phytoplankton, zooplankton and sea water. *J. Marine Biol. Assoc. United Kingdom* **39,** 433–444.

Bishop, C. T., Adams, G. A., and Hughes, E. O. (1954). A polysaccharide from the blue-green alga, *Anabaena cylindrica*. *Can. J. Chem.* **32,** 999–1004.

Bishop, C. T., Anet, E. F. L. J., and Gorham, P. R. (1959). Isolation and identification of the fast-death factor in *Microcystis aeruginosa* NRC–1. *Can. J. Biochem. and Physiol.* **37,** 453–471.

Braarud, T., and Føyn, B. (1930). Beiträge zur Kenntnis des Stoffwechsels im Meere. *Avhandl. Norske Videnskaps. Akad. Oslo. I. Mat. Naturv. Kl.* **14,** 1–24.

Domogalla, B. P., Juday, C., and Peterson, W. H. (1925). The forms of nitrogen found in certain lake waters. *J. Biol. Chem.* **63,** 269–285.

Duursma, E. K. (1960). Dissolved organic carbon, nitrogen and phosphorus in the sea. *Neth. J. Marine Research* **1,** 1–148.

Flint, L. H., and Moreland, C. F. (1946). Antibiosis in the bluegreen algae. *Am. J. Botany* **33,** 218.

Fogg, G. E. (1952). The production of extracellular nitrogenous substances by a blue-green alga. *Proc. Roy. Soc.* **B139,** 372–397.

Fogg, G. E. (1958). Extracellular products of phytoplankton and the estimation of primary production. *J. conseil, Conseil permanent intern. exploration mer, Rappts. et proc.-verb. aux des réunions* **144,** 56–60.

Fogg, G. E., and Boalch, G. T. (1958). Extracellular products in pure cultures of a brown alga. *Nature* **181,** 789–790.

Fogg, G. E., and Miller, J. D. A. (1958). The effect of organic substances on the growth of the freshwater alga *Monodus subterraneus. Verhandel. intern. Ver. Limnol.* **13,** 892–895.

Fogg, G. E., and Westlake, D. F. (1955). The importance of extracellular products of algae in freshwater. *Verhandel. intern. Ver. Limnol.* **12,** 219–232.

Gaffron, H. (1939). Über Anomalien des Atmungsquotienten von Algen aus Zuckerkulturen. *Biol. Zentr.* **59,** 288–302.

Gorham, P. R. (1960). Toxic waterblooms of blue-green algae. *Can. Vet. J.* **1,** 235–245.

Goryunova, S. V. (1950). "Chemical Composition and Kinetics of Excretion in the Blue-green Alga *Oscillatoria splendida* Grew." Academy of Sciences, Moscow. (In Russian.)

Griffin, G. D., and Rees, D. M. (1956). *Anabaena unispora* Gardner and other blue-green algae as possible mosquito control factors in Salt Lake County, Utah. *Proc Utah. Acad. Sci.* **33,** 101–103.

Guillard, R. R. L., and Wangersky, P. J. (1958). The production of extracellular carbohydrates by some marine flagellates. *Limnol. and Oceanog.* **3,** 449–454.

Harder, R. (1917). Ernährungsphysiologische Untersuchungen an Cyanophyceen, hauptsächlich dem endophytischen *Nostoc punctiforme. Z. Botan.* **9,** 145–242.

Harder, R., and Oppermann, A. (1953). Über antibiotische Stoffe bei den Grünalgen *Stichococcus bacillaris* und *Protosiphon botryoides. Arch. Mikrobiol.* **19,** 398–401.

Henriksson, E. (1960). Studies in the physiology of the lichen *Collema.* III. The occurrence of an inhibitory action of the phycobiont on the growth of the mycobiont. *Physiol. Plantarum* **13,** 751–754.

Hughes, E. O., Gorham, P. R., and Zehnder, A. (1958). Toxicity of a unialgal culture of *Microcystis aeruginosa. Can. J. Microbiol.* **4,** 225–236.

Ingram, W. M., and Prescott, G. W. (1954). Toxic fresh-water algae. *Am. Midland Naturalist* **52,** 75–87.

Jørgensen, E. G. (1956). Growth inhibiting substances formed by algae. *Physiol. Plantarum* **9,** 712–726.

Koyama, T., and Thompson, T. G. (1959). Organic acids in sea water. *Communs. Intern. Oceanog. Congr.,* New York.

Krogh, A., Lange, E., and Smith, W. (1930). On the organic matter given off by algae. *Biochem. J.* **24,** 1666–1671.

Lefèvre, M., and Jakob, H. (1949). Sur quelques propriétés des substances actives tirées des cultures d'algues d'eau douce. *Compt. rend. acad. sci.* **229,** 234–236.

Lefèvre, M., Jakob, H., and Nisbet, M. (1952). Auto- et hétéroantagonisme chez les algues d'eau douce *in vitro* et dans les collections d'eau naturelles. *Ann. sta. centre hydrobiol. appl.* **4,** 5–198.

Lewin, R. A. (1956). Extracellular polysaccharides of green algae. *Can. J. Microbiol.* **2,** 665–672.

Lewin, R. A. (1957). Excretion of glycolic acid by *Chlamydomonas. Bull. Japan. Soc. Phycol.* **5,** 74–75. (In Japanese.)

Lewin, R. A. (1958). Vitamin-bezonoj de algoj. *In* "Sciencaj Studoj" (P. Neergaard, ed.), pp. 187–192. Modersmaalet, Haderslev, Copenhagen.

Lucas, C. E. (1947). The ecological effects of external metabolites. *Biol. Revs.* **22,** 270–295.

Lucas, C. E. (1958). External metabolites and productivity. *J. conseil, Conseil. permanent intern. explorations mer, Rappts. et proc.-verb. aux des réunions* **144,** 153–158.

McLaughlin, J. J. A. (1958). Euryhaline chrysomonads: nutrition and toxigenesis in *Prymnesium parvum,* with notes on *Isochrysis galbana* and *Monochrysis lutheri. J. Protozool.* **5,** 75–81.

Miller, J. D. A. (1959). An extracellular enzyme produced by *Monodus. Brit. Phycol. Bull.* **No. 7,** 22.

Miller, J. D. A., and Fogg, G. E. (1958). Studies on the growth of Xanthophyceae in pure culture. II. The relations of *Monodus subterraneus* to organic substances. *Arch. Mikrobiol.* **30,** 1–16.

Muscatine, L., and Hand, C. (1958). Direct evidence for the transfer of materials from symbiotic algae to the tissues of a coelenterate. *Proc. Natl. Acad. Sci. U.S.* **44,** 1259–1263.

Myers, J. (1951). Physiology of the algae. *Ann. Rev. Microbiol.* **5,** 157–180.

Olson, T. A. (1951). Toxic plankton. *In* "Proceedings of Inservice Training Course in Water Works Problems," pp. 86–95. Ann Arbor, Michigan.

Olson, T. A. (1960). Water poisoning—a study of poisonous algae blooms in Minnesota. *Am. J. Public Health* **50,** 883–884.

Pratt, R., and Fong, J. (1940). Studies on *Chlorella vulgaris.* II. Further evidence that *Chlorella* cells form a growth-inhibiting substance. *Am. J. Botany* **27,** 431–436.

Pringsheim, E. G. (1951). Über farblose Diatomeen. *Arch. Mikrobiol.* **16,** 18–27.

Proctor, V. W. (1957). Studies of algal antibiosis using *Haematococcus* and *Chlamydomonas. Limnol. and Oceanog.* **2,** 125–139.

Pütter, A. (1908). Der Stoffhaushalt des Meeres. *Z. allgem. Physiol.* **7,** 321–368.

Reazin, G. H., Jr. (1956). The metabolism of glucose by the alga *Ochromonas malhamensis. Plant Physiol.* **31,** 299–303.

Reich, K., and Rotberg, M. (1958). Some factors influencing the formation of toxin poisonous to fish in bacteria-free cultures of *Prymnesium. Bull. Research Council Israel* **B7,** 199–202.

Rice, T. R. (1954). Biotic influences affecting population growth of planktonic algae. *Fishery Bull. U.S.* **54,** 227–245.

Rose, E. T. (1953). Toxic algae in Iowa lakes. *Proc. Iowa Acad. Sci.* **60,** 738–745.

Saunders, G. W. (1957). Interrelations of dissolved organic matter and phytoplankton. *Botan. Rev.* **23,** 389–409.

Shelubsky, M. (1951). Observations on the properties of a toxin produced by *Microcystis. Proc. Intern. Assoc. Theoret. and Appl. Limnol.* **11,** 362–366.

Shilo, M., and Aschner, M. (1953). Factors governing the toxicity of cultures containing the phytoflagellate *Prymnesium parvum* Carter. *J. Gen. Microbiol.* **8,** 333–343.

Shilo, M., Aschner, M., and Shilo, M. (1953). The general properties of the exotoxin of the phytoflagellate *Prymnesium parvum. Bull. Research Council Israel* **2,** 446.

Sieburth, J. McN. (1959). Antibacterial activity of Antarctic marine phytoplankton. *Limnol.* and *Oceanog.* **4,** 419–424.

Sieburth, J. McN. (1960). Acrylic acid, an "antibiotic" principle in *Phaeocystis* blooms in Antarctic waters. *Science* **132,** 676–677.

Spoehr, H. A., Smith, J., Strain, H., Milner, H., and Hardin, G. J. (1949). Fatty acid antibacterials from plants. *Carnegie Inst. Wash. Publ.* **No. 586,** pp. 1–67

Steemann Nielsen, E. (1952). The use of radio-active carbon (^{14}C) for measuring organic production in the sea. *J. conseil, Conseil permanent intern. exploration mer* **18,** 117–140.

Steemann Nielsen, E. (1955). The production of antibiotics by plankton algae and its effect upon bacterial activities in the sea. *Papers in Marine Biol. and Oceanog.* (*London*) pp. 281–286.

Syrett, P. J. (1958). Fermentation of glucose by *Chlorella vulgaris. Nature* **182,** 1734–1735.

Talling, J. F. (1957). The growth of two plankton diatoms in mixed cultures. *Physiol. Plantarum* **10,** 215–223.

Tolbert, N. E., and Zill, L. P. (1956). Excretion of glycolic acid by algae during photosynthesis. *J. Biol. Chem.* **222,** 895–906.

Tolbert, N. E., and Zill, L. P. (1957). Excretion of glycolic acid by *Chlorella* during photosynthesis. *In* "Research in Photosynthesis" (H. Gaffron, ed.), pp. 228–231. Interscience, New York.

Vallentyne, J. R. (1957). The molecular nature of organic matter in lakes and oceans, with lesser reference to sewage and terrestrial soils. *J. Fisheries Research Board Can.* **14,** 33–82.

von Denffer, D. (1948). Über einen Wachstumshemmstoff in alternden Diatomeenkulturen. *Biol. Zentr.* **67,** 7–13.

Wangersky, P. J. (1952). Isolation of ascorbic acid and rhamnosides from sea water. *Science* **115,** 685.

Wangersky, P. J. (1959). Dissolved carbohydrates in Long Island Sound 1956–1958. *Bull. Bingham Oceanog. Collection* **17,** 87–94.

Wangersky, P. J., and Guillard, R. R. L. (1960). Low molecular weight organic base from the dinoflagellate *Amphidinium carteri. Nature* **185,** 689–690.

Wilkinson, J. F. (1958). The extracellular polysaccharides of bacteria. *Bacteriol. Revs.* **22,** 46–73.

Zavarzina, N. B. (1959). [Substances inhibiting the growth of *Scenedesmus quadricauda.*] *Trudy Vsesoyuz. Gidrobiol. Obshchestva* **9,** 195–205. (In Russian.)

—Part III—

Physiology of Whole Cells and Plants

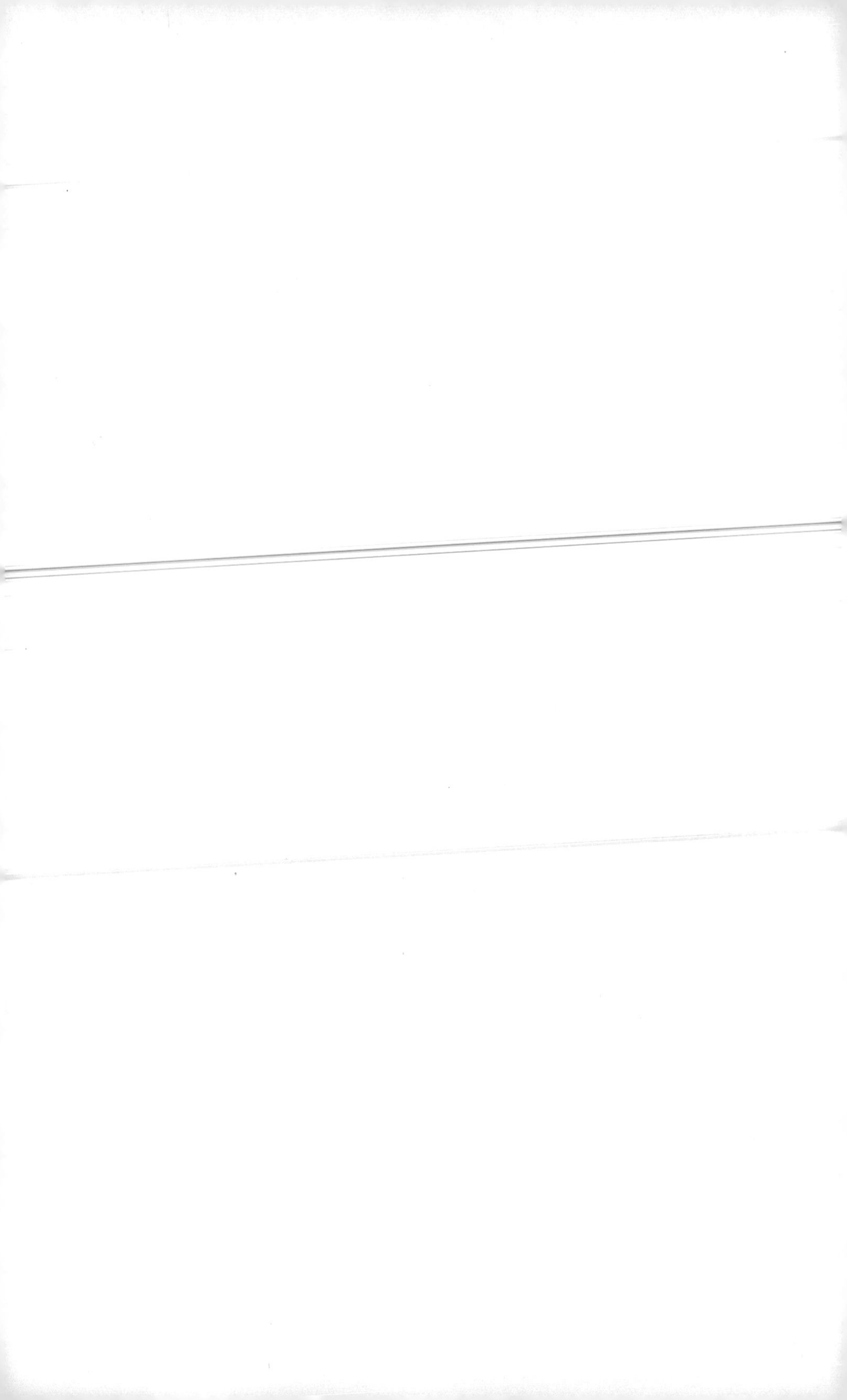

Permeability

E. J. STADELMANN

Institute of Botany, University of Freiburg, Switzerland

I. Introduction

A characteristic feature of living organisms is the formation of membranes differentially permeable to water and solutes. Transport of substances through such a membrane may occur either by *passive transport,* in which the membrane merely acts as a barrier permitting migration of solute or solvent to take place at a reduced rate, or by *active transport,* in which the movement is "driven" by some energy-consuming process taking place in the membrane itself. (Certain aspects of active transport are discussed by Eppley, Chapter 16, this treatise.)

Two types of passive transport may be distinguished:

(1) There may be simple diffusion, a consequence of thermal agitation of molecules in the system, resulting in a net migration of the solute from a region of higher concentration to one of lower concentration (as in experiments on the permeability of non-electrolytes). "Self-diffusion," i.e., the intermixing of identical particles (in practice, similar particles distinguished by isotopes), can now also be studied, as in experiments using heavy water for the determination of water permeability.

(2) The migration may be caused by some "force" located outside the membrane. This may be an osmotic suction force (as employed in experiments on water permeability) or an electric field which causes the migration of charged particles (ions). Its effects will be superimposed on those of simple diffusion and modify the rates of movement and the final distribution of the ions.

Membranes may be found at different levels in the structure of the plant organism. In every cell the protoplast, the nucleus, the plastids, and the mitochondria are surrounded by membranes. In cells with a large central vacuole the protoplasm itself forms a layer lining the interior of the wall, bounded by an outer membrane (the *plasmalemma*) and an inner membrane (the *tonoplast*). Multicellular membranes comprising layers of differentiated cells are often found in higher organisms.

Permeability can be determined only by observing the amount of a substance transported through the membrane. In order to obtain comparable quantitative data, the values must be expressed in standard units of membrane surface, driving force, and time.

For the plant cell, only the permeability of the whole protoplasmic layer between the cell wall and the vacuole has been extensively investigated; a coherent body of data on permeability of other membranes (e.g., nuclear membranes) is not yet available. In tests of non-electrolytes the driving force, i.e., the concentration difference, can readily be determined, but investigations on ion permeability are complicated by the reversible dissociation of salt molecules and often also by active transport ("ion pumps"). As a result, few of the older investigations may be considered to have revealed true permeability values for ions, although by new techniques it may now become possible to overcome these difficulties to some extent (MacRobbie and Dainty, 1958a, b; see also Dainty, 1959). The differential permeability of the outer and inner protoplasmic membranes is also involved in the complex processes of ion uptake and accumulation. Important contributions in this field were made with algal cells by Osterhout (1947), Blinks (1955), and their co-workers. However, only limited information on membrane permeability can be obtained from experiments devoted mainly to the study of salt uptake and accumulation.

More details on the general aspects of permeability and some further references may be found in comprehensive reviews by Collander (1959) and Wartiovaara and Collander (1960), and in Ruhland's "Handbuch der Pflanzenphysiologie" (1956). Polissar (1954) discussed quantitative aspects of the migration of ions through membranes, and Sutcliffe (1959) presented a general review on salt uptake.

Investigations on the permeability of algae were started by the work of Pfeffer (1886/8) on the uptake of aniline dyes by cells of various Zygnematales and of *Oedogonium*. Since then a few species (e.g., of *Chara, Nitella,* and *Spirogyra*) have become standard objects of study in such work, while other classes of algae are rarely investigated because of their less favorable structural features.

In addition to the quantitative data summarized in Tables I to V, qualitative observations are also discussed in this chapter. Most of the information concerns non-electrolytes not normally present in the environment of the cell in the high concentrations used experimentally; only a few data concerning water and ions are available. Such information on the permeability of protoplasmic membranes provides us with a basis for elucidating their structure and composition (cf. Höfler, 1960, 1961).

The problems of water permeability in relation to osmotic regulation in algae are discussed briefly in this treatise by Guillard (Chapter 32).

II. Definition and Methods

A. Definition of Permeability Constant

An absolute measure of permeability is the *permeability constant* (or permeability coefficient), defined as the amount of a substance (dm) passing through a membrane of unit area per unit time (t) when the driving force for the transport has the value of unity. For non-electrolytes the driving force is a concentration difference ΔC, so that for a membrane with a surface area A the permeability constant K becomes:

$$K = \frac{dm}{dt} \cdot \frac{1}{A} \cdot \frac{1}{\Delta C}$$

Its dimension is length $\times$ time^{-1}. In addition, a number of relative measures for permeability have been calculated (cf. Stadelmann, 1956a).

B. Determination of Permeability to Solutes

It is only in experiments with the giant cells of Characeae that the amounts of *solutes* penetrating into the vacuole may be large enough to be determined directly by microchemical analyses (*chemical methods* for determining permeability) or by radioactive tracer techniques. In most other plants the quantity of solute entering or leaving a single cell is too small for chemical analysis, and indirect methods have to be employed to determine the amounts of substances which have penetrated into or escaped from the vacuole. The most important is the *plasmolytic method*. For cells of irregular shape, like those of the majority of algae, only the method of Bärlund (1929; see also Marklund, 1936, and Elo, 1937) permits one to make a good approximation of the permeability constant. Relative permeability values are calculated by using the *recovery-time method* of Hofmeister (1948). For cylindrical cells such as those of *Spirogyra*, the *plasmometric method* gives absolute values for the permeability constant (see Stadelmann 1956b).

C. Determination of Permeability to Water

For measuring the permeability of a membrane to *water*, two different phenomena can be used to bring about the water transport: (i) *osmotic suction forces*, whereby water is drawn through the membrane to a hypertonic solution of a non-permeating solute (*filtration method*); and (ii) *diffusion*, in which no net transport occurs but the rate of exchange is measured by the use of heavy water (HDO) or water labeled with tritium oxide (*diffusion method*). Both have been used in investigations with algae; however, they give different kinds of permeability constants, i.e., osmotic and diffusional, respectively (see Ussing, 1954; Dick, 1959). The value derived by diffusion methods is comparable with permeability constants for dissolved substances.

For further details and for other methods less frequently used with algal cells, see Stadelmann (1956a, b).

III. Permeability of Algal Cells

The results reported here are arranged by algal classes and subdivided in the following sequence: water, non-electrolytes, dyes, and ions. For dyes and ions some information on uptake is also included.

A. Cyanophyta

Cells of the Blue-green Algae differ from those of other algae in a number of special features, including a close association between the protoplasm and the cell wall, an absence of cell vacuoles, and perhaps a higher viscosity of the protoplasm. Consequently, in hypertonic solutions they do not usually exhibit typical plasmolysis, but instead undergo *cytorrhysis*, i.e., shrinkage of the whole cell without its separation from the wall. Filamentous forms exhibit notching or shrinkage of the cells. If plasmolysis occurs, the protoplast usually shows concave plasmolytic forms (Kuchar, 1950; Pernauer, 1958).

Scattered earlier data on permeability (see Drawert, 1949), based on observations of recovery from plasmolysis, are of doubtful value since plasmolysis may seriously damage the cyanophyte cell. A further difficulty arises from the manifestation of differential permeability of the cell wall, and from the fact that the interior of the cell consists of a colloidal system capable of reversible swelling. Thus it is not possible to establish a clear relation between water uptake and osmotic force, as can be done for true solutions.

The only comprehensive measurements of the permeability of Blue-green Algae were made by Elo (1937) on *Oscillatoria princeps* (see Table II) and *O. limosa* with non-electrolytes. The magnitude of the permeability constant is in this case related to the size of the molecule of the solute, the process of permeation here being perhaps mainly one of ultrafiltration through a differentially permeable cell wall. This behavior is unusual among the algae (cf. Höfler, 1942), but it is found in *Beggiatoa. Oscillatoria limosa* occasionally recovered from plasmolysis in glucose solutions after as little as 1 hour (Pernauer, 1958). Prolonged contact of *O. bornetii* with hypertonic urea or glucose solutions resulted in dissolution of the cell wall (Drawert, 1949). Glucose penetrates into cells of *Tolypothrix distorta* much more slowly than into *O. limosa*; glycerol and urea rapidly penetrate cells of both of these algae (Pernauer, 1958).

The permeability of Blue-green Algae to water, dyes, and ions has not yet been examined critically. Brand (1903) produced *plasmoptysis* (i.e., a bursting open of the cells) in *Phormidium* by rapidly transferring filaments, which had been soaked in glycerol for only 30 seconds, into pure water, suggesting a high permeability at least to water. Except for a few cases plasmoptysis has not been produced in other Cyanophyta (Drawert, 1949).

Certain dyes such as methylene blue or toluidine blue enter the trichome of filamentous forms only through the apices and the cross walls, and are unable to penetrate the longitudinal walls. Other dyes penetrate and stain the cells throughout the filaments simultaneously (Drawert, 1949). Staining is often uneven, and some cells of the filaments show more intensive coloration than others. Drawert also tested a few salts (KNO_3, $NaCl$, $CaCl_2$) and showed that these, too, rapidly penetrate the cells. Cytorrhysis is produced by hypertonic aluminum nitrate, possibly as a consequence of an increase in the adhesion of the protoplasm to the cell wall, as is known for other plants (see Stadelmann, 1956c; Höfler, 1958). A high permeability to salts seems to be important for the pronounced osmoregulatory capacity of a brackish-water species of *Oscillatoria* (Pernauer,

1958). *Scytonema julianum* shows considerable variation in the permeability of individual cells even within the same preparation (Schönleber, 1937).

B. Euglenophyta

Since the cells have no rigid cell wall and their shape is unsuitable for the measurement of small changes in volume, these organisms do not lend themselves to quantitative osmotic experiments, and only a few observations on their permeability have been made. The periplast is semipermeable, and transfer of the cells to a hypertonic solution usually results in shrinkage, except in certain genera, e.g., *Phacus* and *Lepocinclis* (Hilmbauer, 1954). Water evidently can penetrate through the periplast of *Euglena, Trachelomonas, Phacus,* and *Lepocinclis* as easily as through many other types of protoplasm which have a low water permeability (Höfler and Höfler, 1952; Hilmbauer, 1954).

C. Pyrrophyta

Diskus (1956) studied the permeability of *Oxyrrhis marina* to various nonelectrolytes, which, when they were able to penetrate into the vacuole, caused the cells to become spherical. Acetamide penetrated most rapidly; methylurea, urea, and glycerol initially caused shrinkage followed by dilation, while concentrated sea water or glucose solutions caused only shrinkage by the abstraction of water. The rapid uptake of water or of acetamide led to plasmoptysis. *Gymnodinium paradoxum* cells are impermeable to urea, but do not shrink in hypertonic solutions of acetamide (Diskus, 1958).

D. Xanthophyta

Experiments with *Tribonema vulgare* and *T. viride* revealed a permeability series (i.e., a list of substances in the sequence of their permeability constants) of the "urea type," though *T. vulgare* was somewhat more permeable to sucrose than would be expected (Lenk, 1956; see Table III).

E. Bacillariophyta

Nitzschia tryblionella showed quite a high permeability for water (Übeleis, 1957); see Table IV.

A variety of non-electrolytes were tested by Marklund (1936) and Elo (1937) on *Melosira* (possibly *M. moniliformis*; see Höfler, 1940) and *Licmophora oedipus*. They present a permeability series quite unlike that usually found for the cells of Green Algae and higher plants. The values for the permeability constants are closer together and higher than those of most plant cells (see Table II), and include surprisingly high rates for the permeability to sugars, which penetrate most other kinds of cells very slowly (Höfler, 1926; Lenk, 1956; Krebs, 1952). A high permeability to glucose and sucrose was also found in certain other diatoms (Höfler, 1940, 1943; Höfler *et al.*, 1956; Übeleis, 1957; see Table I), but this does not seem to be a general characteristic for this group since *Cymbella aspera* and *Caloneis obtusa*, for example, are almost impermeable to sugar (Höfler *et al.*, 1956; Höfler, 1960a).

Follmann (1958a) published some extreme values for the permeability constants of *Melosira arenaria* cells. He also reported that cells of eight diatom species recovered rapidly after plasmolysis in hypertonic sucrose, which he attributed to an active uptake of both water and sugar (cf. Höfler and Url, 1958).

The earlier studies on the uptake of dyes by diatoms were summarized by Hirn (1953a), who herself tested a large number of species with different dyes. She noted that appreciable uptake occurred and the dyes tended to be accumulated in the form of droplets in the cytoplasm or, apparently, even in the vacuole. Some dyes proved harmful, especially to certain species. In *Pinnularia viridis* and some other species, toluidine blue is first absorbed as the cation on the cell walls, and then penetrates the plasmalemma, possibly as undissociated molecules, to accumulate within the cell (Hirn, 1953a; Drawert, 1956). Höfler *et al.* (1956) studied vital staining of *Amphora* with rhodamine B, while Follmann (1958b) reported the effects of various vital dyes on *Coscinodiscus granii.*

The permeability of halophilic diatoms to salts is relatively high (Cholnoky, 1928; Höfler *et al.*, 1956); $CaCl_2$ is able to penetrate into the vacuole of diatom cells, whereas all other plant cells tested appear to be impermeable to this salt. This property may contribute to the high osmoregulatory ability of diatoms, in which, however, active transport seems also to be involved (see Höfler, 1940; Kesseler, 1959).

F. Chlorophyta

1. *Water*

Most studies have been carried out on members of the Zygnematales in which the permeability of the cells to water seems to be very high. The first quantitative data on *Spirogyra*, *Zygnema*, and *Mougeotia* were obtained by Huber and Höfler (1930); see Table IV. Bochsler (1948) obtained wholly different permeability values for the outward and inward movement of water in *Zygnema*. In her studies of *Rhizoclonium*, two species of *Oedogonium*, and some species of Zygnematales, Lenk (1956) found appreciable variations in water permeability, even between different species of *Spirogyra* with similar permeability series for nonelectrolytes (see Table

IV). She observed that zygotes of *Spirogyra mirabilis* can be immersed in hypertonic solutions for relatively long periods before they become plasmolyzed, which in this case is probably attributable to the low permeability of their walls to water. Seemann (1950) examined the relation between the permeability of *Cladophora glomerata* and *Spirogyra* sp. to water (see Table IV) and the pH of the external solution (see below).

The water permeability of the marine coenocyte *Halicystis osterhoutii* was found to decrease when the plants were immersed in sea water diluted more than 50%, due perhaps to a change of the partition coefficient of water at the interface between the protoplasm and the cell sap (Jacques, 1939a).

2. Non-electrolytes

In high concentrations of alcohol, *Hydrodictyon utriculatum* and certain other algae showed plasmoptysis, a result which could also be brought about by a number of other lipid-soluble organic substances (Holdheide, 1932). In more dilute aqueous solutions of alcohol, the cell volume of *Valonia macrophysa* temporarily increased (Laibach, 1932). Such effects are presumably a consequence of the rapid entry of these compounds into the vacuole as compared with water (however, see Collander, 1950). Solutions of small or medium-sized molecules produced plasmolysis in *Hydrodictyon reticulatum*, and some relative values for permeability have been calculated. In sucrose solutions, cytorrhysis may occur, due to the low permeability of the cell wall to substances with so large a molecular volume. Increased adhesion of the protoplasm to the cell wall (Bonte, 1935) seems not to be involved (Horié, 1954). The permeability of some Ulotrichales, Cladophorales, Oedogoniales and Zygnematales was tested by Lenk (1956); see Table III. Hofmeister (1935) and Marklund (1936) established permeability series for *Oedogonium* sp., *Zygnema cyanosporum*, and *Zygnema* sp. (see Table II), which differed somewhat from the standard series obtained with *Chara* cells. The protoplasm of *Oedogonium echinospermum* was found to be unusually permeable to glucose (Lenk, 1956), but other species of this genus are less permeable, and show various degrees of permeability to other nonelectrolytes (see Table III). *Chaetomorpha aerea*, too, is permeable to sugars (Hoffmann, 1932). The permeability of *Pleurotaenium ehrenbergii* and *Closterium dianae* is normal, but other desmids are somewhat more permeable to glycerol (Krebs, 1952), as also is *Ulva pertusa* (Ogata and Takada, 1955).

The first accurate permeability measurements on *Spirogyra* were made by Lepeschkin (1908, 1909), who demonstrated a remarkable variation in the permeability of individual filaments to glycerol, ranging from 1.9×10^{-8} to 6.0×10^{-8} cm.sec.$^{-1}$. In the permeability series established by Elo

(1937) for *Spirogyra* sp. (see Table II), permeability constants of all substances tested were lower than those of *Chara* cells by a fairly constant factor.

3. Dyes

One of the earliest experiments using vital dyes was that of Szücs (1910), who found that the penetration of methyl violet into cells of *Spirogyra* is a passive process. This was later confirmed for many other basic vital dyes by Höfler (1960b).

Two groups of desmids can be distinguished on the basis of their behavior in vital staining. In one group, including *Micrasterias rotata*, *Tetmemorus levis* and *Cosmarium amoenum*, the cells accumulate toluidine blue or brillant cresyl blue, which in highly alkaline media results in a violet coloration of the cell sap. In the second group, which includes *Cosmarium cucurbita*, *C. pyramidatum*, and *Euastrum didelta*, the dye accumulates in the form of small blue-green spherules in the cytoplasm, even at pH values of 7 to 8 (Hirn, 1953b). The ions of certain vital dyes may also stain the cell walls of some, but not all, species of *Closterium* (Höfler and Schindler, 1953). Methyl red, an acidic vital dye, penetrates the cells of *Spirogyra* only as the undissociated molecule, at pH 3.1 to 6.8, and is accumulated solely where tannins are present (Zöttl, 1960). Uptake of methylene blue into the sap of *Valonia ventricosa* was reported by Brooks (cf. 1929). Irwin (cf. 1929) could not verify this result, and attributed Brooks' observation to penetration by the blue dye trimethyl thionin, an oxidation product of methylene blue. 2,6-Dibromophenol indophenol enters *Valonia* coenocytes only when in the reduced form (Brooks, 1926).

4. Ions

Chalaupka (1939) observed that cells of *Bryopsis plumosa*, *Cladophora hutchinsii*, and *Ulva lactuca*, plasmolyzed in concentrated sea water, soon recovered, presumably as a result of the penetration of salts into the vacuole. Osterhout (1911) found that certain species of *Spirogyra* exhibit a higher permeability to NaCl than to $CaCl_2$; Ullrich (1935) showed that the permeability of *Valonia macrophysa* for various anions paralleled their mobility in water. Ion uptake and accumulation in *Valonia* and *Halicystis* vesicles were exhaustively investigated by Osterhout, Jacques, and others (see Collander, 1939; Osterhout, 1947), but few conclusions can be drawn specifically about permeability in these algae. The protoplasm of *Valonia* seems in general to be very impermeable to cations. Using radioactive Na^{22} as a tracer, Brooks (1953) found that sodium penetrated cells of *Valonia* only as far as the protoplasm, whereas in *Halicystis* it also penetrated to the vacuole. The permeability for alkali ions is in the order: K > Na > Rb > Cs (Cooper *et al.*, 1929). Correspondingly, the protoplasm of *Ulva pertusa* is reported to be permeable to K ions, but not to Na or Ca (Ogata and Takada, 1955). For more definite knowledge on the permeability of ions, further investigations of electrochemical equilibria and the action of ion pumps are needed (see Eppley, Chapter 16, this treatise).

5. *Other Solutes*

Guanidine penetrates into *Valonia macrophysa* and reaches 5 to 10% of the external concentration (5 to 50 m*M* in sea water) after 10 hours (Jacques, 1935). Wedding and Erickson (1957) concluded that during the first ten minutes 2,4-dichlorophenoxyacetic acid in aqueous solution penetrates into cells of *Chlorella pyrenoidosa* by simple diffusion, as do alkaloids such as quinine, caffeine, and piperidine into *Spirogyra majuscula*, provided the pH is such as to maintain an appreciable proportion of the undissociated molecules (Tröndle, 1920).

G. Charophyta

Due to its large volume and convenient shape, the characean axial cell is a standard object for permeability experiments.

1. *Water*

Wartiovaara (1944) determined the permeability constant of *Tolypellopsis stelligera* to heavy water (HDO) by using the diffusion method, and obtained a value about one thousand times that of urea. Measurements made with dead cells indicated that the cell wall appreciably reduced the rate of passage of fast penetrating substances such as HDO or ethanol. However, if internal diffusion within the cell sap of these giant cells is rate-limiting, this would complicate the interpretation of such results (Dainty and Hope, 1959; see Table V).

Palva (1939) determined the water permeability of *Tolypellopsis stelligera* by an osmotic method, estimating volume changes by observing changes in the specific gravity of the cell (see Table V). Kamiya and Tazawa (1956), by exact measurements of the speed of water passage through cells of *Nitella flexilis* ("transcellular osmosis"; cf. Osterhout, 1949), presented evidence for differences in the permeability constants for water in exosmosis and endosmosis. However, Dainty and Hope (1959) criticized this evidence and, using *Chara australis*, demonstrated the same basic rates of permeability for water moving in and out of the cell (see Table V).

2. Non-electrolytes

Among the most comprehensive studies of permeability were those of Collander and Bärlund (1933) on *Chara ceratophylla* (see Table II), which established the "lipid sieve theory" for protoplasmic permeability and of Collander (1950) on *Nitella mucronata.* It was found that substances with molecules less than about 4 Å. in diameter penetrated faster than could be accounted for solely on the basis of their lipid solubility (see also Collander, 1949). Permeability of the protoplasm alone was calculated from a comparison of the behavior of living and dead cells of *N. mucronata* (Collander, 1954; see Table II). In these and in similar experiments with substances of high permeability constants, there were no large discrepancies between results obtained with *Nitella mucronata, Chara ceratophylla,* and *Tolypellopsis stelligera* (Wartiovaara, 1942, 1944, 1949; Collander, 1950, 1954; see Table II). However, from experiments with members of a polymer series of polypropylene glycols, it is clear that not only the lipid solubility but also the size of such large molecules influences their ability to penetrate cell membranes (Collander, 1960).

3. Dyes

Collander *et al.* (1943) noted a high rate of penetration of neutral red into protoplasts of *Chara ceratophylla* and *Tolypellopsis stelligera,* evidently attributable to its high lipid solubility. The penetration of brillant cresyl blue, azure B, and other dyes into the vacuole of *Nitella* cells was studied extensively by Irwin (1931a, b).

4. Ions

Characean cells have been frequently selected for studies of the uptake and accumulation of ions from the external medium (see Osterhout, 1947; Hoagland, 1930, 1948; Eppley, Chapter 16, this treatise). In many cases, active transport complicates the interpretation of data on the permeability of the cytoplasmic membranes involved. However, in *Tolypellopsis stelligera* and *Chara ceratophylla* passive ion permeability seems to be very low, since the cations tested passed through the protoplasm less than one-millionth as fast as through a layer of water of the same thickness (Collander, 1939).

The first detailed results on the permeability of algal cells to ions were obtained by MacRobbie and Dainty (1958a), who also discussed the effects of electrochemical potentials and the action of ion pumps. Using a tracer technique with cells of *Tolypellopsis stelligera* in a medium similar to the natural environment, they measured the influx and efflux of Na, K, and Cl ions. Holm-Jensen *et al.* (1944) attempted to calculate the permeability constant of *Tolypellopsis* cells for Na and K ions. MacRobbie and Dainty proposed a three-compartment exchange pattern (cell wall/cytoplasm/vacuole), subsequently adopted also by Diamond and Solomon (1959). In this hypothesis a Cl-ion pump functions at the inner surface of the protoplasm (tonoplast) and a Na pump operates at the plasmalemma. These workers found that the concentrations of K and Na ions in the cell sap did not differ appreciably from those in the protoplasm. The exchange between vacuole and protoplasm was evidently much slower than between protoplasm and external solution. This is in agreement with other observations that, after a brief period in the new medium, the ions penetrated only as far as the protoplasm (Brooks, 1951; Hoagland and Broyer, 1941).

Permeability data for other ions are sparse, and mostly only of a qualitative nature. P^{32}-labeled phosphate ions penetrated only as far as the cytoplasm of *Nitella* sp., and no activity was found in the sap (Brooks, 1951). Nitrate was found to enter the vacuole of *Nitella* sp. very slowly unless the cell was injured (Osterhout, 1922). Chloramphenicol and streptomycin penetrate through the protoplasm of *Nitella clavata,* whereas penicillin does not reach the vacuole. There is evidence that the accumulation of streptomycin in the vacuole is the result of an active transport mechanism (Pramer, 1955, 1956; Litwack and Pramer, 1957).

H. Rhodophyta

Owing to their sensitivity to plasmolytic damage (see Biebl, 1937), few studies of the permeability of Red Algae have been carried out. Only two species, *Nitophyllum punctatum* and *Acrosorium uncinatum,* have been found appropriate for measurements of water permeability; the rates determined were similar to those of other algae (Huber and Höfler, 1930); see Table IV. *Ceramium diaphanum* was found to have permeability properties towards non-electrolytes similar to those of *Chara* (Elo, 1937; see Table II).

Rhodamine B is capable of penetrating the protoplasm of *Polysiphonia* sp., and accumulates in the vacuole. In *Ceramium ciliatum* and *C. diaphanum,* acridine orange produces a yellow coloration in the vacuolar sap (Höfler *et al.,* 1956).

The protoplasm of *Porphyra tenera* appears to be somewhat permeable to K ions, though not to Na, since cells slowly recover after shrinkage in hypertonic solutions of KCl (Ogata and Takada, 1955). MacRobbie and Dainty (1958b), using a tracer technique, measured the fluxes and thus the specific rates of exchange of Na and K ions in *Rhodymenia palmata.* Their results suggest an active transport of Cl ions, and possibly also of K, into the cell vacuoles, and active outward transport of Na ions. Some species which tolerate plasmolysis in concentrated sea water show recovery in a few hours (Biebl, 1937), probably due to permeability of the protoplasm to salts. However, Chalaupka (1939) did not observe recovery of cells of *Ceramium deslongchampsii* under such conditions.

I. Phaeophyta

The permeability of Brown Algae has not yet been tested extensively, and much work needs to be done with this group.

Huber and Höfler (1930) found a high value for the water permeability of *Arthrocladia villosa,* intermediate values for *Heterospora vidovichii* and *Stypocaulon scoparium,* and low ones for *Ectocarpus globifer* and *Sporochnus pedunculatus* (see Table IV). The water permeability of the protoplasmic membrane at the surface of *Fucus vesiculosus* eggs was found to increase with age (Resühr, 1935).

For non-electrolytes a permeability series established for *Pylaiella litoralis* (Marklund, 1936) proved quite similar to that of *Chara* (see Table II). The permeability of *Ectocarpus siliculosus* to urea is of the same order of magnitude as that of many higher-plant cells (Höfler *et al.*, 1956); the permeability of some other marine species to urea was studied by Chalaupka (1939). Resühr tested eggs of *Fucus vesiculosus* (1935) for their permeability to a few non-electrolytes (see Table II).

The extremely rapid uptake of rhodamine B into the physodes of *Ectocarpus siliculosus* cells (see Ogino, Chapter 26, this treatise) indicates a high permeability of both the plasmalemma and the cytoplasm to this dye (Höfler *et al.*, 1956).

Osterhout (1912) found a rapid decrease in the electrical resistance of tissues of *Laminaria* sp. immersed in solutions of NaCl or other alkali salts, which he interpreted as indicating an appreciable permeability to univalent cations. The protoplasm seemed to be much less permeable to alkali earth salts. Chalaupka (1939) observed that all the species of benthonic Brown Algae that she tested remained plasmolyzed in concentrated sea water, but that in NaCl solutions *Sargassum linifolium* recovered from plasmolysis, indicating that increased permeability to NaCl may perhaps be a consequence of membrane damage. However, Biebl (1938) reported a high permeability of certain intertidal Brown Algae to sea salt.

IV. Changes in Permeability

A. Endogenous Factors

The age of cells may considerably influence their permeability. For example, adult *Spirogyra* cells are permeable to urea but young cells are not (Weber, 1931), and fertile filaments of *Oedogonium pringsheimii* bearing oogonia are considerably more permeable than sterile ones (Lenk, 1956). Permeability has also been found to change with the season of the year (see Section V).

B. Physical Factors

The effects of temperature on the rates of penetration of non-electrolytes into *Tolypellopsis stelligera* were tested by Wartiovaara (1942). Permeability increases reversibly if the temperature is raised, with a Q_{10} between 2.5 and 5.5. The Q_{10} was found to be independent of the lipid solubility of the solute but to be somewhat influenced by the size of the molecule.

The effects of light were reviewed by Järvenkylä (1937), who found that illumination tended to increase the permeability of *Chara ceratophylla*, especially to substances with a low lipid solubility, and that blue light was more effective than red. In *Rhodymenia palmata* the rates of exchange of K and Na ions—especially the former—were also accelerated by light, which may be attributed to an increased permeability or to a higher activity of the K transport mechanism (MacRobbie and Dainty, 1958b). Light was reported to promote the entrance of electrolytes into *Halicystis osterhoutii* (Jacques, 1939b), but Laibach (1932) found no obvious effects of light on exosmosis or endosmosis in cells of *Valonia macrophysa*.

Application of an electric current to *Chara* cells did not detectably increase their permeability to the dyes tested as long as they remained undamaged (Suolathi, 1937).

C. Chemical Effects

The direct action of a solute upon the cell membranes and the protoplasm obviously affects measurements of its permeability. For *Spirogyra*, urea is appreciably toxic, causing an increase in permeability (see Höfler, 1934); single salt solutions can also have adverse effects, e.g., on *Nitella* (Osterhout, 1922). Some salts tend to decrease the permeability of algal cells to dyes (Szücs, 1910) or to other salts (Osterhout, 1911, 1912, 1922); in some cases, however, certain concentrations may increase it (Iljin, 1928).

Metabolic inhibitors have been found to decrease the penetration of solutes into the cells of several species of diatoms plasmolyzed in sugar solutions (Follmann, 1958a), and to change the permeability of *Cladophora gracilis* (Follmann and Follmann-Schrag, 1959). The narcotic chloral hydrate, however, was reported to have no appreciable influence on the permeability of *Spirogyra majuscula* to caffeine (Tröndle, 1920).

D. Effects of pH

The permeability of *Spirogyra* and *Cladophora* protoplasm to water reaches a maximum at about pH 7 (Seemann, 1950). At neutrality the rate of diffusion of arsenate into cells of *Valonia macrophysa* is minimum; that of arsenite increases only under alkaline conditions (Brooks, 1925). Contrariwise, uptake of 2,4-dichlorophenoxyacetic acid into *Chlorella pyrenoidosa* is more rapid in more acidic media (Wedding and Erickson, 1957), as would be expected from the theory that such substances penetrate chiefly as the undissociated acids (Simon and Beevers, 1952). A low pH increases the permeability of *Valonia* cells for sugar, thereby promoting its uptake into the vacuole, and accelerates the exosmosis of substances from the cell sap of *Chaetomorpha* (Iljin, 1928). However, Rottenburg (1944) found no relation between the pH of the external solution, within the range 3.5 to 11.1, and the permeability of *Zygnema* cells to urea or glycerol.

V. Comparison of the Permeability of Different Algae

There are no fundamental differences between cells of most algae and those of higher plants in the permeability series for non-electrolytes, except for the high sugar permeability of some diatoms. The Cyanophyta, however, may be somewhat atypical (see Section III.A).

With respect to their permeability, the majority of the algae examined belong to the "urea" type, in which the permeability constant for urea is at least twice as great as that for glycerol (Marklund, 1936). Only a few algae, notably some Chlorophyta, belong to the "glycerol" type. Some species appear to change from the "glycerol" type in winter to the "urea" type in summer (Lenk, 1956). A decrease of the permeability constant in the sequence acetamide > methylurea > urea > glycerol is characteristic of lower plants (Diskus, 1956).

The protoplasm of intertidal Brown Algae is quite permeable to salts, which are readily admitted from concentrated sea water into the vacuole, whereas species occupying zones lower down the shore do not show this degree of permeability to salts. This difference in behavior may be related to ecological conditions (see Biebl, 1938; and Biebl, Chapter 53, this treatise).

VI. Protoplasmic Membranes and Permeability

It is generally assumed that the cytoplasmic membranes constitute the chief barriers to the migration of substances into or out of the protoplasm

(Collander, 1956, 1957; Levitt, 1960; Wartiovaara and Collander, 1960), although Höfler (1960b, 1961) has questioned the unqualified acceptance of this view. The plasmalemma is often considered to be more permeable, especially to ions, than is the tonoplast membrane of the vacuole (Collander, 1957; Walker, 1957). The tonoplast generally exhibits a very low permeability for all ions tested (MacRobbie and Dainty, 1958a; Dainty, 1959). The plasmalemma of *Tolypellopsis stelligera* is apparently much more permeable to K than to Na; that of *Nitella* sp. is almost impermeable to Ca ions (Walker, 1957). The movement of water between the vacuole and the exterior is probably controlled by the whole layer of protoplasma (Höfler, 1950; see Dainty and Hope, 1959; Wartiovaara and Collander, 1960). Other experiments on the resistance of the plasmalemma of various algae to toxic concentrations of salts (Höfler, 1951, 1958; Diannelidis and Mitrakos, 1960; Böhm-Tüchy, 1960) may relate indirectly to their permeability.

The cytoplasm of *Valonia macrophysa* tends to adhere strongly to the cell wall, which presumably accounts for the discrepancies observed between permeability measurements of intact *Valonia* cells and cell walls from which the protoplasmic layer had been removed (Kornmann, 1934, 1935). In other types of cells, however, the cell wall itself is composed of relatively impermeable material and may thus also act as a permeability barrier. In hypertonic media this may lead to shrinkage of the whole cell without plasmolysis, i.e., cytorrhysis, as in Cyanophyta, Euglenophyta, certain Dinophyceae, and the chlorophyte *Hydrodictyon*.

The lipid-sieve theory for protoplasmic membranes, to account for the more rapid penetration of small molecules and of lipid-soluble substances, is generally applicable to algae. Thus for *Tolypellopsis stelligera*, small molecules have a permeability constant about 10 times as high as would be expected on the basis of their lipid solubility, while heavy water (HDO) penetrates a few hundred times more rapidly (Wartiovaara, 1944). Lipid solubility also controls the permeability of the membrane at the surface of *Fucus* eggs (Resühr, 1935).

TABLES I–V APPEAR ON FOLLOWING PAGES

TABLE I

RELATIVE PERMEABILITY CONSTANTS OF DIATOMS FOR SUGARS AND FOR UREA[a]

Species	Sugar	$K_s' \times 10^4$ in sec.$^{-1}$		$\frac{K_s' \text{ (sugar)}}{K_s' \text{ (urea)}}$ [b]	Reference
		Sugar	Urea		
Anomoeoneis bohemica	Sucrose	2.8			Höfler (1940)
	Glucose	3.3			Höfler (1940)
A. sculpta	Glucose	1.8			Höfler (1943)
Pinnularia microstauron	Glucose	1.9			Höfler (1943)
P. sudetica	Glucose	1.0			Höfler (1943)
Anomoeoneis sphaerophora	Sucrose	1.7			Übeleis (1957)
Gyrosigma acuminatum					
(in summer)	Sucrose	2.4	17	0.15	Übeleis (1957)
(in autumn)	Sucrose	1.8			Übeleis (1957)
Licmophora oedipus	Sucrose	3.9	8.1	0.48	Elo (1937)
Melosira cf. *moniliformis*	Sucrose	1.5	11	0.14	Marklund (1936)
Navicula hungarica var. *capitata*	Sucrose	2.7			Übeleis (1957)
N. pygmaea	Sucrose	8.6; 1.9			Übeleis (1957)
N. viridula	Sucrose	2.6			Übeleis (1957)
Nitzschia filiformis	Sucrose	3.3			Übeleis (1957)
N. hungarica	Sucrose	1.4			Übeleis (1957)

N. tryblionella	Sucrose	1.4			Übeleis (1957)
N. tryblionella var. *levidensis*	Sucrose	1.8			Übeleis (1957)
Pinnularia maior	Sucrose	3.1			Übeleis (1957)
P. microstauron	Sucrose	4.3			Übeleis (1957)
For comparison:					
Chara ceratophylla	(Glucose, sucrose)	<0.0028	0.36	<0.0078	Collander and Bärlund (1933)

[a] $K_s' = (dC_i/dt)\,[1/(C_o - C_i)]$, where C_i, C_o are concentrations of the permeating substance, respectively within the vacuole and outside the cell. (Since in calculating this relative permeability constant the surface/volume ratio of the protoplast is not considered, these values can be compared only in cells with the same surface/volume ratio.)

[b] This ratio, being independent of the surface/volume ratio, is equal to the ratio of the absolute permeability constants.

TABLE II

ABSOLUTE PERMEABILITY CONSTANTS ($K_s \times 10^7$ IN CM. SEC.$^{-1}$) FOR SOME NON-ELECTROLYTES*†

Compound	Cyanophyta	Charophyta					Chlorophyta			Bacillariophyta		Rhodophyta	Phaeophyta	
			Nitella mucronata											
	Oscillatoria princeps	*Chara ceratophylla*	*Living cells*[c]	*Dead cells*[c]	*Protoplasm (calculated)*	*Tolypellopsis stelligera*	*Oedogonium* sp.[i]	*Zygnema*	*Spirogyra* sp.[k]	*Melosira* cf. *moniliformis*[i]	*Licmophora oedipus*[k]	*Ceramium diaphanum*	*Pylaiella litoralis*[i]	*Fucus vesiculosus* (eggs)[n]
Methanol	—	≧2800[a]	3230	7500	5700[c] 5600[d]	2400[e] 4700[f]	—	—	—	—	—	—	—	—
Ethanol	—	≧1600[a]	2990	6600	5500[c] 5600[d]	—	—	—	—	—	—	—	—	—
Methylcarbamic acid ethyl ester (= Urethylan)	—	≧1100[a]	—	—	—	1200[g] 2100[h]	—	—	—	—	—			—
Antipyrine	61[l]	610[a]	180	2700	192[c]	—	—	580[i]	—	—	—	—	—	—
Butyramide	100[l]	470[a]	134	4400	139[c]	—	—	—	—	—	—	—	—	—
Trimethyl citrate	31[l]	670[a]	115	2200	121[c]	120[e] 140[f] 150[g] 174[h]	150	210[i]	—	—	—	—		—
Propionamide	110[k]	360[a]	78	5000	79[c]	—	38	110[i]	44	289	170	270[k]	130	94
Acetamide	150[k]	150[a]	66	5800	66[c]	—	—	64[i]	26	47	28	120[k]	28	44
N,N-Diethylurea	72[l]	160[a]	38	3700	39[c]	—	—	—	—	—	—	—	—	—
N,N-Dimethylurea	—	94[a]	15	4400	15[c]	—	—	15[j]	—	—	—	—	—	—
Ethylene glycol	140[k]	120[a]	12	5600	12[c]	27[g]	12	25[i] 21[j]	11	38	11	83[k]	24	14
Thiourea	—	21[a]	3.6	4900	3.6[c]	—	—	1.5[j]	—	—	—	—	—	8.9
Methylurea	83[k]	19[a]	3.2	5000	3.2[c]	—	2.9	1.4[i] 1.6[j]	2.8	12	4.2	11[k]	3.1	7.2
Urea	27[k]	11[a] 1.9[b]	1.3	5700	1.3[c]	2.3[g]	1.4	0.47[i] 0.43[j]	1.6	4.2	1.8	8.9[k]	0.64	1.2

Hexamethylene tetramine	7.8[k]	7.2[a]	0.39	3400	0.39[c]	1.52[g]	0.24	0.075[i]	0.89	3.1	2.3	2.8[m]	—	—
Glycerol	26[k]	2.1[a]	0.032	4300	0.032[c]	0.26[g]	0.17	0.075[i]	—	3.3	1.6	2.0[k]	0.23	<0.53
Malonic diamide	5.6[k]	0.39[a]	—	—	—	—	—	0.075[i] 0.11[j]	—	1.8	1.6	2.5[m]	0.15	—
Erythritol	0.94[k]	0.13[a]	—	—	—	—	—	0.02[i] 0.01[j]	—	1.3	1.3	0.19[k]	0.01	—
Sucrose	—	<0.083[a]	—	—	—	—	—	0.0053[i]	—	0.58	0.83	—	0.0006	—

* $K_s = (dm/dt)\{[1/A(C_o - C_i)]\}$, where A is area of permeable "membrane" (i.e., protoplasmic layer); C_i and C_o are molar concentrations of permeating solute, inside vacuole and outside cell, respectively; and dm/dt is derivative of molar amount of solute in vacuole with respect to time. This value of K_s is an absolute measure of the permeability of the membrane under consideration. It indicates the amount of substance (moles) which pass per unit time (sec.) and per unit area (cm.^{2}) when the concentration difference of the penetrating substance between external solution and cell sap is unity (moles/cm.^{3}).

† REFERENCES: [a] Collander and Bärlund, 1933; [b] Wartiovaara, 1942 (for *Chara* sp.); [c] Collander, 1954; [d] Wartiovaara, 1949; [e] Wartiovaara, 1944 (for the living cell, i.e., cell wall plus protoplasm layer); [f] Wartiovaara, 1944 (calculated for the protoplasm layer alone); [g] Wartiovaara, 1942 (for the living cell, i.e., cell wall plus protoplasm layer); [h] Wartiovaara, 1942 (calculated for the protoplasm layer alone); [i] Marklund, 1936 (values for *Zygnema* refer to *Z. cyanosporum*); [j] Hofmeister, 1935 (for *Zygnema* sp., using the formula

$$K = \left(\frac{dm}{dt}\right)\cdot\left(\frac{V_p}{V_z}\right)\cdot\left[\frac{1}{A(C_o - C_i)}\right]$$

where V_p and V_z are respectively the volume of the protoplast and the total volume enclosed within the cell wall); [k] Elo, 1937; [l] Elo, 1937 (values probably too low, since cell-wall permeability may interfere); [m] Elo, 1937 (values uncertain, probably too high); [n] Resühr, 1935.

TABLE III

RELATIVE PERMEABILITY VALUES ($\Delta G \times 10^4$ IN SEC.$^{-1}$) FOR NONELECTROLYTES[a]

Species	Methyl-urea	Urea	Glycerol	Malonic diamide	Erythritol	Reference
Chlorophyta						
Ulotrichales						
Microspora tumidula	—	0.83–2.2	0.36–0.69	—	—	Lenk (1956)
M. floccosa	—	2.6	5.6	—	—	Lenk (1956)
Hormidium subtile	—	4.2	2.0	—	—	Lenk (1956)
Stigeoclonium cf. *tenue*	—	2.5–11	0.72	—	—	Lenk (1956)
Cladophorales						
Cladophora glomerata	3.8–4.4	0.58–1.6	0.56	0.33	—	Lenk (1956)
Rhizoclonium hieroglyphicum	6.4	1.1	3.3	—	—	Lenk (1956)
Oedogoniales						
Oedogonium echinospermum	8.3	3.1	0.39	0.2	0.083	Lenk (1956)
O. itzigsohnii	13	2.1	0.16	0.78	—	Lenk (1956)
O. pringsheimii	4.4–8.3	5.6–10	0.44–1.4	0.16–0.31	—	Lenk (1956)
O. sp.	7.2	2.7	0.64	0.17	~0.042	Lenk (1956)
Zygnematales						
Closterium dianae	7.5	1.5	0.12	0.049	0.012	Krebs (1952)
Mougeotia scalaris	3.9	0.78	0.39	0.17	—	Lenk (1956)
Mougeotia sp.	—	1.3	0.56	0.14	—	Lenk (1956)
Pleurotaenium ehrenbergii	7.5	1.5	0.12	0.049	0.012	Krebs (1952)
Spirogyra parvula	3.6	0.83	0.33	0.023	—	Lenk (1956)
Sp. varians	7.2	0.47	0.26	0.024	—	Lenk (1956)
Zygnema vaginatum	3.6	0.97	0.67	0.033	—	Lenk (1956)
Zygnema sp.	1.6	0.47	0.58	0.13	0.013	Hofmeister (1935)

Zygnema sp. "RT"	—	5.3	0.92	—	—	Lenk (1956)
Zygnema sp. "S"	1.5	0.75	0.13	—	—	Lenk (1956)
Xanthophyta						
Tribonema viride	6.1	2.2	0.64	—	—	Lenk (1956)
T. vulgare	—	1.0	0.78	—	—	Lenk (1956)
Bacillariophyta						
Caloneis obtusa	3.3	1.1	0.86	—	0.26	Höfler (1960a)

[a] $\Delta G = (G_2 - G_1)/(t_2 - t_1)$; this value can be conveniently determined for cylindrical cells, plasmolyzed in a hypertonic solution of a permeating substance. The degree of plasmolysis G (equal to volume of the protoplast divided by volume of the cell) can be calculated with accuracy, and its change (ΔG) per unit time (t, in seconds) gives a relative measure for permeability. For a given species, the values are approximately in proportion to the absolute permeability constants, K. The relative permeability value ΔG is an experimentally derived measure for permeability. To obtain the relative permeability constant K_s', ΔG should be multiplied by $C_o/(C_o - C_i)$ [which may be estimated to range between 1.6 and 6; see Lenk (1956, p. 183)] in order to take into account the action of the concentration difference.

TABLE IV

Relative Osmotic Permeability Constants (K_{os}', in cm.^{3} sec.$^{-1}$ mole^{-1} and Converted into 10^{-4} atm^{-1} sec^{-1}) for Water[a]

Species	External solute	Concentration (M)	pH	Direction of water flow	Relative osmotic permeability constant (K_{os}') (cm.^{3} × sec.$^{-1}$ × mole^{-1})	(K_{os}') (× 10^4 in atm.$^{-1}$ sec.$^{-1}$)	Reference
Chlorophyta							
Cladophorales							
Cladophora glomerata	Glucose	0.8	4.8	Inward	5.2	2.2	Seemann (1950)
	Glucose	0.8	6	Inward	11	4.6	Seemann (1950)
	Glucose	0.8	7	Inward	13	5.4	Seemann (1950)
	Glucose	0.8	8	Inward	10	4.2	Seemann (1950)
	Glucose	0.8	8.7	Inward	6.7	2.8	Seemann (1950)
Rhizoclonium hieroglyphicum	Glucose	0.8	—	Outward	19	7.9	Lenk (1956)
Oedogoniales							
Oedogonium echinospermum	Glucose	?	—	Outward	12	5.0	Lenk (1956)
Oedogonium sp.	Glucose	0.8	—	Outward	1.2	0.5	Lenk (1956)
Zygnematales							
Spirogyra communis	?	?	—	Outward	6.0	2.5	Lenk (1956)
S. singularis	?	?	—	Outward	21	8.7	Lenk (1956)
S. condensata	?	?	—	Outward	11	4.6	Lenk (1956)
S. gracilis	?	?	—	Outward	20	8.3	Lenk (1956)
S. pseudovarians	?	?	—	Outward	3.0	1.3	Lenk (1956)
S. varians	?	?	—	Outward	5.7	2.4	Lenk (1956)

S. affinis	?	?	—	Outward	6.3	2.6	Lenk (1956)
S. porticalis	?	?	—	Outward	3.7	1.5	Lenk (1956)
S. mirabilis (zygotes)	?	?	—	Outward	1.0	0.42	Lenk (1956)
S. nitida	KNO_3	0.50	—	Outward	9.7	4.0	Huber and Höfler (1930)
Spirogyra sp.	?	?	—	Outward	3.3–31	1.4–13	Lenk (1956)
	KCl	0.50	—	Outward	12	5.0	Huber and Höfler (1930)
Spirogyra sp. "H"	Glucose	0.45	3	Inward	4.8	2.0	Seemann (1950)
	Glucose	0.45	4.3	Inward	9.5	4.0	Seemann (1950)
	Glucose	0.45	4.8	Inward	13	5.4	Seemann (1950)
	Glucose	0.45	6	Inward	16	6.7	Seemann (1950)
	Glucose	0.45	7	Inward	18–25	7.5–10	Seemann (1950)
	Glucose	0.45	8.7	Inward	8.6	3.6	Seemann (1950)
Zygnema sp.	Glucose	0.6–0.8	—	Outward	7.2	3.0	Lenk (1956)
	Glucose	0.5	—	Inward	29	12	Hofmeister (1935)
	?	?	—	Outward	13	5.4	Hofmeister (1935)
Zygnema sp. ("*velox*")	KCl	0.50	—	Outward	120	50	Huber and Höfler (1930)
Zygnema sp. ("*longicellularis*")	$CaCl_2$, KCl	0.50	—	Outward	33	14	Huber and Höfler (1930)
Mougeotia scalaris	Glucose	0.6–0.8	—	Outward	1.7	0.71	Lenk, 1956
Mougeotia sp. 7	KCl	0.40	—	Outward	39	16	Huber and Höfler (1930)
Mougeotia sp. 8	KCl	0.50	—	Outward	129	54	Huber and Höfler (1930)
Bacillariophyta							
Nitzschia tryblionella	Sucrose	0.4	—	Outward	42–265	17–110	Übeleis (1957)

TABLE IV—Continued

Species	External solute	Concentration (M)	pH	Direction of water flow	Relative osmotic permeability constant (K_{os}') (cm.^{3} × sec.$^{-1}$ × mole^{-1})	(K_{os}') (× 10^4 in atm.$^{-1}$ sec.$^{-1}$)	Reference
Rhodophyta							
Nitophyllum punctatum	Sea water	(2.8×)	—	Outward	~17	~7.1	Huber and Höfler (1930)
Acrosorium uncinatum	Sea water	(2.8×)	—	Outward	~8.3	~3.5	Huber and Höfler (1930)
Phaeophyta							
Heterospora vidovichii	Sea water + glucose	1.2	—	Outward	12[b]	4.6	Huber and Höfler (1930)
Ectocarpus globifer	Sea water	(2×)	—	Outward	2.6[b]	1.1	Huber and Höfler (1930)
Arthrocladia villosa	Sea water	(2×)	—	Outward	60[b]	25	Huber and Höfler (1930)

[a] Here

$$K_{os}' = \frac{G_2 - G_1}{t_2 - t_1} \cdot \frac{1}{C_s - C_u}$$

where C_s is the concentration of nonpermeating solute (in mole/cm.^{3}) outside the cell, and C_u is the concentration of all solutes (in mole/cm.^{3}) of the cell sap; other symbols as in Table III. This relative measure for the water permeability can be conveniently used for cylindrical cells plasmolyzed in a solution of a nonpermeating substance (e.g., sugar).

The relative osmotic permeability constant can also be based upon the difference in osmotic suction forces instead of the difference in concentrations. We may easily recalculate it in terms of osmotic suction forces, since 1 $cm.^3$ $mole^{-1}$ corresponds to 4.16×10^{-5} $atm.^{-1}$ (derived from van't Hoff's law for dilute solutions: $P = CRT$, where $R = 82.057$ $cm.^3$ atm. $deg.^{-1}$ $mole^{-1}$ and $T = 20°$ C. = 293° K.) However for concentrations as high as those used here the real relation between osmotic suction forces and concentrations deviates considerably from van't Hoff's law.

[b] Calculated on the assumption that sea water is osmotically equivalent to 1.2 *M* glucose (Huber and Höfler, 1930).

TABLE V

ABSOLUTE PERMEABILITY CONSTANTS (K_d AND K_{os}) FOR WATER[a, b]

Species	External solute	Concentration (M)	Direction of water flow	Diffusional permeability constant, $K_d \times 10^4$ (in cm. sec.$^{-1}$)	Osmotic permeability constants		Reference
					$K_{os} \times 10^6$ (in cm. sec^{-1} atm.$^{-1}$)	$K_{os} \times 10^4$ (in cm. sec^{-1})	
Nitella mucronata							
Living cells	Heavy water	?	No net flow	6.3[a]	—	—	Collander (1954)
Dead cells	Heavy water	?	No net flow	8.5[a]	—	—	Collander (1954)
Protoplasm (calculated)	Heavy water	?	No net flow	~25[a]	—	—	Collander (1954)
Tolypellopsis stelligera							
Living cells	Heavy water	~15–26	No net flow	2.8[a]	—	—	Wartiovaara (1944)
Dead cells	Heavy water	~15–26	No net flow	8.3[a]	—	—	Wartiovaara (1944)
Protoplasm (calculated)	Heavy water	~15–26	No net flow	4.4[a]	—	—	Wartiovaara (1944)
Living cells	Sucrose and NaCl	~0.3	Outward	24[c]	1.8[b]	—	Palva (1939)
Nitella flexilis	Sucrose	0.1–0.5	Inward	410[c]	31[b]	—	Kamiya and Tazawa (1956)
	Sucrose	0.1–0.5	Outward	160[c]	12[b]	—	Kamiya and Tazawa (1956)
Chara australis	Sucrose	0.1–0.5	Inward and outward	120[c]	9.3[b]	—	Dainty and Hope (1959)

Fucus vesiculosus (egg)	Sea water + sucrose	Total ~2	Outward	3.7[c]	0.27[b]	—	Resühr (1935)
Spirogyra communis	?	?	Outward	3.83[e]	—	—	Lenk (1956)
S. singularis	?	?	Outward	11.5[e]	—	—	Lenk (1956)
S. gracilis	?	?	Outward	8.28[e]	—	—	Lenk (1956)
S. affinis	?	?	Outward	2.8[e]	—	—	Lenk (1956)
S. porticalis	?	?	Outward	2.5[e]	—	—	Lenk (1956)
S. nitida	KNO_3	1.78	Outward	4.9[e]	—	—	Huber and Höfler (1930)
Spirogyra sp.	Glucose	0.7	Outward	—	—	0.03[d]	Bochsler (1948)
	Glucose	0.5	Outward	—	—	0.031[d]	Bochsler (1948)
Zygnema sp.	Glucose	0.8	Outward	8.5[e]	—	—	Hofmeister (1935)
Zygnema sp.	?	?	Outward	3.8[e]	—	—	Lenk (1956)
Zygnema sp. ("*Velox*")	Glucose	0.7	Outward	23[e]	—	—	Huber and Höfler (1930)
Zygnema sp.	Glucose	0.855	Outward	—	—	0.039[d]	Bochsler (1948)
	Glucose	0.4–0.6	Outward	—	—	0.072–0.12[d]	Bochsler (1948)

[a] Absolute *diffusional* permeability constant (K_d) for heavy water (outward diffusion):

$$K_d = \frac{dm}{dt} \cdot \frac{1}{A(C_i - C_o)}.$$

(For explanation of symbols, see Table II.)

[b] Absolute *osmotic* permeability constant (K_{os}) for water:

$$K_{os} = \frac{dV_v}{dt} \cdot \frac{1}{A(S_o - S_i)}$$

where V_v is the volume of the vacuole, and $(S_o - S_i)$ is the difference between the osmotic suction forces outside and inside the cell, measured in atm. K_{os} here indicates the volume of water (in cm.^{3}) which passes through the protoplasm layer per unit time (1 sec.) and unit area (1 cm.^{2}) when the difference in osmotic suction force between the external solution (of a nonpermeating substance) and the cell sap is 1 atm.

[c] The values of K_{os} derived as above can be converted into K_d values (in cm. sec^{-1}) if we assume that van't Hoff's law (see legend of Table IV) is also applicable to water transport, i.e.. that the transporting power of a suction-force difference of 1 atm. is the same as

that of a "concentration" difference of water of 4.16×10^{-5} moles cm.$^{-3}$ In addition, the amount of water, dV_v, has to be measured in moles. Therefore the volume dV_v (measured in cm.^{3}) has to be multiplied by 1/18 mole cm.$^{-3}$ and the suction force $(S_o - S_i)$ by 4.16×10^{-5} atm.$^{-1}$ cm.$^{-3}$ mole. It is thus found that K_d (in cm. sec.$^{-1}$) = 1335 atm $\times$ K_{os} (in cm. sec.$^{-1}$ atm.$^{-1}$). The conversion of osmotic permeability constants into diffusion permeability constants involves some theoretical difficulties (cf. Dick, 1959), so that the values so obtained might be affected by some systematic errors.

[d] Osmotic permeability constant (in cm. sec.$^{-1}$)

$$K_{os} = \frac{dC_u}{dt} \cdot \frac{1}{Q(C_s - C_u)}.$$

(This formula is readily derived from the original equation of Bochsler.) Q is the surface/volume ratio of the protoplast; other symbols as in Table IV.

[e] Recalculated from Hofmeister using the original data and applying a conversion formula by Stadelmann. Communicated by courtesy of Hofmeister. [Details in Hofmeister (1962)].

REFERENCES

Bärlund, H. (1929). Permeabilitätsstudien an Epidermiszellen von *Rhoeo discolor*. *Acta Botan. Fennica* **5,** 1–117.

Biebl, R. (1937). Ökologische und zellphysiologische Studien an Rotalgen der englischen Südküste. *Botan. Centr. Beih.* **A, 57,** 381–424.

Biebl, R. (1938). Zur Frage der Salzpermeabilität bei Braunalgen. *Protoplasma* **31,** 518–523.

Blinks, L. R. (1955). Some electrical properties of large plant cells. *In* "Electrochemistry in Biology and Medicine" (T. Shedlovsky, ed.), pp. 187–212. Wiley, New York.

Bochsler, A. (1948). Die Wasserpermeabilität des Protoplasmas auf Grund des Fickschen Diffusionsgesetzes. *Ber. schweiz. botan. Ges.* **58,** 73–123.

Böhm-Tüchy, E. (1960). Plasmalemma und Aluminiumsalz-Wirkung. *Protoplasma* **52,** 108–142.

Bonte, H. (1935). Vergleichende Permeabilitätsstudien an Pflanzenzellen. *Protoplasma* **22,** 209–242.

Brand, F. (1903). Über das osmotische Verhalten der Cyanophyceenzellen. *Ber. deut. botan. Ges.* **21,** 302–309.

Brooks, M. M. (1925). The effects of internal and external pH on the penetration of arsenic from arsenate and arsenite solutions into a living cell. *Am. J. Physiol.* **72,** 222. (Abstr.)

Brooks, M. M. (1926). Studies on the permeability of living cells. VI. The penetration of certain oxidation-reduction indicators as influenced by pH; estimation of the *r*H of *Valonia. Am. J. Physiol.* **76,** 360–379.

Brooks, M. M. (1929). Studies on the permeability of living cells. X. The influence of experimental conditions upon the penetration of methylene blue and trimethyl thionine. *Protoplasma* **7,** 46–61.

Brooks, S. C. (1951). Penetration of radioactive isotopes P^{32}, Na^{24} and K^{42} into *Nitella. J. Cellular Comp. Physiol.* **38,** 83–93.

Brooks, S. C. (1953). The penetration of radioactive sodium into *Valonia* and *Halicystis. Protoplasma* **42,** 63–68.

Chalaupka, I. (1939). Permeabilitätsstudien an Meeresalgen, vornehmlich an Braunalgen. *Thalassia* (*Rovigno d'Istria*) **3** (5), 1–36.

Cholnoky, B. (1928). Über die Wirkung von hyper- und hypotonischen Lösungen auf einige Diatomeen. *Intern. Rev. ges. Hydrobiol. Hydrog.* **19,** 452–500.

Collander, R. (1939). Permeabilitätsstudien an Characeen. III. Die Aufnahme und Abgabe von Kationen. *Protoplasma* **33,** 215–257.

Collander, R. (1949). The permeability of plant protoplasts to small molecules. *Physiol. Plantarum* **2,** 300–311.

Collander, R. (1950). The permeability of *Nitella* cells to rapidly penetrating nonelectrolytes. *Physiol. Plantarum* **3,** 45–57.

Collander, R. (1954). The permeability of *Nitella* cells to non-electrolytes. *Physiol. Plantarum* **7,** 420–445.

Collander, R. (1956). Der Ort des Penetrationswiderstandes. *In* "Handbuch der Pflanzenphysiologie" (W. Ruhland, ed.), Vol. II, pp. 218–229. Springer, Berlin.

Collander, R. (1957). Permeability of plant cells. *Ann. Rev. Plant Physiol.* **8,** 335–348.

Collander, R. (1959). Cell membranes: their resistance to penetration and their capacity for transport. *In* "Plant Physiology" (F. C. Steward, ed.), Vol. II, pp. 3–102. Academic Press, New York.

Collander, R. (1960). The permeation of polypropylene glycols. *Physiol. Plantarum* **13,** 179–185.

Collander, R., and Bärlund, H. (1933). Permeabilitätsstudien an *Chara ceratophylla. Acta Botan. Fennica* **11,** 1–114.

Collander, R., Lönegren, H., and Arhimo, E. (1943). Das Permeationsvermögen eines basischen Farbstoffes mit demjenigen einiger Anelektrolyte verglichen. *Protoplasma* **37,** 527–537.

Cooper, W. C., Jr., Dorcas, M. J., and Osterhout, W. J. V. (1929). The penetration of strong electrolytes. *J. Gen. Physiol.* **12,** 427–433.

Dainty, J. (1959). Ion transport across plant cell membranes. *Proc. Roy. Physical Soc. Edinburgh* **28,** 3–14, 120.

Dainty, J., and Hope, A. B. (1959). The water permeability of cells of *Chara australis* R. Br. *Australian J. Biol. Sci.* **12,** 136–145.

Diamond, J. M., and Solomon, A. K. (1959). Intracellular potassium compartments in *Nitella axillaris. J. Gen. Physiol.* **42,** 1105–1121.

Diannelidis, T., and Mitrakos, K. (1960). Vergleichende Untersuchungen über das Plasmalemma von Meeresalgen. *Protoplasma* **51,** 265–283.

Dick, D. A. T. (1959). Osmotic properties of living cells. *Intern. Rev. Cytol.* **8,** 387–448.

Diskus, A. (1956). Osmoseverhalten und Permeabilität der Gymnodiniale *Oxyrrhis marina. Protoplasma* **46,** 160–169.

Diskus, A. (1958). Das Osmoseverhalten einiger Peridineen des Süsswassers. *Protoplasma* **49,** 187–196.

Drawert, H. (1949). Zellmorphologische und zellphysiologische Studien an Cyanophyceen. I. Mitt. Literaturübersicht und Versuche mit *Oscillatoria borneti* Zukal. *Planta* **37,** 161–209.

Drawert, H. (1956). Die Aufnahme der Farbstoffe. Vitalfärbung. *In* "Handbuch der Pflanzenphysiologie" (W. Ruhland, ed.), Vol. II, pp. 252–289. Springer, Berlin.

Elo, J. E. (1937). Vergleichende Permeabilitätsstudien, besonders an niederen Pflanzen. *Ann. Botan. Soc. Zool. Botan. Fennicae Vanamo* **8,** (6) 1–108.

Follmann, G. (1958a). Über Aufnahme und Bindung von Wasser und Anelektrolyten durch Diatomeen-Zellen. *Planta* **50,** 671–700.

Follmann, G. (1958b). Plasmolyse-Verhalten und Vitalfärbungs-Eigenschaften von *Coscinodiscus granii* Gough. *Bull. inst. océanog.* **55,** (1116) 1–22.

Follmann, G., and Follmann-Schrag, I.-A. (1959). Die Wasserführung plasmolysierter Protoplasten von *Cladophora gracilis* (Griff.) Kuetzing unter dem Einfluss von Atmungsgiften. *Z. Naturforsch.* **14b,** 181–187.

Hilmbauer, K. (1954). Zellphysiologische Studien an Euglenaceen, besonders an *Trachelomonas. Protoplasma* **43,** 192–227.

Hirn, I. (1953a). Vitalfärbung von Diatomeen mit basischen Farbstoffen. *Österr. Akad. Wiss. Math.-naturw. Kl. Sitzber. Abt. I,* **162,** 571–595.

Hirn, I. (1953b). Vitalfärbungsstudien an Desmidiaceen. *Flora (Jena)* **140,** 453–473.

Hoagland, D. R. (1930). The accumulation of mineral elements by plant cells. *In* "Contributions to Marine Biology," pp. 131–144. Stanford Univ. Press, Stanford, California.

Hoagland, D. R. (1948). Lectures on the inorganic nutrition of plants. Chronica Botanica, Waltham, Massachusetts.

Hoagland, D. R., and Broyer, T. C. (1941). Accumulation of salt and permeability in plant cells. *J. Gen. Physiol.* **25,** 865–880.

Höfler, K. (1926). Über die Zuckerpermeabilität plasmolysierter Protoplaste. *Planta* **2,** 454–475.

Höfler, K. (1934). Permeabilitätsstudien an Stengelzellen von *Majanthemum bifolium.* (Zur Kenntnis spezifischer Permeabilitätsreihen, I.) *Österr. Akad. Wiss. Math.-naturw. Kl. Sitzber. Abt. I.* **143,** 213–264.

Höfler, K. (1940). Aus der Protoplasmatik der Diatomeen. *Ber. deut. botan. Ges.* **58,** 97–120.

Höfler, K. (1942). Unsere derzeitige Kenntnis von den spezifischen Permeabilitätsreihen. *Ber. deut. botan. Ges.* **60,** 179–200.

Höfler, K. (1943). Über Fettspeicherung und Zuckerpermeabilität einiger Diatomeen und über Diagonal-Symmetrie im Diatomeenprotoplasten. *Protoplasma* **38,** 71–104.

Höfler, K. (1950). New facts on water permeability. *Protoplasma* **39,** 677–683.

Höfler, K. (1951). Plasmolyse mit Natriumkarbonat. Zur Frage des Plasmalemmas bei Süsswasseralgen und bei Gewebszellen von Landblütenpflanzen. *Protoplasma* **40,** 426–460.

Höfler, K. (1958). Aluminiumsalz-Wirkung auf Spirogyren und Zygnemen. *Protoplasma* **49,** 248–258.

Höfler, K. (1960a). Über die Permeabilität der Diatomee *Caloneis obtusa. Protoplasma* **52,** 5–25.

Höfler, K. (1960b). Permeability of protoplasm. *Protoplasma* **52,** 145–156.

Höfler, K. (1961). Grundplasma und Plasmalemma. Ihre Rolle beim Permeationsvorgang. *Ber. deut. botan. Ges.* **74,** 233–242.

Höfler, K., and Höfler, L. (1952). Osmoseverhalten und Nekroseformen von *Euglena. Protoplasma* **41,** 76–102.

Höfler, K., and Schindler, H. (1953). Vitalfärbbarkeit verschiedener Closterien. *Protoplasma* **42,** 296–311.

Höfler, K., and Url, W. (1958). Kann man osmotische Werte plasmolytisch bestimmen? *Ber. deut. botan. Ges.* **70,** 462–476.

Höfler, K., Url, W., and Diskus, A. (1956). Zellphysiologische Versuche und Beobachtungen an Algen der Lagune von Venedig. *Boll. museo civico storia nat. Venezia* **9,** 63–94.

Hoffmann, C. (1932). Zur Bestimmung des osmotischen Druckes an Meeresalgen. *Planta* **16,** 413–432.

Hofmeister, L. (1935). Vergleichende Untersuchungen über spezifische Permeabilitätsreihen. *Bibl. Botan.* (*Stuttgart*) **113,** 1–83.

Hofmeister, L. (1948). Über die Permeabilitätsbestimmung nach der Deplasmolysezeit. *Österr. Akad. Wiss. Math.-naturw. Kl. Sitzber. Abt. I.* **157,** 83–95.

Hofmeister, L. (1962). Permeabilität für Nicht-Elektrolyte. *In* "Protoplasmatologia" (L. V. Heilbrunn and F. Weber, eds.), Vol. II, C 8 b. Springer, Vienna (in preparation).

Holdheide, W. (1932). Über Plasmoptyse bei *Hydrodictyon utriculatum. Planta* **15,** 244–298.

Holm-Jensen, I., Krogh, A., and Wartiovaara, V. (1944). Some experiments on the exchange of potassium and sodium between single cells of Characeae and the bathing fluid. *Acta Botan. Fennica* **36,** 1–21.

Horié, K. (1954). Permeability of the cell membrane of *Hydrodictyon reticulatum. Cytologia* **19,** 117–129.

Huber, B., and Höfler, K. (1930). Die Wasserpermeabilität des Protoplasmas. *Jahrb. wiss. Botan.* **73,** 351–511.

Iljin, W. S. (1928). Die Durchlässigkeit des Protoplasmas, ihre quantitative Bestimmung und ihre Beeinflussung durch Salze und durch die Wasserstoffionenkonzentration. *Protoplasma* **3,** 558–602.

Irwin, M. (1929). Spectrophotometric studies of penetration. V. Resemblance between the living cell and an artificial system in absorbing methylene blue and trimethyl thionine. *J. Gen. Physiol.* **12,** 407–418.

Irwin, M. (1931a). Studies on penetration of dyes with glass electrode. IV. Penetration of Brillant Cresyl Blue into *Nitella flexilis. J. Gen. Physiol.* **14,** 1–17.

Irwin, M. (1931b). Studies on penetration of dyes with glass electrode. V. Why does Azur B penetrate more readily than methylene blue or crystal violet? *J. Gen. Physiol.* **14,** 19–29.

Jacques, A. G. (1935). The kinetics of penetration. X. Guanidine. *Proc. Natl. Acad. Sci. U. S.* **21,** 488–492.

Jacques, A. G. (1939a). The kinetics of penetration. XVIII. Entrance of water into impaled *Halicystis. J. Gen. Physiol.* **22,** 743–755.

Jacques, A. G. (1939b). The kinetics of penetration. XX. Effect of pH and of light on absorption in impaled *Halicystis. J. Gen. Physiol.* **23,** 41–51.

Järvenkylä, Y. T. (1937). Über den Einfluss des Lichtes auf die Permeabilität pflanzlicher Protoplasten. *Ann. Botan. Soc. Zool.-Botan. Fennica Vanamo* **9,** (3) 1–99.

Kamiya, N., and Tazawa, M. (1956). Studies on water permeability of a single plant cell by means of transcellular osmosis. *Protoplasma* **46,** 394–422.

Kesseler, H. (1959). Mikroskopische Untersuchungen zur Turgorregulation von *Chaetomorpha linum. Kiel. Meeresforsch.* **15,** 51–73.

Kornmann, P. (1934). Osmometer aus lebenden *Valonia*-Zellen und ihre Verwendbarkeit zu Permeabilitätsbestimmungen. *Protoplasma* **21,** 340–350.

Kornmann, P. (1935). Permeabilitätsstudien an *Valonia*-Osmometern. Das Verhalten gegen Neutralsalzlösungen. *Protoplasma* **23,** 34–49.

Krebs, I. (1952). Beiträge zur Kenntnis des Desmidiaceen-Protoplasten. III. Permeabilität für Nichtleiter. *Österr. Akad. Wiss. Math.-naturw. Kl. Sitzber. Abt. I,* **161,** 291–328.

Kuchar, K. (1950). Plasmolyseformverlauf und Trichomzerfall bei zwei Oscillatorien. *Phyton, Ann. rei botan.* **2,** 213–222.

Laibach, F. (1932). Interferometrische Untersuchungen an Pflanzen. II. Die Verwendbarkeit des Interferometers in der Pflanzenphysiologie. *Jahrb. wiss. Botan.* **76,** 218–282.

Lenk, I. (1956). Vergleichende Permeabilitätsstudien an Süsswasseralgen (Zygnemataceen und einigen Chlorophyceen). *Österr. Akad. Wiss. math.-naturw. Kl. Sitzber. Abt. I* **165,** 173–279.

Lepeschkin, W. W. (1908). Über den Turgordruck der vakuolisierten Zellen. *Ber. deut. botan. Ges.* **26a,** 198–214.

Lepeschkin, W. W. (1909). Über die Permeabilitätsbestimmung der Plasmamembran für gelöste Stoffe. *Ber. deut. botan. Ges.* **27,** 129–142.

Levitt, J. (1960). In defense of the plasma membrane-theory of cell permeability. *Protoplasma* **52,** 161–163.

Litwack, G., and Pramer, D. (1957). Absorption of antibiotics by plant cells. III. Kinetics of streptomycin uptake. *Arch. Biochem. Biophys.* **68,** 396–402.

MacRobbie, E. A. C., and Dainty, J. (1958a). Ion transport in *Nitellopsis obtusa. J. Gen. Physiol.* **42,** 335–353.

MacRobbie, E. A. C., and Dainty, J. (1958b). Sodium and potassium distribution and transport in the seaweed *Rhodymenia palmata* (L.) Grev. *Physiol. Plantarum.* **11,** 782–801.

Marklund, G. (1936). Vergleichende Permeabilitätsstudien an pflanzlichen Protoplasten. *Acta Botan. Fennica* **18,** 1–110.

Ogata, E., and Takada, H. (1955). Elongation and shrinkage in thallus of *Porphyra tenera* and *Ulva pertusa* caused by osmotic changes. *J. Inst. Polytech. Osaka City Univ. Ser.* **D 6,** 29–41.
Osterhout, W. J. V. (1911). The permeability of living cells to salts in pure and balanced solutions. *Science* **34,** 187–189.
Osterhout, W. J. V. (1912). The permeability of protoplasm to ions and the theory of antagonism. *Science* **35,** 112–115.
Osterhout, W. J. V. (1922). Direct and indirect determinations of permeability. *J. Gen. Physiol.* **4,** 275–283.
Osterhout, W. J. V. (1947). The absorption of electrolytes in large plant cells. II. *Botan. Rev.* **13,** 194–215.
Osterhout, W. J. V. (1949). Movements of water in cells of *Nitella. J. Gen. Physiol.* **32,** 553–557.
Palva, P. (1939). Die Wasserpermeabilität der Zellen von *Tolypellopsis stelligera. Protoplasma* **32,** 265–271.
Pernauer, S. (1958). Das Verhalten einiger Cyanophyceen bei osmotischen Impulsen. Untersuchungen zur Plasmolysierbarkeit der Cyanophyceen. *Protoplasma* **49,** 262–295.
Pfeffer, W. (1886–1888). Über Aufnahme von Anilinfarben in lebende Zellen. *Untersuch. botan. Inst. Tübingen* **2,** 179–332.
Polissar, M. J. (1954). Diffusion through membranes and transmembrane potentials. *In* "The Kinetic Basis of Molecular Biology" (F. H. Johnson, H. Eyring, and M. J. Polissar, eds.), pp. 515–603. Wiley, New York.
Pramer, D. (1955). Absorption of antibiotics by plant cells. *Science* **121,** 507–508.
Pramer, D. (1956). Absorption of antibiotics by plant cells. II. Streptomycin. *Arch. Biochem. Biophys.* **62,** 265–273.
Resühr, B. (1935). Hydratations- und Permeabilitätsstudien an unbefruchteten Fucus-Eiern (*Fucus vesiculosus L.*) *Protoplasma* **24,** 531–586.
Rottenburg, W. (1944). Die Plasmapermeabilität für Harnstoff und Glycerin in ihrer Abhängigkeit von der Wasserstoffionenkonzentration. *Flora* (*Jena*) **137,** 230–264.
Ruhland, W., ed. (1956). "Handbuch der Pflanzenphysiologie," Vol. II: Allgemeine Physiologie der Pflanzenzelle, 1072 pp. Springer, Berlin.
Schönleber, K. (1937). *Scytonema Julianum.* Beiträge zur normalen und pathologischen Cytologie und Cytogenese der Blaualgen. *Arch. Protistenk.* **88,** 36–68.
Seemann, F. (1950). Zur cH-Abhängigkeit der Wasserpermeabilität des Protoplasmas. *Protoplasma* **39,** 147–175.
Simon, E. W., and Beevers, H. (1952). The effect of pH on the biological activities of weak acids and bases. I. The most usual relationship between pH and activity. *New Phytologist* **51,** 163–190.
Stadelmann, E. (1956a). Mathematische Analyse experimenteller Ergebnisse: Gewinnung der Permeabilitätskonstanten, Stoffaufnahme- und -abgabewerte. *In* "Handbuch der Pflanzenphysiologie" (W. Ruhland, ed.), Vol. II, pp. 139–195. Springer, Berlin.
Stadelmann, E. (1956b). Zur Versuchsmethodik und Berechnung der Permeabilität pflanzlicher Protoplasten. *Protoplasma* **46,** 692–710.
Stadelmann, E. (1956c). Plasmolyse und Deplasmolyse. *In* "Handbuch der Pflanzenphysiologie" (W. Ruhland, ed.), Vol. II, pp. 71–115. Springer, Berlin.
Suolathi, O. (1937). Über den Einfluss des elektrischen Stromes auf die Plasmapermeabilität pflanzlicher Zellen. *Protoplasma* **27,** 496–501.

Sutcliffe, J. F. (1959). Salt uptake in plants. *Biol. Revs. Cambridge Phil. Soc.* **34,** 159–220.

Szücs, J. (1910). Studien über Protoplasmapermeabilität. Über die Aufnahme der Anilinfarben durch die lebende Zelle und ihre Hemmung durch Elektrolyte. *Kaiserl. Akad. Wiss. Wath.-naturw. Kl. Sitzber. Abt. I (Wien)* **119,** 737–773.

Tröndle, A. (1920). Neue Untersuchungen über die Aufnahme von Stoffen in die Zelle. *Biochem. Z.* **112,** 259–285.

Übeleis, I. (1957). Osmotischer Wert, Zucker-und Harnstoffpermeabilität einiger Diatomeen. *Österr. Akad. Wiss. Math.-naturw. Kl. Sitzber. Abt. I,* **166,** 395–433.

Ullrich, H. (1935). Über den Anionendurchtritt bei *Valonia* sowie dessen Beziehungen zum Zellbau. *Planta* **23,** 146–176.

Ussing, H. (1954). Membrane structure as revealed by permeability studies. *Proc. Symposium Colston Research Soc.* **7,** 33–42.

Walker, N. A. (1957). Ion permeability of the plasmalemma of the plant cell. *Nature* **180,** 94–95.

Wartiovaara, V. (1942). Über die Temperaturabhängigkeit der Protoplasmapermeabilität. *Ann. Botan. Soc. Zool. -Botan. Fennicae Vanamo* **16** (1), 1–111.

Wartiovaara, V. (1944). The permeability of *Tolypellopsis* cells for heavy water and methyl alcohol. *Acta Botan. Fennica* **34,** 1–22.

Wartiovaara, V. (1949). The permeability of the plasma membranes of *Nitella* to normal primary alcohols at low and intermediate temperatures. *Physiol. Plantarum* **2,** 184–196.

Wartiovaara, V., and Collander, R. (1960). Permeabilitätstheorien. *In* "Protoplasmatologia" (L. V. Heilbrunn and F. Weber, eds.), Vol. II, C 8 d, pp. 1-98. Springer, Vienna.

Weber, F. (1931). Harnstoff-Permeabilität ungleich alter *Spirogyra*-Zellen. *Protoplasma* **12,** 129–140.

Wedding, R. T., and Erickson, L. C. (1957). The role of pH in the permeability of *Chlorella* to 2,4-D. *Plant Physiol.* **32,** 503–512.

Zöttl, P. (1960). Vitalfärbestudien mit Methylrot. *Protoplasma* **61,** 465–506.

—32—

*Salt and Osmotic Balance**

ROBERT R. L. GUILLARD

Woods Hole Oceanographic Institution, Woods Hole, Massachusetts

I. Introduction

Dissolved salts have two kinds of effects upon aquatic organisms. The first is related to the specific chemical nature of the ions in solution, and to their specific actions on living cells. The second, the osmotic effect, depends essentially on the total number of dissolved particles, and directly influences the movement of water into or out of cells as if the latter were semipermeable vesicles. Freshwater algae must prevent dilution of their cell contents, while algae of saline waters must adapt to what amounts to physiological drought. Algae of estuarine regions, tide pools, mud flats, and the marine littoral zone are often subjected to widely fluctuating osmotic stress.

Organic compounds as well as salts influence the osmotic pressure of solutions and are commonly used in experimental physiological studies; but algae are not usually found in places, like flower nectaries, where a high osmotic pressure is due to organic materials. The osmotic relations of algae living in the tissues of other organisms will not be considered here.

The various waters of the earth differ in their content of dissolved materials. The ionic constituents of inland saline waters are discussed by Hutchinson (1957), those of sea water by Harvey (1955). Mixohaline waters, formed by the mixing of sea and fresh waters, should be described in the terminology of the Venice system (Venice Symposium, 1959). The

* Contribution Number 1071 from the Woods Hole Oceanographic Institution. Partially supported by National Science Foundation Grant 10693.

composition of inland saline waters may differ greatly from that of diluted or concentrated sea water.

There is an enormous literature on the distribution of organisms in waters of all kinds, and on the effects on organisms of changes in salinity. Much is surveyed by Gessner (1959), Remane and Schlieper (1958), and Schwenke (1960); pertinent material is also contained in the Venice Symposium (1959), in Smith (1951), Pearse and Gunter (1957), and Steiner and Eschrich (1958). Smayda (1958) includes a discussion of the distribution of oceanic phytoplankton with respect to salinity, and Provasoli (1958) gives some generalizations on the distribution of algal groups according to chemical characteristics of the water.

II. Osmotic Effects

A. General Considerations

Osmosis may be defined as the diffusion of a solvent through a semipermeable membrane because of a concentration gradient, i.e., from a region of high to one of lower concentration of the *solvent*. For general treatments, see e.g. Giese (1957), and articles in Ruhland (1956). The osmotic pressure (osmotic value, or potential) of a solution may be defined in terms of measurement with an osmometer. The solution, in a vessel having a semipermeable membrane, is in contact through the membrane with pure solvent; the osmotic pressure is equivalent to the hydrostatic pressure which, applied to the solution, just prevents net solvent diffusion through the membrane. A molal solution of an ideal substance (corresponding to an ideal gas) would have an osmotic pressure of 22.4 atm. at 0°C. The osmotic pressure of a solution may be calculated from observation of the vapor pressure or freezing point and knowledge of the corresponding values for the pure solvent.

If the membrane is differentially permeable, so that certain solutes penetrate it but not others, and if the rates of penetration vary, then the equilibrium distribution of the motile solutes and the solvent will in general be influenced. Morales and Shock (1941) point out that relatively large quantities of solvent and solutes can be shifted by the transfer of relatively small amounts of other solutes. A complete mathematical treatment of the distribution of ions and electrostatic potential in a simple Donnan equilibrium for a particular geometry was given by Bartlett and Kromhout (1952). Using a collodion membrane, Meschia and Setnikar (1958) experimentally showed the magnitude of some of these effects.

Biological membranes are differentially permeable; see reviews by Collander (1957), Höfler (1960a), and Stadelmann (Chapter 31 in this treatise). Water penetrates plant cells much faster than do salts or large organic molecules; only certain small organic molecules penetrate with speeds of the same order of magnitude (Collander, 1954). Kamiya and Tazawa (1956) and Dainty and Hope (1959) measured the water permeability of *Nitella flexilis* and *Chara australis* by transcellular osmosis (described first by Osterhout). The latter authors point out that the values of permeability obtained by this method cannot readily be compared with values of "diffusional permeability" determined by plasmolysis or diffusion of heavy water. They also give evidence against the suggestion by Kamiya and Tazawa (1956) that the water permeability of the cell is different in the inward and outward direction.

Osmosis is considered to account for most of the movement of water into or out of algal cells. Nevertheless, the fact that certain phenomena cannot be explained readily on the basis of this mechanism has led to the hypothesis of a "nonosmotic" water uptake that depends on expenditure of energy by the cell. The observations include: (*1*) discrepancies between determinations of osmotic values by different methods (e.g., plasmometric and cryoscopic); (*2*) effects of auxins on water uptake by higher plants; (*3*) the existence of electrical potential differences across cell membranes or across the cytoplasm; and (*4*) effects of respiratory inhibitors on water uptake.

Not all reviewers have accepted the hypothesis of nonosmotic water uptake (see Kramer, 1956; Collander, 1957; Höfler, 1960a). In a study of ion regulation by *Porphyra perforata*, Eppley and Cyrus (1960) found no evidence suggesting active transport of water. Items (*1*) and (*2*) above will not be discussed further (see Kramer, 1956); concerning item (*3*), Blinks and Airth (1957) pointed out that electro-osmosis could account for not more than 1% of either the turgor or water movement of *Nitella clavata* cells.

Questions raised by item (*4*) above—effects of respiratory inhibitors on water uptake—have not been resolved. Many diatoms are known to deplasmolyze relatively quickly in sugar solutions, which effect has generally been interpreted to mean that the cells are unusually permeable to sugars. The discovery that such deplasmolysis of some species was checked by respiratory inhibitors led Bogen and Follmann (1955) to suggest that nonosmotic water uptake was responsible. Since not all diatoms respond to inhibitors in the same way (Follmann, 1956, 1957, 1958; Höfler, 1960b), Follmann distinguished three physiological types of diatoms: those with predominantly osmotic uptake; those with essentially nonosmotic uptake; and many species with both kinds of uptake. Höfler (1960a) suggested that the apparent nonosmotic water uptake might in fact be due to an active sugar transport with concomitant water transfer. Critical determinations of sugar penetration during deplasmolysis have not been made, to this reviewer's knowledge. Follmann (1957) reported that by a cytochemical test he was unable to detect malonamide or erythritol in deplasmolyzed *Melosira nummuloides* auxospores, but stated that such analyses should be done with more refined analytical methods and more material. The absorption of glucose by *Scenedesmus quadricauda* is not by simple diffusion (Taylor, 1960); however, the glucose taken up is metabolized at once and not accumulated. The connection between the active uptake of solutes, the permeability of the membrane, and the motion of water requires further study.

Fischer's (1952) observations suggest that a deplasmolyzing mechanism is not the only adaptation to osmotic stress, because certain mud-flat diatoms, plasmolyzed by concentrated sea water, survived whether or not they manifested a rapid recovery.

B. The Osmotic Pressure of Algal Cells and Response to Osmotic Change

Blinks (1951) summarized much of the available information on this subject, stating that freshwater algae often have an osmotic value of about 5 atm. (roughly 0.1 *M* NaCl). Some new observations can be added,

TABLE I

APPROXIMATE OSMOTIC PRESSURES OF SOME FRESHWATER ALGAE

Algae	Osmotic pressure (atm.)	Molar equivalent (as NaCl)	Reference
Nitella axillaris	8	0.19	Diamond and Solomon (1959)
Nitella flexilis	6.5	0.16	Kamiya and Kuroda (1956)
Blue-green Algae	5	0.12	Pernauer (1958)
Oscillatoria princeps	6.5	0.16	Yamaha and Negoro (1941)
Desmids	6–9	0.14–0.22	Krebs (1952)
Diatoms	4–8	0.1–0.19	Ubeleis (1957)
Scenedesmus quadricauda	6	0.14	Taylor (1960)
Dinoflagellates	5	0.12[a]	Diskus (1958)
Chlamydomonas eugametos	4.6	0.11[b]	Lothring (1941–1942)
Mesotaenium caldariorum	7	0.17[b]	Lothring (1941–1942)
Spirogyra varians	7	0.17[b]	Lothring (1941–1942)
Zygnema sp.	7	0.17[b]	Lothring (1941–1942)

[a] This is the osmotic pressure at which motility ceased, which is probably a fair estimate.

[b] The lowest osmotic pressures, determined by plasmolysis, in a study of the influence of various factors, especially the aging of cultures, on plasmolytic behavior.

from papers not necessarily concerned directly with osmotic pressure (Table I). In addition, Tazawa (1957) described a new method for determining osmotic pressure, involving direct measurements of turgor.

Marine algae, again summarizing from Blinks (1951), vary considerably in osmotic value; some are barely turgid, while others have a value almost double that of sea water. [Sea water of salinity 3.5% has an osmotic pressure of 23.12 atm. Useful tables of osmotic equivalence of sea water, artificial sea water, and various solutions are given by Robinson (1954); also Barnes (1954).] The amounts by which the osmotic values of some marine algae exceed that of sea water are shown in Table II.

Brackish-water algae also tend to maintain an osmotic value higher than that of the surrounding water, which may be illustrated by the green coenocytes. Collander (1936) found the sap concentration of six fresh- and brackish-water species about equally elevated above the values of the surroundings: (Lehtoranta (1956) presented a summary and more data.) The marine genera *Halicystis* and *Valonia* differ greatly, however; the former is barely turgid, the latter has an osmotic value almost double that of sea water; see Table 3 of Steiner and Eschrich (1958).

The capacity to maintain turgor may be surprisingly great; Bünning (1934) observed *Elachista* to do so in a watch glass in which sea water had concentrated to the point of salt crystallization. Kanwisher (1957) found that thalli of *Ulva lactuca*, *Chondrus crispus*, and *Fucus vesiculosus* attained internal chloride concentrations proportional to those in the ambient solutions of concentrated sea water. In a study of the osmotic values of *Enteromorpha clathrata* from California salt ponds, Biebl (1956) observed that over a range of concentration from a tenfold dilution of sea water to a threefold increase, the internal osmotic values of the algae were above

TABLE II

APPROXIMATE EXCESS OF THE OSMOTIC PRESSURE OF SOME MARINE ALGAE OVER THAT OF THE LOCAL SEA WATER

Algae	Osmotic pressure excess (atm.)	Molar equivalent (as NaCl)	Reference
Diatoms	4.5	0.11	Fischer (1952)
Blue-green Algae	11	0.27	Pernauer (1958)
Cladophora sp.	16.5	0.41	Tramèr (1957)
Chaetomorpha sp.	18	0.44	Tramèr (1957)
Griffithsia sp.	16	0.39	Tramèr (1957)
Neomonospora furcellata	12.5	0.31	Tramèr (1959)

those of the water by roughly 20 atm., falling somewhat at both ends of the range.

Very precise measurements were made by Kesseler (1958, 1959) on *Bryopsis plumosa*, *Chara baltica*, and especially *Chaetomorpha linum*, using a new micro-cryoscopic method. The turgor (difference between the cellular osmotic pressure and that of the medium) of the latter alga was almost constant at 14.8–16.5 atm. in the range from fresh water to a salinity of 3.5% in a balanced artificial sea water. In nature, too, turgors were found to be relatively constant. Eppley and Cyrus (1960) found that the content of potassium *plus* sodium of *Porphyra perforata* cells varied almost linearly with salinity in artificial sea water. The sodium content of the brine flagellate *Dunaliella salina* also increased with the concentration of the medium (Marrè *et al.*, 1958; Marrè and Servattaz, 1959). The osmotic excess of the cells was 9 atm. when the external medium had an osmotic value of 67 atm.; it was 25 atm. when the medium had 170 atm. osmotic value. The authors concluded from studies of sodium uptake and exchange that the flagellates were highly permeable, especially to NaCl. They also suggested that salt uptake might be metabolically controlled. On the other hand, the freshwater green flagellate *Chlamydomonas moewusii* is not rapidly permeable to sodium (Ronkin and Buretz, 1960); it does not grow if the osmotic pressure of the medium, maintained by sucrose or NaCl, is ca. 20 atm. or over (Guillard, 1960). It is not known how the cell contents vary with the medium.

The evidence presented so far is consistent with the idea that algae tend to maintain an internal concentration somewhat above that of the medium, and that the latitude of osmotic stress tolerated depends ultimately on the ability of the protoplasm to function when its salt concentration is altered. The ions accumulated are not necessarily those most abundant in the medium.

An entirely different mechanism for adaptation to high salt concentrations was postulated by Pochmann (1959) for a newly discovered non-photosynthetic flagellate, *Choanogaster plattneri*, found in water of salinity 20%. The cells have a complex vacuolar system, producing smaller accessory vacuoles that empty outside the cells and that decrease in activity when the salinity of the medium is lowered. Pochmann believed that the vacuole contains a higher concentration of salt than the medium and serves to excrete salt—thereby conserving water—like the salt glands of marine fishes and birds. This function is opposite to that of contractile vacuoles, which eliminate water from organisms living in a hypotonic medium (see Guillard, 1960). Pochmann suggested that the vacuolar system of *Choanogaster* is comparable to the pusules of marine dinoflagellates, though the function of the latter has not been studied experimentally to this reviewer's knowledge. On the other hand Kofoid (1909), noting that pusules communicate with the exterior through a pore in the sulcus region of the cells (through which food particles are ingested by holozoic species), suggested that the activity of the pusules might lead to the *intake* of liquid, and might thus be concerned with heterotrophic nutrition as in the pinocytosis of leucocytes. The two functions proposed—water conservation and heterotrophic nutrition—are not mutually exclusive, and it is quite reasonable that a structure capable of one of these functions might become adapted to the other. In any event, the possibility that a protistan can maintain an internal concentration lower than that of the external medium is of special interest. Gross (1940) suggested that the planktonic marine diatom *Ditylum brightwellii* might do so, and Gross and Zeuthen (1948) indicated that diatoms might exert a measure of control over their specific gravity by preferential accumulation of lighter materials, such as univalent rather than bivalent ions, or sodium rather than potassium salts. However, there is no proof that *Ditylum* is hypotonic to sea water; other algae known to accumulate specific ions are not hypotonic to the medium.

Guillard (1960) showed that the only irreplaceable function of the contractile vacuoles of *Chlamydomonas moewusii* was the elimination of water. A mutant strain lacking contractile vacuoles survived and grew only if the osmotic pressure of the medium exceeded 1.5 atm. and was provided by compounds such as sucrose or salts that only slowly penetrate the cells. Under such osmotic conditions the vacuoles of the wild strain no longer functioned, and both strains grew rapidly. Six similar mutants, capable of growth in 0.6 atm. osmotic pressure (0.78 gm./liter NaCl) but not in more dilute solutions, were isolated after ultraviolet irradiation of *C. eugametos* by Gowans (1960). Lothring (1941–1942) showed that the cells of *C. eugametos* (= *C. moewusii* Gerloff, Moewus' strain) exhibited obvious plasmolysis at about 4.6 atm.; Guillard found this to be true also of *C. moewusii* (Provasoli's strain). The difference between this value and the 1.5 atm. which is sufficient to stop contractile vacuole activity may be due in part to wall pressure. However, freshwater protozoa, which characteristically lack cell walls, have osmotic values of only about 1.5 atm. (see Prosser, 1950). (The value of 4.6 atm., determined by "plasmolysis" of young *C. moewusii* cells, may be too high.)

Wall pressure undoubtedly plays a role in limiting the uptake of water by osmosis into freshwater algae without contractile vacuoles. If the central vacuole of a green coenocyte is pierced with a hollow glass needle, sap exudes into the tube and water enters the cell through the protoplast membrane (Jacques, 1938). Kamiya and Kuroda (1956) found that in the special case of *Nitella flexilis* cells subjected to trans-cellular osmosis the wall pressure apparently exerted no influence on water movement.

III. Some Effects of Ions and Ion Ratios

Provasoli (1958) reviewed the literature on the influence of the major elements on algal growth, and studies on the effects of total dissolved solids, the predominant anions and cations, the ratio of univalent to bivalent ions, and the ratio of calcium to magnesium. McLachlan (1960) showed that the flagellate *Dunaliella tertiolecta* (also referred to in the literature as *D. euchlora*, and, incorrectly, as "a marine *Chlamydomonas*") has clearly distinguishable osmotic and sodium requirements. In a medium having the ionic proportions of sea water, as the total salinity is decreased by dilution, the growth of *D. tertiolecta* is limited first by the osmotic requirement. If the osmotic value is maintained by the addition of other solutes, the first cation to become limiting on further dilution is sodium. It seems likely, as implied by Droop (1958), that certain other marine algae requiring high salinity have a high nonosmotic sodium requirement. Sodium is involved in "ion pump" systems in a number of algae (see Eppley and Cyrus, 1960; Hope and Walker, 1960; Eppley, Chapter 16, this treatise). Ion transport systems, together with alterations of membrane permeability, are the most obvious mechanisms by which the ionic variation of the medium can influence plant metabolism. It should be profitable to compare the properties of marine algae with those of algae from inland saline waters that have a significantly different ionic composition.

The ability of algae to maintain an osmotic pressure above that of the environment is lowered and finally destroyed by deficiencies of cations in the medium (Kesseler, 1959); resistance to hypotonic stress is also lowered (Schwenke, 1958). Schmitz (1959) observed the influence of salinity on the division rate of the euryhaline diatom *Thalassiosira fluviatilis* in various dilutions of two artificial media, one of which resembled natural river water (mixohaline), while the other had a higher ratio of sulfate to chloride. The maximum growth rate in the first medium was at a salinity of about 1.4%, while in the medium with more sulfate there were three optima, which Schmitz attributed respectively to favorable levels of sulfate, total cations, and total anions. Apparently these three independent effects were revealed by alteration of the composition of the medium.

Large algal coenocytes and many other plants accumulate chloride and discriminate against sulfate; on the other hand *Porphyra perforata* tends to exclude chloride (Eppley, 1958), and the euryhaline *Dunaliella tertiolecta* can grow well with none added to the medium (McLachlan,

1960). Krishna Pillai (1954, 1955) found that Blue-green Algae from saline lagoons discriminated against chloride and accumulated sulfate; in the highest salinities studied they grew only if additional sulfate was provided. The balance of trace metals also became critical under such conditions.

Steemann Nielsen (1954) offered an interesting explanation for the preference of certain freshwater plants, including *Chara* and *Nitellopsis*, for mixohaline water in Finland, pointing out that the fresh waters there were so low in alkalinity that the supply of bicarbonate limited photosynthesis.

References

Barnes, H. (1954). Some tables for the ionic composition of sea water. *J. Exptl. Biol.* **31,** 582–588.

Bartlett, J. H., and Kromhout, R. A. (1952). The Donnan equilibrium. *Bull. Math. Biophys.* **14,** 385–391.

Biebl, R. (1956). Zellphysiologisch-ökologische Untersuchungen an *Enteromorpha clathrata* Roth (Greville). *Ber. deut. botan. Ges.* **69,** 75–86.

Blinks, L. R. (1951). Physiology and biochemistry of algae. *In* "Manual of Phycology" (G. M. Smith, ed.), pp. 263–291. Chronica Botanica, Waltham, Massachusetts.

Blinks, L. R., and Airth, R. L. (1957). Electroosmosis in *Nitella. J. Gen. Physiol.* **14,** 383–396.

Bogen, H. J., and Follmann, G. (1955). Osmotische und nichtosmotische Stoffaufnahme bei Diatomeen. *Planta* **45,** 125–146.

Bünning, E. (1934). Zellphysiologische Studien an Meeresalgen. *Protoplasma* **22,** 444–456.

Collander, R. (1936). Der Zellsaft des Characeen. *Protoplasma* **25,** 201–210.

Collander, R. (1954). The permeability of *Nitella* cells to non-electrolytes. *Physiol. Plantarum* **7,** 420–445.

Collander, R. (1957). Permeability of plant cells. *Ann. Rev. Plant. Physiol.* **8,** 335–348.

Dainty, J., and Hope, A. B. (1959). The water permeability of *Chara australis* R. Br. *Australian J. Biol. Sci.* **12,** 136–145.

Diamond, J. M., and Solomon, A. K. (1959). Intracellular potassium compartments in *Nitella axillaris. J. Gen. Physiol.* **42,** 1105–1121.

Diskus, A. (1958). Das Osmoseverhalten einiger Peridineen des Süsswassers. *Protoplasma* **49,** 187–196.

Droop, M. R. (1958). Optimum relative and actual ionic concentrations for growth of some euryhaline algae. *Verhandl. intern. Ver. Limnol.* **13,** 722–730.

Eppley, R. W. (1958). Sodium exchange and potassium retention by the red marine alga *Porphyra perforata. J. Gen. Physiol.* **41,** 901–911.

Eppley, R. W., and Cyrus, C. C. (1960). Cation regulation and survival of the red alga, *Porphyra perforata,* in diluted and concentrated sea water. *Biol. Bull.* **118,** 55–65.

Fischer, H. (1952). Über das Verhalten einiger Watt-Diatomeen in hypertonischen Lösungen. *Ber. deut. botan. Ges.* **65,** 218–228.

Follmann, G. (1956). Die Permeabilitätsreihe von *Leptocylindrus adriaticus* Schroeder

und das Problem der "Spezifischen Diatomeen-Permeabilität." *Naturwissenschaften* **43,** 306.
Follmann, G. (1957). Die Anelektrolytaufnahme der Auxosporen zentrischer Diatomeen. *Naturwissenschaften* **44,** 567–568.
Follmann, G. (1958). Über Aufnahme und Bindung von Wasser und Anelektrolyten durch Diatomeen-zellen. *Planta* **50,** 671–700.
Gessner, F. (1959). "Hydrobotanik." Vol. II. V.E.B. Deutscher Verlag der Wissenschaften, Berlin.
Giese, A. C. (1957). "Cell Physiology." Saunders, Philadelphia, Pennsylvania.
Gowans, C. S. (1960). Some genetic investigations of *Chlamydomonas eugametos. Z. Vererb.* **91,** 63–73.
Gross, F. (1940). The osmotic relations of the plankton diatom *Ditylum Brightwelli* (West). *J. Marine Biol. Assoc. United Kingdom* **24,** 381–415.
Gross, F., and Zeuthen, E. (1948). The buoyancy of plankton diatoms: a problem of cell physiology. *Proc. Roy. Soc.* **B135,** 382–389.
Guillard, R. R. L. (1960). A mutant of *Chlamydomonas moewusii* lacking contractile vacuoles. *J. Protozool.* **7**(3), 262–268.
Harvey, H. W. (1955). "The Chemistry and Fertility of Sea Waters." Cambridge Univ. Press, London and New York.
Höfler, K. (1960a). Permeability of protoplasm. *Protoplasma* **52,** 145–156.
Höfler, K. (1960b). Über die Permeabilität der Diatomee *Caloneis obtusa. Protoplasma* **52,** 5–25.
Hope, A. B., and Walker, N. A. (1960). Ionic relations of *Chara australis.* III. Vacuolar fluxes of sodium. *Australian J. Biol. Sci.* **13,** 276–291.
Hutchinson, G. E. (1957). "A Treatise on Limnology," Vol. I. Wiley, New York.
Jacques, A. G. (1938). The kinetics of penetration. XV. The restriction of the cellulose cell wall. *J. Gen. Physiol.* **22,** 147–163.
Kamiya, N., and Kuroda, K. (1956). Artificial modification of the osmotic pressure of the plant cell. *Protoplasma* **46,** 423–436.
Kamiya, N., and Tazawa, M. (1956). Studies on water permeability of a single plant cell by means of transcellular osmosis. *Protoplasma* **46,** 394–422.
Kanwisher, J. (1957). Freezing and drying of intertidal algae. *Biol. Bull.* **113,** 275–285.
Kesseler, H. (1958). Eine mikrokryoscopische Methode zur Bestimmung des Turgors von Meeresalgen. *Kiel. Meeresforsch.* **14,** 23–41.
Kesseler, H. (1959). Mikrokryoskopische Untersuchungen zur Turgorregulation von *Chaetomorpha Linum. Kiel. Meeresforsch.* **15,** 51–73.
Kofoid, C. A. (1909). On *Peridinium Steinii* Jörgensen, with a note on the nomenclature of the skeleton of the Peridinidae. *Arch. Protistenk.* **16,** 25–47.
Kramer, P. J. (1956). The uptake of water by plant cells. *In* "Handbuch der Pflanzenphysiologie" (W. Ruhland, ed.), Vol. II, pp. 316–336. Springer, Berlin.
Krebs, I. (1952). Beiträge zur Kenntnis der Desmidiaceen Protoplasten. *Österr. Akad. Wiss. Math.-naturw. Kl. Sitzber. Abt. I* **160,** 578–618.
Krishna Pillai, V. (1954). Growth requirements of a halophilic blue-green alga, *Phormidium tenue* (Menegh). *Indian J. Fisheries* **1,** 130–144.
Krishna Pillai, V. (1955). Observations of the ionic composition of blue-green algae growing in saline lagoons. *Proc. Natl. Inst. Sci. India* **B21,** 90–102.
Lehtoranta, L. (1956). Über den Kationen- und Chlorgehalt des Zellsaftes, insbesonders bei Submersen. *Ann. Botan. Soc. Zool.-Botan. Fennicae Vanamo* **29**(1), 1–164.
Lothring, H. (1941–1942). Beiträge zur Biologie der Plasmolyse. *Planta* **32,** 600–629.

McLachlan, J. (1960). The culture of *Dunaliella tertiolecta* Butcher—a euryhaline organism. *Can. J. Microbiol.* **6,** 367–379.

Marrè, E., and Servattaz, O. (1959). Sul meccanismo di adattamento a condizioni osmotiche estreme in *Dunaliella salina.* II. Rapporto fra concentrazioni del mezzo esterno e composizione del succo cellulare. *Atti accad. nazl. Lincei. Rend. Classe sci. fis. mat. e nat.* **26,** 272–277.

Marrè, E., Servettaz, O., and Albergoni, F. (1958). Sull meccanismo di adattamento a condizioni osmotiche estreme in *Dunaliella salina.* I. Reazioni fisiologiche a variazioni dell'ambiente osmotico. *Atti accad. nazl. Lincei. Rend. Classe sci. fis. mat. e nat.* **25,** 567–574.

Meschia, G., and Setnikar, I. (1958). Experimental study of osmosis through a collodion membrane. *J. Gen. Physiol.* **42,** 429–444.

Morales, M. F., and Shock, N. W. (1941). The general membrane equation: its simple derivation and some of its biological implications. *Bull. Math. Biophys.* **3,** 153–160.

Pearse, A. S., and Gunter, G. (1957). Salinity. *In* "Treatise on Marine Ecology and Paleoecology" (J. W. Hedgpeth, ed.), *Geol. Soc. Am. Mem.* **67,** pp. 129–159.

Pernauer, S. (1958). Das Verhalten einiger Cyanophyceen bei osmotischen Impulsen. *Protoplasma* **49,** 262–295.

Pochmann, A. (1959). Über die Tätigkeit der nichtkontraktilen Importvakuole und den Modus der Osmoregulation bei dem Salzflagellaten *Choanogaster* nebst Bemerkungen über die Funktion der Pusulen. *Ber. deut. botan. Ges.* **72,** 99–108.

Prosser, C. L., ed. (1950). "Comparative Animal Physiology." Saunders, Philadelphia, Pennsylvania.

Provasoli, L. (1958). Nutrition and ecology of protozoa and algae. *Ann. Rev. Microbiol.* **12,** 279–308.

Remane, A., and Schlieper, C. (1958). Die Biologie des Brackwasser. *In* "Die Binnengewässer" (A. Thienemann, ed.), Vol. 22, 348 pp. E. Schweizerbart'sche Verlagsbuchhandlung (Nägele u. Obermiller), Stuttgart, Germany.

Robinson, R. A. (1954). The vapour pressure and osmotic equivalence of sea water. *J. Marine Biol. Assoc. United Kingdom* **33,** 449–455.

Ronkin, R. R., and Buretz, K. M. (1960). Sodium and potassium in normal and paralyzed *Chlamydomonas. J. Protozool.* **7,** 109–113.

Ruhland, W., ed. (1956). "Handbuch der Pflanzenphysiologie," Vol. II: Allgemeine Physiologie der Pflanzenelle; Vol. III: Pflanze und Wasser. Springer, Berlin.

Schmitz, W. (1959). Zur Frage der Klassification der binnenländischen Brackwasser. *In* "Symposium on the Classification of Brackish Waters." *Arch. Oceanog. Limnol. (Suppl.)* **11,** 179–226.

Schwenke, H. (1958). Über einige zellphysiologische Faktoren der Hypotonieresistenz mariner Rotalgen. *Kiel. Meeresforsch.* **14,** 130–150.

Schwenke, H. (1960). Neuere Erkenntnisse über die Beziehungen zwischen den Lebensfunktionen mariner Pflanzen und dem Salzgehalt des Meer- und Brackwassers. *Kiel. Meeresforsch.* **16,** 28–47.

Smayda, T. J. (1958). Biogeographical studies of marine phytoplankton. *Oikos* **9,** 158–191.

Smith, G. M., ed. (1951). "Manual of Phycology." Chronica Botanica, Waltham, Massachusetts.

Steemann Nielsen, E. (1954). On the preference of some fresh water plants in Finland for brackish waters. *Botan. Tidskkr.* **51,** 242–247.

Steiner, M., and Eschrich, W. (1958). Die osmotische Bedeutung der Mineralstoffe.

In "Handbuch der Pflanzenphysiologie" (W. Ruhland, ed.), Vol. IV, pp. 334–354. Springer, Berlin.

Taylor, F. J. (1960). The absorption of glucose by *Scenedesmus quadricauda*. I. Some kinetic aspects. *Proc. Roy. Soc.* **B151,** 400–418.

Tazawa, M. (1957). Neue Methode zur Messung des osmotischen Wertes einer Zelle. *Protoplasma* **48,** 342–359.

Tramèr, P. O. (1957). Zur Kenntnis der Saugkraft des Meerwassers und einiger Hydrophyten. *Ber. schweiz. botan. Ges.* **67,** 411–419.

Tramèr, P. O. (1959). Über osmotische Zustandsgrössen einiger Hydrophyten. *Ber. schweiz. botan. Ges.* **69,** 323–341.

Übeleis, I. (1957). Osmotischer Wert, Zucker-und-Harnstoffpermeabilität einiger Diatomeen. *Österr. Akad. Wiss. Math.-naturw. Kl. Sitzber. Abt. I.* **166,** 395–433.

Venice Symposium, Societas Internationalis Limnologiae. (1959). "Symposium on the Classification of Brackish Waters." *Arch. Oceanog. Limnol.* (*Suppl.*) **11,** 248 pp. (Comitato talassografico, Italy.)

Yamaha, G., and Negoro, K. (1941). Über die Plasmolyse bei Zyanophyzeen. Ein osmotischer Versuch bei *Oscillatoria princeps* Vauch. *Sci. Repts. Tokoyo Kyôiku Daigaku, Sect.* **B5,** 261–295.

—33—

Temperature

ERASMO MARRÈ

Institute of Plant Sciences, University of Milan, Italy

I. Introduction[1]

The influence of temperature on living organisms may be considered from two points of view: (i) that of the ecologist, who is interested in this factor primarily as it affects the distribution of species (see Biebl, Chapter 53, this treatise); and (ii) that of the physiologist, striving towards an understanding of the biochemical and biophysical mechanisms which permit certain species to prosper under conditions incompatible with the survival of others. An understanding of the physiological aspects of the problem appears fundamental for a rational approach to the ecological aspect. Algae, as a group, provide interesting material for studies of the relation between temperature and biological activity, because of the extremely wide range of thermal environments which they occupy and the high degree of adaptation shown by various species to their particular environments.

II. The Upper Limit of Mesothermal Species

Few common freshwater algae grow at temperatures above 25–30°C. Interest in the commercial applications of algal mass culture stimulated searches for thermophilic species and studies of their metabolism. In par-

[1] Abbreviations used in this chapter: TPN and TPNH, triphosphopyridine nucleotide and its reduced form; DPN and DPNH, diphosphopyridine nucleotide and its reduced form; FMN, flavin mononucleotide; ATPase, adenosine triphosphatase.

ticular, Blue-green Algae with temperature optima for growth at 35–40°C. and with short generation times (g.t.) have been studied, e.g., *Anacystis nidulans* (g.t. = 2 hours; Kratz and Myers, 1955), and *Oscillatoria subbrevis* (g.t. = 3 hours; Moyse *et al.*, 1957).

Sorokin (1959a; see also Sorokin and Myers, 1953) isolated a thermophilic *Chlorella* (strain 7–11–06) with a temperature maximum for growth at 42°C., and compared several of its physiological reactions with those of *Chlorella pyrenoidosa*, which closely resembles it but which does not grow at temperatures above 29°C. Its temperature optima, not only for growth (38–39°C.) but also for photosynthesis (40–42°C.) and respiration (40–42°C.), considerably exceeded those of *C. pyrenoidosa* (25–26°, 32–35°, and 30°C., respectively).

The availability of pure cultures of phagotrophic algae, the cells of which are permeable to a great variety of dissolved as well as particulate nutrients, opens the way for experimental analyses of the metabolic breakdowns which might be responsible for setting upper limits to the temperature range for growth. *Ochromonas malhamensis* ordinarily stops growing when the temperature is raised to about 36°C. However, by supplying extra amounts of metals, amino acids, vitamin B_{12}, and folic acid (or products of folic acid metabolism such as purines and thymidine), this limit can be pushed to 38°C. (Hutner *et al.*, 1957). In a similar manner, the upper growth threshold of *O. danica* can be raised from 35° to 38°C., or even higher if certain lipids are supplied (Frank *et al.*, 1960).

At these high temperatures, the integrity of the plastid in *Ochromonas* spp. is retained. Contrariwise, when certain cultures of *Euglena gracilis* are maintained in organic media at 36°C., permanent apochlorosis ("bleaching") results; growth then becomes obligately heterotrophic. At 37°C. or somewhat higher, there emerge nutritional deficiencies for as yet unidentified growth factors (Baker *et al.*, 1955).

III. Adaptations to Extreme Temperatures

A. Stenothermy and Eurythermy

Adaptations to adverse temperature conditions are found at the one extreme among the snow algae, and at the other in the Blue-green Algae of hot springs. In both cases, the adaptation generally appears to be highly specialized in that a change of a few degrees above or below the optimum temperature for growth and reproduction may impair biological activity. Organisms with such narrow limits of tolerance are referred to as *stenothermal*. At somewhat more equable temperatures, species are found showing a variety of temperature requirements, ranging from the algae of the warmer seas, which fail to survive temperatures several degrees above zero (Biebl, 1939a, b), to intertidal algae, exposed to and surviving wide diurnal and seasonal temperature variations, and to such eurythermal diatoms as *Pinnularia appendiculata,* which is reported to extend from the cold waters of mountain lakes to thermal waters at temperatures as high as 70°C. (Sprenger, 1930). However, records of diatoms "living" in water above 50°C. should be accepted with reservations (cf. Copeland, 1936, p. 210). Perhaps the most remarkable cases of *eurythermy* occur among the encrusting algae (mainly Chlorophyta and Cyanophyta) and the lichens of bare mountain rocks in polar regions, where the surface temperatures may vary within a few hours from $-60°$ to $+10°C.$ or higher.

In this chapter, particular attention will be given to some cases of adaptation to extreme temperatures, notably the algae of snow and of hot springs.

B. Snow Algae

Several algae, belonging to different groups, have been reported to grow on the surface of snow, and even to be ecologically restricted to this habitat, e.g., *Scotiella* (Chodat, 1922), *Raphidonema* (Lagerheim, 1892), *Ancylonema nordenskioldii, A. meridionale, Chlamydomonas nivalis,* and *C. flavo-virens* (Fritsch, 1935). Active life at such a low temperature poses two separate problems: (i) the ability of these forms to survive freezing during the extremely cold nights; and (ii) their capacity to carry on the metabolic reactions required for growth and assimilation when the day temperature is in the neighborhood of 0°C. The first problem is common to many plants, and is usually defined as that of "frost resistance". According to most recent views (e.g., Levitt, 1956; Kanwisher, 1957), frost resistance is attributable to an abnormal capacity of the protoplasm to survive mechanical damage caused by the freezing and thawing of intracellular liquids. An additional mechanism of defense against low temperatures could probably lie in a high osmotic pressure of the cell sap, correlated with a lowering of its freezing point. For example, it has been observed that *Dunaliella salina,* a flagellate capable of growing in concentrated saline media, can under such conditions survive without freezing, and even maintain some motility, at temperatures as low as $-15°C.$ The osmotic concentration of the cell contents of this alga has experimentally been found to be equal to or slightly higher than that of the medium (Marré and Servettaz, 1959). However, a high internal osmotic value does not always result in "cold tolerance"; indeed in *Griffithsia,* an intertidal Red Alga, the higher osmotic pressure of the apical cells is associated with a higher sensitivity to damage by cold (Biebl, 1939b).

On the other hand, the main handicap to active growth and metabolism at temperatures close to 0°C. lies in the slowness of chemical reactions under these conditions, which could be overcome only by a correspondingly higher efficiency of all essential enzymatic mechanisms. This interesting aspect of the problem has been largely unexplored. It is also possible that the biological activity of snow algae may be confined to periods of high light intensity, when part of the radiant energy could be employed to raise the temperature of the cells above that of the environment. In this connection it should be noted that many snow algae are characterized by large amounts of carotenoid pigments, usually red, which might be expected to aid in effecting such an energy conversion.

C. Algae from Hot Springs

1. Ecological Observations

The ecology, physiology, and biochemistry of the many algae of hot springs present several extremely interesting problems, since for most forms of life this habitat is incompatible not only with an orderly course of biological functions, but also with the maintenance of the integrity of cell structure, and even of the structure of such fundamental molecules as proteins and nucleic acids. Most hot-spring algae belong to the Cyanophyta; the primitive phylogenetic position of this group may be significant in regard to this type of adaptation.

The distribution of many thermal algae in relation to temperature has been discussed by Copeland (1936) and Yoneda (1952), and more thoroughly reviewed by Vouk (1950). Vouk points out that, as a rule, the higher the temperatures tolerated by such algae, the narrower is the range of temperature under which they flourish. This indicates that the problem is concerned with a high degree of specialization rather than with a general adaptability to such abnormal environmental conditions.

According to Bünning and Herdtle (1946), temperatures between 50° and 70°C. are as a rule only tolerated in nature under conditions of high light intensity and of high environmental CO_2 content, where the latter, measured in the atmosphere immediately above the source, ranges around 2 to 5%. Many hot-spring algae can be grown experimentally at lower temperatures, though in nature they tend to be rapidly overgrown by mesothermal species. Thus Inman (1940) showed that certain Blue-green Algae taken from a 65°C. hot spring were capable of active photosynthesis at temperatures as low as 20°C. In the following paragraphs detailed attention will be given to some of the special physiological and biochemical aspects of "heat tolerance".

2. Physicochemical Considerations

To understand heat tolerance, one has to analyze both the inhibitory effects of high temperature upon life in normal organisms and the mechanisms by which heat-resistant organisms overcome these obstacles, bearing in mind that: (i) thermophilic organisms not only survive, but assimilate and grow, at high temperatures; and (ii) for most of these species such high temperatures are an absolute requirement for active life.

A high temperature may affect life through its action either on the external medium, or directly on the cells, or on both. The effect of temperature on the solubility of CO_2 and O_2 could be important, since the two solubility coefficients (α) are reduced, between 25° and 60°C., from 0.76 to 0.36 (for CO_2), and from 0.028 to 0.019 (for O_2). Thus in hot-spring waters the availability of oxygen for respiration, or of CO_2 for photosynthesis, is appreciably diminished. This could partly explain the generally slow growth of thermal species. However, since many heat-labile organisms are known to respire and to photosynthesize actively at O_2 and CO_2 levels even lower than those of hot-spring waters, the reduced solubility of these gases cannot be a major limiting factor for life at high temperatures.

Inasmuch as it acts directly on the cell, a high temperature may induce death by shifting the equilibria between different reaction sequences with different Q_{10} values,[2] a consideration equally important at lower temperatures. Since respiration has a higher Q_{10} value than has photosynthesis, it has been suggested that higher temperatures could be less favorable for growth of plants because they raise the compensation point. (Some experiments on this question are discussed below.) However, this argument could be applied to any species adapted to a given temperature range, such as the algae of the warm seas (Biebl, 1939b). Moreover, we now know of physiological mechanisms which control oxidative metabolism by coupling electron-transfer to the storage of high-energy phosphate bonds; also important is the capacity of cells to equilibrate different functions by substrate-induced changes of enzyme synthesis. It thus seems improbable that a difficulty in harmonizing different processes could play a major role in establishing high temperature limits to life. In the search for the main limiting factor, one is thus brought back to that already suggested by Sachs, the thermolability of proteins and other macromolecules. The capacity of hot-spring organisms to endure heat should therefore be dependent on some peculiar biochemical characteristic of their protoplasm.

[2] Q_{10} is the factor by which the velocity is increased on raising the temperature by 10°C.

3. Physiological Effects of Temperature Changes

The mechanism of extreme heat tolerance in algae has hitherto been the object of a rather limited number of investigations *in vivo* and *in vitro*, which nevertheless suffice to establish certain peculiar features of their metabolism.

Experiments carried out by Bünning and Herdtle (1946), largely with *Oscillatoria geminata* isolated from a hothouse, indicated that tolerance of these algae to heat may be attributed to a combination of the following features:

(*a*) the ability to carry out photosynthesis efficiently and continuously at temperatures (30 to 40°C.) somewhat higher than normal;

(*b*) the low Q_{10} of the respiratory processes at these temperatures, with the consequence that the respiratory rate does not accelerate to a lethal catabolism.

Essentially similar results were obtained by Prat and Kubin (1956) with *Mastigocladus laminosus*, *Oscillatoria animalis*, and *Symploca* sp. Possibly, as Bünning and Herdtle suggested, the low water content of the cells results in the limitation of biochemical processes by physical diffusion, which might account for the low Q_{10} values found.

The photosynthesis and respiration of *Aphanocapsa thermalis* as a function of temperature were studied by Marrè and Servettaz (1956). They found that temperatures several degrees lower than that of acclimatization (the temperature to which the growing alga is adapted) strongly depressed both processes, that the effect on photosynthesis was much more marked than on respiration, and that both effects were completely reversible. Temperatures a few degrees above that of acclimatization induced first a rapid, irreversible decline of photosynthesis, and then a similar decline of respiration. The margin between the acclimatization temperature and the irreversible inactivation temperature was much narrower for algae which had been acclimatized at the highest temperatures. These data suggest: (*a*) that the main factor in heat resistance is the heat stability of protoplasmic structures; (*b*) that the possibility of more active metabolism at higher temperatures is fully exploited by these thermal algae, which live close to the upper limit of heat tolerance for their biochemical systems; and (*c*) that a single species may show different degrees of thermal adaptation, depending on the temperature at which the alga had been grown and adapted. For instance, Löwenstein (1903) showed that filaments of *Mastigocladus laminosus*, which had originally grown at 52°C., apparently lost their ability to withstand temperatures in excess of 40°C. after they had been maintained for 5 months at 5° to 8°C. A similar indication of de-adaptation was noted in *Oscillatoria geminata* (Bünning and Herdtle, 1946).

In another alga from hot springs, the paradoxical *Cyanidium caldarium** which can be cultured at temperatures as high as 55°C. (Allen, 1959), the photosynthetic mechanism likewise becomes more sensitive than respiration as either the upper or the lower limit of the temperature range is approached (Fukuda, 1957). Both the saturating CO_2 concentration and the saturating light intensity for photosynthesis are consistently higher in this species than in a typical Green Alga such as *Chlorella*, a result with interesting phylogenetic implications.

4. *Enzymes from Hot-Spring Algae*

Various enzymes corresponding to those of normal algae have been demonstrated in hot-spring species. Harvey (1924) found a reductase and a peroxidase in a *Phormidium* species acclimatized at 73°C. Marré and Servettaz (1957) demonstrated, in acetone-powder preparations from *Aphanocapsa*, the activity of a TPNH- and DPNH-oxidase, a TPNH-oxidized glutathione reductase, and a TPNH-cytochrome-*c* reductase. The cytochrome reductase from this thermal alga showed a considerably higher heat stability than that extracted from *Anabaena cylindrica*, with optimum growth at ca. 25–30°C. The enzyme from the thermal alga retained full activity after heating for 5 minutes at 85°C., whereas that from *Anabaena* was almost completely inactivated by such treatment. Moreover, the requirement for FMN as a cofactor appeared much lower for the *Aphanocapsa* enzyme, especially after partial denaturation by heat. It was shown experimentally that the heat-resistance of the cytochrome reductase from *Aphanocapsa* could not be due to the presence of a dialyzable substance with an unspecific protective action, since extracts of *Aphanocapsa* exerted no protective action on the *Anabaena* enzyme preparation. It was concluded that special protein configurations, possibly less dependent than usual on hydrogen bonds for their integrity, must be the main factor responsible for heat tolerance in hot-spring algae. In agreement with this hypothesis, it was found that the cytochrome reductase and catalase isolated from *Aphanocapsa* were considerably more resistant to treatment by acetamide and by urea, reagents which dissociate hydrogen bonds, than were the corresponding enzymes from *Anabaena* (Marrè *et al.*, 1958).

5. *Mechanism of Heat Tolerance*

It appears that heat tolerance in hot-spring algae is a consequence primarily of the singular capacity of their proteins to endure without denaturation abnormally high temperatures, thereby enabling them to flourish in environments which tend to exclude competition from other species. The requirement for a high-temperature habitat might then be an

* Regarding the taxonomy of this organism, see Note 11 of Appendix A in this treatise by Silva.

evolutionary consequence of biological selection for enzymes with higher temperature optima.

The indications of an unusual protein structure in heat-resistant algae are in agreement with similar data from thermophilic bacteria. Militzer *et al.* (1950) found that heat-resistant enzymes can be extracted from certain bacteria which grow best at ca. 55°C. Cell proteins and flagellar proteins from several thermophilic bacteria can be subjected to temperatures above 60°C. without significant denaturation (Koffler and Sale, 1957). In these cases too, the evidence indicates that the heat tolerance does not depend on the presence of protective agents (Adye *et al.*, 1957), but rather on the relative stability of the intramolecular bonds. Thus, in the heat-stable ATPase isolated by Marsh *et al.*, only about half as many hydrogen bonds were broken by heat during a given period of time as in the homologous enzyme from yeast (Marsh and Militzer, 1956). Likewise the viscosities of solutions of flagellar proteins from thermophilic bacteria were unchanged by reagents which break hydrogen bonds (Mallett and Koffler, 1957). Evidence summarized by Fogg (1956) suggests that the molecular structures of the proteins of Blue-green Algae may be more rigid than those of other organisms, perhaps being cross-linked by primary bonds rather than solely by hydrogen bonds.

An interesting problem arises at this point. If all the proteins of thermophilic organisms differ in structure from the proteins playing corresponding roles in ordinary heat-labile organisms, how could such profound differences have developed? A simultaneous change of all genes controlling the synthesis of single proteins seems, of course, highly improbable; the hypothesis of a series of mutations over an extended period of time does not appear much more satisfactory. It may be that in heat-resistant organisms there has arisen some single key factor acting on all proteins and protoplasmic structures and controlled by only a small number of genes. In this connection, it may be noted that the extraction of enzymes from thermal Cyanophyta requires a much more extended preliminary treatment with fat solvents (such as isobutanol) than in the case of heat-labile forms. Possibly certain lipophilic groups may act to replace hydrogen bonds for intramolecular stabilization, as postulated by Mallet and Koffler (1957). The fact that the thermophilic species able to withstand extremely high temperatures are found almost exclusively among so-called primitive organisms, the Cyanophyta and the bacteria, suggests that these species are survivors from a period when the common environment was characterized by high temperatures and high CO_2 pressures, and that thermophily may be a primitive, rather than an adaptive, character. If this is so, then the study of thermophiles would be of additional interest by providing an insight into the biochemistry and the physiology of the earliest forms of cellular life on our planet.

References

Adye, J., Koffler, H., and Mallet, G. E. (1957). The relative thermostability of flagella of thermophilic bacteria. *Arch. Biochem.* **67,** 251–253.

Allen, M. B. (1959). Studies with *Cyanidium caldarium,* an anomalously pigmented chlorophyte. *Arch. Mikrobiol.* **32,** 270–277.

Baker, H., Hutner, S. H., and Sobotka, H. (1955). Nutritional factors in thermophily; a comparative study of bacilli and *Euglena. Ann. N.Y. Acad. Sci.* **62,** 349–376.

Biebl, R. (1939a). Protoplasmatische Oekologie der Meeresalgen. *Ber. deut. botan. Ges.* **57,** 79–90.

Biebl, R. (1939b). Über die Temperaturresistenz von Meeresalgen verschiedener Klimazonen und verschieden tiefer Standorte. *Jahrb. wiss. Botan.* **88,** 389–420.

Bünning, E., and Herdtle, H. (1946). Physiologische Untersuchungen an thermophilen Blaualgen. *Z. Naturforsch.* **1,** 93–99.

Chodat, R. (1922). Materiaux pour l'histoire des algues de la Suisse. *Bull. soc. botan. Genève* **13,** 66–114.

Copeland, J. J. (1936). Yellowstone thermal Myxophyceae. *Ann. N.Y. Acad. Sci.* **36,** 1–229.

Crozier, W. J., and Federighi, H. (1924–1925). Critical thermal increment for the movement of *Oscillatoria. J. Gen. Physiol.* **7,** 137–150.

Fogg, G. E. (1956). The comparative physiology and biochemistry of the blue-green algae. *Bacteriol. Revs.* **20,** 148–165.

Frank, O., Hutner, S. H., Baker, H., Cox, D., Packer, E., Siegel, S., Aaronson, S., and Amsterdam, D. (1960). Sugar- and glycerol-free media for *Ochromonas danica. J. Protozool.* **7,** 13.

Fritsch, F. E. (1935). "The Structure and Reproduction of the Algae," Vol. I. Cambridge Univ. Press, London and New York.

Fukuda, I. (1957). Physiological studies on a thermophilic blue-green alga, *Cyanidium caldarium* Geitler. *Botan. Mag. (Tokyo)* **71,** 79–86.

Harvey, R. B. (1924). Enzymes of thermal algae. *Science* **60,** 481–482.

Hutner, S. H., Baker, S., Aaronson, S., Nathan, H. A., Rodriguez, E., Lockwood, S., Sanders, M., and Petersen, R. A. (1957). Growing *Ochromonas malhamensis* above 35°. *J. Protozool.* **4,** 259–269.

Inman, O. L. (1940). Studies on the chlorophylls and photosynthesis of thermal algae from Yellowstone Park, California and Nevada. *J. Gen. Physiol.* **23,** 661–666.

Kanwisher, J. (1957). Freezing and drying in intertidal algae. *Biol. Bull.* **113,** 275–285.

Kratz, W. A., and Myers, J. (1955). Nutrition and growth of several blue-green algae. *Am. J. Botany* **42,** 282–287.

Koffler, H., and Sale, G. O. (1957). The relative thermostability of proteins from thermophilic bacteria. *Arch. Biochem. Biophys.* **67,** 249–251.

Lagerheim, G. (1892). Die Schneeflora des Pichincha. *Ber. deut. botan. Ges.* **10,** 517–534.

Levitt, J. (1956). "The Hardiness of Plants," 278 pp. Academic Press, New York.

Löwenstein, A. (1903). Über die Temperaturgrenzen des Lebens bei der Thermalalge *Mastigocladus laminosus* Cohn. *Ber. deut. botan. Ges.* **21,** 317–323.

Mallett, G. E., and Koffler, H. (1957). Hypotheses concerning the relative stability of flagella from thermophilic bacteria. *Arch. Biochem. Biophys.* **67,** 254–256.

Marrè, E., and Servettaz, O. (1956). Ricerche sull adattamento proteico in organismi termoresistenti. I. Sul limite di resistenza all'inattivazione termica dei sistemi respiratorio e fotosintetico di alghe termali. *Atti accad. nazl. Lincei Rend. Classe sci. fis. mat. e nat.* **20,** 72–79.

Marrè, E., and Servettaz, O. (1957). Ricerche sull'adattamento proteico in organismi termoresistenti. II. Sulla termoresistenza in vitro del sistema citocromo riduttasico di cianoficee termali. *Atti accad. nazl. Lincei Rend. Classe sci. fis. mat. e nat.* **22,** 91–98.

Marrè, E., and Servettaz, O. (1959). Sul meccanismo di adattamento a condizioni osmotiche estreme. II. Concentrazione del mezzo esterno e composizione del succo cellulare. *Atti accad. nazl. Lincei Rend. Classe sci. fis. mat. e nat.* **26,** 272–278.

Marrè, E., Albertario, M., and Vaccari, E. (1958). Ricerche sull 'adattamento proteico in organismi termoresistenti. III. Relativa insensibilità di enzimi di cianoficee termali a denaturanti che agiscono rompendo i ponti di idrogeno. *Atti accad. nazl. Lincei Rend. Classe sci. fis. mat. e nat.* **24,** 349–353.

Marsh, C., and Militzer, W. (1956). Thermal enzymes. VIII. Properties of a heat-stable inorganic pyrophosphatase. *Arch. Biochem. Biophys.* **60,** 439–451.

Militzer, W., Sonderegger, T. B., Georgi, C. E., and Tuttle, L. C. (1950). Thermal enzymes. II. Cytochromes. *Arch. Biochem.* **26,** 299–306.

Moyse, A., Couderc, D., and Garner, J. (1957). L'influence de la température sur la croissance et la photosynthèse d'*Oscillatoria subbrevis* (Cyanophycée). *Rev. cytol. e biol. végétales* **18,** 293–304.

Prat, S., and Kubin, S. (1956). [Photosynthesis and respiration of thermophilic Blue-green Algae.] *Fiziol. Rasteni: Akad. Nauk S.S.S.R.* **3,** 505–515. (In Russian.)

Rochleder, O. (1958). "Chemie und Physiologie der Pflanzen," p. 145.

Sorokin, C. (1959a). Tabular comparative data for the low- and high-temperature strains of *Chlorella. Nature* **184,** 613–614.

Sorokin, C. (1959b). Kinetic studies of temperature effects on the cellular level. *Biochim. et Biophys. Acta* **38,** 197–204.

Sorokin, C., and Myers, J. (1953). A high-temperature strain of *Chlorella. Science* **117,** 330–331.

Sprenger, W. J. (1930). Bacillariales aus Thermen der Umgebung von Karlsbad. *Arch. Protistenk.* **71,** 502–542.

Vouk, V. (1950). "Grundriss zu einer Balneobiologie der Thermen." Birkhäuser, Basel, Switzerland.

Yoneda, Y. (1952). A general consideration of the thermal Cyanophyceae of Japan. *Mem. Coll. Agr. Kyoto Univ.* **62,** 1–20.

—34—

Invisible Radiations

M. B. E. GODWARD

Department of Botany, Queen Mary College, University of London, England

I. Introduction and Definitions

The term *radiation* in the present context refers generally to those kinds of electromagnetic or particle radiation that can cause extensive damage to living tissue. Ultraviolet light, x-rays, γ-rays, and beams of fast electrons, protons, α-particles and neutrons all fall within this category. The types of damage differ significantly from one form of radiation to another. Thus ultraviolet light (U.V.) (wavelength range approximately 2000 to 3000 Å) produces excitation of an electron in an atom or molecule, raising it to a state of higher energy. This may lead to chemical change, but the effect is less drastic than ionization. Ionizing radiations include x-rays (wavelength range approximately 0.05 to 10 Å.), γ-rays (wavelength 0.05 Å. or less) and β-rays or high-energy electrons. These last three kinds of ionizing radiation usually produce roughly the same kind of damage, whereas that produced by densely ionizing particles, such as neutrons or α-particles, tends to be extremely concentrated in the region of the particle's path.

The dose of x-rays and γ-rays is usually defined in roentgens (*r.*). One r. is that quantity of x- or γ-radiation which produced 2.1×10^9 ion pairs in 1 ml. of air at 0°C. and 760 mm. pressure. Under standard conditions one r. represents an energy absorption of 0.111 ergs per milliliter, or 87 ergs per gram of air. Other units in use include the rad, which can be employed for all radiations including neutrons; 1 rad corresponds to an absorption of energy of 100 ergs per gram.

A short section on infrared light (nonionizing radiations with wavelengths exceeding 7500 Å.) is appended.

Lea (1956) divided radiation damage into two general types: (i) direct effects due to molecular damage occurring in the molecule where the energy has been absorbed (this has been visualized in terms of a particle hitting a target); (ii) indirect effects, in which molecular damage is brought about by the chemical reactions of free radicals produced (for example, in the water of the cell) as a primary effect of the radiation.

The effects of the experimental irradiation of microorganisms are usually evaluated by determining the percentage of organisms surviving as a function of the magnitude of the dose. Less frequently, studies have been made on the comparative effects of different kinds of radiations, and of varying intensity or dose-rate. Valuable information on the mechanisms by which radiation produces its effects has been provided by fractionation of the dose (cf. Wolff, 1959). Fundamental properties of survival curves were worked out by Crowther (1926) and Lea (1956) in relation to the number of "hits" (e.g., by photons or ionizing particles) needed to cause inactivation of a cell. Thus in the simplest case, where one hit in a sensitive region is sufficient for inactivation (e.g., of certain bacteriophages), the curve of survival, plotted against radiation dose, is exponential. Nonexponential curves may be interpreted as a consequence of the need for at least two hits, indicating that there is not just one single target which can be effectively destroyed by a single hit. Surviving cells have presumably accumulated a sublethal amount of damage. In algae, nonexponential survival curves have been found in *Chlamydomonas* spp. by Nybom (1953) and Jacobson (1957) and in *Dunaliella salina* by Ralston (1939; see also Forssberg, 1933a, b). The nonexponential form of some survival curves (e.g., of *Stichococcus* sp.) may be perhaps correlated with approximate polyploidy (Zirkle and Tobias, 1953).

The term "ultimate survival" has been used by the author in referring to cultures which, after receiving a nearly lethal dose and after being maintained for a period of several weeks under conditions favorable for growth, prove to have contained at least one surviving cell which retained the ability finally to multiply at a normal rate [*Chlorella* sp. (Bonham and Palumbo, 1951); *C. pyrenoidosa* (Nizam and Godward, unpublished)]. This concept has some empirical use though it lacks absolute significance, the chance of one or more cells surviving in the irradiated culture being obviously a function of the initial population.

A dose which leaves 50% of the population alive is referred to as LD 50[1];

[1] LD = lethal dose.

where an exponential inactivation curve is desired the value LD 37 may be more useful, because it represents the dose for which each sensitive unit has received, on the average, one hit (Lea, 1956). However, LD 50 does not necessarily have a comparable meaning for different organisms, because experimentally one uses different criteria for death and assesses it at different time intervals after irradiation; e.g, mammals, 30 days (Hollaender, 1954–1956), *Chlorella,* 10 to 12 days (Bonham and Palumbo, 1951); *Chlamydomonas,* a "few" days (Jacobson, 1957). Some other arbitrary definition of survival may be preferable (Howard and Horsley, 1960); alternatively a standard reduction of growth-rate may be used (Gray and Scholes, 1951).

Although the LD 50 may be relatively easily determined when one irradiates a bacteriophage (in which viability is assessed solely by plaque formation), in the case of an alga this value is not so readily determined because not all fatally damaged cells can be observed to die at once. "*Immediate*" *death* and *death on division* may be distinguished (Halberstaedter and Back, 1942; Nybom, 1953). Among the first effects of irradiation are aberrant mitoses with failure of cytokinesis, resulting in the formation of giant cells which may die within the following 2 or 3 weeks, presumably due in some measure to chromosome damage which produces mechanical difficulties at mitosis. Such chromosome damage includes the loss of acentric fragments which fail to move to the poles of the mitotic spindle, with the frequent result that the dividing cell or daughter cells die. In higher plants such cytological effects have been precisely studied (Lea, 1956; Revell, 1955). Some of these effects can be distinguished following a dose of only 50 to 100 r.; 500 r. may be destructive of a meristem; a resistant seed or pollen grain may survive, though exhibiting aberrations, after receiving 50,000 to 100,000 r. Although in algae observation of chromosome changes is more difficult, fragmentation and bridge-formation have been observed and studied quantitatively by the present author and her co-workers, Dodge, Patel, Prasad, and Rayns. Algal chromosomes may be atypical in centromeric organization, e.g., in the Zygnematales (Conjugales), *Euglena,* and some Dinophyta (Dodge and Godward, unpublished), so that radiation-produced fragments are not lost and the nucleus is not thereby inactivated, as shown for *Spirogyra* by Godward (1954a, b). Other algae such as *Cladophora, Oedogonium, Volvox,* and *Pandorina,* with chromosomes of the conventional plant type, may be expected to show cytological effects similar to those of higher plants. Lethal genetic damage may also be sustained without the production of any cytologically recognizable aberrations, as demonstrated in *Euglena* (Leedale, unpublished).

During the delay in cell division that is generally one of the first consequencies of irradiation, much of the original damage is repaired. For example, restitution of primary chromosome breaks presumably occurs in a proportion of cases which, though not directly observable, can be calculated from the target theory. If no such repair occurred, the effect of successive doses would be strictly additive, which is seldom found to be the case (Wolff, 1959).

It should be recalled throughout this chapter that since cells of the Cyanophyta have not been shown to possess nuclei or chromosomes like those of other plants, no radiation damage has been assessed cytologically. There is likewise no information on cytological damage in many unicellular algae such as *Chlorella*, in which the chromosomes are very small.

II. Effects of Specific Radiations

A. α-Particles

Holweck and Lacassagne (1931a, b) estimated the target areas of sensitive organelles in *Polytoma uvella* by irradiating cells on an agar surface in air, 15 mm. from a polonium source, with an estimated density of 56 α-particles per micron squared per minute. Their observations are summarized in Table I. By irradiating filaments of *Zygnema* sp. with α-particles from a similar source, Petrová (1942) showed that to kill a cell by irradiation of the cytoplasm alone required 700 times as much energy as was needed to incapacitate the nucleus.

TABLE I

SOME EFFECTS OF α-PARTICLE IRRADIATION OF *Polytoma uvella*[a]

Damage[b] in terms of impaired:				Estimated target		
Cell enlargement	Flagellar motility	Cell division	Survival (5 days)	Number	Size (μ)	Possible organelle damaged
−	−	−	+	2	2.3	Nucleus
−	−	+	+	1	0.25	Centrosome
−	+[c]	+	+	2 (?)	0.15	Basal granules of flagella
+	+	+	+			

[a] After Holweck and Lacassagne (1931a, b). + = observed damage or impairment; − = no observed adverse effects.

[b] Doses increased from top row of table to bottom row.

[c] Cells capable only of rotation, apparently due to immobilization of one of the paired flagella, were also noted.

TABLE II

SOME CYTOLOGICAL AND OTHER EFFECTS OF β-RAYS (10^6 ELECTRON VOLTS) ON VARIOUS ALGAE[a]

Effect	Dose (in rads × 10^3)	Alga
Continuing nuclear division with "ultimate survival" (as defined in text)	10	*Chaetomorpha melagonium*
	100	*C. melagonium* (zoospores)
	15	*Spirogyra crassa*
	40	*Spirogyra subechinata*
	20 to 50	*Mougeotia* sp.
	20 to 50	*Zygnema cylindricum*
	20 to 50	*Cosmarium subtumidum*
LD 50 after 2 days	1000	*Chlorella pyrenoidosa*
LD 50 after 10 days	200	*Chlorella pyrenoidosa*
LD 90 after 18 days	1000	*Chlorella pyrenoidosa*
"Ultimate survival" (as defined in text)	1000	*Chlorella pyrenoidosa*
	100 to 200	*Anabaena* sp.
	20 to 50	*Eudorina elegans*
Continuing nuclear division, with 50% anaphase bridges at 1st division	30	*Eudorina elegans*
Death on misdivision (anaphase bridges)	3	*Prorocentrum micans*
"Immediate" death	20	*Prorocentrum micans*
	500	*Eudorina elegans*

[a] Data of Godward *et al.* (unpublished).

B. β-Rays

Accelerated β-rays, i.e., high-energy electrons (10^6 e.v.), were used in the following work, none of which has yet been published (see Table II).

1. *Effects on Growth and Viability*

The observed effects of 1,000,000 rads upon cultures of *Chlorella pyrenoidosa* subsequently kept in continuous light began with a fall in the viable cell count, which continued to decrease more slowly for 18 days after irradiation, and then increased again. A second dose of the same magnitude produced similar effects. The recovered cells were somewhat larger than normal, but their growth rate was equal to that of the control culture. In comparable tests at 10,000 rads no difference was observed between the effects of x-rays and high-energy β-rays.

2. *Effects on Cytoplasm and Nucleus*

Giant cells, in which the nucleus was abnormally large, probably polyploid, and apparently unable to divide further, were formed in *Chlorella pyrenoidosa* and *Mougeotia* sp. after irradiation (see Table II). All such cells died after a few weeks. After irradiation of *Anabaena* sp., a transverse rupture in the heterocyst walls was noted, which persisted as a constant feature of the surviving culture for over a year; after 3 years, however, it disappeared.

Chromosomes were examined in all the irradiated algae listed in Table II, though in *Chlorella* they are so minute that no details of changes could be observed. In *Euglena gracilis,* in which the normal chromosomes are quite atypical, no change was seen in the chromosomes or in the course of mitosis even after lethal doses had been administered. The observed effects of irradiation included a temporary loss of motility (Leedale and Godward, unpublished). In *Prorocentrum micans,* anaphase bridges of increased complexity were formed with increased dose, and ultimately there was extensive fragmentation of all chromosomes. The fragments remained together, but mitosis was blocked (Dodge and Godward, unpublished). In *Chaetomorpha melagonium,* nuclei with more than one or two fragments were nonviable. Fragmentation of chromosomes in *Eudorina elegans* was seen in all divisions of cells which had received a sublethal dose of irradiation; however, in the cultures which ultimately survived the chromosomes appeared normal.

3. *Recovery*

Photoreactivation—a partial restoration of viability in cells illuminated shortly after receiving a dose of radiation (in this case β-rays)—has been demonstrated in *Eudorina elegans* (Rayns and Godward, unpublished).

C. γ-Rays

γ-Rays from Co^{60} were applied to *Chlorella pyrenoidosa* by Zill and Tolbert (1958). At doses which depressed the rate of fixation of CO_2, the evolution of O_2 continued normal, indicating that these processes have separate sites of radiation sensitivity. Fixation of CO_2 declined exponentially with increasing dose; calculation showed that the target volume is smaller than that of the chloroplast. Inhibition of CO_2 fixation was greatest immediately after irradiation; recovery, possibly by some form of photoreactivation, followed within 5 hours after exposure to light. Gailey and Tolbert (1958) noted that 10 hours after irradiation, with a dose of 150,000 r., chlorophyll formation declined, perhaps as a consequence of the inhibition of enzyme synthesis.

Eudorina elegans seems to be slightly more sensitive to γ-rays than to high-energy β-rays. In our experiments 20,000 rads caused rapid death, whereas a small proportion of the cells survived a dose of 15,000 rads (Rayns and Godward, unpublished). A dose of 150,000 rads was just sublethal to *Euglena gracilis* (Leedale, personal communication), while some Blue-green Algae proved capable of surviving more than 10^6 rads when administered at dose rates of about 200,000 r. per hour (Shields *et al.*, 1961).

D. X-rays

1. *Effects on Growth and Viability*

In Table III certain effects of different doses on various algae are tabulated. Although Forssberg (1933a, b) found in x-irradiated suspensions of *Chlorella* sp. that the number of undivided cells present 3 days after they had received 5000 r. was dependent on the original intensity, i.e., on the dose rate, and although some other authors have also studied dose-rate effects, such data are omitted here since no systematic investigation of dose-rate has yet been made. Algae are nearly all more resistant than higher plants. Possibly this is correlated in some cases with the atypical or minute size of their chromosomes or the condition of polyploidy. The genus *Oedogonium*, with chromosomes of the same order of size as those of many higher plants and with typical localized centromeres, is the most sensitive alga so far examined.

2. *Effects on Cytoplasm and Nucleus*

Acetabularia lends itself especially well to studies on the differential effects of radiations on the cytoplasm and the nucleus, since the cells are so easily enucleated. Some of the results obtained with *A. mediterranea* by Six (1956b, 1958) are shown in Table I. In other algae, the cytological effects of x-rays on cytoplasm are most easily seen in cells in which the nucleus normally does not divide, as in the internodal cells of *Chara* (Moutschen, 1957), or wl ere it has become unable to divide as a result of radiation damage, e.g., *Spirogyra crassa* after receiving 20,000 r. (Godward, unpublished). The cell size and the volume of cytoplasm increased in both of these cases, effects which Moutschen attributed to an abnormal release of enzymes in cytoplasm damaged by irradiation. Effects on chromosomes were observed in *Spirogyra crassa* (Godward, 1954a, b), in which fragmentation increased with dose; the fragments survived, and many of the new karyotypes could be maintained in culture. There was certainly some rejoining of fragments, but it remains uncertain to what extent this may have been responsible for recovery of the cells. In *Oedogonium*, A. Howard (unpublished) observed the usual lesions.

TABLE III

Some Cytological and Other Effects of X-rays on Various Algae

Effect	Dose (in roentgens $\times 10^3$)	Alga	Reference
None	2	*Spirogyra crassa*	Godward (1954a, b)
	3	*Pandorina morum*	Halberstaedter and Back (1942)
	10	*Chlorella pyrenoidosa*	Nizam and Godward (unpublished)
	10	*Chara vulgaris*	Moutschen and Dahmen (1956)
Abnormal cells	10	*Chlorella vulgaris*	Pietschmann (1937)
18% genetic death	40	*Astasia longa*	Schoenborn (1954)
50% death	0.75	*Oedogonium* sp. (zoospores)	Howard and Horsley (1960)
	8 (hard)	*Chroococcus* sp.	Bonham and Palumbo (1951)
	11	*Ankistrodesmus* sp.	Bonham and Palumbo (1951)
	12	*Dunaliella salina*	Ralston (1939)
	16	*Mesotaenium caldariorum*	Langendorff *et al.* (1933)
	18	*Chlorella* sp.	Bonham and Palumbo (1951)
	28	*Chroococcus* sp.	Bonham and Palumbo (1951)
	40 (soft)	*Chlorella* sp.	Bonham and Palumbo (1951)
	45	*Chlamydomonas reinhardtii*	Nybom (1953); Jacobson (1957)
		C. eugametos	Jacobson (1957)
		C. moewusii	Jacobson (1957)
	>100	*Synechococcus* sp.	Bonham and Palumbo (1951)
Death on division	5 to 300	*Pandorina morum*	Halberstaedter and Back (1942)
Viable gamete formation suppressed	40	*Acetabularia mediterranea*	Six (1958)
Regeneration suppressed	400	*Acetabularia mediterranea*	Six (1958)

Hat formation (in enucleate portions) suppressed	50 to 100	*Acetabularia mediterranea*	Six (1958)
"Ultimate survival" (as defined in text)	15	*Spirogyra crassa*	Godward (1954a, b)
	100	*Chlorella* sp.	Bonham and Palumbo (1951)
	600	*Chara vulgaris*	Moutschen (1957)
"Immediate" death	10	*Chlamydomonas reinhardtii*	Nybom (1953)
	600	*Pandorina morum*	Halberstaedter and Back (1942)
	750	*Acetabularia mediterranea*	Bacq *et al.* (1955)

3. *Effects on Photosynthesis and Respiration*

No published data have been found.

4. *Recovery and Protection*

Cells which would otherwise have died have clearly been shown to be resuscitated by certain postirradiation treatments (Nybom, 1953; Jacobson, 1957). Recovery in *Chlamydomonas reinhardtii* is enhanced at a temperature of 30°C., especially if the dose is fractionated (i.e., delivered in several short doses separated by periods without irradiation), and recovery is inhibited by a low temperature (5°C) or by depletion of assimilable nitrogen in the medium (Jacobson, 1957). Wolff and Luippold (1956), discussing the general question of such recovery, postulated that damaged chromosomes might rejoin and that this process might be blocked by agents which inhibit respiration and oxidative phosphorylation. Jacobson invoked a similar mechanism to account for recovery in *Chlamydomonas.*

Protection from x-radiation damage is afforded by the presence of reducing substances such as thiourea and cysteine in the medium, and by anaerobiosis produced by bubbling nitrogen gas through it (Nybom, 1953). Damage is increased by irradiating the cells in the light, an effect possibly related to the production of oxygen by photosynthesis.

E. Ultraviolet Light

1. *Effects on Growth and Viability*

The intense mercury-vapor emission line at 253.7 mμ is the most frequently used source of U.V. for biological purposes, since the purine and pyrimidine bases of nucleic acids absorb strongly in this region. Doses in the range between 10^2 and 10^3 μw./cm.^{2} (1 μw. = 10 ergs/sec.) generally produce genetic damage. In a continuous culture system with a constant cell density, artificial light, and 5% CO_2 in air, viability of *Chlorella pyrenoidosa,* measured indirectly by growth-curve effects, was found to fall exponentially with increasing doses (Redford and Myers, 1951). In *Chlamydomonas* spp., however, the survival curves were not linear (Nybom, 1953). Phytoplankton species of deeper water appeared to be more sensitive than those of shallow water (Gessner and Diehl, 1951). *Chlorella vulgaris* was found to be more resistant than *Ankistrodesmus* sp. and *Scenedesmus quadricauda* (Gessner and Diehl, 1951); *C. pyrenoidosa* was more resistant than *Eudorina elegans* (Rayns and Godward, unpublished).

The effects of other wavelengths upon survival were investigated by Meier (1932) and her co-workers. In *Chlorella vulgaris,* peaks of sensitivity were found at 240.0 mμ and 260.0 mμ; bleaching effects were noted at

253.6, 265.2, 269.9, 275.3, 289.4, 296.7 and 302.2 mμ. Meier (1939) reported that at 297.7 mμ two-thirds of the lethal dose seemed to stimulate growth; she also found some stimulation of *Stichococcus bacillaris* at 235.2, 248.3, 265.2, and 296.7 mμ (1941).

2. *Effects on Cytoplasm and Nucleus*

Whereas the lethal effects of x-rays on *Chlamydomonas* tend to be delayed, partly as a consequence of abortive nuclear division, those of U.V. irradiation are immediate, possibly because they result from protein denaturation or photo-oxidation of enzymes (Nybom, 1953). Errera and Vanderhaeghe (1957) found the cytoplasm of *Acetabularia mediterranea* to be more sensitive to U.V. than the nucleus, and noted that damage to the nucleus could result from irradiation of the cytoplasm only. Ultraviolet light of short wavelengths (254 mμ) had more profound morphogenetic effects on enucleate *Acetabularia* cells than did that of longer wavelengths (281 or 297 mμ), suggesting that purine or pyrimidine bases are involved in morphogenesis (Six, 1956a). Holweck and Lacassagne (1931b) reported that the function of motility and the ultimate viability of cells of *Polytoma uvella* were equally susceptible to damage by U.V. irradiation. According to Meier (1940) the effects of U.V. irradiation on *Stichococcus* included a progressive shortening and "weakening" of the cells, although irradiated cells were of a deeper green color and reproduced more rapidly.

There are few records of U.V.-induced chromosome changes. Dodge and Godward (unpublished) found a low percentage of anaphase bridges in cells of *Prorocentrum micans* subjected to sublethal doses.

3. *Induction of mutants*

Morphological mutants were obtained in *Cosmarium turpinii* by Korn (1959), in *Chlamydomonas* spp. by Lewin (1952), Nybom (1953) and Gowans (1960), and in *Euglena gracilis* by Lewin (1960). Biochemical mutants, obtained largely by U.V. irradiation, have been reviewed by Ebersold (Chapter 49 of this treatise).

4. *Effects on Photosynthesis and Respiration*

Arnold (1933) showed that the rate of photosynthesis in *Chlorella pyrenoidosa* is reduced by U.V. irradiation. Inactivation was found to behave initially as a first-order process, but later it became one of the second or a higher order (Holt *et al.*, 1951; Redford and Myers, 1951); i.e., as the U.V. dose increased, the photosynthetic rate fell at first exponentially, and then faster. Inactivation of photosynthesis was a function not merely of dose but also of intensity, i.e., it was dose-rate dependent. It was not affected by the nutritional state of the cells, by the presence of glucose, or by illumination with visible light during U.V. irradiation. The evidence indicated that U.V. inhibits the "Hill reaction." The photosynthetic mechanism of *Scenedesmus* (strain D_1) appears to be less resistant to U.V. than that of *Chlorella pyrenoidosa* (Holt *et al.*, 1951); likewise photosynthesis in *Scenedesmus quadricauda, Phaeodactylum tricornutum, Skeletonema costatum, Dunaliella euchlora, Platymonas subcordiformis,* and an unknown species of *Chlorella* proved more sensitive than in *Botryococcus braunii, Ankistrodesmus falcatus,* and *Chlorella vulgaris*. The sensitivity of these algae tended to be low in the lag phase, to rise to a peak in the log phase, and to decline again in the stationary phase of their growth (McLeod and McLachlan, 1959). Higher doses of short-wave U.V. destroy chlorophyll (Meier, 1934; Gessner and Diehl, 1951; McLeod and McLachlan, 1959).

Biebl (1952) asserted that certain marine algae have a "constitutional" protoplasmic resistance to damaging effects of short-wave U.V., as contrasted with their environmentally conditioned resistance to damage by visible light. He obtained U.V. absorption spectra with small peaks between 290 and 360 mμ for several species, and postulated that some of the energy absorbed at these wavelengths might be used for photochemical reactions. It was found in fact that in *Chlorella, Anacystis, Porphyridium, Dunaliella, Monochrysis, Phaeodactylum,* and *Coscinodiscus* chlorophyll luminescence, which can be detected in the dark 0.2 second after excitation, is induced by U.V. as well as by visible light (McLeod, 1958). Action spectra of the luminescence of chlorophyll and the accessory pigments showed a distinct rise in the U.V. at 2500 Å. McLeod recalled earlier speculations on the possible activity of U.V. in photosynthesis, and suggested that it might be due to the absorption of energy by proteins.

Though endogenous respiration of *Chlorella vulgaris* is not affected by moderate doses of U.V., the oxidation of exogenous glucose or acetate is severely inhibited. Glucose oxidation is more sensitive to U.V. than is photosynthesis (Holt *et al.*, 1951; Redford and Myers, 1951). In agreement with these observations, Zill and Tolbert (1958) found that with increasing U.V. doses the photosynthetic quotient of *C. pyrenoidosa* remained close to unity until photosynthesis was completely stopped, apparently as a consequence of the destruction of chlorophyll.

5. *Protection from Radiation Damage*

Substances such as cysteine, thiourea, and glutathione may protect cells of *Chlamydomonas* spp. against damage by U.V.; anaerobiosis, however, is ineffective (Nybom, 1953). Since these reducing substances all absorb U.V.

in the region between 240 and 280 mμ (Schoenborn, 1956), their action is readily explicable and could be unrelated to their reducing action. However, similar protection against other forms of ionizing radiation has also been reported.

Photoreactivation is a potent means of promoting the recovery of cells damaged by U.V. radiation. Light of 10,000 lux, applied immediately after irradiation of various *Chlamydomonas* species, was increasingly effective for illumination periods up to 160 minutes (Nybom, 1953). Photoreactivation has also been demonstrated in U.V. irradiated *Eudorina elegans* (Rayns and Godward, unpublished). Fractionation of the U.V. dose also permitted slight recovery of *Chlorella pyrenoidosa* in the short intervals between irradiation (Nizam and Godward, unpublished), possibly because visible light promotes the re-formation of chlorophyll bleached by U.V.

F. Infrared Light

Féher and Frank (1936, 1939) claimed that certain soil algae are able to use some of the infrared absorbed by chlorophyll for carbon assimilation. Although Baatz (1939) refuted this possibility on experimental grounds, and there is no other evidence in support of it, Féher and Frank (1940) later reaffirmed their position.

G. Summary and Conclusion

Some distinction may be drawn between: (*a*) the action of U.V., which may produce genetic or lethal damage, is not markedly influenced by external oxygen, may be stimulatory in small doses, and has various other photochemical effects; and (*b*) the action of ionizing radiation, which has at least partly "indirect" effects, is markedly influenced by the presence of external oxygen and reducing substances, and is destructive only. Visible light, if administered sufficiently soon after irradiation, may partially reverse the effects of U.V., and has even some effect on x-irradiated cells.

In their metabolism and in their chromosome organization the classes of algae differ as profoundly from one another as they do from other groups of organisms such as fungi, bacteria, and animals. The great variety in their responses to radiation, which is perhaps to be expected, makes attempts at generalization difficult. Sensitivity to radiation damage is in general low; in some cases it may be comparable with that of higher plants (where it seems to be correlated with a similar chromosome organization), but in others it is so slight that some cells may survive 2,000,000 rads and continue to grow with vigor indefinitely. Such a degree of resistance, or capacity for recovery, is approached but not equaled by certain fungal spores and highly polyploid ciliates.

ACKNOWLEDGMENTS

The author expresses her gratitude to Mr. Don Moore at the Medical Research Council Unit, Hammersmith Hospital, London, for making available the van de Graaff machine as a source of high-energy β-rays and x-rays, and for carrying out the irradiations; to the Atomic Energy Authority, Wantage, Berks., for providing a source of γ-rays and irradiating the material; and to the Department of Scientific and Industrial Research for provision of constant-temperature chambers for algal culture.

REFERENCES

Arnold, W. (1933). The effect of ultraviolet light on photosynthesis. *J. Gen. Physiol.* **17,** 135–143.

Baatz, I. (1939). Über das Verhalten von Bodenalgen in kurzwelligen Infrarot. *Arch. Mikrobiol.* **10,** 508–514.

Bacq, Z. M., Damblon, J., and Herve, A. (1955). Radiorésistance d'une algue, *Acetabularia mediterranea* Lamour. *Compt. rend. soc. biol.* **149,** 1512–1519.

Biebl, R. (1952). Ultraviolett Absorption der Meeresalgen. *Ber. deut. botan. Ges.* **65,** 36–40.

Bonham, K., and Palumbo, R. (1951). Effects of x-rays on snails, crustacea, and algae. *Growth* **15,** 155–188.

Chase, F. M. (1941). Increased stimulation of the alga *Stichococcus bacillaris* by successive exposures to short wave lengths of the ultraviolet. *Smithsonian Inst. Publs. Misc. Collections* **99,** Paper No. 17, Publ. No. 3603, 16 pp.

Crowther, J. A. (1926). The action of x-rays on *Colpidium colpoda. Proc. Roy. Soc.* **B100,** 390–404.

Errera, M., and Vanderhaeghe, F. (1957). Effets des rayons U.V. sur *Acetabularia mediterranea. Exptl. Cell Research* **13,** 1–11.

Fehér, D., and Frank, M. (1936). Untersuchungen über die Lichtökologie der Bodenalgen. I. *Arch. Mikrobiol.* **7,** 1–10.

Fehér, D., and Frank, M. (1939). Untersuchungen über die Lichtökologie der Bodenalgen. II. Der unmittelbare Beweis des autotrophen Algenwachstums beim Abschluss des sichtbaren Anteils der strahlenden Energie. *Arch. Mikrobiol.* **10,** 247–264.

Fehér, D., and Frank, M. (1940). Ergänzende Bemerkungen zu unseren Arbeiten über die Lichtökologie der Bodenalgen. *Arch. Mikrobiol.* **11,** 80–84.

Forssberg, A. (1933a). Der Zeitfaktor in der biologischen Wirkung von Röntgenstrahlen. II. Untersuchungen an Algen und Drosophila-Puppen. *Acta Radiol.* **14,** 399–407.

Forssberg, A. (1933b). Der Zeitfaktor in der biologischen Wirkung von Röntgenstrahlen. III. *Acta Radiol.* **14,** 407–499.

Gailey, F. B., and Tolbert, N. E. (1958). Effect of ionizing radiation on the development of photosynthesis in etiolated wheat leaves. *Arch. Biochem. Biophys.* **76,** 188–195.

Gessner, F., and Diehl, A. (1951). Die Wirkung natürlicher Ultraviolettstrahlung auf die Chlorophyllzerstörung von Planktonalgen. *Arch. Mikrobiol.* **15,** 434–439.

Godward, M. B. E. (1954a). Cytotaxonomy of *Spirogyra. Congr. intern. botan.,* **8ᵉ** *Congr., Paris, 1954* Sect. 17, p. 17. (Abstr.)

Godward, M. B. E. (1954b). Irradiation of *Spirogyra* chromosomes. *Heredity* **8,** 293. (Abstr.)

Gowans, C. S. (1960). Some genetic investigations on *Chlamydomonas eugametos. Z. Vererbungslehre* **91,** 63–73.

Gray, L. H., and Scholes, M. E. (1951). Growth rate studies on broad bean root. *Brit. J. Radiol.* **24,** 82.

Halberstaedter, L., and Back, A. (1942). The effects of x-rays on single colonies of *Pandorina. Brit. J. Radiol.* **15,** 124–128.

Hollaender, A., ed. (1954–1956). "Radiation Biology," 3 Vols. McGraw Hill, New York.

Holt, A. S., Brooks, I. A., and Arnold, W. A. (1951). Effects of 2537 Å on green algae and chloroplast preparations. *J. Gen. Physiol.* **34,** 627–645.

Holweck, F., and Lacassagne, A. (1931a). Action des rayons α sur *Polytoma uvella.* Détermination des "cibles" correspondant aux principales lésions observées. *Compt. rend. soc. biol.* **107,** 812–814.

Holweck, F., and Lacassagne, A. (1931b). Essai d'interprétation quantique des diverses lésions produites dans les cellules par les radiations. *Compt. rend. soc. biol.* **107,** 814–817.

Howard, A., and Horsley, R. J. (1960). Filamentous green algae for radiobiological study. *Intern. J. Radiation Biol.* **2,** 319–330.

Jacobson, B. S. (1957). Evidence for recovery from x-ray damage in *Chlamydomonas. Radiation Research* **7,** 394–407.

Korn, R. W. (1959). Ultraviolet-induced morphological mutations in *Cosmarium turpinii* Bréb. *Proc. Intern. Botan. Congr.* **2,** 199. (Abstr.)

Langendorff, H., Langendorff, M., and Reuss, A. (1933). Über die Wirkung von Röntgenstrahlen verschiedener Wellenlange auf biologische Objekte. *Strahlentherapie* **46,** 655–662.

Lea, D. E. (1956). "Actions of Radiations on Living Cells" (L. H. Gray, ed.), 2nd ed. Cambridge Univ. Press, London and New York.

Lewin, R. A. (1952). U.V. induced mutations in *Chlamydomonas moewusii. J. Gen. Microbiol.* **6,** 233–248.

Lewin, R. A. (1960). A device for obtaining mutants with impaired motility. *Can. J. Microbiol.* **6,** 21–25.

McLeod, G. C. (1958). Delayed-light action spectra of several algae in visible and U.V. light. *J. Gen. Physiol.* **42,** 243–250.

McLeod, G. C., and McLachlan, J. (1959). The sensitivity of several algae to U.V. irradiation of 2537 Å. *Physiol. Plantarum* **12,** 306–309.

Meier, F. E. (1932). Lethal action of ultra-violet light on a unicellular green alga. *Smithsonian Inst. Publs. Misc. Collections* **87,** Paper No. 10, Publ. No. 3173, 11 pp.

Meier, F. E. (1934). Colonial formation of unicellular algae under various light conditions. *Smithsonian Inst. Publs. Misc. Collections* **92,** Paper No. 5, Publ. No. 3256, 14 pp.

Meier, F. E. (1939). Stimulative effect of short wave lengths in the ultraviolet on the alga *Stichococcus bacillaris. Smithsonian Inst. Publs. Misc. Collections* **98,** Paper No. 23, Publ. No. 3549, 19 pp.

Moutschen, J. (1957). Action du rayonnement X sur la croissance des cellules internodales de l'algue *Chara vulgaris* L. *Experientia* **13,** 240.

Moutschen, J., and Dahmen, M. (1956). Spermiogénèse de *Chara vulgaris* L. *Rev. cytol. et biol. végétales* **17,** 3–4.

Nybom, N. (1953). Some experiences from mutation experiments in *Chlamydomonas. Hereditas* **39,** 317–324.

Petrová, J. (1942). Über den Vergleich der α-Strahlenempfindlichkeit von Kern und Plasma. *Ber. deut. botan. Ges.* **60,** 148–151.

Pietschmann, K. (1937). Über Bestrahlung von *Chlorella vulgaris* mit Röntgenstrahlen. *Arch. Mikrobiol.* **8,** 180–206.

Ralston, H. J. (1939). Immediate and delayed action of x-rays on *Dunaliella salina. Am. J. Cancer* **37,** 288–297.

Redford, E. L., and Myers, J. (1951). Some effects of ultraviolet radiations on the metabolism of *Chlorella. J. Cellular Comp. Physiol.* **38,** 217–243.

Revell, S. H. (1955). A new hypothesis for "chromatid" changes. *In* "Radiobiology Symposium, 1954" (Z. M. Bacq and P. Alexander, eds.), pp. 243–253. Academic Press, New York.

Schoenborn, H. W. (1954). Lethal effects of ultraviolet and x-radiation on the protozoan flagellate *Astasia longa. Physiol. Zoöl.* **26,** 312–319.

Schoenborn, H. W. (1956). Protection against lethal damage induced by ultraviolet radiation. *J. Protozool.* **3,** 97–99.

Shields, L. M., Durrell, L. W., and Sparrow, A. H. (1961). Preliminary observations on radiosensitivity of algae and fungi from soils of the Nevada test site. *Ecology* **42,** 440–441.

Six, E. (1956a). Die Wirkung von Strahlen auf *Acetabularia.* I. Die Wirkung von u-v Strahlen auf kerlose Teile von *Acetabularia mediterranea. Z. Naturforsch.* **11b,** 463–470.

Six, E. (1956b). Die Wirkung von Strahlen auf *Acetabularia.* II. Die Wirkung von Röntgenstrahlen auf kernlose Teile von *Acetabularia mediterranea. Z. Naturforsch.* **11b,** 598–603.

Six, E. (1958). III. Die Wirkung von Röntgenstrahlen und ultravioletten Strahlen auf kernhaltige Teile von *Acetabularia mediterranea. Z. Naturforsch.* **13b,** 6–14.

Wolff, S. (1959). Interpretation of induced chromosome breakage and rejoining *Radiation Research Suppl.* **1,** 453–469.

Wolff, S., and Luippold, H. E. (1956). The biochemical aspects of chromosome rejoining. *In* "Progress in Radiobiology" (J. S. Mitchell, B. E. Holmes, and C. L. Smith, eds.) Oliver & Boyd, Edinburgh.

Zill, L. P., and Tolbert, N. E. (1958). The effect of ionizing and ultraviolet radiations on photosynthesis. *Arch. Biochem. Biophys.* **76,** 196–203.

Zirkle, R. E., and Tobias, C. A. (1953). Effects of ploidy and linear energy transfer on radiobiological survival curves. *Arch. Biochem. Biophys.* **47,** 282–306.

—35—

Intracellular Movements

W. HAUPT

Botanical Institute, University of Tübingen, Germany

I. Introduction

Movements of the cytoplasm or protoplasmic particles such as chloroplasts can be grouped into two divisions: (*a*) autonomous movements going on more or less continuously, at least under certain conditions, without changing the basic structure of the cell, e.g., agitation or rotation (see Section II); (*b*) movements which bring about new distributions of protoplasmic material and thus change the basic cellular structure. After the latter dislocation, movement ceases, either more or less irreversibly, leading to morphogenesis (see Section III below; also Nakazawa, Chapter 43, this treatise), or until environmental conditions change (e.g., orientation of chloroplasts to light; see Section IV). Obviously such classifications can be only a matter of convenience, since our knowledge of the mechanisms is still very incomplete (see Section V). Only those movements will be discussed here which are typical of some particular algae or for the study of which certain algae have provided the material of choice.

II. Typical Cytoplasmic Streaming

There are cytoplasmic streamings of different degrees of regularity. The less ordered forms, such as agitation and circulation, are to be found in many algae, such as desmids and *Spirogyra* (Kamiya, 1959; Küster, 1956); but rotation in its most typical form, with constancy both in orientation and in speed, is of greater interest in that here we have some preliminary ideas as to the mechanism of this phenomenon.

One of the best known cases of rotation is that in the internodal cells of Characeae, e.g., *Chara* and *Nitella.* Since the most important papers in this field are reviewed by Kamiya (1959), we shall restrict ourselves here to some fundamental facts. There are two different cytoplasmic layers, the stationary outer ectoplasm, and the inner endoplasm. The latter is in constant motion, which continues in the absence of external stimuli; it is referred to as "primary streaming." Since the endoplasm moves down one side and up the other side of the cell, there must be two stationary strips where these pathways adjoin. These lines are marked by special structures of the wall and probably also of the cytoplasm. The direction of the streaming is fixed, in that upward movement takes place on the outer side of a "leaf" cell or on that side of an internodal cell which gives rise to the oldest "leaf"; see reviews by Kamiya (1959) and Küster (1956). The relation between helical streaming and spiral growth of the cell has been discussed by Probine and Preston (1958) and Green (1959). In *Acetabularia, Codium, Bryopsis,* and *Caulerpa* (Kamiya, 1959; Küster, 1956) streaming takes place in a number of definite cytoplasmic tracks, separated from each other by stationary zones. Corresponding striations occur in the ectoplasm, and can still be seen when the whole endoplasm has been concentrated by centrifugation to one end of the cell (Takata, 1958). According to Hämmerling (1931), in growing cells of *Acetabularia* the direction of streaming appears to be random, whereas it becomes strongly oriented acropetally when the formation of cysts begins (see below).

In *Nitella* and *Chara,* unlike angiosperms and some other algae, the majority of chloroplasts are fixed in the ectoplasm, and only an occasional chloroplast is carried with the moving endoplasm. Most of the chloroplasts are somewhat elongated, but the few to be found along the stationary strips are often round (Jarosch, 1956).

III. Movements of Nuclei

The spontaneous rotation of nuclei in isolated drops of Characeae protoplasm, observed by Valkanov (1934) and Fetzmann (1959), clearly shows the possibility of active nuclear movement within the cytoplasm.

One of the most interesting instances of nuclear migration in morphogenesis can be seen in *Acetabularia* (Hämmerling, 1931, 1953). Here, the single cell nucleus is situated in one of the rhizoids and remains there without movement during the whole vegetative development of the plant, growing enormously without increase of chromosomal material. There are no other evident changes until the reproductive cap characteristic of this genus has reached its maximum size. At this stage the giant nucleus divides repeatedly to form several thousand secondary nuclei, which migrate for several millimeters up the stalk into the cap where they induce cyst formation (Hămmerling, 1931, 1953, 1955). Whether or not this movement is exclusively a passive one due to the oriented cytoplasmic streaming, as Hämmerling suggests, cannot be stated with certainty at the moment, but this seems probable. Hämmerling (1939, 1953) showed by interspecific grafts that the cytoplasmic factors responsible for the initiation of nuclear divisions, and therefore also for the induction of nuclear migration, are not species-specific. However, the nature of these factors is still unknown.

Another example of nuclear movement has been described in pennate diatoms by Geitler (1932, 1953). When an auxospore of *Cocconeis* begins to germinate into the primordial vegetative cell, two new valves must be formed. In this process the nucleus moves first to one side of the auxospore, where it divides into a normal nucleus and a pycnotic one. At the same time cell-wall formation is initiated, leading to the production of the first valve, the epitheca. The nucleus then moves to the opposite side of the auxospore, where the same events are repeated and the second valve, the hypotheca, is formed.

Other instances of nuclear migration are those associated with the processes of fertilization. The movement of gamete nuclei across intercellular bridges in *Spirogyra*, and, in Red Algae, the movement of the male nucleus down the trichogyne and the complex migration pattern of diploid nuclei in the carpogonium, are among the better known examples which would certainly merit physiological investigation.

IV. Orientation Movements of Chloroplasts

The chloroplasts of many plants are able to orient to light. For example, those of *Vaucheria* move to the front and back walls when illuminated from one side with light of low or medium intensity, and thus absorb as much light as possible, whereas in strong light they move to the side walls and are thereby protected against damage. This phenomenon has been observed in a number of algae of widely different forms of thallus; as a general rule, the chloroplasts tend to move to those places which are most strongly or most weakly illuminated (Senn, 1908). These movements are usually referred to as positive and negative phototaxis of the chloroplasts, implying that the effective light absorption takes place in the chloroplasts themselves. However, an alternative hypothesis should not be overlooked. As early as 1880, Stahl suggested that light absorption might take place in the cytoplasm, leading to the formation of a gradient which in turn could direct the orientation of the chloroplasts. Some facts in favor of this hypothesis are discussed below; see also Haupt (1959a).

An outstanding type of chloroplast "phototaxis" occurs in *Mougeotia*, in which the plank-shaped chloroplast turns inside the cell so as to expose its face or its profile to dim or bright light respectively (Senn, 1908; see also Lewis, 1898). But whereas in algae such as *Vaucheria* the response is at least superficially phototactic, in that orienting movements take place only during the period of the light stimulus, the responses of *Mougeotia* are more complex, since induction can also occur in brief periods of illumination, the orientation movements taking place later in the dark (Lewis, 1898). Haupt (1959b) confirmed this observation "low-intensity movements," but for "high-intensity movements" this aftereffect could not be demonstrated; nor was it possible to confirm Lewis' assertion that the angle of turning depends on the illumination time, with the full turn of 90 degrees being completed only after a certain duration of induction. For low-intensity movements, if a chloroplast or chloroplast region responds at all, it completes the orientation movement, though among the chloroplasts within a filament there is some variation in the threshold stimulus required to bring about a response (Haupt, 1959b).

The question now arises whether the type of chloroplast movement observed in *Mougeotia* is based upon the same physiological events as that of *Vaucheria*. At first this seems unlikely since the action spectrum in *Mougeotia* exhibits the well-known red:far-red antagonism phenomenon (Haupt, 1959b), whereas in *Vaucheria* and similar cases only blue light is effective (Senn, 1908). However, since the red:far-red reaction system is responsible only for the low-intensity plastid movements of *Mougeotia*, whereas the high-intensity movements are induced only by blue and violet light (Mosebach, 1958; Haupt and Schönbohm, 1962), the difference between the *Mougeotia* and *Vaucheria* systems may not be so important. From experiments on *Mougeotia*, there is increasing evidence that the light stimulus is perceived by the cytoplasm rather than by the chloroplast (Bock and Haupt, 1961); this appears to be true also for the red:far-red system of other plants (Butler *et al.*, 1959). Moreover, in *Mougeotia* the pigment molecules in question may be oriented, since polarized light is effective only when the plane of vibration is perpendicular to the cell axis. This has been well established for the low-intensity reaction (Haupt, 1960a); however in the high-intensity reaction, too, there seems to be considerable dependence on the polarization plane (Haupt and Schönbohm, 1962).

A third type of chloroplast movement is exemplified by the plastids of the diatom *Nitzschia palea*, which reversibly round up in darkness. This contraction, like the re-expansion which takes place in the light, occupies only a few hours; moreover it is perhaps significant that in illumination above the threshold level the rate of recovery appears to be independent of intensity (von Denffer, 1949). Geitler (1960, 1961) reported a completely different type of chloroplast movement in the hair cells of *Coleochaete*. Here the single chloroplast rotates autonomously without any visible participation of the cytoplasm, and is apparently not readily influenced by environmental changes.

V. Some Remarks on Mechanisms

In attempting to explain all protoplasmic movements on a common basis, Kamiya (1959) suggested that the motive forces are always produced at sol–gel interfaces. In favor of this suggestion is the velocity distribution in a *Nitella* cell, where the highest speeds are observed just at the boundary between ectoplasm and endoplasm. More detailed observations by Jarosch (1960) point to the same interpretation. The observation that very near to a rotating chloroplast there are minute local streamlets in the cytoplasm, the direction of which is opposite to that of the chloroplast (Jarosch, 1956), suggests that the movements of nuclei and chloroplasts may also be explained in terms of such sol–gel interrelationships. The participation in all protoplasmic movements of actomyosin-like contractile fibrils, located in the plasmagel at the gel–sol interfaces of ectoplasm, chloroplast, or nucleus, may explain all the known facts (Kamiya, 1959; cf. Jarosch, Chapter 36, this treatise). From observations on foraminifera, Jahn and Rinaldi (1959) came to a similar hypothesis, except that the shearing forces which they postulated could be established between two protoplasmic portions in a gel state, and not merely between gel and sol portions. An alternative hypothesis, based upon metabolic gradients which act like drag-forces (Rashevsky; cited in Stewart and Stewart, 1959a, b) does not seem adequate to explain such phenomenon as the rotation in Characeae (Haupt, 1960b).

References

Bock, G. und Haupt, W. (1961). Die Chloroplastendrehung bei *Mougeotia*. III. Die Frage der Lokalisierung des Hellrot-Dunkelrot-Pigmentsystems in der Zelle. *Planta* **57,** 518–530.

Butler, W. L., Norris, K. H., Siegelman, H. W., and Hendricks, S. B. (1959). Detection, assay, and preliminary purification of the pigment controlling photoresponsive development of plants. *Proc. Natl. Acad. Sci. U.S.* **45,** 1703–1708.

Fetzmann, E. L. (1959). Über rotierende Eigenbewegungen der Zellkerne und Plastiden von *Chara foetida*. *Protoplasma* **49,** 549–556.

Geitler, L. (1932). Der Formwechsel der pennaten Diatomeen (Kieselalgen). *Arch. Protistenk.* **78,** 1–226.

Geitler, L. (1953). Das Auftreten zweier obligater, metagamer Mitosen ohne Zellteilung während der Bildung der Erstlingsschalen bei den Diatomeen. *Ber. deut. Botan. Ges.* **66,** 222–227.

Geitler, L. (1960). Spontane Rotation des Chromatophors, lokalisierte Plasmaströmung und elektive Vitalfärbung in den Haarzellen von Coleochaeten. *Österr. Botan. Z.* **107,** 45–79.

Geitler, L. (1961). Spontaneous partial rotation and oscillation of the protoplasm in *Coleochaete* and other Chlorophyceae. *Am. J. Botan.* **48,** 738–741.
Green, P. B. (1959). Wall structure and helical growth in *Nitella. Biochim. et Biophys. Acta* **36,** 536–538.
Hämmerling, J. (1931). Entwicklung und Formbildungsvermögen von *Acetabularia mediterranea. Biol. Zentr.* **51,** 633–647.
Hämmerling, J. (1939). Über die Bedingungen der Kernteilung und der Zystenbildung bei *Acetabularia mediterranea. Biol. Zentr.* **59,** 158–193.
Hämmerling, J. (1953). Nucleo-cytoplasmic relationships in the development of *Acetabularia. Intern. Rev. Cytol.* **2,** 475–498.
Hämmerling, J. (1955). Über mehrkernige Acetabularien und ihre Entstehung. *Biol. Zentr.* **74,** 420–427.
Haupt, W. (1959a). Chloroplastenbewegung. *In* "Handbuch der Pflanzenphysiologie" (W. Ruhland, ed.), Vol. 17, Part 1, pp. 278–317. Springer, Berlin.
Haupt, W. (1959b). Die Chloroplastendrehung bei *Mougeotia.* I. Über den quantitativen und qualitativen Lichtbedarf der Schwachlichtbewegung. *Planta* **53,** 484–501.
Haupt, W. (1960a). Die Chloroplastendrehung bei *Mougeotia.* II. Die Induktion der Schwachlichtbewegung durch linear polarisiertes Licht. *Planta* **55,** 465–479.
Haupt, W. (1960b). Bewegungen. *Fortschr. Botan.* **22,** 372–393.
Haupt, W., und Schönbohm, E. (1962). Das Wirkungsspektrum der "negativen Phototaxis" des *Mougeotia*-Chloroplasten. *Naturwissenschaften* **49,** 42.
Jahn, T. L., and Rinaldi, R. A. (1959). Protoplasmic movement in the foraminiferan, *Allogromia laticollaris*; and a theory of its mechanism. *Biol. Bull.* **117,** 100–118.
Jarosch, R. (1956). Plasmaströmung und Chloroplastenrotation bei Characeen. *Phyton (Buenos Aires)* **6,** 87–107.
Jarosch, R. (1960). Die Dynamik im Characeen-Protoplasma. *Phyton (Buenos Aires)* **15,** 43–66.
Kamiya, N. (1959). Protoplasmic streaming. *In* "Protoplasmatologia—Handbuch der Protoplasmaforschung." (L. V. Heilbrunn and F. Weber, eds.), Vol. VIII, 3a, Springer, Vienna.
Küster, E. (1956). "Die Pflanzenzelle," 3rd ed. Fischer, Jena, Germany.
Lewis, I. J. (1898). The action of light on *Mesocarpus. Ann. Botany (London)* **12,** 418–421.
Mosebach, G. (1958). Zur Phototaxis von *Mougeotia (Mesocarpus). Planta* **52,** 3–46.
Probine, M. C., and Preston, R. D. (1958). Protoplasmic streaming and wall structure in *Nitella. Nature* **182,** 1657–1658.
Senn, G. (1908). "Die Gestalts- und Lageveränderungen der Pflanzen-Chromatophoren." Engelmann, Leipzig.
Stahl, E. (1880). Über den Einfluss von Richtung und Stärke der Beleuchtung auf einige Bewegungserscheinungen im Pflanzenreich. *Botan. Ztg.* **38,** 297 ff.
Stewart, P. A., and Stewart, B. T. (1959a). Protoplasmic movement in slime mold plasmodia. *Exptl. Cell Research* **17,** 44–58.
Stewart, P. A., and Stewart, B. T. (1959b). Protoplasmic streaming and the fine structure of slime mold plasmodia. *Exptl. Cell Research* **18,** 374–377.
Takata, M. (1958). Protoplasmic streaming in *Acetabularia calyculus.* (In Japanese.) *Kagaku (Tokyo)* **28,** 142. (Cited by Kamiya, 1959.)
Valkanov, A. (1934). Über die kinetische Energie einiger Zellbestandteile. *Protoplasma* **20,** 20–30.
von Denffer, D. (1949). Die planktische Massenkultur pennater Grunddiatomeen. *Arch. Mikrobiol.* **14,** 159–202.

—36—

Gliding

R. JAROSCH

Biological Laboratory, Österreichische Stickstoffwerke A.G., Linz/Donau, Austria

I. Definition

"Gliding" is defined as the active movement of an organism in contact with a solid substratum where there is neither a visible organ responsible for the movement nor a distinct change in the shape of the organism.

II. Occurrence

Gliding is exhibited by many Protista. Among the Blue-green Algae, the hormogones of Nostocales and the trichomes of such genera as *Oscillatoria, Spirulina, Phormidium, Anabaena,* and *Cylindrospermum* generally exhibit this movement. Cells of several colorless cyanophytes (e.g., *Beggiatoa Thiothrix, Leucothrix*), Myxobacteriales, and certain Spirochaetales (*Spirochaeta plicatilis*) glide; see the comparative review by Weibull (1960). Unicellular Blue-green Algae sometimes exhibit small jerking movements (Geitler, 1936). Gliding is further found among those pennate diatoms which possess a true raphe, and among certain species of Euglenophyta (especially those of the *Euglena intermedia* group), the aplanospores of Red Algae, and the protozoan Gregarinae and Labyrinthulae. Gliding with the aid of a flagellum in contact with a solid substratum is sometimes observed in flagellate cells (see Brokaw, Chapter 38, this treatise). Desmids, especially *Closterium*, show a different kind of movement (see Section VII).

III. Relation between Cell Surface and Mode of Movement

The movement may either be a simple gliding (e.g., *Oscillatoria rubescens*, and species of *Anabaena*, *Cylindrospermum*, *Gloeotrichia*, *Isocystis*, *Phormidium*), or it may be combined with a dextral or sinistral rotation around the longitudinal axis, as in other species of *Oscillatoria* (Correns, 1897; Schmid, 1923), and may be correspondingly associated with a fine pattern of markings which run either straight or helically around the axis. In *Euglena* the periplast or pellicle pattern is helical, in Gregarinae it ranges from helical to straight. *Oscillatoria* and *Beggiatoa* may exhibit helical trichome patterns (Geitler, 1936), which can also sometimes be revealed by indirect means, e.g., by observation of the arrangement of negative nodes of plasmolysis (Kuchar 1950; Pernauer, 1958).

Though in most diatoms the raphe is straight, in *Cylindrotheca gracilis* it is helical. The path of motile diatoms seems to be essentially dependent on the shape of the raphe. Nultsch (1956) distinguished at least three types: (i) the *Navicula*-type, with a straight movement; (ii) the *Amphora*-type, in which the path is usually curved; and (iii) the *Nitzschia*-type, which always exhibits curved pathways with two different radii. In addition, Nultsch differentiated between *Schwenkbewegung* (swaying movement), in which the cells turn around one end, and *Drehbewegung* (rotating movement), in which they pivot around the center.

IV. Rate and Relation to the Substrate

Table I shows the rates of gliding for different organisms observed on glass slides at room temperatures. *Nitzschia palea* is able to penetrate into 2% agar and to move inside this medium. A less solid substrate diminishes the ability for gliding; thus the rate of movement of *Nitzschia putrida* is 2.7 μ/sec. on glass, but only 0.8 μ/sec. on agar (Wagner, 1934).

V. Shifting of Adhering Particles

When gliding organisms are immersed in suspensions of particles of carmine, India ink, or indigo (Siebold, 1849), the adhering particles can be seen to migrate on the surface of the cells at a rate often comparable to that of gliding, and sometimes considerably faster, as in some Chlamydobacteriales.

TABLE I

GLIDING RATES OF ALGAE AND OTHER ORGANISMS ON GLASS SLIDES AT ROOM TEMPERATURE

Species	Average rate (μ/sec.)	Maximum rate (μ/sec.)	Reference
Diatoms			
Amphora montana	2–4	5	Nultsch (1956)
Bacillaria paradoxa	7–8[a]	—	Jarosch (1958)
Cymatopleura solea	4–6	—	Jarosch (1958)
Navicula buderi	4–6	8	Nultsch (1956)
Navicula radiosa	14	20	Heidingsfeld (1943)
Nitzschia closterium	5	9	Peteler (1940)
Nitzschia communis	5–7	8.5	Nultsch (1956)
Nitzschia palea	8–10	14	Nultsch (1956)
Nitzschia putrida	—	2.7	Wagner (1934)
Nitzschia stagnorum	5–7	9	Nultsch (1956)
Other organisms			
Beggiatoa sp.	6–8	—	Jarosch (1958)
Oscillatoria limosa	2–3	—	Jarosch (1958)
Spirochaeta plicatilis	1–5	(irregular)	Jarosch (1958)
Euglena deses	2–3	—	Jarosch (1958)
Gregarina polymorpha	5–6	—	Jarosch (1958)

[a] Relative shifting of a living cell against a dead one. The relative movement of two living cells is twice as fast.

A species of *Beggiatoa*, for instance, shows a gliding speed of 6 to 8 μ/sec. and a particle movement of 30 μ/sec. relative to the surface (Jarosch, 1958). The relation between particle movement and gliding in *Oscillatoria* was investigated by Hosoi (1951), Schulz (1955), and Dodd (1960). The particles show no motion if the trichome glides forward, but if the trichome is stopped, the particles move backward at a similar rate. If there is a surface pattern of lines, particles move along them; likewise, on organisms which rotate during gliding, the particles follow a spiral course. Such adhering particles frequently rotate, probably as a consequence of their contact with surface pathways moving at different speeds or in opposite directions. In fact, on the raphes of large diatoms (e.g., *Nitzschia sigmoidea*), particles may sometimes be observed to move in opposite directions simultaneously.

VI. Polarity, Rhythm, and Dependence on External Factors

A. Direction

Among the gliding organisms the Euglenaceae and Gregarinae possess a differentiated anteroposterior axis, and always glide in one direction unless damaged (Günther, 1927; Kümmel, 1958). The other gliding organisms can move in either direction, though a tendency toward an axial polarity has been reported in the movement of certain diatoms (e.g., *Surirella*).

B. Reversing Rhythm and the Influence of Light

Many diatoms exhibit autonomous to-and-fro or "shunting" movements (Heidingsfeld, 1943), in which the direction may alternate at intervals of the order of a minute. Sometimes, under conditions of unilateral illumination, for instance, a topotaxis or phototaxis may be a consequence of more movement in one direction than in the other. On occasion, the reverse movement may be arrested altogether, so that the organism progresses by a series of jerks (Nultsch, 1956). Similar rhythmic reversals of direction have been observed in *Oscillatoria*, though the periods are somewhat longer, about 10 to 20 minutes (Drews, 1957, 1959). In *O. limosa* the speed of the cells gradually decreases before the reversal, after which they accelerate rapidly (Burkholder, 1934).

In unilateral illumination, some organisms tend to migrate in groups, as observed in diatoms by Wagner (1934) and in the trichomes of Blue-green Algae (e.g., *Anabaena variabilis*, *Cylindrospermum licheniforme*) by Drews (1957, 1959).

C. *Bacillaria paradoxa*

The curious diatom *B. paradoxa* forms colonies in which the cells are associated by their valve faces, which are parallel to one another, and shift by sliding against one another with a constant autonomous rhythmic period of about 80 sec. at 20°C. At room temperature the cells may slide past one another at 10 μ/sec. The rhythms of the cells in each colony are generally synchronous, so that the colony extends and contracts as a whole. However, in large colonies synchrony is less accurate; the cells at the ends may be moving in opposite directions relative to their neighbors, so that the colony has a zigzag shape. The period remains constant in strong light, in solutions of altered osmotic concentration, and in the presence of narcotics, but the amplitude of shifting decreases; the period decreases if the temperature is raised. The substance which holds the cells together must be very elastic since an extending colony, suddenly released from a mechanical constraint, springs rapidly into the extended position.

D. Further External Factors

Legler and Schindler (1939) observed that diatom cells in which the cell content had been displaced to one end by centrifugation tended to glide with this end as the anterior pole. The cells often show negative thigmotaxis (Wagner, 1934; Hofmeister, 1940; see Halldal, Chapter 37, this treatise). According to von Denffer (1949), the motility of *Nitzschia palea* is dependent on light. In liquid cultures the cells tend to agglutinate into spherical clumps, loosely held together by mucilage, and to disperse when transferred to a glass slide. These movements do not take place in darkness. Damage by strong light or other adverse chemical or physical agents frequently results in short, jerking movements, whereas plasmolysis generally results in a partial or complete arrest of the gliding movement of diatoms (Höfler, 1940) and *Oscillatoria* (Schmid, 1923; Prat, 1925), which may be reversible (Höfler, 1940; Pernauer, 1958). Brackish-water species, such as *Navicula ostrearia* (Höfler *et al.*, 1956) and *Oscillatoria veneta* (Pernauer, 1958), may continue to exhibit gliding movements even when plasmolyzed. Some effects of temperature on the speed of gliding have been reported by Crozier and Federighi (1924), Crozier and Stier (1926), Heidingsfeld (1943), and Drews (1959). In studies on the effects of pH and temperature on the rate of movement of *Oscillatoria*, Burkholder (1934) found that the mechanism was temperature-dependent and conformed to the Arrhenius equation. The rate can be retarded by narcotics (Jarosch, 1955), and may even be accelerated in certain diatoms by neutral red (Wagner, 1934).

VII. Secretion of Mucilage

Secretions of mucilage can be made visible in high concentrations of Indian ink or carmine. Desmids (especially *Closterium*) clearly progress by the secretion of mucilage through pores at one cell pole (Klebs, 1885–1886). As the volume of this mucilage increases, apparently by swelling, the cell is pushed along (Jarosch, 1958). During true gliding movement, however, mucilage is apparently secreted all over the surface and, though actively shifted parallel to the cell, may not itself be the cause of the cell movement. An *Oscillatoria* filament leaves behind it a mucilage trace interrupted by "windows" (Niklitschek, 1934). A lump of mucilage has been observed to accumulate at the back end of a gliding *Euglena* and to be trailed along with the cell for a while before it is left behind. On the free raphes of large diatoms similar aggregations of mucilage can sometimes be observed in Indian ink preparations (Jarosch, 1958).

In *Phaeodactylum tricornutum*, in which two cell forms occur, respectively oval and biradiate, only the former, which possess a silica valve and raphe, are capable of gliding movements and of producing a gelatinous capsule (Lewin *et al.*, 1958). However, when secretion of mucilage leads to the complete enclosure of the cell within a capsule, as occurs in *Navicula pelliculosa* under certain conditions, the ability to move is lost (Lewin, 1955). According to Niklitschek (1934), certain mucilage-free Oscillatoriaceae can glide only over one another or along old mucilage paths; *Euglena* was also observed to follow such paths (Günther, 1927).

On the surfaces of actively gliding cells of *Euglena*, diatoms, and gregarines, small threads and granules can sometimes be stained with dilute cresyl violet acetate (Cresyl Echtviolett, 100 p.p.m.) (Diskus, 1955; Jarosch, 1958). They are distinguishable from the mucilage proper, and may be in organic connection with the cellular protoplasm.

VIII. Mechanism

Among the numerous hypotheses for the mechanism of gliding, the most firmly established is the concept of "contraction waves" (Schmid 1918, 1923), which have been reportedly observed in *Oscillatoria* and *Beggiatoa* by Phillips (1904), Crozier and Federighi (1924), and Johnson and Baker (1947). Ullrich (1926, 1929) described in *O. sancta* both longitudinal waves, with a wavelength of 25 μ and a period of oscillation of about 2.9 sec., and barely perceptible transverse waves, with amplitudes of 0.24 to 0.30 μ. Similar waves have been reported or suspected in other gliding organisms, e.g., *Euglena* (Günther, 1927; Kamiya, 1939), myxobacteria (Knaysi, 1951; Bisset, 1952; Bisset and Grace, 1954), and gregarines (Richter, 1959), although many authors (e.g., Schulz, 1955) do not believe in their existence.

Gliding of diatoms has been explained by extramembranous activity of the protoplasm (Schultze, 1865; Engelmann, 1879; Müller, 1893–1909), though it exhibits no essential difference from that of other organisms. The shifting of adhering particles, in particular, is similar to the movement of particles in the cell protoplasm, and cannot be explained by a swelling of mucilage, osmotic streaming, etc. Since such particles can sometimes be observed to move in the same direction as the gliding of the cells, Martens (1940) considered that extramembranous activity alone is inadequate to explain the movement of diatoms. However, this discrepancy may be explained by the presence in the same raphe of two streams—not necessarily both visible—moving in opposite directions (Hofmeister, 1940).

Extensive investigations by the author (Jarosch, 1958, 1960) have led to the following hypothesis. The protein fibrils of the protoplasm form a dynamic system of oscillations, like the "polyrhythmic system" of Kamiya (1953, 1959). Submicroscopic transverse waves in a single fibril (arrow W, Fig. 1) cause it to glide in the opposite direction (arrow P). Such waves would be able to shift a great quantity of protoplasm if the fibrils were fixed to the cell wall, e.g., in the protoplasmic circulation in cells of Characeae (Fig. 2). Fibrils fixed to the surface could cause the shifting of a secreted mucilage or a gliding of the cell over the substratum (Fig. 3). Figure 4 shows

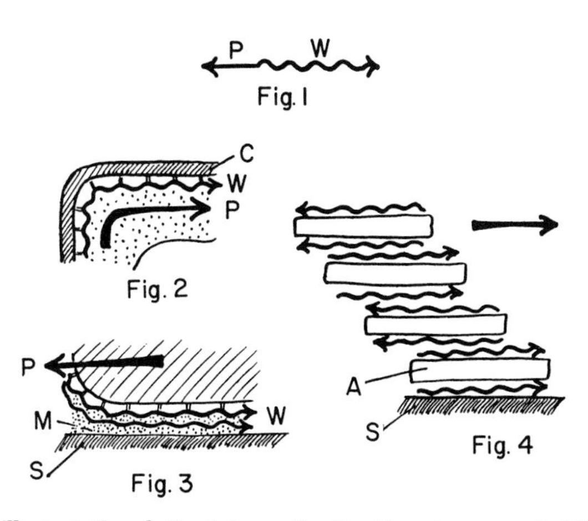

Figures 1 to 4 illustrate the relation between the direction of movement of the submicroscopic transverse waves in the hypothetical protoplasmic fibrils and the consequent gliding movements.

FIG. 1. A single free fibril: the waves running in one direction (W) cause the shifting of the fibril in the opposite direction (P).

FIG. 2. Longitudinal section through a *Chara* cell: the undulations (W) of the fibrils, which are fixed to the cell wall (C), cause the shifting of the inner layer of protoplasm (P).

FIG. 3. Longitudinal section through a gliding organism: (W) direction of waves in extramembranous surface-fixed fibrils; (P) direction of gliding; (M) mucilage; (S) substrate.

FIG. 4. Direction of waves in extramembranous fibrils during a contraction of *Bacillaria paradoxa*: (A) an individual cell; (S) substrate.

the postulated direction of waves in the extramembranous fibrils of *Bacillaria paradoxa* during a contraction of the colony. Although such waves as are visible under the microscope may not themselves be responsible for gliding movements, they could be interpreted as a result of the superposition of submicroscopic waves. It is hoped that such a theory may be ultimately subjected to an experimental test.

REFERENCES

Bisset, K. A. (1952). "Bacteria." Livingstone, Edinburgh and London.

Bisset, K. A., and Grace, J. B. (1954). The nature and relationships of autotrophic bacteria. *In* "Autotrophic Micro-organisms" (B. A. Fry and J. L. Peel, eds.), pp. 28–53. Cambridge Univ. Press, London and New York.

Burkholder, P. R. (1934). Movement in the Cyanophyceae. *Quart. Rev. Biol.* **9**, 438–459.

Correns, C. (1897). Über die Membran und die Bewegung der Oscillatorien. *Ber. deut. botan. Ges.* **15,** 139–148.

Crozier, W. J., and Federighi, H. (1924). Critical thermal increment for the movement of *Oscillatoria. J. Gen. Physiol.* **7,** 137–150.

Crozier, W. J., and Stier, T. J. B. (1926). Temperature characteristics for speed of movement of thiobacteria. *J. Gen. Physiol.* **10,** 185–193.

Diskus, A. (1955). Färbestudien an den Schleimkörperchen und Schleimausscheidungen einiger Euglenen. *Protoplasma* **45,** 460–477.

Dodd, J. D. (1960). Filament movement in *Oscillatoria sancta* (Kütz) Gomont. *Trans. Am. Microscop. Soc.* **79,** 480–485.

Drews, G. (1957). Die phototaktischen Reaktionen einiger Cyanophyceen. *Ber. deut. botan. Ges.* **70,** 259–262.

Drews, G. (1959). Beiträge zur Kenntnis der phototaktischen Reaktionen der Cyanophyceen. *Arch. Protistenk.* **104,** 389–430.

Engelmann, T. W. (1879). Über die Bewegung der Oscillatorien und Diatomeen. *Botan. Zeitung* **37,** 49–56.

Geitler, L. (1936). Schizophyceen. *In* "Handbuch der Pflanzenanatomie" (K. Linsbauer, ed.), pp. 1–138. Borntraeger, Berlin.

Günther, F. (1927). Über den Bau und die Lebensweise der Euglenen. *Arch. Protistenk.* **60,** 511–590.

Heidingsfeld, I. (1943). Phototaktische Untersuchungen an *Navicula radiosa.* Dissertation, University of Breslau, Germany.

Höfler, K. (1940). Aus der Protoplasmatik der Diatomeen. *Ber. deut. botan. Ges.* **58,** 97–120.

Höfler, K., Url, W., and Diskus, A. (1956). Zellphysiologische Versuche und Beobachtungen an Algen der Lagune von Venedig. *Boll. museo civico Venezia* **9,** 63–94.

Hofmeister, L. (1940). Mikrurgische Studien an Diatomeen. *Z. wiss. Mikroskop.* **57,** 259–273.

Hosoi, A. (1951). Secretion of the slime substance in *Oscillatoria* in relation to its movement. *Botan. Mag. (Tokyo)* **64,** 14–17.

Jarosch, R. (1955). Untersuchungen uber Plasmastromung. Dissertation, University of Vienna.

Jarosch, R. (1958). Zur Gleitbewegung der niederen Organismen. *Protoplasma* **50,** 277–289.

Jarosch, R. (1960). Die Dynamik im Characeen-Protoplasma. *Phyton (Buenos Aires)* **15,** 43–66.

Johnson, F. H., and Baker, R. F. (1947). The electron and light microscopy of *Beggiatoa. J. Cellular Comp. Physiol.* **30,** 131–146.

Kamiya, N. (1939). Die Rhythmik des metabolischen Formwechsels der Euglenen. *Ber. deut. botan. Ges.* **57,** 231–240.

Kamiya, N. (1953). The motive force responsible for protoplasmic streaming in the myxomycete plasmodium. *Ann. Rept. Sci. Works, Fac. Sci., Osaka Univ.* **1,** 53–83.

Kamiya, N. (1959). Protoplasmic streaming. *In* "Protoplasmatologia" (L. V. Heilbrunn und F. Weber, eds.), Vol. VIII, Part 3a. Springer, Vienna.

Klebs, G. (1885–1886). Über Bewegung und Schleimbildung der Desmidiaceen. *Biol. Zentr.* **5,** 353–367.

Knaysi, G. (1951). "Elements of Bacterial Cytology." Cornell Univ. Press (Comstock), Ithaca, New York.

Kuchar, K. (1950). Plasmolyseverlauf und Trichomzerfall bei zwei Oscillatorien *Phyton (Graz, Austria)* **2,** 213–222.

Kümmel, G. (1958). Die Gleitbewegung der Gregarinen. Elektronenmikroskopische und experimentelle Untersuchungen. *Arch. Protistenk.* **102,** 501–522.

Legler, F., und Schindler, H. (1939). Zentrifugierungsversuche an Diatomeenzellen. *Protoplasma* **33,** 469–473.

Lewin, J. C. (1955). The capsule of the diatom *Navicula pelliculosa. J. Gen. Microbiol.* **13,** 162–169.

Lewin, J. C., Lewin, R. A., and Philpott, D. E. (1958). Observations on *Phaeodactylum tricornutum. J. Gen. Microbiol.* **18,** 418–426.

Martens, P. (1940). La locomotion des diatomées. *La Cellule* **48,** 277–306.

Müller, O. (1893, 1894, 1896, 1897, 1908, 1909). Die Ortsbewegung der Bacillariaceen. *Ber. deut. botan. Ges.* **11,** 571–576. **12,** 136–143. **14,** 54–64, 111–128. **15,** 70–86. **26a,** 676–685, **27,** 27–43.

Niklitschek, A. (1934). Das Problem der Oscillatorienbewegung. *Beih. botan. Zentr. Abt.* **A 52,** 205–257.

Nultsch, W. (1956). Studien über die Phototaxis der Diatomeen. *Arch. Protistenk.* 101, 1–68.

Pernauer, S. (1958). Das Verhalten einiger Cyanophyceen bei osmotischen Impulsen. *Protoplasma* **49,** 262–295.

Peteler, K. (1940). Bewegungserscheinungen und Gruppenbildungen bei *Nitzschia closterium. Ber. oberhess. Ges. Natur-u. Heilk. Giessen, Naturw. Abt.* **19,** 122–161.

Phillips, O. P. (1904). A comparative study of the cytology and movements of the Cyanophyceae. *Contrib. Botan. Lab. Univ. Pennsylvania* **2,** 237–335.

Prat, S. (1925). Beitrag zur Kenntnis der Organisation der Cyanophyceen. *Arch. Protistenk.* **52,** 142–165.

Richter, J. E. (1959). Bewegungsphysiologische Untersuchungen an polycystiden Gregarinen unter Anwendung des Mikrozeitrafferfilms. *Protoplasma* **51,** 197–241.

Schmid, G. (1918). Zur Kenntnis der Oscillarienbewegung. *Flora (Jena)* **111,** 327–379.

Schmid, G. (1923). Das Reizverhalten künstlicher Teilstücke, die Kontraktilität und das osmotische Verhalten der *Oscillatoria jenensis. Jahrb. wiss. Botan.* **62,** 328–419.

Schultze, M. (1865). Über die Bewegung der Diatomeen. *Arch. mikroskop. Anat.* **1,** 376.

Schulz, G. (1955). Bewegungsstudien sowie elektronenmikroskopische Membranuntersuchungen an Cyanophyceen. *Arch. Mikrobiol.* **21,** 335–370.

Siebold, C. T. (1849). Über einzellige Pflanzen und Tiere. *Z. wiss. Zool.* **1,** 270.

Ullrich, H. (1926). Über die Bewegung von *Beggiatoa mirabilis* und *Oscillatoria jenensis. Z. wiss. Biol. Abt. E. Planta* **2,** 295–324.

Ullrich, H. (1929). Über die Bewegungen der Beggiatoaceen und Oscillatoriaceen. *Planta* **9,** 144–194.

von Denffer, D. (1949). Die planktische Massenkultur pennater Grunddiatomeen. *Arch. Mikrobiol.* **14,** 159–202.

Wagner, J. (1934). Beiträge zur Kenntnis der *Nitzschia putrida* Benecke, insbesonders ihrer Bewegung. *Arch. Protistenk.* **82,** 86–113.

Weibull, C. (1960). Movement. *In* "The Bacteria" (I. C. Gunsalus and R. Y. Stanier, eds.), pp. 180–188. Academic Press, New York.

—37—

Taxes

PER HALLDAL

Department of Plant Physiology, University of Lund, Sweden

I. Introduction

Taxes are movements of whole organisms or of organelles, induced by external stimuli. Different taxes (*phototaxis*, *geotaxis*, etc.) are distinguished according to the type of stimulus. A taxis is *positive* when the stimulus attracts the organisms, and *negative* when it repels them. Tactic movements may be further divided into two types, namely *topic* and *phobic*. In *topotaxis* movement is directed towards or away from the source of the external stimulus, either by an active steering mechanism as in most flagellates and in some Blue-green Algae (Nostocaceae; Drews 1957, 1959), or as a result of "trial and error" motion, as in some diatoms (e.g., *Navicula*; Nultsch, 1956) and Oscillatoriaceae (Drews, 1957). In phobotaxis accumulation or repulsion results from a series of movements dependent on the specific morphology of the organism, under the influence of sudden changes or gradients of stimuli.

Taxes in algae have been reviewed by Haupt (1959; phototaxis, geotaxis, and chemotaxis), Umrath (1959; galvanotaxis), and Bendix (1960; phototaxis).

II. Phototaxis

A. Cyanophyta

Among this group examples of both phobo-phototaxis and topo-phototaxis have been distinguished by Drews (1957, 1959), who showed that the topic response is not the result of a series of isolated phobotactic movements, as may be assumed for flagellates (see below). Both negative and positive responses have been reported for both reaction types, as well as reversals of direction induced by sudden changes in light intensity (Harder, 1920; Burkholder, 1934). *Oscillatoria jenensis* is peculiar in that only negative reactions have been observed in its responses to light (Schmid, 1921).

The phobo-phototactic responses at different wavelengths of light exhibit wide variations from one species to another. Drews (1957, 1959) reported positive responses for *Oscillatoria mougeotii* between 515 and 670 mμ, and for *Phormidium uncinatum* between 610 and 780 mμ. Blue light was ineffective for both species. *Anabaena variabilis* and *Cylindrospermum licheniforme* did not show a phobic response at any wavelengths, though they reacted topotactically to light.

Considerable interspecific variation is also shown in the spectral sensitivity of topo-phototaxis, which has been reported over the whole spectral range of visible light. In experiments by Drews (1957, 1959) *Anabaena variabilis* and *Cylindrospermum licheniforme* reacted positively to yellow and orange-red light. *Phormidium uncinatum* reacted positively between 400 and 610 mμ, and *Oscillatoria mougeotii* between "near ultraviolet" and 730 mμ. Although these data cannot be interpreted as action spectra, such analyses can be expected to reveal a complicated pattern of pigment mediation for both types of reactions. Clearly the environmental conditions, such as radiant energy, temperature, and the composition and pH of the medium before and during such experiments, must be carefully controlled if general conclusions are to be drawn.

B. Flagellates

1. The Stigma (Eyespot)

It is now certain that the stigma is *not* the photoreceptor in phototaxis, since both phobo-phototaxis and topo-phototaxis occur in algae lacking this organelle (e.g., Luntz, 1931; Gössel, 1957; Hartshorne, 1953). It is further assumed, but not proved, that the photoreceptor in *Euglena* is the thickening near the flagellar base, shown by Vávra (1956) to be an independent organ. The stigma is now considered as an auxiliary body which may aid in the orientation of some motile forms, e.g., *Euglena*, possibly by periodically shading the photoreceptor during cell rotation. Theories on the possible mechanisms of different types of stigma have been reviewed by Haupt (1959).

2. Reaction Types

Mast (1911), Jost (1913), Jennings (1910) and others assumed that apparently topic movements of flagellates result from a series of isolated phobic reactions which thereby orient the cells in relation to the light path. Bancroft (1913) and Buder (1919), however, considered phobic and topic responses to be quite distinct, a view also accepted by Bünning and Schneiderhöhn (1956), Gössel (1957) and Haupt (1959). Halldal (1959) showed that motile cells of a number of different algae, when exposed to a gradient of light, ultimately accumulated at a certain light intensity, possibly as a result of phobotaxis. In *Platymonas subcordiformis* either a positive or a negative response was obtained, according to the wavelengths of light (Halldal, 1960). The action spectra for an induced phototactic response, presumably phobotaxis, differ considerably from the action spectrum of topo-phototaxis in this species (see below). It must therefore be concluded that the two reaction types are distinct.

Bünning and Tazawa (1957) observed microscopically the behavior of *Euglena gracilis* on a slide under a coverslip when the cells were subjected to sudden external changes in light intensity or when they swam across a sharp boundary between light and darkness. In experiments performed in this way, phobic responses could be studied without interference from topic reactions, and these workers concluded that the taxis was purely phobic. They postulated that the positive phototaxis could be controlled by the periodic shading of a photoreceptive spot by some more opaque organelle such as the stigma. Gössel (1957), working with a chlorophyll-free strain of this species which still possessed both a stigma and a photoreceptor, found that for some unknown reason the cells exhibited only negative phototaxis under all conditions tested. However, both positive and negative phototaxis occurred in a chlorophyll-free form without a stigma.

3. Action Spectra

In *Euglena*, Bünning *et al.* determined action spectra of negative phobo-phototaxis. That of a chlorophyll-free strain exhibited a maximum at 415 mμ (Gössel, 1957), while tactic activity at 430 mμ was much less. For the normal green strain Bünning and Schneiderhöhn (1956) similarly recorded a maximum at 415 mμ, but as the wavelength increased high activity was retained up to 500 mμ.

Action spectra of topo-phototaxis have been determined for a number of species. Positive phototaxis in an apochlorotic, stigma-less *Euglena* showed a maximum at 415 mμ, as for negative phobotaxis (Gössel, 1957). The action spectrum for positive phototaxis in a green strain of *Euglena* had a maximum around 490 mμ (Bünning and Schneiderhöhn, 1956), in the same region as that obtained by Mast (1917) for both positive and negative phototaxis in *Euglena viridis*. Halldal (1958) determined action spectra of both positive and negative topo-phototaxis for a number of Volvocales and Dinophyceae. The Volvocales all had maxima around 490 mμ. Among the Dinophyceae, *Gonyaulax catenella* and *Peridinium trochoideum* had a maximum at 475 mμ, whereas that of *Prorocentrum micans* was at 570 mμ. Action-spectra determinations in the ultraviolet strongly indicate that the photoreceptive pigment involved in phototaxis is a carotenoprotein (Halldal, 1961).

4. *Mode of Response*

Some of the factors affecting tactic responses in certain Volvocales, in particular *Platymonas subcordiformis*, were examined by Halldal. A positive taxis generally occurred at low light intensities, a negative one in bright light (Halldal, 1959). The balance of the cations Ca, Mg, and K is evidently of importance (Halldal, 1957, 1959), since in a certain combination of these three cations an apparently random motion of the cells was obtained, the population being divided equally into positively and negatively tactic individuals. In such balanced populations, blue light (400 to 540 mμ, max. 430 mμ) or red light (660 to 700 mμ, max. 680 mμ) induced a 90 to 100% positive response, whereas yellow light (560 to 630 mμ, max. 590 mμ) had the reverse effect. All observations were made at a radiant energy of 21,000 ergs per $cm.^2$ per second. At the most effective wavelengths, the induction time for these responses was about 3 minutes (Halldal, 1960). The mode of response and swimming activity were shown to be independent of the total and partial gas pressures of CO_2 and O_2 in the medium (Blum and Fox, 1932; Halldal, 1959).

Halldal (1959) reported that the action of 2,4-dinitrophenol (5×10^{-4} *M* to 10^{-3} *M*) on *Platymonas* reversed the phototaxis from negative to positive, providing indirect evidence that flagellar motility may be dependent on high-energy phosphorylation (Links, 1955; Mayer and Poljakoff-Mayber, 1959; see Brokaw, Chapter 38, this treatise).

Many colony-forming Volvocales, such as *Volvox* and *Pandorina*, and the zoospores and gametes of many other algae, exhibit phototaxis. Rothert (1903) reported that in *Gonium pectorale* and *Pandorina morum* the phototactic responses are sensitive to narcosis by ether or chloroform, though this is not the case in *Chlamydomonas reinhardtii*. If this is so, it might indicate a specific action of the narcotics on the intercellular transmission of flagellar coordination, which would be involved in the tactic movements of colonial flagellates but not in a unicellular alga.

For an extensive survey of this subject the reader is referred to Haupt (1959).

C. Desmids

Cells of desmids move slowly over solid substrates, presumably by the extrusion of slime (cf. Jarosch, Chapter 36, this treatise). The speed of cells of *Micrasterias denticulata*, which may be directed by light, was found to be about 200 to 400 μ per hour (Hygen, 1938), but for *Closterium acerosum* Klebs (1885) recorded speeds up to 112 μ in 30 seconds. In studies of the phototactic behavior of *Closterium moniliferum*, Stahl (1880) noted that at a low light intensity the cells first oriented with the long axis in the direction of the light, and then began to move towards its source. At intermediate intensities, the cells oriented perpendicular to the direction of the light without further movement; while in light of a high intensity (direct sunlight) they reacted negatively. Other desmids behaved similarly, and these results were later confirmed and extended by Aderhold (1888), Pringsheim (1912a), Steinecke (1926), and Kol (1927).

Bendix (1957), who studied the behavior of the desmid *Micrasterias rotata* var. *evoluta* in a projected spectrum, was able to distinguish two types of movement. In one the cells moved in a path perpendicular to the beam, from the long-wavelength end of the spectrum into the green as far as 570 mμ. In the other type they moved in the plane of the beam, and at longer wavelengths this movement was correlated with the red absorption maximum of chlorophyll *a*, implicating this pigment in the response (see Bendix, 1960, p. 178).

D. Diatoms

Positive and negative phobic and topic reactions have been reported to occur in pennate diatoms and have been studied under various qualities of light and environmental conditions (Richter, 1903; Heidingsfeld, 1943; Nultsch, 1956).

All the species studied by Nultsch showed positive topotaxis to "white" light, but for *Navicula radiosa* Heidingsfeld recorded exclusively negative topotaxis at all light intensities. In his spectral sensitivity analyses, Nultsch recorded positive topotaxis between 400 and 550 mμ. The topotactic responses to blue and green light were in agreement with earlier observations by Verworn (1889) and Heidingsfeld (1943).

Using "white" light, Nultsch (1956) recorded positive phobotactic responses in four species; a fifth, *Nitzschia stagnorum*, reacted positively at low light intensities (10 to 1000 lux), and negatively in bright light (5000 to 10,000 lux). *Navicula radiosa*, which in topotactic analyses was negative, showed positively phobotaxis at a high light intensity (2500 to 10,000 lux); a somewhat indistinct reaction at intermediate intensities (100 to 1000 lux), and a slight negative response at low intensities (30 to 50 lux). Spectral sensitivity analyses of phobo-phototaxis reveal a complicated picture. In short-time experiments, *Navicula radiosa* responded negatively to blue light and positively to red, though even in red light after two and a half hours the reaction became negative (Heidingsfeld, 1943). In Nultsch's experiments, *Navicula buderi* exhibited positive phototaxis over the whole spectral range from 400 to 700 mμ; while *Nitzschia palea*, *N. communis*, and *Amphora montana* were positive between 400 and 510 mμ, but usually negative between 550 and 700 mμ. Nultsch assumed that oxygen evolved during photosynthesis, which is proportionately more active in red light, may have distorted the results. Such indirect effects of light and other environmental factors are certainly important in such experiments; a re-investigation of diatom phototaxis with this consideration in mind would be desirable.

III. Geotaxis

Only a few observations on geotaxis are available. Oltmanns (1917) and Mast (1911) recorded positive responses in *Volvox*. Cells of *Platymonas* may gather at the bottom of a cuvette or at the top, depending on environmental conditions (Halldal, unpublished). In the movements of *Closterium* on vertical glass walls, Klebs (1885) and Gerhardt (1913) saw evidence for negative geotaxis, not determined by light (as stated by Aderhold, 1888).

IV. Galvanotaxis

Galvanotaxis has been investigated chiefly with flagellates (see Umrath, 1959). The response is said to be negative or cathodic when the cells tend to move towards the cathode, and positive or anodic when they move to the anode. Both modes of response may occur within the same species, depending on internal and on other environmental factors. Prolonged exposure to the current may reverse the reaction type (Carlgren, 1900).

In *Polytoma uvella*, *Chlamydomonas variabilis*, and *Carteria ovata*, Gebauer (1930) recorded positive galvanotaxis at high partial pressures of O_2 and negative at low. At an intermediate O_2 pressure, the response was positive at low pH values and negative at high, unlike that to be expected of ampholyte colloid particles. The behavior of *Eudorina elegans*, however, was contrary to that of the other flagellates. In *Volvox*, darkness and dim light induced a positive galvanotaxis, whereas in direct sunlight it was negative (Terry, 1909; Bancroft, 1907). Terry also showed an effect of preillumination with different colors upon the galvanic response. Further, Mast (1927) reported that positively phototactic *Volvox* colonies were negatively galvanotactic, and vice versa.

V. Chemotaxis

Chemotaxis means the attraction or repulsion of organisms in concentration gradients of chemical substances. Though it is usually detected under the microscope by the use of a capillary pipette filled with a solution of the agent in question and introduced into a suspension of reactive cells or organisms, this method is subject to certain variations (see Tsubo, 1957). One should also recall that it is easy to simulate a positive chemotactic response by the action of a toxic agent which, since it arrests the movements of the organisms, causes them to accumulate in its presence without actually attracting them (Clayton, 1957).

Among the flagellates, phobo-chemotaxis is the most common and possibly the only existing type (Links, 1955), though topotaxis may be exhibited by fern spermatozoids (Metzner, 1923). Phobic responses have been reported for *Chilomonas* and *Euglena* by Jennings (1910); for zoospores of several different algae by Ulehla (1911); and for *Polytoma* by Links (1955), who questioned Pringsheim's (1923) report of topotaxis in this genus. To account for phobo-chemotaxis, Links postulated that the initial shock reaction results from a rapid decrease in energy available to the motor apparatus, and that any substance or condition which can suddenly and reversibly increase the quantity or consumption of the energy-supplying substance (probably adenosine triphosphate) may thereby induce a chemotactic response.

Pringsheim and Mainx (1926) and Pringsheim (1937) showed that *Polytoma uvella* may exhibit chemotactic reactions to several saturated and unsaturated monobasic organic acids, primary alcohols, esters, aldehydes, and ketones, but not to sugars or ether. In addition, the cells respond chemotactically to oxygen (Pringsheim, 1912b, p. 274) and to a number of inorganic ions, in particular K, Ca, and Mg (see Links, 1955).

In some algae, chemotaxis is an important part of their sexual behavior. Pascher (1931–1932) observed that sperms of *Sphaeroplea* gathered around a cotton thread which had been soaked in egg water from the same species. Cook *et al.* (1948) and Cook and Elvidge (1951) demonstrated that the sperms of *Fucus vesiculosus*, *F. serratus*, and *F. spiralis* were nonspecifically attracted by some volatile agent in *Fucus*-egg water, probably *n*-hexane or a similar hydrocarbon. According to Tsubo's experiments (1957, 1959, and 1961) one of the mating types of an isogamous heterothallic *Chlamydomonas* produces a volatile substance chemotactic for cells of the other mating type. Although the natural substance has not been identified, the gametes of this organism, like those of *C. moewusii* and *C. eugametos*, are attracted by aqueous solutions of coal gas or of simple gases such as ethylene or acetylene.

Hagen-Seyfferth (1959), who carefully reinvestigated the experiments on chemotaxis in *Chlamydomonas eugametos* described by Moewus (1939), questioned the value of the method he employed for studying chemotaxis, and by variations in the experimental procedure was generally able to refute his results. The effect of crocin were not tested, but Hagen-Seyfferth expressed considerable doubt regarding Moewus' claims of the action of this substance upon motility.

VI. Thermotaxis

Thermotaxis is induced by temperature differences and occurs commonly among the Chlorophyta (Reimers, 1927). The reaction is usually negative. *Euglena proxima* and *Astasia klebsii* also show a pronounced thermotactic response, but no thermotaxis has been observed in *Euglena viridis*.

VII. Thigmotaxis

Thigmotaxis is induced by contact; the response is usually negative, as has been reported for *Nitzschia putrida* by Wagner (1934).

References

Aderhold, R. (1888). Beitrag zur Kenntnis richtender Kräfte bei der Bewegung niederer Organismen. *Jena. Z. Naturw.* **22,** 310–342.

Bancroft, F. W. (1907). The mechanism of the galvanotropic orientation in *Volvox. J. Exptl. Zool.* **4,** 157–163.

Bancroft, F. W. (1913). Heliotropism, differential sensitility and galvanotropism in *Euglena. J. Exptl. Zool.* **15,** 383–428.

Bendix, S. W. (1957). Phototaxis in the desmid *Micrasterias rotata* var. *evoluta*. Ph.D. Thesis, University of California.

Bendix, S. W. (1960). Phototaxis. *Botan. Rev.* **26,** 145–208.

Blum, H. F., and Fox, D. L. (1932). Light response of the brine flagellate *Dunaliella salina* with respect to wavelength of light. *Univ. Calif. (Berkeley) Publs. Physiol.* **8,** 21–30.

Buder, J. (1919). Zur Kenntnis der phototaktischen Richtungsbewegungen. *Jahrb. wiss. Botan.* **58,** 105–220.

Bünning, E., and Schneiderhöhn, G. (1956). Über das Aktionsspektrum der phototaktischen Reaktionen von *Euglena*. *Arch. Mikrobiol.* **24,** 80–90.

Bünning, E., and Tazawa, M. (1957). Über die negativ-phototaktische Reaktion von *Euglena*. *Arch. Mikrobiol.* **27,** 306–310.

Burkholder, P. R. (1934). Movement in the Cyanophyceae. *Quart. Rev. Biol.* **9,** 438–459.

Carlgren, O. (1900). Über die Entwiklung des constanten galvanischen Stromes auf niedere Organismen. *Arch. Anat. u. Physiol. Physiol. Abt.*, pp. 49–76.

Clayton, R. K. (1957). Patterns of accumulation resulting from taxes and changes in motility of micro-organisms. *Arch. Mikrobiol.* **27,** 311–319.

Cook, A. H., and Elvidge, J. A. (1951). Fertilization in the *Fucaceae*: investigations on the nature of the chemotactic substance produced by eggs of *Fucus serratus* and *F. vesiculosus*. *Proc. Roy. Soc.* **B138,** 97–114.

Cook, A. H., Elvidge, J. A., and Heilbron, I. (1948). Fertilization, including chemotactic phenomena in the *Fucaceae*. *Proc. Roy. Soc.* **B135,** 293–301.

Drews, G. (1957). Die phototaktischen Reaktionen einiger Cyanophyceen. *Ber. deut. botan. Ges.* **70,** 259–262.

Drews, G. (1959). Beiträge zur Kenntnis der phototaktischen Reaktionen der Cyanophyceen. *Arch. Protistenk.* **104,** 389–430.

Gebauer, H. (1930). Zur Kenntnis der Galvanotaxis von *Polytoma uvella* und einigen anderen Volvocineen. *Beitr. Biol. Pflanz.* **18,** 463–498.

Gerhardt, K. (1913). Beitrag zur Physiologie von *Closterium*. Thesis, University of Jena, Germany. (Cited from Hygen, 1938.)

Gössel, I. (1957). Über das Aktionsspektrum der Phototaxis chlorophyllfreier Euglenen und über die Absorption des Augenflecks. *Arch. Mikrobiol.* **27,** 288–305.

Hagen-Seyfferth, M. (1959). Zur Kenntnis der Geisseln und der Chemotaxis von *Chlamydomonas eugametos* Moewus (*C. moewusii* Gerloff). *Planta* **53,** 376,

Halldal, P. (1957). Importance of calcium and magnesium ions in phototaxis of motile green algae. *Nature* **179,** 215.

Halldal, P. (1958). Action spectra of phototaxis and related problems in Volvocales, Ulva-gametes and Dinophyceae. *Physiol. Plantarum* **11,** 118–153.

Halldal, P. (1959). Factors affecting light response in phototactic algae. *Physiol. Plantarum* **12,** 742–752.

Halldal, P. (1960). Action spectra of induced phototactic response changes in *Platymonas*. *Physiol. Plantarum* **13,** 726–735.

Halldal, P. (1961). Ultraviolet action spectra of positive and negative phototaxis in *Platymonas subcordiformis*. *Plant Physiol.* **14,** 133–139.

Harder, R. (1920). Über die Reaktionen freibeweglicher pflanzlicher Organismen auf plötzliche Änderungen der Lichtintensität. *Z. Botan.* **12,** 353.

Hartshorne, J. N. (1953). The function of the eyespot in *Chlamydomonas*. *New Phytol.* **52,** 292–297.

Haupt, W. (1959). Die Phototaxis der Algen. *In* "Handbuch der Pflanzenphysiologie," (W. Ruhland, ed.), Vol. XVII, Part 1, pp. 318–370. Springer, Berlin.

Heidingsfeld, I. (1943). Phototaktische Untersuchungen an *Navicula radiosa*. Thesis, University of Breslau, Germany. (Unpublished; cited in Nultsch, 1956.)

Hygen, G. (1938). Über die Fähigkeit phototaktischer Bewegung bei den Desmidiaceen. *Bergens Museums Årbok, Naturv. Rekke* **9,** 1–35.

Jennings, H. S. (1906). The behacior of lower organisms. Nordwood, Mass., USA.

Jost, L. (1913). "Vorlesungen über Pflanzenphysiologie" S. Tischer, Jena, Germany.

Klebs, G. (1885). Über Bewegung und Schleimbildung der Desmidiaceen. *Biol. Zentr.* **5,** 353–367.

Kol, E. (1927). Über die Bewegung durch Schleimbildung einiger Desmidiaceen aus der Hohen Tátra. *Folia Cryptogamica* **1,** 435–522. (German summary.)

Links, J. (1955). Een hypothese over het mechanisme van de (phobo-)chemotaxis. Ph.D. Thesis, University of Leyden, Netherlands.

Luntz, A. (1931). Untersuchungen über die Phototaxis. I. *Z. vergleich. Physiol.* **14,** 68–92.

Mast, S. O. (1911). "Light and the Behavior of Organisms." Wiley, New York.

Mast, S. O. (1917). The relation between spectral color and stimulation in the lower organisms. *J. Exptl. Zool.* **22,** 471–528.

Mast, S. O. (1927). Response to electricity in *Volvox* and the nature of galvanic stimulation. *Z. vergleich. Physiol.* **5,** 739–761.

Mayer, A. M., and Poljakoff-Mayber, A. (1959). The phototactic behaviour of *Chlamydomonas snowiae. Physiol. Plantarum* **12,** 8–14.

Metzner, P. (1923). Studien über die Bewegungsmechanik der Spermatozoiden. *Beitr. allgem. Botan.* **2,** 435–499.

Moewus, F. (1939). Über die Chemotaxis von Algengameten. *Arch. Protistenk.* **92,** 485–526.

Nultsch, W. (1956). Studien über die Phototaxis der Diatomeen. *Arch. Protistenk.* **101,** 1–68.

Oltmanns, F. (1917). Über Phototaxis. *Z. Botan.* **9,** 257–338.

Pascher, A. (1931–1932). Über Gruppenbildung und "Geschlechtswechsel" bei den Gameten einer Chlamydomonadine (*Chlamydomonas paupera*). *Jahrb. wiss. Botan.* **75,** 551–580.

Pringsheim, E. G. (1912a). Kulturversuche mit chlorophyllführenden Mikroorganismen. I. Die Kultur von Algen in Agar. *Beitr. Biol. Pflanz.* **11,** 305–333.

Pringsheim, E. G. (1912b). "Die Reizbewegungen der Pflanzen." Springer, Berlin.

Pringsheim, E. G. (1923). Zur Physiologie saprophytischer Flagellaten. *Beitr. allgem. Botan.* **2,** 88–137.

Pringsheim, E. G. (1937). Beiträge zur Physiologie saprophytischer Algen und Flagellaten. *Planta* **26,** 665–691.

Pringsheim, E. G., and Mainx, F. (1926). Untersuchungen an *Polytoma uvella* Ehrb. *Planta* **1,** 583–623.

Reimers, H. (1927). Über die Thermotaxis niederer Organismen. *Jahrb. wiss. Botan.* **61,** 242–290.

Richter, O. (1903). Reinkulturen von Diatomeen. *Ber. deut. botan. Ges.* **21,** 493–506.

Rothert, W. (1903). Über die Wirkung des Aethers und Chloroforms auf die Reizbewegungen der Mikroorganismen. *Jahrb. wiss. Botan.* **39,** 1–70.

Schmid, G. (1921). Beiträge zur Kenntnis der Oscillarienbewegung. *Jahrb. wiss. Botan.* **60,** 572–627.

Stahl, E. (1880). Über den Einfluss von Richtung und Stärke der Beleuchtung auf einige Bewegungserscheinungen im Pflanzenreiche. Bewegung der Desmidieen. *Botan. Z.* **38,** 393–400.

Steinecke, F. (1926). Über das Vorkommen von Gipskristallen bei den Desmidiaceen. *Botan. Arch.* **14,** 312–318.

Terry, O. P. (1909). Galvanotropism of *Volvox. Am. J. Physiol.* **15,** 235–243.

Tsubo, Y. (1957). On the mating reaction of a *Chlamydomonas*, with special references to clumping and chemotaxis. *Botan. Mag.* (*Tokyo*) **70,** 327–334.

Tsubo, Y. (1959). Chemotactic behaviour of the gametes of *Chlamydomonas. Proc. Intern. Botan. Congr., 9th Congr.,* Vol. II, Abstr. 404.

Tsubo, Y. (1961). Chemotaxis and sexual behavior in *Chlamydomonas*. *J. Protozool.* **8,** 114–121.

Ulehla, V. (1911). Ultramikroskopische Studien über Geisselbewegung. *Biol. Zentr.* **31,** 645–654.

Umrath, K. (1959). Galvanotaxis. *In* "Handbuch de Pflanzenphysiologie" (W. Ruhland, ed.), Vol. XVII, Part 1, pp. 164–167. Springer, Berlin.

Vávra, J. (1956). Ist der Photoreceptor eine unabhängige Organelle der Eugleniden? *Arch. Mikrobiol.* **25,** 223–225.

Verworn, M. (1889). "Psycho-physiologische Protisten-Studien." Fischer, Jena, Germany.

Wagner, J. (1934). Beiträge zur Kenntnis der *Nitzschia putrida* Benecke, insbesondere ihrer Bewegung. *Arch. Protistenk.* **82,** 86–113.

—38—

Flagella

C. J. BROKAW

Zoology Department, University of Minnesota, Minneapolis, Minnesota

I. Introduction: Morphology of Flagella and Flagellar Movement[1]

Movement by means of flagella occurs in the algae in diverse situations, ranging from essentially uniflagellate cells (e.g., *Euglena*) to multiflagellate colonies (e.g., *Volvox*). There are no general distinctions between the flagellar movement of algae and that of flagellated protozoa and flagellated cells of higher plants and animals.

The typical flagellum is a long, flexible cylinder, 0.2 μ in diameter and 10 to 40 μ or more in length. The distal end may be thinner or slightly tapered, perhaps only by abrasion. Thin, hairlike, lateral appendages, about 1 μ long, are found on some flagella. These "flimmer" are especially well shown in photographs published by Manton *et al.* (1953). Their functional significance is unknown, although under some circumstances such projections could considerably influence or even reverse the propulsive effect of waves passing along the flagellum (Taylor, 1952).

Studies of flagellar ultrastructure by electron microscopy have shown that a "9 + 2" pattern of longitudinal fibrils is common to all cilia and flagella (Bradfield, 1955; Manton, 1956). Essentially, they all consist of a cylinder of 9 double fibrils surrounding a central pair (see Fig. 1). The structure of the surrounding matrix or sheath is much less clear. In *Euglena,* which has a flagellum bearing flimmer, longitudinal striations have been occasionally seen in the sheath (Brown, 1945; Wolken, 1956). In some of the studies carried out by Manton and her collaborators on flagellated cells of algae it

[1] Abbreviations used throughout this chapter: ADP, adenosine diphosphate; ATP, adenosine triphosphate.

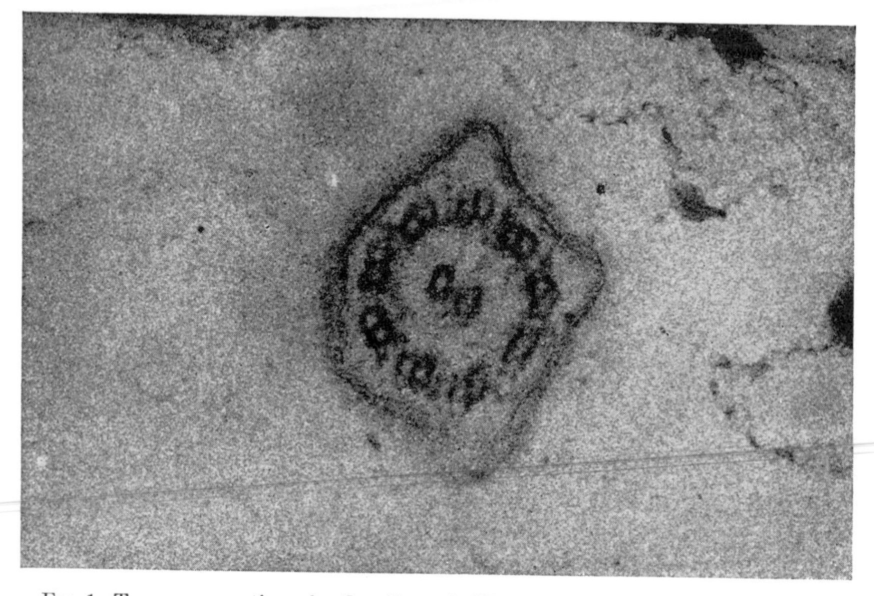

FIG. 1. Transverse section of a flagellum of *Chlamydomonas reinhardtii* (× 120,000).

has been possible to define the plane of orientation of the central pair of filaments with reference to other features of the cell body. In studies of ciliated epithelium (Fawcett and Porter, 1954) it was possible to state that the plane of beat was at right angles to the line joining the two central fibrils of the cilium; but similar information on algal flagella is not yet available.

Flagella are usually attached to a basal granule or blepharoplast (see Fig. 2). In algae there is little evidence that the basal granule is concerned with control of flagellar activity; presumably its major significance is in the synthesis of the flagellum.

No analysis of patterns of flagellar movement in algae has been made with definitive photographic techniques such as those used by Gray (1955) in his study of the movement of the sea-urchin sperm flagellum. In the antherozoid of *Fucus vesiculosus*, where the motility of the cell is primarily due to a posteriorly directed flagellum, the flagellar movement appears similar to that of the sea urchin sperm tail, with uniform sinusoidal waves passing along the flagellum. In algae with paired anterior flagella, such as *Chlamydomonas* and *Polytoma*, the movement has been described as a "breast-stroke" type (Lewin, 1952a), implying asymmetry of beat as in cilia. It may instead be a symmetrical movement similar to that of the sea-urchin sperm tail, with the modification that the flagella are bent posteriorly close to their emergence from the cell (Lowndes, 1941; Brokaw, 1961). In other cases, such as *Euglena*, the flagellar movement is considerably more

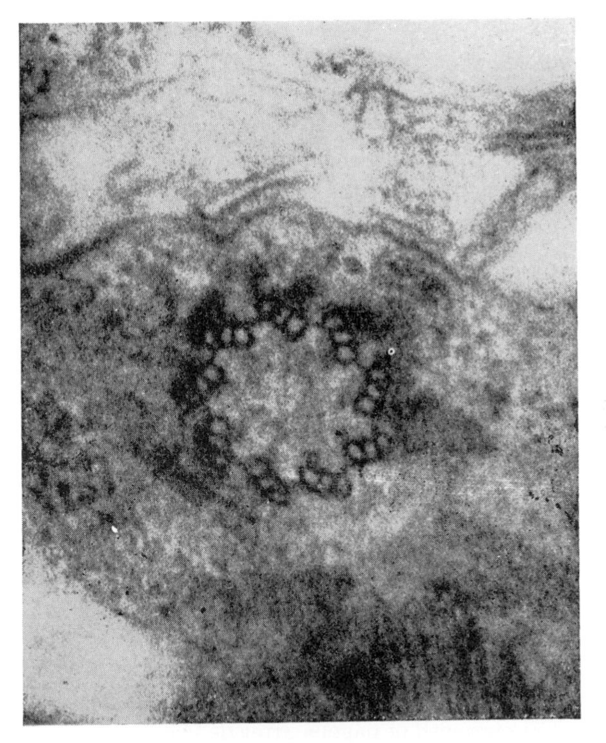

FIG. 2. Transverse section of a flagellar basal body of *Chlamydomonas reinhardtii* (× 120,000). (Both figures are hitherto unpublished electron micrographs kindly provided by Dr. I. R. Gibbons, Biological Laboratories, Harvard University.)

complex, although probably the passage of some form of flectional wave from the base of the flagellum towards the tip is universal (Lowndes, 1941; Brown, 1945).

II. The Movement of Flagellated Algae

The net result of flagellar movement is propulsion, usually in a helical path, at speeds of 50 to 300 μ/second. The efficiency of flagellar movement is low, as has been shown in the case of sperm movement (Gray and Hancock, 1955). In *Polytoma*, only about 10% of the work expended is effective in pushing the cell body through the medium, the remainder being necessary to sustain the lateral oscillation of the flagella against the viscous resistance of the medium (Brokaw, 1961). Variations in the pattern of movement, such as "shock reactions," have been observed in many motile algae and are thought to be the basis for some forms of tactic behavior. The physiological mechanisms responsible for these variations have not been clarified.

Creeping over surfaces by means of the flagella has also been described;

it can be seen particularly clearly in otherwise paralyzed mutants of *Chlamydomonas* (Lewin, 1952b).

Measurements of the swimming velocity of flagellated cells provide quantitative results which can be correlated with alterations in the cellular environment. For instance, studies with *Euglena gracilis* var. *bacillaris* have related the speed of forward swimming to the pH of the medium, revealing optima at pH 5.8 and 7.0 (Lee, 1954a). The swimming speed of cells in acetate-tryptone medium, pH 7.4, after growth at 23°C., had a temperature optimum at 30°C. (Lee, 1954b). The rate could be increased by illumination; the action spectrum of this photokinetic effect suggested absorption by a carotenoid pigment (Wolken and Shin, 1958). However, since these measurements could indicate environmental effects on general cell metabolism or a complex combination of cellular systems, or indirectly operated adaptive-behavior mechanisms, they may have little direct bearing on the physiology of flagellar movement.

More lasting alterations in motility are provided by various culture conditions, and in some cases flagella may be absent. For example, *Chlamydomonas moewusii* loses its flagella during division, and does not form flagella at all when cultured below pH 6 (Lewin, 1952a). When grown on agar, its flagella gradually become shorter after several days in the dark, but they regain their original length in 1 to 2 hours after illumination or after suspension in distilled water (Lewin, 1953). Hagen-Seyfferth (1959) obtained repeated regrowth of flagella after deflagellation by brief exposure to dilute acetic acid or high temperatures. Flagellar growth after suspension in distilled water was independent of illumination, was most rapid in the pH range 5.6 to 8.2, and did not occur below pH 4.6 or above pH 10. Fastest flagellar growth (about 0.3 to 0.4 μ/minute) occurred at 30°C. The biochemistry of flagellar synthesis remains to be studied.

Permanently paralyzed mutants of *C. moewusii* were isolated after ultraviolet irradiation (Lewin, 1952b; see Ebersold, Chapter 49, this treatise). These mutants have flagella which appear morphologically normal (Gibbs *et al.*, 1959), except that some are abnormally short. Flagellar movement is absent or limited to a slight quivering of the distal end. Several distinct mutants have been obtained, and "dikaryons," formed by the sexual union of two different mutants, recovered motility before completing cell fusion (Lewin, 1954). Ronkin (1959) found that the oxygen consumption of paralyzed cells of *C. moewusii* was only about 90% that of normal cells, and suggested that this decrease would be expected if no energy was being used for locomotion.

III. Studies on Isolated Algal Flagella

Study of the biochemistry and physiology of isolated flagella allows a clear distinction to be made between the physiology of the flagella themselves and the physiology of the cells which bear them. Such work would be of limited value if flagella were merely inert filaments operated by an active organelle imbedded in the cell, but the evidence indicates that flagella are themselves independent contractile organelles. Persistence of movement for brief intervals after detachment of flagella from some algae has been known for a long time (Fischer, 1894). Photographic and mathematical analysis of the shape of the contractile wave of the sea-urchin sperm tail has shown conclusively that mechanical energy must be generated along the length of the flagellum (Gray, 1955; Machin, 1958).

Tibbs (1958) removed flagella from *Polytoma uvella* by shaking the cells in distilled water containing a few drops of chloroform. The flagellar material was sedimented and analyzed. It contained about 20% lipid, 6.6 to 8.4% hexose, and negligible amounts of bound phosphorus. The cysteine content of the flagellar protein (0.4%) was significantly lower than that characteristic of actin or myosin [about 1.4% (Bailey, 1954)], but ATPase (adenosine triphosphatase) activity was present. In an earlier paper Tibbs (1957), using less pure flagellar preparations collected by acid precipitation rather than by sedimentation, found that the ATPase activity was maximally activated by 0.006 *M* Mg at pH 7.4.

Jones and Lewin (1960) removed flagella from *Chlamydomonas moewusii* by cooling concentrated cell suspensions to −12°C. for 40 minutes; 60–65% of the dry weight consisted of protein, 6–8% was carbohydrate, and the remainder was presumed to be lipid. They surmised that the insolubility of flagella in typical protein-solubilizing reagents might be attributable to complexed lipids. The amino acid composition roughly resembled that of actin or myosin (Bailey, 1954), except that only a trace of cysteine could be detected.

Brokaw (1961) removed flagella from *Polytoma uvella* by stirring the cells in 50% glycerol containing magnesium ions and thioglycolate, at pH 7.8, and maintaining the preparations at −20°C. This procedure made it possible to recover intact flagella which could be reactivated when placed in solutions containing ATP and suitable concentrations of magnesium and potassium ions. The beat of the reactivated flagella was qualitatively similar to that of normal flagella, with contractile waves passing along the flagellum from the base towards the tip, but it had either a significantly lower frequency or a reduced amplitude, depending on the ATP concentration. Forward progression at speeds of 10 to 20 μ/second was commonly observed; but an undamaged detached flagellum might be expected to swim faster than 135 μ/second, the normal swimming speed of *Polytoma*.

Flagella from *Polytoma* were also collected by high-speed centrifugation for biochemical study. The ATPase activity of the isolated flagella was equal to that calculated to be necessary to sustain the normal movement of the cell, so that the efficiency of the movement of the isolated flagella was only a small fraction of its normal value. An adenylate kinase responsible for dephosphorylation of ADP and movement in ADP solutions was also demonstrated, and the nucleotide specificity of the ATPase activity was studied.

Similar preparations of isolated flagella were prepared from *Chlamydomonas moewusii* (Brokaw, 1960). Flagella from a paralyzed mutant could not be activated in ATP solutions, and had only 30–40% of the ATPase activity of flagella from normal cells, indicating that the lesion responsible for paralysis affected the conversion of chemical energy into mechanical work actually within the flagella.

Mintz and Lewin (1954) isolated flagella from *C. moewusii* by treatment with hydrochloric acid at pH 3.0 and used them to prepare antibodies, but no serological differences could be found between flagella from paralyzed and normal cells.

IV. The Mechanism of Flagellar Motility

Current ideas of the mechanism of flagellar motility involve the assumption that a flagellum contains contractile elements in some respects similar to muscle, although the evidence available shows no more similarity between flagellar protein and muscle protein than might be expected simply because both use ATP as an energy source. The contractile elements are presumed to go through a spontaneous cycle of contraction and relaxation using energy derived from the dephosphorylation of ATP which is somehow supplied from the cell body. The cycle may be mechanically triggered, as in the rapid oscillation of insect flight muscle (Pringle, 1957). Its biochemical basis is unknown, but it may perhaps be revealed eventually by continued studies of isolated algal flagella.

References

Bailey, K. (1954). Structure proteins. II. Muscle. *In* "The Proteins" (H. Neurath and K. Bailey, eds.), Vol. II, pp. 951–1055. Academic Press, New York.

Bradfield, J. R. G. (1955). Fibre patterns in animal flagella and cilia. *Symposia Soc. Exptl. Biol.* **9,** 306–334.

Brokaw, C. J. (1960). Decreased adenosine triphosphatase activity of flagella from a paralyzed mutant of *Chlamydomonas moewusii*. *Exptl. Cell Research* **19,** 430–431.

Brokaw, C. J. (1961). Movement and nucleoside polyphosphatase activity of isolated flagella from *Polytoma uvella*. *Exptl. Cell Research* **22,** 151–162.

Brown, H. P. (1945). On the structure and mechanics of the protozoan flagellum. *Ohio J. Sci.* **45,** 247–301.

Fawcett, D. W., and Porter, K. R. (1954). A study of the fine structure of ciliated epithelium. *J. Morphol.* **94,** 221–281.

Fischer, A. (1894). Über die Giesseln einiger Flagellaten. *Jahrb. wiss. Botan.* **26,** 187–235.
Gibbs, S. P., Lewin, R. A., and Philpott, D. E. (1959). The fine structure of the flagellar apparatus of *Chlamydomonas moewusii. Exptl. Cell Research* **15,** 619–622.
Gray, J. (1955). The movement of sea-urchin spermatozoa. *J. Exptl. Biol.* **32,** 775–801.
Gray, J., and Hancock, G. J. (1955). The propulsion of sea-urchin spermatozoa. *J. Exptl. Biol.* **32,** 802–814.
Hagen-Seyfferth, M. (1959). Zur Kenntnis der Geisseln und der Chemotaxis von *Chlamydomonas eugametos* Moewus (*C. moewusii* Gerloff). *Planta* **53,** 376–401.
Jones, R. F., and Lewin, R. A. (1960). The chemical nature of the flagella of *Chlamydomonas moewusii. Exptl. Cell Research* **19,** 408–410.
Lee, J. W. (1954a). The effect of pH on forward swimming in *Euglena* and *Chilomonas. Physiol. Zoöl.* **27,** 272–275.
Lee, J. W. (1954b). The effect of temperature on forward swimming in *Euglena* and *Chilomonas. Physiol. Zoöl.* **27,** 275–280.
Lewin, R. A. (1952a). Studies on the flagella of algae. I. General observations on *Chlamydomonas moewusii* Gerloff. *Biol. Bull.* **103,** 74–79.
Lewin, R. A. (1952b). Ultraviolet induced mutations in *Chlamydomonas moewusii* Gerloff. *J. Gen. Microbiol.* **6,** 233–248.
Lewin, R. A. (1953). Studies on the flagella of algae. II. Formation of flagella by *Chlamydomonas moewusii* in light and in darkness. *Ann. N. Y. Acad. Sci.* **56,** 1091–1093.
Lewin, R. A. (1954). Mutants of *Chlamydomonas moewusii* with impaired motility. *J. Gen. Microbiol.* **11,** 358–363.
Lowndes, A. G. (1941). On flagellar movement in unicellular organisms. *Proc. Zool. Soc.* (*London*) **111A,** 111–134.
Machin, K. E. (1958). Wave propagation along flagella. *J. Exptl. Biol.* **35,** 796–806.
Manton, I. (1956). Plant cilia and associated organelles. *In* "Cellular Mechanisms in Differentiation and Growth" (D. Rudnick, ed.), pp. 61–72. Princeton Univ. Press, Princeton, New Jersey.
Manton, I., Clarke, B., and Greenwood, A. D. (1953). Further observations with the electron microscope on spermatozoids in the brown algae. *J. Exptl. Botany* **4,** 319–329.
Mintz, R. H., and Lewin, R. A. (1954). Studies on the flagella of algae. V. Serology of paralyzed mutants of *Chlamydomonas. Can. J. Microbiol.* **1,** 65–67.
Pringle, J. W. S. (1957). Myogenic rhythms. *In* "Recent Advances in Invertebrate Physiology" (B. T. Scheer, ed.), pp. 99–115. University of Oregon, Eugene, Oregon.
Ronkin, R. R. (1959). Motility and power dissipation in flagellated cells, especially *Chlamydomonas. Biol. Bull.* **116,** 285–293.
Taylor, G. I. (1952). Analysis of the swimming of long and narrow animals. *Proc. Roy. Soc.* **A214,** 158–183.
Tibbs, J. (1957). The nature of algal and related flagella. *Biochim. et Biophys. Acta* **23,** 275–288.
Tibbs, J. (1958). The properties of algal and sperm flagella, obtained by sedimentation. *Biochim. et Biophys. Acta* **28,** 636–637.
Wolken, J. J. (1956). Molecular morphology of *Euglena gracilis* var. *bacillaris. J. Protozool.* **3,** 211–221.
Wolken, J. J., and Shin, E. (1958). Photomotion in *Euglena gracilis.* I. Photokinesis. II. Phototaxis. *J. Protozool.* **5,** 39–46.

—39—

Laboratory Cultures

JACK MYERS

Departments of Botany and Zoology, University of Texas, Austin, Texas

I. Introduction

Liquid cultures of algae may be managed for many different purposes and accordingly the rationale of design varies between wide limits. Beyond the problem of maintenance, as in stock cultures, there are commonly two categories of objectives: (1) the production of algae in such quantities as may be desired for further study of their cytological, physiological, or biochemical characteristics; and (2) the study of growth characteristics of the algae.

II. Growth Characteristics

Presented diagrammatically in Fig. 1 is a typical growth curve for a batch culture of an alga growing with an adequate supply of CO_2 and nutrient salts, at an appropriate temperature, and under any specified illumination and form of the culture vessel. In spite of its apparent similarity to growth curves of other microorganisms, the important factor of illumination governs the shape of the curve for photosynthetic algae. Throughout the growth curve the increasing quantity of cells causes mutual shading to increase, so that the effective illuminance per cell, I_c, becomes progressively lower than the incident illuminance, I_0.

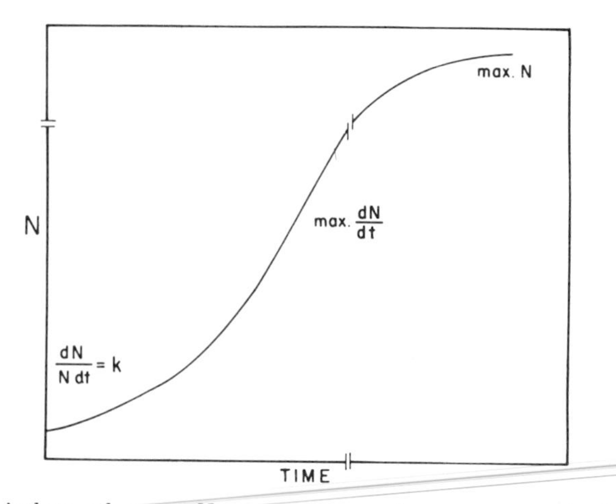

FIG. 1. A typical growth curve. N represents some chosen measure of cell quantity.

The consequences of mutual shading may be interpreted from Fig. 2, which shows the dependence of the specific growth rate on the effective illuminance, I_c. Figure 2 is also drawn as a generalized curve based upon experimental data for very sparse cultures in which I_0 and I_c were almost identical (Phillips and Myers, 1954; Sorokin and Krauss, 1958). A compensation point for growth, I_b, has not been demonstrated experimentally, but is inferred with confidence from characteristics of algal metabolism.

Three portions of the growth curve (Fig. 1) may be interpreted in terms of the effective illumination, as follows. At low cell concentrations, as long as mutual shading is negligible, the increase in cell quantity is *exponential* and the specific growth rate is constant: $dN/Ndt = k$. For a high incident illuminance the exponential phase may continue in spite of mutual shading

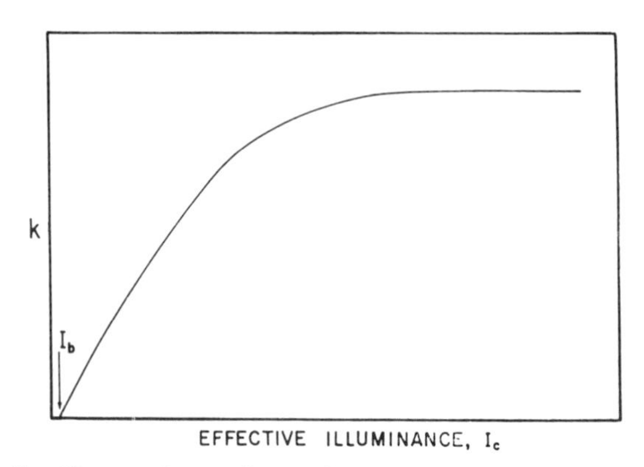

FIG. 2. Specific growth rate, k, as a function of effective illuminance, I_c.

as long as all cells are maintained above light saturation. As the cell concentration increases and the fraction of light absorbed approaches 100%, the approximately constant energy income of the culture is reflected in a *linear* increase in cell quantity, i.e., dN/dt is constant. Thereafter, any further increase in cell quantity merely increases the overhead demands of basal or endogenous metabolism. Whatever the incident illumination, the effective illumination, I_c, will eventually approach a value I_b (Fig. 2) and the quantity of cells in the culture will approach a maximum.

It will be recognized that the above discussion is abbreviated and ignores several different second-order effects. More complete treatments of the kinetics of algal growth in relation to illumination have been presented by Myers (1953), van Oorschot (1955), Talling (1960), Tamiya (1951), and Tamiya *et al.* (1953a).

The three labeled portions of the growth curve of Fig. 1, providing three different kinds of information about an algal culture, may be drastically altered by changing the specified parameters. The following general remarks may be made about the three different criteria of growth.

The *specific growth rate*, $k = dN/Ndt$, provides an intrinsic measure of the rate of total metabolism leading to cell synthesis. It is a first-order reaction constant with dimensions of time^{-1}. Some authors use a statement of generation time or doubling time, $t_2 = 0.69/k$, apparently preferred as a measure more readily interpreted. Other authors have used a logarithmic basis since the integrated form $\log_{10}(N/N_0) = k't$, in which $k' = k/2.3$, allows one conveniently to follow exponential growth as $\log_{10}N$ plotted against time ($\log_e 2 = 0.69$; $\log_e 10 = 2.3$).

The specific growth rate provides a useful measure of growth in response to such factors as light intensity (Fig. 2), temperature, or added organic substrates. Furthermore, it can provide a criterion of the adequacy of inorganic nutrition; any algal species requires an optimum medium to sustain a maximum specific growth rate.

A second criterion of growth, which might be labeled *production*, is the rate of increase in cell quantity with time, dN/dt. It has little general theoretical value, since it describes the total culture including such empirical factors as the shape of the culture vessel and its illumination; it may be of practical importance whenever the objective is simply the maximum production of algal cells, as in large-scale mass cultures. As noted by Ketchum *et al.* (1949), and confirmed incidentally in subsequent work of others, the maximum production in a culture never occurs during exponential growth while the specific growth rate is maximum, but is reached at some cell density at which the product of cell quantity and specific growth rate (kN) is maximum.

A third criterion of growth, *yield*, represents the maximum quantity of cells achievable in a specific set of circumstances. Under the conditions set for Fig. 1, the yield for a given alga depends on the incident illumination and the form of the vessel. Higher yields can be obtained when thinner layers of cell suspension are illuminated, as is predictable from Beer's Law and as was demonstrated, at least in crude fashion, by Myers *et al.* (1951). Under other culture conditions the yield may be a measure of the nutrient requirements of a given alga. It has been used to establish the essentiality of certain trace elements, e.g., the molybdenum requirement of *Scenedesmus* (Arnon *et al.*, 1955), and to estimate the efficiency of cell synthesis from glucose by *Chlorella* in the dark (Samejima and Myers, 1958).

III. Production of Experimental Material

For critical biochemical or physiological studies of an alga, one should start with experimental material grown under definable conditions. For some problems, as in taxonomic or cytological work, the quantity required is small, but for others, as in biochemical work on nucleic acids, the requirement may be large. There is always a necessity for defining the conditions under which the cells were developed. Most algae are not notably mutable, but in many forms the cell composition, particularly the pigmentation, is highly variable and sensitive to culture conditions. The author is, perhaps, particularly aware of this problem because of his own early bitter experiences. However, the increasing sophistication of experimental work places increasing emphasis on culture conditions.

Attention should be called to the limitations of the procedure, fortunately no longer common, of using algae from an agar culture for experimental work. When a layer of algae of any thickness has developed on an agar surface, the lowermost cells are drastically shaded while those of the top layer are deficient in nutrients as a result of diffusion limitations. The culture conditions for cells washed off an agar surface are thus not definable.

With reference to the preceding discussion of growth characteristics, questions can be raised as to the most desirable point of harvest on the growth curve (Fig. 1). As a rule one should choose some point during exponential growth during which period the specific growth rate, k, is constant, and only N, the number of cells present, is changing. Unfortunately, as previously noted, maximum production is not achieved under these conditions, and therefore the time of harvest is usually chosen at a later period of greater production. In any one laboratory, a standardized regimen of inoculation and harvesting can be made to yield practically reproducible results. A far more elegant procedure is the establishment of steady-state conditions, to be discussed below, which can be set up to provide material harvested from any chosen point on the growth curve.

For the production of cells deficient in some specified nutrient, a harvest at or near the point of maximum yield (the so-called stationary phase) may be desired. For such purposes it must be confirmed that the maximum yield is established by no other factor than the limiting supply of the nutrient in question. This technique has been used extensively, notably by Pirson (1955), to describe physiological effects of nutrient deficiency. Unless the limitation determining the maximum yield can be clearly specified, or if no such limitation is desired, it is inadvisable to harvest the cells at the upper portion of the growth curve.

IV. Methods of Culture

A. Media

Innumerable formulas exist for media, variously selected by different laboratories and then carried down through the years. For some exacting algae nutrient requirements have been carefully determined (Provasoli and Pintner, 1956). More commonly, a previous recipe is modified empirically until it satisfies some test of desirability for a newly investigated alga. The result is that most published media meet the test that "they work," but for few can one assert that all of the components are really required in the concentrations specified. A further difficulty is that media such as "Chu No. 10" (Chu, 1942), originally developed from ecological considerations, may admirably promote growth at low cell concentrations but may exhibit severe deficiencies when used to develop such higher cell densities as may be desired for experimental work.

A résumé of the various algal media would have only doubtful significance, but some general observations on the problems of medium construction are possible. These may be resolved into the following component problems: (1) major elements and total salt concentration; (2) nitrogen source; (3) pH; (4) trace elements; (5) organic growth factors; and (6) stability. The major elements or ions always comprise potassium, magnesium, sulfate, and phosphate, and sometimes include sodium and calcium. Lower values of pH are preferred since they tend to prevent the precipitation of calcium, magnesium, and iron phosphates in particular. Ammonia, nitrate, and urea are widely but not universally used by algae as nitrogen sources. Nitrate uptake is attended by an increase in pH; ammonia uptake, by a pH decrease. The use of urea is often advantageous, though it is partially hydrolyzed by autoclaving. The requirement for nitrogen is greater than that for any other element except carbon. Since most unicellular algae contain about 8% nitrogen, each gram/per liter (dry weight) of cells produced requires 80 mg./liter of nitrogen, i.e., 580 mg./liter of KNO_3.

Trace elements have usually been provided in mixtures found empirically to be effective; however, the necessity for each individual component has not always been proved. Iron, zinc, manganese, sodium, calcium, molybdenum, copper, cobalt, and vanadium have been demonstrated as essential for one or more algae. However copper, cobalt, and vanadium are required only in concentrations measured in micrograms per liter, and demonstration of their requirement demands extreme purification of the water and of all the nutrient salts employed; their routine addition may be justified as a precaution, but not as a practical necessity. An important aid to the stabilization in solution of trace elements is the use of chelating agents, which form easily reversible complexes and therefore effectively buffer most of the essential cations. Citrate has been used commonly. Ethylenediamine tetra-acetic acid has the advantage that it is metabolically almost inert for all organisms and does not support the growth of contaminants in non-sterile media (Hutner *et al.*, 1950). Many algae have growth-factor requirements, notably for thiamine and cobalamin (see Droop, Chapter 8, this treatise).

B. Culture Vessels

Vessels for algal cultures have ranged from test tubes to concrete tanks. A common design for growing quantities in the laboratory has been based on an Erlenmeyer flask equipped with inlet and outlet tubes for aeration. Frequently the flasks are held at a constant temperature in a water bath, shaken to prevent settling, and illuminated from below. A mechanical arrangement for such a system has been described by Fogg *et al.* (1959). The use of large, squat flasks with provisions for continuous culture has been recommended by Benson *et al.* (1949). Arnon *et al.* (1955) have used Roux culture bottles laid horizontally in order to expose only a thin layer of algae to vertical illumination. The author prefers the simplicity of test tubes of 25 to 250 ml. capacity immersed in a glass-walled water bath and illuminated from the side (Myers, 1950). A glass tube, containing a cotton air filter and inserted through a large cotton plug, provides aeration and thereby sufficient agitation to prevent the cells from settling.

Larger vessels such as carboys have been used for the production of larger quantities (Scott, 1943; Spoehr *et al.*, 1949; Myers *et al.*, 1951). Krauss and Thomas (1954) used an open 300-liter vat of 20 cm. fluid depth. In such vessels the extremely thick layer of suspension severely limits the cell concentrations which can be obtained. Vertical glass columns 10 cm. in diameter have also been used (Reisner and Thompson, 1956). A very thin layer of cell suspension can be illuminated in an all-glass apparatus patterned essentially upon a Liebig condenser (Myers and Clark, 1944). Although originally designed for an automatic steady-state culture system, such vessels can be managed by periodic manual harvesting and dilution. For example, a "sleeve" chamber, 66 cm. long and 6 cm. outside diameter, with an effective volume of 400 ml., illuminated by four 40-watt fluorescent lamps, can produce 2.5 gm. dry weight of *Chlorella* cells per day.

C. Illumination

For the growth of algae in the laboratory, illumination is commonly provided by either fluorescent or tungsten lamps. The former have the advantage of greater efficiency for visible light output. Tungsten lamps have the advantage of a "cleaner" emission spectrum; in fluorescent lamps the mercury lines persist in addition to the spectral output of the phosphor. Some questions of unfavorable effects of fluorescent lamps have been raised (e.g. Algéus, 1951), and it is possible that sufficient of the shorter wavelengths of the mercury lines (e.g., 365 mμ) may penetrate the glass envelope to retard algal growth; but in the author's laboratory no such unfavorable effects of fluorescent lamps have been observed. The "soft white" phosphors with higher energy at longer wavelengths appear to be more efficient, as measured by algal production per watt of electrical input.

In order to minimize the apparent thickness of an algal suspension and thereby to obtain better penetration by light, it is desirable in practice to use multidirectional illumination. For example, a culture in a test tube or vertical column may be illuminated by a surrounding bank of tubular lamps. In such a case the illuminance or irradiance received by the culture, measured in any absolute units such as lux or μwatts/cm.^{2}, is not easily specified, since most light-measuring devices are designed for unilateral illumination. Even a single bank of fluorescent lamps does not give strictly unidirectional (i.e., parallel) light; a photocell and a culture vessel may "see" the illumination quite differently. Insufficient attention has been devoted to this problem, with the result that light intensities as measured in different laboratories are not necessarily comparable.

D. Provision of Carbon

The convenient microbiological practice of providing no other aeration than gaseous exchange by diffusion through cotton plugs is often inadequate for the culture of algae. Provision of sufficient oxygen by diffusion from the normal high concentration in air is far more easily achieved than the provision of carbon dioxide. Carbon dioxide limitation is not always important, but aeration by air enriched with carbon dioxide is usually advantageous. The production of 1 mg. of algal dry weight requires an uptake of about 0.5 mg. of carbon, which is equivalent to 1.8 mg. or about 1.0 ml. of CO_2 under laboratory conditions.

In the cultivation of algae for the study of photosynthesis, aeration with about 5% CO_2 in air has been the almost universal practice following the original suggestion of Warburg. In *Chlorella* the rate of growth is CO_2-saturated, or nearly so, at a dissolved-CO_2 concentration in equilibrium with that in air (Steemann Nielsen, 1955). However, in an illuminated algal suspension the medium is not in equilibrium with the gas phase even when aerated vigorously; there is a CO_2 gradient between the gas and liquid phases, proportional to the rate of CO_2 uptake by the algae. Although 5% CO_2 in the aerating gas mixture was arbitrarily chosen to ensure a continually adequate diffusion gradient, for most purposes 0.5% CO_2 would be equally satisfactory. In a batch culture aerated at a constant rate, the effective concentration of CO_2 in solution decreases as the algal production rate increases. A tacit assumption in the use of 5% CO_2 is that characteristics of the cells are independent of the CO_2 concentration in the medium. Convincing evidence to the contrary is lacking, although Steemann Nielsen (1955) has questioned the advisability of using higher CO_2 concentrations.

In laboratory studies designed to obtain data applicable to ecological conditions, the investigator faces an apparent dilemma with respect to CO_2 just as serious as that with respect to nutrient concentrations. If he chooses aeration with 5% CO_2 he faces criticism for his choice of an unnaturally high concentration. If, on the other hand, he chooses aeration with air and permits the development of measurable cell concentrations, the CO_2 concentration may be reduced to levels below those commonly encountered in natural waters, where cell concentrations are typically low and CO_2 is rarely a limiting factor. It would appear, therefore, that air enriched with CO_2 would be preferable for laboratory cultures even when the experiments are directed toward ecological problems.

For most algae carbon may also be provided as organic solutes such as glucose or acetate in the medium, rather than as carbon dioxide in the gas phase. Technically the simplest method for producing *Chlorella* in large quantities requires only large cotton-stoppered flasks, the provision of glucose, light (not necessarily bright), and aeration or mechanical agitation. The widely used vitamin B_{12}-assay procedures developed by Hutner and his co-workers (e.g., Hutner *et al.*, 1956) achieve simplicity in the use of 10 ml. or 25 ml. glass-capped flasks for cultures of *Euglena gracilis*.

E. Measurement of Cell Quantity

In the preceding discussion, dimensions for the measurement of cell quantity have been purposely left undefined. Under steady-state growth conditions all possible indices of cell quantity should yield equivalent results. However, in a batch culture the character of the growth curve may vary considerably, depending on the index used (cf. Arnon *et al.*, 1955). The choice of an index therefore depends on the kind of question asked.

In early work on algal physiology the *cell number*, as determined by counting cells with a hemocytometer, was the preferred index of quantity. The first excellent analysis of algal growth rates by Bristol-Roach (1926) was based on such cell counts and on estimates of cell size in *Scenedesmus*. Subsequently other indexes, such as cell volume and dry weight, came into use, partly because they proved to be less laborious though of equal precision. Recent attention to the phases of the life cycle and the possibilities of synchronized cell division (see below) has placed new importance on direct determinations of cell size and number.

The most direct index of cell quantity is the *dry weight*. This may be conveniently determined by centrifuging an aliquot of suspension containing 5 to 50 mg. of cells, washing the cells once in water to remove occluded salts, transferring them in a minimum volume of water to a tared dish, and drying them at 100°C to constant weight. (Preliminary drying at 90°C will prevent loss by spattering.) For certain fragile algae, such as *Ochromonas*, the washing must be eliminated and appropriate corrections made for contaminant salts.

Determinations of *cell nitrogen*, often used as an index of cell quantity for other microorganisms, might be a preferred method for algae such as *Porphyridium cruentum*, the cells of which have large capsular sheaths.

Cell volume, although not favored in work with other microorganisms, has been a commonly used index of algal cell quantity. Centrifuge tubes with a capacity of about 5 ml., and with the bottom end of precision-bore capillary tubing calibrated to an internal volume of about 0.050 ml., are convenient for this purpose. For dilute suspensions a large aliquot may be centrifuged in ordinary vessels and the cells then transferred quantitatively to the capillary tube. Sufficient centrifugal force and time must be employed to obtain a constant and minimum packed cell volume; for *Chlorella pyrenoidosa* one hour of centrifugation at a relative centrifugal force of 2500 is adequate. The relation between dry weight and cell volume for many unicellular algae falls in the range of 0.22 to 0.27 gm./ml.

The *optical density* of an algal suspension, as determined by any precise colorimeter, can be used as a routine measure of relative cell quantity. Depending upon the design of the instrument, the observed optical density is determined principally either by absorption (colorimetric) or by scattering (turbidometric) (cf. Mestre, 1935). When the cell pigmentation and size vary, as they commonly do during the lifetime of a batch culture, the optical density is not a constant function of cell volume or dry weight, even for any one instrument. However, under steady-state conditions, even as they are approached during early exponential growth in a batch culture, the optical density is generally a reliable relative measure of cell quantity. For many purposes, convenient culture vessels are test tubes accepted by instruments such as the Evelyn or the Bausch & Lomb colorimeters. With sufficient attention to detail, such as the selection of optically uniform test tubes, a plot of the logarithm of optical density against time gives the expected linear relation within the range of the instrument. The method is adapted to the development of an optimum medium (Kratz and Myers, 1955) and to studies of the gross effects of temperature (Sorokin and Myers, 1953) or light intensity (Sorokin and Krauss, 1958); it may even be used to provide a crude estimate of the number of viable cells (Redford and Myers, 1951).

V. Reproducibility and Uniformity

In Section I, attention was directed to the changing environmental conditions associated with the growth of cells in a batch culture. The possibility of stabilizing the environmental conditions is presented by a steady-state culture device. The earliest form of such a device as applied to algae was that of Myers and Clark (1944). In principle such a continuous-culture apparatus holds a culture at some chosen point on its growth curve by dilution of the suspension controlled by a photometric system. In its original form the culture vessel was a vertical sleeve-shaped chamber illuminated by vertical tubular lamps in such a way that the effective illumination was independent of the total volume of the suspension. The cells were manually harvested at intervals, leaving a small volume of suspension as inoculum. The apparatus has been applied successfully to the culture of species of *Chlorella, Scenedesmus, Euglena, Anabaena, Anacystis,* and other microscopic algae. Such cultures can serve for (1) the production of reproducible cell material day after day, and (2) the study of cellular characteristics as a function of some single environmental factor which is purposely varied (Myers, 1946).

The controlled-dilution principle has also been applied to a smaller culture vessel with a constant-level overflow (Phillips and Myers, 1954). This apparatus is more easily managed and simpler to use for the study of specific growth rates as a function of illuminance.

A kind of inverted steady-state device is the chemostat of Novick and Szilard (1950) and Monod (1950), which employs a constant dilution rate. In the chemostat, operation is stable only if illuminance or the concentration of some component of the medium limits the specific growth rate. It would be useful for producing algae with a constant nutrient deficiency. It has been used to study the efficiency of cell production by *Chlorella* under illuminance equivalent to sunlight (Myers and Graham, 1959).

Steady-state culture devices solve the problem of providing reproducible experimental material, since the harvested suspensions contain a constant frequency distribution of cells at different stages in the life cycle. The cells are not homogeneous with respect to "age," however, and may not be homogeneous in metabolic characteristics, as demonstrated by Neeb (1952) and by Pirson and Doring (1952) for *Hydrodictyon*, by Tamiya *et al.* (1953b) for *Chlorella ellipsoidea*, and by Sorokin and Myers (1957) for a high-temperature strain of *Chlorella*. A discussion of synchronous cultures and of phenomena attending cell division is presented by Hase (see Chapter 40, this treatise). Reproducibility and uniformity of experimental material are clearly not identical problems. It is probable that a judicious compromise between the techniques of synchronous culture and steady-state culture, if this can be achieved, will provide a common solution.

References

Algéus, S. (1951). Studies on the cultivation of algae in artificial light. *Physiol. Plantarum* **4,** 742–753.

Arnon, D. I., Ichioka, P. S., Wessel, G., Fujiwara, A., and Woolley, J. T. (1955). Molybdenum in relation to nitrogen metabolism. I. Assimilation of nitrate nitrogen by *Scenedesmus*. *Physiol. Plantarum* **8,** 538–551.

Benson, A. A., Calvin, M., Haas, S., Aronoff, S., Hall, A. G., Bassham, J. A., and Weigl, J. W. (1949). C^{14} in photosynthesis. *In* "Photosynthesis in Plants" (J. Franck and W. E. Loomis, eds.), pp. 381–401. Iowa State Coll. Press, Ames, Iowa.

Bristol-Roach, B. M. (1926). On the relation of certain soil algae to some soluble carbon compounds. *Ann. Botany (London)* **40,** 149–201.

Chu, S. P. (1942). The influence of the mineral composition of the medium on the growth of planktonic algae. *J. Ecol.* **30,** 284–325.

Fogg, G. E., Smith, W. E. E., and Miller, J. D. A. (1959). An apparatus for the culture of algae under controlled conditions. *J. Biochem. Microbiol. Technol. Eng.* **1,** 59–76.

Hutner, S. H., Provasoli, L., Schatz, A., and Haskins, C. P. (1950). Some approaches to the role of metals in the metabolism of microorganisms. *Proc. Am. Phil. Soc.* **94,** 152–170.

Hutner, S. H., Bach, M. K., and Ross, G. I. M. (1956). A sugar-containing basal medium for vitamin B_{12}-assay with *Euglena*; application to body fluids. *J. Protozool.* **3,** 101–112.

Ketchum, B. H., Lillick, L., and Redfield, A. C. (1949). The growth and optimum yields of algae in mass culture. *J. Cellular Comp. Physiol.* **33,** 267–280.

Kratz, W. A., and Myers, J. (1955). Nutrition and growth of several blue-green algae. *Am. J. Botany* **42,** 282–287.

Krauss, R. W., and Thomas, W. H. (1954). The growth and inorganic nutrition of *Scenedesmus obliquus* in mass culture. *Plant Physiol.* **29,** 205–214.

Mestre, H. (1935). A precision photometer for the study of suspensions of bacteria and other microorganisms. *J. Bacteriol.* **30,** 335–358.

Monod, J. (1950). La technique de culture continue; théorie et applications. *Ann. inst. Pasteur* **79,** 390–401.

Myers, J. (1946). Influence of light intensity on cellular characteristics of *Chlorella. J. Gen. Physiol.* **29,** 419–427.

Myers, J. (1950). The culture of algae for physiological research. *In* "The Culturing of Algae" (J. Brunel, G. W. Prescott, and L. H. Tiffany, eds.), pp. 45–51. Antioch Press, Yellow Springs, Ohio.

Myers, J. (1953). Growth characteristics of algae in relation to the problems of mass culture. *In* "Algal Culture: From Laboratory to Pilot Plant" (J. S. Burlew, ed.), *Carnegie Inst. Wash. Publ. No.* **600,** pp. 37–54.

Myers, J., and Clark, L. B. (1944). An apparatus for the continuous culture of *Chlorella. J. Gen. Physiol.* **28,** 103–112.

Myers, J., and Graham, J. R. (1959). On the mass culture of algae. II. Yield as a function of cell concentration under continuous sunlight irradiance. *Plant Physiol.* **34,** 345–352.

Myers, J., Phillips, J. N., and Graham, J. R. (1951). On the mass culture of algae. *Plant Physiol.* **26,** 539–548.

Neeb, O. (1952). *Hydrodictyon* als Objekt einer vergleichenden Untersuchung physiologischer Grössen. *Flora (Jena)* **139,** 39–95.

Novick, A., and Szilard, L. (1950). Description of the chemostat. *Science* **112,** 715–716.

Phillips, J. N., and Myers, J. (1954). Measurement of algal growth under controlled steady-state conditions. *Plant Physiol.* **29,** 148–161.

Pirson, A. (1955). Functional aspects in mineral nutrition of plants. *Ann. Rev. Plant Physiol.* **6,** 71–114.

Pirson, A., and Doring, H. (1952). Induzierte Wachstumperioden bei Grünalgen. *Flora (Jena)* **139,** 314–328.

Provasoli, L., and Pintner, I. J. (1956). Ecological implications of *in vivo* nutritional requirements of algal flagellates. *Ann. N. Y. Acad. Sci.* **56,** 839–851.

Redford, E. L., and Myers, J. (1951). Some effects of ultraviolet radiations on the metabolism of *Chlorella. J. Cellular Comp. Physiol.* **38,** 217–244.

Reisner, G. S., and Thompson, J. F. (1956). The large scale laboratory culture of *Chlorella* under conditions of microelement deficiency. *Plant Physiol.* **31,** 181–185.

Samejima, H., and Myers, J. (1958). On the heterotrophic growth of *Chlorella pyrenoidosa. J. Gen. Microbiol.* **18,** 107–117.

Scott, G. T. (1943). The mineral composition of *Chlorella pyrenoidosa* grown in culture media containing varying concentrations of calcium, magnesium, potassium, and sodium. *J. Cellular Comp. Physiol.* **21,** 327–338.

Sorokin, C., and Krauss, R. W. (1958). The effects of light intensity on the growth rates of green algae. *Plant Physiol.* **33,** 109–113.

Sorokin, C., and Myers, J. (1953). A high temperature strain of *Chlorella. Science* **117,** 330–331.

Sorokin, C., and Myers, J. (1957). The course of respiration during the life cycle of *Chlorella* cells. *J. Gen. Physiol.* **40,** 579–592.

Spoehr, H. A., Smith, J. H. C., Strain, H. H., Milner, H. W., and Hardin, G. J. (1949). Fatty acid antibacterials from plants. *Carnegie Inst. Wash. Publ. No.* **586,** pp. 1–67.

Steemann Nielsen, E. (1955). Carbon dioxide as carbon source and narcotic in photosynthesis and growth of *Chlorella pyrenoidosa. Physiol. Plantarum* **8,** 317–335.

Talling, J. F. (1960). Comparative laboratory and field studies of photosynthesis by a marine planktonic diatom. *Limnol. Oceanog.* **5,** 62–77.

Tamiya, H. (1951). Some theoretical notes on the kinetics of algal growth. *Botan. Mag. (Tokyo)* **64,** 167–173.

Tamiya, H., Hase, E., Shibata, K., Mituya, A., Iwamura, T., Nihei, T., and Sasa, T. (1953a). Kinetics of the growth of *Chlorella. In* "Algal Culture: From Laboratory to Pilot Plant" (J. S. Burlew, ed.), *Carnegie Inst. Wash. Publ. No.* **600,** pp. 204–232.

Tamiya, H., Iwamura, T., Shibata, K., Hase, E., and Nihei, T. (1953b). Correlation between photosynthesis and light-dependent metabolism in the growth of *Chlorella. Biochim. et Biophys. Acta* **12,** 23–40.

van Oorschot, J. L. P. (1955). Conversion of light energy in algal culture. *Mededel. Landbouwhogeschool Wageningen* **55,** 225–276.

Cell Division

EIJI HASE

Institute of Applied Microbiology, University of Tokyo, and Tokugawa Institute for Biological Research, Tokyo, Japan

I. Introduction[1]

Recent developments in the technique of synchronous culture of unicellular algae have opened a new avenue of approach to the biochemical and physiological control of cell division in algae. Interesting and important phenomena, which had been missed with ordinary methods of investigation, have been brought to light in this way.

Since it proliferates only by asexual means, *Chlorella* may be regarded as exhibiting the simplest type of algal life cycle. All of the studies reported here relate to species of this genus, which all behave more or less similarly. In the method developed for *C. ellipsoidea* by Tamiya and his co-workers (Tamiya *et al.*, 1953; Iwamura *et al.*, 1955; Hase *et al.*, 1957; Morimura, 1959) a synchronous culture is started from a homogeneous population of small, young, and photosynthetically most active cells, referred to as "dark" cells or D_a-cells, obtained by culturing the alga first in bright light at an intensity of 10,000 lux and later at a lower intensity of 800 lux. Another method of cell synchronization was developed by several workers (Lorenzen, 1957, 1959; Lorenzen and Ruppel, 1960; Sorokin and Myers, 1957; Bongers, 1958; Sueoka, 1960; Bernstein, 1960) who synchronized cultures of various species of *Chlorella* and *Chlamydomonas* by subjecting them to alternating periods of light and darkness. Padilla and James (1960) employed alternat-

[1] Abbreviations used in this chapter: DNA, deoxyribose nucleic acid; RNA, ribose nucleic acid.

ing periods of 15° and 25°C. for synchronizing cell division in cultures of the colorless euglenid *Astasia longa.*

II. Various Stages in the Life Cycle of *Chlorella*

Using the technique of synchronous culture developed by Tamiya and his co-workers, Morimura (1959) showed that the course of the life cycle of *Chlorella* is appreciably influenced by temperature in the range of 9 to 25°C, and by light intensity between 400 and 25,000 lux. At a constant temperature of 16°C., D_a-cells grew faster in brighter light, but the time at which the cells began to divide simultaneously was independent of the light intensity. Consequently at the moment of incipient cell division the cells grown at higher light intensities were larger, and the average number (n) of daughter cells produced from a single mother cell was correspondingly larger, being $n = 4.2$ at 10 to 25 kilolux, $n = 3.7$ at 5 kilolux, $n = 3.1$ at 2.5 kilolux, and $n = 2$ at 1 kilolux. On the other hand, at a constant saturating light intensity of 10 kilolux, growth was faster at higher temperatures, and cell division set in earlier and proceeded more rapidly to completion. Temperature, unlike light, did not affect the average number of daughter cells produced, which was constant at $n = 4$ within the range of 9 to 25°C. The following four phases in the life cycle were distinguished: (I) the phase of "growth," which represents the period of active increase in cell mass; (II) the phase of "early ripening," during which various synthetic processes occur in preparation for the formation of daughter cells; (III) the phase of "late ripening" of the incipient mother cells; and (IV) the phase of "autospore liberation." This sequence is illustrated schematically in Fig. 1. The

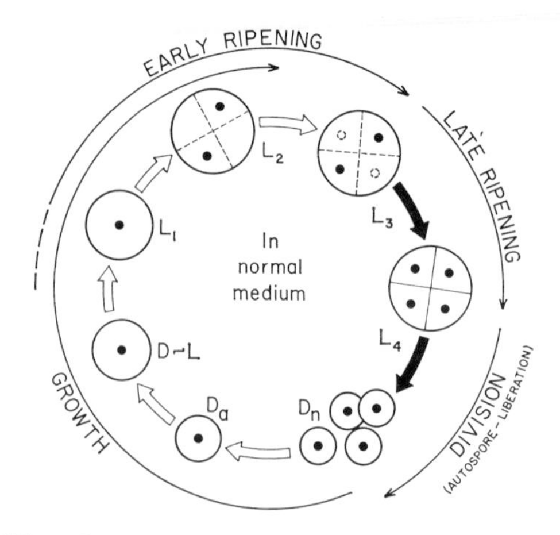

FIG. 1. Normal life cycle of *Chlorella ellipsoidea.* The solid spots represent nuclei.

phases of "growth" and "early ripening" are both light- and temperature-dependent processes, whereas those of "late ripening" and "autospore liberation" are temperature-dependent but not light-dependent. Distinction of three different stages of "light" cells (L_1, L_2, and L_3) in the early ripening phase was made on the basis of their different contents of DNA and different capacities for division when incubated in the dark (Iwamura *et al.*, 1955). When cells at a later stage of "early ripening" (L_3-cells) are transferred to the dark under aerobic conditions, they proceed through the "late ripening" phase to become L_4-cells, and then liberate daughter cells (designated as D_n-cells) which under photosynthesizing conditions develop into D_a-cells. When the life cycle is in operation under continuous illumination, however, D_n-cells as such are only rarely liberated from the mother cells; instead, they enlarge to some extent inside the mother cells and then emerge in the form of D_a-cells, or sometimes as $D \sim L$-cells, i.e., cells at the stage just prior to ripening.

Since at a constant temperature cell division occurs at a definite time independent of light intensity, one may postulate the existence of some temperature-dependent factor which, largely independently of the degree of growth and ripening, induces the onset of division, or switches the process of "early ripening" to "late ripening" and "autospore liberation."

In their study of *Chlorella pyrenoidosa* (Emerson strain), Lorenzen and Ruppel (1960) distinguished three successive developmental phases, I, II, and III, which seem to correspond, respectively, to the "growth," "early ripening," and "late ripening" phases proposed by Tamiya and his associates. Pirson and Lorenzen (1958), using the technique of regularly alternating light-and-dark incubation, observed that cell division occurred at a definite time after the beginning of the light period, and that the onset of cell division was independent of both the intensity (2–15 kilolux) and the duration of the light periods, but slightly dependent on temperature. They assumed the existence of an "endogenous time-factor," independent of light intensity and temperature, controlling the onset of cell division. The temperature-independence of this factor is open to question, however, since the temperature range tested was limited between 20° and 30°C. As shown by Morimura (1959), in *C. ellipsoidea* the temperature dependence of the onset of cell division was more striking at lower (9–16°C.) than at higher (16–25°C.) temperatures.

III. Effects of Some Chemicals and Nutrient Deficiency on the Course of Cell Division

The four developmental phases in the life cycle of *Chlorella ellipsoidea* were affected differently by various antimetabolites or antibiotics, as well as by a deficiency of essential nutrient elements from the culture solution.

Morimura and Tamiya (unpublished) found that chloramphenicol, at a concentration of $3 \times 10^{-3}M$, completely inhibited growth, although it allowed the cells to divide into two smaller daughter cells ($n = 2$). This indicates that, at least for one round of the life cycle starting from D_a-cells, synthesis of new protein is not essential for cell division. Some antimetabolites, such as 2-thiocytosine ($10^{-2}M$; $n = 8$), 6-azathymine ($10^{-5}M$; $n = 6$), and maleic hydrazide ($10^{-3}M$; $n = 8$), apparently stimulated the ripening phase, and thereby increased the n-value, without affecting the other developmental phases of growth, whereas 8-azaguanine ($10^{-3}M$) and 8-azaxanthine ($10^{-3}M$) completely suppressed cell division without affecting growth. Kinetin ($10^{-3}M$), gibberellin ($10^{-2}M$), and colchicine ($10^{-2}M$) had no apparent effect on any phase of cell development; the permeability of the cell wall to these substances remains to be examined.

Hase *et al.* (1957) found that, in media lacking nitrate, phosphate, K or Mg, the D_a-cells could perform one cycle of cell division, giving rise to different numbers of daughter cells which were unable to grow normally (control: $n = 6.5$; N-free: $n = 2.4$; P-free: $n = 3.5$; K-free: $n = 5.1$; Mg-free: $n = 6.4$). In the case of sulfur deficiency, however, the cells grew apparently normally only until the earlier stage of ripening, whereupon development was arrested and division inhibited. More or less similar suppression of division was observed in the presence of certain antagonists to sulfur-containing amino acids, such as $3 \times 10^{-5}M$ selenomethionine (Shrift, 1954, 1959), $10^{-2}M$ ethionine, or $10^{-2}M$ allylglycine (Morimura and Tamiya, unpublished). With synchronous cultures in a normal culture medium it was found that sulfur is assimilated most actively in the ripening phases (Hase *et al.*, 1957). These observations point to an essential role of sulfur in the processes of cell division.

IV. Cellular Metabolism Associated with Cell Division

The photosynthetic activity of *Chlorella ellipsoidea* was found to be greater during the growing phase than in the ripening phase (Nihei *et al.*, 1954). The capacity of light-induced "dark" CO_2-fixation, measured by the tracer technique using C^{14}, showed a similar trend, being almost zero at the stage of the ripening phase and increasing rapidly during subsequent phases of the cycle. The photosynthetic quotient (CO_2/O_2) was found to be approximately unity during the growing and late ripening phases, but it declined to 0.3–0.5 during the early ripening phase (see below).

The endogenous respiration was appreciably higher during the ripening phases than in the growing phase, suggesting the requirement of respiratory energy for the processes of ripening and cell division (Nihei *et al.*, 1954). In fact, when cells at the early ripening phase (L_3) were kept under anaerobic conditions, they were completely unable to divide (Tamiya *et al.*, 1953).

Somewhat different results were reported by other workers (Sorokin, 1957; Sorokin and Myers, 1957; Bongers, 1958; Lorenzen, 1959). According to Lorenzen, the photosynthetic activity of *C. pyrenoidosa* was apparently influenced directly by the light-dark treatment and was almost independent of the stage in the life cycle. Sorokin and Myers (1957) observed that both endogenous and glucose respiration of a high-temperature strain of *Chlorella* became most active shortly after the beginning of the light period, and that respiratory activity decreased slowly during the following phases of the cycle. In the process of cell division, however, the rate of respiration declined rapidly to its original low level. It would be of value to reinvestigate these systems, taking into account the differences in the species employed and in the methods of cell synchronization adopted in different laboratories.

According to Nihei (1955), the abnormally low value of the photosynthetic quotient (CO_2/O_2) observed in the early ripening phase was due to surplus O_2 liberated by a photochemical process at least outwardly similar to the Hill reaction. The nature of the intracellular hydrogen acceptor was not determined. However, this photochemical process was found to be associated with the assimilation of orthophosphate into energy-rich polyphosphate, which accumulates in large quantities in the cells during the early ripening phase. Conceivably the energy of this phosphorus compound is utilized, in some way as yet unknown, in various endergonic reactions involved in the processes of ripening and division.

The variation of nucleic acids during the course of the life cycle was studied by Iwamura (1955; Iwamura *et al.*, 1955) and by Lorenzen and Ruppel (1960), who obtained essentially the same results. The amount of DNA per cell increased during the early ripening phase (II), whereas RNA increased from the beginning of the growth phase (I), almost keeping pace with the increase of cell mass or of total protein. In detailed investigations of changes in the base composition of RNA and DNA during the life cycle, Iwamura and Myers (1959; Iwamura, 1960) found two types of DNA which differed in their intracellular distribution as well as in the rate of phosphorus turn-over (see Chapter 13 in this treatise). Lorenzen (1958) observed that the DNA, which is ordinarily stainable with the Feulgen reagent, became unstainable at the end of the growth phase (I), which may be associated with some physical change in the state of the DNA molecules.

Hase *et al.* (1958) showed that, by controlling the supply of S- and N-sources in a synchronous culture of *Chlorella ellipsoidea*, the processes of nuclear and cell division could be experimentally separated (Fig. 2), indicating the importance of sulfur in ripening and division (i.e., in the formation of DNA, nuclear division and cell division). By performing similar experiments with S^{35} as a tracer, and by thereby following the distribution of S-compounds in various fractions of cell material, it was shown that certain S-containing peptide-nucleotide compounds, in a cold-acid-soluble fraction, accumulated just before and during nuclear division, and that certain unidentified S-compounds in a hot-acid-soluble fraction appeared prior to and during the process of cell division (Hase *et al.*, 1959a, b, c, 1960a, b, 1961).

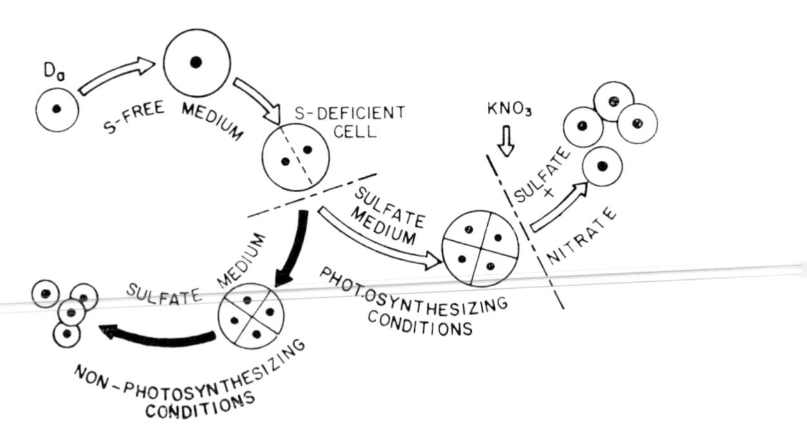

FIG. 2. Life cycle of *Chlorella ellipsoidea*, as modified by limited supplies of sulfur or nitrogen. The broken lines indicate arrested development.

As illustrated in Fig. 2, S-starved cells which have been arrested at an early stage of ripening have different nutritional requirements, for the resumption of nuclear and cellular division, under photosynthesizing and nonphotosynthesizing conditions. In the latter case the addition of sulfate alone is sufficient to induce maturation and cell division, whereas under photosynthesizing conditions nitrate as well as sulfate is necessary to bring this about. Sulfur-starved cells presumably also require for their further maturation and division both S- and N-sources; but whereas under nonphotosynthesizing conditions the nitrogen is supplied from some endogenous source, under photosynthesizing conditions such endogenous N-substances would be depleted by the synthesis of some material not utilizable for the processes of maturation and division. There is evidence that under nonphotosynthesizing conditions the utilization of endogenous N-substances and the incorporation of sulfur lead to the formation of the hot-acid-soluble S-compounds mentioned above (Hase *et al.*, 1960b, 1961).

These results raise the hope that further studies using the technique of synchronous culture, in combination with modern biochemical methods, will yield important information on the physiological control of ripening and cell division in algae.

REFERENCES

Bernstein, E. (1960). Synchronous division in *Chlamydomonas moewusii*. *Science* **131,** 1528–1529.

Bongers, L. H. J. (1958). Changes in photosynthetic activity during algal growth and multiplication. *Mededeel. Landbouwhogeschool Wageningen* **58,** 1–10.

Hase, E., Morimura, Y., and Tamiya, H. (1957). Some data on the growth physiology of *Chlorella* studied by the technique of synchronous culture. *Arch. Biochem. Biophys.* **69,** 149–165.

Hase, E., Morimura, Y., Mihara, S., and Tamiya, H. (1958). The role of sulfur in the cell division of *Chlorella*. *Arch. Mikrobiol.* **32,** 87–95.

Hase, E., Mihara, S., Otsuka, H., and Tamiya, H. (1959a). New peptide-nucleotide compounds obtained from *Chlorella* and yeasts. *Biochim. et Biophys. Acta* **32,** 298–300.

Hase, E., Mihara, S., Otsuka, H., and Tamiya, H. (1959b). Sulfur-containing peptide-nucleotide complex isolated from *Chlorella* and yeast cells. *Arch. Biochem. Biophys.* **83,** 170–177.

Hase, E., Otsuka, H., Mihara, S., and Tamiya, H. (1959c). Role of sulfur in the cell division of *Chlorella,* studied by the technique of synchronous culture. *Biochim. et Biophys. Acta* **35,** 180–189.

Hase, E., Mihara, S., and Tamiya, H. (1960a). Sulfur-containing deoxypentose polynucleotides obtained from *Chlorella* cells. *Biochim. et Biophys. Acta* **39,** 381–382.

Hase, E., Mihara, S., and Tamiya, H. (1960b). Role of sulfur in the cell division of *Chlorella,* with special reference to the sulfur compounds appearing during the process of cell division. I. *Plant & Cell Physiol.* **1,** 131–142.

Hase, E., Mihara, S., and Tamiya, H. (1961). Role of sulfur in the cell division of *Chlorella,* with special reference to the sulfur compounds appearing during the process of cell division. II. *Plant & Cell Physiol.* **2,** 9–24.

Iwamura, T. (1955). Change of nucleic acid content in *Chlorella* cells during the course of their life-cycle. *J. Biochem.* **42,** 575–589.

Iwamura, T. (1960). Distribution of nucleic acids among subcellular fractions of *Chlorella*. *Biochim. et Biophys. Acta* **42,** 161–163.

Iwamura, T., and Myers, J. (1959). Changes in the content and distribution of the nucleic acid bases in *Chlorella* during the life cycle. *Arch. Biochem. Biophys.* **84,** 267–277.

Iwamura, T., Hase, E., Morimura, Y., and Tamiya, H. (1955). Life cycle of the green alga *Chlorella* with special reference to the protein and nucleic acids contents of cells in successive formative stages. *Ann. Acad. Sci. Fennicae Ser. A. II.* **No. 60,** 89–102.

Lorenzen, H. (1957). Synchrone Zellteilungen von *Chlorella* bei verschiedenen Licht-Dunkel-Wechseln. *Flora* **144,** 473–496.

Lorenzen, H. (1958). Periodizität von Nuklealreaktion und Kernteilung in *Chlorella*. *Ber. deut. botan. Ges.* **71,** 89–97.

Lorenzen, H. (1959). Die photosynthetische Sauerstoffproduktion wachsender *Chlorella* bei langfristig intermittierender Belichtung. *Flora* **147,** 382–404.

Lorenzen, H., and Ruppel, H. G. (1960). Versuche zur Gliederung des Entwicklungsverlaufs der *Chlorella*-Zelle. *Planta* **54,** 394–403.

Morimura, Y. (1959). Synchronous culture of *Chlorella*. I. Kinetic analysis of the life cycle of *Chlorella ellipsoidea* as affected by changes of temperature and light intensity. *Plant & Cell Physiol.* **1,** 49–62.

Nihei, T. (1955). A phosphorylative process, accompanied by photochemical liberation of oxygen, occurring at the stage of nuclear division in *Chlorella* cells. *J. Biochem.* **42,** 245–256.

Nihei, T., Sasa, T., Miyachi, S., Suzuki, K., and Tamiya, H. (1954). Change of photosynthetic activity of *Chlorella* cells during the course of their normal life cycle. *Arch. Mikrobiol.* **21,** 155–164.

Padilla, G. M., and James, T. W. (1960). Synchronization of cell division in *Astasia longa* on a chemically defined medium. *Exptl. Cell Research* **20,** 401–415.

Pirson, A., and Lorenzen, H. (1958). Ein endogener Zeitfaktor bei der Teilung von *Chlorella. Z. Botan.* **46,** 53–67.

Shrift, A. (1954). Sulfur-selenium antagonism. II. Antimetabolite action of selenomethionine on the growth of *Chlorella vulgaris. Am. J. Botany* **41,** 223–230.

Shrift, A. (1959). Nitrogen and sulfur changes associated with growth uncoupled from cell division in *Chlorella vulgaris. Plant Physiol.* **34,** 505–512.

Sorokin, C. (1957). Changes in photosynthetic activity in the course of cell development in *Chlorella. Physiol. Plantarum* **10,** 659–666.

Sorokin, C., and Myers, J. (1957). The course of respiration during the life cycle of *Chlorella* cells. *J. Gen. Physiol.* **40,** 579–592.

Sueoka, N. (1960). Mitotic replication of deoxyribonucleic acid in *Chlamydomonas reinhardi. Proc. Natl. Acad. Sci. U.S.* **46,** 83–91.

Tamiya, H., Iwamura, T., Shibata, K., Hase, E., and Nihei, T. (1953). Correlation between photosynthesis and light-independent metabolism in the growth of *Chlorella. Biochim. et Biophys. Acta* **12,** 23–40.

—41—

Cell Expansion

PAUL B. GREEN

Division of Biology, University of Pennsylvania, Philadelphia, Pennsylvania

I. Introduction

Plants increase in volume by the expansion of cells. In some regions the cell expansion is coupled with cell division to produce new volume in the form of new cells. In others the expansion may proceed unaccompanied by cell division to produce relatively large cells. Among certain algae this latter type of expansion produces some of the largest plant cells or syncytia: the vesicle of *Valonia*, the internode of the Charales, the branching tube cells of *Vaucheria* and the siphonous Chlorophyta. Since these cells are not embedded in tissues, they are especially suitable for the experimental investigation of the mechanism of cell expansion.

II. General Mechanism of Cell Expansion

A. Internal Pressure

The expansion of a plant cell may be regarded as a result of the yielding of the cell wall to pressure of the cytoplasm and vacuole contained within it (Heyn, 1931). This pressure derives in large part from the osmotic potential of the solutes in the vacuole, and in freshwater algae amounts to approximately 5 atm. or roughly 5 kg./cm.^{2} (Blinks, 1951). When this pressure is translated into terms of stress on the cell wall, following the equations of

Castle (1937), the values for the stresses are in the range of 10 to 30 kg./mm.[2] (assuming a cell radius of 0.5 mm. and a wall thickness of 1 μ, as in *Nitella axillaris* internodes).

B. The Yielding Cell Wall

In the Brown, Green, and Red Algae the stresses caused by internal pressure are generally borne by cell walls which consist of more or less crystalline microfibrils embedded in an amorphous matrix. Both the microfibrils and the matrix are of a carbohydrate nature. The microfibrils have been shown in many cases to consist of native or of "mercerized" cellulose (Preston, 1952; Northcote, 1958; Myers and Preston, 1959). In certain marine algae they contain xylose as well as glucose; in *Porphyra umbilicalis* the basic structural unit is mannose (Cronshaw *et al.*, 1958). Microfibrils always lie in the plane of the cell surface. The process of cell expansion is accompanied by the synthesis of many components within the cell, and it is quite clear that the yielding cell wall itself increases in mass during the process. This is obvious in large-celled algae, such as *Valonia* and *Nitella*, where the cell wall undergoes an increase in area of several thousand-fold while retaining a firm texture. In the case of *Nitella axillaris* the increase in wall mass during elongation is relatively so rapid that the wall thickness doubles as the internode elongates from 1 to 40 mm. in length (Green, 1958a). This increase in total wall mass (though not necessarily in wall thickness) is also characteristic of cell expansion in some higher plants (Roelofsen, 1959).

III. The Site of Addition of New Wall Materials

The mode of the addition of new mass to the wall has long been a problem. Two possibilities, not mutually exclusive, are (i) apposition, where new material is added only at the inner wall surface, and (ii) intussusception, where new material is added throughout the wall's thickness. As with the question of mass increase in the growing wall, the answer was almost self-evident in certain algal forms. Strasburger observed that in the lamellar wall of the apical cell of *Bornetia* the outermost lamellae appeared to burst and to be replaced by new lamellae apposed on the inner wall surface (see Küster, 1956). The addition of new lamellae in this way was also evident in the development of the *Valonia* vesicle as studied by means of the electron microscope (Steward and Mühlethaler, 1953). The experimental approach to the problem also dates from the last century and relates to the algae. Noll applied the stain Berlin blue to living *Caulerpa* and *Derbesia*; after a period of growth in the absence of the dye, he observed that colorless wall material had been deposited interior to the stained portion (see Küster, 1956). Since the cell walls of a number of plants contain nitrogenous substances, presumed to be protein, it was still thought possible that cytoplasm might penetrate at least the inner cell wall and add new material by intussusception. The question has been studied with a variety of techniques, as outlined below.

The growing cylindrical cell wall in most higher plant tissues (coleoptile-parenchyma, root cortex, etc.) and certain algae [*Hydrodictyon* (Diehl *et al.*, 1939); *Nitella* (Correns, 1893); *Bryopsis* (Green, 1960a)] is characterized by a predominantly transverse arrangement of microfibrils, as indicated by the fact that such walls are negatively birefringent (Frey-Wyssling, 1959). This orientation persists throughout the phase of enlargement and is believed to be responsible for elongation of the cell as a cylinder. The strong microfibrils are assumed to prevent rapid increase in girth and to permit predominant extension along the cell axis. Roelofsen and Houwink (1953) examined replicas of wall surfaces of a variety of cells by electron microscopy, and observed that the inner surface showed the expected transverse arrangement of microfibrils whereas the outer surface showed the microfibrils in either random or axial array. They proposed that the observed structures represented the two ends of a gradient of microfibrillar arrangement, from transverse on the inside to random or axial on the outside. Such a structure would result from the apposition of transverse microfibrils upon the inner wall surface. After a given group of such microfibrils had been covered by new appositions, it would undergo passive reorientation to the axial direction. This model for cylindrical wall elongation is called "multi-net growth."

Support for the above view that the cell wall extends passively, in that it increases in mass only at its inner surface, has come from work on *Nitella*. During elongation in tritiated water the cell was found to incorporate the isotope tritium (H^3) into the dry matter of the wall (Green, 1958b). The location of the labeled material was deduced by comparisons of radioactivity from the inner and outer surfaces of the same piece of cell wall. After short periods of growth most of the isotope incorporated was found to be localized at or near the inner surface, as would be expected from growth by apposition.

The *Nitella* cell wall is also suitable for investigations of the gradient of microfibrillar arrangement postulated by multi-net growth. A study of this subject (Green, 1960b) involved the analysis of single thicknesses of wall which had been torn so as to present wedge-like areas tapering gradually to a vanishing thickness. With the inner wall surface facing upward it was shown that, at a point on the wedge where only half the normal wall thickness remained, the action on polarized light (negative retardation) was much less than half the action of the entire wall thickness. This indicated that there were many more transverse fibrils in the inner half of the wall thickness than in the outer half. Positive birefringence of the thinnest parts of the tapering wedge indicated a predominantly axial arrangement of microfibrils in the outermost part of the wall. It was concluded that the *Nitella axillaris* cell wall has the multi-net structure. The internodal cell of this alga shows a twisting or helical growth (Green, 1954), and the microfibrils therefore would tend to be drawn out into the direction of maximum surface expansion which is nearly axial. Probine and Preston (1958) found a flat helical position of major extinction for *N. opaca* which was correlated with the lines of flow of the protoplasmic stream. This indicates that a slight spirality of the wall is associated with the helical or "twisting" growth of this cell.

Other cylindrical cells of freshwater algae (e.g., *Spirogyra, Oedogonium*) have been investigated, but apparently not to the extent of the characeous internodal cell.

IV. Walls with Crossed Fibrillar Texture

While the above account suggests a high degree of similarity between the walls of Green Algae and the primary cell wall of higher plants, there remain many marine Chlorophyta and a few freshwater species with a wall structure which is obviously different from the typical multi-net structure and resembles more the secondary wall of higher plants. In these algae the wall consists of individual lamellae each with a nearly parallel arrangement of cellulose microfibrils. From one lamella to the next the direction of this alignment is changed through an angle of 60 to 100 degrees. The "crossed fibrillar texture" (Preston, 1952) which occurs in the vesicle wall of *Valonia* has been extensively investigated by Preston and co-workers by x-ray diffraction methods. This pioneering work was followed by the extensive investigations of Steward and Mühlethaler (1953), Wilson (1951, 1955), Cronshaw and Preston (1958) and others, so that now this algal cell wall is the most thoroughly investigated of all. In brief, the *Valonia* wall appears to consist of a series of microfibrillar lamellae with the microfibrils in one layer forming a steep Z-spiral while those in the next run up the vesicle in a more slowly ascending S-spiral. Many layers of each type are found in the wall. In occasional lamellae there is a third microfibrillar direction, bisecting the angle between the other two. The convergence of all microfibrillar directions at the poles of the vesicle, as indicated in the early x-ray diagrams, has also been shown by optical methods (Wilson, 1951, 1955).

The cylindrical cell walls of *Chaetomorpha* and *Cladophora* show a crossed fibrillar texture, the visible striations apparently corresponding to microfibrillar orientations (see references in Roelofsen, 1959). Although these cells are cylindrical, their wall structure is much like that of *Valonia*, though there are only two microfibrillar directions (near-axial and near-transverse) and, in *Chaetomorpha*, there are occasional layers of noncellulosic materials between the alternating lamellae (Nicolai and Frey-Wyssling, 1938). The early stages of wall development in these algae were described by Nicolai and Preston (1960), and further details on wall expansion were given by Frei and Preston (1961). Good general accounts of the work on *Valonia* and other algal cell walls are given by Preston (1952, 1959) and Roelofsen (1959).

V. Relation of Cell Shape to Wall Structure in Vesicles and Cylinders

If we consider a model system in which cell expansion results from a yielding of the wall to the outward pressure of the vacuole, then we may expect the direction of expansion to be controlled by the structure of the wall. Though the crossed fibrillar texture of the vesicles of *Valonia* and *Dictyosphaeria* exhibits an arrangement which is far from random, the various layers of the cell wall are so diversely oriented that one would not expect any strong preference in the direction of yielding.

The predominant transverse or nearly transverse arrangement of microfibrils in most higher-plant cells and in some species of *Hydrodictyon*, *Nitella*, etc., could well explain the preferential yielding in the direction of the cell axis. In those cases in which the crossed fibrillar texture is present in cylindrical cells, such as those of *Chaetomorpha*, *Cladophora*, and related algae, the presence of a great many longitudinal fibrils leads to the question whether the wall could be sufficiently anisotropic in structure to account for the cylindrical elongation. In *Chaetomorpha linum* there is evidence that longitudinally and transversely oriented microfibrillar lamellae occur together at all levels in the wall. The ends of the cell show negative birefringence, however, and the middle is either isotropic or positively birefringent. If expansion were localized at the ends of the cell, the model would fit this type of alga, too. Preston (1952) pointed out the difficulty of explaining cylindrical elongation in a cell such as that of *Cladophora*, where the x-ray diagram and striation pattern indicate a crossed fibrillar texture like that in *Valonia*. Correns (1893) reported that the cell walls of *Cladophora* were either isotropic or positively birefringent in surface view, the latter observation indicating that longitudinal microfibrils may even predominate. The fine structure in the growing part of the cell wall, however, has not been fully elucidated.

VI. Wall Structure in Cells Showing a Complex Pattern of Expansion

A correlation between cell wall structure and cell shape has been found in the pinnate coenocyte of *Bryopsis* (Green, 1960a). The walls of both the main cylindrical axis and the lateral cylindrical protrusions are negatively birefringent, and thus presumably have a transverse arrangement of microfibrils. The site of the origin of lateral processes is initially marked by an apparent arrangement of microfibrils (high index of refraction) in a concentric pattern. The center of the disk-like area protrudes to produce the lateral branch; the periphery of the area persists as a lateral-base.

The cells of desmids have a strong bilateral symmetry. At each division two new semicells are built in the region of the old isthmus, so that each new cell consists of an old semicell and a new one which usually develops as its mirror image. As a result of displacing the nucleus by centrifugation, Waris (1951) obtained enucleate semicells which nevertheless started to produce the companion semicell. In *Micrasterias* such an enucleate semicell showed little synthesis of either protein or cellulose in the wall, although a considerable amount of pectin was laid down. If the cells were kept in a simple mineral medium, the new lobes were smaller and far less complex in outline than those of a normal semicell. The addition of various auxins, metabolic intermediates such as xanthine and thymine, or inhibitors such as sodium iodoacetate and potassium cyanide, often increased the size of the new semicell, particularly in the lateral lobes; but the resulting form was no less bizarre. This might indicate that the specificity of pattern of the cell outline is normally associated with the cellulose component, which in such abnormal semicells is absent. In these desmids the presence of the nucleus is evidently necessary for the development of the normal pattern and shape of the cell, although this does not seem to be the case in the regeneration of enucleate stalks of *Acetabularia* (Hämmerling, 1953; see also Richter, Chapter 42, this treatise).

The changes of shape observed when certain algae are grown under unusual conditions may reflect disturbances of an orientation mechanism possibly involving the cell wall. The normally cylindrical cells of *Chara vulgaris* become ovoid when grown in colchicine (Delay, 1957), and *Bryopsis* grown in dilute agar media shows abnormal dimensions at the growing tip (Weihe, 1960).

VII. Summary

Cell expansion may be viewed as a result of the yielding of the cell wall to pressure of the expanding vacuole. During this process the wall increases in mass as well as in area. The addition of mass apparently occurs at the inner wall surface and involves the deposition of microfibrils embedded in an amorphous matrix. The physical properties—particularly the microfibrillar orientations—of the accumulated depositions appear to be important in the control of both the rate and the direction of cell expansion.

REFERENCES

Blinks, L. R. (1951). Physiology and biochemistry of algae. *In* "Manual of Phycology" (G. M. Smith, ed.), pp. 263–284. Chronica Botanica, Waltham, Massachusetts.

Castle, E. S. (1937). Membrane tension and orientation of structure in the plant cell wall. *J. Cellular Comp. Physiol.* **10,** 113–121.

Correns, C. (1893). Zur Kenntnis der inneren Struktur einiger Algenmembranen. *Zimmermann's Beitr. Morphol. u. Physiol. Pflanzen* **1,** 260–305.

Cronshaw, J., and Preston, R. D. (1958). A re-examination of the fine structure of the walls of vesicles of the green alga *Valonia. Proc. Roy. Soc.* **B148,** 137–148.

Cronshaw, J., Myers, A., and Preston, R. D. (1958). A chemical and physical investigation of the cell walls of some marine algae. *Biochim. et Biophys. Acta* **27,** 89–103.

Delay, C. (1957). Action de la colchicine sur la croissance et la différentiation de l'appareil végétatif de *Chara vulgaris* L. *Compt. rend. acad. sci.* **244,** 485–487.

Diehl, J. M., Gorter, C. J., van Iterson, G., Jr., and Kleinhoonte, A. (1939). The influence of growth hormone on hypocotyls of *Helianthus* and the structure of their cell walls. *Rec. trav. botan. néerl.* **36,** 709–798.

Frei, E., and Preston, R. D. (1961). Cell wall organization and wall growth in the filamentous green algae *Cladophora* and *Chaetomorpha*. I. *Proc. Roy. Soc.* **B154,** 70–94.

Frey-Wyssling, A. (1959). "Die pflanzliche Zellwand." Springer, Berlin.

Green, P. B. (1954). The spiral growth pattern of the cell wall in *Nitella axillaris. Am. J. Botany* **41,** 403–409.

Green, P. B. (1958a). Structural characteristics of developing *Nitella* internodal cell walls. *J. Biophys. Biochem. Cytol.* **4,** 505–516.

Green, P. B. (1958b). Concerning the site of the addition of new wall substances to the elongating *Nitella* cell wall. *Am. J. Botany* **45,** 111–116.

Green, P. B. (1960a). Wall structure and lateral formation in the alga *Bryopsis. Am. J. Botany* **47,** 476–481.

Green, P. B. (1960b). Multinet growth in the cell wall of *Nitella. J. Biophys. Biochem. Cytol.* **7,** 289–296.

Hämmerling, J. (1953). Nucleo-cytoplasmic relationships in the development of *Acetabularia. Intern. Rev. Cytol.* **2,** 475–498.

Heyn, A. N. J. (1931). Der Mechanismus der Zellstreckung. *Rec. trav. botan. néerl.* **28,** 133–244.

Küster, E. (1956). "Die Pflanzenzelle." Fisher, Jena, Germany.

Myers, A., and Preston, R. D. (1959). Fine structure in the red algae. III. A general survey of cell-wall structure in the red algae. *Proc. Roy. Soc.* **B150,** 456–459.

Nicolai, E., and Frey-Wyssling, A. (1938). Über den Feinbau der Zellwand von *Chaetomorpha. Protoplasma* **30,** 401.

Nicolai, E., and Preston, R. D. (1960). Cell wall studies in the Chlorophyceae. III. Differences in structure and development in the *Cladophoraceae. Proc. Roy. Soc.* **B151,** 39–47.

Northcote, D. H. (1958). The cell walls of higher plants; their composition, structure, and growth. *Biol. Revs.* **33,** 53–102.

Preston, R. D. (1952). "The Molecular Architecture of Plant Cell Walls." Chapman and Hall, New York.

Preston, R. D. (1959). Wall organization in plant cells. *Intern. Rev. Cytol.* **8,** 33–60.

Probine, M. C., and Preston, R. D. (1958). Protoplasmic streaming and wall structure in *Nitella. Nature* **182,** 1657–1658.

Roelofsen, P. A. (1959). "The Plant Cell Wall." Borntraeger, Berlin.

Roelofsen, P. A., and Houwink, A. L. (1953). Architecture and growth of the primary wall in some plant hairs and the *Phycomyces* sporangiophore. *Acta Botan. Neerl.* **2,** 218–225.

Steward, F. C., and Mühlethaler, K. (1953). The structure and development of the cell-wall in the Valoniaceae as revealed by the electron microscope. *Ann. Botany* (*London*) **17,** 295–316.

Waris, H. (1951). Cytophysiological studies on *Micrasterias*. III. *Physiol. Plantarum* **4,** 387–409.

Weihe, K. V. (1960). Über Formänderungen der *Bryopsis*-Zellspitzen. *Arch. Mikrobiol.* **35,** 44–52.

Wilson, K. (1951). Observations on the structure of the cell wall of *Valonia ventricosa* and of *Dictyosphaeria favulosa*. *Ann. Botany* (*London*) **15,** 279–288.

Wilson, K. (1955). The polarity of the cell wall of *Valonia*. *Ann. Botany* (*London*) **19,** 289–292.

—42—

Nuclear–Cytoplasmic Interactions

GERHARD RICHTER

Botanical Institute, University of Tübingen, Germany

I. Introduction[1]

In theory one can investigate the relationship between the nucleus and the cytoplasm of a cell by observing how various cellular processes are affected by removal of the nucleus. In practice, however, very few organisms possess cells with features suitable for this technique. The principle was used for plants as early as 1887, when Klebs succeeded in producing anucleate portions of cells of *Spirogyra* and *Zygnema*, and of the moss *Funaria*, by means of plasmolysis. He observed that they remained viable for at least two months, and that, in light, starch accumulated in the chloroplasts. In a more intensive study of the same problem, Van Wisselingh (1909) showed that anucleate portions of the cytoplasm of *Spirogyra* cells, separated by centrifugation, not only survived for several weeks but remained capable of growth, photosynthesis, cytoplasmic streaming, plastid formation, and the accumulation of fat and tannin. More recent experiments by Yoshida (1956) showed that anucleate fragments of plasmolyzed *Elodea* cells can synthesize starch and even chlorophyll. Waris (1951), working with the desmid *Micrasterias*, obtained anucleate semicells by centrifugation. Their

[1] Abbreviations used in this chapter: ATP = adenosine triphosphate; RNA = ribonucleic acid.

life span was rather limited under normal conditions, but could be prolonged from 1 to 10 days in adequate calcium concentrations at pH 4, and under these conditions new pectic membranes were formed around the new semi-cells.

Since 1931 the unicellular and uninucleate Green Algae, *Acetabularia* spp., have become favorite objects for studies of the nuclear control of morphogenesis and of other cytoplasmic processes. The fundamental work of Hämmerling and his co-workers revealed that anucleate cells and cell portions of *A. mediterranea* can not only survive for some months but also retain a remarkable capacity for morphogenesis controlled by the specific genetic characteristics of the original nucleus. Grafting experiments between nucleate and anucleate plants, or between several nucleate cell portions, disclosed other interesting and important facts; during recent years, studies in different laboratories have dealt with various biochemical aspects of the problem. These investigations form the subject of this chapter. Similar investigations on the protozoa *Amoeba* and *Stentor* have been reviewed by Brachet and Lang (1961).

II. General Morphology and Life Cycle of *Acetabularia*

All members of the family Dasycladaceae (order Dasycladales) are littoral Green Algae of subtropical and tropical seas. The three species chiefly dealt with here are *Acetabularia mediterranea,** *A. crenulata,* and *Acicularia schenkii.* All are unicellular and uninucleate organisms during the vegetative phase of their development (for details see Hämmerling, 1931, 1944). A mature plant consists of a basal, lobed rhizoid, containing the single nucleus with an extraordinarily well-developed nucleolus; a cylindrical stalk several centimeters long; and a discoid or umbrella-shaped cap composed of several radially arranged compartments. A transverse section through the stalk shows that the chloroplasts lie in a rather thin layer of cytoplasm between the cell wall and the central vacuole. Several whorls of apical hairs are borne on young plants, and similar appendages persist above the cap, and sometimes below, as the so-called *corona superior* and *corona inferior.* In *A. mediterranea* only one fertile cap is developed, whereas in *A. crenulata* the stalk tip continues to grow and to produce a succession of caps. The number, shape, and degree of adhesion of the cap rays, the form of the coronae, and the presence of hairs on the latter are the chief specific characters in the genus.

During the vegetative phase of development the nucleus does not divide, but undergoes an enormous enlargement correlated with the growth of the

* Regarding the taxonomy of this organism, see Silva, this treatise, Appendix A, Note 44.

cell. When the cap is almost completely differentiated, this primary nucleus divides into a great number (10,000–15,000 in *A. mediterranea*) of secondary nuclei, which within a few hours migrate up the stalk into the cap rays. Here cyst formation occurs as each nucleus becomes surrounded by a small portion of cytoplasm and a membrane. Further nuclear divisions follow, and with the eventual breakdown of the cap the cysts are released into the water. Meiosis then takes place, and flagellated isogametes are ultimately liberated through a preformed opening in the cyst membrane. They copulate, and the resulting zygotes begin to grow almost at once. Under laboratory conditions, the life cycle (see Fig. 1) can be completed in about 6 months; in nature it takes at least a year. (For details, see Schulze, 1939).

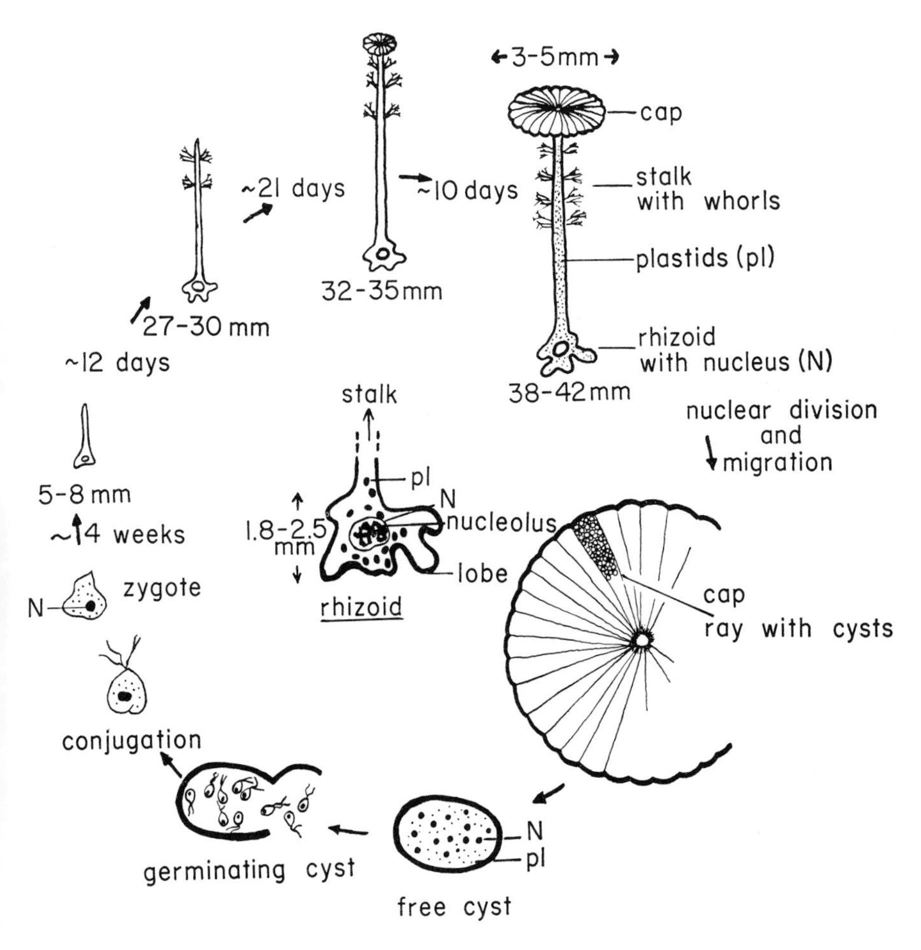

FIG. 1. Life cycle and general anatomy of *Acetabularia mediterranea*.

III. Morphogenetic Studies

A. Cell Regeneration

As the single nucleus is situated in the rhizoid, it can easily be removed by amputation (Fig. 2). Anucleate "cells" so obtained display a remarkable capacity for regeneration (Hämmerling, 1932, 1934a, b, c). The stalk can continue to grow, producing several whorls of hairs and finally a normal cap, or, in case of *A. crenulata*, several caps (Hämmerling 1943a). Many anucleate cells also form additional hair whorls and a cap at the basal end. When a cell is cut into several portions (Fig. 2), these may retain some morphogenetic capacity, depending on their original position in the cell. Apical stalk portions tend to elongate and form a cap at one or both ends; but the anucleate basal portions, especially if derived from older cells, seldom produce a new stalk and never produce a normal cap.

Nucleate basal portions begin to regenerate a complete cell after a few days, and can be made to do this repeatedly by successive amputations of the newly formed stalks. An isolated rhizoid, or even just the lobe containing the nucleus, is sufficient for regeneration of a normal plant. According to Hämmerling (1934a):

(1) The capacity of a cell for morphogenesis is determined by its content of "morphogenetic substances." These are continuously produced by the nucleus, or under its close control; they accumulate in the cytoplasm, and may control morphogenesis even after removal of the nucleus.

(2) Since an apical anucleate portion has more "morphogenetic capacity" than a basal one of similar length, a concentration gradient of morphogenetic substances may be postulated.

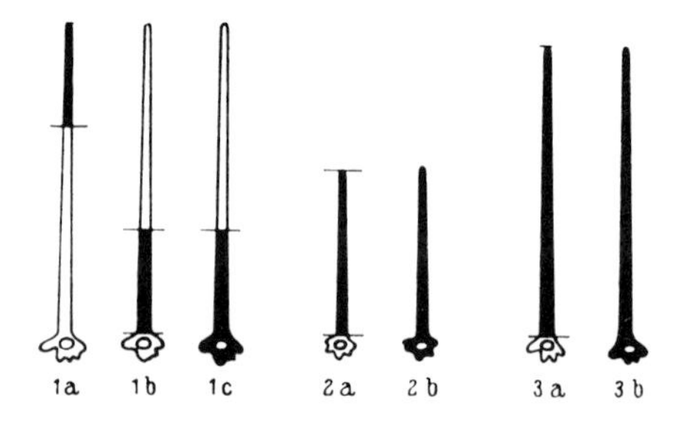

Fig. 2. Diagram showing (in black) the different types of nucleate and anucleate cells and cell portions most frequently used for the investigations discussed in this chapter: (*1a*) apical or anterior, (*1b*) basal or posterior, and (*1c*) rhizoid portion; (*2a*) and (*3a*) anucleate cells, (*2b*) and (*3b*) normal nucleate cells.

B. Intercellular Grafts

In one of the first experiments of Hämmerling (1934a), an anucleate, decapitated portion was grafted on a rhizoid by inserting its apical end into the cut end of the latter. After a few days the original basal end of the stalk started to grow and to form a cap, indicating a reversal of polarity. Similarly a basal section, grafted on a decapitated distal portion of another stalk, was induced to form a cap, presumably under the influence of morphogenetic substances which migrated into it (Sato, unpublished).

Once *Acetabularia* had been found suitable for grafting experiments, new approaches to the problem of morphogenesis were possible. The experiments described below were done with the following species: *A. mediterranea* (Hämmerling, 1932, 1934a, b, 1943b); *A. wettsteinii* (Hämmerling, 1934b, 1946); *A. crenulata* (Hämmerling, 1943a, b; Maschlanka, 1946; Werz, 1955); and *Acicularia schenkii* (Beth, 1943a, b; Maschlanka, 1943).

(*a*) Uninucleate systems result from grafting an anucleate stalk portion on a rhizoid base. Regeneration occurs at the free end of the stalk portion. A segment from another cell may be interposed to join the cut ends of apical and rhizoid section.

(*b*) Binucleate grafts are produced by joining the cut ends of the nucleate portions of two cells. A new stalk usually develops at the point of fusion, and later may produce whorls and one or more caps.

(*c*) A trinucleate system can easily be obtained by removing the tip of

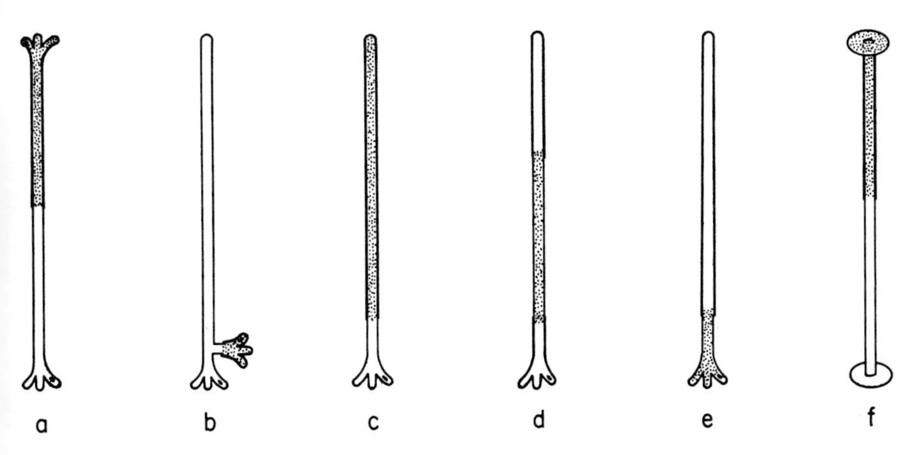

FIG. 3. Diagram of interspecific grafts between *cren.* (stippled) and *med.* portions: (*a*) two nucleate rhizoid portions; (*b*) *med.* rhizoid portion + *cren.* rhizoid; (*c*) *med.* rhizoid + *cren.* anucleate portion; (*d*) *cren.* anucleate portion joining a *med.* rhizoid and a *med.* apical portion; (*e*) *cren.* rhizoid + *med.* anucleate portion; (*f*) anucleate *med.* tip + anucleate *cren.* tip. (Redrawn, after Hämmerling, 1943b.)

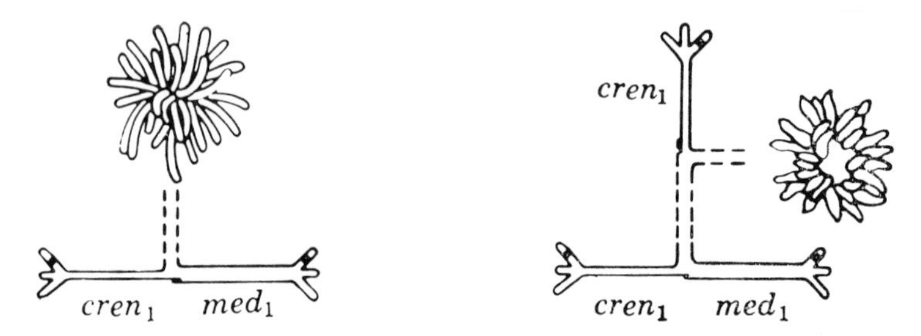

FIG. 4. Binucleate and trinucleate *cren.* + *med.* grafts and the resulting caps. Regenerated stalk indicated by broken lines. For further explanations, see Fig. 6. (From Hämmerling, 1953.)

the first regenerate stalk from a binucleate graft and adding a third rhizoid portion.

(*d*) Similarly, by combining two binucleate grafts a quadrinucleate system results.

Figures 3 and 4 show the various types of grafts so obtained.

It should be mentioned that in these experiments not only the nucleus but also a certain amount of the surrounding cytoplasm is transplanted. By means of another technique, however, it is possible successfully to "implant" single nuclei into anucleate cells or cell portions (Hämmerling, 1955; Richter, 1959a; see Section IV.A below).

In interspecific grafting experiments between *A. mediterranea* and *A. crenulata* (abbreviated below as "*med.*" and "*cren.*"), only certain differences in the structure of the mature caps were initially considered. Each species is characterized by the number of cap rays, by the presence or absence of hairs on the corona superior. Some of the experimental results obtained by Hämmerling or his co-workers from the different combinations outlined above may be summarized here, as follows:

(1) Control grafts containing one to four nuclei from the same species produce either pure *med.*- or pure *cren.*-type caps.

(2) A graft between two anucleate, distal, *med.* and *cren.* portions gives rise to a characteristic cap intermediate in character between the two species (Hämmerling, 1943b).

(3) In systems where an anucleate portion is grafted on a nucleate one of another species, the first-formed caps from the regenerated stalk may exhibit certain features of the anucleate distal partner, and a more or less intermediate type of cap is formed. If the first cap is amputated, the graft may produce a second cap with characters determined only by the species of the rhizoid nucleus (Fig. 5).

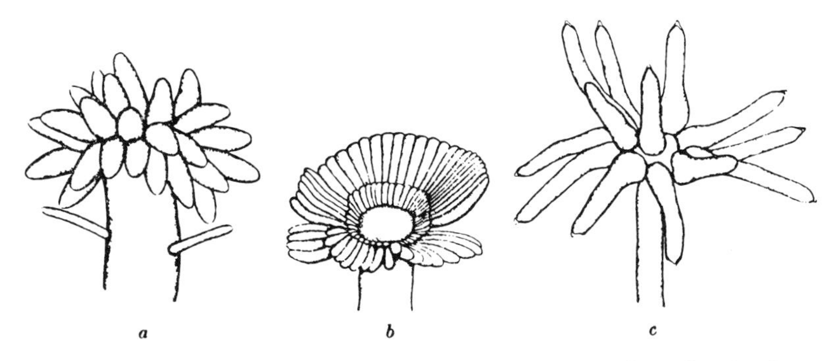

FIG. 5. Cap types produced by uninucleate 0 *cren.* + 1 *med.* grafts: (*a*) first cap, intermediate type, rays *cren.*-like, but without spurs; (*b*) second cap, formed after amputation of the first, *med.* type; (*c*) for comparison) *cren.* cap, large isolated rays with spurs. (From Hämmerling, 1934b.)

(4) Uninucleate grafts in which an anucleate *cren.* segment has been inserted between a *med.* tip and a *med.* base (Fig. 3) tend to produce intermediate-type caps, indicating that some morphogenetic influence has apparently migrated from the inserted *cren.* section into the *med.* tip.

(5) Trinucleate grafts with two *cren.* nuclei and one *med.* nucleus produce intermediate caps which tend towards the pure *cren.* type, although no spur is developed (Fig. 6).

A close correlation is thus indicated between the specific types and proportions of the nuclei and the final form of the cap. The postulated morphogenetic substances, produced under the control of the nucleus, are species-specific, and in binucleate and trinucleate grafts their effects appear to be additive. The evidence indicates also that anucleate portions retain some of the morphogenetic substance, and that longer segments contain more than shorter ones and therefore exert more profound effects on the ultimate form of the cap in interspecific grafts. Once a cap is formed, however, the local supply is depleted, and when replenished from the nucleus the morphogenetic substance carries only the specific imprint of the latter. Hence in uninucleate grafts between two species the second cap is no longer intermediate in character, but conforms to the species type of the nucleus present. Moreover, the supply can apparently be depleted even without cap formation, since if anucleate basal segments of one species are cultivated for 3 weeks before being grafted on nucleate rhizoidal portions of another, the caps ultimately formed are purely of the latter type (Richter, 1959b).

Another series of grafting experiments (Beth, 1943a, b, c) indicated the complexity of the morphogenetic substances. These grafts involved *Acetabularia mediterranea* and a strain of *Acicularia schenkii* in which, even under

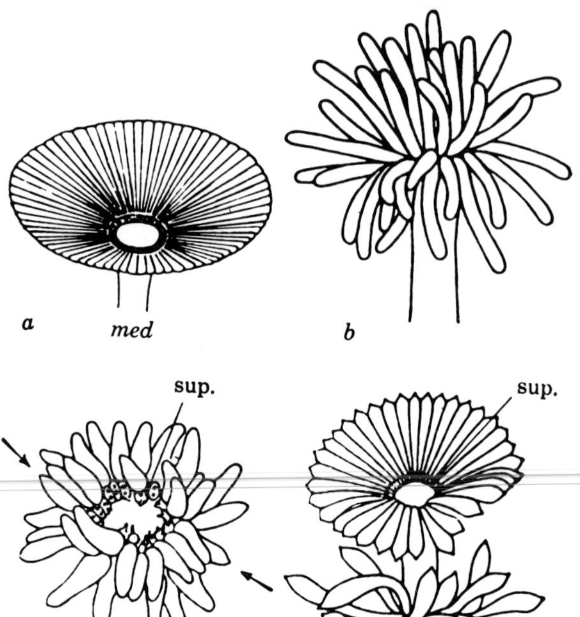

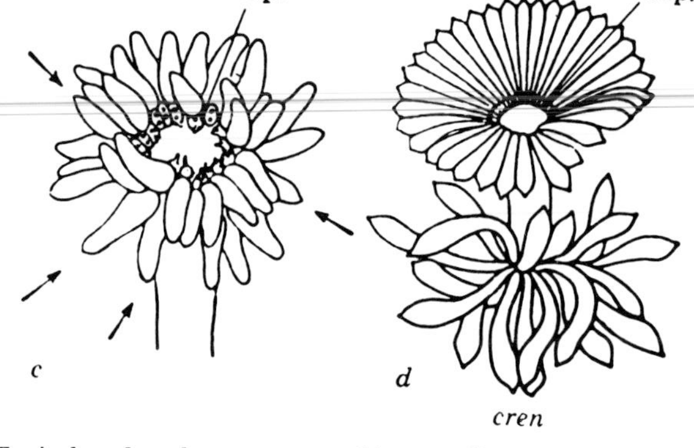

FIG. 6. Typical *med.* and *cren.* caps, and intermediate caps produced by interspecific grafts: (*a*) *med.* cap with 72 rays, no spurs; (*b*) intermediate cap from 1 *cren.* + 1 *med.* with 33 slim isolated rays without spurs; (*c*) "*cren.*-like" intermediate cap from 2 *cren.* + 1 *med.* with 34 rays, some pointed; (*d*) two *cren.* caps, the lower with isolated rays, the upper with coherent rays, each ray in both cases bearing a spur. "sup." = corona superior. (From Hämmerling, 1953.)

favorable conditions, cap formation occurs less frequently than in all the *Acetabularia* species examined. Among regenerating nucleate parts of *Acicularia* only about one-sixth produce caps; but grafts between nucleate or anucleate portions of *Acicularia* and nucleate portions of *Acetabularia mediterranea* regularly form caps, which usually exhibit typical *Acicularia* characters. Conversely, grafts between anucleate portions of *A. mediterranea* and nucleate portions of *Acicularia* generally fail to produce caps. Presumably the morphogenetic substances formed by *Acicularia* do not become fully functional unless combined with an *A. mediterranea* nucleus. Thus according to Hämmerling (1946) and Beth (1943b) we have to deal with two types of morphogenetic substances, respectively initiating and controlling the form of the caps. The former appear not to be species-specific, the latter are at least partially so.

Werz (1955) investigated the regeneration products of multinucleate grafts by examining the following interspecific combinations of nuclei: 3 *cren.* + 1 *med.*; 2 *cren.* + 1 *med.*; 2 *cren.* + 2 *med.*; 1 *cren.* + 1 *med.*; 1 *cren.* + 2 *med.*; 1 *cren.* + 3 *med.* Multinucleate control grafts of the same species, like 2 *cren.*; 3 *cren.*; 4 *cren.* and 2 *med.*; 3 *med.*; 4 *med.*, showed normal development and form of the caps. Generally only the features of the first cap formed in each graft were examined. The graft was then cut into its component nucleate portions; if they showed normal and complete regeneration, Werz concluded that their nuclei had been alive and active in the original graft. His results are summarized below.

(1) The earlier results of Hämmerling and Beth could be confirmed, in that the character of the cap depends on the nuclear balance of the two species in the graft. There is an uninterrupted sequence of intermediate types leading from the one species to the other as the proportionate number of the nuclei of the latter increases.

(2) In interspecific grafts, different features of the cap are influenced to different degrees. For example, in the degree of serration of the corona inferior and the shape of the outer margin of the rays, the *cren.* characteristic, a spur, is readily suppressed (Table I). The young initials of the cap rays, on the other hand, strongly reflect the *cren.* type in the combination 1 *cren.* + 2 *med.*, though this is suppressed by the presence of a third *med.* nucleus (i.e., 1 *cren.* + 3 *med.*). To account for these observations one might postulate a third type of morphogenetic substance, responsible, for instance, for the inhibitory action of the *med.* nuclei on the formation of *cren.* spurs.

(3) The rates of growth and of net protein synthesis are similar in uninucleate and multinucleate systems.

TABLE I

SPUR FORMATION ON *Acetabularia* CAP RAYS IN INTERSPECIFIC GRAFTS[a]

Number of nuclei	Apiculate	Acute	Spatulate	Obtuse	Emarginate
4 *cren.*	100				
3 *cren.* + 1 *med.*	40.9	23.7	13.7	21.7	
2 *cren.* + 1 *med.*	11.9	15.9	20.7	51.5	
2 *cren.* + 2 *med.*		0.1	14.3	85.4	
1 *cren.* + 2 *med.*			1.4	91.6	8.0
1 *cren.* + 3 *med.*				69.8	30.2
4 *med.*					100

[a] Percentage of ray tips classified in the five predominant categories.

(4) Nuclear division occurs synchronously and is normal in the grafts: 4 *med.*; 4 *cren.*; 3 *cren.* + 1 *med.*; 1 *cren.* + 1 *med.*; 1 *cren.* + 3 *med.* From the 4 *cren.* and 4 *med.* grafts, respectively, 75% and 91% of the caps formed normal cysts; in all interspecific combinations, however, the cysts degenerated. Evidently incompatibility is manifest only at the end of the vegetative phase.

Since these interspecific grafts involve both the nucleus and cytoplasm of both species, they do not present information on whether the morphogenetic substances of one species can act as well on the cytoplasm of another species as on its own. Experiments in which isolated nuclei, virtually free from cytoplasm, were implanted into anucleate cell portions of another species suggested that morphogenetic interactions can indeed occur (Hämmerling, 1955; also unpublished results).

C. Integrated Influence of the Nucleus and of External Factors

Acetabularia cultures grow well in illumination of 2,500 lux. When *A. mediterranea* is cultivated in relatively weak light (500 lux), no caps are formed, though the cells are capable of limited growth. If such cells are enucleated by amputation of the rhizoid and then transferred to bright light, normal cap formation is resumed. Beth (1953a) therefore considered that only the precursors of morphogenetic substances are formed in weak light, and that for their conversion into active compounds strong light is needed. Apparently this reaction is independent of nuclear control, since it can proceed in anucleate cells. This is taken to indicate that the determination of morphogenetic capacity in the cell ceases with the removal of the nucleus, although its speed of growth and differentiation is still influenced by light and temperature. Beth (1953b) showed that cells enucleated at an advanced stage of development produce caps at the same rate and of the same form as normal cells, but those enucleated at an earlier stage tend to form caps prematurely and of abnormal character. According to Beth, the presence of a rapidly growing nucleus in younger cells may retard early cap formation; consequently, enucleation would overcome this inhibition.

Further information on the relations between the nucleus and the production of the morphogenetic substances was provided by Hämmerling and Hämmerling (1959). Young cells of *A. mediterranea* were first kept in darkness for 11 days; the rhizoids were then amputated and the resulting anucleate portions transferred to normal illumination. In such portions growth, regeneration, and protein synthesis were appreciably more active than in controls which had not been subjected to the preliminary period of darkness. Evidently, though growth and other synthetic processes had ceased during the dark period, the nucleus had continued to produce the precursors of morphogenetic substances, which accumulated in the cytoplasm. These became effective in the light and were responsible for the subsequent high rates of growth, regeneration, and protein synthesis.

When normal plants were treated with 2,4-dinitrophenol or acriflavine (trypaflavine) the results obtained were in this respect similar to those from cells incubated in darkness (Brachet *et al.*, 1955; Werz, 1957a).

Evidently in *Acetabularia* the production of morphogenetic substances occurs under the intimate control of the nucleus. It proceeds even under conditions, such as insufficiency of light, in which many other synthetic processes do not function. The precursors formed under these conditions, however, must be activated by light before they become effective, a process which is independent of the nucleus. Once morphogenesis has started, there is no further requirement for strong light.

D. Influence of the Cytoplasm on the Nucleus

All of the experiments described so far were conducted primarily to investigate nuclear control of cytoplasmic activities. However, we must bear in mind that the cytoplasm can in turn influence the nucleus. Thus Hämmerling (1939, 1953, 1958) found that removal of the mature cap from an *Acetabularia* cell shortly before division of the primary nucleus postpones mitosis until after a new cap has been regenerated. By repeating this operation one can keep cells in their vegetative phase almost indefinitely. Conversely, when the stalk of an old plant with a mature cap is grafted on the rhizoid of a young plant, the primary nucleus starts to divide prematurely.

Darkness, treatment with inhibitors of oxidative phosphorylation, or localized ultraviolet irradiation reduces the size of both the nucleus and its nucleolus in a normal *Acetabularia* cell, indicating that the condition of the nucleus depends upon the level of cytoplasmic energy production (Stich, 1951, 1956; Brachet, 1952; Stich and Hämmerling, 1953; Brachet *et al.*, 1955; Hämmerling and Stich, 1956a, b; Brachet and Olszewska, 1960).

IV. Biochemical Studies

A. RNA and Protein Synthesis

Research on *Acetabularia* has more recently been concerned mainly with the relations between the nucleus and cellular metabolism. In 1953–1955 Brachet and his co-workers studied the incorporation of labeled $C^{14}O_2$ and orotic acid-C^{14} into the ribonucleic acid (RNA) fraction of *Acetabularia* cells. Since the positive results they obtained were mostly due to turnover or exchange, they did not thereby prove net synthesis of RNA, and it thus became important to follow the total RNA content of anucleate cells as a function of time. By means of a C^{14}-adenine isotope-dilution method, Brachet and his co-workers (1955) found that the increase of RNA in regenerating anucleate cells during the first few days after cutting was significantly more rapid than in their nucleate counterparts. Later, the synthesis of RNA became much more sluggish in the anucleate segments, though it continued to be detectable for several weeks. On the other hand Richter (1957, 1959c), using a modification of the Ogur and Rosen method for the extraction of RNA, was unable to confirm these observations of Brachet *et al.* In his experiments with both apical and basal segments, under conditions in which nucleate cells exhibited a constant linear increase of RNA and protein, net synthesis of RNA ceased immediately after removal of the nucleus although, as demonstrated by Vanderhaeghe (1954), protein synthesis remained active for about 21 days before slowing down. In nucleate basal portions, which possess the capacity for complete regeneration, there was a close correlation between regeneration and net synthesis of RNA (Figs. 7–9). These results were later confirmed with

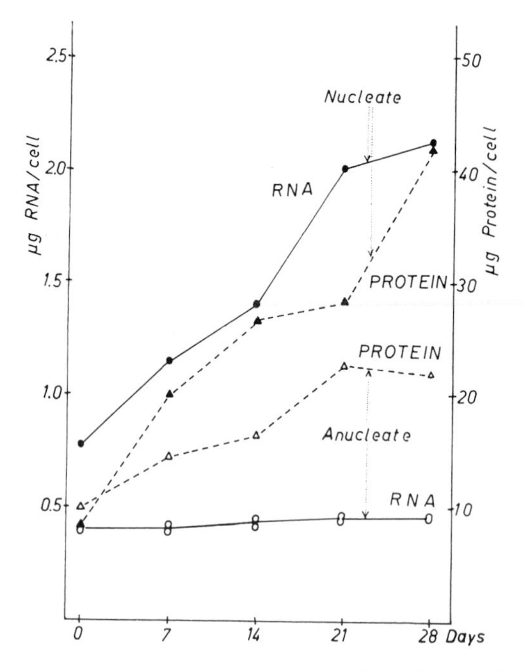

FIG. 7. RNA and protein synthesis in nucleate and anucleate cells of the same stalk length. The rhizoids of the nucleate cells were removed just before the determinations. (From Richter, 1959c.)

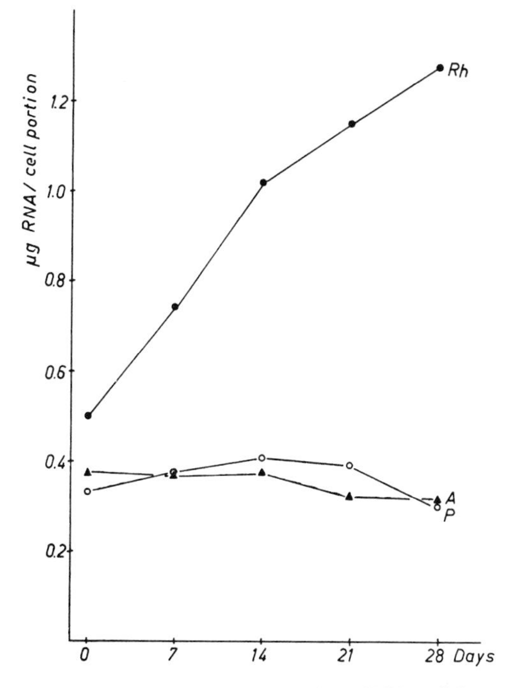

FIG. 8. RNA synthesis in nucleate rhizoid portions (Rh) and in anucleate apical (A) and basal (P) portions, all derived from plants of the same culture and initially all of equal length (12 mm.). (From Richter, 1959c.)

Hämmerling and Richter's cultures in Brachet's laboratory by the isotope-dilution method (Naora *et al.*, 1959).

In another series of experiments by Richter (1959a, b) "aged" anucleate basal portions of *Acetabularia mediterranea* were supplied with an active nucleus, either by grafting the portion on a nucleate rhizoid, or by direct implantation of an isolated nucleus (Fig. 10). Both *med.* and *cren.* nuclei were used, with similar results. There was a steady increase in the content of RNA and protein in the newly formed regenerate, which thus resembled a normal, growing cell, whereas the original anucleate stalk portion showed no change in form or color, and its RNA and protein content remained constant (Fig. 11). Removal of the nucleus from such a system by amputation of the rhizoid had the same effect as in normal plants: net synthesis of RNA ceased immediately, while active protein synthesis in the new stalk continued for at least three weeks.

It is therefore clear that the nucleus plays an important role in the synthesis of cytoplasmic RNA. However, one should not ignore the possibility

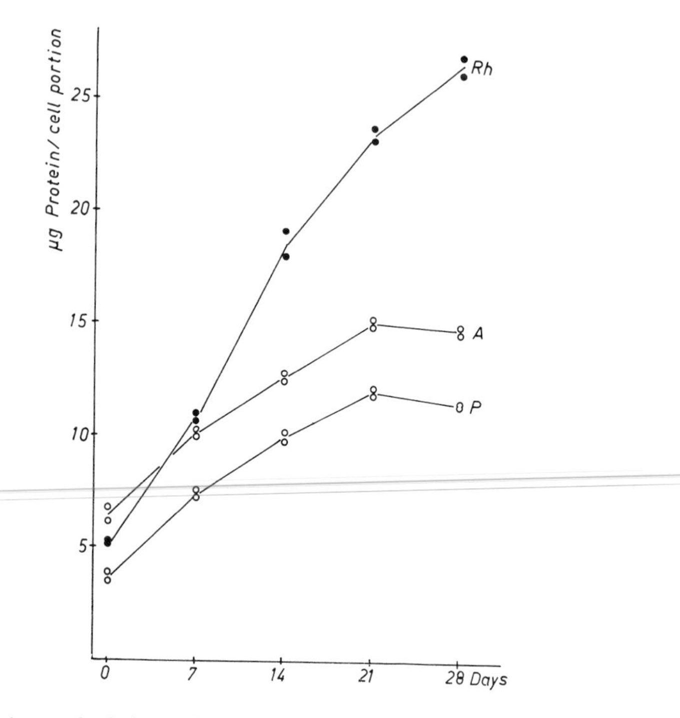

FIG. 9. Protein synthesis in nucleate rhizoid portions (Rh) and in anucleate apical (A) and basal (P) portions; c.f. Fig. 8. (From Richter, 1959c.)

that the apparent constancy of the RNA content in anucleate systems may result from simultaneous synthesis and breakdown in the cytoplasm, conceivably of different kinds of RNA. In fact, in anucleate cells of *Acetabularia mediterranea* subjected to ultraviolet irradiation, the total RNA decreased markedly for 16 days and then showed an appreciable increase (Richter, 1959d), thereby supporting the view that some RNA can be synthesized even in the absence of the nucleus. The effects of ultraviolet irradiation on whole cells resembled those of enucleation, in that the net synthesis of RNA was arrested while some protein synthesis still continued.

In experiments with C^{14}-labeled substrates, Naora *et al.* (1960) confirmed that no net synthesis of RNA takes place in anucleate portions, though these can readily incorporate adenine or orotic acid as well as CO_2. An important fact which emerged from these studies was that some RNA could still be synthesized in the chloroplasts (see Section IV.C), presumably at the expense of other cytoplasmic components. Further investigations of RNA synthesis in different cell fractions are needed to test this hypothesis.

The rate of net protein synthesis in anucleate cells and cell portions is highest in growing or regenerating parts; protein is also synthesized, how-

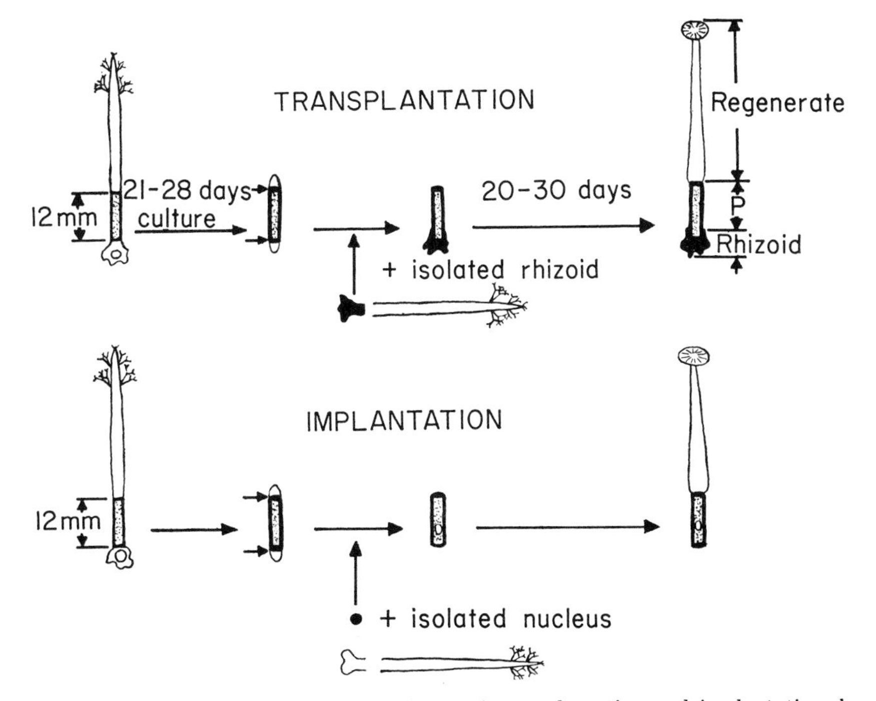

FIG. 10. Diagrams illustrating techniques of transplantation and implantation, by which anucleate "aged" cell portions are supplied with an active nucleus. (From Richter, 1959a.)

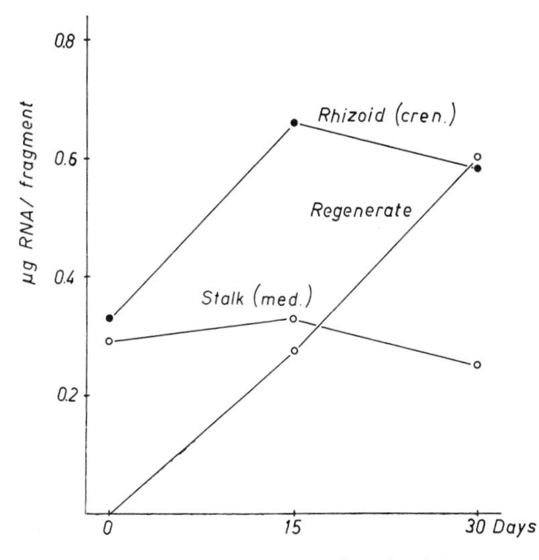

FIG. 11. RNA synthesis in three components of a mixed 1 *cren.* + 0 *med.* transplant, comprising a *cren.* rhizoid, an "aged" anucleate *med.* stalk portion, and the *cren.*-type regenerate. (From Richter, 1959b.)

ever, in anucleate basal portions, which exhibit practically no regenerative growth (Clauss, 1958; Hämmerling and Werz, 1958; Richter, 1959c; Werz and Hämmerling, 1959). Growth is thus not merely a consequence of protein synthesis, but is also dependent on morphogenetic substances (Hämmerling *et al.*, 1958). The actual limiting factor of protein synthesis is the degree of growth defined as volume increase (Werz and Hämmerling, 1959).

B. Enzymes

Among the proteins still synthesized in anucleate cells and cell portions, the enzyme systems are clearly affected to different degrees. Thus after removal of the nucleus the activity of phosphorylase and invertase increases considerably (Keck and Clauss, 1958; Clauss, 1959), that of aldolase increases only slightly (Brachet *et al.*, 1955; Baltus, 1959), while acid phosphatase activity soon begins to decrease (Keck and Clauss, 1958). It is not certain, of course, that these changes reflect the amounts of the enzymes themselves, but from the present evidence this cannot be denied.

C. Chloroplasts and Chloroplast Pigments

An increase in the number of chloroplasts in anucleate cells was reported by Hämmerling (1934b); details of this process, however, are still lacking. The ratios of chlorophyll *a*: chlorophyll *b*: carotenoids in *Acetabularia* cells remain unchanged after enucleation, while the quantities continue to increase appreciably for as long as 30 days after removal of the nucleus (Clauss, 1958; Richter, 1958). There is also a significant increase in the protein content of the plastids (Clauss, 1958). The steady increase in the number of these organelles, which can be shown to function normally, indicates that chloroplast RNA is synthesized even in the absence of a nucleus (Richter, 1958; see IV.A above).

D. Auxin

Thimann and Beth (1959) found that indoleacetic acid, at optimum concentrations of 10^{-5} to $10^{-6}M$, doubled the stalk elongation of normal cells of *Acetabularia mediterranea*. It had a marked influence on cap formation, too, though the optimum concentration required was higher, $10^{-4}M$. Naphthaleneacetic acid, at $10^{-5}M$, had a similar effect. Anucleate cells likewise responded to the addition of these substances by an increase of stalk growth and cap formation, but in general higher concentrations of auxin were needed. The auxin antagonist, 2,3,5-triiodobenzoic acid, inhibited cap formation by anucleate cells.

Clearly auxin is not one of the morphogenetic substances; since it does not increase cap formation in older plants which have already initiated this process, it cannot be the cap-producing substance either. Its role lies in promoting one or more of the processes involved in stalk growth and cap formation, possibly involved with cell-wall extension.

E. Miscellaneous

The effects of several metabolites and antimetabolites on growth and morphogenesis in normal and anucleate cells of *Acetabularia mediterranea* were studied by Brachet (1958, 1959). Other investigations of the metabolic activities of anucleate cells of *Acetabularia* included those on: the synthesis of soluble sugars (Clauss and Keck, 1959) and of cell wall material (Werz, 1957b; Clauss and Keck, 1959); photosynthesis, respiration, and ATP synthesis (Brachet *et al.*, 1955); formation of polyphosphates (Stich, 1955, 1956); and ammonia and phosphorus metabolism (Bremer and Schweiger, 1960; Schweiger and Bremer, 1960). All of these processes continue in anucleate cells, which in the synthesis of sugars and ammonia apparently surpass the capacity of their nucleate counterparts.

By cytochemical methods, Werz (1959, 1960a) identified in cells of *Acetabularia mediterranea* certain proteins and a highly polymerized RNA which showed a specific reaction with the dye azocarmine B. These were absent in cells or cell portions which did not grow or regenerate, and disappeared after enucleation or transfer of growing cells to darkness, suggesting some relationship with the morphogenetic substances postulated by Hämmerling. With a basic dye, azure I, Werz (1960b) demonstrated a specific reticulate structure in the growing zone of *Acetabularia* cells which had been predigested with trypsin to remove interfering proteins. The reticulum was apparently composed of an acidic polysaccharide, possibly a polyuronide. According to Werz it can always be detected in regions of cells where growth or morphogenesis is being initiated. The structure remains unchanged after about 30 days in darkness.

The importance of sulfur-containing proteins has also been demonstrated. Methionine accumulates at the distal tip of the cell when cap formation begins (Olszewska and Brachet, 1960), and agents like β-mercaptoethanol, which reduce —S—S— bonds in proteins, completely inhibit cap formation (Brachet, 1959). Furthermore, localized irradiation of the tip of the stalk with ultraviolet inhibits not only morphogenesis and net RNA synthesis, but also suppresses the formation of sulfur-containing proteins (Brachet and Olszewska, 1960). Evidently such proteins, like RNA, are synthesized in the nucleus, transferred to the cytoplasm, and accumulated in the growing tip, where they play an important role in morphogenesis.

References

Baltus, E. (1959). Evolution de l'aldolase dans les fragments anucléées d'*Acetabularia*. *Biochim. et Biophys. Acta* **33**, 337–339.

Beth, K. (1943a). Entwicklung und Regeneration von *Acicularia schenkii*. *Z. induktive Abstammungs-u. Vererbungslehre* **81**, 252–270.

Beth, K. (1943b). Ein- und zweikernige Transplantate zwischen *Acetabularia mediterranea* und *Acicularia schenkii*. *Z. induktive Abstammungs- u. Vererbungslehre* **81**, 271–312.

Beth, K. (1943c). Ein- und zweikernige Transplantate verschiedener *Acetabulariaceen*. *Naturwissenschaften* **31**, 206–207.

Beth, K. (1953a). Experimentelle Untersuchungen über die Wirkung des Lichtes auf die Formbildung von kernhaltigen und kernlosen *Acetabularia*-Zellen. *Z. Naturforsch.* **8b**, 334–342.

Beth, K. (1953b). Über den Einfluss des Kerns auf die Formbildung von *Acetabularia*. *Z. Naturforsch.* **8b**, 771–775.

Brachet, J. (1952). Quelques effets des inhibiteurs des phosphorylations oxydatives sur des fragments nucléés et énucléés d'organismes unicellulaires. *Experientia* **8**, 347–349.

Brachet, J. (1958). The effects of various metabolites and antimetabolites on the regeneration of fragments of *Acetabularia mediterranea*. *Exptl. Cell Research* **14**, 650–651.

Brachet, J. (1959). The role of sulfhydryl groups in morphogenesis. *J. Exptl. Zool.* **142**, 115–139.

Brachet, J., and Lang, A. (1961). Differentiation and development. *In* "Handbuch der Pflanzenphysiologie" (H. Ruhland, ed.), Vol. XV, in press. Springer, Berlin.

Brachet, J., and Olszewska, M. (1960). Influence of localized ultra-violet irradiation on the incorporation of adenine-8-^{14}C and D,L-methionine-^{35}S in *Acetabularia mediterranea*. *Nature* **187**, 945–955.

Brachet, J., Chantrenne, H., and Vanderhaeghe, F. (1955). Recherches sur les interactions biochimiques entre le noyau et le cytoplasm chez les organismes unicellulaires. II. *Acetabularia mediterranea*. *Biochim. et Biophys. Acta* **18**, 544–563.

Bremer, H. J., and Schweiger, H. G. (1960). Der NH_3-Gehalt kernhaltiger und kernloser Acetabularien. *Planta* **55**, 13–21.

Clauss, H. (1958). Über quantitative Veränderungen der Chloroplasten- und cytoplasmatischen Proteine in kernlosen Teilen von *Acetabularia mediterranea*. *Planta* **52**, 334–350.

Clauss, H. (1959). Das Verhalten von Phosphorylase in kernhaltigen und kernlosen Teilen von *Acetabularia mediterranea*. *Planta* **52**, 534–542.

Clauss, H., and Keck, K. (1959). Über die löslichen Kohlenhydrate der Grünalge *Acetabularia mediterranea* und deren quantitative Veränderungen in kernhaltigen und kernlosen Teilen. *Planta* **52**, 543–553.

Hämmerling, J. (1931). Entwicklung und Formbildungsvermögen von *Acetabularia mediterranea*. I. Die normale Entwicklung. *Biol. Zentr.* **51**, 633–647.

Hämmerling, J. (1932). Entwicklung und Formbildungsvermögen von *Acetabularia mediterranea*. II. Das Formbildungsvermögen kernhaltiger und kernloser Teilstücke. *Biol. Zentr.* **52**, 42–61.

Hämmerling, J. (1934a). Über formbildende Substanzen bei *Acetabularia mediterranea*, ihre räumliche und zeitliche Verteilung und ihre Herkunft. *Arch. Entwicklungsmech. Organ.* **131**, 1–81.

Hämmerling, J. (1934b). Regenerationsversuche an kernhaltigen und kernlosen Zellteilen von *Acetabularia wettsteinii*. *Biol. Zentr.* **54**, 650–665.

Hämmerling, J. (1934c). Entwicklungsphysiologische und genetische Grundlagen der Formbildung bei der Schirmalge *Acetabularia. Naturwissenschaften* **22,** 829–836.

Hämmerling, J. (1939). Über die Bedingungen der Kernteilung und Cystenbildung bei *Acetabularia mediterranea. Biol. Zentr.* **59,** 158–193.

Hämmerling, J. (1943a). Entwicklung und Regeneration von *Acetabularia crenulata. Z. induktive Abstammungs-u. Vererbungslehre* **81,** 84–113.

Hämmerling, J. (1943b). Ein- und zweikernige Transplantate zwischen *Acetabularia mediterranea* und *A. crenulata. Z. induktive Abstammungs-u. Vererbungslehre* **81,** 114–180.

Hämmerling, J. (1944). Zur Lebensweise, Fortpflanzung und Entwicklung verschiedener Dasycladaceen. *Arch. Protistenk.* **97,** 7–56.

Hämmerling, J. (1946). Neue Untersuchungen über die physiologischen und genetischen Grundlagen der Formbildung. *Naturwissenschaften* **33,** 337–342, 361–365.

Hämmerling, J. (1953). Nucleocytoplasmic interactions in the development of *Acetabularia. Intern. Rev. Cytol.* **2,** 475–498.

Hämmerling, J. (1955). Neuere Versuche über Polarität und Differenzierung bei *Acetabularia. Biol. Zentr.* **74,** 545–554.

Hämmerling, J. (1958). Über wechselseitige Abhängigkeit von Zelle und Kern. *Z. Naturforsch.* **13b,** 440–448.

Hämmerling, J., and Hämmerling, C. (1959). Kernaktivität bei aufgehobener Photosynthese. *Planta* **52,** 516–527.

Hämmerling, J., and Stich, H. (1956a). Einbau und Ausbau von ^{32}P im Nucleolus (nebst Bemerkungen über intra- und extranucleäre Proteinsynthese). *Z. Naturforsch.* **11b,** 158–161.

Hämmerling, J., and Stich, H. (1956b). Abhängigkeit des ^{32}P-Einbaues in den Nucleolus vom Energiezustand des Cytoplasmas, sowie vorläufige Versuche über Kernwirkungen während der Abbauphase des Kernes. *Z. Naturforsch.* **11b,** 162–165.

Hämmerling, J., and Werz, G. (1958). Über den Wuchsmodus von *Acetabularia. Z. Naturforsch.* **13b,** 449–454.

Hämmerling, J., Clauss, H., Keck, K., Richter, G., and Werz, G. (1958). Growth and protein synthesis in nucleated and enucleated cells. *Exptl. Cell Research Suppl.* **6,** 210–226.

Keck, K., and Clauss, H. (1958). Nuclear control of enzyme synthesis in *Acetabularia. Botan. Gaz.* **120,** 43–49.

Klebs, G. (1887). Über den Einfluss des Kernes in der Zelle. *Biol. Zentr.* **7,** 161–168.

Maschlanka, H. (1943). Zweikernige Transplantate zwischen *Acetabularia crenulata* und *Acicularia schenkii. Naturwissenschaften* **31,** 549.

Maschlanka, H. (1946). Kernwirkungen in artgleichen und artverschiedenen *Acetabularia*-Transplantaten. *Biol. Zentr.* **65,** 167–176.

Naora, H., Richter, G., and Naora, H. (1959). Further studies on the synthesis of RNA in enucleate *Acetabularia mediterranea. Exptl. Cell Research* **16,** 134–136.

Naora, H., Naora, H., and Brachet, J. (1960). Studies on independent synthesis of cytoplasmic ribonucleic acids in *Acetabularia mediterranea. J. Gen. Physiol.* **43,** 1083–1102.

Olszewska, M., and Brachet, J. (1960). Incorporation de la D,L-méthionine-^{35}S dans l'algue *Acetabularia mediterranea. Arch. intern. physiol. et biochem.* **68,** 693–694.

Richter, G. (1957). Zur Frage der RNS-Synthese in kernlosen Teilen von *Acetabularia. Naturwissenschaften* **44,** 520–521.

Richter, G. (1958). Das Verhalten der Plastidenpigmente in kernlosen Zellen und Teilstücken von *Acetabularia mediterranea. Planta* **52,** 258–275.

Richter, G. (1959a). Die Auslösung kerninduzierter Regeneration bei gealterten kernlosen Zellteilen von *Acetabularia* und ihre Auswirkungen auf die Synthese von RNS und Cytoplasmaproteinen. *Planta* **52,** 554–564.

Richter, G. (1959b). Regeneration und RNS-Synthese bei der Einwirkung eines artfremden Zellkerns auf gealterte kernlose Zellteile von *Acetabularia mediterranea. Naturwissenschaften* **45,** 629–630.

Richter, G. (1959c). Die Auswirkungen der Zellkernentfernung auf die Synthese von RNS und Cytoplasma-Proteinen bei *Acetabularia mediterranea. Biochim. et Biophys. Acta* **34,** 407–419.

Richter, G. (1959d). Das Verhalten von RNS und löslichen Cytoplasma-Proteinen in u.v.-bestrahlten kernhaltigen und kernlosen Zellen von *Acetabularia. Z. Naturforsch.* **14b,** 100–104.

Schulze, K. L. (1939). Zytologische Untersuchungen an *Acetabularia mediterranea* und *Acetabularia wettsteinii. Arch. Protistenk.* **92,** 179–225.

Schweiger, H. G., and Bremer, H. J. (1960). Das Verhalten verschiedener P-Fraktionen in kernhaltigen und kernlosen *Acetabularia mediterranea. Z. Naturforsch.* **15,** 395–400.

Stich, H. (1951). Experimentelle karyologische und cytochemische Untersuchungen an *Acetabularia mediterranea. Z. Naturforsch.* **6b,** 319–326.

Stich, H. (1955). Synthese und Abbau der Polyphosphate von *Acetabularia* nach autoradiographischen Untersuchungen des ^{32}P-Stoffwechsels. *Z. Naturforsch.* **10b,** 281–284.

Stich, H. (1956). Änderungen von Kern und Polyphosphaten in Abhängigkeit von dem Energiegehalt des Cytoplasmas von *Acetabularia. Chromosoma* **7,** 693–707.

Stich, H., and Hämmerling, J. (1953). Der Einbau von ^{32}P in die Nucleolarsubstanz des Zellkerns von *Acetabularia mediterranea. Z. Naturforsch.* **8b,** 329–333.

Thimann, K. V., and Beth, K. (1959). Action of auxins on *Acetabularia* and the effect of enucleation. *Nature* **183,** 946–948.

Vanderhaeghe, F. (1954). Les effets de l'énucléation sur la synthèse des proteines chez *Acetabularia mediterranea. Biochim. et Biophys. Acta* **15,** 281–287.

Van Wisselingh, C. (1909). Zur Physiologie der *Spirogyra*-Zelle. *Botan. Centr. Beih.,* **24,** 1. Abt., 133–210.

Waris, H. (1951). Cytological studies on *Micrasterias.* III. Factors influencing the development of enucleate cells. *Physiol. Plantarum* **4,** 387–409.

Werz, G. (1955). Kernphysiologische Untersuchungen an *Acetabularia. Planta* **46,** 113–153.

Werz, G. (1957a). Die Wirkung von Trypaflavin auf Kern und Cytoplasma von *Acetabularia mediterranea. Z. Naturforsch.* **12b,** 559–563.

Werz, G. (1957b). Membranbildung bei kernlosen wachsenden und nicht wachsenden Teilen von *Acetabularia. Z. Naturforsch.* **12b,** 739–740.

Werz, G. (1959). Über polare Plasmaunterschiede bei *Acetabularia. Planta* **53,** 502–517.

Werz, G. (1960a). Anreicherung von Ribonucleinsäure in der Wuchszone von *Acetabularia mediterranea. Planta* **55,** 22–37.

Werz, G. (1960b). Über Strukturierungen der Wuchszonen von *Acetabularia mediterranea. Planta* **55,** 38–56.

Werz, G., and Hämmerling, J. (1959). Proteinsynthese in wachsenden und nicht wachsenden kernlosen Zellteilen von *Acetabularia. Planta* **53,** 145–161.

Yoshida, Y. (1956). On the senescence of chloroplast in the presence of nucleus in plasmolysed *Elodea* cells. *J. Fac. Sci. Niigata Univ., Ser.* **II,** 73–78.

—43—

Polarity

S. NAKAZAWA

Biology Department, Yamagata University, Yamagata, Japan

I. Introduction

Polarity, or axial differentiation, is a feature common to the growth of almost all cells, organs, and individuals. Among the algae, eggs of the Fucaceae and the stalks of *Acetabularia* have been studied most extensively in this connection because, in their relative simplicity, they are convenient subjects for experimental morphogenetic analysis. There have been a few studies of polarity in other algae, notably *Caulerpa* (Zimmermann, 1929; Dostál, 1945), *Bryopsis* (Steinecke, 1925; Jacobs, 1951), *Sphacelaria* (Zimmermann and Heller, 1956), and *Enteromorpha* (Weber, 1959); reference to these may be found in the general reviews by Bloch (1945), Bünning (1958) and Nakazawa (1960b).

II. Eggs of Fucoids

A. Early Observations

The *Fucus* egg is spherical, 60 to 90 μ in diameter, with no visible polar differentiation before or just after the fertilization. Upon germination, a protuberance arises at one side of the egg which elongates to form the primary rhizoid, while the rest of the egg gives rise to the main part of the embryo. Thus the determination of the point of origin of the rhizoid indicates the axis of polarity of the egg. The number of primary rhizoids varies according to the species, being generally 1 in *Fucus* and *Pelvetia*, 4 in

Cystoseira and *Cystophyllum crassipes*, 8 in *Coccophora* and *Sargassum confusum*, 16 in *Sargassum enerve* and other species, and 32 in *S. tortile*, *Cystophyllum sisymbrioides*, etc. Since Rosenvinge's (1889) famous experiments demonstrating that the rhizoid of the *Fucus* egg originated on the side away from the light, many investigations on this object have been reported. Most of the classical ones were reviewed in 1940 by Whitaker. The polarity of various fucoid eggs may be determined by (*a*) unilateral white light, (*b*) unilateral ultraviolet irradiation, (*c*) concentration gradients of auxin and, more unequivocally, of dinitrophenol, (*d*) pH gradients, (*e*) "group effects," (*f*) temperature gradients, (*g*) an electric field, (*h*) mechanical deformation, (*i*) centrifugal redistribution of the cell contents, and (*j*) the point of entry of the spermatozoid. All of these apparently diverse effects were explained by Whitaker's unified CO_2–pH gradient theory, according to which the rhizoid originates in the region where the auxin activity is optimum, this being a function of the pH which in turn is determined by the concentration of CO_2 evolved in respiration.

B. Effects of Light

The determination of polarity by unilateral light was reexamined by Child (1941), who noted that bipolar embryos could be obtained by reversing the direction of the light at hourly intervals. The minimum duration of exposure for the determination of an axial gradient was tested by Haupt (1958), who found that it could not be reduced by an increase in the light intensity; thus determination required at least 60 minutes even under 14,000 lux of white light. A similar determination of polarity by light was recognized by Subrahmanyan (1957) in *Pelvetia canaliculata*.

Following up the observations by Child (1941), Jaffe (1956) found a striking polarotropic response in *Fucus* eggs. If the eggs are exposed to plane-polarized white light coming both from above and below, they tend to germinate horizontally in the plane of vibration, forming two rhizoids in opposite directions. This implies that the polarity of the *Fucus* egg arises epigenetically, rather than being determined by the rotation of a preformed asymmetric structure. If the eggs are exposed to unilateral polarized light, the rhizoids are also formed on opposite sides, their regions of origin being restricted to the subequatorial zone, 90 to 135 degrees away from the light source (Jaffe, 1958). The mechanism Jaffe proposed is illustrated in Fig. 1, which shows that, when the egg is illuminated from above, only the subequatorial zone *A* remains dark, while zone *B* receives light considerably reduced by reflection. Thus, the rhizoids evidently originate in the darkest region. The mechanism of their bipolar origin, in the plane of polarization of incident light, is indicated in Fig. 2. It is postulated

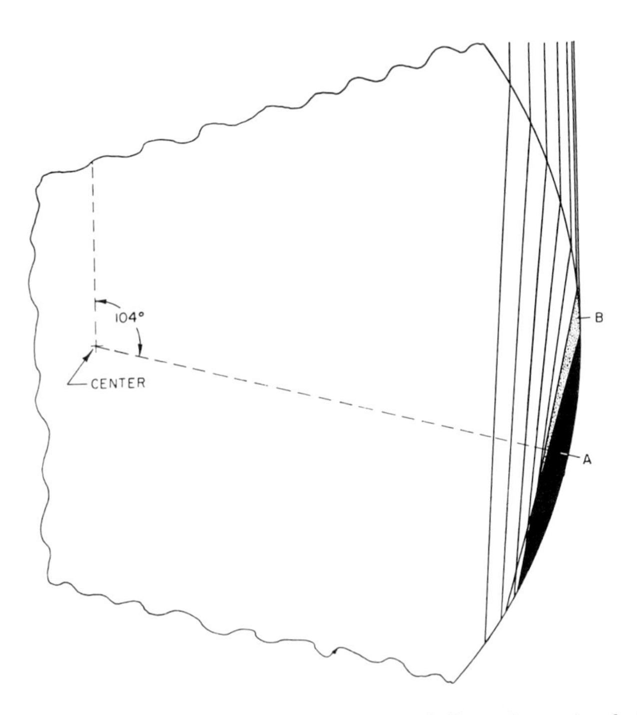

FIG. 1. Model of *Fucus* egg, unilaterally illuminated. Zone *A* remains dark, while in zone *B* at least half of the light is lost by reflection. (From Jaffe, 1958).

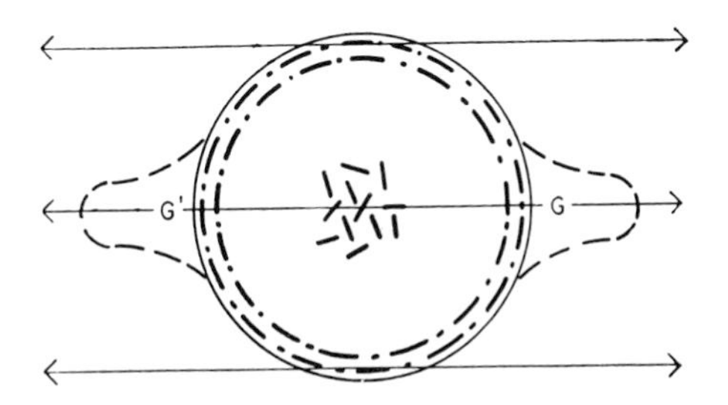

FIG. 2. Model of *Fucus* egg showing that the photoreceptor molecules in the germinating regions (*G* and *G'*) absorb least light if they are periclinally oriented near the cell surface. Light enters perpendicular to plane of page, vibrating as shown by double-headed arrows. Dashes and dots represent the photoreceptor molecules. (From Jaffe, 1958.)

that the photoreceptor molecules of the egg are peripheral, with their axes of maximum absorption arranged periclinally, so that those on opposite sides of the egg (*G* and *G'*) absorb least light, and it is therefore in these regions that the rhizoids originate. When *Fucus* eggs are irradiated unilaterally with monochromatic ultraviolet light (2804 Å.), the primary rhizoid ultimately originates at the shaded side. If, however, such ultraviolet-irradiated eggs are immersed in a hypertonic sucrose solution, plasmolysis occurs on the non-irradiated side (Reed and Whitaker, 1941), which implies that rhizoid formation is preceded by differentiation in physical properties at the periphery of the cytoplasm. If, according to Whitaker's theory, the origin of the rhizoid is in fact determined by a local optimum concentration of auxin, then, Bünning (1958) suggested, the effect observed by Reed and Whitaker may be a consequence of the photochemical destruction of auxin by ultraviolet light absorbed by riboflavin. According to Bünning (1958) the absorption spectrum of riboflavin actually exhibits three peaks between 250 and 500 mμ, corresponding to three of the four wavelengths found most effective for the induction of polarity in *Fucus* eggs, and thus providing indirect evidence for this hypothesis.

C. Other External and Internal Factors

The CO_2–pH gradient theory of Whitaker was questioned by Jaffe (1955), who discovered that the tendency of clustered *Fucus* eggs to produce their rhizoids towards each other (i.e., the "group effect") persisted despite the replacement of respiratory CO_2 emission by photosynthetic CO_2 uptake, and was not influenced by a 500-fold change in the buffer capacity of the medium.

Eggs of *Coccophora*, *Fucus*, and *Sargassum* are stained vitally with neutral red, brilliant green, and some other basic dyes. The staining is uniform before the determination of a polar axis, but tends to begin selectively at the rhizoid pole when this has become morphologically differentiated as a bulge, indicating that permeability to the dye is highest at the rhizoid pole. Neither this polar difference of permeability nor the consequent site of origin of the rhizoid is altered if the contents of the egg are stratified by centrifuging at 50,000 *g* (Nakazawa, 1957, 1959a). When immersed in a hypertonic solution of NaCl, the *Coccophora* egg at anaphase or telophase of the first nuclear division undergoes a peculiar plasmolysis in the equatorial zone as well as at the prospective rhizoid pole. This plasmolysis pattern, too, is stable even if the egg at that stage is strongly centrifuged (Nakazawa, 1960a). These facts suggest that the polarity of the egg is based on differentiation in the cortical cytoplasm, which is almost unaffected by centrifugation. On the other hand, no sign of polarity was detected in the peripheral cytoplasm by Levring (1947, 1949, 1952), who observed that the peripheral cytoplasm of the unfertilized egg of *Fucus* and some related algae uniformly exhibits double refraction, positive in the radial direction, indicating the presence of rod-shaped molecules arranged perpendicularly to the surface. After fertilization, however, the birefringence becomes negative in the radial direction when, presumably as the cell wall develops, elongate molecules are laid down parallel with its surface.

In *Coccophora* and *Sargassum* the rhizoid originates at the point of entry of the spermatozoid (Abe, 1941). A fertilization cone arises as the spermatozoid approaches the egg; the spermatozoid then enters the egg cell through the cone, which later bulges out to become the primary rhizoid. In *Coccophora langsdorfii* (Fig. 3) the jelly coat on one side of the egg is elongated to form a stalk, and this differential thickening of the mucilaginous envelope affects the polarity axis of the egg in two ways. In the first place, it tends to obstruct the approach of spermatozoids, so that there is a tendency for the first sperm cell to reach the egg on the side away from its stalk and to initiate the first rhizoid in that region. The observed frequency of rhizoids arising distally is 1.3 times the frequency of proximal rhizoids (Nakazawa, 1959b). In the second place, the polar axis of the egg of *Coccophora* or

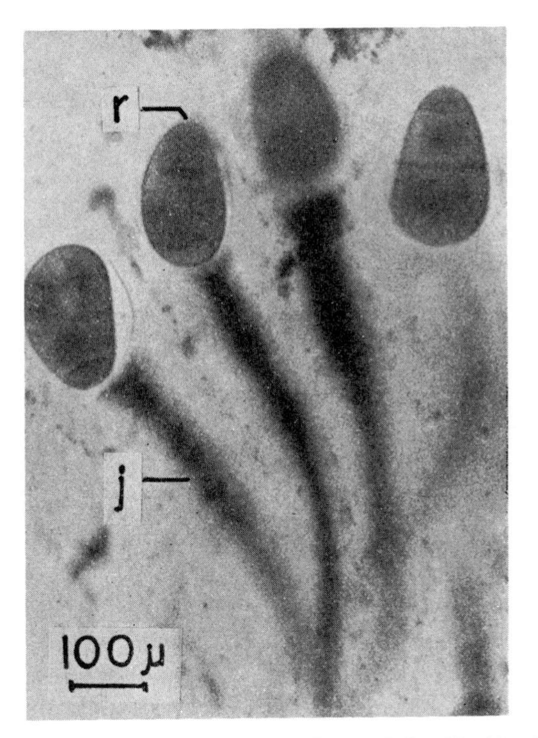

FIG. 3. Photograph showing the evident tendency of the rhizoid pole (*r*) of eggs of *Coccophora langsdorfii* to be oriented away from the jelly stalk (*j*). Latter is stained with Congo red. (From Nakazawa, 1959b.)

Sargassum is also determined to some extent by the shape of the cell, as was shown experimentally in the case of *Fucus* by Whitaker (1940). The form of the egg is in turn chiefly determined by the jelly coat, a small egg being completely spherical whereas a larger one tends to be elongated in the axis of its stalk (Nakazawa, 1957). In such elongated eggs, the rhizoid usually originates at either end of this axis; it only rarely arises laterally.

It should be borne in mind that polarity in an egg of *Sargassum* arises even if germination by parthenogenesis is induced by means of hypertonic treatment (Hiroe and Inoh, 1955a). Mature but unfertilized eggs of *Sargassum*, stored for two days in normal sea water, eventually produce a rhizoidal bulge, though in the absence of spermatozoids they do not develop further (Hiroe and Inoh, 1955b). This seems to show that even before fertilization an axis of polarity is inherent in the egg, though it is subject to modification by later events.

III. Other Algae

The polarity in the germination of *Vaucheria* zoospores was shown to be affected by unilateral visible light, gravity, and a "group effect" (Weber, 1958). Here, too, the mere displacement of the intracellular material by gentle centrifugation did not affect the determination of the polarity by other factors, indicating that the polar differentiation of the *Vaucheria* spore resides in its peripheral cytoplasm.

In the giant cell of *Acetabularia*, prior to the actual formation of the cap, there appears an apicobasal gradient, postulated by Hämmerling (1934) to be determined by a morphogenetic substance released from the nucleus. Morphogenetic substances accumulated in the apical zone are invoked to account for the fact that the rate of protein synthesis in an apical, enucleated piece is twice as high as in a piece of the same length cut out from the basal region (Hämmerling *et al.*, 1959); see Fig. 4. Gradients of an unidentified material which stains with azocarmine B were presumed to be closely related to that of the morphogenetic substance (Werz, 1959). From studies of regeneration of the cap in enucleated *Acetabularia* sections, Hämmerling (1959) concluded that the gradients of morphogenetic substance are less steep in the dark and therefore are not controlled by the polar structure of the protoplasm alone. However, it seems certain that polar gradients must always have their basis in the natural polarity of the protoplasm itself (Nakazawa, 1960b). The protoplasmic polarity may in turn be attributed to an orderly arrangement of polar molecules in the cortical layer of the cytoplasm (Bünning, 1958; Nakazawa, 1960b). Thus the spiral arrangements of plastids in *Euglena, Chara,* etc., may be considered to indicate the spiral nature of the peripheral layers of protoplasm

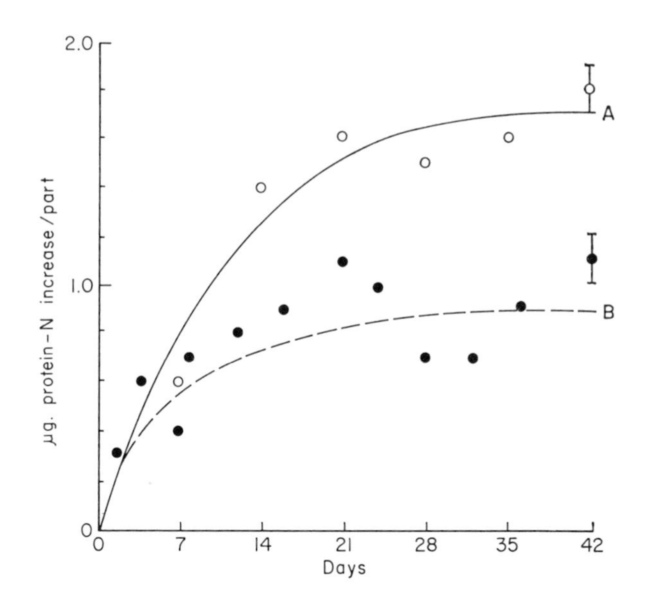

FIG. 4. Increase of protein nitrogen of 10 mm. segments of *Acetabularia mediterranea* in light: (*A*) Enucleated anterior portions; (*B*) Enucleated posterior portions. (From Hämmerling *et al.*, 1959.)

(Bünning, 1958). Since the axis of the spiral corresponds with the growth axis of the alga, it is presumably determined by the inherent polarity of the cell. Similarly, cell walls of *Valonia* reveal under the polarization microscope a fine spiral structure with double refraction (Wilson, 1955), indicating that the micelles of cellulose are arranged spirally around the polar axis of the cell.

REFERENCES

Abe, K. (1941). Weitere Untersuchungen über die Befruchtung von *Coccophora* und *Sargassum*. *Sci. Repts. Tôhoku Imp. Univ. Fourth Ser.* **16,** 441–444.

Bloch, R. (1945). Polarity in plants. *Botan. Rev.* **9,** 261–310.

Bünning, E. (1958). Polarität und inäquale Teilung des pflanzlichen Protoplasten. *In* "Protoplasmatologia—Handbuch der Protoplasmaforschung" (L. V. Heilbrun and F. Weber, eds.), Vol. 8: Physiologie des Protoplasmas, Part 9a, pp. 1–86. Springer, Vienna.

Child, C. M. (1941). "Patterns and Problems of Development." University of Chicago Press, Chicago, Illinois.

Dostál, R. (1945). Morphogenetic studies on *Caulerpa prolifera*. *Česká akad. véd a umĕrí v Praze—Bull. intern. (sci., math. et nat.)* **46,** 133–149.

Hämmerling, J. (1934). Über formbildende Substanzen bei *Acetabularia mediterranea*, ihre räumliche und zeitliche Verteilung und ihre Herkunft. *Wilhelm Roux' Arch. Entwicklungsmech. Organ.* **131,** 1–81.

Hämmerling, J. (1959). Über Bildung und Ausgleich des Polaritätsgefälles bei *Acetabularia*. *Planta* **53,** 522–531.

Hämmerling, J., Clauss, H., Keck, K., Richter, G., and G. Werz (1959). Growth and protein synthesis in nucleated and enucleated cells. *In* "The Relationship Between Nucleus and Cytoplasm" (Symposium held at l'Université Libre de Bruxelles, Belgium, June 9–13, 1958). *Exptl. Cell Research Suppl.* **6,** 210–226.

Haupt, W. (1958). Beobachtungen zur Polaritätsinduktion bei keimenden *Fucus*eiern. Publications from the Biological Station, Espegrend (H. Brattstrom, ed.), No. 24, 9 pp. *In* "Universitetet I Bergen Årbok 1958. Naturvitenskapelig rekke No. 13." John Griegs Boktrykkeri, Bergen, Norway.

Hiroe, M., and Inoh, S. (1955a). Artificial parthenogenesis in *Sargassum piluliferum* C. Ag. *Botan. Mag. (Tokyo)* **67,** 271–274.

Hiroe, M., and Inoh, S. (1955b). Some experiments on the eggs of *Sargassum piluliferum* C. Ag. *Biol. Bull. Okayama Univ.* **2,** 85–94.

Jacobs, W. P. (1951). Studies on cell differentiation: the role of auxin in algae, with particular reference to rhizoid-formation in *Bryopsis*. *Biol. Bull.* **101,** 300–306.

Jaffe, L. (1955). Do *Fucus* eggs interact through a CO_2-pH gradient? *Proc. Natl. Acad. Sci. U.S.* **41,** 267–270.

Jaffe, L. (1956). Effects of polarized light on polarity of *Fucus*. *Science* **123,** 1081–1082.

Jaffe, L. (1958). Tropistic responses of zygotes of the Fucaceae to polarized light. *Exptl. Cell Research* **15,** 282–299.

Levring, T. (1947). Remarks on the surface layers and the formation of the fertilization membrane in *Fucus* eggs. *Medd. Götenborgs Botan. Trädgård* **17,** 97–105.

Levring, T. (1949). Fertilization experiments with *Hormosira banksii* (Turn.) Dcne. *Physiol. Plantarum* **1,** 45–55.

Levring, T. (1952). Remarks on the submicroscopical structure of eggs and spermatozoids of *Fucus* and related genera. *Physiol. Plantarum* **5,** 528–539.

Nakazawa, S. (1957). Developmental mechanics of Fucaceous algae. VI. A unified theory on the polarity determination in *Coccophora*, *Fucus* and *Sargassum* eggs. *Sci. Repts. Tôhoku Univ., Fourth Ser.* **23,** 119–130.

Nakazawa, S. (1959a). General mechanism of the polarity determination in some fucoid eggs. *Naturwissenschaften* **46,** 333–334.

Nakazawa, S. (1959b). Developmental mechanics of Fucaceous algae. XIII. Polarity determination in *Coccophora* eggs relating to the position of jelly stalk. *Sci. Repts. Tôhoku Univ. Fourth Ser.* **25,** 231–238.

Nakazawa, S. (1960a). Developmental mechanics of Fucaceous algae. XVI. Plasmolytic patterns in ultracentrifuged *Coccophora* eggs. *Phyton (Buenos Aires)* **15,** 129–136.

Nakazawa, S. (1960b). Nature of the protoplasmic polarity. *Protoplasma* **52,** 274–294.

Reed, E. A., and Whitaker, D. M. (1941). Polarized plasmolysis of *Fucus* eggs with particular reference to ultraviolet light. *J. Cellular Comp. Physiol.* **18,** 329–338.

Rosenvinge, K. L. (1889). Influence des agents extérieurs sur l'organisation polaire et dorsiventrale des plants. *Rev. gén. botan.* **1,** 53.

Steinecke, F. (1925). Über Polarität von *Bryopsis*. *Botan. Arch.* **12,** 97–118.

Subrahmanyan, R. (1957). Observations on the anatomy, cytology, development of the reproductive structures, fertilization and embryology of *Pelvetia canaliculata* Dcne. et Thun. III. Liberation of reproductive bodies, fertilization and embryology. *J. Indian Botan. Soc.* **36,** 374–395.

Weber, W. (1958). Zur Polarität von *Vaucheria*. *Z. Botan.* **46,** 161–198.
Weber, W. (1959). Das Polaritätsverhalten der *Enteromorpha*zelle bei Verzweigung und Schwärmerbildung. *Z. Botan.* **47,** 251–257.
Werz, G. (1959). Über polare Plasmaunterschiede bei *Acetabularia*. *Planta* **53,** 502–521.
Whitaker, D. M. (1940). Physical factors of growth. *Growth* (*Suppl.*) pp. 75–90.
Wilson, K. (1955). The polarity of the cell-wall of *Valonia*. *Ann. Botany* **19,** 289–292.
Zimmermann, W. (1929). Experimente zur Polarität von *Caulerpa* und zum allgemeinen Polaritätsproblem. *Wilhelm Roux' Arch. Entwicklungsmech. Organ.* **116,** 669–688.
Zimmermann, W., and Heller, H. (1956). Polarität und Brutknospenentwicklung bei *Sphacelaria*. *Pubbl. staz. zool. Napoli* **28,** 289–304.

—44—

Growth Substances[1]

HERBERT M. CONRAD AND PAUL SALTMAN[2]

Department of Biochemistry, School of Medicine, University of Southern California, Los Angeles, California

I. Characterization of Endogenous Growth Substances[3]

There is circumstantial evidence for growth hormones in many multicellular or filamentous algae, e.g., apical dominance and the abscission of old fruiting laterals in *Fucus* (Knight, 1947; Burrows, 1956), and the phototropic responses observed in members of the Dasycladaceae (Hämmerling, 1952; Puiseux-Dao, 1957), in *Bryopsis* (Darsie, 1939) and in *Vaucheria* (Weber, 1958); nevertheless, objective data are few.

Janke (1939) reviewed some of the early studies of auxins in algae. Van der Weij (1933), using the standard *Avena* curvature test, found auxin activity in the cell wall and attached protoplasm of *Valonia macrophysa*,

[1] This work is supported in part by a research grant from the National Science Foundation.

[2] Senior Research Fellow, United States Public Health Service, National Institutes of Health.

[3] Abbreviations used in this article:

Abbreviation	Compound	A 10^{-3} M solution contains:
IAA	3-Indoleacetic acid	175 mg./liter
IBA	3-Indolebutyric acid	203 mg./liter
IPA	3-Indolepropionic acid	189 mg./liter
NAA	Naphthaleneacetic acid	186 mg./liter
PAA	Phenylacetic acid	136 mg./liter
GA	Gibberellic acid	346 mg./liter

and, in much lower concentrations, in the vacuolar sap, but could detect none in *Gracilaria acicularis*, *Cystoseira barbata*, or *Codium tomentosum*. Du Buy and Olson (1937) with the same technique detected the presence of auxin in various parts of *Fucus vesiculosus*, the highest concentrations being in the egg cells and sperm, and the lowest in the thalli and fruiting tips. Van Overbeek (1940a, b) found auxin activity in *Macrocystis pyrifera* ranging from 0.5 μg. equivalents of IAA per kilogram fresh weight in the terminal blades down to about 0.15 μg. in the bladders and stipe, and in other Brown Algae in concentrations ranging from 0.05 μg./kilogram in *Desmarestia aculeata* to 0.5 μg./kilogram in *Fucus evanescens*. The absolute amounts recorded in all such extractions are subject to question, since Thimann and Skoog (1940), working with higher plants, showed that the amount of auxin obtained is a function of the time of extraction with ether, apparently due to a slow enzymatic liberation of auxin which may continue for a considerable time, in the case of *Ulva lactuca* (kept at 1°C.) for as long as 8 months (Thimann *et al.*, 1942). Williams (1949) demonstrated in *Laminaria agardhii* the presence of a growth substance similar to or identical with IAA.

Auxin or auxin-like substances may be synthesized by many algae. There are numerous reports of growth-stimulating agents in washings, extracts, or decoctions of algae (e.g., De Valera, 1940; Kylin, 1942; Pringsheim, 1946). Such evidence alone is unsatisfactory unless precautions are taken to separate, by chromatography or other means, the auxins from other potential growth factors, antagonists, or inactivating agents which might be present. Bentley (1958), however, critically examined by solvent extraction and chromatography the naturally occurring auxins of a variety of algae including *Chlorella pyrenoidosa*, *Oscillatoria* sp., and *Anabaena cylindrica*. One substance, which she separated by paper chromatography in ammoniacal isopropanol, moved with an R_F similar to that of IAA,[4] was active in the *Avena* coleoptile straight-growth test, and gave color reactions characteristic of indole compounds. She also obtained evidence for an unstable, water-soluble complex, which breaks down under certain conditions to yield a product similar to IAA, but which was not fully characterized.

II. Physiological Responses of Algae to Growth Substances

A. Studies with Auxins

The first report of an effect of a known growth factor on an alga was by Yin (1937). Using *Chlorella vulgaris*, he found no increase in the rate of cell division, but about a 20% increase in the final cell size after incubation

[4] R_f = distance the compound travels relative to the solvent front.

with 10 mg./liter IAA for 12 days. In view of the negative result of Pratt (1937), and observations on the cell size of *Chlorella* by later workers (see Hase, Chapter 40, this treatise), the specificity of this effect is questionable. In the same year a report on the effect of IAA on *Chlorella miniata, C. pyrenoidosa, Cystococcus cohaerans, Oocystis naegelii,* and *Scenedesmus flavescens* was published by Leonian and Lilly (1937), who concluded that it was stimulatory only at concentrations greater than 33.3 mg./liter. Pratt (1937) studied the stimulatory effects of IAA, IPA, and IBA on *Chlorella vulgaris,* and found the optimum concentration for each to be 50 mg./liter. After a 25-day incubation period he observed a 20-fold increase in cell number due to IAA, a 12-fold increase with IPA, and an 11-fold increase with IBA. However, probably because Pratt included sugar in this growth medium, he was unable to confirm Yin's statement that IAA causes an increase in cell size.

Brannon and Bartsch (1939) likewise studied the effects of IAA, NAA, and PAA on *Chlorella,* determining growth after 12 days by turbidometric methods and by direct cell counts. In concentrations of 3–30 mg./liter the yields were slightly increased. Brannon and Sell (1945) found a 4-fold increase in dry weight of *Chlorella pyrenoidosa* after a 30-day growth period in media containing 10 mg./liter IAA, although this concentration was found to be inhibitory to *Chlorella vulgaris* by Manos (1945). However, since Brannon *et al.* dissolved the substances to be tested in ethanol, their results should be discounted for reasons indicated by Bach and Fellig (1958) and Street *et al.* (1958). (See below.)

In an extensive study of the effects of auxins on the growth of Chlorophyta, Algéus (1946) noted great variations in the response of several species of *Scenedesmus* to growth hormones. For *S. obliquus,* IAA was the most potent growth-promoting substance of the several tested, while NAA, PAA, IPA, IBA, and phenylpropionic acid followed in decreasing order of activity. In a mineral medium containing 0.1 mg./liter of IAA, *S. obliquus* exhibited a 3-fold increase in cell number, without any increase in average cell size, whereas *S. dimorphus, S. acuminatus, S. acutiformis,* and *S. quadricauda* were either unaffected or were inhibited by the IAA. Of the two varieties of *Chlorella vulgaris* tested, one (strain A) responded with a 3-fold increase in cell number at 1.0 mg./liter IAA, while the other was severely inhibited. Roborgh and Thomas (1948), also using a variety of *C. vulgaris,* observed an increase in cell size, but not in number, in a medium containing 1 mg./liter IAA. These observations emphasize the high degree of variability which may be found in the responses of different species or of varieties of the same species.

Elliot (1939) noted a moderate relative increase after 15 days in the cell number of *Euglena gracilis* due to IAA at concentrations of 0.1 to 1.0 mg./

liter, but observed no increase in cell size. Ondratschek (1940) reported that IAA at a concentration of 10 mg./liter stimulated the development of several mixotrophic algae; his work was severely criticized by Lwoff (1947).

Kylin (1942) found that 100 mg./liter IAA stimulated cell division in *Ulva* and *Enteromorpha*, although Levring (1943) detected no effect of this agent on the growth of *Ulva*. Contrary to observations with higher plants, the cell size was not altered. Kylin also reported stimulation of growth with thiamine at an optimum concentration of 10 mg./liter, and glucose and ascorbic acid at 100 to 1000 mg./liter, indicating that the effects of IAA were not specific.

In experiments by Jacobs (1951), *Bryopsis plumosa* shoots produced no rhizoids in the controls, whereas in most cases rhizoids were formed in sea water containing 100 mg./liter IAA, an unusually high concentration for physiological action. Later Davidson (1952), in studies of the effects of growth substances on the increase in dry weight of *Rhizoclonium hieroglyphicum*, found that, after 16 days, IAA at concentrations ranging from 10^{-6} to $10^{-3}M$ stimulated growth by 37%, IBA at $10^{-9}M$ increased growth by 53%, and NAA at 10^{-9} M caused a 20% increase in growth. Since these experiments were carried out with ethanol solutions of the growth substances, the interpretation of the results may be questioned (see below).

In experiments with *Codium decorticatum*, Williams (1952) noted that after 31 days 5 mg./liter IAA induced an increase of approximately 30%, compared with controls, in the length of apical segments, but was without effect on segments from the base of the thalli. King (1954) obtained a 2- to 3-fold increase in the diameter of algal masses of *Phycopeltis hawaiiensis* grown in a medium containing 1 mg./liter IAA under continuous fluorescent light or natural sunlight. Dao (1956) reported a 4- to 5-fold increase in the growth of *Acetabularia mediterranea* in the presence of 1 mg./liter IAA, the effects of which varied with the seasons. A 3-fold increase was noted in the presence of 5 mg./liter of tryptophan; other amino acids were unfortunately not tested. Such studies, carried out in nonsterile sea water, are subject to serious experimental errors. Thimann and Beth (1959) reported a 2-fold stimulation in the growth of rhizoids of *A. mediterranea* after incubation for 25 days with IAA at 10^{-5} or $10^{-6}M$; a 4-fold increase in the rate of formation of caps (reproductive organs) was also observed. NAA was effective at concentrations about one-tenth those of IAA. Hustede (1957) reported that IAA promotes the formation and germination of zoospores in *Vaucheria*, and that tryptophan induces a similar effect.

Most of the studies mentioned so far have related to Green Algae; responses of a few Brown Algae to auxins have also been reported. Williams (1949) showed that concentrations of applied IAA above 0.1 mg./liter produced injurious effects on meristematic tissue of *Laminaria agardhii*. The following differential effects were noted: at 0.1 mg./liter, disintegration of transition zone between stipe and lamina; at 5.0 mg./liter, disintegration between middle and distal third of lamina; at 10.0 mg./liter, 30% loss of weight in the distal third and thin periphery of lamina. Davidson (1950) noted that the fronds of *Ascophyllum nodosum* responded to 10^{-9} to $10^{-4}M$ IAA with a distinct increase in apical growth compared to the control; the greatest response was about 50% with 10^{-5} M IAA dissolved in ethanol. Fronds treated with $10^{-7}M$ NAA responded with a similar increase; above $10^{-6}M$ NAA, growth was inhibited. IBA was inhibitory at all concentrations tested greater than $10^{-9}M$. There was a 15% increase in the apical growth of fronds of *Fucus evanescens* at the following concentrations: $10^{-5}M$ IAA; $10^{-8}M$ NAA; and $10^{-8}M$ IBA.

Harder and von Witsch (1943) observed no enhancement of the growth of freshwater pennate diatoms by IAA, but Bentley (1958) found that concentrations of 1 to 10 mg./liter not only resulted in an increase in the chain length of *Skeletonema* but also doubled the cell number, in relation to auxin-free controls grown for the same period of time. IAA produced no significant effect on either growth or cell size of *Monodus subterraneus* (Miller and Fogg, 1958).

B. Studies Involving Other Growth Substances

Provasoli (1957, 1958) found kinetin and IAA to favor initiation of filaments and elongation of sporelings of *Ulva lactuca*. He reported that the optimum concentration of IAA was 0.05 mg./liter for filament initiation, whereas 0.3 mg./liter proved inhibitory; 0.05 mg./liter IAA together with 0.1 mg./liter kinetin gave a far more pronounced response, although 0.1 mg./liter kinetin alone induced rhizoids without affecting elongation; GA at 0.1 mg./liter induced maximum elongation. Kinoshita and Teramoto (1958) reported that 0.01 mg./liter gibberellin induced an increase of 28% in the growth of sections of *Porphyra* fronds; they also observed accelerated cell division at 10–100 mg./liter. Griffin (1958) reported an increase in cell size of *Euglena* at 1000 mg./liter of GA after 48 hours; but since no effects were noticed at lower concentrations, this result should be accepted with caution. Thimann and Beth (1959) found no action of kinetin or GA on vegetative growth or cap formation in *Acetabularia*.

Conrad *et al.* (1959) observed that the net increase in wet weight of *Ulothrix subtilissima* grown for 15 days in a medium containing 3 μg./liter IAA was 13 times greater than the net increase observed in the same medium in the absence of IAA. The use of ethanol in this study was deliberately avoided (see below). Higher concentrations of IAA were inhibitory. The same alga gave a 7-fold increase in a medium containing 0.05 mg./liter GA. Here too, higher concentrations were inhibitory. The increase in wet weight was directly correlated with increased cell division. There was no interaction between IAA and GA when tested together.

III. General Experimental Considerations

The transport of IAA into algal cells has received some attention. Albaum *et al.* (1937) observed that the penetration of 0.0036*M* IAA into *Nitella* is more rapid at low pH values, with a maximum at about pH 4.0. The kinetics of penetration can be interpreted by a first-order equation indicative of diffusion or of a monomolecular reaction of the undissociated auxin. Algéus (1946) drew a similar conclusion from studies with *Chlorella vulgaris* and *Scenedesmus obliquus*, and also noted an increased promotion of cell division by IAA at low pH.

The critical work of Bach and Fellig (1958) and of Street *et al.* (1958) on the stimulation of the growth of *Chlorella vulgaris* by ethanol demonstrated that many of the previous studies on the apparent effects of growth substances on algae involved a serious experimental error, since the growth factors tested had been dissolved in alcoholic solutions. Street *et al.* (1958) found that, in the absence of ethanol, IAA produced only a 12% increase in cell number over controls, whereas in the presence of 0.4 ml./liter of ethanol there was a 150% enhancement of yield. Such effects, attributed merely to the nonspecific nutritional value of the alcohol, were noted both in cultures grown in the dark and in light. Bach and Fellig (1958) likewise demonstrated a stimulation of growth by alcohol in several algal species, including species of *Chlorella, Coccomyxa, Scenedesmus,* and *Euglena,* but in no case were they able to detect a significant growth-promoting effect of any of a wide variety of auxins tested. In the experiments of Algéus (1946), Thimann and Beth (1959), and Conrad *et al.* (1959) the use of ethanol was purposely avoided.

It is important to realize that an alga with a vitamin requirement, growing in a vitamin-deficient medium, would respond to organic supplements in a manner which might not be easily distinguishable from its response to a growth-substance. For this reason a careful analysis of both yield and growth rate should be made for any system tested. The profound morphological and physiological changes induced by growth substances in higher plants are far more readily characterized.

The use of algae in investigations of the mechanisms of action of plant growth substances has not been fully exploited. Some algae show a remarkable response in a comparatively short time. With proper precautions and adequate consideration of experimental artifacts there are many advantages to be gained by studying such simple plant systems, uncomplicated by the extensive morphological differentiation of higher plants.

Editorial note

[Though the English terminology is unsatisfactory, a general distinction can be made between *growth substances,* as discussed in this article, and *growth factors* or vitamins, as reviewed by Droop in Chapter 8. Agents of an auxin nature tend to change the form or the rate of growth, but growth can take place in media from which they are absent. Vitamins, on the other hand, are by definition essential growth factors, and in their absence no growth can occur, or growth is limited by the rate of synthesis of a vitamin. For micro-organisms, the distinction is usually clear from an examination of growth curves of cell densities plotted against time, with and without the supplement. If the growth *rates* are similar but the final *yield* is increased, we are dealing with an essential or auxiliary nutrient. If the final *yields* are similar, but the growth rate is accelerated, then we may consider the supplement as a growth factor. Confusion arises chiefly in those cases where a growth factor (vitamin) is synthesized at a rate so low that it limits growth, i.e., where the requirement is not absolute.]

References

Albaum, H. G., Kaiser, S., and Nestler, H. A. (1937). Relation of pH to penetration of IAA into *Nitella* cells. *Am. J. Botany* **24,** 513–518.

Algéus, S. (1946). Untersuchungen über die Ernährungsphysiologie der Chlorophyceen. *Botan. Notiser* pp. 129–280.

Bach, M. K., and Fellig, J. (1958). Effect of ethanol and auxins on the growth of unicellular algae. *Nature* **182,** 1359–1360.

Bentley, J. A. (1958). Role of plant hormones in algal metabolism and ecology. *Nature* **181,** 1499–1501.

Brannon, M. A., and Bartsch, A. F. (1939). Influence of auxin on the growth of *Chlorella vulgaris. Am. J. Botany* **26,** 271–279.

Brannon, M. A., and Sell, H. M. (1945). The effect of IAA on the dry weight of *Chlorella pyrenoidosa. Am. J. Botany* **32,** 257–258.

Burrows, E. M. (1956). Growth control in the Fucaceae. *Abstr. Intern. 2nd Symposium, Seaweed Symposium, Trondheim,* **1955** pp. 163–170.

Conrad, H., Saltman, P., and Eppley, R. (1959). Effects of auxin and gibberellic acid on growth of *Ulothrix. Nature* **184,** 556–557.

Dao, S. (1956). À propos de l'action du tryptophane sur l'*Acetabularia mediterranea. Compt. rend. acad. sci.* **243,** 1552–1554.

Darsie, M. L. (1939). Certain aspects of phototropism, growth and polarity in the single celled marine alga, *Bryopsis.* Thesis, Stanford University, Stanford, California.

Davidson, F. F. (1950). The effect of auxin on the growth of marine algae. *Am. J. Botany* **37,** 501–510.

Davidson, F. F. (1952). Effects of growth substances on *Rhizoclonium hieroglyphicum*. *Am. J. Botany* **39**, 700–704.
de Valera, M. (1940). Differences in growth of *Enteromorpha* in various culture media. *Kgl. Fysiograf. Sällskap. i Lund Förh.* **10**, 52–58.
du Buy, H. G., and Olson, R. A. (1937). The presence of growth regulators during early development of *Fucus*. *Am. J. Botany* **24**, 609–611.
Elliot, A. M. (1939). Effects of phytohormones on *Euglena* in relation to light. *Trans. Am. Microscop. Soc.* **58**, 385–388.
Griffin, D. N. (1958). The effect of gibberellic acid upon *Euglena*. *Proc. Oklahoma Acad. Sci.* **38**, 14–15.
Hämmerling. J. (1952). Über die Fortpflanzung von *Cymopolia*, nebst Bemerkungen über *Neomeris* und *Dasycladus*. *Biol. Zentr.* **71**, 1–10.
Harder, R., and von Witsch, H. (1943). Über Massenkultur von Diatomeen. *Ber. deut. botan. Ges.* **60**, 146–152.
Hustede, H. (1957). Untersuchungen über die stoffliche Beeinflussung der Entwicklung von *Stigeoclonium falklandicum* und *Vaucheria sessilis* durch Tryptophanabkömmlinge. *Biol. Zentr.* **76**, 555–556.
Jacobs, W. P. (1951). Studies in cell-differentiation: the role of auxin in algae, with particular reference to rhizoid formation in *Bryopsis*. *Biol. Bull.* **101**, 300–306.
Janke, A. (1939). The growth factor question in microbiology. *Zentr. Bakteriol., Parasitenk. Abt. II.* **100**, 409–459.
King, J. W. (1954). An investigation of haematochrome accumulation in *Phycopeltis hawaiiensis*. *Pacific Sci.* **8**, 205–208.
Kinoshita, S., and Teramoto, K. (1958). On the efficiency of gibberellin on the growth of *Porphyra*-frond. *Bull. Japan. Soc. Phycol.* **6**, 85–88.
Knight, M. (1947). Biological study of *Fucus vesiculosus* and *Fucus serratus*. *Proc. Linnean Soc. London* **159**, 87–90.
Kylin, H. (1942). Influence of glucose, ascorbic acid, and heteroauxin on the seedlings of *Ulva* and *Enteromorpha*. *Kgl. Fysiograf. Sällskap. i Lund Förh.* **12**, 135–148.
Leonian, L. H., and Lilly, V. G. (1937). Is heteroauxin a growth promoting substance? *Am. J. Botany* **24**, 135–139.
Levring, T. (1943). Some culture experiments with *Ulva* and artificial sea water. *Kgl. Fysiograf. Sällskap. i Lund. Förh.* **16**, 45–56.
Lwoff, A. (1947). Some aspects of the problem of growth factors for protozoa. *Ann. Rev. Microbiol.* **1**, 101–112.
Manos, E. (1945). The effect of heteroauxin on the growth of *Chlorella*. *Biol. Revs. City Coll. New York* **7**, 11–15.
Miller, J. D. A., and Fogg, G. E. (1958). Studies on the growth of Xanthophyceae in pure culture. II. The relation of *Monodus subterraneus* to organic substances. *Arch. Mikrobiol.* **30**, 1–16.
Ondratschek, K. (1940). Experimentelle Untersuchungen über den Einfluss von Wirkstoffen auf die Vermehrung einiger mixotropher Algen. *Arch. Mikrobiol.* **11**, 89–117.
Pratt, R. (1937). Influence of auxins on growth of *Chlorella vulgaris*. *Am. J. Botany* **25**, 498–501.
Pringsheim, E. G. (1946). "Pure Cultures of Algae." Cambridge Univ. Press, London and New York.
Provasoli, L. (1957). Effect of plant hormones on seaweeds. *Biol. Bull.* **113**, 321.
Provasoli, L. (1958). Effect of plant hormones on *Ulva*. *Biol. Bull.* **114**, 375–384.
Puiseux-Dao, S. (1957). Comportement de fragments anuclées de *Batophora Oerstedi*

J. Ag. (Dasycladacée) dans l'eau de mer contenant, soit un acide aminé, soit une auxine, *Comptes rend. acad. sci.* **245,** 2371–2374.

Roborgh, J. R., and Thomas, J. B. (1948). The synthesis of growth substances. I. The action of heteroauxin on the growth of *Chlorella vulgaris* and the influence of light on the production of growth substances. *Koninkl. Ned. Akad. Wetenschap. Proc.* **B51,** 87–99.

Street, H. E., Griffith, D. J., Thresher, C. L., and Owens, M. (1958). Ethanol as a carbon source for the growth of *Chlorella vulgaris. Nature* **182,** 1360–1361.

Thimann, K. V., and Beth, K. (1959). Action of auxins on *Acetabularia* and the effect of enucleation. *Nature* **183,** 946–948.

Thimann, K. V., and Skoog, F. (1940). The extraction of auxin from plant tissues. *Am. J. Botany* **27,** 951–960.

Thimann, K. V., Skoog, F., and Byer, A. C. (1942). The extraction of auxin from plant tissues. II. *Am. J. Botany* **27,** 598–606.

van der Weij, H. G. (1933). Occurrence of growth substances in marine algae. *Proc. Koninkl. Akad. Wetenschap., Amsterdam* **36,** 759–760.

van Overbeek, F. (1940a). Auxin in marine algae. *Plant Physiol.* **15,** 291–299.

van Overbeek, F. (1940b). Auxin in marine plants. *Botan. Gaz.* **101,** 940–947.

Weber, W. (1958). Zur Polarität von *Vaucheria. Z. Botan.* **46,** 161–198.

Williams, L. G. (1949). Growth regulating substances in *Laminaria agardhii. Science* **110,** 169–170.

Williams, L. G. (1952). Effects of IAA on growth in *Codium. Am. J. Botany* **39,** 107–109.

Yin, H. C. (1937). Effect of auxin on growth of *Chlorella vulgaris. Proc. Natl. Acad. Sci. U. S.* **23,** 174–179.

—45—

Inhibitors

R. W. KRAUSS

Department of Botany, University of Maryland, College Park, Maryland

I. Introduction

The occasional need to limit the growth of algae certainly antedates any efforts to promote it. Historically, attempts to remove algae from natural or man-made reservoirs of waters have led to a search for chemical agents more or less toxic to algae. In only rare cases has the total destruction of all life in a given body of water been the objective. Selective toxicity—leading to algal deterioration without detriment to the potability or the usefulness of the water to support other plant or animal life—has been the conscious goal of countless experiments. In view of the current attitude concerning the "unity" of biochemistry, achievement of that goal might appear remote or even theoretically impossible. Some success, however, has been achieved. Agents have been found which at low concentrations are lethal to algae but which are harmless, or only slightly toxic, to other forms of life. Conspicuous in this regard is copper sulfate, which has traditionally served as an eradicant of algae in lakes and ponds (Bartsch, 1954; McVeigh and Brown, 1954), although comparative studies have shown that susceptibility varies widely with the species. If the criterion for selectivity is taken to be toxicity for all algae, and for algae alone, then the success of the search must be considered conditional, though continued efforts are being made to find agents active against a broader spectrum of species. A justification of such exploration, especially in the field of plant-growth regulation, can be found in the excellent examples of the relatively higher toxicity of 2,2-dichloropropionic acid for monocotyledons than for dicotyledons, and the reverse situation for 2,4-dichlorophenoxyacetic acid and its derivatives.

TABLE I

MINIMUM CONCENTRATIONS OF ANTIBIOTICS INHIBITING GROWTH OF ALGAE IN INORGANIC MEDIA

Agent	Alga	Concentration (p.p.m.)[1]	Reference
Acti-dione	*Ankistrodesmus falcatus*	1	Zehnder and Hughes (1958)
	Gomphonema parvulum	2	Palmer and Maloney (1955)
	Haematococcus lacustris	1	Zehnder and Hughes (1958)
	Nitzschia palea	2	Palmer and Maloney (1955)
	Raphidonema longiseta	1	Zehnder and Hughes (1958)
	Scenedesmus obliquus	2	Palmer and Maloney (1955)
	Scenedesmus obliquus	1	Zehnder and Hughes (1958)
	Tribonema aequale	1	Zehnder and Hughes (1958)
Bacitracin	*Anabaena variabilis*	100	Galloway and Krauss (1959a)
	Chlorella pyrenoidosa	1000	Galloway and Krauss (1959a)
	Chlorella pyrenoidosa	5000	Tomisek *et al.* (1957)
	Scenedesmus obliquus	1000	Galloway and Krauss (1959a)
Carbomycin	*Chlorella pyrenoidosa*	100	Tomisek *et al.* (1957)
Chloromycetin	*Anabaena variabilis*	10	Galloway and Krauss (1959a)
	Chlorella pyrenoidosa	100	Galloway and Krauss (1959a)
Erythromycin	*Chlorella pyrenoidosa*	725	Tomisek *et al.* (1957)
Gliotoxin	*Chlamydomonas*	31–62	Foter *et al.* (1953)
	Chlorella pyrenoidosa	2	Tomisek *et al.* (1957)
	Gomphonema	8	Foter *et al.* (1953)
	Nitzschia	8	Foter *et al.* (1953)
	Nostoc	125–250	Foter *et al.* (1953)
	Phormidium	8–16	Foter *et al.* (1953)
	Scenedesmus	125–250	Foter *et al.* (1953)
Gramidicin	*Anabaena variabilis*	1000	Galloway and Krauss (1959a)
	Chlorella pyrenoidosa	1000	Galloway and Krauss (1959a)
	Scenedesmus obliquus	1000	Galloway and Krauss (1959a)

[1] p.p.m. = parts per million = mg. per liter

TABLE I—Continued

Agent	Alga	Concentration (p.p.m.)	Reference
Neomycin	*Chlamydomonas*	4	Foter *et al.* (1953)
	Gomphonema	4	Foter *et al.* (1953)
	Nitzschia	4	Foter *et al.* (1953)
	Nostoc	4	Foter *et al.* (1953)
	Phormidium	4	Foter *et al.* (1953)
	Scenedesmus	16–32	Foter *et al.* (1953)
Neothiolutin	*Chlorella pyrenoidosa*	10	Tomisek *et al.*, (1957)
Netropsin	*Chlorella pyrenoidosa*	10	Tomisek *et al.* (1957)
Penicillin G	*Anabaena variabilis*	0.1	Galloway and Krauss (1959a)
	Chlorella pyrenoidosa	1000	Galloway and Krauss (1959a)
	Microcystis aeruginosa	2	Palmer and Maloney (1955)
	Scenedesmus obliquus	1000	Galloway and Krauss (1959a)
Pleocidin	*Chlorella pyrenoidosa*	5	Tomisek *et al.* (1957)
Polymyxin A	*Chlamydomonas*	10–20	Foter *et al.* (1953)
	Chlorella pyrenoidosa	5	Tomisek *et al.* (1957)
	Gomphonema	5–10	Foter *et al.* (1953)
	Nitzschia	10–20	Foter *et al.* (1953)
	Nostoc	10–20	Foter *et al.* (1953)
	Phormidium	20–40	Foter *et al.* (1953)
	Scenedesmus	80	Foter *et al.* (1953)
Polymyxin B	*Anabaena variabilis*	5	Galloway and Krauss (1959a)
	Chlorella pyrenoidosa	5	Galloway and Krauss (1959a)
	Chlorella variegata	2	Foter *et al.* (1953)
	Cylindrospermum licheniforme	2	Foter *et al.* (1953)
	Gomphonema parvulum	2	Foter *et al.* (1953)
	Microcystis aeruginosa	2	Foter *et al.* (1953)
	Nitzschia palea	2	Foter *et al.* (1953)
	Scenedesmus obliquus	40	Galloway and Krauss (1959a)
Puromycin	*Chlorella pyrenoidosa*	0.5	Tomisek *et al.* (1957)

TABLE I—Continued

Agent	Alga	Concentration (p.p.m.)	Reference
Streptomycin	*Anabaena variabilis*	0.1	Galloway and Krauss (1959a)
	Chlamydomonas	9–18	Foter *et al.* (1953)
	Chlorella pyrenoidosa	10	Galloway and Krauss (1959a)
	Chlorella pyrenoidosa	581	Tomisek *et al.* (1957)
	Coccomyxa ellipsoidea	25	Zehnder (1951)
	Cylindrospermum licheniforme	2	Palmer and Maloney (1955)
	Cystococcus placodii	10	Zehnder (1951)
	Gomphonema	4–9	Foter *et al.* (1953)
	Microcystis aeruginosa	2	Palmer and Maloney (1955)
	Nitzschia	2–4	Foter *et al.* (1953)
	Nostoc	2	Foter *et al.* (1953)
	Phormidium	2	Foter *et al.* (1953)
	Scenedesmus	18–36	Foter *et al.* (1953)
	Scenedesmus obliquus	1	Galloway and Krauss (1959a)
Terramycin	*Anabaena variabilis*	10	Foter *et al.* (1953)
	Chlamydomonas	233	Foter *et al.* (1953)
	Chlorella pyrenoidosa	1000	Galloway and Krauss (1959a)
	Gomphonema	233	Foter *et al.* (1953)
	Microcystis aeruginosa	2	Palmer and Maloney (1955)
	Nitzschia	233	Foter *et al.* (1953)
	Nostoc	233	Foter *et al.* (1953)
	Phormidium	58–117	Foter *et al.* (1953)
	Scenedesmus	233	Foter *et al.* (1953)
	Scenedesmus obliquus	1000	Galloway and Krauss (1959a)
Tetracycline	*Chlorella pyrenoidosa*	1000	Galloway and Krauss (1959a)
	Scenedesmus obliquus	1000	Galloway and Krauss (1959a)
Thiolutin	*Chlorella pyrenoidosa*	5	Tomisek *et al.* (1957)

That the two subclasses of the angiosperms can be so differentiated toxicol-gically offers promise of similar achievements with the algae.

In general, no toxic compounds have been found that differ markedly in their action on the three groups of plant microorganisms—algae, fungi, and bacteria. Moreover, there is no clear-cut pattern of difference in the resistance or susceptibility of the major classes of algae to poisons, although a few compounds, such as 2,3-dichloronaphthoquinone (Phygon), have proved to be especially toxic to all tested species of Blue-green Algae (Fitzgerald *et al.*, 1952). In fact, some marked differences have occasionally been detected in the response of even closely related algae. As an example, polymyxin B is much more toxic to *Chlorella pyrenoidosa* than to *C. vulgaris* (Galloway and Krauss, 1959a).

Among the more extensive studies on potential inhibitors of algal growth, especially in fresh water, were those of Fitzgerald *et al.* (1952; see also Fitzgerald, 1957), Palmer and Maloney (1955), Jakob and Nisbet (1955) and Galloway and Krauss (1959a). Inhibitory concentrations of some toxins of biological origin (i.e., of antibiotics) are listed in Table I, and agents inhibitory at known metabolic sites, in Table II. The data given in these tables relate only to the effects of the various compounds in defined media, at specific pH values, under specified culture conditions. That the pH can drastically affect the activity of the agents has been well demonstrated by Wedding *et al.* (1954), Erickson *et al.* (1955) and Tomisek *et al.* (1957). The presence of organic carbon sources has generally been found to protect the algae and consequently to raise the minimum lethal levels above those in strictly inorganic media (Geoghegan, 1957; Galloway and Krauss, 1959b). In some cases the action of antibiotics (Provasoli *et al.*, 1948) and respiratory inhibitors (Kandler, 1955) in the light differs from that in the dark.

Not only do the pH and composition of the medium influence the effectiveness of a given agent, but the density of the cell population used in the initial inoculum or in the test culture often influences the lethal range. This may be due entirely to differences in the mechanism of accumulation of the agent in the organism. Even if the concentration of a poison in the medium is low, the cells may die quickly because of an efficient accumulating mechanism, either actively metabolic or purely adsorptive in nature. The local concentration on or within the cell may be considerably greater than that in the medium. Consequently the true toxic level from the standpoint of the individual cell is difficult to determine without analyses to reveal the rates of accumulation. Studies with *Nitella clavata* indicated that streptomycin and chloramphenicol were absorbed by the cells, possibly by a binding carrier system requiring respiratory energy, but that penicillin G was not (Pramer, 1955, 1956; Litwack and Pramer, 1957). In the field of antibiosis, investigators have demonstrated the relation between the permea-

TABLE II

THE ACTION OF SELECTIVE METABOLIC INHIBITORS ON ALGAE

Compound	Action	Concentration (μM)	Organism	Reference
Aminotriazole	Inhibits chlorophyll synthesis.	20	*Euglena gracilis* *Ochromonas* spp.	Aaronson and Scher (1960)
Azaserine	Inhibits transamination.	20	*Scenedesmus obliquus*	Bassham and Calvin (1957)
Calcium dipicrylamine	Inhibits photosynthesis.	1–10	*Chlorella* sp.	Bierhuizen (1957)
$CdCl_2$	Blocks —SH groups. Forms mercaptide. Blocks Si uptake.	100–1000	*Navicula pelliculosa*	Lewin (1954)
4,6-Dinitrocresol	Prevents respiratory formation of high-energy phosphate bonds. Inhibits photosynthesis.	33	*Ulva lactuca*	Scott and Hayward (1953)
2,4-Dinitrophenol	Uncouples photosynthetic and respiratory phosphorylation.	3000–7000 20–2000	*Chlorella pyrenoidosa* *Ankistrodemus braunii*	Holzer (1951) Kessler (1955)
HCN	Blocks anion absorption, not respiration.	3	*Scenedesmus quadricauda*	Osterlind (1951)
Phenylurethane	Inhibits photosynthesis and respiration.	1000	*Ulva lactuca*	Scott and Hayward (1953) Lewin and Mintz (1955)
Polymyxin-B	Blocks glucose-6-phospho-isomerase.	250–500	*Chlorella pyrenoidosa*	Galloway and Krauss (1959b)
Seleno-methionine	Blocks normal methionine absorption and incorporation.	10	*Chlorella vulgaris*	Shrift (1954)
Sodium arsenate	Permits bypass of iodoacetate inhibition of 3-phosphoglyceraldehyde dehydrogenase.	5000	*Ulva lactuca*	Scott and Hayward (1953)
Sodium arsenite	Blocks —SH groups. Forms mercaptide. Blocks Si uptake.	1000	*Navicula pelliculosa*	Lewin (1954)

Sodium fluoroacetate	Inhibits aconitase.	250–2500	*Chilomonas paramecium*	Holz (1954)
	Causes citric acid accumulation.	100,000–400,000	*Chlorella vulgaris*	Merrett and Syrett (1960)
		100,000	*Chilomonas paramecium*	Holz (1954)
Sodium (or potassium) iodoacetate	Causes citric acid accumulation. Blocks respiration.	100	*Navicula pelliculosa*	Lewin (1954)
	Blocks 3-phosphoglyceraldehyde dehydrogenase. Causes K loss.	1000	*Ulva lactuca*	Proctor (1957)
Sodium malonate	Blocks succinic dehydrogenase.	1000–50,000	*Chilomonas paramecium*	Holz (1954)
Streptomycin	Bleaches chloroplasts. Inhibits chlorophyll synthesis.	10–1700	*Euglena gracilis*	Provasoli *et al.* (1951a)
		20	*Ochromonas* spp.	Aaronson and Scher (1960)
Sulfanilamide	Blocks pteroylglutamic acid synthesis from *p*-aminobenzoic acid.	100	*Chlamydomonas moewusii*	Lewin (1950)
			Scenedesmus obliquus	Krauss and Thomas (1954)

bility of the cells to an agent and its minimum lethal concentration (see Stadelmann, Chapter 31, this treatise).

II. Toxic Agents for Control Purposes

A number of extensive studies have been directed to the determination of concentrations at which certain compounds are toxic to algae. The two chief purposes have been: (i) to discover agents potentially useful for the control of algae in natural waters; and (ii) to determine concentrations which might be safely employed for eliminating contaminants in the purification of algal cultures.

The results of attempts to control algal populations in lakes and reservoirs do not always agree with laboratory data. Aside from considerations of cost, the different amounts of colloidal material and the different acidities of natural waters often render substances, promising on the basis of laboratory results, unsuitable in field use. The presence of natural or synthetic chelating agents has also been shown to reduce dramatically the toxicity of heavy metals such as copper, rendering otherwise toxic levels ineffective. However, the converse may sometimes be true, for algae grown in the laboratory under ideal conditions may be more resistant to poisons than cells of the same species in a natural environment (Moore and Kellerman, 1905).

At present the most commonly employed compound for the control of algae is copper sulfate, at concentrations near 1 p.p.m. (Bartsch, 1954; Palmer and Maloney, 1955; Maloney and Palmer, 1956; Vaulina, 1957; Mackenthun, 1959). Some algae, however, have been reported to tolerate concentrations considerably higher than this (Galloway and Krauss, 1959a). Many, though not all, common algae are susceptible to 1 or 2 p.p.m. of copper sulfate, whereas most fish and other animals are uninjured. Dichloronaphthaquinone has also been successfully employed against Blue-green Algae in field trials (Fitzgerald and Skoog, 1954). Sodium arsenite in concentrations of about 1 p.p.m. has been recommended for use in lakes, but the general cumulative toxicity of the arsenicals has naturally restricted their use.

Knowledge of toxicity comes not only from the search for algicides, but even more from two other fields, those of algal purification and of specific metabolic inhibitors. The first of these has involved attempts to develop ways of isolating algae in pure culture through the use of bactericidal agents. Attempts to eliminate bacteria and fungi, which are common contaminants, have naturally led to the use of antibiotics (Goldzweig-Shelubsky, 1951; Provasoli *et al.*, 1951a; Spencer, 1952). Though their use can reduce the number of bacteria and so increase the efficiency of routine purification procedures, the toxicity of the antibiotics towards algae varies unpredictably with the species and may render them unsuitable for this purpose (see Table I). Conspicuous in this regard has been the bleaching of *Chlorella* and *Euglena* by streptomycin(Dubé,1952; Provasoli *et al.*,1951a) and the inhibition of algal isomerases by polymyxin (Krauss and Galloway, 1956; Galloway and Krauss, 1959a).

III. Toxic Agents in Metabolic Investigations

Albert (1960) has stated that comparative biochemistry is the master key for the logical discovery of selectively toxic agents. The reverse is also true: the employment of inhibitors provides valuable clues which aid us to elucidate metabolic pathways and to reveal specific differences in biochemical capabilities. In fact, such differences in the responses of organisms to various poisons are often due to the susceptibility of a given enzyme or protein peculiar to the species. The inhibition may be of the competitive-analog-block type, or the configuration of the enzyme protein may be especially susceptible to chemical denaturation. Regardless of the mechanism by which it inhibits, there is no doubt that selective toxicity operates at the molecular level. By establishing blocks in reaction sequences, it is often possible to shunt metabolism into alternative pathways which are normally little used or which are adaptive only under periods of stress. It is also possible to build up for detection and analysis sufficient amounts of certain transitory metabolic intermediates that would otherwise be overlooked. In such ways selective inhibitors have served to develop our knowledge of inorganic nutrition, photosynthesis, and respiration (Woolley, 1959).

Table II lists compounds that have been used as selective inhibitors for certain metabolic reactions in the algae. Such an agent should be employed with caution, however, since it is not always certain that it blocks only one metabolic pathway. The toxicity is often relative, acting on one system or one enzyme somewhat more than on others. Moreover a given inhibitor, which blocks a terminal step in a reaction sequence, may behave differently when different compounds initiate the sequence. For example, Merrett and Syrett (1960) showed that in *Chlorella vulgaris* fluoroacetate and malonate inhibit the oxidation of acetate more strongly than that of glucose, although the inhibitor would be expected to block the reaction sequence equally, regardless of substrate. They suggested that this may be due to the different susceptibilities of physiologically separate"pools" of intermediates within the cell. A similar situation exists in the well-known fact that HCN only partially inhibits endogenous respiration, although it completely blocks glucose respiration (Österlind, 1951). The site of action of HCN is believed to be cytochrome oxidase, presumably the terminal oxidase for both respiratory sequences (Syrett, 1951). It appears that either HCN is not as specific as has been thought, or independent terminal oxidase systems exist in some algae. (See James, 1953; and Gibbs, Chapter 4, this treatise).

The inadvisability of trusting too much to specificity is further exemplified by the failure of CO to inhibit respiration in *Anabaena*, although HCN, azide, and *o*-phenanthroline are effective respiratory inhibitors (Webster and Frenkel, 1953). Additional difficulties arise from the marked differences in the characteristic pH optima, reaction rates, and specificities, etc., of enzymes catalyzing the same reactions in different organisms. Owing to difficulties in preparing enzymes from algal cells, little has been done with cell-free systems to confirm results obtained from intact cells (Krauss, 1958; see Jacobi, Chapter 7, this treatise).

The demonstration of inhibition by a specific poison or antimetabolite can be taken only as presumptive evidence for the existence of the normally susceptible enzyme. Although considerable progress has been made in investigations on the action of many sulfanilamides and antibiotics on bacteria (Burkholder, 1959), little attention has yet been paid to similar investigations with algae. However, the investigation of the action of polymyxin B on *Chlorella* spp. provided suggestive evidence for an alternative pathway for the breakdown of glucose and galactose through galactose-6-phosphate and tagatose phosphate, apart from the Embden-Myerhoff route (Galloway and Krauss, 1959a, b).

IV. Other Inhibitors of Algal Growth

There are many reports on the effects of one alga on another in mixed cultures (Lefèvre *et al.*, 1952; McVeigh and Brown, 1954; Jakob, 1954a, b; Jørgensen, 1956). Such observations suggest that inhibitors may be secreted by algae, but unfortunately no specifically active agents have been isolated or identified either from laboratory cultures or from natural populations (Talling, 1957). Although Proctor (1957) identified a fatty acid from cultures of *Chlamydomonas reinhardtii* which appears to suppress growth of *Haematococcus pluvialis*, identification of the active compound was only tentative. The discovery of an auto-inhibitor, chlorellin, in old cultures of *Chlorella vulgaris* (Oneto and Pratt, 1954; Pratt, 1940; Pratt and Fong, 1940; Pratt *et al.*, 1945) stimulated a study by Spoehr *et al.* (1949) of other species of *Chlorella*, but these authors, too, found only fatty acids with slight antibiotic properties.

The inhibitory action of radioactive compounds on algae has been noted (e.g., Peterson, 1947), but little has been done to identify the precise effects of the radiation, though it has been observed that cell enlargement associated with suppression of cell division is a common symptom. Porter and Watson (1954) showed that tritium oxide at 5 millicuries per milliliter almost completely inhibits growth of *Chlorella pyrenoidosa*, although cultures can be grown in media in which 99% of the water is replaced by the nonradioactive D_2O (Moses *et al.*, 1958; Katz, 1960). The isotopes P^{32}, Sr^{90}, Y^{90}, and S^{35} inhibit growth even at levels below 200 microcuries per milliliter (Porter and Knauss, 1954).

References

Aaronson, S., and Scher, S. (1960). Effect of amino-triazole and streptomycin on multiplication and pigment production of photosynthetic microorganisms. *J. Protozool.* **7,** 156–158.

Albert, A. (1960). "Selective Toxicity," 2nd ed. Wiley, New York.

Bartsch, A. F. (1954). Practical methods for control of algae and water weeds. *Public Health Repts.* **69,** 749–757.

Bassham, J. A., and Calvin, M. (1957). "The Path of Carbon in Photosynthesis." Prentice-Hall, Englewood Cliffs, New Jersey.

Bierhuizen, J. F. (1957). Inhibition of growth and metabolism of *Chlorella* and some other plant types by Ca dipicryl amine and other poisons. *Mededeel. Landbouwhogeschool Wageningen* **57,** 1–59.

Burkholder, P. R. (1959). Antibiotics. *Science* **129,** 1457–1465.

Dubé, J. F. (1952). Observations on a chlorophyll-deficient strain of *Chlorella vulgaris* obtained after treatment with streptomycin. *Science* **116,** 278–279.

Erickson, L. C., Wedding, R. T., and Brannaman, B. L. (1955). Influence of pH on 2,4-dichlorophenoxyacetic and acetic acid activity in *Chlorella*. *Plant Physiol.* **30,** 69–74.

Fitzgerald, G. P. (1957). The control of the growth of algae with CMU. *Trans. Wisconsin Acad. Sci.* **46,** 281–294.

Fitzgerald, G. P., and Skoog, F. (1954). Control of blue-green algae blooms with 2,3-dichloronaphthoquinone. *Sewage and Ind. Wastes* **26,** 1136–1140.

Fitzgerald, M. J., Gerloff, G. C., and Skoog, F. (1952). Studies on chemicals with selective toxicity to blue-green algae. *Sewage and Ind. Wastes* **24,** 888–896.

Foter, M. J., Palmer, C. M., and Maloney, T. E. (1953). Antialgal properties of various antibiotics. *Antibiotics & Chemotherapy* **3,** 505–508.

Galloway, R. A., and Krauss, R. W. (1959a). The differential action of chemical agents, especially polymyxin B, on certain algae, bacteria, and fungi. *Am. J. Botany* **46,** 40–49.

Galloway, R. A., and Krauss, R. W. (1959b). Mechanisms of action of polymyxin B on *Chlorella* and *Scenedesmus*. *Plant Physiol.* **34,** 380–389.

Geoghegan, M. J. (1957). The effect of some substituted methylureas on the respiration of *Chlorella vulgaris* var. *viridis*. *New Phytologist* **56,** 71–80.

Goldzweig-Shelubsky, M. (1951). The use of antibiotic substances for obtaining monoalgal bacteria-free cultures. *Palestine J. Botany Jerusalem Ser.* **5,** 129–131.

Holz, G. G. (1954). The oxidative metabolism of a cryptomonad flagellate, *Chilomonas paramecium*. *J. Protozool.* **1,** 114–120.

Holzer, H. (1951). Photosynthese und Atmungskettenphosphorylierung. *Z. Naturforsch.* **6b,** 424–430.

Jakob, H. (1954a). Compatibilités et antagonismes entre algues du sol. *Compt. rend. acad. sci.* **238,** 928–930.

Jakob, H. (1954b). Sur les propriétés antibiotiques énergiques d'une algue du sol: *Nostoc muscorum* Ag. *Compt. rend. acad. sci.* **238,** 2018–2020.

Jakob, H., and Nisbet, M. (1955). Action de certains derivés d'ammonium quaternaire sur quelques poissons et algues d'eau douce. *Verhandl. intern. Ver. Limnol.* **12,** 726–735.

James, W. O. (1953). The use of respiratory inhibitors. *Ann. Rev. Plant Physiol.* **4,** 59–90.

Jørgensen, E. G. (1956). Growth inhibiting substances formed by algae. *Physiol. Plantarum* **9,** 712–726.

Kandler, O. (1955). Über die Beziehungen zwischen Phosphathaushalt und Photosynthese. III. Hemmungsanalyse der lichtabhängigen Phosphorylierung. *Z. Naturforsch.* **10b,** 38–46.

Katz, J. J. (1960). The biology of heavy water. *Sci. American* **203,** 106–116.

Kessler, E. (1955). Über die Wirkung von 2,4 DNP auf Nitratreduktion und Atmung von Grünalgen. *Planta* **45,** 94–105.

Krauss, R. W. (1958). Physiology of the freshwater algae. *Ann. Rev. Plant Physiol.* **9,** 207–244.

Krauss, R. W., and Galloway, R. A. (1956). The biological action of polymyxin B. *Science* **124,** 939.

Krauss, R. W., and Thomas, W. H. (1954). The growth and inorganic nutrition of *Scenedesmus obliquus* in mass culture. *Plant Physiol.* **29,** 205–214.

Lefèvre, M., Jakob, H., and Nisbet, M. (1952). Auto- et hétéroantagonisme chez les algues d'eau douce *in vitro* et dans les collections d'eau naturelles. *Ann. Sta. Centrale Hydrobiol. Appl. Paris* **4,** 5–198.

Lewin, J. C. (1954). Silicon metabolism in diatoms. I. Evidence for the role of reduced sulfur compounds in silicon utilization. *J. Gen. Physiol.* **37,** 589–599.

Lewin, R. A. (1950). Induced vitamin-requiring mutants of *Chlamydomonas. Nature* **166,** 196.

Lewin, R. A., and Mintz, R. H. (1955). Inhibitors of photosynthesis in *Chlamydomonas. Arch. Biochem. Biophys.* **54,** 246–248.

Litwack, G., and Pramer, D. (1957). Absorption of antibiotics by plant cells. III. Kinetics of streptomycin uptake. *Arch. Biochem. Biophys.* **68,** 396–402.

Mackenthun, K. M. (1959). "The Chemical Control of Aquatic Nuisances." Committee on Water Pollution, Madison, Wisconsin.

McVeigh, I., and Brown, W. (1954). *In vitro* growth of *Chlamydomonas chlamydogama* Bold and *Haematococcus pluvialis* Flotow em. Wille, in mixed cultures. *Bull. Torrey Botan. Club* **81,** 218–233.

Maloney, T. E., and Palmer, C. M. (1956). Toxicity of six chemical compounds in thirty cultures of algae. *Water & Sewage Works* **103,** 509–513.

Merrett, M. J., and Syrett, P. J. (1960). The relationship between glucose oxidation and acetate oxidation in *Chlorella vulgaris. Physiol. Plantarum* **13,** 237–249.

Moore, G. T., and Kellerman, K. F. (1905). Copper as an algicide and disinfectant in water supplies. *U. S. Dept. Agr. Bur. Plant Ind. Bull.* **No. 76.**

Moses, V., Holm-Hansen, O., and Calvin, M. (1958). Response of *Chlorella* to a deuterium environment. *Biochim. et Biophys. Acta* **28,** 62–70.

Oneto, J. F., and Pratt, R. (1954). Studies on *Chlorella vulgaris.* X. Influence of the age of the culture on the accumulation of chlorellin. *Am. J. Botany* **32,** 405.

Österlind, S. (1951). Anion absorption by an alga with cyanide resistant respiration. *Physiol. Plantarum* **4,** 528–534.

Palmer, C. M., and Maloney, T. E. (1955). Preliminary screening for potential algicides. *Ohio J. Sci.* **55,** 1–8.

Peterson, B. O. (1947). The effect of radium and uranium on the growth of *Skeletonema costatum* Grev. *Acta Horti Gotoburgensis* **17,** 107–112.
Porter, J. W., and Knauss, H. J. (1954). Inhibition of growth of *Chlorella pyrenoidosa* by beta-emitting radioisotopes. *Plant Physiol.* **29,** 60–63.
Porter, J. W., and Watson, M. S. (1954). Gross effects of growth inhibiting levels of tritium oxide on *Chlorella pyrenoidosa. Am. J. Botany* **41,** 550–555.
Pramer, D. (1955). Absorption of antibiotics by plant cells. *Science* **121,** 507–508.
Pramer, D. (1956). Absorption of antibiotics by plant cells. II. Streptomycin. *Arch. Biochem. Biophys.* **62,** 265–273.
Pratt, R. (1940). Influence of the size of inoculum on the growth of *Chlorella vulgaris* in freshly prepared culture medium. *Am. J. Botany* **27,** 52–56.
Pratt, R., and Fong, J. (1940). Studies on *Chlorella vulgaris.* II. Further evidence that *Chlorella* forms a growth-inhibiting substance. *Am. J. Botany* **27,** 431–436.
Pratt, R., Oneto, J. F., and Pratt, J. (1945). Studies on *Chlorella vulgaris.* X. Influence of the age of the culture on the accumulation of chlorellin. *Am. J. Botany* **32,** 405–408.
Proctor, V. W. (1957). Studies of algal antibiosis using *Haematococcus* and *Chlamydomonas. Limnol. and Oceanog.* **2,** 125–139.
Provasoli, L., Hutner, S. H., and Schatz, A. (1948). Streptomycin-induced chlorophyll-less races of *Euglena. Proc. Soc. Exptl. Biol. Med.* **69,** 279–282.
Provasoli, L., Hutner, S. H., and Pintner, I. J. (1951a). Destruction of chloroplasts by streptomycin. *Cold Spring Harbor Symposia Quant. Biol.* **16,** 113–120.
Provasoli, L., Pintner, I., and Packer, L. (1951b). Use of antibiotics in obtaining pure cultures of algae. *Proc. Am. Soc. Protozool.* **2,** 6.
Scott, G. T., and Hayward, H. R. (1953). The influence of iodoacetate on the sodium and potassium content of *Ulva lactuca* and the prevention of its influence by light. *Science* **117,** 719–721.
Shrift, A. (1954). Sulfur-selenium antagonism. II. Antimetabolic action of selenomethionine on the growth of *Chlorella vulgaris. Am. J. Botany* **41,** 345–352.
Spencer, C. P. (1952). On the use of antibiotics for isolating bacteria-free cultures of marine phytoplankton organisms. *J. Marine Biol. Assoc. United Kingdom* **31,** 97–106.
Spoehr, H. A., Smith, J. H. C., Strain, H. H., Milner, H. W., and Hardin, G. J. (1949). Fatty acid antibacterials from plants. *Carnegie Inst. Wash. Publ.* **No. 586,** 1–67.
Syrett, P. J. (1951). The effects of cyanide on the respiration and oxidative assimilation of glucose by *Chlorella vulgaris. Ann. Botany (London)* **15,** 473–492.
Talling, J. F. (1957). The growth of two plankton diatoms in mixed cultures. *Physiol. Plantarum* **10,** 215–223.
Tomisek, A., Reid, M. R., Short, W. A., and Skipper, H. E. (1957). Studies on the photosynthetic reaction. III. The effects of various inhibitors upon growth and carbonate-fixation in *Chlorella pyrenoidosa. Plant Physiol.* **32,** 1–10.
Vaulina, Z. N. (1957). [The effect of copper on soil algae.] *Botan Zhur.* **42,** 1097–1099. (In Russian.)
Webster, G. C., and Frenkel, A. W. (1953). Some respiratory characteristics of the blue-green alga *Anabaena. Plant Physiol.* **28,** 63–69.
Wedding, R. T., Erickson, L. C., and Brannaman, B. L. (1954). Effect of 2,4-dichlorophenoxyacetic acid on photosynthesis and respiration. *Plant Physiol.* **29,** 64–69.
Woolley, D. W. (1959). Antimetabolites. *Science* **129,** 615–621.
Zehnder, A. (1951). Über den Einfluss antibiotischer Stoffe auf das Wachstum von Grünalgen. *Experientia* **7,** 99–100.
Zehnder, A., and Hughes, E. O. (1958). The antialgal activity of acti-dione. *Can. J. Microbiol.* **4,** 399–408.

—46—

Rhythms*

B. M. SWEENEY
University of California, La Jolla, California
and
J. W. HASTINGS
University of Illinois, Urbana, Illinois

I. Introduction

In common with other organisms, the algae have evolved in a periodic environment, where night alternates with day and the tides inexorably rise and fall. As is well known, the physiological processes of these plants may show concomitant periodicities. These fall into two major categories, the exogenous and the endogenous rhythms. They may be further classified according to the length of the period as diurnal (period about 24 hours), tidal (period about $12\frac{1}{2}$ hours) and semilunar and lunar (period about 14.8 and 29 days).

Exogenous rhythms are those which occur as a direct consequence of the environmental changes. Predictions concerning these responses may be made on the basis of our understanding of algal physiology and biochemistry. An example of a rhythm in this category is illustrated in Fig. 1. The pH of the culture medium containing a photosynthetic alga, *Gonyaulax*

* Studies reported here have been supported by grants to the authors from the National Science Foundation.

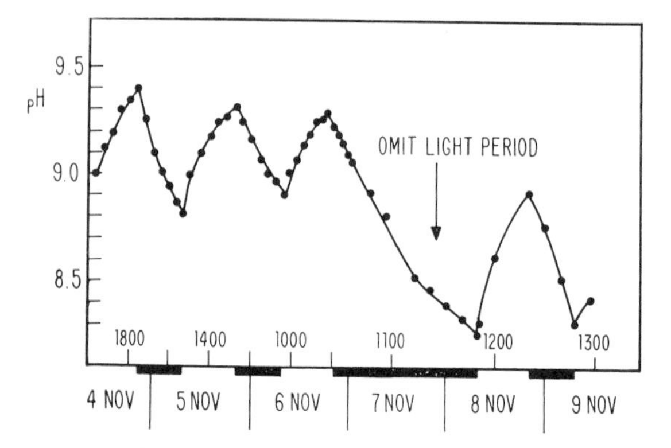

FIG. 1. An exogenous rhythm: the change in pH of a culture of *Gonyaulax polyedra* exposed to alternating light and dark periods as indicated on the abscissa. The pH (recorded automatically) is directly dependent on the prevailing light conditions, increasing when the culture is illuminated and decreasing as long as the culture is in the dark. This point is illustrated in the experiment where, when one light period was omitted, the pH continued to fall. (Original.)

polyedra, varies as a result of carbon dioxide fixation in the light and its release by respiration in darkness. The pH continues to vary rhythmically only as long as light and darkness alternate at regular intervals.

Endogenous rhythms are those which can be shown experimentally or statistically to be to some extent independent of environmental changes, and which thus evidently involve innate (i.e., endogenous) physiological processes. An example of an endogenous rhythm with some evidence supporting this supposition is shown in Fig. 2.

II. Endogenous Rhythms in Algae

A. Diurnal Rhythms

Various endogenous rhythms have been described in algae, including rhythms of respiration, photosynthetic capacity, phototaxis, luminescence, and cell division. The rhythms discussed hereafter are considered to be endogenous, although in some cases adequate experimental proof for this contention is lacking.

Rhythmicity in photosynthesis may be quite general. Doty and Oguri (1957) and Yentsch and Ryther (1957) observed that C^{14} fixation varies diurnally in naturally occurring phytoplankton populations. Photosynthetic capacity shows a diurnal rhythm in cultures of *Chlorella pyrenoidosa* (Pirson and Lorenzen, 1958a) and *Gonyaulax polyedra* (Hastings and Astrachan, 1959; Hastings, 1959). In normally alternating periods of light and darkness, the maximum occurs at about midday in *Chlorella* and

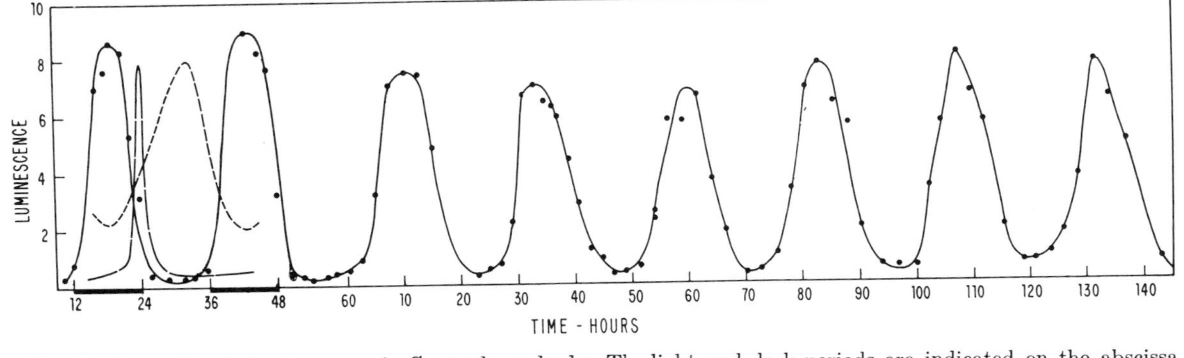

FIG. 2. The rhythm in luminescence in *Gonyaulax polyedra*. The light and dark periods are indicated on the abscissa. When the cells are no longer exposed to dark periods (continuous light, 1000 lux), the rhythm persists, indicating that the rhythm is endogenous. The phase relationships of two other endogenous rhythms in *Gonyaulax*, that of cell division and that of photosynthesis, are indicated in one cycle. These rhythms continue in continuous dim light as does the rhythm in luminescence. (Luminescence in arbitrary units. Original; some data taken from previous publications.)

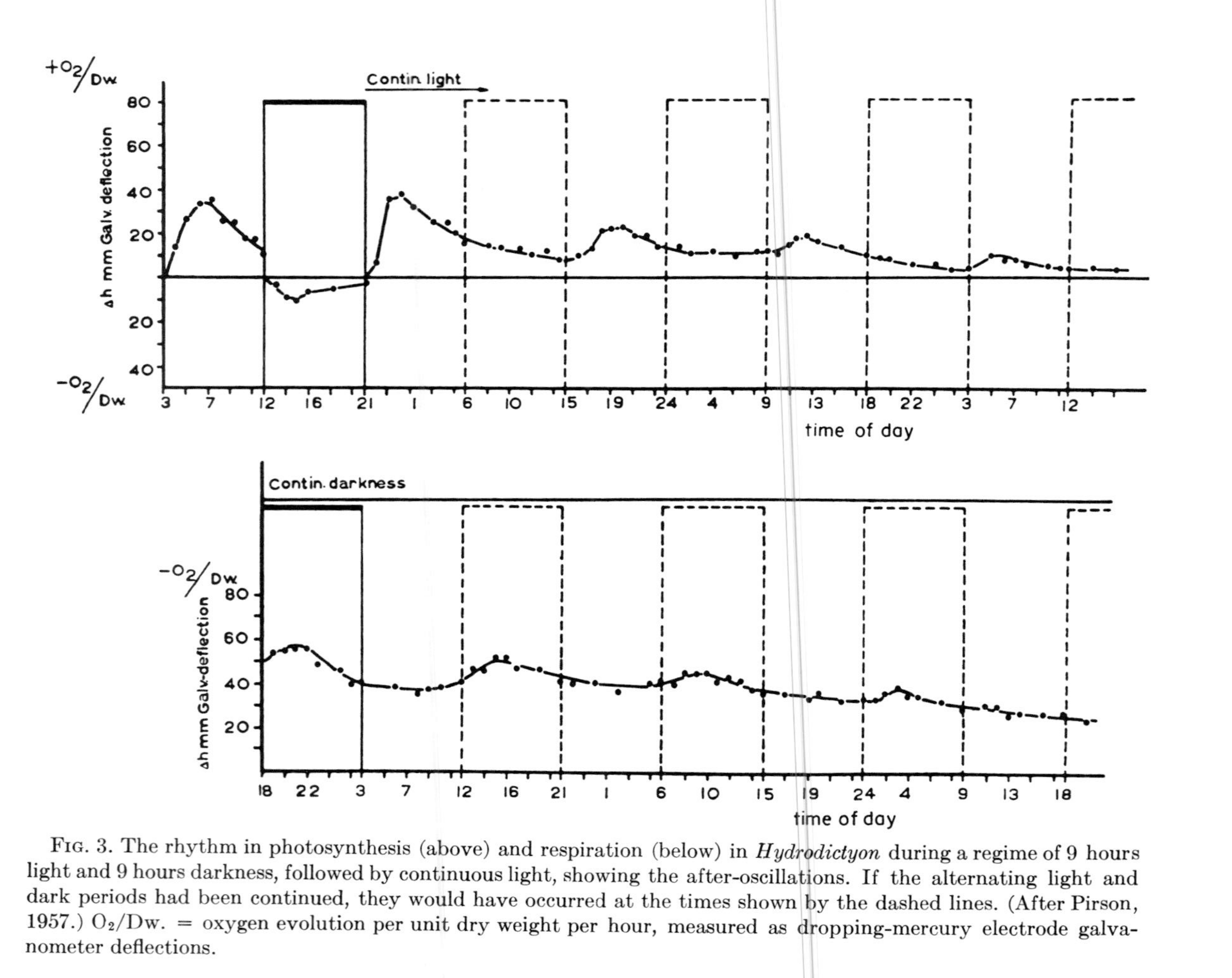

FIG. 3. The rhythm in photosynthesis (above) and respiration (below) in *Hydrodictyon* during a regime of 9 hours light and 9 hours darkness, followed by continuous light, showing the after-oscillations. If the alternating light and dark periods had been continued, they would have occurred at the times shown by the dashed lines. (After Pirson, 1957.) O_2/Dw. = oxygen evolution per unit dry weight per hour, measured as dropping-mercury electrode galvanometer deflections.

somewhat after midday in *Gonyaulax* (Fig. 2). In *Gonyaulax*, the rhythm has been demonstrated in cells maintained in continuous light of low intensity, and thus has been shown to be endogenous.

In *Hydrodictyon* grown in alternating light and darkness, photosynthesis, respiration, and growth rate all exhibit periodic fluctuations that persist for a short while after the alga has been transferred to constant conditions (Fig. 3) (Pirson *et al.*, 1954; Schön, 1955; Pirson and Schön, 1957; Pirson, 1957). In fact, Pirson *et al.* reported that this alga is unable to survive in continuous light for any length of time.

The diurnal phototactic rhythm in *Euglena*, described by Pohl (1948), was investigated in detail by Bruce and Pittendrigh (1956, 1958). Phototactic activity is greatest during the day. This rhythm may persist for long periods in darkness interrupted only by the test light, which was shown to be too weak to act as a timing signal. No rhythm is manifest in cultures grown on organically enriched media.

In members of the genus *Gonyaulax* (*G. polyedra*, *G. monilata*, *G. catenella*), bioluminescence shows a marked diurnal rhythmicity (Fig. 2), which continues in cultures kept in continuous light of low intensity (500 to 1500 lux), but cannot be demonstrated in cultures maintained in continuous bright light (Massart, 1893; Moore, 1909; Sweeney and Hastings, 1957). Although Massart (1893) reported a diurnal rhythm of luminescence in the heterotrophic dinoflagellate *Noctiluca miliaris*, Nicol (1958) found no evidence of this. We have also studied *Noctiluca* and have found no diurnal rhythm in either the frequency or the intensity of luminescence flashes.

At least part of the diurnal change in light emission may be attributed to changes in the activities of components of the luminescent system. Extracts prepared from cells during the night have enzyme (luciferase) activities from 3 to 8 times as high as those prepared during the day (Hastings and Sweeney, 1957a). Similar diurnal fluctuations in the concentration of the substrate (luciferin) have also been found, indicating that changes of enzyme activity are not alone responsible for the diurnal rhythm of luminescence.

Under natural conditions, cell division in some dinoflagellates occurs only at night (Gough, 1905; Jörgensen, 1911). This periodicity was confirmed experimentally for several photosynthetic dinoflagellates: *Peridinium triquetrum* (Braarud and Pappas, 1951); *Gonyaulax polyedra* (Sweeney and Hastings, 1958); and *Gymnodinium splendens*, *Prorocentrum micans*, and *Gonyaulax sphaeroidea* (Sweeney, 1959) (Fig. 4). Maximum numbers of dividing cells were found at different times in different species. A diurnal rhythmicity of mitosis occurs in *Spirogyra* (Karsten, 1918) and *Oedogonium* (Bühnemann, 1955d) but not in the colorless dinoflagellate *Oxyrrhis marina* (Nozawa, 1940).

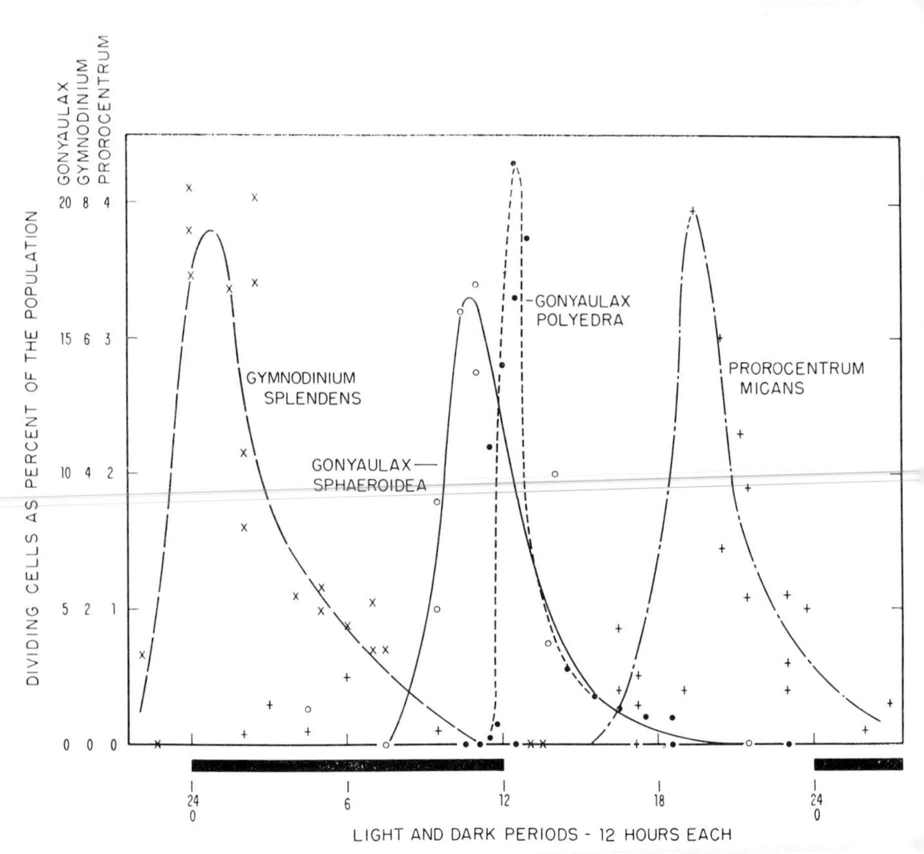

FIG. 4. The time at which cell division occurs in various dinoflagellates growing with the 12-hour light:12-hour dark cycle indicated on the abscissa. KEY: (×) *Gymnodinium splendens;* (○) *Gonyaulax sphaeroidea;* (●) *Gonyaulax polyedra;* (+) *Prorocentrum micans.* (Original.)

In *Gonyaulax,* the cell-division rhythm continues in constant dim light, with maxima in the number of dividing cells approximately every 24 hours, although the generation time under these conditions is 3 to 4 days (Sweeney and Hastings, 1958). We have also observed a similar persistent cell division rhythm in *Gymnodinium,* the "natural" period in this organism being about 20 hours at 20°C.

The several green *Euglena* species studied by Leedale (1959) divide only during the evening hours. The endogenous nature of this rhythm is not clearly established but is suggested by the fact that shortening the day does not hasten division. Like the rhythm in phototaxis, the mitotic rhythm in *Euglena gracilis* disappears when the organism is grown in media containing organic supplements. In the Euglenophyta, as in the dinoflagellates, colorless species show no rhythmicity.

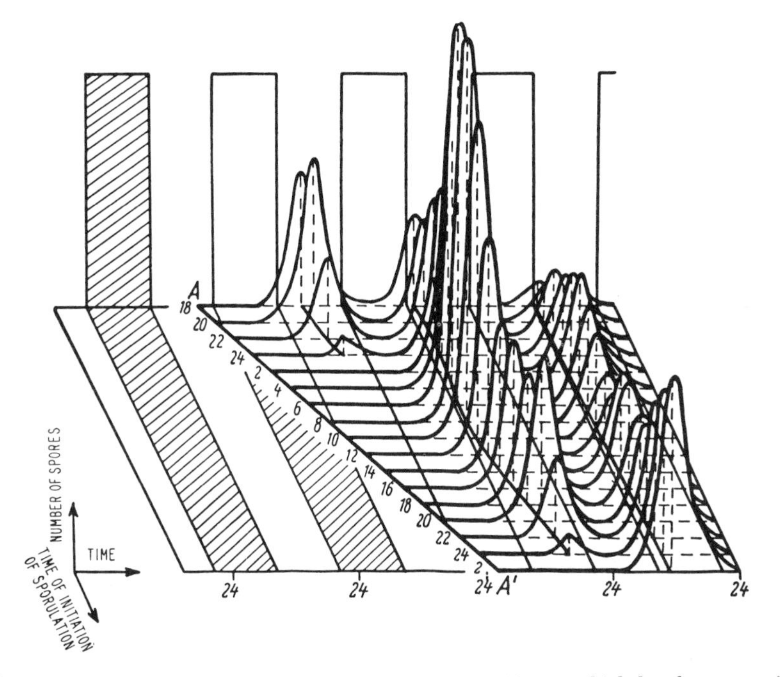

FIG. 5. The rhythm in sporulation in *Oedogonium cardiacum* which has been growing vegetatively on a 12-hour light:12-hour dark cycle. To induce and record sporulation, cells are transferred to new medium in continuous light at *A–A'*. Sporulation is maximum at midday, independent of the time of this transfer. (After Bühnemann, 1955a.)

Asexual reproduction in certain algae is rhythmic. Sporulation in *Oedogonium cardiacum* occurs only at midday, and the rhythm continues for 3 to 6 days in constant light and temperature (Fig. 5) (Bühnemann, 1955a). In *Chlorella*, too, the increase in cell number by the release of autospores takes place only at a certain time, determined by the beginning of the preceding light period (Lorenzen, 1957, 1958; Pirson and Lorenzen, 1958b). Proof of a true endogenous rhythm is lacking in this case, however.

B. Tidal Rhythms

A tidal rhythm in phototaxis, observed in certain organisms inhabiting tide flats, may be viewed as an adaptation that prevents the organisms from being washed out to sea at high tide. *Euglena limosa* becomes negatively phototactic and buries in the mud just before the time of high tide, but is positively phototactic at other times during the daylight hours (Bracher, 1932). The diatoms *Pleurosigma* (Fauvel and Bohn, 1907) and *Hantzschia* (Fauré-Fremiet, 1951) come to the surface of the beach or mud flat only at low tide; similar observations on littoral diatoms and dinoflagellates were reported by Ganapati *et al.* (1959). All these organisms continue to show their tidal rhythmic behavior for several days when observed in the laboratory away from the influence of the tide.

C. Lunar Rhythms

Lunar rhythms in the release of gametes and spores have been observed in the Brown Algae *Sargassum* (Tahara, 1909), *Nemoderma* (Kuckuck, 1912) and *Dictyota* (Williams, 1905; Hoyt, 1927), and in the Green Algae *Halicystis* (Hollenberg, 1936) and *Ulva* (Smith, 1947). In *Ulva*, which grows in tide pools, release of gametes and spores has been shown to occur at the time of low tide and only at the especially low tides of the spring series, as shown in Fig. 6. The simultaneous release of large numbers of gametes into relatively shallow water must increase the probability of fertilization. Many of the plants mentioned above have been observed to retain lunar periodicity in the laboratory.

Diurnal and lunar periodicities in the respiration of *Fucus* in constant light and temperature have been reported by Brown *et al.* (1955). The overt rhythms in this case are not pronounced, being apparent only after statistical treatment of the data. It would be most useful to extend our knowledge concerning these rhythms in *Fucus*.

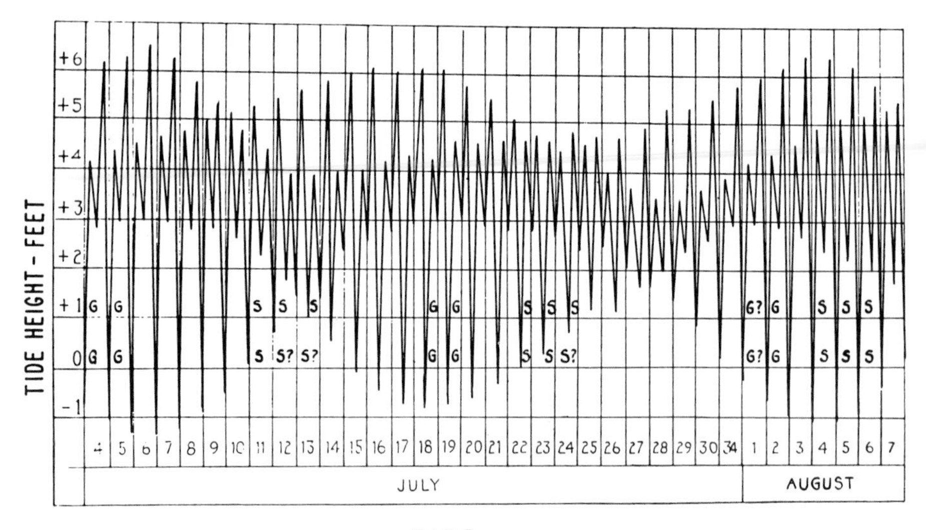

FIG. 6. Chart showing the daily range of tides from July 4 to August 7, 1944. Days on which gametes were liberated from gametophytes of *Ulva lobata* growing at mean low-tide level and at the 1-foot level at Cypress Point, California, are labeled G. Days on which sporophytes liberated zoospores are labeled S. A period of 14 days is evident in the release of both gametes and spores. (After Smith, 1947.)

III. Susceptibility to the Influence of Experimental Factors

A. Illumination Regime

The properties of diurnal rhythms appear to be very similar in a wide variety of plants and animals, including certain of the algae which have been discussed here (Bruce and Pittendrigh, 1957; Bünning, 1958). Moreover, there are many interesting similarities between the properties of these rhythms and the properties of mechanical or electronic oscillators (Pittendrigh and Bruce, 1959).

For example, under constant conditions the rhythms show "natural periods" which, in every organism so far investigated, may be somewhat different from 24 hours. The biological rhythms may be "entrained" or forced to match in frequency some external oscillator like the 24-hour light-dark cycle of day and night. Furthermore, one may entrain rhythms to periods different from 24 hours. In *Gonyaulax*, for example, one may entrain to periods as short as 14 hours by exposing the cells to alternating light and dark periods of 7 hours each. After the entraining program is stopped, even when it has been maintained for as long as a year, the rhythm at once reverts to its natural period (Hastings and Sweeney, 1958).

Moreover, no "learning" or "training" is involved. In arhythmic *Gonyaulax* cells which have been maintained in constant light for as long as 3 years, a single change in the light intensity is all that is necessary to initiate the characteristic 24-hour rhythm of luminescence.

When light-dark cycles are present, the phase bears a fixed relationship to this entraining schedule. The phase of the biological rhythm changes when the phase of the entraining schedule is changed. However, Bruce and Pittendrigh (1958) showed that an external cycle is not necessary in order to change the phase of a biological rhythm. Exposure to a "single perturbation," i.e., a single change in the light intensity, is quite adequate to shift phase. This is comparable to the behavior of any mechanical oscillator, such as a pendulum; only a single disturbance, not a new entrainment, is required to shift the phase.

B. Temperature

A feature of the biological rhythms which has attracted much interest is their independence of temperature, as shown by the fact that the natural period of diurnal rhythms is very nearly the same at temperatures which may be as much as 10° or 20°C. apart. Only relatively small effects of temperature upon period have been reported for rhythms in a variety of organisms, including the algae *Oedogonium* (Bühnemann, 1955c), *Euglena* (Bruce and Pittendrigh, 1956), and *Gonyaulax* (Hastings and Sweeney, 1957b). As pointed out by Pittendrigh and Bruce (1957), reasonable accuracy is a crucially important feature of any functionally useful clock, which must be temperature-independent if it is to operate under a variety of environmental conditions.

However, in *Gonyaulax* the system is relatively inaccurate and not strictly temperature-independent: thus at 16°C. the period is 22.8 hours, while at 26°C. it is 26.5 hours. In *Oedogonium*, too, the Q_{10} for the process is less than one. The fact that the "clock" runs faster at lower temperatures would be explicable on the basis of a biochemical temperature-compensating mechanism.

Twenty-four hour temperature cycles, with alternating warm and cool periods, have been shown to "entrain" rhythms in a manner similar to that described for light-dark cycles, although the effect is less marked (Sweeney and Hastings, 1960). The phase of the rhythm usually bears a fixed relation to the temperature cycle, the cooler phase corresponding to darkness.

C. Inhibitors

Several studies on the effects of inhibitors and other biochemical agents on various systems showing an endogenous rhythm have indicated that both the phase and the period are remarkably insensitive to chemical control. One of the most carefully documented studies in this connection is that of Bühnemann (1955b) on the rhythm of sporulation in *Oedogonium*. He found essentially no effect of copper sulfate, cyanide, fluoride, arsenate, 2,4-dinitrophenol, iodoacetate, indoleacetate, quinine, cocaine, adenosine triphosphate, or riboflavin. Bünning (1958) has reviewed these and similar experiments with inhibitors by other workers.

In studies with *Gonyaulax*, the present authors utilized both specific and nonspecific inhibitors in attempts to modify phase and/or period, employing slightly different techniques from those of previous workers, the major difference being that the cells were subjected to inhibitors intermittently rather than continuously. In such tests, several compounds were found to be ineffective, including 5′-fluoro-2′-deoxyuridine, chloramphenicol, *p*-fluorophenylalanine, mono- and dichlorophenyl dimethylurea, gibberellin, kinetin, urethane, and chlorpromazine (Hastings, 1960). The "accuracy" of the rhythm was apparently decreased in the presence of a variety of inhibitors, which however had no specific effect upon phase. On the other hand, sodium arsenite and chloromercuribenzoate (both dithiol reagents) caused a marked shifting of phase, the amount and direction depending upon the time of day when the cells were exposed to the inhibitor. It is evident that chemically induced phase shifts may occur, but the mechanism and specificity of the reaction are not yet clear. Continuous exposure of cells to chloramphenicol or colchicine produced small changes in period. It is hoped that a more detailed analysis of these effects may yield information concerning the biochemistry of the rhythmic mechanism.

The rhythm of phototaxis in *Euglena* is also susceptible to chemical influence. It was noted that the rhythm could be apparently suppressed by the presence of unidentified organic substances present in "enriched" media. Moreover, cells grown in a mineral medium compounded with pure heavy water (D_2O) exhibited, after acclimatization, a natural period which differed by several hours from cells kept in H_2O (Bruce and Pittendrigh, 1960).

IV. Discussion

The evidence indicating that rhythmic phenomena are dependent on a cellular clock-like mechanism has been reviewed by Bünning (1958), Pittendrigh (1958), and Hastings (1959). There is little evidence at the present time, however, to suggest the actual physicochemical nature of this rhythmic mechanism. In fact, there is distressingly little information concerning any biochemical aspects of rhythmic processes in living organisms. There may be difficulty in distinguishing between the biochemical system which constitutes a "chemical clock" and systems which are merely clock-controlled, although a variety of distinctions are possible. It was shown, for example, that the luminescent system of *Gonyaulax* should not be considered an autonomous biochemical oscillator, being apparently coupled to an independent timing mechanism (Hastings and Sweeney, 1958).

The fact that temperature independence has evolved leads one to the suggestion that an accurate biological clock must somehow be of importance to the economy of cells and organisms (Pittendrigh, 1954). This is supported by experiments in which it has been found that certain plants survive poorly or not at all under constant conditions (Pirson *et al.*, 1954; Highkin and Hanson, 1954; Hillman, 1956). To some organisms, such as those exhibiting tidal phototactic periodicity, the "usefulness" of the rhythm is evident. In other cases, it is clear that the clock serves to time a biochemical or physiological process (e.g., photosynthesis, luminescence) to operate most actively at the appropriate time of day.

The importance in other cases may be more obscure. Possibly timing has adaptive significance in terms of economy to the cell whereby biosyntheses occur maximally at a time when the products are best utilized. It is conceivable that this timing could act at the level of gene function, or be regulated thereby. It is also possible that gene action might be discontinuous in time, so that the action of different genes would be staggered. These are at present mere speculations, introduced only to suggest the possible relevance of biological clocks to the general pattern of cellular function.

References

Braarud, T., and Pappas, I. (1951). Experimental studies on the dinoflagellate *Peridinium triquetrum* (Ehrb.) Lebour. *Avhandl. Norske Videnskap Akad. Oslo.* I. *Mat.-Naturv. Kl.* **2,** 1–23.

Bracher, R. (1932). The light relations of *Euglena limosa* Gard. I. The influence of intensity and quality of light on phototaxy. *J. Linnean Soc. London, Botany* **51,** 23–42.

Brown, F. A., Jr., Freeland, R. O., and Ralph, C. L. (1955). Persistent rhythms of O_2-consumption in potatoes, carrots, and the seaweed, *Fucus*. *Plant Physiol.* **30,** 280–292.

Bruce, V. G., and Pittendrigh, C. S. (1956). Temperature independence in a unicellular "clock." *Proc. Natl. Acad. Sci. U.S.* **42,** 676–682.

Bruce, V. G., and Pittendrigh, C. S. (1957). Endogenous rhythms in insects and microorganisms. *Am. Naturalist* **91,** 179–195.

Bruce, V. G., and Pittendrigh, C. S. (1958). Resetting the *Euglena* clock with a single light stimulus. *Am. Naturalist* **92,** 295–306.

Bruce, V. G., and Pittendrigh, C. S. (1960). An effect of heavy water on the phase and period of the circadian rhythm in *Euglena*. *J. Cellular Comp. Physiol.* **56,** 25–31.

Bühnemann, F. (1955a). Die rhythmische Sporenbildung von *Oedogonium cardiacum* Wittr. *Biol. Zentr.* **74,** 1–54.

Bühnemann, F. (1955b). Das endodiurnale System der Oedogoniumzelle. II. Der Einfluss von Stoffwechselgiften und anderen Wirkstoffen. *Biol. Zentr.* **74,** 691–705.

Bühnemann, F. (1955c). Das endodiurnale System der *Oedogonium* Zelle. III. Über den Temperatureinfluss. *Z. Naturforsch.* **10b,** 305–310.

Bühnemann, F. (1955d). Das endodiurnale System der *Oedogonium* Zelle. IV. Die Wirkung verschiedener Spektralbereiche auf die Sporulation-und Mitoserhythmik. *Planta* **46,** 227–255.

Bünning, E. (1958). "Die physiologische Uhr," 105 pp. Springer, Berlin.

Doty, M. S., and Oguri, M. (1957). Evidence for a photosynthetic daily periodicity. *Limnol. and Oceanog.* **2,** 37–40.

Fauré-Fremiet, E. (1951). The tidal rhythm of the diatom *Hantzschia amphioxys*. *Biol. Bull.* **100,** 173–177.

Fauvel, P., and Bohn, G. (1907). Les rhythmes des marées chez les Diatomées littorales. *Compt. rend. soc. biol.* **62,** 121–123.

Ganapati, P. N., Rao, M. V. Lakshmana, and Rao, D. V. Subba. (1959). Tidal rhythms of some diatoms and dinoflagellates inhabiting the intertidal sands of the Visakhapatnam Beach. *Current Sci. (India)* **28,** 450–451.

Gough, L. E. (1905). Report on the plankton of the English Channel in 1903. *In* "First Report on Fishery and Hygrographical Investigations in the North Sea and Adjacent Waters (Southern Area); 1902–1903." *Marine Biol. Assoc. United Kingdom Intern. Fishery Invest.*, pp. 325–377.

Hastings, J. W. (1959). Unicellar clocks. *Ann. Rev. Microbiol.* **13,** 297–309.

Hastings, J. W. (1960). Biochemical aspects of rhythms: phase shifting by chemicals. *Cold Spring Harbor Symposia Quant. Biol.* **25,** 131–143.

Hastings, J. W., and Astrachan, L. (1959). A diurnal rhythm of photosynthesis. *Federation Proc.* **18,** 65.

Hastings, J. W., and Sweeney, B. M. (1957a). The luminescent reaction in extracts of the marine dinoflagellate, *Gonyaulax polyedra*. *J. Cellular Comp. Physiol.* **49,** 209–226.

Hastings, J. W., and Sweeney, B. M. (1957b). On the mechanism of temperature independence in a biological clock. *Proc. Natl. Acad. Sci. U. S.* **43,** 804–811.

Hastings, J. W., and Sweeney, B. M. (1958). A persistent diurnal rhythm of luminescence in *Gonyaulax polyedra*. *Biol. Bull.* **115,** 440–458.

Highkin, H. R., and Hanson, J. B. (1954). Possible interaction between light-dark cycles and endogenous daily rhythms on the growth of tomato plants. *Plant Physiol.* **29,** 301–302.

Hillman, W. S. (1956). Injury of tomato plants by continuous light and unfavorable photoperiodic cycles. *Am. J. Botany* **43,** 80–96.

Hollenberg, G. J. (1936). A study of *Halicystis ovalis*. II. Periodicity in the formation of gametes. *Am. J. Botany.* **23,** 1–3.

Hoyt, W. D. (1927). The periodic fruiting of *Dictyota* and its relation to the environment. *Am. J. Botany* **14,** 592–619.

Jörgensen, E. (1911). Die *Ceratien*. Eine kurze Monographie der Gattung *Ceratium* Shrank. *Intern. Rev. ges. Hydrobiol. Hydrog. Biol. Suppl. Ser.* **II,** p. 1–124.

Karsten, G. (1918). Über die Tagesperiode der Kern- und Zellteilungen. *Z. Botan.* **10,** 1–20.

Kuckuck, P. (1912). Neue Untersuchungen über *Nemoderma* Schousboe. *Wiss. Meeresuntersuch. Abt. Helgoland. N. F.* **5,** 117–152.

Leedale, G. F. (1959). Periodicity of mitosis and cell division in the Euglenineae. *Biol. Bull.* **116,** 162–174.

Lorenzen, H. (1957). Synchrone Zellteilungen von *Chlorella* bei verschiedenen Licht-Dunkel-Wechseln. *Flora (Jena)* **144,** 473–496.

Lorenzen, H. (1958). Periodizität von Nuklealreaktion und Kernteilung in *Chlorella*. *Ber. deut. botan. Ges.* **71,** 89–97.

Massart, J. (1893). Sur l'irritabilité des Noctiluques. *Bull. Sci. France et Belg.* **25,** 59–76.

Moore, B. (1909). Observations on certain organisms of (a) variations in reaction to light and (b) a diurnal periodicity of phosphoresence. *Biochem. J.* **4,** 1–29.

Nicol, J. A. C. (1958). Observations on luminescence in *Noctiluca*. *J. Marine Biol. Assoc. United Kingdom* **37,** 535–549.

Nozawa, K. (1940). Problem of the diurnal rhythm in the cell division of the dinoflagellate, *Oxyrrhis marina*. *Annotationes Zool. Japan.* **19,** 170–174.

Pirson, A. (1957). Induced periodicity of photosynthesis and respiration in *Hydrodictyon*. *In* "Research in Photosynthesis" (H. Gaffron and others, eds.), pp. 490–499. Interscience, New York.

Pirson, A., and Lorenzen, H. (1958a). Photosynthetische Sauerstoffentwicklung von *Chlorella* nach Synchronisation durch Licht-Dunkel-Wechsel. *Naturwissenschaften* **20,** 497–498.

Pirson, A., and Lorenzen, H. (1958b). Ein endogener Zeitfaktor bei der Teilung von *Chlorella*. *Z. Botan.* **46,** 53–67.

Pirson, A., and Schön, W. J. (1957). Versuche zur Analyse der Stoffwechselperiodik bei *Hydrodictyon*. *Flora (Jena)* **144,** 447–466.

Pirson, A., Schön, W. J., and Döring, H. (1954). Wachstums- und Stoffwechselperiodik bei *Hydrodictyon*. *Z. Naturforsch.* **9b,** 349–353.

Pittendrigh, C. S. (1954). On temperature independence in the clock system controlling emergence time in *Drosophila*. *Proc. Natl. Acad. Sci. U. S.* **40,** 1018–1029.

Pittendrigh, C. S. (1958). Perspectives in the study of biological clocks. *In* "Perspectives in Marine Biology" (A. A. Buzzati-Traverso, ed.), pp. 239–268. Univ. of California Press, Berkeley and Los Angeles, California.

Pittendrigh, C. S., and Bruce, V. G. (1957). An oscillator model for biological clocks. *In* "Rhythmic and Synthetic Processes in Growth" (D. Rudnick, ed.), pp. 75–109. Princeton Univ. Press, Princeton, New Jersey.

Pittendrigh, C. S., and Bruce, V. G. (1959). Daily rhythms as coupled oscillator systems and their relation to thermoperiodism and photoperiodism. *In* "Photoperiodism and Related Phenomena in Plants and Animals" (R. B. Withrow, ed.) pp. 475–505. A.A.A.S., Washington, D. C.

Pohl, R. (1948). Tagesrhythmus in phototaktischen Verhalten der *Euglena gracilis*. *Z. Naturforsch.* **3b,** 367–374.

Schön, W. J. (1955). Periodische Schwankungen der Photosynthese und Atmung bei *Hydrodictyon*. *Flora (Jena)* **142,** 347–380.

Smith, G. M. (1947). On the reproduction of some Pacific coast species of *Ulva*. *Am. J. Botany* **34,** 80–87.

Sweeney, B. M. (1959). Endogenous diurnal rhythms in marine dinoflagellates. Preprints Intern. Oceanog. Congr., August–September, 1959 (Mary Sears, ed.), pp. 204–207. Am. Assoc. Advancement Sci., Washington, D. C.

Sweeney, B. M., and Hastings, J. W. (1957). Characteristics of the diurnal rhythm of luminescence in *Gonyaulax polyedra*. *J. Cellular Comp. Physiol.* **49,** 115–128.

Sweeney, B. M., and Hastings, J. W. (1958). Rhythmic cell division in populations of *Gonyaulax polyedra*. *J. Protozool.* **5,** 217–224.

Sweeney, B. M., and Hastings, J. W. (1960). Effects of temperature upon diurnal rhythms. *Cold Spring Harbor Symposia Quant. Biol.* **25,** 87–104.

Tahara, M. (1909). On the periodic liberation of the oospheres in *Sargassum*. *Botan. Mag. (Tokyo)* **23,** 151–153.

Williams, J. L. (1905). Studies in the Dictyotaceae. III. The periodicity of the sexual cells in *Dictyota dichotoma*. *Ann. Botany* **19,** 531–560.

Yentsch, C. S., and Ryther, J. H. (1957). Short-term variations in phytoplankton chlorophyll and their significance. *Limnol. and Oceanog.* **2,** 140–142.

—47—

Sporulation

K. ERBEN

Botanical Institute, University of Cologne, Germany

I. Introduction

The term "spore" is somewhat loosely used in phycology. It embraces (*a*) various kinds of cells involved in vegetative propagation and dispersal (*aplanospores* and *zoospores*), and (*b*) specialized resting cells, usually with thickened or sculptured walls (*hypnospores, akinetes*, etc.). The former may be produced by mitosis or following a reduction division (*gonospores*); the latter by mitosis or following the fusion of gametes (*zygospores* and *oospores*). Resting spores are common among terrestrial and freshwater Green and Blue-green Algae in habitats frequently liable to desiccation. They also occur among certain centric diatoms with a limited planktonic season. *Auxospores* are a characteristic developmental stage of diatom zygotes, whereby the cells overcome the increasing physical constraints of a rigid siliceous wall that becomes progressively smaller at each cell division. Finally, we may include a reference to *heterocysts*, which occur, usually singly, in many filamentous Blue-green Algae. They have been considered to be degenerate reproductive organs (Geitler, 1921) or "enzyme reservoirs" (Canabaeus, 1929), but without more experimental evidence their function remains obscure (see Fogg, 1951; Drawert and Tischer, 1956; de Puymaly, 1957).

Spores of one sort or another occur in the life cycles of the majority of algae, and many incidental observations have been published concerning their formation and development. By comparison with the extensive studies of the spores of fungi or bacteria, however, there have been very few critical investigations of the physical and biochemical factors involved in algal sporulation and germination. The present discussion is therefore limited to a small number of more carefully studied cases.

II. Vegetative Spores

A fundamental study of the physiology of asexual reproduction in algae was made by Klebs (1896; cf. Oltmanns, 1923; Jost, 1934). In a great variety of experiments he found that alteration of one or more of the following six major factors would usually induce the production of zoospores:

(*1*) Lowering of the concentration of the culture medium led to zoospore formation in species of *Vaucheria*, *Protosiphon*, *Hydrodictyon*, *Bumilleria*, etc. Similar results were obtained by an increased concentration in the case of some species of *Oedogonium* and, together with an increased intensity of light, in *Vaucheria clavata* and *V. repens*.

(*2*) Transfer from moving to still water was found effective for *Vaucheria clavata*, *Oedogonium pluviale*, and species of *Ulothrix*, *Draparnaldia*, etc.

(*3*) Darkness, or a decrease in the intensity of light, was effective for *Vaucheria*, *Protosiphon*, and *Oedogonium capillare*.

(*4*) A reduction in temperature from 13–17° to 5–6°C. induced *Bumilleria sicula* to form zoospores. *Oedogonium* reacted similarly to a rise of temperature from 10 to 15°C.

(*5*) Increased humidity, or transfer to an aqueous medium, induced certain terrestrial algae, such as *Vaucheria repens*, *Protosiphon*, and *Botrydium*, to sporulate, provided the change was sufficiently abrupt.

(*6*) The addition of certain organic compounds to cultures maintained in darkness stimulated sporulation in some cases.

Many observations by other workers have confirmed these results of Klebs; for example, in species of *Cladophora*, where changes in the pH of the medium are effective (Ulehla, 1923), and in *Ulothrix* (Gross, 1931), *Tribonema* (Hawlitschka, 1932), *Oedogonium* (Freund, 1908, 1928), *Stigeoclonium subspinosum* (Jüller, 1937), *Tetraëdron bitridens* (Starr, 1954), and *Cladophora* (Lück, 1957). The main objective from a physiologist's point of view is to determine which systems within the cell are influenced by the changed external conditions. Klebs himself tended to the opinion that all of the various factors which induce sporulation in a given species act by affecting one specific system within the cell, which in turn controls spore formation—a logical but still unproven hypothesis.

More recently, several workers have approached the problem with more refined methods. By increasing the light intensity, Neeb (1952) successfully induced formation of zoospores in old cultures of *Hydrodictyon reticulatum* which had been standing for some weeks in dim light. He noted a brief burst of photosynthetic activity about 30 hours before the zoospores became motile; this was followed by a steady decline in photosynthetic activity, during which time consumption of oxygen increased markedly. The final division of the cytoplasm to form spores, however, occurred without any further detectable change in metabolic activity.

In the freshwater alga *Oedogonium cardiacum*, Bühnemann (1955a, b, c, d) demonstrated a daily maximum of sporulation correlated with an endogenous diurnal rhythm. The length of the light-dark cycle determines the timing which persists for 4 to 6 days after transfer to continuous light. Somewhat surprisingly, at higher temperatures the cycle is extended, e.g., from 20 hours at 17.5°C. to 25 hours at 27.5°C, but it is not appreciably influenced by sublethal concentrations of metabolic poisons such as cyanide, dinitrophenol, or fluoride. According to Ruddat (1961), in cultures of *O. cardiacum* that are maintained in constant light (thus exhibiting no regular fluctuation in zoospore formation), a single period of 6 hours or more in darkness or at a temperature below 20°C, is sufficient to initiate rhythmic sporulation. A light-controlled diurnal periodicity of sporulation has also been reported in other species, e.g., *Vaucheria sessilis* (Rieth, 1959), which thus present new material for research in the field of reproductive cycles. At least in such unicellular algae as *Chlorella* (Tamiya *et al.*, 1953; Pirson and Lorenzen, 1958; Lorenzen, 1959; Soeder, 1960) and *Chlamydomonas* (Scherbaum, 1960), cycles of illumination have been successfully used to synchronize cell divisions (see Hase, Chapter 40, this treatise). Since in both of these genera vegetative reproduction occurs exclusively by the production of zoospores or aplanospores, data derived from the study of such systems may help to elucidate some of the endogenous factors involved in the mechanism of sporulation. Seasonal or lunar rhythms of spore formation, especially in marine algae, have been known for some time (see Czaja, 1929; Smith, 1947; Sweeney and Hastings, Chapter 46, this treatise); they are probably induced by cyclic ecological changes which have not yet been analyzed in adequate detail.

A further point arises from the observation of von Denffer and Hustede (1955) that in *Vaucheria sessilis* the proportion of developing antheridia which redifferentiate into zoosporangia can be increased from 7% (control) to 49% by the addition of 10^{-3} mg/liter indolyl-3-acetic acid. Hustede (1957) further showed that the formation of zoospores in *V. sessilis* is stimulated by tryptophan, tryptamine, tryptophol (indolyl-3-ethanol), indolyl-3-acetic acid, indolyl-3-propionic acid, and indolyl-3-glyoxylic acid, whereas indolyl-3-aldehyde and indolyl-3-carboxylic acid inhibit the process. Tryptophan even induced sporulation under a light intensity of 5000 lux, which normally suppresses it. In view of the instability of many tryptophan derivatives in the light, these experiments suggest an alternative possibility for the role of light intensity in spore formation, which had been considered by League and Greulach (1955), on the basis of their experiments with *V. sessilis*, to be a consequence of limited photosynthesis.

Certain unidentified substances which accumulate in culture media have been reported to be responsible for the inhibition of sporulation in such species as *Enteromorpha compressa* (Moewus, 1948), *Oedogonium cardiacum* (Bühnemann, 1955e), *Stigeoclonium falklandicum*, and *Vaucheria sessilis* (Hustede, 1957).

III. Gonospores

Gonospores are more or less direct products of meiotic division; their formation therefore depends on the induction and completion of normal meiosis. A knowledge of factors that control the development of gonospores is thus of considerable interest, particularly for genetic investigations, because this is the point in the life cycle which usually constitutes the main obstacle in experimental studies, especially with otherwise suitable haploid freshwater species.

Some information is available from experiments on zygote germination. By the use of a method developed for *Chlamydomonas moewusii* by Lewin (1949, 1957), it was found that gonospore formation could also be regularly induced in *C. reinhardtii* (Levine and Ebersold, 1958; Ebersold and Levine, 1959; Levine and Folsome, 1959). The zygotes, after illumination for 18 hours followed by a 6-day maturation period in darkness at 25 °C., release gonospores within 15 to 24 hours (depending on the genotype) after being transferred to a fresh agar medium in the light. *C. eugametos* (Gowans, 1960) and *Gonium pectorale* (Stein, 1958) respond similarly. Starr (1949) demonstrated a high percentage of germination in *C. chlamydogama* zygotes after incubation for 48 hours at 37 °C. *Dunaliella salina* zygotes apparently require a more complex treatment, involving desiccation and resuspension in media of increasing salinity (Lerche, 1937). Apart from these examples among the Volvocales, there are few data on successful methods for the induction of gonospore formation from resting zygotes (see Lewin, 1954). Starr (1955, 1959) effected considerable germination of aged zygotes of the desmids *Cosmarium turpinii* and *C. formosulum*, which had been previously dried and frozen, by immersing them in a fresh medium. But even simulated natural conditions of this sort failed in other cases, such as *Oedogonium* spp. (Mainx, 1931; Gussewa, 1931). The factors responsible for initiating the reduction division and its associated metabolic changes can only rarely be identified. Once meiosis has started, however, the course of subsequent events generally proceeds without the need for further external stimuli.

IV. Auxospores of Diatoms

Whereas conditions for zygospore formation in most algae have long been known to be related to factors evoking sexually active gametes, the situation regarding the auxospores of diatoms was less clear until the early 1950's. This was mainly because of the lack of knowledge concerning the sexual process in centric diatoms and its relation to auxospores, which are by no means resting stages (compare the extensive discussions in, e.g., Persidsky, 1929; Geitler, 1932; Fritsch, 1935). It now seems generally established that in all diatoms meiosis is restricted to cells within a certain size range, and that auxospores usually develop from newly formed zygotes (Wiedling, 1943, 1948; von Stosch, 1951, 1954, 1955, 1956; Geitler, 1952; Ermolaeva, 1953; Patrick, 1954). Consequently, auxospore formation can take place only when cells of the necessary size are present. Since no reliable case of asexual auxospore formation is known, their initiation must depend chiefly on metabolic conditions associated with a certain cell size or, more specifically, on the physiological state of the young zygotes.

For technical reasons, most workers have studied the effects of external factors, such as light, temperature, and substrate, upon the whole period of sexual activity which culminates in auxospore production (Karsten, 1899; Bachmann, 1904; Müller, 1906; Schreiber, 1931; Geitler, 1932; Braarud, 1944; Ermolaeva, 1953). It appears that the cells of diatoms, in the course of their natural reduction in size, pass through a certain size range in which the tendency for meiosis and gamete production is greatest (Erben, 1959). The limits of this size range, and the proportion of suitably sized cells which in fact undergo meiosis, are features subject to environmental influence (von Stosch, 1954; Bruckmayer-Berkenbusch, 1955). In cultures of *Melosira nummuloides* it was not found possible to inhibit auxospore formation altogether, or to initiate it in noncompetent cells, by varying the concentration of nitrate or phosphate within the range which permitted mitotic growth, or by modifying the metabolism with sublethal concentrations of 2,4-dinitrophenol, iodoacetic acid, or *o*-phenanthroline (Erben, 1959). On the other hand, treatment with continuous light of 3400 lux intensity at 24°C. inhibited sporulation completely, although it permitted intensive vegative growth leading to cells with the smallest diameter known for this species (Bruckmayer-Berkenbusch, 1955). Here, as usual, it remains unknown which of the necessary events leading to the mature auxospore—meiosis, gamete fusion, or zygote development—is actually influenced. However, some indication is provided by an observation by von Stosch (1954) that in cultures of *Lithodesmium* sp. containing cells of suitable size, only oogonia were produced under continuous light, whereas low light intensity in a natural light-dark cycle led exclusively to the formation of antheridia.

V. Resting Spores

McKater and Burroughs (1926) discovered that in *Polytomella citri* encystment took place only when the cells had accumulated starch, and that maximum encystment was apparently a consequence of optimum growth conditions. Droop (1955) suggested that in *Haematococcus pluvialis* a necessary condition for encystment is an excess of carbon assimilation over dissimilation, as occurs in cultures in which carbon assimilation and accumulation of organic material in the light continue after cell division has been arrested. In fact, in those of his experiments which were performed in the dark with sodium acetate as the main carbon source, the acetate concentration had to be higher for encystment than for growth.

Conditions for the development of resting zygotes in Volvocales may be similar to those for encystment in *Haematococcus*. Thus the motile zygotes of *Brachiomonas submarina* develop into thick-walled spores only if adequately illuminated (Droop, 1955). According to Lewin (1957), the formation of viable resting zygotes in *Chlamydomonas moewusii* appears to require normal photosynthesis. Zygotes maintained in the absence of carbon dioxide or receiving less than 14 hours illumination (with CO_2 present) did not reach a stage capable of germination. Light is also necessary for the corresponding phase of development in *C. reinhardtii*, *C. eugametos*, and *Gonium pectorale*, as mentioned above.

VI. Heterocysts

Fogg (1949) found that in *Anabaena cylindrica* the formation of heterocysts is favored by a low light intensity, or by the addition of glucose or succinic acid to the medium. Nitrate, glycine, or asparagine, on the other hand, temporarily delays their appearance; the ammonium ion inhibits heterocyst production as long as it is present in the medium.

VII. Conclusion

Although descriptive studies have elucidated details of the life cycles of a great variety of algal species, the physiology of most of the functional stages involved remains completely obscure. At present no generalization on the basic factors controlling the phenomena of sporulation can be more than speculative. Our present knowledge of metabolism and biochemistry provides little information which might help us understand these determinative steps in cell development because we still have only crude observations of their correlation with certain relatively unspecific external factors. However, new concepts and techniques are now becoming available, especially with the experimental induction of synchronously sporulating cultures, by which it should eventually be possible to identify specific metabolic processes underlying the production of spores.

References

Bachmann, H. (1904). *Cyclotella bodanica* var. *lemanica* O. Mül. im Vierwaldstätter See und ihre Auxosporenbildung. *Jahrb. wiss. Botan.* **39,** 106–133.

Braarud, T. (1944). Experimental studies on marine plankton diatoms. *Avhandl. Norske Videnskaps-Akad. Oslo. I. Mat.-Naturv. Kl.* **10,** 1–16.

Bruckmayer-Berkenbusch, H. (1955). Die Beeinflussung der Auxosporenbildung von *Melosira nummuloides* durch Aussenfaktoren. *Arch. Protistenk.* **100,** 183–211.

Bühnemann, F. (1955a). Die rhythmische Sporenbildung von *Oedogonium cardiacum* Wittr. *Biol. Zentr.* **74,** 1–54.

Bühnemann, F. (1955b). Das endodiurnale System der *Oedogonium*zelle. II. Der Einfluss von Stoffwechselgiften und anderen Wirkstoffen. *Biol. Zentr.* **74,** 691–705.

Bühnemann, F. (1955c). Das endodiurnale System der *Oedogonium*zelle. III. Über den Temperatureinfluss. *Z. Naturforsch.* **10b,** 305–310.

Bühnemann, F. (1955d). Das endodiurnale System der *Oedogonium*zelle IV. Die Wirkung verschiedener Spektralbereiche auf die Sporulations- und Mitoserhythmik. *Planta* **46,** 227–255.

Bühnemann, F. (1955e). Untersuchungen über einen Sporulations- und Wachstumshemmstoff in *Oedogonium*-Kulturen. *Arch. Mikrobiol.* **23,** 14–27.

Canabaeus, L. (1929). Über die Heterocysten und Gasvakuolen der Blaualgen und ihre Beziehung zueinander. *Pflanzenforschung* **13,** (R. Kolkwitz, ed.) Fischer, Jena.

Czaja, A. T. (1929). Periodizität. *In* "Tabulae Biologicae" (C. Oppenheimer and L. Pincussen, eds.) Vol. V, pp. 362–509. Junk, Berlin.

de Puymaly, A. (1957). Les hétérocystes des algues bleues: leur nature et leur rôle. *Le botaniste* **46,** 209–270.

Drawert, H., und Tischer, I. (1956). Über Redox-Vorgänge bei Cyanophyceen unter besonderer Berücksichtigung der Heterocysten. *Naturwissenschaften* **43,** 132.

Droop, M. R. (1955). Some factors governing encystment in *Haematococcus pluvialis. Arch. Mikrobiol.* **21,** 267–272.

Ebersold, W. T., and Levine, R. P. (1959). A genetic analysis of linkage group I of *Chlamydomonas reinhardi. Z. Vererbungslehre* **90,** 74–82.

Erben, K. (1959). Untersuchungen über Auxosporenentwicklung und Meioseauslösung an *Melosira nummuloides* (Dillw.) C.A. Agardh. *Arch. Protistenk.* **104,** 165–210.

Ermolaeva, L. M. (1953). [Über die Auxosporenbildung bei der Alge *Cyclotella Meneghiniana* Ktz.] (In Russian.) *Doklady Acad. Nauk S.S.S.R.* **91,** 165–168; cf. *Ber. wiss. Biol.* **89,** 329.

Fogg, G. E. (1949). Growth and heterocyst production in *Anabaena cylindrica* Lemm. II. In relation to carbon and nitrogen metabolism. *Ann. Botany (London)* **13,** 241–259.

Fogg, G. E. (1951). Growth and heterocyst production in *Anabaena cylindrica* Lemm. III. The cytology of heterocysts. *Ann. Botany (London)* **15,** 23–35.

Freund, H. (1908). Neue Versuche über die Wirkung der Aussenwelt auf die ungeschlechtliche Fortpflanzung der Algen. *Flora (Jena)* **98,** 41.

Freund, H. (1928). Über die Bedingungen des Wachstums von *Oedogonium pluviale.* Ein Beitrag zur Frage des Stickstoff- und Phosphoretiolements. *Planta* **5,** 520–548.
Fritsch, F. E. (1935). "The Structure and Reproduction of the Algae," Vol. I. Cambridge Univ. Press, London and New York.
Geitler, L. (1921). Versuch einer Lösung des Heterocystenproblems. *Sitzber. Akad. Wiss. Wien Math. naturw. Kl. Abt. I,* **130,** 223–245.
Geitler, L. (1932). Der Formwechsel der pennaten Diatomeen (Kieselalgen). *Arch. Protistenk.* **78,** 1–224.
Geitler, L. (1952). Oogamie, Mitose, Meiose und metagame Teilung bei der zentrischen Diatomee *Cyclotella. Österr. Botan. Z.* **99,** 506–520.
Gowans, C. S. (1960). Some genetic investigations on *Chlamydomonas eugametos. Z. Vererbungslehre* **91,** 63–73.
Gross, I. (1931). Beiträge zur Entwicklungsgeschichte der Protophyten. VII. Entwicklungsgeschichte, Phasenwechsel und Sexualität bei der Gattung *Ulothrix. Arch. Protistenk.* **73,** 207–234.
Gussewa, K. (1931). Über die geschlechtliche und ungeschlechtliche Fortpflanzung von *Oedogonium capillare Ktz.* im Licht der sie bestimmenden Verhältnisse. *Planta* **12,** 293–326.
Hawlitschka, E. (1932). Die Heterokonten-Gattung *Tribonema. Pflanzenforschung* **No. 15.** (R. Kolkwitz, ed.) Fischer, Jena.
Hustede, H. (1957). Untersuchungen über die stoffliche Beeinflussung der Entwicklung von *Stigeoclonium falklandicum* und *Vaucheria sessilis* durch Tryptophanabkömmlinge. *Biol. Zentr.* **76,** 555–595.
Jost, L. (1934). Physiologie der Fortpflanzung. *In* "Handwörterbuch der Naturwissenschaften" Vol. IV, 2nd ed., pp. 435–450. Fischer, Jena.
Jüller, E. (1937). Der Generations- und Phasenwechsel bei *Stigeoclonium subspinosum. Arch. Protistenk.* **89,** 55-93.
Karsten, G. (1899). Die Diatomeen der Kieler Bucht. *Wiss. Meeresunters. Kiel* **4,** 17–205.
Klebs, G. (1896). "Die Bedingungen der Fortpflanzung bei einigen Algen und Pilzen." Fischer, Jena. (2nd ed., 1928.)
League, E. A., and Greulach, V. A. (1955). Effects of daylength and temperature on the reproduction of *Vaucheria sessilis. Botan. Gaz.* **117,** 45–51.
Lerche, W. (1937). Untersuchungen über Entwicklung und Fortpflanzung in der Gattung *Dunaliella. Arch. Protistenk.* **88,** 236–268.
Levine, R. P., and Ebersold, W. T. (1958). Gene recombination in *Chlamydomonas reinhardi. Cold Spring Harbor Symposia Quant. Biol.* **23,** 101–109.
Levine, R. P., and Folsome, C. E. (1959). The nuclear cycle in *Chlamydomonas reinhardi. Z. Vererbungslehre* **90,** 215–222.
Lewin, R. A. (1949). Germination of zygospores in *Chlamydomonas. Nature* **164,** 543-544.
Lewin, R. A. (1954). Sex in unicellular algae. *In* "Sex in Microorganisms" (D. H. Wenrich, ed.), pp. 100-133. Am. Assoc. Advancement Sci., Washington, D. C.
Lewin, R. A. (1957). The zygote of *Chlamydomonas moewusii. Can. J. Botan.* **35,** 795–804.
Lorenzen, H. (1959). Die photosynthetische Sauerstoffproduktion wachsender *Chlorella* bei langfristig intermittierender Belichtung. *Flora (Jena)* **147,** 382–404.
Lück, H. B. (1957). Au sujet d'un cas de formation d'aplanospores dans le genre *Cladophora. Bull. inst. océanog.* **No. 1105,** 4 pp.

McKater, J., and Burroughs, R. D. (1926). The cause and nature of encystment in *Polytomella citri. Biol. Bull.* **50,** 38.

Mainx, F. (1931). Physiologische und genetische Untersuchungen an *Oedogonium.* I. *Z. Botan.* **24,** 481–527.

Moewus, F. (1948). Zur Genetik und Physiologie der Kern- und Zellteilung. I. Die Apomiktosis von *Enteromorpha*-Gameten. *Biol. Zentr.* **67,** 277–293.

Müller, O. (1906). Pleomorphismus, Auxosporen und Dauersporen bei *Melosira*-Arten. *Jahrb. wiss. Botan.* **43,** 49–88.

Neeb, O. (1952). *Hydrodictyon* als Objekt einer vergleichenden Untersuchung physiologischer Grössen. *Flora (Jena)* **139,** 39–95.

Oltmanns, F. (1923). "Morphologie und Biologie der Algen." Vol. III, 2nd ed. Fischer, Jena.

Patrick. R. (1954). Sexual reproduction in diatoms. *In* "Sex in Microorganisms" (D. H. Wenrich, ed.), pp. 82–99. Am. Assoc. Advancement Sci., Washington, D. C.

Persidsky, B. M. (1929). "The Development of Auxospores in the Group of *Centricae* (Bacillariaceae)." Moscow (published privately). Abstracted in *Botan. Zentr.* **159,** 235.

Pirson, A., and Lorenzen, H. (1958). Ein endogener Zeitfaktor bei der Teilung von *Chlorella. Z. Botan.* **46,** 53–66.

Rieth, A. (1959). Periodizität beim Ausschlüpfen der Schwärmsporen von *Vaucheria sessilis. Flora (Jena)* **147,** 35–42.

Ruddat, M. (1961). Versuche zur Beeinflussung und Auslösung der endogenen Tagesrhythmik bei *Oedogonium cardiacum* Wittr. *Z. Botan.* **49,** 23–45.

Scherbaum, O. H. (1960). Synchronous division of microorganisms. *Ann. Rev. Microbiol.* **14,** 283–310.

Schreiber, E. (1931). Über Reinkulturversuche und experimentelle Auxosporenbildung bei *Melosira nummuloides. Arch. Protistenk.* **73,** 331.

Smith, G. M. (1947). On the reproduction of some Pacific Coast species of *Ulva. Am. J. Botany* **34,** 80–87.

Soeder, C. J. (1960). Studien zur Entwicklungsphysiologie von *Chlorella pyrenoidosa* Chick unter besonderer Berücksichtigung der Salzkonzentration im Medium. *Flora (Jena)* **148,** 489–516.

Starr, R. C. (1949). A method of effecting zygospore germination in certain Chlorophyceae. *Proc. Natl. Acad. Sci. U. S.* **35,** 453–456.

Starr, R. C. (1954). Reproduction by zoospores in *Tetraëdron bitridens. Am. J. Botany* **41,** 17–20.

Starr, R. C. (1955). Zygospore germination in *Cosmarium Botrytis* var. *subtumidum. Am. J. Botany* **42,** 577–581.

Starr, R. C. (1959). Sexual reproduction in certain species of *Cosmarium. Arch. Protistenk.* **104,** 155–164.

Stein, J. R. (1958). A morphologic and genetic study of *Gonium pectorale. Am. J. Botany* **45,** 664–672.

Tamiya, H., Shibata, K., Sasa, T., Iwamura, T., and Morimura, Y. (1953). Effect of diurnally intermittent illumination on the growth and some cellular characteristics of *Chlorella. In* "Algal Culture From Laboratory to Pilot Plant" (J. S. Burlew, ed.) *Carnegie Inst. Wash. Publ.* **No. 600,** pp. 76–84.

Ulehla, V. (1923). Über CO_2- und pH-Regulation des Wassers durch einige Süsswasseralgen. *Ber. deut. botan. Ges.* **41,** 20–31.

von Denffer, D., and Hustede, H. (1955). Wuchsstoffbedingte Umstimmung von der sexuellen zur vegetativen Entwicklung bei *Vaucheria sessilis*. *Flora (Jena)* **142,** 489–492.

von Stosch, H. A. (1951). Entwicklungsgeschichtliche Untersuchungen an zentrischen Diatomeen. I. Die Auxosporenbildung von *Melosira varians*. *Arch. Mikrobiol.* **16,** 101–135.

von Stosch, H. A. (1954). Die Oogamie von *Biddulphia mobiliensis* und die bisher bekannten Auxosporenbildungen bei den Centrales. *Congr. intern. botan. 8e Congr. Paris,* **1954,** Sect. 17, 58–68.

von Stosch, H. A. (1955). Pennate Diatomeen. *Z. Botan.* **43,** 89–99.

von Stosch, H. A. (1956). Entwicklungsgeschichtliche Untersuchungen an zentrischen Diatomeen. II. Geschlechtszellenreifung, Befruchtung und Auxosporenbildung einiger grundbewohnender Biddulphiaceen der Nordsee. *Arch. Mikrobiol.* **23,** 327–365.

Wiedling, S. (1943). Die Gültigkeit der MacDonald-Pfitzer'schen Regel bei der Diatomeengattung *Nitzschia*. *Naturwissenschaften* **31,** 115.

Wiedling, S. (1948). Beiträge zur Kenntnis der vegetativen Vermehrung der Diatomeen. *Botan. Notiser* **3,** 322.

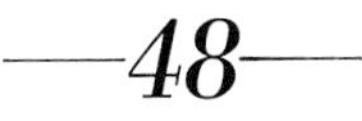

Sexuality*

ANNETTE WILBOIS COLEMAN

Biology Department and McCollum-Pratt Institute, Johns Hopkins University, Baltimore, Maryland

I. Introduction[1]

Sexual reproduction has been described in all major classes of algae except the Euglenophyta (but see Biecheler, 1937) and the Cyanophyta.

* This review was completed in 1960 during the tenure of National Science Foundation Post-doctoral Fellowship No. 49005. The author is indebted to Dr. Lutz Wiese for Figs. 3–5.

[1] The publications of the late Dr. Franz F. Moewus could be mentioned in nearly all sections of this chapter. However, in view of the reported difficulties in repeating much of his work (Ryan, 1955; Hartmann, 1955; Renner, 1958), as well as theoretical objections to his conclusions (Philip and Haldane, 1939; Thimann, 1940), such references have been

Nonetheless, most algae are also capable of unlimited multiplication by one or more means of vegetative reproduction, and strains of many microscopic species can be cultivated in the laboratory indefinitely without expressing any tendency for sexual reproduction. The means evolved for genetic recombination range from simple fusion of motile isogametes to the union of more elaborate but immotile gametes of Red Algae and the apparent autogamy found in some diatoms. For the most part, the original descriptions of sexuality were based on observations of collections from nature in which mating occurred spontaneously, or, particularly among marine forms, on the examination of specimens collected at known times of sexual reproduction, often correlated with the season or the stage of the tide (see Sweeney and Hastings, Chapter 46, this treatise). From such collections the life cycles of a great many algae were deciphered, yet the causes of the onset of sexuality at specific times remained unknown.

The first extensive investigation of the relationship between growth and sexual reproduction in algae was made by Klebs (1896). Several of the species which he maintained in the laboratory formed zygotes under conditions of high light intensity and nutrient salt deficiency, showing that sexual reproduction could be induced at will by changes in the external environment. Nevertheless, he could find no uniform set of conditions which would evoke sexuality in all the species he examined, and many remained in an asexual phase despite a variety of treatments tested.

II. Genetic Control of Sexuality

Soon after the turn of the century, new techniques and new ideas became available to students of algal sexuality. Of greatest importance were the development of chemically defined culture media for the growth of bacteria-free strains, and the discovery of the role of genetics in the control of sexuality. Clonal cultures (populations of cells derived by mitosis from a single original isolate) came to be recognized as necessary for experiments attempting to distinguish between genotypic and phenotypic variation.

A. Self-sterility

Among sexually reproducing organisms, clonal cultures may be of two kinds: (i) those which can form fertile zygotes by themselves; and (ii) those which can do so only when mixed with another sexually compatible clone. The former condition, self-compatibility, has come to be equated

omitted—primarily for the sake of brevity, for the validity of each would require discussion. Suffice it to say that probably no other man has done more to stimulate work in the field of algal genetics and sexuality, albeit by the controversial nature of his own results.

with the term *homothallism* or monoecie, and the latter condition, self-sterility, with *heterothallism* or dioecie. In those heterothallic algae which are isogamous—i.e., in which the gametes are morphologically indistinguishable—the initial designations of the two mating types as "male and "female," or as *plus* and *minus*, have been arbitrary.

The precise definitions of these terms have been discussed, with reference to the fungi, by Whitehouse (1949). Since these terms were not originally applied to algae, and since the genetic basis of the conditions they describe is usually unknown, difficulties arise in their more general usage. To complicate matters further, in some strains which have been shown to be homothallic, e.g., *Vaucheria* spp., the mechanism of marked sexual dimorphism, *per se*, cannot be equated with the cause of self-sterility.

B. Mating Types

Self-sterility may not be the only genetic block to mating among various heterothallic clones of a species. Frequently, such clones can be classified into two or more mating types on the basis of their ability to form zygotes. At least two possible breeding systems might be expected in such heterothallic species:

(*1*) *Bipolar*, in which all clones may be classified as one or the other of two mating types. More than one *syngen* or sexually isolated subspecies (see Sonneborn, 1957) may exist within the species, but each syngen contains only one pair of mating types.

(*2*) *Multipolar*, in which any clone will mate with any other clone; within a syngen, there is no theoretical limit to the number of mating types.

Although there are few cases where more than two strains of a species have been examined, so far only the bipolar breeding system has been found in the algae studied: *Chlamydomonas eugametos* (= *C. moewusii*) (Bernstein and Jahn, 1955); *Gonium pectorale* (Stein, 1958b); and *Pandorina morum* (Coleman, 1959). In all three cases, multiple syngens have been reported. A multipolar breeding system has not yet been found in any cultivated alga, though it occurs in some Protozoa. Tetrapolarity, where any strain in a syngen crosses with all others but two, is apparently unique to the Basidiomycetes.

A system of *relative sexuality* was proposed by Hartmann (1925) to explain some of the results he obtained with *Ectocarpus siliculosus*. This system differs empirically from multipolar breeding in that reactions between different strains are of differing, though constant, intensities. All strains can be classified in a linear array according to the intensity of their mating reaction, adjacent strains crossing weakly if at all. Hartmann's theory has not been confirmed using pure cultures of any alga under constant conditions.

There is no reason to assume that all strains of any widely distributed species will exhibit the same breeding system, particularly since a number of species containing both heterothallic and homothallic strains are already known. Further genetic as well as physiological studies are needed, since we still know very little about the manner in which the genetic constitution controls mating behavior.

III. Environmental Control of Sexuality

A. Induction of Sexuality

There is ample evidence that the appearance of sexuality is largely controlled by nutritional factors, and that only when an alga reaches a certain physiological state is it susceptible to the stimuli inducing gametogenesis. This sexually mature condition is easy to recognize in algae with marked sexual differentiation. Even among species in which there is no clear cellular differentiation before gametogenesis, the "mature" state appears to be correlated with a characteristic cell size which arises in the course of growth of a culture. This is particularly evident among the diatoms (reviewed by Patrick, 1954; Geitler, 1957a) where gamete size has been used as a guide to mating type, though sometimes without justification (Geitler, 1957b).

For several freshwater Green Algae, there is more precise knowledge of the nutritional changes leading to sexuality. The following discussion on the effects of light, temperature, nutrition, and pH on the induction of the sexual process draws heavily on investigations of a single genus, *Chlamydomonas*. This is because (*a*) the cells of many species can be easily and rapidly grown; (*b*) bacteria-free clones of several species are available; (*c*) they can be maintained in chemically defined media; and (*d*) since some of these species are heterothallic, mating can be initiated at a time determined experimentally by mixing the mating types. Many reports on a wide variety of other algae confirm the importance of light, nutrient depletion, and temperature in the induction of mating activity (e.g., Hustede, 1957).

B. Nutrient Supply

For investigations of the nutritional control of mating, cells have generally been harvested from a growth medium and suspended for various periods of time in distilled water or in dilute media. However, observations on the appearance of gametes in relation to the growth curve of a culture often indicate that sexually active cells first appear toward the end of the exponential phase. This is true for *Chlamydomonas reinhardtii* (Sager and Granick, 1953), *Chlamydomonas* sp. 24* (Tsubo, 1956), and *Pandorina morum* (Wilbois, 1958).

Sager and Granick showed clearly that, of the various components of the medium, only the concentration of the nitrogen source directly influenced sexuality. Depletion of the available nitrogen resulted in the formation of active gametes; conversely, the addition of a nitrogenous solute led to a rapid loss of sexual activity. This led Sager and Granick to suggest that the triggering of sexuality by a lowered available nitrogen concentration might be a mechanism common to many algae which form resting zygotes. Such behavior would aid the survival of the species during periods of depleted nutrients, and would thus presumably be favored by selection. The empirical value of this discovery has already been shown in research on *Chlamydomonas chlamydogama* (Trainor, 1958), *Chlamydomonas* sp. 24* (Tsubo, 1956), and *Pandorina morum* (Wilbois, 1958), where experimentally lowering the concentration of the nitrogen source aids in obtaining gamete suspensions. Correspondingly, at least in the latter two cases, gamete activity appears to be depressed by added nitrogen sources.

On the other hand, although nitrate as well as ammonia destroys gametic ability (Sager and Granick, 1953), Lewin could not grow *C. reinhardtii* on nitrate as a sole nitrogen source (personal communication), which suggests that perhaps the inhibition of gamete activity by added salts is a phenomenon only indirectly related to gamete induction by nitrogen depletion. Unlike the species mentioned above, *Chlamydomonas moewusii* failed to respond significantly to a lowered nitrogen content of the growth medium (Bernstein and Jahn, 1955), and the activity of gametes of this species was unaffected by an added nitrogen source, at least for several hours (Lewin, 1956). A comparative study by Trainor (1959) showed that this difference could not be attributed merely to different experimental conditions, and confirmed that in *C. moewusii*, unlike the other three *Chlamydomonas* species which he examined, sexual activity was apparently unaffected by the nitrogen content of the suspending medium.

If, as Sager and Granick suggested, the selection of a trigger mechanism for sexuality is related to the type of zygote formed, then an abundance of nutrients might be expected to induce sexuality in species with zygotes which germinate immediately. Such an alga is the marine form, *Chaetomorpha aerea*, for which Köhler (1956) reported that the best method of obtaining gametes is to flood the culture with fresh nutrient medium. In this case, however, there was no indication of which constituents of the medium might be responsible for the primary stimulus.

Among other solutes which may affect mating, a definite role has been

* = *C. moewusii* var. *rotunda* (Tsubo, 1961).

assigned to the calcium ion in the case of *Chlamydomonas moewusii*. Lewin (1954a) reported that washed gametes resuspended in distilled water were unable to mate unless calcium was added to the suspension. He also found that 0.2 *M* citrate inhibited the pairing of gametes, presumably by chelating calcium ions, and that this effect could be reversed by raising the calcium concentration.

C. pH

Within the bounds of physiological neutrality, there appear to be no striking effects of pH on the mating reaction of algal gametes so far investigated. Mating has been reported in cultures ranging in pH from 4 to 8.5. Several authors mention an increased zygote yield correlated with a slight rise in the pH value of the medium, but the effect is rarely as much as two-fold.

D. Temperature

Mating can occur over approximately as wide a temperature range as that which allows growth. In *Chlamydomonas* spp., a rise in temperature may, under some conditions, increase the proportion of gametes in a cell population (Förster, 1957), as well as the rate of their appearance when cultures are first placed in the light (Lewin, 1956; Förster, 1957). In both *Chlamydomonas eugametos* and *C. moewusii*, when gametes are placed in the dark the rate of loss of gametic activity is dependent upon the temperature, the loss of sexuality being approximately exponential (Lewin, 1956). Temperatures up to 35°C. apparently enhance the total zygote formation in *C. chlamydogama* (Trainor, 1960), but the interpretation of these results is complicated by the fact that the zygote yields were not determined in the samples until 4 days after they had been removed from conditions of light and high temperature.

E. Light

Probably the most universal inducer of sexuality in algae is light. Incident light energy of less than 1000 lux may limit the sexual reproduction of a culture, and experimental work has been done mainly in the range of 3000 to 6000 lux. For many cultivated species, light signals the onset of sexuality; in cultures of *Pandorina morum* subjected to a cycle of 16 hours light and 8 hours darkness, mating occurs only during a certain portion of the light period each day (Wilbois, 1958). Lewin (1956) found that *Chlamydomonas moewusii* gametes, which require light for the induction of sexual activity, remain active in darkness for a period of several minutes, the length of this "dark lag" being proportional to the length of the previous light period. Ultimately sexual activity is lost. When such inactive gametes are reilluminated, sexual activity rises rapidly after a brief "light lag" period, which is more or less constant in length.

In several species of *Chlamydomonas*, sexual reproduction can occur in the absence of light, although usually only under special cultural conditions and with a relatively low yield of zygotes (Smith, 1946; Sager and Granick, 1953). With *C. reinhardtii*, Sager and Granick (1954) succeeded in growing cells from the first inoculum through complete zygote formation entirely in the dark by using a medium low in nitrogen and high in acetate. They suggested that light, in evoking sexuality, acts through photosynthetic assimilation which leads to the depletion of the available nitrogen supply in the medium. If this is so, then in light-requiring forms the action spectrum of the induction of gamete formation should correspond to the action spectrum of chlorophyll and thus resemble the absorption spectrum of the cell. This was found to be approximately true for a mixed suspension of the two mating types of *C. moewusii*, previously sexually inactivated in the dark, the action spectrum of which exhibited maxima at about 450 mμ and 675 mμ; see Fig. 1 (Lewin, 1956).

Dissimilar results were obtained by Förster (1957), who carried out more critical experiments with *C. eugametos*. He used light of various wavelengths to irradiate suspensions of male cells, which normally require light for activation, and then mixed them with female gametes, sexually active in darkness. The action spectrum which he obtained in this way was nearly the reverse of the absorption spectrum of the cells, with a major peak at 590 mμ and a minor one at 460 mμ (Fig. 2). With *C. moewusii*, in which both mating types require light, he obtained similar action spectra, considerably different from that published by Lewin, though the same strains were used (Förster, 1959).

An explanation for this discrepancy was suggested by Stifter (1959), who pointed out that sometimes the proportion of gametes in a *C. eugametos* suspension can be raised by illumination with red light at 680 mμ for several hours. This effect presumably results from photosynthesis, since it requires the presence of CO_2. Stifter suggested that Lewin's action spectrum was determined by the use of cells whose mating potential could be raised by photosynthesis, and for this reason it paralleled the absorption spectrum of chlorophyll. On the other hand, since in most of the cultures used by Förster the mating potential could not be raised by a period of photosynthesis, the action spectra he obtained presumably reflected some action of light other than its role in photosynthesis. This second action of light, Stifter suggested, may nevertheless be linked to

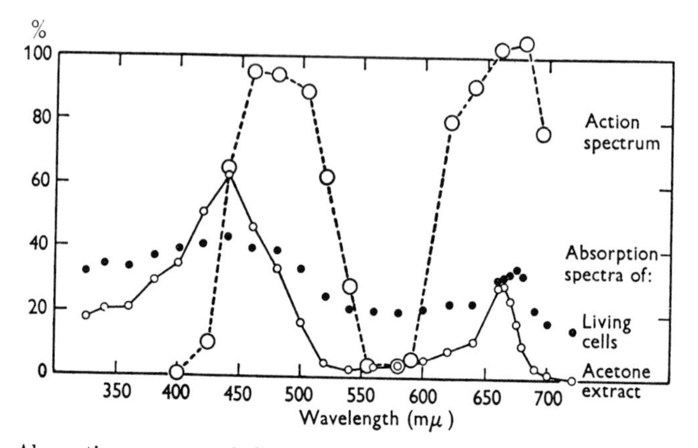

FIG. 1. Absorption spectra of *Chlamydomonas moewusii* cell suspension (dots) and extracted pigments (solid line). Action spectrum for induction of pairing in a mixed gamete suspension of *C. moewusii* (broken line) relative to that obtained after 5 minutes of exposure to bright white light. After Lewin (1956).

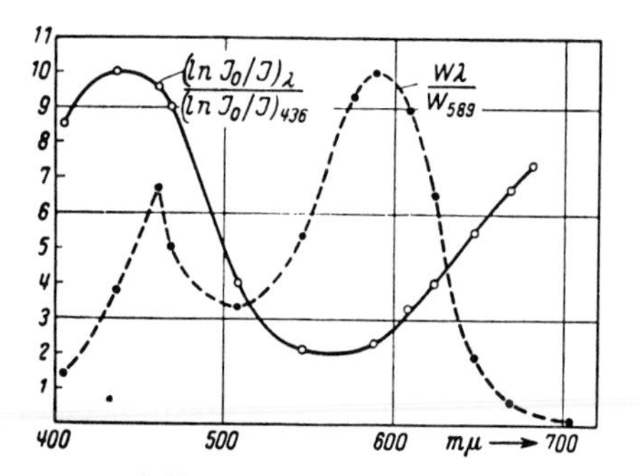

FIG. 2. Absorption spectrum of *Chlamydomonas eugametos* cell suspension (solid line), referred to $\lambda = 436$. Action spectrum for gamete formation of male *C. eugametos* strain (broken line). W_λ/W_{589} represents the mating enhancement relative to that obtained where $\lambda_{589} = 10$; I_0 is intensity of radiation entering medium; I, intensity after traversing medium. After Förster (1957) and Stifter (1959).

photosynthesis, since it can be demonstrated only in cells whose photosynthetic requirement is already satisfied. For this and other reasons, she proposed that the second action of light, that related solely to the activation of gametes, could involve some respiratory enzyme, linked to the photosynthetic cycle, and with an absorption spectrum characterized by maxima at 460 and 590 mμ.

F. Oxygen

Both Stifter (1959), with the male strain of *Chlamydomonas eugametos*, and Lewin (1956), with *C. moewusii*, found that production and maintenance of gametes in the light requires the presence of oxygen, and that gametic activity is lost more rapidly in the dark under anaerobic conditions than in the presence of air. Under anaerobic conditions in the dark, the female *C. eugametos* strain is sexual, as shown by its immediate clumping ability, although no gynogamone is detectable in the medium (see Section V.D). Oxygen is thus required for the mating reaction, or at least enhances it, in all cases so far examined.

Excess oxygen does not adversely affect mating (Stifter, 1959).

G. Carbon Dioxide

A CO_2 requirement for mating was not directly demonstrable in *C. moewusii* (Lewin, 1956); however, in the case of male gametes of *C. eugametos*, the removal of CO_2 from either a brightly or a dimly illuminated cell suspension halved the proportion of active gametes within an hour (Stifter, 1959). Addition of normal air soon restored the original level of gametic activity. According to Stifter, excess CO_2, even as little as 5% CO_2 in air, within 10 to 20 minutes drastically reduced the proportion of gametes in a suspension.

With male *Chlamydomonas eugametos*, the CO_2 requirement for the maintenance of a high gametic level is not apparent when previously darkened cells are first exposed to low light intensity. In this case, suspensions deprived of CO_2 become sexually active more rapidly than suspensions provided with CO_2, at least for the first few minutes in the light.

If this apparently inhibitory effect of CO_2 on initial gametogenesis is related to the role of CO_2 in photosynthesis, then it should perhaps be pointed out that Förster used similar cells and conditions to determine the action spectrum of gametogenesis, i.e., previously darkened cells and very low light intensities for 4 to 8 minutes. Thus, his action spectra would reflect not only absorption of light by a receptor specific for mating but also the inhibition of gametogenesis by photosynthesis. It would be of particular interest to determine under such conditions the action spectrum of gamete formation in the absence of CO_2. It should also be borne in mind that there are pathways of CO_2 fixation other than photosynthesis, and that these may also be concerned in reproductive activity, as has been shown for the differentiation of resistant sporangia in the fungus *Blastocladiella* (Cantino, 1956).

IV. The Course of the Mating Reaction

A. Mechanisms

Although little is known of the underlying physiology of gametogenesis, the various stages of the mating process itself have been found more amenable to study. Of the great variety of mechanisms among algae for bringing together two nuclei in syngamy, one has been examined in far greater detail than any other. This is the system in which both sexes produce motile gametes, as found in various genera of the Chlorophyta, Xanthophyta, Pyrrophyta (Dinophyceae), and Phaeophyta. The basic pattern of events appears to be common to all, i.e., an aggregation of gametes in clumps, and a pairing of cells of complementary mating type followed by their fusion to form a single cell, the zygote. Sooner or later, during the maturation of the zygote, the two nuclei fuse.

The time required for this process (excluding nuclear fusion) varies among different algae, from a total of 5 minutes in *Chaetomorpha aerea* to as long as several hours in *Chlamydomonas moewusii*. The manner of clumping may also differ: thus in *Ectocarpus siliculosus* several male gametes have been seen to cluster around a single female cell, their flagella adherent to the female cell surface; whereas in many Green Algae the clumps usually comprise several gametes of both sexes clustered by the agglutination of their flagellar tips. Variation is also encountered in the mode of cell fusion. In most cases this appears to begin with contact between specialized areas of the two gametes and to proceed rapidly to completion. However, gametes of *Chlamydomonas moewusii* (Lewin, 1954a) and *Chlamydomonas* sp. 24 (Tsubo, 1956) initially form only a narrow bridge between their apices and swim about in pairs for some hours before the completion of cell fusion.

B. Phases

If complementary strains of a heterothallic species are mixed, so that the mating reaction begins and the stages proceed more or less in synchrony, the course of the reaction can be followed stage by stage, and the following phases can be recognized: (*1*) physiological and morphological differentiation leading to gamete formation; (*2*) clumping; (*3*) pairing; (*4*) cell fusion; and (*5*) nuclear fusion.

1. Physiological and Morphological Differentiation

In some genera, e.g., *Chaetomorpha* and *Eudorina*, the first recognizable stage of sexual reproduction is a characteristic cell cleavage which results in the formation of a number of gametes from each cell. In other algae, such as *Chlamydomonas eugametos* and *Pandorina morum*, no special cleavage is apparent; cells indistinguishable from vegetative cells are released and act directly as gametes, as indicated by their ability to clump immediately in appropriate combinations. In the colonial green flagellates, this clumping ability is present even before gametic cells have been released from the colonies, as shown by the clumping of whole colonies upon mixing of complementary mating types (Stein, 1958a, b; Coleman, 1959).

2. Clumping

The intensity of the clumping reaction, as indicated by the number of gametes per clump, is proportional to the number of gametes present in the reaction mixture (Geitler, 1931; Köhler, 1957; Tsubo, 1957). Lewin (1956) briefly discussed the problem of different degrees of gametic activity, but no experimental evidence is yet available on this point. The initial site of agglutination is the tip of the flagellum either in one (Brown Algae) or in both sexes (many Green Algae). Gametes attached to the clump bob and swing, as their flagella flex, in a characteristic manner which helps the investigator to distinguish sexual clumping from other types of cell aggregation. Gametes may become detached from one clump, swim away, and perhaps join another. There has been no case reported in which true clumping occurred between wild strains which could not also complete zygote formation, suggesting that interstrain mating specificity is expressed in the agglutination phenomenon or earlier. Considerable progress has been made in the chemical elucidation of this phenomenon in *Chlamydomonas* (see below).

Several investigators have been able to obtain clumping reactions in *Tetraspora* and *Chlamydomonas* spp. between living gametes of one mating type and gametes of the complementary type killed by gentle heating, ultraviolet irradiation, or osmic acid (Geitler, 1931; Hutner and Provasoli, 1951; Tsubo, 1957), but with *Chaetomorpha* gametes similar attempts, using the same treatments, were unsuccessful (Köhler, 1956).

Köhler found that the clumping ability of *Chaetomorpha* gametes is not affected by a variety of proteases, but that after treatment with 1% citrate flagellar activity is impaired and clumping is decreased to the point where gametes which clump fall apart almost immediately. This may be related to the chelation of calcium ions, since Lewin (1954a) showed that calcium is required for clumping in *C. moewusii*.

3. Pairing

Pairing appears to be a consequence of the clumping reaction. Typically, among Green Algae, pairs of gametes, attached by their flagella, soon become detached from the clumps, and experiments using marked gametes of heterothallic strains have invariably shown that each pair comprises one gamete of each mating type (Lerche, 1937; Lewin, 1950; Sager, 1955; Tsubo, 1957). The possibility that stage *2* can be separated experimentally from stage *3* is supported by the observation of Trainor (1958) that under limiting-light conditions clumping in *Chlamydomonas chlamydogama* may occur without the consequent appearance of pairs, and the finding of *C. moewusii* mutants which can clump but which do not form pairs (Lewin, personal communication).

The physiological basis for the change-over from clumping to pairing remains a mystery.

4. Cell Fusion

The next step, cell fusion, appears to involve a mechanism quite different from that of flagellar agglutination. Concentrations of citrate which inhibit the agglutination of gamete flagella in *Chlamydomonas moewusii* have no adverse effect on cell fusion if applied after pairing has begun (Lewin, 1954a). From the many published observations it seems probable that fusion can be initiated only at a particular region of the cell surface, usually at the apex. Since the apices of two compatible gametes may touch without fusion at other times, before pairing begins, competence to fuse, presumably with some differentiation of the apical surface, may develop only under special circumstances. Chemical inhibitors of cell fusion differ from those which have been reported to inhibit clumping or pairing. *Chaetomorpha* gametes treated with a bacterial protease, subtilisin, though they retain their ability to clump, do not go on to form zygotes (Köhler, 1956). Similarly, treatment of *Chlamydomonas eugametos* gametes with reagents which "block" SH groups suppresses cell fusion but not clumping (Wiese, personal communication).

The efficiency of the pairing process generally ensures that no more than two gametes fuse, but there are in the literature a few references to possible triple fusion (Geitler, 1931; Sager, 1955). Genetic evidence presented by Levine and Ebersold (1960) suggested that in *Chlamydomonas reinhardti* such triploid zygotes can be fertile. Their normal incidence remains low, hoever, perhaps because immediately after gamete fusion is initiated the flagella of paired cells lose their ability to agglutinate and thereby prevent further fusions (Lewin, 1954a; Coleman, 1959).

5. Nuclear Fusion

Most freshly formed resting zygotes of freshwater Green Algae require a period of light before they become capable of germination. It is not known whether nuclear fusion is connected with this light requirement, or whether illumination is required solely for growth and the production of storage materials.

V. Hormonal Substances

The literature contains numerous references to hormone-like activities in cell-free centrifugates or filtrates which evidently play a role in the mating reactions of algae (see Raper, 1952, 1957). There has been no exact chemical identification of any of the substances involved, but recent advances in chemical techniques, which have led to the successful identification of chemotactic substances in slime molds (Heftmann *et al.*, 1959) and in moths (Butenandt, 1959; Jacobson *et al.*, 1960), hold great promise for advances in this field. Four phenomena have been credited to the action of specific secreted substances: (A) the chemotaxis of gametes; (B) the release of gametes from cell walls; (C) the strengthening of the mating potential of "weak" gametes; and (D) the clumping of gametes.

A. Chemotaxis

A number of investigators, after careful observation of the mating reaction in various isogamous algae, have stated that they observed no evidence of chemotaxis (Hutner and Provasoli, 1951; Köhler, 1956; Coleman, 1959). The association of the gametes could be explained satisfactorily as a consequence of phototaxis, by which the gametes tend to congregate in illuminated regions, followed by random contact of flagella, resulting in clumps. There is at least one report to the contrary, however. From observations of the reactions between filtrates or killed gametes of one mating type and live gametes of the other, Tsubo (1957) presented evidence that in *Chlamydomonas* sp. 24, *minus* cells produce a substance attractive to *plus* cells. However, he did not establish final proof of chemotaxis, as distinct from the promotion of cell clumping.

Among oogamous algae, as might be expected, suggestions of chemotaxis are quite numerous. Hoffman (1960) reported that the egg substances attractive to spermatozoids of *Oedogonium* are species-specific. In the only chemical study of such a substance, Cook and Elvidge (1951) presented striking evidence for the presence of a chemotactic substance given off by the unfertilized egg cells of *Fucus serratus* and attractive to the spermatozoids. The substance is not species-specific, since it attracts spermatozoids of at least two other *Fucus* species. Extensive but incomplete chemical analysis suggested that the active factor is a volatile substance, probably a short-chain hydrocarbon.

B. Gametogenesis

In many algae, some known to be heterothallic, gametes are released regardless of the presence or absence of sexually compatible cells of the species. Others appear to require a stimulus from the complementary mating type before the gametes are shed. Diwald (1938) found that in the dinoflagellate *Glenodinium lubiniensiforme* gametes appeared only among cells treated with culture fluid of a sexually compatible culture. Likewise, in clonal cultures of the colonial flagellate *Pandorina morum* the gametes, though capable of sexual clumping, are not normally released from the colonial matrix unless colonies of the complementary mating type are present. Cell-free supernatant solutions also exhibit this gamete-inducing action, but the chemical nature of the active agent(s) is unknown (Coleman, 1959).

C. Enhancement of Sexual Activity

A type of extracellular substance with an altogether different role in the mating reaction was reported by Jollos (1926) for *Dasycladus clavaeformis*. Filtrates of strongly reactive gametes of one type increased the sexual activity of weakly reactive gametes of the *same* mating type. Köhler (1957), who reinvestigated this alga more recently, was unable to repeat these results, and suggested that Jollos had been led astray since at that time knowledge of clumping reactions was so incomplete.

D. Substances Inducing Clumping

The cell contact preliminary to fertilization in algae always involves some variety of superficial agglutination: cell to cell (as in the Red Algae and the chlorophyte order Zygnematales); or flagellum to cell (e.g., *Fucus*); or flagellum to flagellum (e.g., *Chlamydomonas*). Thus, it may not be surprising that, of the various reports of hormone-like phenomena in sexual reproduction among the algae, the most frequently described effect is cell clumping induced by a cell-free culture fluid. In such cases (Geitler, 1931; Lerche, 1937; Förster and Wiese, 1954a, b) it has been found that the addition of an active filtrate from one mating type to gametes of the other may induce *auto-agglutination*, which cannot readily be distinguished from normal clumping but which does not result in pairing or zygote formation (Figs. 3, 4, and 5). The clump-inducing substance is mating-type specific, and is consumed in the reaction.

Extensive work by Förster and co-workers (1954a, b, 1956) has characterized the clump-inducing substances of both male and female mating types of *Chlamydomonas eugametos* as glycoproteins of high molecular weight. The female substance, or *gynogamone*, is relatively stable to temperatures below 100°C. It is found associated with a colloidal fraction which is 95% homogeneous by ultracentrifugation and electrophoresis, and which contains approximately 6% nitrogen and 3% sulfur. Two pentoses, two hexoses, and twelve amino acids have been found in hydrolyzates by paper chromatography. Gynogamone induces clumping at a concentration as low as 1 mg./liter. Less is known of the analogous substance produced by the male strain of *C. eugametos*. It is readily inactivated by temperatures as low as 50°C., but in composition it appears to be similar to gynogamone.

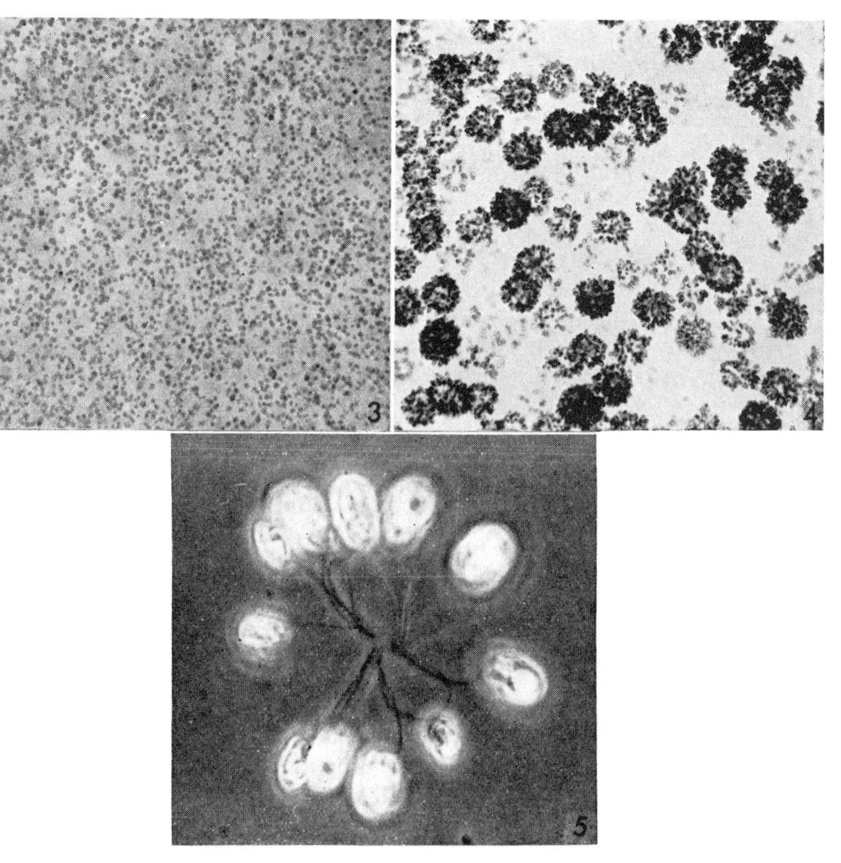

Fig. 3. Male gametes of *Chlamydomonas eugametos* (×89).
Fig. 4. Clumping of male gametes in the presence of gynogamone (×89).
Fig. 5. A single clump of male gametes induced by the addition of gynogamone (×1000).

One can demonstrate clump-inducing substances in cell-free media from both mating types of two other commonly used species of *Chlamydomonas*, *C. moewusii* (Förster, 1959), and *C. reinhardtii* (Förster and Wiese, 1955). Those from the *C. moewusii* and *C. eugametos* strains act interchangeably, in agreement with the genetic compatibility of these races. They have no observable effect on *C. reinhardtii* gametes, and correspondingly *C. reinhardtii* filtrates have no effect on *C. eugametos* or *C. moewusii* gametes. The demonstration of such substances in these strains is particularly interesting in view of similar but unsuccessful attempts by Smith (1946), Bernstein and Jahn (1955), and Hutner and Provasoli (1951), using these same algal strains; this indicates that cultural conditions are of utmost importance in the preparation of active filtrates. The substances which induce clumping are apparently liberated from the flagella; as Förster pointed out, there are also undoubtedly strain differences which may underly the tightness of attachment of these substances to the flagellar surface. Support for this suggestion comes from the observations (on *C. moewusii* and *Chlamydomonas* sp. 24) that killed gametes of one mating type could induce clumping among gametes of the other, although centrifugal supernatants were ineffective (Hutner and Provasoli, 1951; Tsubo, 1957). It remains to be seen whether flagella isolated from such cells are capable of inducing clumping.

The clumping substance of the marine alga *Chaetomorpha aerea* may be of a somewhat different nature from those reported for freshwater algae. Köhler (1956) was unable to obtain a clumping reaction with either cell-free filtrates or gametes killed in various ways. He also found that treatment with a protease did not prevent clumping, whereas Förster *et al.* (1956) reported that proteases destroy the gynogamone of *Chlamydomonas eugametos.*

The appearance of the clumping reaction, and its apparent species-specificity in every case investigated, suggest a similarity to the phenomenon of agglutination *via* an antigen-antibody reaction. However, antisera prepared against cells or flagella of *C. moewusii* (Hutner and Provasoli, 1951; Mintz and Lewin, 1954) or against purified gynogamone of *C. eugametos* (Wiese, personal communication) failed to distinguish between the two mating types, although a species-specific antibody may have been present. Antisera prepared against *Pandorina morum* flagella appear to exhibit syngen specificity (Coleman and Coleman, unpublished).

E. Intracellular Substances

Lewin (1954b) presented circumstantial evidence for the existence of an intracellular substance controlling one feature of the mating reaction in *Chlamydomonas moewusii.* In this species, paired cells form a narrow cytoplasmic bridge between their apices, and remain motile for some hours before completion of gamete fusion. Lewin observed that, after the bridge was formed, the flagella of the *minus* gamete ceased to beat while those of the *plus* gamete continued to do so, suggesting that a substance inhibitory to *minus* cell motility had crossed the bridge after pairing.

VI. Conclusion

The algae offer unique characteristics for the study of the physiology of mating. Homothallic and heterothallic species of several different genera are available in pure culture; they can be grown on chemically defined media; their mating reactions can be induced at will and are easily observed, and the various stages can often be separated by chemical or physical agents; results can be expressed quantitatively in terms of pair numbers, in some cases, or zygote numbers. At least three types of extracellular chemical compounds specifically associated with the mating reaction have been found. With the wealth of new techniques now available for synchronizing cultures, for obtaining cell particulates, and for identifying unknown biological compounds, it should be possible to examine the role of intermediary pathways of metabolism in mating reactions. Electron microscopy, both of flagella and of fusing cells, may yet reveal characteristics unique to gametes. With these studies should come some understanding of the genetic actions which underly the phenomena of self-sterility, mating-type specificity, and sexual differentiation.

References

Bernstein, E., and Jahn, T. L. (1955). Certain aspects of the sexuality of two species of *Chlamydomonas*. *J. Protozool.* **2,** 81–85.

Biecheler, B. (1937). Sur l'existence d'une copulation chez une euglène verte et sur les conditions globale qui la déterminent. *Compt. rend. soc. biol.* (*Paris*) **124,** 1264–1266.

Butenandt, A. (1959). Über den Sexual-Lockstoff des Seidenspinners *Bombyx mori*. *Z. Naturforsch.* **14b,** 283–284.

Cantino, E. C. (1956). The relation between cellular metabolism and morphogenesis in *Blastocladiella*. *Mycologia* **48,** 225–240.

Coleman, A. W. (1959). Sexual isolation in *Pandorina morum*. *J. Protozool.* **6,** 249–264.

Cook, A. H., and Elvidge, J. A. (1951). Fertilization in the Fucaceae: investigations on the nature of the chemotactic substance produced by eggs of *Fucus serratus* and *F. vesiculosus*. *Proc. Roy. Soc.* **B138,** 97–114.

Diwald, K. (1938). Die ungeschlechtliche und geschlechtliche Fortpflanzung von *Glenodinium lubiniensiforme* spec. nov. *Flora* (*Jena*) **132,** 174–192.

Förster, H. (1957). Das Wirkungsspektrum der Kopulation von *Chlamydomonas eugametos*. *Z. Naturforsch.* **12b,** 765–770.

Förster, H. (1959). Die Wirkungsstärken einiger Wellenlängen zum Auslösen der Kopulation von *Chlamydomonas moewusii*. *Z. Naturforsch.* **14b,** 479–480.

Förster, H., and Wiese, L. (1954a). Untersuchungen zur Kopulationsfähigkeit von *Chlamydomonas eugametos*. *Z. Naturforsch.* **9b,** 470–471.

Förster, H., and Wiese, L. (1954b). Gamonwirkungen bei *Chlamydomonas eugametos*. *Z. Naturforsch.* **9b,** 548–550.

Förster, H., and Wiese, L. (1955). Gamonwirkung bei *Chlamydomonas reinhardi*. *Z. Naturforsch.* **10b,** 91–92.

Förster, H., Wiese, L., and Braunitzer, G. (1956). Über das agglutinierend wirkende Gynogamon von *Chlamydomonas eugametos*. *Z. Naturforsch.* **11b,** 315–317.

Geitler, L. (1931). Untersuchungen über das sexuelle Verhalten von *Tetraspora lubrica*. *Biol. Zentr.* **51,** 173–187.

Geitler, L. (1957a). Die sexuelle Fortpflanzung der pennaten Diatomeen. *Biol. Rev.* **32,** 261–295.

Geitler, L. (1957b). Über die Paarung bei *Navicula seminulum* und die Ausprägung der Geschlechtsmerkmale bei pennaten Diatomeen. *Ber. deut. botan. Ges.* **70,** 45–48.

Hartmann, M. (1925). Über relative Sexualität bei *Ectocarpus siliculosus. Naturwissenschaften* **13,** 975–980.

Hartmann, M. (1955). Sex problems in algae, fungi and protozoa. *Am. Naturalist* **89,** 321–346.

Heftmann, E., Wright, B., and Liddel, G. (1959). Identification of a sterol with acrasin activity in a slime mold. *J. Am. Chem. Soc.* **81,** 6525–6526.

Hoffman, L. (1960). Chemotaxis of *Oedogonium* sperms. *The Southwestern Naturalist* **5,** 111–116.

Hustede, H. (1957). Untersuchungen über die stoffliche Beeinflussung der Entwicklung von *Stigeoclonium falklandicum* und *Vaucheria sessilis* durch Tryptophan Abkömmlinge. *Biol. Zentr.* **76,** 555–595.

Hutner, S. H., and Provasoli, L. (1951). The phytoflagellates. *In* "Biochemistry and Physiology of Protozoa" (A. Lwoff, ed.), pp. 27–128. Academic Press, New York.

Jacobson, M., Beroza, M., and Jones, W. A. (1960). Isolation, identification and synthesis of the sex attractant of Gypsy moth. *Science* **132,** 1011.

Jollos, V. (1926). Untersuchungen über die Sexualitätsverhältnisse von *Dasycladus clavaeformis. Biol. Zentr.* **46,** 279–295.

Klebs, G. (1896). "Die Bedingungen der Fortpflanzung bei einigen Algen und Pilzen." Fischer, Jena, Germany.

Köhler, K. (1956). Entwicklungsgeschichte, Geschlechtsbestimmung, und Befruchtung bei *Chaetomorpha. Arch. Protistenk.* **101,** 223–268.

Köhler, K. (1957). Neue Untersuchungen über die Sexualität bei *Dasycladus* and *Chaetomorpha. Arch. Protistenk.* **102,** 209–218.

Lerche, W. (1937). Untersuchungen über Entwicklung und Fortpflanzung in der Gattung *Dunaliella. Arch. Protistenk.* **88,** 236–268.

Levine, R. P., and Ebersold, W. T. (1960). The genetics and cytology of *Chlamydomonas. Ann. Rev. Microbiol.* **14,** 197–216.

Lewin, R. A. (1950). Gamete behaviour in *Chlamydomonas. Nature* **166,** 76.

Lewin, R. A. (1954a). Sex in unicellular algae. *In* "Sex in Microorganisms" (D. Wenrich, ed.), pp. 100–133. Am. Assoc. Advancement Sci., Washington, D. C.

Lewin, R. A. (1954b). Mutants of *Chlamydomonas moewusii* with impaired motility. *J. Gen. Microbiol.* **11,** 358–363.

Lewin, R. A. (1956). Control of sexual activity in *Chlamydomonas* by light. *J. Gen. Microbiol.* **15,** 170–185.

Mintz, R. H., and Lewin, R. A. (1954). Studies on the flagella of algae. V. Serology of paralyzed mutants of *Chlamydomonas. Can. J. Microbiol.* **1,** 65–67.

Patrick, R. (1954). Sexual reproduction in diatoms. *In* "Sex in Microorganisms" (D. Wenrich, ed.), pp. 82–99. Am. Assoc. Advancement Sci., Washington, D. C.

Philip, U., and Haldane, J. B. S. (1939). Relative sexuality in unicellular algae. *Nature* **143,** 334.

Raper, J. R. (1952). Chemical regulation of sexual processes in the thallophytes. *Botan. Rev.* **18,** 447–545.

Raper, J. R. (1957). Hormones and sexuality in lower plants. *Symposia Soc. Exptl. Biol.* **No. 11,** 143–165.

Renner, O. (1958). Auch etwas über F. Moewus, *Forsythia* und *Chlamydomonas. Z. Naturforsch.* **13b,** 399–403.

Ryan, F. J. (1955). Attempt to reproduce some of Moewus' experiments on *Chlamydomonas* and *Polytoma. Science* **122,** 470.

Sager, R. (1955). Inheritance in the green alga *Chlamydomonas reinhardi*. *Genetics* **40,** 476–489.

Sager, R., and Granick, S. (1953). Nutritional studies with *Chlamydomonas reinhardi*. *Ann. N. Y. Acad. Sci.* **56,** 831–838.

Sager, R., and Granick, S. (1954). Nutritional control of sexuality in *Chlamydomonas reinhardi*. *J. Gen. Physiol.* **37,** 729–742.

Smith. G. M. (1946). The nature of sexuality in *Chlamydomonas*. *Am. J. Botany* **33,** 625–630.

Sonneborn, T. M. (1957). Breeding systems, reproductive methods, and species problems in Protozoa. *In* "The Species Problem" (E. Mayr, ed.), pp. 155–324. Am. Assoc. Advancement Sci., Washington, D. C.

Stein, J. R. (1958a). A morphological study of *Astrephomene gubernaculifera* and *Volvulina steinii*. *Am. J. Botany* **45,** 388–397.

Stein, J. R. (1958b). A morphologic and genetic study of *Gonium pectorale*. *Am. J. Botany* **45,** 664–672.

Stifter, I. (1959). Untersuchungen über einige Zusammenhänge zwischen Stoffwechsel- und Sexualphysiologie an dem Flagellaten *Chlamydomonas eugametos*. *Arch. Protistenk.* **104,** 364–388.

Thimann, K. V. (1940). Sexual substances in the algae. *Chronica Botanica* **6,** 31–32.

Trainor, F. R. (1958). Control of sexuality in *Chlamydomonas chlamydogama*. *Am. J. Botany* **45,** 621–626.

Trainor, F. R. (1959). A comparative study of sexual reproduction in four species of *Chlamydomonas*. *Am. J. Botany* **46,** 65–70.

Trainor, F. R. (1960). Mating in *Chlamydomonas chlamydogama* at various temperatures under continuous illumination. *Am. J. Botany* **47,** 482–484.

Tsubo, Y. (1956). Observations on sexual reproduction in a *Chlamydomonas*. *Botan. Mag. (Tokyo)* **69,** 1–6.

Tsubo, Y. (1957). On the mating reaction of a *Chlamydomonas*, with special references to clumping and chemotaxis. *Botan. Mag. (Tokyo)* **70,** 299–340.

Tsubo, Y. (1961). Chemotaxis and sexual behavior in *Chlamydomonas*. *J. Protozool.* **8,** 114–121.

Whitehouse, H. L. K. (1949). Heterothallism and sex in the fungi. *Biol. Revs. Cambridge Phil. Soc.* **24,** 411–447.

Wilbois, A. D. (1958). Sexual isolation in *Pandorina morum* Bory. Ph.D. Thesis, Indiana University, Bloomington, Indiana.

—49—

Biochemical Genetics

W. T. EBERSOLD

Department of Botany, University of California, Los Angeles, California

I. Introduction

Until recently there were no studies on algae which combined biochemical and genetic methods. There are two primary reasons for the absence of this type of investigation. In those species which have received most attention from a physiological or biochemical standpoint, present techniques for genetic analysis are inadequate or sexuality is unknown. On the other hand, in those which have been investigated genetically, the types of characteristics used do not lend themselves readily to biochemical analysis. A notable exception is the genus *Chlamydomonas*. Much of the following discussion will therefore be concerned with those studies of *Chlamydomonas* species from which both biochemical and genetic information has been obtained.

The first reported genetic investigations using *Chlamydomonas* were those of Pascher (1918), who studied the segregation of morphological characters in crosses between two unidentified *Chlamydomonas* species. The subsequent publications of Moewus (e.g. 1950) will not be discussed here, since they have been severely criticized and many attempts to confirm his reported findings have been unsuccessful (c.f. Pätau, 1941; Harte, 1948; Gowans, 1960). More recent studies were initiated by Lewin (1953), who found that mutant strains of *C. moewusii* having flagellar abnormalities, or requiring for growth an organic carbon source, thiamine, or *p*-aminobenzoic acid, represented single-gene mutations at different loci. He also studied the physiological character of some non-heterotrophic mutants of *C. dysosmos* (a homothallic species) and of *C. debaryana* (heterothallic), but did not examine these mutations from a genetic standpoint (Lewin, 1954a). Subsequently, the single-gene nature of mutations affecting growth requirements for an organic substrate (acetate), nicotinamide, purines, thiamine, and *p*-aminobenzoic acid was demonstrated by Gowans (1960), using a strain of "*C. eugametos*" (which is sexually compatible with *C. moewusii* and, in the opinion of Gerloff (1940), should be thus renamed). Strains of *C. reinhardtii* have also been found with mutations affecting: (*a*) movement of flagella; (*b*) resistance to streptomycin, sulfanilamide, and methionine sulfoximine; (*c*) pigments; and (*d*) growth requirements for an organic carbon source or for arginine, thiamine, *p*-aminobenzoic acid, or nicotinamide (Ebersold, 1956; Eversole, 1956; Levine, 1960; Sager, 1955; Weaver, 1952; Wetherell and Krauss, 1957). The genetic behavior and linkage relationships of the loci represented by these mutations, including those of *C. moewusii* and *C. eugametos*, have been reviewed by Levine and Ebersold (1960).

II. Carbon-source Mutants

Mutant strains of *C. reinhardtii* requiring acetate for growth in the light may provide excellent material for investigating the genetic control of the process of photosynthesis. More than 100 mutants of this type have now been isolated (Levine, personal communication). These strains are incapable of photosynthetic growth, and several have been shown to be unable to fix carbon dioxide. This condition, as pointed out by Levine (1960), is subject to three possible explanations: (i) their pigments may be affected either qualitatively or quantitatively; (ii) they may be impaired in some other way in their ability to carry out certain light reactions; or (iii) they may be unable to carry out certain closely related dark reactions. In one mutant strain (*ac*-21), which was studied in some detail, the rate of carbon-dioxide fixation was approximately one-tenth that of wild-type cells, although the mutant was able to carry out the Hill reaction at almost the normal rate. Both the mutant and wild-type cells were found to fix carbon dioxide in darkness in the presence of ribulose-1,5-diphosphate, indicating the presence of an active ribulose-1,5-diphosphate carboxylase system, but the mutant was unable to carry out the process of photosynthetic phosphorylation. It has thus been shown that at least one of the steps in photosynthesis is controlled by a single genetic locus.

III. Pigment Mutants

No significant qualitative or quantitative difference between the pigment content of the mutant *ac*-21 and that of wild-type cells could be detected.

However, mutant strains distinguishable from wild-type by qualitative or quantitative pigment differences are of common occurrence among *C. reinhardtii* cells following ultraviolet irradiation. This type of mutant opened the possibility of investigating the genetic control of the biosynthetic pathway of chlorophyll synthesis. Based on diverse mutant strains of *Chlorella vulgaris* which accumulate presumed chlorophyll precursors, Granick (1951) formulated a scheme for the biosynthesis of chlorophyll (see also Dubé, 1952; Allen *et al.*, 1960). The nature of the mutations involved is unknown, since in *Chlorella* it is not yet possible to distinguish a cytoplasmic defect from a nuclear gene mutation.

Genetic blocks in chlorophyll synthesis have been found in four independently isolated mutants of *Chlamydomonas reinhardtii* (Eversole, 1956; Sager, 1955; Weaver, 1952). Three of these are apparently unable to convert protochlorophyll into chlorophyll when grown in darkness on a medium containing acetate; the characters of the fourth, a yellow-brown mutant which accumulates red-brown chlorophyll precursors, are attributable to a genetic block preceding protochlorophyll (see Bogorad, Chapter 23, this treatise). The specific step blocked in this mutant is not known, but protoporphyrin 9 has been identified among the precursors which accumulate. These results indicate that in *Chlamydomonas* the steps involved in the biosynthesis of chlorophyll may be similar to those described for *Chlorella*.

Several mutant strains of *Chlorella pyrenoidosa* (Allen *et al.*, 1960) and *C. vulgaris* (Claes, 1954) are characterized by the absence or reduced content of α- and β-carotene and xanthophylls, the primary carotenoid constituents of wild-type cells. Instead, these mutants accumulate carotenoids such as phytoene, phytofluene, and ζ-carotene, which are not found in the wild-type strain, but apparently accumulate in the mutants because of blocks in biosynthetic pathways leading to normal end-products.

A mutant of *Chlamydomonas reinhardtii* was found which possesses only a small amount of carotenoid pigments, but which, unlike the *Chlorella* mutants mentioned above, does not accumulate either phytoene or phytofluene. This mutant, described as *pale-green*, has been studied in some detail (Chance and Sager, 1957; Sager and Zalokar, 1958). A comparison of the wild-type strain with the mutant showed that, when grown in darkness, *pale-green* cells had approximately 1/15 the chlorophyll content (moles chlorophyll per gram protein) and less than 1/200 the carotenoid content of wild-type cells. Unlike the wild-type cells, in which six carotenoid pigments were demonstrated, mutant cells were found to contain only α- and β-carotene. Although the absolute chlorophyll content of the mutant was low, the chlorophyll/carotenoid ratio was much higher than that of the wild-type strain, and the photosynthetic capacity (moles carbon dioxide taken up per mole chlorophyll per second) was higher in mutant than in wild-type cells. These findings were interpreted as supporting the hypothesis that carotenoids, except possibly in small amounts, are not essential for photosynthesis in green plants.

IV. Vitamin and Amino-Acid Mutants

In *C. reinhardtii*, seven independently isolated mutant strains requiring thiamine for growth can be divided into five categories on the basis of their specific growth-factor requirements, linkage relationships, and susceptibility to growth inhibition by analogs of thiamine (Ebersold and Levine, 1959; Ebersold, unpublished). The mutants *thi*-3, *thi*-4, *thi*-4a, and *thi*-4b use thiamine or vitamin thiazole (4-methyl-5-β-hydroxyethyl thiazole); *thi*-8 uses thiamine or vitamin pyrimidine (2-methyl-4-amino-5-ethoxymethyl pyrimidine); *thi*-1 can use only thiamine; and *thi*-2 requires either thiamine or thiazole *plus* pyrimidine. Among the mutants that can utilize thiazole, the growth of *thi*-4, *thi*-4a, and *thi*-4b is not inhibited by the analogs oxythiamine or pyrithiamine or by a combination of both; genetically these three mutations (designated *thi*-4, 5, and 6 by Levine and Ebersold, 1960) are either allelic or very closely linked in linkage group IV. On the other hand, *thi*-3 is inhibited by a combination of both analogs, and the mutant gene is located in linkage group I. These four mutations probably represent blocks at two different steps in the biosynthesis of thiazole, each controlled by a distinct genetic locus. Nutritionally, *thi*-2 resembles two thiamine mutants of the fungus *Neurospora crassa* (56501 and 17084, designated *thi*-1 by Barratt *et al.*, 1954), which also require both thiazole and pyrimidine for growth. Mutants of this type are difficult to explain on the basis of the scheme proposed by Tatum and Bell (1946), who concluded that thiazole and pyrimidine condense to form thiamine. Harris (1955) proposed that the double requirement of these mutants can best be explained by assuming that pyrimidine condenses preferentially with a precursor of thiazole, this compound then being converted into thiamine. When the latter step is blocked (by the gene *thi*-1 in the case of *Neurospora*), exogenous thiazole might condense with pyrimidine to form thiamine. The possibility also exists that *thi*-2 is impaired in its ability to utilize thiamine (B. M. Eberhart, personal communication). As yet, no direct biochemical investigation of the thiamine mutants of *Chlamydomonas* has been carried out, although they should provide suitable material for studying the biosynthesis of thiamine.

The single-gene nature of several mutant strains of *C. reinhardtii* requiring nicotinamide for growth was demonstrated by Eversole (1956). Four of these mutants show a positive growth response to substances which are known precursors of nicotinic acid in *Neurospora* (Bonner, 1951). In addition to the mutant strains isolated by Eversole, several nicotinamide-requiring mutants have subsequently been found (R. P. Levine, personal communication). In the biosynthesis of nicotinamide the specific steps that are blocked in these mutants have not been determined.

Only two amino-acid-requiring mutants of *Chlamydomonas* have been found, and both are unable to synthesize arginine. Mutant *arg*-1 (Ebersold, 1956) uses ornithine, citrulline, or arginine, whereas *arg*-2 (Eversole, 1956) uses arginine only. Interestingly, *arg*-1 cells become white when grown on limiting amounts of arginine. These mutants are under investigation, in the laboratory of R. P. Levine at Harvard University, in order to determine the location of the metabolic block in each mutant and the nature of the relationship between bleaching and the arginine requirement of *arg*-1.

From among x-irradiated cells of *C. eugametos*, Wetherell and Krauss (1957) isolated several nutritional mutant strains that did not behave in the usual manner. These mutants were apparently unique in that growth was enhanced by any one of a number of different vitamins. Crosses of two of these mutant strains with wild-type yielded tetrads with one normal:one mutant:two dead cells. Because of the unconventional growth response and genetic behavior of these mutant strains, Wetherell and Krauss investigated the genetic nature of *C. eugametos* as compared with haploid and diploid strains of *C. reinhardtii*. From a comparison of x-ray survival curves, and from a comparison of the fraction of subnormal-sized colonies surviving x-radiation, they postulated that there may be a considerable amount of duplication in the genome of *C. eugametos*, and that the mutant strains they isolated may represent deletions of one portion of the genetic material which is normally duplicated elsewhere. Their data might be interpreted to indicate that in haploid *C. reinhardtii*, too, portions of the genome may be duplicated, but to a lesser extent than in *C. eugametos*. Genetic studies by other workers have not yet produced supporting evidence for this hypothesis.

As mentioned above, only two stable mutant strains of *Chlamydomonas* have been found to require an amino acid for growth. Although there have been no studies primarily concerned with the isolation of amino-acid mutants, it is surprising that they have been recovered so rarely in comparison with vitamin or carbon-source mutations. The manifest absence of amino-acid mutant strains may be attributable to the lack of a proper selective medium or to a basic difference between the amino-acid metabolism of *Chlamydomonas* and that of other microorganisms. Another possible explanation is that a part of the genetic material is duplicated in haploid cells, as proposed by Wetherell and Krauss, in portions of the genome specifically concerned with amino-acid synthesis.

V. Flagella Mutants

A series of ultraviolet-induced mutants of *Chlamydomonas moewusii* was obtained in which the flagella were absent, or short, or of normal length but with impaired activity (Lewin, 1949, 1952). Among the mutants examined, no differences from wild-type strain were detected either in their flagellar fine structure (Gibbs *et al.*, 1959) or in their serological reaction (Mintz and Lewin, 1954). Many paralysis mutations were shown genetically to be attributable to single genes; several cases of allelism or apparent recurrences of the same mutation were detected. At least nine distinct loci are concerned with flagellar movement; three of these are closely linked to their respective centromeres, and one is sex-linked. One partial suppressor was identified which proved to be specific for a single paralysis locus (R. A. Lewin, personal communication).

In certain cases paralyzed cells recovered normal motility when paired in so-called "dikaryons" with wild-type gametes, indicating the transfer of a substance or substances essential for normal flagellar activity. In such "recoverable" strains, dikaryons between two different mutants likewise recovered motility, providing evidence for the existence of several dissimilar agents which could be exchanged reciprocally between copulants (Lewin, 1954b). Although presumably there must be biochemical differences between paralyzed mutants and wild-type cells, attempts to demonstrate motility factors in cell extracts have proved unsuccessful. However, Brokaw (1960) discovered that the detached flagella of one mutant (M. 1002), unlike those from wild-type cells, do not exhibit autonomous movement in the presence of adenosine triphosphate, and that they have only 30% of the apyrase activity of normal flagella. Possible clues to the metabolic impairment of paralyzed mutants may also be found in the reports by Ronkin (1959) and Ronkin and Buretz (1959) that the respiratory rate of M. 1001 was only 86% that of wild-type cells, and that the rate of potassium uptake by mutant cells was lower than normal.

By the use of a simple selection device, paralyzed mutants of *Euglena gracilis* were obtained (Lewin, 1960); but since recombination could not be detected, they are not at present amenable to genetic study.

VI. Eyespot Mutants

A mutant of *Chlamydomonas reinhardtii* with the eyespot reduced or absent was isolated by Hartshorne (1953, 1955). The cells retained their sensitivity to light, but tactic movements appeared to be less organized. Genetic studies indicated that the *eyeless* phenotype was controlled by a single mutant gene. The data suggested in addition that the wild-type allele of *eyeless* suppressed the effect of certain genes for reduced colony size.

VII. Cytoplasmic Inheritance

In *Chlamydomonas reinhardtii* a system has been found in which the interaction of chromosomal and non-chromosomal factors can be investigated.

Mutant strains resistant to streptomycin were shown to be controlled by two different mechanisms (Sager, 1954). In one class of mutant, resistant to 100 mg./liter streptomycin (*sr*-100, originally *sr*-1), resistance was inherited in Mendelian fashion, while mutants of the second class, resistant to 500 mg./liter streptomycin (*sr*-500, originally *sr*-2), showed uniparental transmission of resistance. When *sr*-500 was crossed with a wild-type strain, all progeny were streptomycin-resistant if the *sr*-500 gene was carried by the *plus* mating-type parent, whereas from the reciprocal type of cross all progeny were streptomycin-sensitive. Subsequently it was found that resistance to 1500 mg./liter streptomycin (*sr*-1500) and streptomycin dependence (*sd*) were transmitted in the same uniparental fashion (Sager, 1960; Sager and Tsubo, 1960). In addition it was shown that a chromosomal gene modified the phenotypes of both *sr*-100 and *sr*-500 by increasing their levels of resistance to more than 2000 mg./liter streptomycin, although it had no effect on the streptomycin-sensitive wild-type cells. For a variety of reasons Sager (1960) favored the hypothesis that the phenotype expressed by *sr*-500, *sr*-1500, or *sd* cells is controlled by cytoplasmic factors (cytogenes) whose replication and segregation are somehow integrated with cell division.

VIII. Conclusion

This chapter presents a brief review of several approaches used in the study of the biochemical genetics of algae. It is hoped that in the future further significant contributions will be made in demonstrating the genetic control of the biochemical steps involved in photosynthesis, in pigment formation, and in other features of algal metabolism.

References

Allen, M. B., Goodwin, T. W., and Phagpolngarm, S. (1960). Carotenoid distribution in certain naturally occurring algae and in some artificially induced mutants of *Chlorella pyrenoidosa*. *J. Gen. Microbiol.* **23,** 93–104.

Barratt, R. W., Newmeyer, D., Perkins, D. D., and Garnjobst, L. (1954). Map construction in *Neurospora crassa*. *Advances in Genet.* **6,** 1–93.

Bonner, D. M. (1951). Gene-enzyme relationships in *Neurospora*. *Cold Spring Harbor Symposia Quant. Biol.* **16,** 143–157.

Brokaw, C. J. (1960). Decreased adenosine triphosphatase activity of flagella from a paralyzed mutant of *Chlamydomonas moewusii*. *Exptl. Cell Research* **19,** 430–432.

Chance, B., and Sager, R. (1957). Oxygen and light-induced oxidations of cytochrome, flavoprotein, and pyridine nucleotide in a *Chlamydomonas* mutant. *Plant Physiol.* **32,** 548–560.

Claes, H. (1954). Analyse der biochemischen Synthesekette für Carotinoide mit Hilfe von *Chlorella*-Mutanten. *Z. Naturforsch.* **9b,** 461–469.

Dubé, J. F. (1952). Observations on a chlorophyll-deficient strain of *Chlorella vulgaris* obtained after treatment with streptomycin. *Science* **116,** 278–279.

Ebersold, W. T. (1956). Crossing over in *Chlamydomonas reinhardi. Am. J. Botany* **43,** 408–410.

Ebersold, W. T., and Levine, R. P. (1959). A genetic analysis of linkage group I of *Chlamydomonas reinhardi. Z. induktive Abstammungs-u. Vererbungslehre* **90,** 74–82.

Eversole, R. A. (1956). Biochemical mutants of *Chlamydomonas reinhardi. Am. J. Botany* **43,** 404–407.

Gerloff, J. (1940). Beiträge zur Kenntnis der Variabilität und Systematik der Gattung *Chlamydomonas. Arch. Protistenk.* **94,** 311–502.

Gibbs, S. P., Lewin, R. A., and Philpott, D. E. (1959). The fine structure of the flagellar apparatus of *Chlamydomonas moewusii. Exptl. Cell Research* **15,** 619–622.

Gowans, C. S. (1960). Some genetic investigations on *Chlamydomonas eugametos. Z. induktive Abstammungs-u. Vererbungslehre* **91,** 63–73.

Granick, S. (1951). Biosynthesis of chlorophyll and related pigments. *Ann. Rev. Plant Physiol.* **2,** 115–144.

Harris, D. L. (1955). Alternative pathways in thiamine biosynthesis in *Neurospora. Arch. Biochem. Biophys.* **57,** 240.

Harte, C. (1948). Das Crossing-over bei *Chlamydomonas. Biol. Zentr.* **67,** 504–510.

Hartshorne, J. N. (1953). The function of the eyespot in *Chlamydomonas. New Phytologist* **52,** 292–297.

Hartshorne, J. N. (1955). Multiple mutation in *Chlamydomonas reinhardi. Heredity* **9,** 239–248.

Levine, R. P. (1960). Genetic control of photosynthesis in *Chlamydomonas reinhardi. Proc. Natl. Acad. Sci. U. S.* **46,** 972–978.

Levine, R. P., and Ebersold, W. T. (1960). The genetics and cytology of *Chlamydomonas. Ann. Rev. Microbiol.* **14,** 197–216.

Lewin, R. A. (1949). Genetics of *Chlamydomonas*—paving the way. *Biol. Bull.* **97,** 243–244.

Lewin, R. A. (1952). Ultraviolet induced mutations in *Chlamydomonas moewusii* Gerloff. *J. Gen. Microbiol.* **6,** 233–248.

Lewin, R. A. (1953). The genetics of *Chlamydomonas moewusii* Gerloff. *J. Genet.* **51,** 543–560.

Lewin, R. A. (1954a). The utilization of acetate by wild-type and mutant *Chlamydomonas dysosmos. J. Gen. Microbiol.* **11,** 459–471.

Lewin, R. A. (1954b). Mutants of *Chlamydomonas moewusii* with impaired motility. *J. Gen. Microbiol.* **11,** 358–363.

Lewin, R. A. (1960). A device for obtaining mutants with impaired motility. *Can. J. Microbiol.* **6,** 21–25.

Mintz, R. H., and Lewin, R. A. (1954). Studies on the flagella of algae. V. Serology of paralyzed mutants of *Chlamydomonas. Can. J. Microbiol.* **1,** 65–67.

Moewus, F. (1950). Sexualität und Sexualstoffe bei einem einzelligen Organismus (*Chlamydomonas*). *Z. Sexualforsch.* **1,** 1–25.

Pätau, K. (1941). Eine statistische Bemerkung zu Moewus' Arbeit "Die Analyse von 42 erblichen Eigenschaften der *Chlamydomonas-eugametos* Gruppe. III." *Z. induktive Abstammungs-u. Vererbungslehre* **79,** 317–319.

Pascher, A. (1918). Über die Kreuzung einzelliger, haploider Organismen: *Chlamydomonas. Ber. deut. botan. Ges.* **36,** 163–168.

Ronkin, R. R. (1959). Motility and power dissipation in flagellated cells, especially *Chlamydomonas. Biol. Bull.* **116,** 285–293.

Ronkin, R. R., and Buretz, K. M. (1959). Sodium and potassium in normal and paralyzed *Chlamydomonas. J. Protozool.* **7,** 109–114.

Sager, R. (1954). Mendelian and non-mendelian inheritance of streptomycin resistance in *Chlamydomonas reinhardi. Proc. Natl. Acad. Sci. U. S.* **40,** 356–363.

Sager, R. (1955). Inheritance in the green alga *Chlamydomonas reinhardi. Genetics* **40,** 476–489.

Sager, R. (1960). Genetic systems in *Chlamydomonas. Science* **132,** 1459–1465.

Sager, R., and Tsubo, Y. (1960). Inheritance of streptomycin resistance and dependence in *Chlamydomonas. Bacteriol. Proc.* (*Soc. Am. Bacteriologists*) **60,** 184.

Sager, R., and Zalokar, M. (1958). Pigments and photosynthesis in a carotenoid-deficient mutant of *Chlamydomonas. Nature* **182,** 98–100.

Tatum, E. L., and Bell, T. T. (1946). *Neurospora.* III. Biosynthesis of thiamine. *Am. J. Botany* **33,** 15–20.

Weaver, E. C. (1952). A chlorophyll deficient mutant of *Chlamydomonas reinhardi.* M.A. Thesis, Stanford University, Stanford, California.

Wetherell, D. F., and Krauss, R. W. (1957). X-ray induced mutations in *Chlamydomonas eugametos. Am. J. Botany* **44,** 609–619.

—Part IV—

Physiological Aspects of Ecology

Appendixes

—50—

Freshwater Algae

J. F. TALLING

Freshwater Biological Association, Ambleside, England

I. Introduction

In their responses to the vicissitudes of a natural environment, freshwater algae illustrate behavior of unique physiological and ecological interest. The populations of these predominantly microscopic organisms, particularly the planktonic forms, are often comparatively easily sampled and enumerated, show large oscillations of numbers resulting from short generation times, and inhabit relatively homogeneous media in water-bodies usually more sharply delimited and accessible than those that surround their marine counterparts. In recent years knowledge of their ecology has benefited from an immensely stimulating though occasionally overzealous transfer of information from general algal physiology, based largely upon cultured material. This chapter will consider features of physiological interest in the behavior of natural populations. Planktonic species have received much more attention than attached forms in quantitative field surveys, experiments and studies with cultures, and so provide most of the illustrations here.

II. Sources of Information

This subject can be approached from observations of algal distribution in space and in time, and from the properties of population samples and their behavior in field and laboratory experiments.

Differences in the algal floras of separate water-bodies can rarely be used to identify the response to a single factor, as many significant factors are likely to vary simultaneously, and their separation by an analysis of multiple correlations (Deevey, 1940) must be far from complete. Such comparisons have given valuable general information on the tolerances of many species or groups; examples include the salinity tolerance of diatoms (e.g. Kolbe, 1932; Gessner, 1959), and differences in the floras of unproductive and productive lakes (Pearsall, 1921, 1932), which can be much greater than is suggested by analysis of the major inorganic nutrients (Lund, 1957). Within the same water-body, differences of algal development at different depths can illustrate a response to a strong light gradient (e.g., Oberdorfer, 1928), but may involve other factors such as sinking rates (Lund 1959) or resistance to desiccation (Kann, 1933).

The quantitative measurement of changes in algal abundance with time, in relation to environmental factors, is one of the most direct methods of studying algal growth and its limitation in nature. The considerable literature on seasonal periodicity is largely concerned with qualitative aspects of species and community succession, or changes in some measure of total algal density such as chlorophyll content. The detailed dynamics of individual species can be particularly informative for phases of recolonization following the periodic return of favorable conditions.

More specific information concerning response to environment can be derived from samples of algae exposed *in situ* or in the laboratory; they may be obtained directly from a natural population or by culture. Field experiments have been used chiefly to measure rates of growth and photosynthesis at different depths, in relation to varying illumination and temperature. A much wider range of behavior can be followed under laboratory conditions, especially with cultured material, but its application to events in nature is usually by indirect inference or analogy. Simultaneous laboratory and field experiments with the same algal material (e.g., Talling, 1960) are largely unexploited, as are direct comparisons of the constitution and behavior of material from natural and cultured populations of the same species (e.g., Gerloff and Skoog, 1954, 1957a, b; Mackereth, 1953; Lund, 1950; Talling, 1957a).

III. Growth Rates and Population Kinetics

Exponential rather than linear increase is the fundamental property of algal growth, of which the specific or relative growth rate is the appropriate measure. Under favorable conditions, values of 1 to 3 cell divisions per day are typical of many freshwater species in the exponential phase of growth.

The equivalents of this and other phases of growth in natural populations are less easily recognized, owing largely to the continual loss of individuals by sinking, grazing, etc, and to difficulties of representative sampling in populations with uneven distribution. Unfortunately it is not usual to plot algal numbers on a logarithmic scale to illustrate exponential growth and density-dependent factors (e.g., Scourfield, 1897; Hutchinson, 1944, Fig. 6; Pearsall *et al.*, 1946; Lund, 1949). Long periods of apparently exponential growth are not infrequent, at least in the plankton; such constancy in specific rates of population increase has aroused comment (Pearsall *et al.*, 1946; Lund, 1949). The spring growth of many planktonic diatoms provides examples (e.g., Lund, 1949, 1950; Verduin, 1951, 1952). The range between initial population minima and resultant maxima can be as large as 1 to 10,000, as observed in the *Asterionella* cycles studied by Lund, with specific rates of population increase close to 1 cell division per week (cf. also Verduin, 1951, 1952).

The suppression of exponential increase, giving rise to other phases of a growth cycle, is less easily explained. Although of varied origins (see Sections IV and V), some integration is implied in any stable annual cycle. In natural populations, identification of the lag phases of growth, and many reports of "physiological dormancy," are largely speculative (cf. Lund, 1949; Fogg, 1958). Population maxima may form apparently stationary phases, but these are rarely long-lived and a dynamic balance between active growth and losses can often be suspected (e.g., Gessner, 1948; Grim, 1952). Population decline is frequently exponential (e.g., Lund, 1950), as in experimental cultures; its severity often controls the number of survivors available for subsequent growth, and possibly even influences the whole seasonal succession. Thus Lund (1949, 1950, 1954, 1955b) demonstrated two contrasting, yet often coexisting, types of annual population cycle in the plankton diatoms *Asterionella formosa* and *Melosira italica.* The *Asterionella* cycle is marked by periods of prolonged exponential growth and rapid declines with high mortality. Much of the seasonal increase and decrease of *Melosira* in the euphotic zone arises from the vertical redistribution of cells by turbulence and sinking, and the smaller mortality of *M. italica* is accompanied by less vigorous growth.

IV. Light and Temperature Requirements; Photosynthesis

In freshwater habitats, conditions of light and temperature are often closely associated, as in seasonal and latitudinal relationships to solar radiation and distribution with depth; consequently much debate has centered on their relative importance. Seasonal changes of temperature tend to lag behind corresponding changes in illumination, so that a combination of supposed light and temperature tolerances has been suggested to determine seasonal occurrence (Findenegg, 1943; cf. also Rodhe, 1948; Vollenweider, 1950). Experimental work is lacking for most of the species discussed, but within the usual ecological range of 0 to 25°C. a sufficiently narrow temperature tolerance can rarely be demonstrated in cultures (e.g., Lund, 1950, 1954). Possibly the strongest indications of stenothermy are found among "snow algae" (e.g., Kol, 1944; see Marré, Chapter 33, this treatise) and in the chrysophyte *Hydrurus foetidus*, which characteristically occurs in cold streams (cf. Bursa, 1934).

In lakes, the effects of light are most clearly shown by the depth distribution of populations and their activities. Light penetration into different fresh waters is extremely varied (Sauberer and Ruttner, 1941), but the depth with 1% of the surface intensity usually marks the lower limit of the euphotic zone in which there can be appreciable growth and photosynthesis. Some characteristics of communities of littoral algae growing near this limit were reviewed by Fritsch (1931). Intensive and often reddish pigmentation is common, but complementary chromatic adaptation (Sauberer and Ruttner, 1941, pp. 133–144) appears less evident than in the marine littoral. Field experiments show a direct relationship between the exponential decrease of light intensity and the photosynthetic rates at greater depths, but near the surface the mechanisms of photosynthesis tend to become saturated or often even inhibited by light (e.g., Manning and Juday, 1941; Talling, 1957a, b). These features can be interpreted by the well-known rate-intensity curves of photosynthesis. Growth rates often vary with depth in a similar manner (Lund, 1949; Talling, 1955; Cannon *et al.*, 1961), but light saturation is less evident in growth on submerged glass slides (Godward, 1937; Pearsall *et al.*, 1946). A differentiation of "sun"- and "shade"-characteristics in photosynthesis may develop in algae from different depths (Gessner, 1949; and personal observations on *Asterionella formosa*). Changes of light intensity at the water surface also cause modifications of depth profiles of growth and photosynthesis (e.g., Talling, 1957b, c; Edmondson, 1956). Low light intensities in winter often prevent the increase of plankton populations except in the shallowest lakes (e.g., Lund, 1950, 1955a), although the growth of algae attached near the surface may continue (e.g. Knudson, 1957).

A vertical stratification of temperature and density (see Hutchinson, 1957) has important indirect effects on algal distribution and growth, by limiting the vertical transfer of dissolved substances and phytoplankton (e.g., Lund, 1954, 1955b, 1959; Gessner, 1955). It can develop diurnally, as in shallow waters, or seasonally, as in most lakes, or may persist indefinitely in meromictic lakes.

V. Nutrient Requirements and Assimilation

Culture studies are required to establish the absolute nutrient requirements for each species. In their major ionic composition some particularly successful culture media resemble many natural waters (Chu, 1942, 1943, 1949; Rodhe, 1948; Provasoli and Pintner, 1953). Concentration requirements established in cultures for nutrients, particularly nitrate and phosphate, are often much greater than those known to permit growth in nature. This feature has been related to the larger volumes of natural media (Gerloff and Skoog, 1954; Provasoli, 1958), but for many of the more evenly distributed populations of plankton this seems implausible. The comparatively high cell densities and rapid growth rates usual in cultures may be partly responsible. Evidence of nutrient requirements in specific natural populations is provided by correlations between algal growth and water chemistry, estimates of the quantities of nutrients removed by the crop (e.g., Gardiner, 1941; Lund, 1950), comparisons of the chemical composition of naturally grown cells with that known to develop in cultures under limiting conditions (e.g., Mackereth, 1953; Gerloff and Skoog, 1954, 1957b), and enrichment experiments (e.g., Strøm, 1933; Potash, 1956; Fish, 1956; Gerloff and Skoog, 1957b).

A comparison of the chemical composition of most fresh waters with that of algal cells suggests that silicon, nitrogen, and phosphorus are among the elements most likely to be depleted by algal growth. A considerable depletion of silicon during diatom maxima has often been observed (e.g., Einsele and Vetter, 1938; Lund, 1950; Komarovsky, 1953; Jørgensen, 1957), and in some examples it clearly limits the maximum algal densities attained (Lund, 1950). Jørgensen (1957) claimed that epiphytic diatoms might be able to utilize silicon associated with leaves of their host, the reed *Phragmites*, during periods of silicon depletion by planktonic diatoms. Elsewhere, however, such periods were usually unfavorable even to the growth of *Tabellaria flocculosa* epiphytic on *Phragmites* (Knudson, 1957).

Studies of nitrogen utilization are complicated by the occurrence of several possible inorganic sources: NO_3^-, NH_4^+, and N_2. The first two can be utilized by most algae, although varied preferences have been described (Proctor, 1957a; Provasoli, 1958). In most well-aerated and unpolluted fresh waters, nitrate predominates over ammonia, but lowered nitrate concentrations are often associated with warmer conditions (Hutchinson, 1957). In one tropical water-body, concentrations below 0.02 mg. nitrate-N/liter may have limited the growth of a plankton diatom (Prowse and Talling, 1958), though elsewhere algae have been reported to utilize inorganic nitrogen at concentrations below 0.1 mg. N/liter (e.g., Lund, 1950).

The quantitative significance of nitrogen fixation by Blue-green Algae in fresh waters is still not clear (see Fogg, 1956; also, in this treatise, Chapter 9). Although *Anabaena cylindrica* has been shown unequivocally to fix nitrogen, and there are circumstantial data indicating a similar faculty in planktonic species of *Anabaena* (Hutchinson, 1957, p. 847; Prowse and Talling, 1958), no bloom-forming species of this genus has experimentally been shown to do so. Indirect evidence for nitrogen fixation by planktonic Blue-green Algae has also been presented by Aleev and Mundretsova (1937), Hutchinson (1941a), Sawyer (1952), and Kusnezow (1959). Nevertheless, nitrogen fixation could not be demonstrated in pure cultures of such abundant planktonic species as *Microcystis* (*Diplocystis*) *aeruginosa* and *Aphanizomenon flos-aquae* (Williams and Burris, 1952), which are often common in summer when inorganic nitrogen is low (e.g., Pearsall, 1930, 1932; Gerloff and Skoog, 1954, 1957b).

It is widely believed that in many bodies of fresh water growth of phytoplankton tends to be limited by the supply of inorganic phosphate. Concentrations of phosphate-P are often below 10 μg./liter, and in some lakes they may be reduced to 1 μg./liter or less by uptake during the growth of algae (Hutchinson, 1957; Gardiner, 1941). Nevertheless such an uptake of phosphorus may reflect a storage by algal cells much in excess of their immediate growth requirements, yet available to be utilized in later growth when the external supply of this element is depleted (Rodhe, 1948; Mackereth, 1953). At least one species has been shown capable of P-uptake from lake waters, but not from artificial media, containing as little as 1 μg./liter of phosphate-P, suggesting a phosphate-sparing factor in the lake waters (Rodhe, 1948; Mackereth, 1953). A rapid turnover of phosphate in lakes is also likely (Hayes *et al.*, 1952; Rigler, 1956).

Reduction by algae of the concentrations of major inorganic ions is usually slight, although many (Ca, Mg, Na, K, HCO_3, SO_4) are known nutrients. Sulfate deficiency has been deemed important in some African waters (Beauchamp, 1953), but the most direct evidence for this is based on differences of algal yields in dense enrichment cultures (Fish, 1956). Despite some attention to the ratios of total monovalent to divalent cations (Pearsall, 1932; Provasoli *et al.*, 1954) and of Ca to Mg (Vollenweider, 1950; Provasoli *et al.*, 1954), there are apparently no well-established ecological consequences in fresh waters. The chief importance of bicarbonate for algal cells lies in its role in the transport of carbon dioxide. Its removal by photosynthesis causes a rise of pH and a consequent shift in the equilibrium of the residual ions towards carbonate, but the growth and survival of many algae seem to be endangered only at pH values exceeding 10 (e.g., Steemann Nielsen, 1955; Jenkin, 1936; Gerloff *et al.*, 1950, 1952).

Little is known of the natural availability of iron and trace metals such as Mn, Mo, Co, Zn, V, and Cu. Only very low concentrations of iron can exist in true solution in well-oxygenated neutral or basic waters (e.g., Cooper, 1948). A loose combination with carriers or chelating substances may greatly increase the availability of iron and some trace metals (Provasoli *et al.*, 1957). Various forms of organic matter (reviewed by Saunders, 1957; Vallentyne, 1957; Fogg, Chapter 30, this treatise) may be involved, including the yellow organic acids of lake waters (Shapiro, 1957, 1958) and various extracellular algal products, both soluble (Fogg and Westlake, 1955) and insoluble. Otherwise there is little direct ecological evidence for the utilization of organic substances, either as sources of nitrogen or phosphorus (cf. Mackereth, 1953; Hutchinson, 1957; Gessner, 1959) or as energy sources (but see Rodhe, 1955), though some are valuable constituents of culture media; possibly suitable compounds seldom occur in nature in significant concentration. Ecological aspects of organic micronutrients are discussed in Chapter 8 of this treatise by Droop.

There is evidence that some nutrients may show toxicity even at relatively low concentrations which can be reached in nature. Examples are iron (Uspenski, 1927), manganese (Guseva, 1941, 1952; Gerloff and Skoog, 1957a), phosphate (Rodhe, 1948), and copper (Hutchinson, 1957, p. 816).

VI. Species Differentiation and Interactions

Few examples of floristic differences and species succession can be related to known physiological differences between algae. Succession is as likely to result from inherent differences in specific growth rate, and variability of the initial "inocula" (e.g., Prowse and Talling, 1958), as from different responses to environmental factors during active growth. Competitive relationships complicate the problem (e.g., Proctor, 1957b) so that a stable equilibrium may rarely be reached (Hutchinson, 1941b, 1944); but they do so by contributing to, rather than superseding, the immediate influences of environmental physics and chemistry upon algal behavior.

The frequent growth of planktonic *Dinobryon* spp. after the subsidence of an algal maximum has been discussed by Pearsall (1932) and Hutchinson (1944). Rodhe (1948) believed that the reduction of unfavorable concentrations of phosphate was probably involved (see also Bamforth, 1958), although *Dinobryon* has been cultured in phosphate-rich media (Rosenberg, 1938; personal observations). Complex interactions involving several diatoms (*Asterionella, Tabellaria*) and a Blue-green Alga (*Oscillatoria*) in the lake Esthwaite Water were described by Canter and Lund (1951); here silica depletion, shading effects, and selective parasitism by chytrids were involved. Contrasts between the growth cycles of *Asterionella* and *Melosira* spp. (see Section III) are governed largely by differences in their respective sinking rates and consequent responses to thermal stratification

(Lund, 1950, 1954, 1955b); an unusual tolerance of anaerobic conditions is also important for *Melosira* spp. (Lund, 1954; Talling, 1957d). Species interactions may develop from specific inhibitory or stimulatory substances produced by algae (see Fogg, Chapter 30, this treatise). There are few recorded instances of physiological differentiation within a species, such as might result from differences in previous light conditions. For example, in some mixed populations in the Nile, two forms of *Melosira granulata* showed different responses to unfavorable conditions (Prowse and Talling, 1958). Some populations of *Asterionella* have been considered as distinct ecotypes (Ruttner, 1937).

VII. Production and Productivity

In this section, quantitative aspects of organic synthesis in freshwater algal communities are discussed. Although the terms are often used loosely, *productivity* best denotes the *rate* of synthesis, and *production* the resulting quantity or yield (cf. MacFadyen, 1948). Because of their ultimate dependence on solar energy, both measures are normally also related to unit area. Overall restrictions are imposed by specific rates of growth and photosynthesis, and by the supply of available nutrients and energy. The increases observed in natural populations can rarely be taken as measures of productivity, owing to continuous depletion of the populations which is usually difficult to measure; however, see Thomas (1950) and Grim (1950). Consequently a variety of less direct approaches have been used (e.g., Lund and Talling, 1957). In favorable examples, changes in the composition of the medium during algal growth provide useful quantitative estimates; examples include the depletion of silicon (e.g., Grim, 1939; Gardiner, 1941), and the diurnal liberation of oxygen or removal of carbon dioxide by photosynthesis (e.g., Vinberg and Yarovitzina, 1939; Juday *et al.*, 1943; Talling, 1957c). Phosphate depletion, much utilized in oceanography as a measure of production (see Yentsch, Chapter 52, this treatise), is of dubious value (Section V). Photosynthetic changes occur in a definite distribution within the euphotic zone, and can be estimated in enclosed samples suspended at various levels or exposed in the laboratory. Such measurements, and their integration with depth to give estimates per unit area, account for most of the present intensive work on phytoplankton productivity.

The contributions of population density and synthetic activity per unit population can be distinguished.

A. Population Density

Densities of phytoplankton per unit volume are usually very low by comparison with laboratory cultures, and rarely exceed a few milligrams of dry weight, micrograms of chlorophyll, or cubic millimeters of cell volume per liter. In nature, however, phytoplankton populations may extend to considerable depths, so that the quantity of chlorophyll per unit area can reach a magnitude comparable with that of closed terrestrial communities (Gessner, 1949, 1959). More significant for productivity is the plant population of the euphotic zone. Its maximum value per unit area depends on the extinction of light associated with the population density; several sources of data suggest a maximum of about 300 mg. chlorophyll per square meter (Steemann Nielsen, 1957). The size of the euphotic zone population per unit area shows less variation than do population densities per unit volume, since low densities are often associated with deep euphotic zones.

B. Population Activity

This may be assessed by specific growth rates (see Section III) or by rates of photosynthesis per unit of population. Maximum values of both are known for many freshwater algae; 1 to 3 cell divisions per day and 4 to 12 mg. CO_2/mg. chlorophyll/hour are typical (e.g., Manning and Juday, 1941; Gessner, 1943, 1944, 1949; Ichimura 1958). Average rates in a natural population, however, are often considerably reduced by limitations of light and temperature. Such quantitative restrictions of growth and of photosynthesis have been discussed by Lund (1949) and Talling (1955), and by Clarke (1939), Verduin (1956), and Talling (1957b), respectively. Average daytime rates of photosynthesis for populations in the euphotic zone are often about half of the maximum, light-saturated rates (Steemann Nielsen, 1952; Verduin, 1956; Rodhe *et al.*, 1958), and on a 24-hour basis night introduces a further limitation. There is evidence for some diurnal variation in photosynthetic capacity (Ohle, 1958, 1961; cf. Yentsch, Chapter 52, this treatise). Photosynthetically inactive populations seem surprisingly uncommon, even in late stages of growth or under conditions of nutrient deficiency (Talling, 1957a, c; cf. also Rodhe, 1958a, Fig. 12). The extent of respiration losses in nature is largely conjectural.

C. Areal Productivity

Estimates of phytoplankton productivity from photosynthesis determinations are generally expressed in gm. C per square meter per day, and may be transformed into the percentage utilization of incident radiation (cf. Vinberg, 1948). The four classes of values used by Steemann Nielsen (1954) to summarize oceanic data appear also to be generally applicable to fresh waters; very unproductive waters yield values less than 0.1 gm. C/m.^{2}/day, and the most productive waters may yield 0.5 to 3 gm. C/m.^{2}/day (e.g., Rodhe, 1958b). Indeed, the highest values approach those which can be obtained in algal mass cultures (Tamiya, 1957), but most natural populations show much lower rates, due mainly to their low densities.

Areal rates are usually calculated by the summation of many rates per unit volume determined at different depths in the euphotic zone. The depth distribution of photosynthetic activity is influenced by the surface-light intensity, by temperature, and by underwater light penetration, as well as by the distribution of the algae themselves (e.g., Manning and Juday, 1941; Talling, 1957b; Rodhe *et al.*, 1958) During daylight hours, the areal productivity is relatively insensitive to changes of surface intensity, due to saturation effects (Steemann Nielsen, 1954, 1958; Talling, 1957b). Very dense populations have high photosynthetic rates within the euphotic zone, but at the same time greatly reduce the depth of the latter by self-shading effects (Steemann Nielsen, 1957). This restriction tends to set an upper limit to the productivity per unit area of natural waters, however densely they may be populated.

References

Aleev, B. S., and Mudretsova, K. A. (1937). [The role of the phytoplankton in the nitrogen dynamics in the water of a 'blossoming' pond.] *Mikrobiologiya* **6,** 329–338; *Chem. Abstr.* **33,** 5446 (1939). (In Russian.)

Bamforth, S. S. (1958). Ecological studies on the planktonic protozoa of a small artificial pond. *Limnol. and Oceanog.* **3,** 398–412.

Beauchamp, R. S. A. (1953). Sulphates in African inland waters. *Nature* **171,** 769–771.

Bursa, A. (1934). *Hydrurus foetidus* Kirch. in der polnischen Tatra. Oekologie, Morphologie. I. *Bull. acad. polon. sci.* B I, pp. 69–84.

Cannon, D., Lund, J. W. G., and Sieminska, J. (1961). The growth of *Tabellaria flocculosa* (Roth) Kütz. var. *flocculosa* (Roth) Knuds. under natural conditions of light and temperature. *J. Ecol.* **49,** 277–288.

Canter, H. M., and Lund, J. W. G. (1951). Studies on plankton parasites. III. Examples of interaction between parasitism and other factors determining the growth of diatoms. *Ann. Botany (London)* **15,** 361–371.

Chu, S. P. (1942). The influence of the mineral composition of the medium on the growth of planktonic algae. I. Methods and culture media. *J. Ecol.* **30,** 284–325.

Chu, S. P. (1943). The influences of the mineral composition of the medium on the growth of planktonic algae. II. The influence of the concentration of inorganic nitrogen and phosphate phosphorus. *J. Ecol.* **31,** 109–148.

Chu, S. P. (1949). Experimental studies on the environmental factors influencing the growth of phytoplankton. *Sci. & Technol. China* **2,** 37–52.

Clarke, G. L. (1939). The utilisation of solar energy by aquatic organisms. *In* "Problems of Lake Biology" (F. R. Poulton, ed.) *Publ. Am. Assoc. Advance. Sci.* **No. 10,** 27–38.

Cooper, L. H. N. (1948). Some chemical considerations on the distribution of iron in sea water. *J. Marine Biol. Assoc. United Kingdom* **27,** 314–321.

Deevey, E. S (1940). Limnological studies in Connecticut. V. A contribution to regional limnology. *Am. J. Sci.* **238,** 717–741.

Edmondson, W. T. (1956). The relation of photosynthesis by phytoplankton to light in lakes. *Ecology* **37,** 161–174.

Einsele, W., and Vetter, H. (1938). Untersuchungen über die Entwicklung der physikalischen und chemischen Verhältnisse im Jahreszyklus in einem mässig eutrophen See (Schleinsee bei Langenargen). *Intern. Rev. Hydrobiol.* **36,** 285–324.

Findenegg, I. (1943). Untersuchungen über die Ökologie und die Produktionsverhältnisse des Planktons im Kärntner Seengebiete. *Intern. Rev. Hydrobiol.* **43,** 368–429.

Fish, G. R. (1956). Chemical factors limiting growth of phytoplankton in Lake Victoria. *E. African Agr. J.* **21,** 152–158.

Fogg, G. E. (1956). The comparative physiology and biochemistry of the blue-green algae. *Bacteriol. Revs.* **20,** 148–165.

Fogg, G. E. (1958). Relationships between metabolism and growth in plankton algae. *J. Gen. Microbiol.* **16,** 294–297.

Fogg, G. E., and Westlake, D. F. (1955). The importance of extracellular products of algae in freshwater. *Verhandl. intern. Ver. Limnol.* **12,** 219–232.

Fritsch, F. E. (1931). Some aspects of the ecology of the freshwater algae. (With special reference to static waters.) *J. Ecol.* **19,** 233–272.

Gardiner, A. C. (1941). Silicon and phosphorus as factors limiting development of diatoms. *J. Soc. Chem. Ind. (London)* **60,** 73–78.

Gerloff, G. C., and Skoog, F. (1954). Cell contents of nitrogen and phosphorus as a measure of their availability for growth of *Microcystis aeruginosa. Ecology* **35,** 348–353.

Gerloff, G. C., and Skoog, F. (1957a). Availability of iron and manganese in southern Wisconsin lakes for the growth of *Microcystis aeruginosa. Ecology* **38,** 551–556.

Gerloff, G. C., and Skoog, F. (1957b). Nitrogen as a limiting factor for the growth of *Microcystis aeruginosa* in southern Wisconsin lakes. *Ecology* **38,** 556–561.

Gerloff, G. C. Fitzgerald, G. P., and Skoog, F. (1950). The mineral nutrition of *Coccochloris peniocystis. Am. J. Botany* **37,** 835–840.

Gerloff, G. C., Fitzgerald, G. P., and Skoog, F. (1952). The mineral nutrition of *Microcystis aeruginosa. Am. J. Botany* **39,** 26–32.

Gessner, F. (1943). Die assimilatorische Leistung des Phytoplanktons, bezogen auf seinen Chlorophyllgehalt. *Z. Botan.* **38,** 414–424.

Gessner, F. (1944). Der Chlorophyllgehalt der Seen als Ausdruck ihrer Produktivität. *Arch. Hydrobiol.* **40,** 687–732.

Gessner, F. (1948). The vertical distribution of phytoplankton and the thermocline *Ecology* **29,** 386–389.

Gessner, F. (1949). Der Chlorophyllgehalt im See und seine photosynthetische Valenz als geophysikalische Problem. *Schweiz. Z. Hydrol.* **11,** 378–410.

Gessner, F. (1955). "Hydrobotanik," Vol. I: Energiehaushalt. VEB Deutscher Verlag des Wissenschaften, Berlin.

Gessner, F. (1959). "Hydrobotanik," Vol. II: Stoffhaushalt. VEB Deutscher Verlag der Wissenschaften, Berlin.

Godward, M. B. E. (1937). An ecological and taxonomic investigation of the littoral algal flora of Lake Windemere. *J. Ecol.* **25,** 496–568.

Grim, J. (1939). Beobachtungen am Phytoplankton des Bodensees (Obersee) sowie deren rechnerische Auswertung. *Intern. Rev. Hydrobiol.* **39,** 193–315.

Grim. J. (1950). Versuche zur Ermittelung der Produktionskoeffizienten einiger Planktophyten in einem flachen See. *Biol. Zentr.* **69,** 147–174.

Grim, J. (1952). Vermehrungsleistungen planktischer Algenpopulationen in Gleichgewichtsperioden. *Arch. Hydrobiol. Suppl.* **20,** 238–260.

Guseva, K. A. (1941). Die Wasserblüte im Utscha-Stausee. (In Russian; German summary.) *Trudi Zoologicheskogo Instituta Akad. Nauk S.S.S.R.* **7,** 89–121.

Guseva, K. A. (1952). [Studies of rivers and water reservoirs. Blooming waters, their cause, and associated measures of competition.] (In Russian.) *Trudy Vsesoyuz. Gidrobiol. Obshchestva* **4,** 3–92.

Hayes, F. R., MacCarter, J. A., Cameron, M. L., and Livingstone, D. A. (1952). On the kinetics of phosphorus exchange in lakes. *J. Ecol.* **40,** 202–216.

Hutchinson, G. E. (1941a). Limnological studies in Connecticut. IV. The mechanism of intermediary metabolism in stratified lakes. *Ecol. Monographs* **11,** 21–60.

Hutchinson, G. E. (1941b). Ecological aspects of succession in natural populations. *Am. Naturalist* **65,** 406–418.

Hutchinson, G. E. (1944). Limnological studies in Connecticut. VII. A critical examination of the supposed relationship between phytoplankton periodicity and chemical changes in lake waters. *Ecology* **25,** 3–26.

Hutchinson, G. E. (1957). "A Treatise on Limnology," Vol. I: Geography, Physics and Chemistry. Wiley, New York.

Ichimura, S. (1958). On the photosynthesis of natural phytoplankton under field conditions. *Botan. Mag. (Tokyo)* **71,** 110–116.

Jenkin, P. M. (1936). Reports on the Percy Sladen expedition to some Rift Valley lakes in Kenya in 1929. VII. Summary of the ecological results, with special reference to the alkaline lakes. *Ann. Mag. Nat. Hist.* [10] **18,** 133–181.

Jørgensen, E. G. (1957). Diatom periodicity and silicon assimilation. Experimental and ecological investigations. *Dansk Botan. Arkiv* **18,** 1–54.

Juday, C., Blair, J. M., and Wilda, E. F. (1943). The photosynthetic activities of the aquatic plants of Little John Lake, Vilas county, Wisconsin. *Am. Midland Naturalist* **30,** 426–446.

Kann, E. (1933). Zur Ökologie des litoralen Algenaufwuchses im Lunzer Untersee. *Intern. Rev. Hydrobiol.* **28,** 172–227.

Knudson, B. M. (1957). Ecology of the epiphytic diatom *Tabellaria flocculosa* (Roth.) Kütz. var. *flocculosa* in three English lakes. *J. Ecol.* **45,** 93–112.

Kol, E. (1944). Vergleich der Kryovegetation der nördlichen und südlichen Hemisphäre. *Arch. Hydrobiol.* **40,** 835–846.

Kolbe, R. W. (1932). Grundlinien einer allgemeinen Ökologie der Diatomeen. *Ergeb. Biol.* **8,** 221–348.

Komarovsky, B. (1953). A comparative study of the phytoplankton of several fish ponds in relation to some of the essential chemical constituents of the water. *Bull. Research Council Israel* **2,** 379–410.

Kusnezow, S. I. (1959). "Die Rolle der Mikroorganismen im Stoffkreislauf der Seen" (Russian translation by A. Pochman). Deutscher Verlag der Wissenschaften, Berlin.

Lund, J. W. G. (1949). Studies on *Asterionella formosa* Hass. I. The origin and nature of the cells producing seasonal maxima. *J. Ecol.* **37,** 389–419.

Lund, J. W. G. (1950). Studies on *Asterionella formosa* Hass. II. Nutrient depletion and the spring maximum. *J. Ecol.* **38,** 1–35.

Lund, J. W. G. (1954). The seasonal cycle of the plankton diatom *Melosira italica* Kütz. subsp. *subarctica* O. Müll. *J. Ecol.* **42,** 151–179.

Lund, J. W. G. (1955a). The ecology of algae and waterworks practice. *Proc. Soc. Water Treatment Exam.* **4,** 83–109.

Lund. J. W. G. (1955b). Further observations on the seasonal cycle of *Melosira italica* (Ehr.) Kütz. subsp. *subarctica* O. Müll. *J. Ecol.* **43,** 91–102.

Lund, J. W. G. (1957). Chemical analysis in ecology illustrated from Lake District tarns and lakes. II. Algal differences. *Proc. Linnean Soc. London* **167,** 165–171.

Lund, J. W. G. (1959). Buoyancy in relation to the ecology of the freshwater phytoplankton. *Brit. Phycol. Bull* **1**(7), 1–17.

Lund, J. W. G., and Talling, J. F. (1957). Botanical limnological methods with special reference to the algae. *Botan. Rev.* **23,** 489–583.

MacFadyen, A. (1948). The meaning of productivity in biological systems. *J. Animal Ecol.* **17,** 75–80.

Mackereth, F. J. (1953). Phosphorus utilization by *Asterionella formosa* Hass. *J. Exptl. Botany* **4,** 296–313.

Manning, W. M., and Juday, R. E. (1941). The chlorophyll content and productivity of some lakes in north-eastern Wisconsin. *Trans. Wisconsin Acad. Sci.* **33,** 363–393.

Oberdorfer, E. (1928). Lichtverhältnisse und Algenbesiedlung im Bodensee. *Z. Botan.* **20,** 465–568.

Ohle, W. (1958). Diurnal production and destruction rates of phytoplankton in lakes. *Rappts. Conseil permanent intern. exploration mer.* **144,** 129–131.

Ohle, W. (1961). Tagesrhythmen der Photosynthese von Planktonbiocoenosen. *Verhandl. intern. Ver. Limnol.* **14,** 113–119.

Pearsall, W. H. (1921). The development of vegetation in the English Lakes, considered in relation to the general evolution of glacial lakes and rock basins. *Proc. Roy. Soc.* **B92,** 259–284.

Pearsall, W. H. (1930). Phytoplankton in the English lakes. I. The proportions in the waters of some dissolved substances of biological importance. *J. Ecol.* **18,** 306–320.

Pearsall, W. H. (1932). Phytoplankton in the English lakes. II. The composition of the phytoplankton in relation to dissolved substances. *J. Ecol.* **20,** 241–262.

Pearsall, W. H., Gardiner, A. C., and Greenshields, F. (1946). Freshwater biology and water supply in Britain. *Sci. Publs. Freshwater Biol. Assoc. England* **No. 11.**

Potash, M. (1956). A biological test for determining the potential productivity of water. *Ecology* **37,** 631–639.

Proctor, V. W. (1957a). Preferential assimilation of nitrate by *Haematococcus pluvialis*. *Am. J. Botany* **44,** 141–143.

Proctor, V. W. (1957b). Some controlling factors in the distribution of *Haematococcus pluvialis*. *Ecology* **38,** 457–462.

Provasoli, L. (1958). Nutrition and ecology of protozoa and algae. *Ann. Rev. Microbiol.* **12,** 279–308.

Provasoli, L., and Pintner, I. J. (1953). Ecological implications of *in vitro* nutritional requirements of algal flagellates. *Ann. N. Y. Acad. Sci.* **56,** 839–851.

Provasoli, L., McLaughlin, J. J. A., and Pintner, I. J. (1954). Relative and limiting concentrations of major mineral constituents for the growth of algal flagellates. *Trans. N. Y. Acad. Sci.* **16,** 412–417.

Provasoli, L., McLaughlin, J. J. A., and Droop, M. R. (1957). The development of artificial media for marine algae. *Arch. Mikrobiol.* **25,** 392–428.

Prowse, G. A., and Talling, J. F. (1958). The seasonal growth and succession of plankton algae in the White Nile. *Limnol. and Oceanog.* **3,** 222–238.

Rigler, F. H. (1956). A tracer study of the phosphorus cycle in lake water. *Ecology* **37,** 550–562.

Rodhe, W. (1948). Environmental requirements of freshwater plankton algae. *Symbolae Botan. Upsalienses* **10,** 1–149.

Rodhe, W. (1955). Can plankton production proceed during winter darkness in subarctic lakes? *Verhandl. intern. Ver. Limnol.* **12,** 117–122.

Rodhe, W. (1958a). Aktuella problem inom Limnologien. *In* "Svensk Naturvetenskap 1957–1958" Vol. 11, pp. 55–107. Haeggströms, Stockholm, Sweden.

Rodhe, W. (1958b). The primary production in lakes: some results and restrictions of the ^{14}C method. *Rappts. Conseil permanent intern. exploration mer* **144,** 122–128.

Rodhe, W., Vollenweider, R. A., and Nauwerck, A. (1958). The primary production and standing crop of phytoplankton. *In* "Perspectives in Marine Biology" (A. A. Buzzati Traverso, ed.), pp. 299–322. Univ. of California Press, Berkeley and Los Angeles.

Rosenberg, M. (1938). The culture of algae and its applications. *Repts. Freshwater Biol. Assoc. Brit. Empire* **6,** 43–46.

Ruttner, F. (1937). Ökotypen mit verschiedener Vertikalverteilung im Plankton der Alpenseen. *Intern. Rev. Hydrobiol.* **35,** 7–34.

Sauberer, F., and Ruttner, F. (1941). Die Strahlungsverhältnisse der Binnengewässer. *In* "Probleme der kosmischen Physik" Vol. XXI. Akademische Verlagsges., Leipzig.

Saunders, G. W. (1957). Interrelations of dissolved organic matter and phytoplankton. *Botan. Rev.* **23,** 389–410.

Sawyer, C. N. (1952). Some new aspects of phosphates in relation to lake fertilization. *Sewage and Ind. Wastes* **24,** 768–776.

Scourfield, D. J. (1897). The logarithmic plotting of certain biological data. *J. Queckett Microscop. Club* **6,** 419–423.

Shapiro, J. (1957). Chemical and biological studies on the yellow organic acids of lake water. *Limnol. and Oceanog.* **2,** 161–179.

Shapiro, J. (1958). Yellow acid-cation complexes in lake water. *Science* **127,** 702–704.

Steemann Nielsen, E. (1952). The use of radioactive carbon (C^{14}) for measuring organic production in the sea. *J. conseil, Conseil permanent intern. exploration mer* **18,** 117–140.

Steemann Nielsen, E. (1954). On organic production in the oceans. *J. conseil, Conseil permanent intern. exploration mer* **19,** 309–328.

Steemann Nielsen, E. (1955). The production of organic matter by the phytoplankton in a Danish lake receiving extraordinarily great amounts of nutrient salts. *Hydrobiologia* **7,** 68–74.

Steemann Nielsen, E. (1957). The chlorophyll content and the light utilization in communities of plankton algae and terrestrial higher plants. *Physiol. Plantarum* **10,** 1009–1021.

Steemann Nielsen, E. (1958). Light and the organic production in the sea. *Rappt. Conseil permanent intern. exploration mer* **144,** 141–148.

Strøm, K. M. (1933). Nutrition of algae. *Arch. Hydrobiol.* **25,** 38–47.

Talling, J. F. (1955). The relative growth rates of three plankton diatoms in relation to underwater radiation and temperature. *Ann. Botany (London)* **19,** 329–341.

Talling, J. F. (1957a). Photosynthetic characteristics of some freshwater plankton diatoms in relation to underwater radiation. *New Phytologist* **56,** 29–50.

Talling, J. F. (1957b). The phytoplankton population as a compound photosynthetic system. *New Phytologist* **56,** 133–149.

Talling, J. F. (1957c). Diurnal changes of stratification and photosynthesis in some tropical African waters. *Proc. Roy. Soc.* **B147,** 57–83.

Talling, J. F. (1957d). Some observations on the stratification of Lake Victoria. *Limnol. and Oceanog.* **2,** 213–221.

Talling, J. F. (1960). Comparative laboratory and field studies of photosynthesis by a marine planktonic diatom. *Limnol. and Oceanog.* **5,** 62–77.
Tamiya, H. (1957). Mass culture of algae. *Ann. Rev. Plant Physiol.* **8,** 309–334.
Thomas, E. A. (1950). Beitrag zur Methodik der Produktionsforschung in Seen. *Schweiz. Z. Hydrol.* **12,** 25–37.
Uspenski, E. E. (1927). Eisen als Faktor für die Verbreitung niederer Wasserpflanzen. *Pflanzenforschung* **9,** 1–104.
Vallentyne, J. R. (1957). The molecular nature of organic matter in lakes and oceans, with lesser reference to sewage and terrestrial soils. *J. Fisheries Research Board Can.* **14,** 33–82.
Verduin, J. (1951). Comparison of spring diatom crops of western Lake Erie in 1949 and 1950. *Ecology* **32,** 662–668.
Verduin, J. (1952). Photosynthesis and growth rates of two diatom communities in western Lake Erie. *Ecology* **33,** 163–168.
Verduin, J. (1956). Energy fixation and utilisation by natural communities in western Lake Erie. *Ecology* **37,** 40–49.
Vinberg, G. G. (1948). The efficiency of utilization of solar-radiation by plankton. (In Russian.) *Priroda* **12,** 29–35.
Vinberg, G. G., and Yarovitzina, L. I. (1939). Daily changes in the quantity of dissolved oxygen as a method for measuring the value of primary production. (*In* Russian; English summary.) *Arb. Limnol. Sta. Kossino* **22,** 128–143.
Vollenweider, R. A. (1950). Ökologische Untersuchungen von planktischen Algen auf experimenteller Grundlage. *Schweiz. Z. Hydrol.* **12,** 193–262.
Williams, A. E., and Burris, R. H. (1952). Nitrogen fixation by blue-green algae and their nitrogenous composition. *Am. J. Botany* **39,** 340–342.

Soil Algae

J. W. G. LUND

Freshwater Biological Association, Ambleside, England

I. Introduction

Wherever there is soil there are algae on or near the surface. They have no unique physiological or biochemical features, but their characteristics reflect the severity of conditions at the air–soil interface.

II. The Physical Environment

A. Moisture

Soil algae grow best on damp soil. Waterlogging produces semiaquatic conditions and communities; hence the use of liquid enrichment cultures may give a wholly false impression of the quantitative composition of the soil flora (as in Fehér, 1936, 1948). They may grow poorly in water (Vischer, 1945) but are less sensitive to submergence than are aerial algae; on the other hand, they are unable to grow actively in the absence of liquid water.

Many species can withstand prolonged and severe drought. Longevity is discussed by Bristol (1919), Lipman (1941) and Becquerel (1942), and survival under desert conditions by Fletcher and Martin (1948), Basilevich *et al.* (1952), Vogel (1955), Degopik (1956), Shields *et al.* (1957), and Sdobnikova (1958). Some of the published data are contradictory. Thus, though *Nitzschia palea* (Kütz) W. Sm. is found abundantly only in moist soils, according to Brendemühl (1949), Zauer (1956), and Shtina (1959), and was killed when dried in air in von Denffer's (1949) experiments, it was experimentally shown to be capable of withstanding prolonged periods of drying (Evans, 1958, 1959), and was the only diatom which developed from soils preserved air-dry for 70 to 98 years (Bristol, 1919; Becquerel, 1942). Such contradictions may arise from incorrect identifications or from the way in which the experiments and observations are made. The conditions prevailing before or during drying are probably important (e.g., Füchtbauer, 1957a, b; Petit, 1877).

Resistance to drought, high temperatures, and high salt concentrations tend to go together and to be attributable to the character of the walls and underlying protoplasm (Fritsch and Haines, 1923) rather than to the osmotic pressure of the sap, high though this may be (de Puymaly 1924a, b; see review by Stocker, 1956).

That diatoms and Xanthophyta are generally the groups most susceptible to severe or prolonged drought is suggested by the results of longevity tests and by the predominance of other groups in deserts (Bristol Roach, 1927a). Yet Petersen (1935) found no such differences during a three-year investigation. In general, the numbers of viable cells fall markedly during the process of drying, particularly if this is rapid, and thereafter more slowly during storage (Bristol Roach, 1927a; personal observations).

B. Light

The commonly observed aggregation of algae on the surface of soil may seem proof that they can withstand full sunlight and sudden changes in light intensity. The former is not confirmed experimentally (e.g., Delarge, 1937; von Denffer, 1949), but there is no good evidence that the light intensity in nature ever reaches lethal intensities except in direct sunlight. However, the precise position of the various algal cells in relation to irregularities of the soil surface has not been studied. In some deserts superficial crusts are common (see references in 1.A.), which may provide some measure of protection.

Xanthophyta have been considered to be "shade" forms because they are most abundant a few centimeters below the surface (Fritsch, 1936; John, 1942; Flint, 1958), but at these depths there is no light (Sauberer and Härtel, 1959) except in fissures. The great abundance of some species on old feces (Bristol Roach 1927a, p. 575; personal observation) suggests a predilection for organic matter.

Fehér and Frank (1936, 1939) put forward evidence for autotrophic growth of soil algae in unknown spectral regions beyond the infrared and ultraviolet, far outside the known limits for photosynthesis. Baatz (1939) contested their evidence, and Fehér and Frank's (1940) rejoinder does not remove all the obvious criticisms.

C. Temperature

There is little information on the temperatures tolerated by soil algae. Short periods at 60° to 90°C. did not kill certain Cyanophyta when wet, and they could apparently tolerate heating to 95° to 100°C. when dry (Glade, 1914), but 110°C. seems fatal to all soil algae (Booth, 1946). Spores of *Cylindrospermum* are much more resistant than vegetative cells (Glade, 1914), indicating the importance of previous conditions in the soil which may control the formation of spores (Demeter, 1956). In liquid cultures of many algae, growth inhibition or death occurs above 30° to 50°C. (Smith, 1944; Clendenning *et al.*, 1956; Miller and Fogg, 1957). These temperatures are near the observed maxima of temperate soils (Geiger, 1950), but below those in many arid or tropical ones. As with aerial algae, the temperature sensitivity of soil algae seems to increase with rising relative humidity (Izerott, 1937).

The few species investigated (Glade, 1914; Kärcher, 1931; Petersen, 1935; Becquerel, 1936) were so resistant to cold that the abundance of algae in polar regions is not surprising (e.g., Dorogostaĭskaya, 1959). However, the reduction in the numbers of viable cells during wintry weather is probably partly a result of the cold (Lund, 1947; Shtina, 1959). Höfler's (1951) work on bog algae showed that the rate of freezing is important and that death may be related to water loss; it is therefore not surprising that some species are resistant both to severe drought and to cold.

III. The Chemical Environment

A. pH and Salt Tolerance

Most soil algae are found over a fairly wide pH range, but towards the extremes the relationship between pH and the composition of the flora is generally more marked (Gollerbach, 1936; Shelhorn, 1936; John, 1942; Lund, 1945–1946, 1947). Nearly neutral or slightly alkaline soils have the most species, but not necessarily the largest algal crops. Most Cyanophyta prefer neutral or alkaline soils. Chlorophyta may abound at almost any natural pH, but certain species preponderate in highly acid soils where Cyanophyta and diatoms are usually rare or absent. Numerous errors in identification, the common misuse of liquid cultures, and the consequent alteration of the natural soil pH therein, make it difficult to discover the pH limits tolerated in nature.

The factors underlying the importance of pH are unknown, though naturally there is often a correlation with calcium. *Monodus subterraneus* Boye Pet. grows well at pH 8.9 provided the ratio of uni- to bivalent cations is high (Miller and Fogg, 1957). However, many algae, notably Cyanophyta, are abundant in salty, arid soils of widely diverse ionic composition (see references in Section I.A). Many species are almost ubiquitous in soils and, therefore, seem to be tolerant of wide ranges and ratios of salt concentrations. Gistl (1932) grew several species of soil algae in sea water; indeed, a high salt content of the medium enhances the growth of *Porphyridium* (Pringsheim and Pringsheim, 1956).

B. Nutrition

1. Inorganic Nutrients

Little is known about the specific nutritional needs of soil forms, though much of the vast literature on the physiology of algae, notably of *Chlorella*, must refer to such algae. Their productivity, in nonarid soils, seems to follow that of higher plants, and they have been used for the bioassay of mineral nutrients in soil (Tchan, 1959).

The smaller size of certain freshwater diatoms when growing in soil has been related to nutrition (e.g., Gistl, 1932), but this is doubted (Lund, 1945). The same applies to the feeble silicification of some species (Hustedt, 1942; Lewin, 1957).

Phosphorus and nitrogen are most frequently cited as the main elements whose addition will stimulate increased growth of algae in soil, the views about other elements being more contradictory (Stokes, 1940a, b; Lund, 1945–46, 1947; Okuda and Yamagouchi, 1956; Knapp and Lieth, 1952; Chodat and Chastain, 1957; De and Mandal, 1956). Certain soil algae, such as *Prasiola crispa*, may be nitrophilous (Petersen, 1928, 1935; Knebel, 1936; Barkman, 1958; personal observations), but factors such as pH and total salt content are just as likely to be responsible for their observed patterns of distribution. In view of the complex interrelation between ions (Gistl, 1933; Pratt, 1941; Miller and Fogg, 1957) and the chelating action of humic acid and other soil components (Prát, 1955), little progress can be made without detailed studies of pure cultures.

2. Fixation of Atmospheric Nitrogen

About one-third of the Cyanophyta which have been shown to fix nitrogen, including the much-studied *Nostoc muscorum* (see Fogg, 1956b; Eyster, 1958), are certainly true soil algae, while several others, originally isolated from waterlogged fields or liquid cultures inoculated with soil, probably also belong to this category. Since such algae grow well in alkaline culture media containing adequate amounts of phosphorus and molybdenum (Bortels, 1940; see review by Fogg and Wolfe, 1954), additions of lime, phosphates and possibly molybdenum to rice fields often stimulate their growth (De and Sulaiman, 1950b; De and Mandal, 1956; Hernandez, 1956; Okuda and Yamaguchi, 1956).

It is not known to what extent these algae fix nitrogen in normal soils (see Stokes, 1940b). Most Cyanophyta flourish only in fertile soils, generally rich in nitrates (e.g., Lund, 1947) which tend to depress fixation by species with this ability (De, 1939). When ammonium salts are present, they need little or no molybdenum (De, 1939; Wolfe, 1954a, b; Fogg and Wolfe, 1954). In many soils, however, Blue-green Algae only predominate after samples have been kept moist in the laboratory for a long time (John, 1942) and presumably have lost most of their available nitrogen (Lund, 1947). In nature, likewise, the surface of undisturbed soils, if rich in algae, may become so depleted of available nitrogen that only nitrogen-fixing species can grow there.

3. Organic Nutrients

Most soil algae seem to be completely autotrophic, in that they are able to manufacture all their essential metabolites (e.g., Gäumann and Jaag, 1950; Provasoli, 1958). However, many are facultative heterotrophs (mixotrophs), capable of utilizing organic compounds in the light or in the dark (Lewin, 1953; Fogg, 1953). A few are obligate phototrophs (Fogg, 1953, 1956a; Fogg and Miller, 1958; Lewin, 1950).

The classic investigations of Bristol Roach (1926, 1927a, b, 1928a, b) suggested that heterotrophic growth of algae within the soil is ecologically important. Later work indicated that in nature such growth is, at the most, slight even in the presence of substances such as glucose (Petersen, 1935; Fritsch, 1936; Stokes, 1940a; De and Sulaiman, 1950a; Tchan and Whitehouse, 1953). Though many algal cells are washed to considerable depths by rain, with or without the assistance of burrowing animals (Petersen, 1935), probably little growth occurs in the absence of light.

Fresh organic manures are inimical to most algae except certain Euglenophyta, but this may be because of the competition for organic nutrients from the vast multiplication of heterotrophic microorganisms (Stokes, 1940a; personal observations). Later, as in sewage lagoons in which mineralization has progressed, large algal crops may develop on the surface of manured soils.

IV. The Biological Environment

Since the physiological needs of soil algae and of higher plants are similar, there may be competition between them for nutrients, moisture and light (Piercy, 1917; Knapp and Lieth, 1952; Bolyshev, 1952). However, the growth of algae in the soil is usually so slight that they are unlikely to be of direct economic importance in this respect.

Stimulation or inhibition of other organisms by soil algae has occasionally been found in laboratory experiments, but its significance in nature is uncertain. Several soil algae by the liberation of thiamine stimulate the growth of fungi needing this vitamin or its components. Some algae and fungi grow together without ill effects (Gäumann and Jaag, 1950); others may not (Litvinov, 1956). The production of substances inhibiting other algae, microorganisms, and higher plants has been recorded (Flint, 1947; Bolyshev, 1952; Harder and Oppermann, 1953; Jakob, 1954a, b). Shtina (1954, 1956, 1959) found no indications of inhibition of roots by soil algae. Conversely, the roots of some plants stimulate algal growth, whereas those of conifers depress it. It has been reported that algae in water cultures may protect roots from fungal attack (Engle and McMurtrey, 1940). A relationship between the growth of rice and Blue-green Algae is well attested (see above). The algae, some of which fix nitrogen, seem in turn to benefit from the CO_2 supplied by live and decaying roots (De and Sulaiman, 1950b).

Modern work has neither proved nor disproved early accounts of a close relationship between soil algae and the nitrogen-fixing *Azotobacter* (Waksman, in Martin, 1940; Stokes, 1940b; Sulaiman, 1944; Russel, 1950; Szolnoki, 1951; Shtina, 1959). A few algae live more or less symbiotically with higher plants, being thus exceptions to the rule that there is very little heterotrophic growth within the soil. Examples are provided by species of *Nostoc* in *Gunnera* (Harder, 1917; also personal observations) and in the root nodules of *Trifolium* (Bhaskaran and Venkataraman, 1958).

V. Effects on the Physics and Chemistry of Soil

Algae are often abundant on heathy, arid, eroded, or primitive soils. They may change the physical structure of the soil, and, by helping to conserve moisture and by adding humus, colloids, and potential fertilizers, promote its colonization by higher plants. However, soil algae flourish particularly under conditions in which higher plants are sparse or absent. These matters are discussed by Treub (1888), Backer (1929), Booth (1941), Drew (1942), Fletcher and Martin (1948), Bolyshev (1952), Vogel (1955), Ponomareva (1956), and Shields *et al.* (1957).

Algae do not alter the structure or texture of normal soils significantly, though they may add appreciable amounts of organic matter (Lewin, 1956; Shtina, 1959), as has been demonstrated for takyr crusts by Bolyshev and Evdokimova (1944). In arid and saline soils, however, periodic massive growth of algae may cause considerable changes in the balance of inorganic factors such as carbonates, pH, silica, iron, and aluminum (Proshkina-Lavrenko, 1942; Bolyshev, 1952; Bazilevich and Shelyakina, 1956; Degopik, 1956; Bazilevich *et al.*, 1952; Shields *et al.*, 1957; Shtina, 1959).

References

[References covering a wide field or citing much of the literature are preceded by an asterisk.]

Baatz, I. (1939). Über das Verhalten von Bodenalgen im kurzwelligen Infrarot. *Arch. Mikrobiol.* **10,** 508–514.

Backer, C. A. (1929). "The Problem of Krakatao as Seen by a Botanist." The Hague, Netherlands.

Barkman, J. J. (1958). "Phytosociology and Ecology of Cryptogamic Epiphytes." Van Gorcum, Assen, Netherlands.

Bazilevich, N. I., Gollerbach, M. M., Litvinov, M. A., Rodin, L. E., and Shteinberg, D. M. (1952). [Concerning the rôle of biological factors in the formation of the takyrs on the route of the main Turkmensk Canal.] *Botan. Zhur.* **38,** 3–30. (In Russian.)

Bazilevich, N. I., and Shelyakina, O. A. (1956). [The movement of mineral substances in takyrs.] *In* "Takyry Zapadnoi Turkmenii Puti Sel'skokhozyaĭstvennogo Osvoeniya," pp. 483–488. Moscow. (In Russian.)

Becquerel, P. (1936). La vie latente de quelques algues et animaux inférieurs aux basses températures et la conservation de la vie dans l'univers. *Compt. rend. Acad. Sci.* **202,** 978.

Becquerel, P. (1942). Revivescence et longévité de certaines algues en vie latente dans les terres desséchés des plantes des vieux herbiers. *Compt. rend. Acad. Sci.* **214,** 986–988.

Bhaskaran, S., and Venkataraman, G. S. (1958). Occurrence of blue-green alga in the nodules of *Trifolium alexandrinum. Nature* **181,** 277–278.

Bolyshev, N. N. (1952). [The origin and evolution of the soils of takyrs.] *Pochvovedenie* pp. 403–417. (In Russian.)

Bolyshev, N. N., and Evdokimova, T. I. (1944). [The nature of takyr crusts.] *Pochvovedenie* pp. 345–352. (In Russian.)

Booth, W. E. (1941). Algae as pioneers in plant succession and their importance in erosion control. *Ecology* **22,** 38–46.

Booth, W. E. (1946). The thermal death point of certain soil inhabiting algae. *Proc. Montana Acad. Sci.* **5/6,** 21–23.

Bortels, H. (1940). Über die Bedeutung des Molybdäns für stickstoffbindende Nostocaceen. *Arch. Mikrobiol.* **11,** 155–186.

Brendemühl, I. (1949). Über die Verbreitung der Erddiatomeen. *Arch. Mikrobiol.* **14,** 407–449.

Bristol, B. M. (1919). On the retention of vitality by algae from old stored soils. *New Phytologist* **18,** 92–107.

Bristol Roach, B. M. (1926). On the relation of certain soil algae to some soluble carbon compounds. *Ann. Botany (London)* **40,** 149–201.

Bristol Roach, B. M. (1927a). On the algae of some normal English soils. *J. Agr. Sci.* **17,** 563–588.

Bristol Roach, B. M. (1927b). On the carbon nutrition of some algae isolated from soil. *Ann. Botany (London)* **41,** 509–517.

Bristol Roach, B. M. (1928a). On the influence of light and glucose on the growth of a soil algae. *Ann. Botany (London)* **42,** 317–345.

* Bristol Roach, B. M. (1928b). The present position of our knowledge of the distribution and functions of algae in the soil. *Proc. Papers Intern. Congr. Soil Sci., 1st Congr., Washington* **3,** 30–38.

Chodat, F., and Chastain, A. (1957). Recherches sur le potentiel algologique des sols. *Bull. soc. botan. France* **104,** 437–451.

Clendenning, K. A., Brown, T. E., and Eyster, H. C. (1956). Comparative studies of photosynthesis in *Nostoc muscorum* and *Chlorella pyrenoidosa*. *Can. J. Botany* **34,** 943–966.

De, P. K. (1939). The role of blue-green algae in nitrogen fixation in rice-fields. *Proc. Roy. Soc.* **B127,** 121–139.

De, P. K., and Mandal, L. N. (1956). Fixation of nitrogen by algae in rice fields. *Soil Sci.* **81,** 453–458.

De, P. K., and Sulaiman, M. (1950a). Influence of algal growth in the rice fields on the yield of crop. *Indian J. Agr. Sci.* **20,** 327–342.

De, P. K., and Sulaiman, M. (1950b). Fixation of nitrogen in rice soils by algae as influenced by crop, CO_2, and inorganic substances. *Soil Sci.* **70,** 137–151.

Degopik, I. Ya. (1956). [The hydrochemical regime of the superficial waters on takyrs.] *Doklady Acad. Nauk. S.S.S.R.* **106,** 1083–1086. (In Russian.)

Delarge, L. (1937). Recherches sur la culture d'une Schizophycée, *Phormidium uncinatum* Gom. *Mém. soc. roy. sci. Liège, Collection in 4e ser.* **2,** 3–38.

Demeter, O. (1956). Über Modifikationen bei Cyanophyceen. *Arch. Mikrobiol.* **24,** 105–133.

de Puymaly, A. (1924a). Sur le vacuome des algues vertes adaptées à la vie aérienne. *Compt. rend. Acad. Sci.* **178,** 958–960.

* de Puymaly, A. (1924b). Recherches sur les algues vertes aériennes. Bordeaux, France. (Dissertation, Paris, 1925.)

Dorogostaĭskaya, E. V. (1959). [Concerning the soil algal flora of the spotted tundras of the far north.] *Botan. Zhur.* **44,** 312–321. (In Russian.)

Drew, W. B. (1942). The revegetation of abandoned cropland in the Cedar Creek area, Boone and Callaway Counties, Missouri. *Missouri Agr. Expt. Sta. Research Bull.* **No. 344,** 1–52.

Engle, H. B., and McMurtrey, J. E. (1940). Effect of algae in relation to aeration, light and sources of phosphorus on growth of tobacco in solution cultures. *J. Agr. Research* **60,** 487–502.

Evans, J. H. (1958). The survival of freshwater algae during dry periods. I. An investigation of the algae of five small ponds. *J. Ecol.* **46,** 149–167.

Evans, J. H. (1959). The survival of freshwater algae during dry periods. II. Drying experiments. III. Stratification of algae in pond margin litter and mud. *J. Ecol.* **47,** 55–81.

Eyster, C. (1958). The micro-element nutrition of *Nostoc muscorum*. *Ohio J. Sci.* **58,** 25–33.

Fehér, D. (1936). Untersuchungen über die regionale Verbreitung der Bodenalgen. *Arch. Mikrobiol.* **7,** 439–76.

Fehér, D. (1948). Researches on the geographical distribution of soil microflora. II. The geographical distribution of soil algae. *Erdészeti Kísérletek* **48,** 57–93.

Fehér, D., and Frank, M. (1936). Untersuchungen über die Lichtökologie der Bodenalgen. *Arch. Mikrobiol.* **7,** 1–31, 490.

Fehér, D., and Frank, M. (1939). II. Der unmittelbare Beweis des autotrophen Algenwachstum beim Abschluss des sichtbaren Anteils der strahlenden Energie. *Arch. Mikrobiol.* **10,** 247–264.

Fehér, D., and Frank, M. (1940). Ergänzende Bemerkungen zu unserem Arbeiten über die Lichtökologie der Bodenalgen. *Arch. Mikrobiol.* **11,** 80–84.

Fletcher, J. E., and Martin, W. P. (1948). Some effects of algae and molds in the rain-crust of desert soils. *Ecology* **29,** 95–100.

Flint, E. A. (1958). Biological studies of some tussock grassland. IX. Algae. Preliminary observations. *New Zealand J. Agr. Research* **1,** 991–997.

Flint, L. H. (1947). Antibiotic activity in the genus *Hapalosiphon. Proc. Louisiana Acad. Sci.* **10,** 30–31.

* Fogg, G. E. (1953). "The Metabolism of Algae." Methuen, London.

* Fogg, G. E. (1956a). The comparative physiology and biochemistry of the blue-green algae. *Bacteriol. Revs.* **20,** 148–165.

Fogg, G. E. (1956b). Nitrogen fixation by photosynthetic organisms. *Ann. Rev. Plant Physiol.* **7,** 51–70.

Fogg, G. E., and Miller, J. D. A. (1958). The effect of organic substances on the growth of the freshwater alga *Monodus subterraneus. Verhandl. intern. Ver. Limnol.* **13,** 892–895.

Fogg, G. E., and Wolfe, M. (1954). The nitrogen metabolism of the blue-green algae (Myxophyceae). *In* "Autotrophic Micro-organisms" (B. A. Fry and J. L. Peel, eds.), pp. 99–125. Cambridge Univ. Press, London and New York.

* Fritsch, F. E. (1936). The rôle of the terrestrial alga in nature. *In* "Essays in Geobotany in Honor of W. A. Setchell," pp. 195–217. Univ. of California Press, Berkeley, California.

Fritsch, F. E., and Haines, F. M. (1923). The moisture-relations of terrestrial algae. II. The changes during exposure to drought and treatment with hypertonic solutions. *Ann. Botany (London)* **37,** 683–728.

Füchtbauer, W. (1957a). Trockenresistenzsteigerung nach osmotischer Adaptation bei *Saccharomyces* und *Chlorella. Arch. Mikrobiol.* **26,** 209–230.

Füchtbauer, W. (1957b). Über den Zusamennhang der Trockenresistenz einiger Einzeller mit ihrem Retentionswasser und Mineralstoffgehalt. *Arch. Mikrobiol.* **26,** 231–253.

Gäumann, E., and Jaag, O. (1950). Bodenbewohnende Algen als Wuchsstoffspender für bodenbewohnende pflanzenpathogene Pilze. *Phytopathol. Z.* **17,** 218–228.

* Geiger, R. (1950). "The Climate Near the Ground" (Engl. translation by M. N. Stewart *et al.*) Harvard Univ. Press, Cambridge, Massachusetts.

Gistl, R. (1932). Zur Kenntnis der Erdalgen. *Arch. Mikrobiol.* **3,** 634–649.

Gistl, R. (1933). Erdalgen und Düngung. *Arch. Mikrobiol.* **4,** 348–378.

Glade, E. (1914). Zur Kenntnis der Gattung *Cylindrospermum. Beitr. Biol. Pflanz.* **12,** 295–343.

* Gollerbach, M. M. (1936). [Concerning the composition and distribution of algae in soils.] *Trudy Akad. Nauk S.S.S.R. (Leningrad), Inst. Botan. Ser.* **2,** 99–301. (In Russian; French resumé.)

Harder, R. (1917). Ernährungsphysiologische Untersuchungen an Cyanophyceen, hauptsächlich dem endophytischen *Nostoc punctiforme. Z. Botan.* **9,** 145–242.

Harder, R., and Oppermann, A. (1953). Antibiotische Stoffe bei *Stichococcus bacillaris* and *Protosiphon botryoides*. *Arch. Mikrobiol.* **19,** 398–401.

Hernandez, S. C. (1956). Studies on soil fertility in Hyderabad, India. *J. Soil Sci. Soc. Philippines* **8,** 19–22.

Höfler, K. (1951). Zur Kälteresistenz einiger Hochmooralgen. *Verhandl. zool.-botan. Ges. Wien* **92,** 234–241.

Hustedt, F. (1942). Aerophile Diatomeen in der nordwestdeutschen Flora. *Ber. deut. botan. Ges.* **60,** 55–72.

Itzerott, H. (1937). Untersuchungen zum Wasserhaushalt von *Prasiola crispa*. *Jahrb. wiss. Botan.* **84,** 254–275.

Jakob, H. (1954a). Compatabilités et antagonismes entre algues du sol. *Compt. rend. Acad. Sci.* **238,** 928–930.

Jakob, H. (1954b). Sur les propriétés antibiotiques énergiques d'une algue du sol: *Nostoc muscorum* Ag. *Compt. rend. Acad. Sci.* **238,** 2018–2020.

John, R. P. (1942). An ecological and taxonomic study of the algae of British soils. I. Distribution of the surface-growing algae. *Ann. Botany (London)* **6,** 323–349.

Kärcher, H. (1931). Über die Kälteresistenz einiger Pilze und Algen. *Planta* **14,** 515–517.

Knapp, R., and Lieth, H. (1952). Über Ursachen des verstärkten Auftretens von erdbewohnenden Cyanophyceen. *Arch. Mikrobiol.* **17,** 292–299.

Knebel, G. (1936). Monographie der Algenreihe der Prasiolales, inbesondere von *Prasiola crispa*. *Hedwigia* **75,** 1–120.

Lewin, J. C. (1950). Obligate autotrophy in *Chlamydomonas Moewusii* Gerloff. *Science* **112,** 652–653.

Lewin, J. C. (1953). Heterotrophy in diatoms. *J. Gen. Microbiol.* **9,** 305–313.

Lewin, J. C. (1957). Silicon metabolism in diatoms. IV. Growth and frustule formation in *Navicula pelliculosa*. *Can. J. Microbiol.* **3,** 427–433.

Lewin, R. A. (1956). Extracellular polysaccharides of green algae. *Can. J. Microbiol.* **2,** 665–672.

Lipman, C. B. (1941). The successful revival of *Nostoc commune* from a herbarium specimen eighty-seven years old. *Bull. Torrey Botan. Club* **68,** 664–666.

Litvinov, M. A. (1956). [Biocoenoses of soil microscopic fungi of takyrs.] *In* "Takyry zapadnoi Turkmenii Puti sel'skokhozyaĭstvennogo Osvoeniya." pp. 55–74. Moscow (In Russian.)

* Lund, J. W. G. (1945–1946). Observations on soil algae. I. The ecology, size and taxonomy of British soil diatoms. Part I. *New Phytologist* **44,** 196–219; *ibid.* Part II, **45,** 56–110.

* Lund, J. W. G. (1947). Observations on soil algae. II. Notes on groups other than diatoms. *New Phytologist* **46,** 35–60.

Martin, T. L. (1940). The occurrence of algae in some arid soils of Utah. *Intern. Congr. Microbiol., 3rd Congr., New York Rept. Proc.* pp. 697–698.

Miller, J. D. A., and Fogg, G. E. (1957). Studies on the growth of Xanthophyceae in pure culture. I. The mineral nutrition of *Monodus subterraneus* Peterson. *Arch. Mikrobiol.* **28,** 1–17.

Okuda, A., and Yamaguchi, M. (1956). Distribution of nitrogen-fixing microorganisms in paddy soils in Japan. *Rept. Intern. Congr. Soil Sci., 6th Congr., Paris* pp. 521–526.

Petersen, J. B. (1928). The aerial algae of Iceland. *In* "The Botany of Iceland" Vol. II, pp. 327–447 Copenhagen.

* Petersen, J. B. (1935). Studies on the biology and taxonomy of soil algae. *Dansk Botan. Arkiv; Res. Botan.* **8,** 1–183.

Petit, M. P. (1877). La désiccation fait-elle périr les diatomées? *Bull. soc. botan. France* **24,** 367–369.

Piercy, A. (1917). The structure and mode of life of a form of *Hormidium flaccidum*. *Ann. Botany* (*London*) **31,** 513–537.

Ponomareva, V. V. (1956). [The humus of takyrs.] *In* "Takyry zapadnoi Turkmenii Puti sel'skokhozyaĭstvennogo Osvoeniya," pp. 411–438. Moscow. (In Russian.)

Prát, S. (1955). The influence of humus substances (capucines) on algae. *Folia Biol.* **1,** 321–326.

Pratt, R. (1941). Studies on *Chlorella vulgaris*. IV. Influence of the molecular proportions of KNO_3, KH_2PO_4 and $MgSO_4$ in the nutrient solution on the growth of *Chlorella*. *Am. J. Botany* **28,** 492–497.

Pringsheim, E. G., and Pringsheim, O. (1956). Kleine Mitteilungen über Flagellaten und Algen. IV. *Porphyridium cruentum* und *Porphyridium marinum*. *Arch. Mikrobiol.* **24,** 169–173.

Proshkina-Lavrenko, A. I. (1942). [The diatomaceous algae of depressions in relation to the origin of solods.] *Pochvovedenie,* pp. 38–43. (In Russian.)

* Provasoli, L. (1958). Nutrition and ecology of protozoa and algae. *Ann. Rev. Microbiol.* **12,** 279–308.

* Russel, E. J. (1950). "Soil Conditions and Plant Growth." (8th ed., revised by E. W. Russell). Longmans, Green, New York.

* Sauberer, F., and Härtel, O. (1959). Pflanze und Strahlung. *In* "Probleme der Bioklimatologie," Vol. V. Leipzig.

Sdobnikova, N. V. (1958). [On the algal flora of the takyrs in the northern part of the Turansk Plain.] *Botan. Zhur.* **43,** 1675–1681. (In Russian.)

Shelhorn, M. (1936). Zur Ökologie und Biologie der Erdalgen. *Naturw. Landwirtsch.* **18,** 1–54.

Shields, L. M., Mitchell, C., and Drouet, F. (1957). Alga- and lichen-stabilized surface crusts as soil nitrogen sources. *Am. J. Botany* **44,** 489–498.

Shtina, E. A. (1954). [The effect of agricultural plants on the algal flora of soil.] *Trudy Kirovskii sel'skokhoz. Inst.* **10,** 59–69. (In Russian.)

Shtina, E. A. (1956). [On the interrelation of soil algae with higher plants.] *Vestnik Moskov. Univ. 6 Ser. Fiz. Mat. i Estestven. Nauk.* **No. 1,** 93–98. (In Russian.)

* Shtina, E. A. (1959). Algae solorum caespitoso- podzolensium regionis Kirovskensis. *Trudy Botan. Inst. im. V. L. Komarova Akad. Nauk S.S.S.R., Ser. II, Sporovye Rasteniya* **12,** 36–141. (In Russian.)

Smith, F. B. (1944). Occurrence and distribution of algae in soils. *Proc. Florida Acad. Sci.* **7,** 44–49.

* Stocker, O. (1956). Wasseraufnahme und Wasserspeicherung bei Thallophyten. *In* "Handbuch der Pflanzenphysiologie" (W. Ruhland, ed.), Vol. III, pp. 160–172. Springer, Berlin.

Stokes, J. L. (1940a). The influence of environmental factors upon the development of algae and other microorganisms in the soil. *Soil Sci.* **49,** 171–184.

Stokes, J. L. (1940b). The rôle of algae in the nitrogen cycle of the soil. *Soil Sci.* **49,** 265–275.

Sulaiman, M. (1944). Effect of algal growth on the activity of *Azotobacter* in rice fields. *Indian J. Agr. Sci.* **14,** 277–282.

Szolnoki, J. (1951). [Data on the nitrogen fixation of *Azotobacter oxyidans*.] *Ann. Inst. Biol.* (*Tihany*) *Hung. Acad. Sci.* **20,** 245–247. (In Hungarian; Russian and English resumés.)

Tchan, Y. T. (1959). Study of soil algae. III. Bioassay of soil fertility by algae. *Plant and Soil* **10,** 220–231.

Tchan, Y. T., and Whitehouse, J. A. (1953). Study of soil algae. II. The variation of algal population in sandy soils. *Proc. Linnean Soc. N. S. Wales* **78,** 160–170.

Treub, M. (1888). Notice sur la nouvelle flore de Krakatau. *Ann. Jard. botan. Buitenzorg.* **7,** 213–223.

Vischer, W. (1945). Heterokonten aus alpinen Böden, speziell dem schweizerischen Nationalpark. *Ergeb. Wiss. Untersuch. Schweiz Natl. Parks* **1,** 481–511.

Vogel, S. (1955). Niedere 'Fensterplanzen' in der südafrikanischer Wüste. Eine ökologische Schilderung. *Beitr. Biol. Pflanz.* **31,** 45–135.

von Denffer, D. (1949). Die planktische Massenkultur pennater Grunddiatomeen. *Arch. Mikrobiol.* **14,** 149–202.

Wolfe, M. (1954a). The effect of molybdenum upon the nitrogen metabolism of *Anabaena cylindrica.* I. A study of the molybdenum requirement for nitrogen fixation and for nitrate and ammonia assimilation. *Ann. Botany (London)* **8,** 299–308.

Wolfe, M. (1954b). The effect of molybdenum upon the nitrogen metabolism of *Anabaena cylindrica.* II. A more detailed study of the action of molybdenum in nitrate assimilation. *Ann. Botany (London)* **18,** 309–325.

* Zauer, L. M. (1956). [To the knowledge of the algae of the plant associations of the Leningrad district.] *Trudy Botan. Inst. im. V. L. Komarova Akad. Nauk S.S.S.R., Ser. II, Sporovye Rasteniya* **10,** 33–174. (In Russian.)

—52—

Marine Plankton

CHARLES S. YENTSCH[1]

Woods Hole Oceanographic Institution, Woods Hole, Massachusetts

I. Introduction

The growth of interest in the physiology of marine phytoplankton has largely resulted from a closer examination of the environmental factors influencing plant metabolism in the sea. Much of our present knowledge comes from indirect sources, such as laboratory experiments which have

[1] Contribution No. 925 from the Woods Hole Oceanographic Institution. This work was in part supported by contract AT-(30-1)-1918 with the U.S. Atomic Energy Commission. The author expresses sincere thanks to Dr. B. H. Ketchum for helpful criticisms.

utilized marine cultures, or studies of general plankton ecology, in which primary production is emphasized.

Under some circumstances the environmental physiologist can conduct experiments at sea using natural populations, but the investigator is then usually faced with the problem of having to work with low concentrations of plant material and with physiological techniques that may be too insensitive for adequate experimentation.

To answer questions stimulated by field observations, some of the best results have been obtained by combining observational and experimental approaches. To some degree this approach is limited, for it is not easy to duplicate, in a controlled experiment, such characteristic factors of the ocean environment as light, temperature, nutrient composition and concentration, and turbulence.

II. Illumination in the Euphotic Zone

In the oceans only the upper layer of 100 meters or less is illuminated with sufficient intensity to permit photosynthesis. Radiation within this zone varies diurnally, seasonally, and geographically.

Light intensity in the euphotic layer decreases logarithmically with depth. In temperate zones, at the level where the light intensity is reduced to approximately 1% of that at the surface, photosynthesis is on the average just sufficient to satisfy respiratory needs. This is termed the *compensation depth*, and the light intensity here is termed the *compensation intensity*.

The vertical extent of the euphotic zone chiefly depends on: (i) the radiation coming from the sun and sky; (ii) reflection and back-scattering from the sea surface; and (iii) attenuation throughout the water column.

A. Penetration of Light

Pure water is more transparent to light of the visible region of the spectrum than to ultraviolet or infrared radiation, which is more rapidly attenuated as it passes through the water (see Holmes, 1957). The data of James and Birge (see Hutchinson, 1957) show that maximum transmission is between 470 and 490 mμ, and that transparency decreases sharply toward the longer wavelengths and gradually toward the shorter. According to Kalle (1938), the selective attenuation of short wavelengths in sea water is due to dissolved yellow substances. However, Yentsch (1960) observed that the blue absorption band of phytoplankton pigments in natural concentrations is of sufficient magnitude to reduce the transmission of blue light. Dense concentrations of phytoplankton organisms possessing only chlorophylls and carotenoid pigments may color the water from green to yellow. Red water is caused by plants having, in addition, orange carotenoids or phycobilins which shift the wavelengths of maximum transmission into the red end of the spectrum.

Of possible importance to many biological processes in surface waters

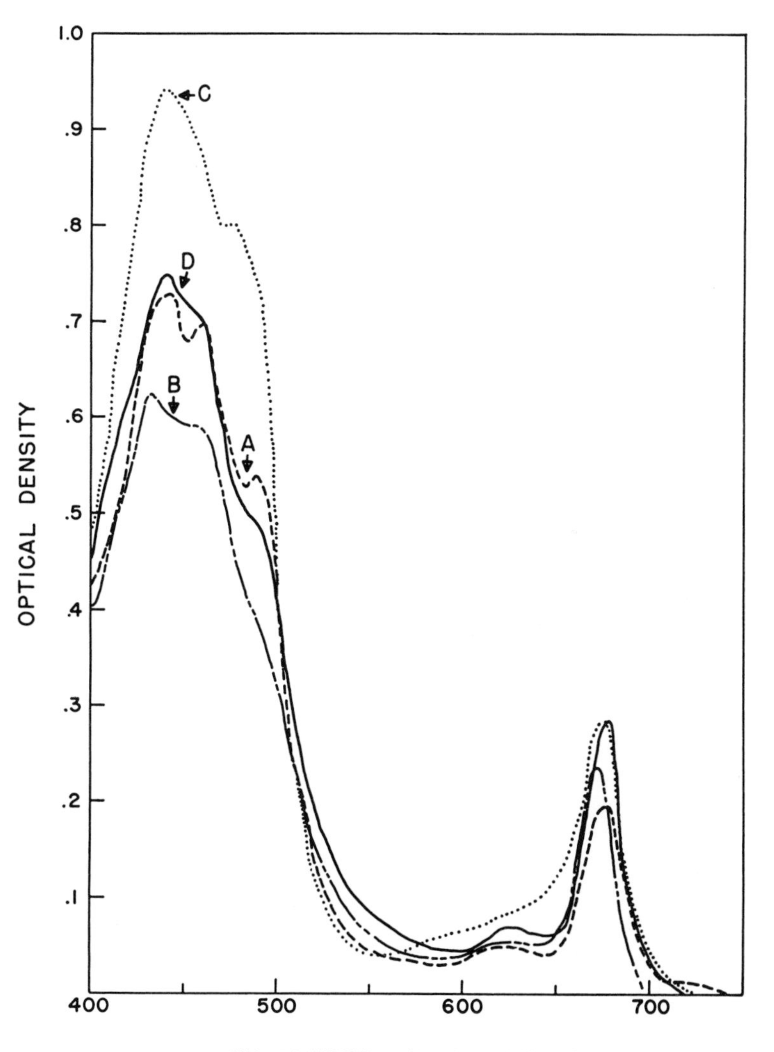

Fig. 1. Absorption spectra of extracts of plankton algae: (A) a diatom, *Cyclotella sp.*; (B) a dinoflagellate, *Amphidinium sp.*; (C) a green flagellate, *Chlamydomonas*; (D) a natural population sampled from Woods Hole waters. (From Yentsch, 1960.)

is the penetration of near-ultraviolet radiation. These wavelengths are severely attenuated in coastal waters; nevertheless, in the open waters of tropical oceans Jerlov (1951) observed that only 14% of the radiation at wavelengths of about 310 mμ is absorbed in the uppermost meter of water. Many workers have assigned biologically inhibitory roles to these short wavelengths (e.g., Steemann Nielsen, 1952). Midday inhibition of photosynthesis has been suggested to be the result of ultraviolet radiation, although, as will be shown below, inhibition of photosynthesis can occur when intensities of visible light alone are high. Moreover, McLeod (1958) observed active photosynthesis, in excess of respiration, in cultures of marine diatoms illuminated only by radiation at 310 mμ.

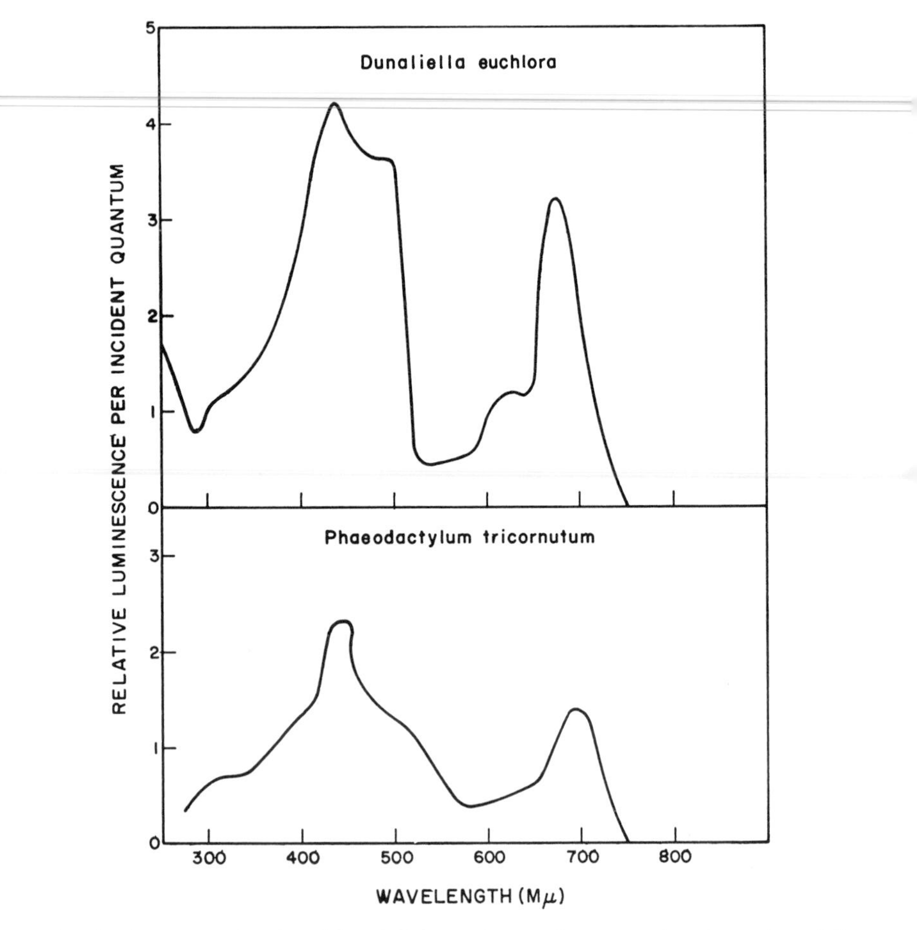

FIG. 2. Action spectra for delayed light emission of a green flagellate, *Dunaliella euchlora*, and of a diatom, *Phaeodactylum tricornutum*. (From McLeod, 1958.)

B. Absorption and Utilization of Light by Phytoplankton

The ability of the plant cell to absorb light is dependent upon its pigments. Our knowledge of the pigments of phytoplankton has been obtained mainly from studies of the large seaweeds and of cultured microscopic algae. In most phytoplankton populations, light absorption by chlorophylls and carotenoids greatly exceeds any effect of phycobilins because of the usual dominance of diatoms and dinoflagellates. Figure 1 shows an *in vivo* absorption spectrum of a natural phytoplankton population taken in waters off Woods Hole, compared with absorption curves for a diatom, a dinoflagellate, and green flagellate. These curves all exhibit a chlorophyll absorption maximum in the red region and a second peak, resulting from the combined absorptions of chlorophylls and carotenoids, in the blue region. The additional absorption by chlorophyll *b* is clearly visible in the region of 630–645 mμ in the case of the green flagellate, but is not apparent in the natural population.

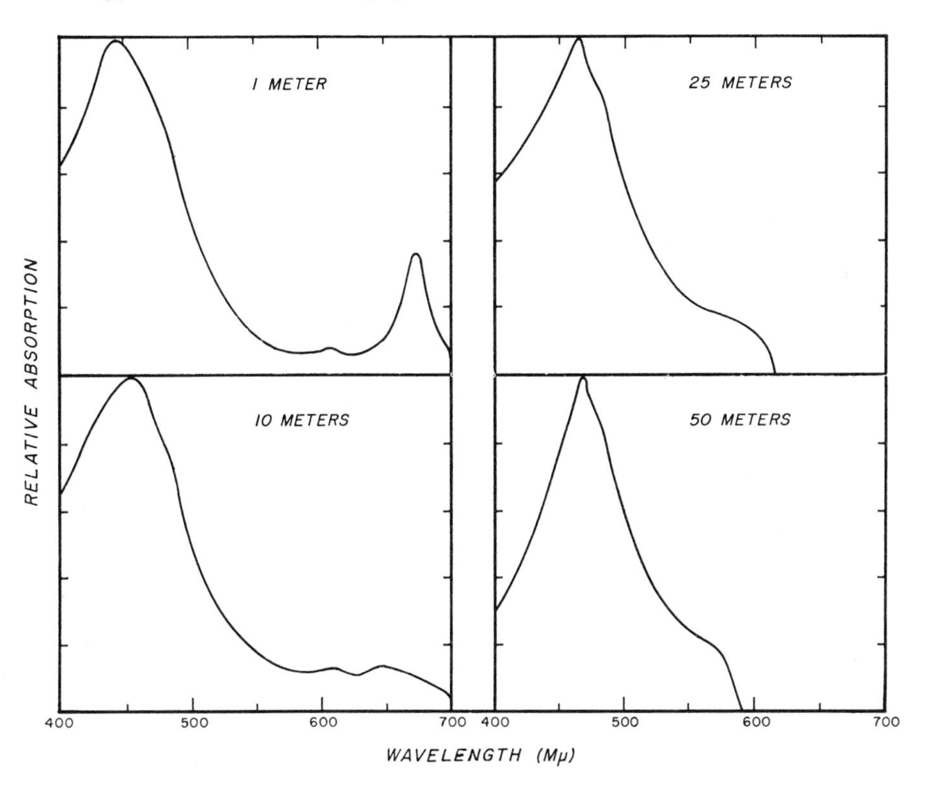

FIG. 3. Relative absorption by phytoplankton of light available at different depths in clear water.

Action spectra of photosynthesis have yet to be measured in natural populations. McLeod (1958) showed that the action spectra of cultures of two marine phytoplanktonic organisms (Fig. 2) resemble closely an *in vivo* absorption spectrum of natural phytoplankton. McLeod's data, like those of previous workers, show active participation of the carotenoids as well as of the chlorophylls in photosynthesis.

It used to be thought that the pigment composition of marine phytoplankton was comparatively unsuitable for the efficient utilization of the prevailing bluish-green submarine light. But from measurements by Jerlov (1951) it is now apparent that clear ocean water transmits more blue light than was previously supposed. More accurate measurements of the *in vivo* absorption spectra of phytoplankton show that its pigment composition is in fact well suited for absorption of the wavelengths of maximum transmission in clear sea water. From data on the penetration of light and its absorption by the pigments of phytoplankton, action spectra at different depths can be estimated (Fig. 3). The red peak of chlorophyll cannot be active in photosynthesis below 10 meters, while the blue light absorbed by chlorophyll is active only to 50 meters. Below 50 meters carotenoids are the principal absorbers. Since the wavelengths of maximum penetration in the oceans fall within the range of carotenoid absorption, cells with carotenoids can absorb available light at any depth; the utilization of this light in photosynthesis depends, of course, on its intensity.

III. Effect of Light Intensity on Photosynthesis

A. Photosynthesis as a Function of Light and Depth

From experiments with cultures of different algae isolated from Woods Hole plankton, Ryther (1956) observed that photosynthesis was light-

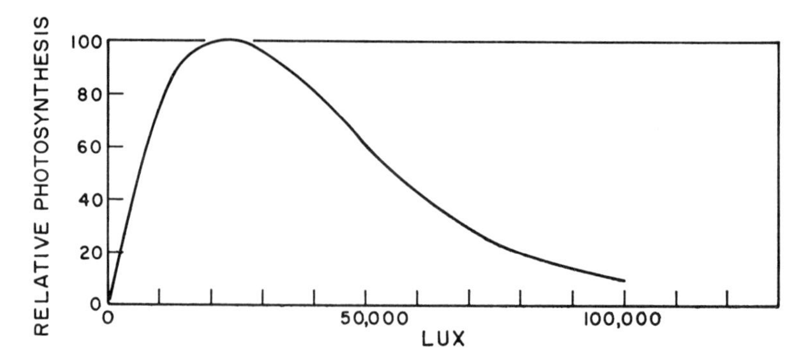

FIG. 4. Relative net photosynthesis (C^{14} uptake) of cultured marine phytoplankton at different light intensities. (From Ryther, 1956.)

saturated at intensities of 5000 to 7500 lux for green algae, 10,000 to 20,000 lux for diatoms, and 25,000 to 30,000 lux for dinoflagellates. Inhibition of photosynthesis was noted in all groups at intensities exceeding the saturation values by 10,000 lux or more. Ryther concluded that the "typical" photosynthesis:light curve for natural populations consisting mostly of diatoms should be of the form shown in Fig. 4. Similar curves had previously been obtained by studies using cultures of diatoms suspended at different depths throughout the euphotic zone (Jenkin, 1937) and in the open ocean (Steemann Nielsen and Aabye Jensen, 1957).

B. Adaptation

It turns out, however, that Ryther's "typical" curve for natural populations does not apply to all environmental conditions in the sea. Adaptation to ambient light intensities, either high or low, modifies the photosynthesis: light ratios (Steemann Nielsen and Hansen, 1959a; Ryther and Menzel, 1959). If, for example, the surface of the euphotic zone is not actively stirred by the wind, the phytoplankton in various layers is exposed to

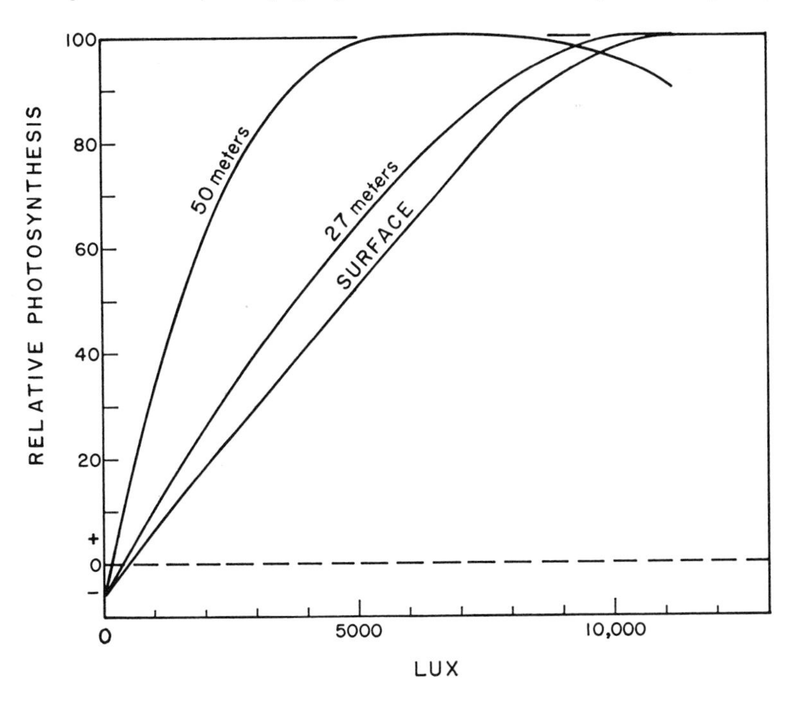

FIG. 5. Relative photosynthesis at different light intensities. Arctic summer phytoplankton from various depths, as indicated. (From Steemann Nielsen and Hansen, 1959a.)

various constantly high or low light levels for a period, and adaptation to ambient light intensities occurs. This results in cells near the surface becoming so-called "sun" plants (Fig. 5), with a high rate of photosynthesis at high light intensities, while phytoplankton in the lower layers of the euphotic zone develop "shade" characteristics, whereby maximum photosynthesis occurs at relatively lower light intensities.

IV. Photosynthesis:Respiration (PS:R) Ratios

A. Constant PS:R Ratios

Whereas measurements of the reduction of $C^{14}O_2$ *in situ* have provided a means for the direct determination of photosynthesis by phytoplankton, no comparable method has been developed for the direct measurement of respiration. Some of the discrepancies observed between different methods of measuring net photosynthesis depend upon differences in the observed rates of respiration.

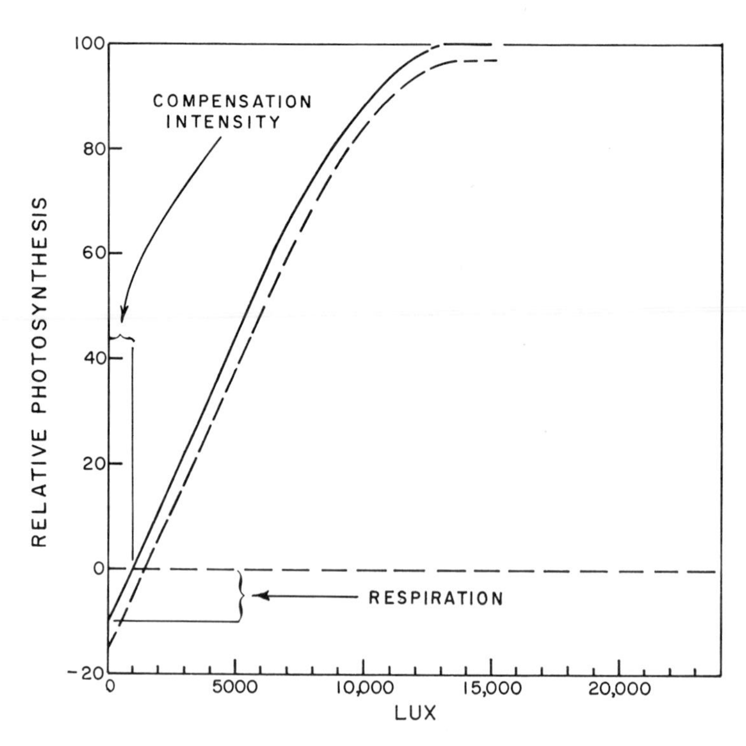

FIG. 6. Relative net photosynthesis at different light intensities. Surface phytoplankton, Dana station 10561. KEY: ——, observed values; ----, corrected for short-term measurements. (From Steemann Nielsen and Hansen, 1959b.)

In introducing the C^{14} technique, Steemann Nielsen assumed that the maximum rate of photosynthesis in natural populations was about 10 times that of respiration. Over a 24-hour period the respiration by phytoplankton throughout the euphotic zone would be about 40% of the total photosynthesis. However, Ryther's (1954b) experiments with the marine flagellate *Dunaliella* showed that, with the onset of nutrient deficiency, photosynthesis declined more rapidly than respiration, and that when cells had ceased dividing, the two processes were carried out at approximately equal rates. Ryther further demonstrated that at the compensation intensity, determined by measurements of dissolved oxygen in bottles exposed to light and darkness, uptake of C^{14} could not be detected. He therefore concluded that the C^{14} method directly measured net photosynthesis, and that there was no convenient means for estimating respiration on a daily basis.

Steemann Nielsen (1955) and Steemann Nielsenn and Al Kholy (1956) claimed that bactericidal substances produced by algae in the light bottles, together with the generally bad experimental techniques employed by Ryther *et al.*, were responsible for the discrepancies between the oxygen determinations of the latter and Steeman Nielsen's radiocarbon measurements of photosynthesis. However, the production of bactericidal sub-

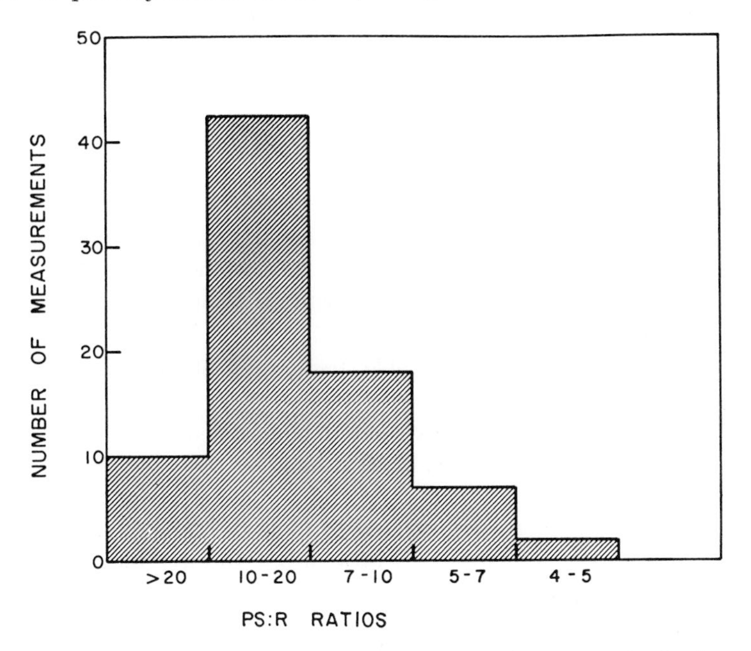

FIG. 7. Distribution of values for the ratio of photosynthesis to respiration in samples of marine phytoplankton. (From Steemann Nielsen and Hansen, 1959b.)

stances by marine phytoplankton has not been clearly demonstrated as a general phenomenon; on the contrary, it appears that bacteria and healthy phytoplankton usually coexist cooperatively. McLachlan and Yentsch (1959) observed no obvious deleterious effects of *Dunaliella euchlora* on the growth of associated bacteria. S. W. Watson (personal communication), who surveyed more than 100 different cultures of phytoplanktonic organisms for antibacterial activity against *Escherichia coli*, found only slight antibiosis in extremely few cases, and mostly no antibiosis at all.

To substantiate their original claims of a constant PS:R ratio of about 10 for light-saturated phytoplankton, Steemann Nielsen and Hansen (1959b) invoked a straight-line relationship between short-term C^{14} observations at different light intensities (Fig. 6), assuming the difference between zero and the intercept on the y-axis to represent the organic carbon oxidized on respiration in darkness. In most of their observations made on the Dana expedition (Fig. 7), respiration was between 6 and 10% of maximum photosynthesis. The reason given by Steemann Nielsen and Hansen for the constant PS:R ratio was that populations of nutrient-deficient cells capable of only slow reproduction tend to disappear from the euphotic zone, being rapidly removed by sinking or grazing. They concluded that in long-term experiments the C^{14} technique provides a measure of net photosynthesis, but that in short-term experiments some value between net and gross photosynthesis is obtained.

B. Variable PS:R Ratios

It is nevertheless unlikely that the value of the PS:R ratio could be constant for all phytoplankton populations. Differences in the availability of nutrients certainly account for the relative abundance of phytoplankton throughout the oceans, and undoubtedly cells in the oligotrophic surface waters must at times be deficient in nutrients and thereby reduced in their capacity for photosynthesis (see below).

Ryther (1956) discussed possible factors affecting the PS:R ratio throughout the euphotic zone during the extremes of seasonal radiation. He considered a population distributed homogeneously throughout the euphotic zone with a PS:R ratio of 10 at light saturation. This population, when exposed to low radiation at a depth of 38 meters, would have a daily PS:R ratio of 0.2, whereas if exposed to high radiation at a depth of 95 meters it would have a daily PS:R ratio of 2.6. In phytoplankton populations of an environment comparable to the waters off Newport, Rhode Island (Fig. 8), a PS:R ratio of 10 to 15 would be expected, to account for the low net productivity during winter months. If the PS:R ratio were

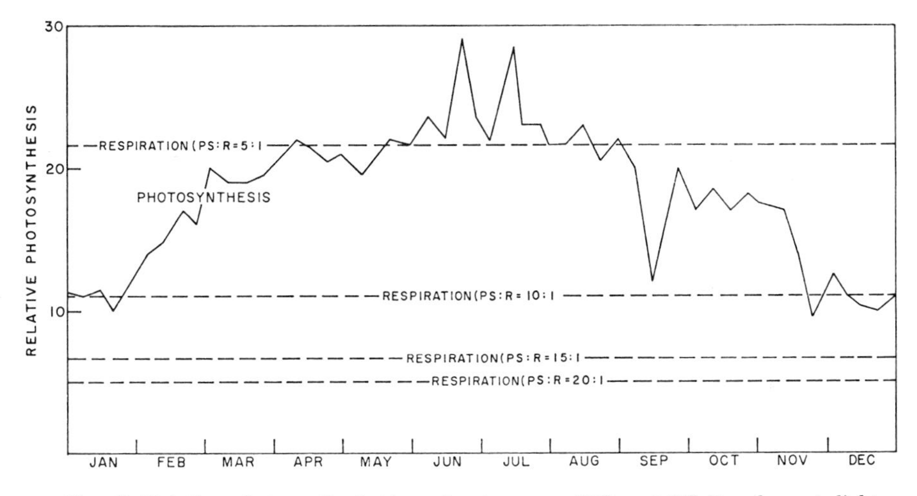

FIG. 8. Relative photosynthesis throughout a year. Different PS:R values at light saturation indicated, for phytoplankton from Newport, Rhode Island. (From Ryther, 1956.)

reduced to 5, significant net productivity would be apparent only during June and July when there is most light.

V. Photosynthesis-Pigment Relationships in Natural Populations

A. Assimilation in Natural Populations

Using measurements of net oxygen evolution, Ryther and Yentsch (1957) measured the ratios of total photosynthesis to chlorophyll *a* at light saturation in natural phytoplankton populations collected from Woods Hole waters and in cultures of marine algae. Although their data, like others reported in the literature, varied considerably, it was calculated that in natural populations 3.7 grams of carbon per hour is assimilated per gram of phytoplankton chlorophyll. Further investigation of populations of phytoplankton on the continental shelf off New York (Ryther and Yentsch, 1958) revealed that, despite wide variation in the chlorophyll content, this parameter of chlorophyll activity showed no obvious correlation with any of the environmental conditions examined (e.g., temperature or the availability of nutrients). Such a correlation may have existed, though obscured by experimental errors. As shown in Fig. 9, most of the values obtained by the C^{14} method were below 3.7, whereas values obtained by the oxygen method ranged between 1.0 and 4.0, and were evenly distributed about the mean. The values obtained by the C^{14} method support

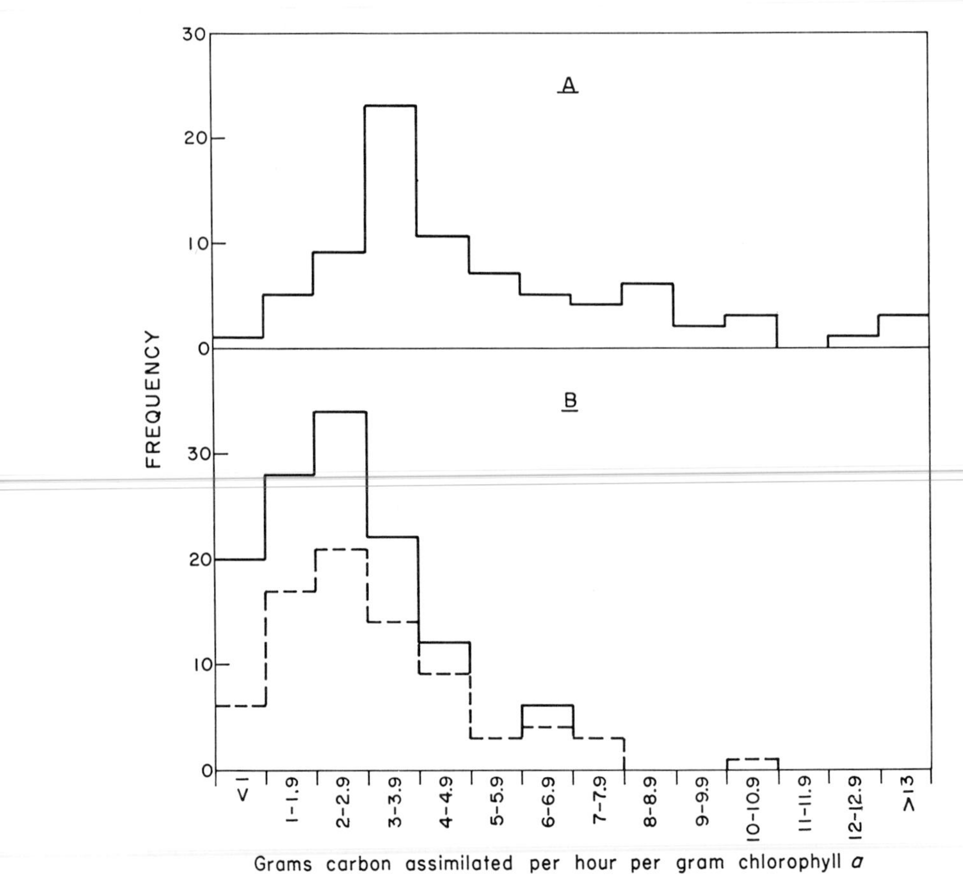

FIG. 9. Frequency distribution of chlorophyll efficiency values (grams of carbon assimilated per hour per gram of chlorophyll *a* at light saturation) measured in phytoplankton samples from a depth of 10 meters in the Atlantic Ocean off New York: (A) oxygen method; (B) C^{14} method; broken line represents values taken simultaneously with those shown in (A). (From Ryther and Yentsch, 1958.)

Ryther's contention that in light-saturated natural populations the PS:R ratio may be quite variable.

B. Factors Affecting Chlorophyll Activity

Steemann Nielsen and Hansen (1959a) presented evidence that in "shade" phytoplankton from 60° N. latitude, the chlorophyll-activity parameter did not show much variation at different light intensities. Populations at higher latitudes, are of course, exposed to much lower temperatures, and Steemann Nielsen and his colleagues attributed the maintenance of high rates of photosynthesis in relation to chlorophyll to a greater concentration of photosynthetic enzymes in cells of northern populations. However, although they assumed that the chlorophyll content per unit of organic matter is the same at different latitudes, no data exist to substantiate this.

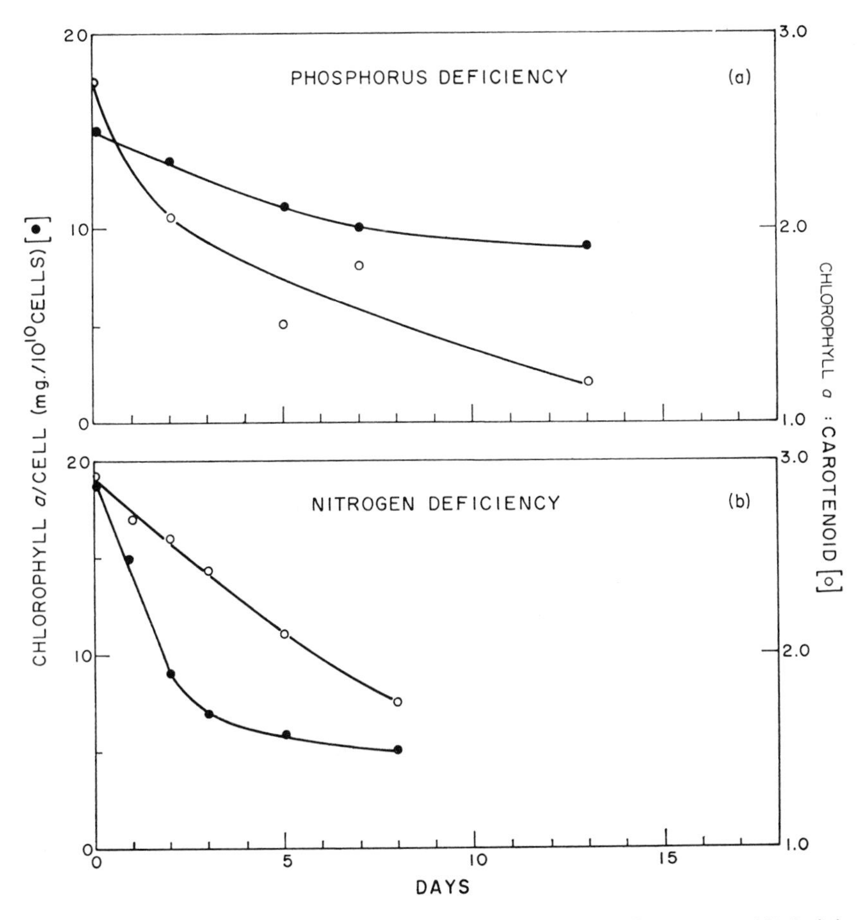

FIG. 10a. Chlorophyll *a* per cell (●) and ratio of chlorophyll *a*:carotenoid[3] (○) in cultures of *Dunaliella euchlora* grown with limiting phosphorus. (From Ketchum *et al.*, 1958a.)

FIG. 10b. Chlorophyll *a* per cell (●) and ratio of chlorophyll *a*:carotenoid[3] (○) in cultures of *Dunaliella euchlora* grown with limiting nitrogen. (From Ketchum *et al.*, 1958a.)

[3] Note that this ratio is the reciprocal of that discussed in the text of this chapter, where the ratios are given as carotenoid:chlorophyll *a*.

English (1959), who examined the relation between photosynthesis and chlorophyll in arctic waters by the C^{14} method, arrived at a value of 0.6 for the parameter of chlorophyll activity, which is considerably lower than the high 3.7 value obtained by Ryther and Yentsch, using light and dark bottles, for the waters off New York. English attributed this difference to the obvious effects of the low temperature (−1.5°C.) of arctic waters on

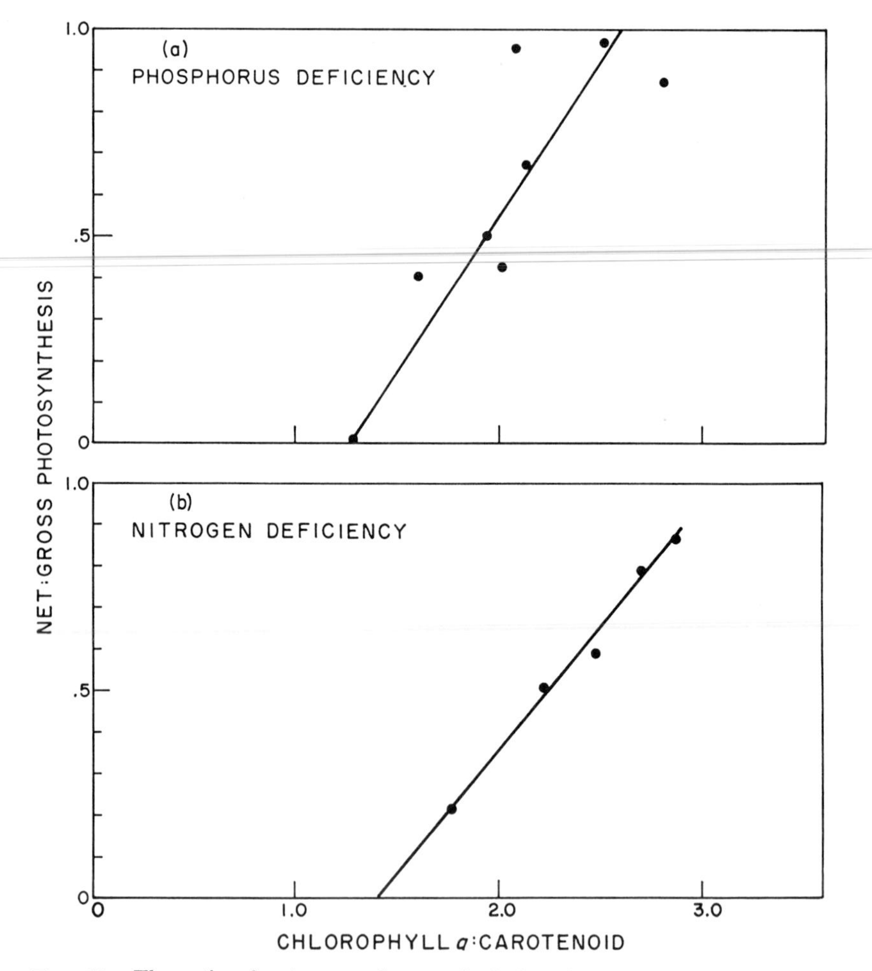

FIG. 11a. The ratio of net:gross photosynthesis in relation to the chlorophyll *a*: carotenoid[3] ratio in cultures of *Dunaliella euchlora* grown with limiting phosphorus. (From Ketchum *et al.*, 1958a.)

FIG. 11b. The ratio of net:gross photosynthesis in relation to the chlorophyll *a*: carotenoid[4] ratio in cultures of *Dunaliella euchlora* grown with limiting nitrogen. (From Ketchum *et al.*, 1958a.)

[4] See footnote 3 on p. 783.

the "dark" (i.e., chemical) reactions of photosynthesis. On the basis of C^{14} determinations, the chlorophyll-activity parameter for New York waters is lower, about 1.5 (Ryther and Yentsch, 1958), but still considerably above the value of 0.6 obtained by English.

Typical light adaptation with respect to depth is found only when the euphotic zone is thermally stabilized and vertical mixing thereby reduced. In surface phytoplankton with "sun" characteristics (Steemann Nielsen and Hansen, 1959a), one finds a high rate of photosynthesis per unit chlorophyll and a relatively high value for the light intensity for compensation (600 to 1200 lux). Phytoplankton near the bottom of the euphotic zone exhibits "shade" characteristics with a lower rate of photosynthesis per unit of chlorophyll and a lower compensation intensity (200 to 450 lux).

Cultures of the marine flagellate *Dunaliella euchlora* exhibit a greatly reduced capacity for net photosynthesis in media containing limiting concentrations of nitrogen and phosphorus (Ketchum *et al.*, 1958a). Symptomatic of the deficiency are changes in the pigment composition (Figs. 10a and 10b), notably a decline in the chlorophyll content and a consequent decrease in the ratio of chlorophyll to carotenoid. These changes can be directly correlated with a decrease in photosynthetic capacity (Figs. 11a and 11b).

It is frequently claimed that the accuracy of quantitative determinations of the pigments of natural plankton, as an indication of photosynthetic capacity, may be seriously reduced by the presence of dead cellular debris containing chlorophyll (Krey *et al.*, 1957; Banse, 1957). Such debris might account for the large variations observed in chlorophyll-activity parameters and perhaps also in chlorophyll:protein ratios. However, as indicated above, there is no reason to expect that such parameters should be constant in all natural populations. Similarly, measurements of organic nitrogen and chlorophyll in marine algae cultured under various nutrient conditions showed no constancy in the chlorophyll:nitrogen ratio (Yentsch and Vaccaro, 1958).

It is our general experience with land plants and cultures of planktonic algae that pigments tend to be stabilized when associated with protein within the cells. On the other hand, pigments in solutions, collodial smears, and cellular fragments tend to break down rapidly in the presence of light and oxygen. The euphotic zone is therefore an unsuitable environment for the preservation of undegraded pigments. Absorption curves of acetone extracts of particulate matter from deep-water samples (Fig. 12) show a disappearance of the red peak of chlorophyll *a* below 300 meters. The remaining ether-soluble pigments comprise mainly carotenoids and phaeophytins (Yentsch and Ryther, 1959). Whether these pigments are contained in viable cells or in fragments of plant debris was not determined, though

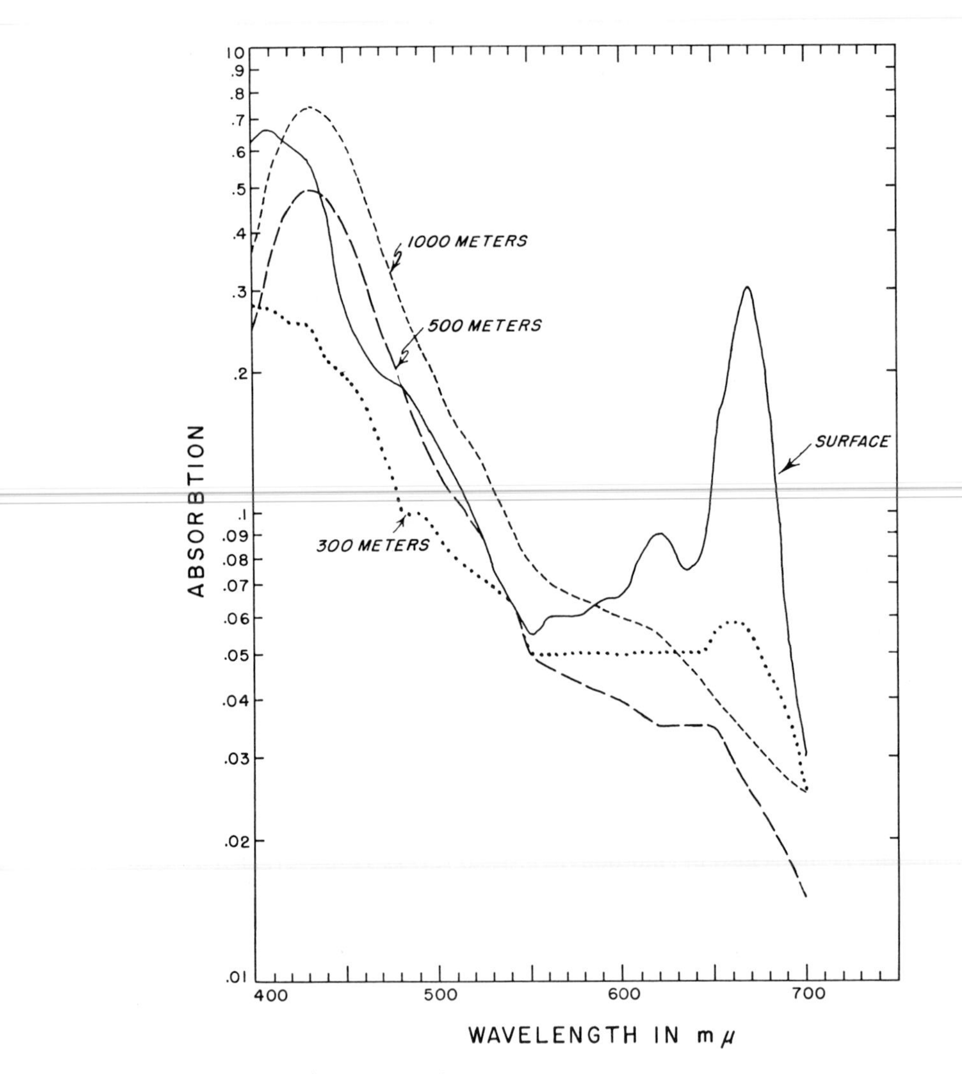

FIG. 12. Absorption spectra of acetone extracts of particulate matter from various depths in the Sargasso Sea. (From Yentsch and Ryther, 1959.)

the presence of appreciable quantities of pheophytin provides evidence that cell disintegration has occurred.

C. Diurnal Fluctuations

The photosynthetic capacity of phytoplankton at light saturation varies diurnally. Doty and Oguri (1957) found that populations sampled from sur-

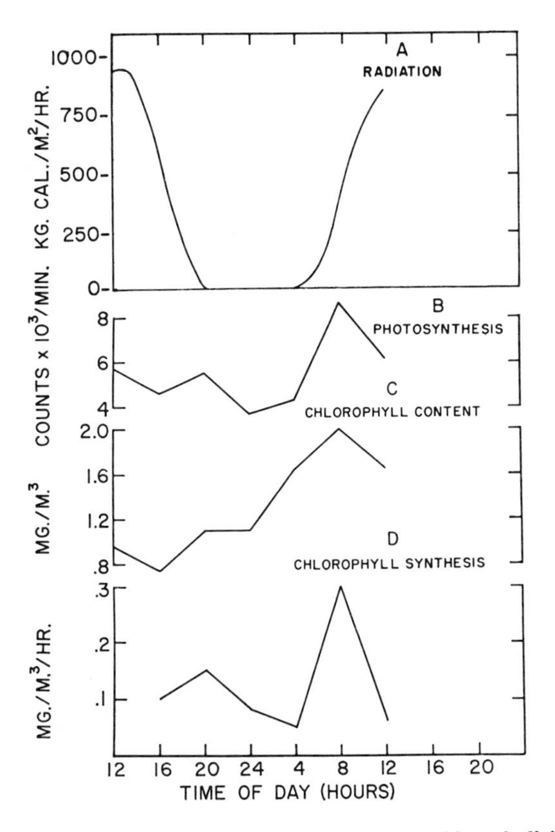

FIG. 13. Diurnal variation of light, photosynthesis and chlorophyll in Woods Hole waters: (A) intensity of solar radiation at the surface (kg. calories/m.²/hour); (B) relative photosynthesis at 20,000 lux, as measured by 15-minute C^{14} experiments (counts $\times$ 10^3/minute); (C) chlorophyll *a* content (mg./m.³); (D) chlorophyll *a* synthesis at 20,000 lux during preceding 4-hour period (mg./m.³/hour). (From Yentsch and Ryther, 1957.)

face waters throughout the day and illuminated by light at a saturating intensity exhibit maximum photosynthetic activity at 0800 hours, at a rate approximately 7 times as great as that measured under similar conditions at midnight. Yentsch and Ryther (1957) noted a similar rhythm in phytoplankton of Woods Hole waters, although in their experiments the ratio of maximum to minimum photosynthetic capacities was only 2 (Fig. 13).

Yentsch and Ryther (1957) and Shimada (1958) further observed that these fluctuations of the photosynthetic capacity at light saturation could be directly correlated with *in situ* changes in the chlorophyll *a* content of the phytoplankton per unit volume of water. Yentsch and Ryther postulated that a diurnal photosynthetic rhythm might result from the reduction in the concentration of chlorophyll by bleaching during the periods of highest light intensity and its resynthesis when intensities were lower. Measurements of pigment content of the cells during the course of a day supported this hypothesis (Fig. 13).

Marked diurnal changes in the amounts of phytoplankton pigments per cells were also observed by Yentsch and Scagel (1958). The lowest concentrations of chlorophylls and carotenoids were found at midday and at midnight, whereas the highest concentrations occurred during the morning and in the late afternoon, when the light intensities at the surface were most favorable for photosynthesis. The phenomenon was marked in plankton samples from deeper layers, whereas cells collected close to the surface showed only small diurnal fluctuations of pigmentation.

VI. Nutrition

A. Seasonal Changes in Fixed Nitrogen and Phosphorus

The growth of marine phytoplankton is probably never seriously arrested by the lack of inorganic carbon. Also in excess of normal requirements are the elements sodium, calcium, potassium, bromine, boron, magnesium, and sulfur. On the other hand, since phosphorus and nitrogen are essential plant nutrients normally present in only small amounts, these elements have often been indicated as factors limiting the growth of plankton. Their concentrations decrease markedly in surface waters as a result of phytoplankton growth, but they are periodically replenished by decomposition of organic matter, and by diffusion and turbulent mixing from waters below the euphotic zone. For a general review of the chemistry and fertility of sea water, see Harvey (1955).

B. Nitrogen:Phosphorus Ratios

In the English Channel the concentrations of nitrogen and phosphorus in the surface water are reduced at similar rates (Harvey, 1926). The ratio in which these compounds are depleted from sea water—i.e., $([N]_2 - [N]_1):([P]_2 - [P]_1)$—is generally comparable to the atomic ratio of these elements ($N/P = \sim 20$) in marine phytoplankton organisms (Redfield, 1934; Fleming, 1940). Harris and Riley (1956) obtained an average value of 16.7 for the N:P ratio in net plankton, including some zooplankton, from surface waters of Long Island Sound. Phifer and Thompson (1937) and Cooper (1937a, 1938) reported somewhat lower N:P ratios in the water itself, while the ratios obtained by Riley (1956) and Ketchum *et al.* (1958b) for coastal and continental-slope surface waters varied from near zero during the summer to 10 in midwinter; nevertheless, the

$([N]_2 - [N]_1) : ([P]_2 - [P]_1)$ ratio did not vary greatly. The extremely low N:P ratios of the cells indicate that the phytoplankton continues to grow and multiply for a time, but the intracellular content of the limiting element, nitrogen, is reduced to a minimum value (Ketchum, 1939a, b).

C. Phosphorus Utilization

When all components of the phosphorus cycle in the sea are known, including the particulate fraction as well as dissolved organic and inorganic phosphorus, it becomes possible to evaluate the phosphorus utilization by marine phytoplankton in nature (Redfield *et al.*, 1937). Assuming an approximately constant ratio of carbon to phosphorus in living organisms, one can indirectly estimate the phosphorus utilized by measuring the photosynthetic activity in a given water mass, and can then compare this value with the actual observed changes in phosphate concentration. If one ignores the effects of horizontal and vertical mixing, then the difference between these two values reflects the amount of phosphorus which is being recycled in the course of phytoplankton reproduction. By this method, Redfield *et al.* concluded that phosphate was recycled by phytoplankton between 2 and 16 times within a period of 2 months. Although their data did not show any consistent seasonal trends, they illustrated the importance of regeneration and recycling in the maintenance of phytoplankton productivity.

Measurements of cellular nitrogen and phosphorus in the natural environment are complicated by the presence of particulate detritus and by the adsorption of otherwise soluble compounds on cells, organic debris, and suspended particles of silt and clay. As much as 30% of the insoluble phosphorus may be rapidly released by acidifying with hydrochloric acid (1:200); though some of this may come from cell sap, it is considered unlikely that organic phosphorus compounds would be hydrolyzed by such dilute acid (Ketchum *et al.*, 1958a). In cells from cultures of marine plankton algae, the ratio of atoms of phosphorus to molecules of chlorophyll usually falls between the limits of 0.035 and 0.07 and shows little variation between species. In particulate matter from sea water, however, this ratio is lower and varies considerably, reaching 0.018 to 0.035 only when the chlorophyll concentrations are high. Inshore waters usually contain relatively more particulate phosphorus than do offshore waters, which tends to increase the ratio, while high ratios offshore suggest that in dying and dead cells the breakdown of chlorophyll is more rapid than the liberation of inorganic phosphorus from cellular material.

D. Nitrogen Utilization

Nitrogen is available to plants in the oceans as inorganic salts of ammonium, nitrite, and nitrate. Although inorganic phosphorus is almost always the major fraction of the total phosphorus, inorganic nitrogen is usually a relatively small fraction of the total fixed nitrogen in the euphotic zone. Of the various inorganic nitrogenous components, nitrate is usually the most abundant, though it shows marked seasonal variation.

There is conflicting information on the question of which nitrogen source is most readily utilized by phytoplankton. *Nannochloris* and *Stichococcus* grew well in media containing nitrate, nitrite, or ammonia. The diatom "*Nitzschia*" (subsequently identified as *Phaeodactylum tricornutum*)[2] grew poorly in media with ammonia (Ryther, 1954a), yet Harris (1959) found ammonia to be far superior to nitrate or nitrite in promoting chlorophyll synthesis in natural populations, presumably composed largely of diatoms and dinoflagellates. Nitrite was more effective than nitrate in the promotion of chlorophyll production (Harris, 1959), providing support for the assertion by Harvey (1953) that the absorption of nitrite by phytoplankton may be qualitatively different from that of nitrate. Under some conditions phytoplankton algae may themselves produce and excrete detectable quantities of nitrite, as was demonstrated in cultures of *Skeletonema costatum*, *Lauderia* sp. and *Isochrysis* sp. (Vaccaro and Ryther, 1960). The production of nitrite occurs especially at low light intensities, and is maintained only as long as nitrate is available, presumably due to the fact that the reduction of nitrate proceeds more rapidly than the further reduction of nitrite to amino compounds. Vaccaro and Ryther concluded that considerable concentrations of nitrite can be expected in the sea, particularly near the bottom of the euphotic zone, if nitrate is present in sufficient quantity.

Marine animals excrete nitrogen largely as ammonia, urea, uric acid, trimethylamine, and amino acids (see Harvey, 1955; Cooper, 1937b), and sea water may contain considerable quantities of these substances. A number of workers have demonstrated the utilization of organic forms of nitrogen by plankton algae. Harvey (1940) found that a natural population of diatoms could use urea, uric acid, or any of a number of amino acids as a nitrogen source. According to Ryther (1954a), *Nannochloris* and *Stichococcus* grew well with urea, uric acid, cystine, asparagine, or glycine, but "*Nitzschia*" (=*Phaeodactylum*) grew poorly with such organic nitrogen compounds.

Chlorophyll synthesis and decomposition in phytoplankton bear a close relationship to nitrogen metabolism (Harvey, 1953; Yentsch and Vaccaro, 1958). Under conditions of nitrogen deficiency, chlorophyll synthesis slows down and the cell content begins to decline, while carotenoid pigments may continue to be synthesized for a longer period and then

[2] Regarding the taxonomy of this organism, see Note 22 of the Appendix to this treatise by Silva.

decline at a much lower rate than does the chlorophyll. A high carotenoid: chlorophyll ratio in cultures of algae tends to reflect a nutrient deficiency, which may be correlated with a reduction in the photosynthetic capacity of the cells.

The ratio of organic nitrogen to chlorophyll was not constant in cultures, but depended on the degree of chlorosis of the cell, which could be represented quantitatively by the carotenoid:chlorophyll ratio (Yentsch and Vaccaro, 1958) (Fig. 14). It was assumed that similar ratios would be obtained for a variety of other species under similar conditions, and hence that this ratio would be of significance also in natural communities. Ryther *et al.* (1958), noting that in diatom blooms the ratio of carotenoid:chlorophyll *a* varied diurnally and could be directly correlated with incident light intensity, concluded that the decrease of chlorophyll at high light intensities was due to the exhaustion of nutrients resulting from active photosynthesis and growth. Similar observations were made by Yentsch and Scagel (1958) in the surface waters of East Sound, Washington, while further data obtained by Yentsch *et al.* (unpublished) showed a close relation between the availability of fixed nitrogen and the pigment ratio.

The presence of Blue-green Algae such as *Trichodesmium* and *Katagnymene* in the offshore surface waters of tropical oceans, where the concentration of dissolved nitrogen compounds is extremely low, has led to speculation on the possible capacity of these plants to fix elementary nitrogen. Although most members of the Oscillatoriaceae are evidently

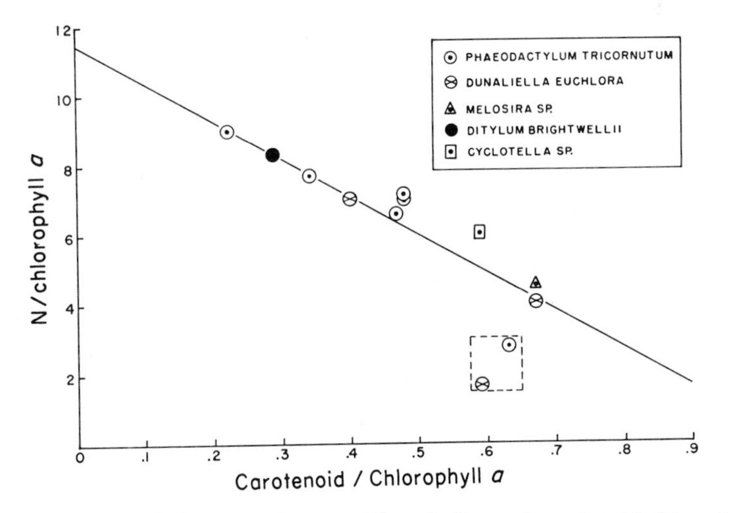

FIG. 14. Relationship between nitrogen:chlorophyll *a* and carotenoid:chlorophyll *a* in five cultures of marine phytoplankton. Points in square were observed in cultures grossly deficient in nitrogen. (From Yentsch and Vaccaro, 1958.)

unable to carry out this reaction (Fogg and Wolfe, 1954; see also Fogg, Chapter 9, this treatise), fixation of N^{15} was demonstrated in *Trichodesmium thiebautii* from the Sargasso Sea by Dugdale *et al.* (1961).

E. Silicon, Iron, Manganese, and Vitamin B_{12}

The silicate requirement by diatoms and the low concentrations of this ion in natural waters have suggested that silicate may often be a limiting factor for diatom growth. Diatoms growing in water low in silicate tend to develop thin walls (see Lewin, Chapter 27, this treatise). Atkins (1928) computed plankton productivity in the sea near Plymouth, England, from the uptake of phosphorus, carbon dioxide, and silicate, respectively. Since the value computed on the basis of phosphorus changes exceeded that from silicate changes, he concluded that much of the productivity was due to plants which did not require silicate. Later information on the prevalence of unsilicified nannoplankton supported this conclusion. In a series of samples of Atlantic surface waters, the relative depletions of N and Si, like those of N and P (see above), were found to reflect the average content of these elements in plankton (Richards, 1958).

The availabilities of iron and manganese have been considered as occasionally limiting to open-ocean populations. Both elements are found predominantly in the particulate state, since the cations are readily adsorbed on organic surfaces; it is conceivable that the euphotic zone could thereby be completely depleted of either element or of both (Harvey, 1955). Harvey (1937) demonstrated that marine phytoplankton can directly utilize ferric hydroxide adsorbed on the cell walls. Goldberg (1952) obtained experimental data which he interpreted to indicate that the diatom *Asterionella japonica* could utilize only particulate or colloidal ferric hydroxide, whereas iron complexed by organic matter was not available to them for growth. Manganese requirements by phytoplankton, too, may be met by adsorption and utilization of the hydrated oxides.

Evidence that a natural deficiency of these elements may limit plankton growth was presented by Harvey (1955), who demonstrated that the addition of Fe or Mn could result in an increase in growth beyond that observed in sea water enriched with only nitrogen and phosphorus (Harvey, 1955). Similar experiments of Ryther and Guillard (1959), using uptake of C^{14} as an index of growth, pointed to the same conclusion.

It has been established by a number of workers that manganese is needed in nitrate reduction (see, in this treatise, Syrett, Chapter 10, and Wiessner, Chapter 17); with ammonia present in adequate amounts, no Mn is required. The presence of sufficient ammonia in sea water would thus eliminate a requirement for manganese, but it is doubtful if this ever occurs in nature. Since iron is essential for all living systems containing cytochrome, the requirement for this element cannot be by-passed as can that for manganese.

Many phytoplankton organisms are now known to require vitamin B_{12}. The ecological implications of this are discussed in Chapter 8 of this treatise by Droop.

VII. Temperature and Salinity

Circumstantial observations on temperature and salinity tolerance ranges, based on field data alone, must be accepted with caution. More significant data have been obtained for a few species, e.g., Pacific littoral or Baltic dinoflagellates, which were found experimentally to grow at temperatures between 5° and 25° or 30°C. and in a salinity range between 1.0 and 4.0% (Barker, 1935; Braarud, 1951; Nordli, 1957). A discussion of osmotic tolerance in algae is presented by Guillard (Chapter 32, this treatise).

VIII. Flotation and Diffusion Problems

As the cells of the majority of species in marine populations of phytoplankton are non-motile and more dense than the surrounding sea water, they tend slowly to sink. Because of the disastrous consequences of sinking from illuminated areas into the dark depths, it has often been suggested that phytoplankton organisms must have developed adaptive features to retard sinking. Many features such as spines, mucilage envelopes, and oil globules have been considered as means to this end. Since the resistance of a sinking cell is a function of its surface area, it is presumably of advantage to reduce the cell size or in other ways to increase the surface:volume ratio. Munk and Riley (1952), however, challenged the importance of some of the morphological devices supposedly evolved for the retardation of sinking, pointing out that the formation of cell chains in diatoms, especially in centric forms, actually reduces the surface area and hence would be expected to increase sinking rates. Since the absorption of nutrients also involves surface:volume relations, it may be that the small size or elongated form of many plankton cells is rather an adaptation to conditions of low nutrient concentrations. For large planktonic cells, however, one must seek other mechanisms, such as a reduction in the ionic content of the cell sap, demonstrated in *Ethmodiscus rex* by Beklemishev *et al.* (1961).

Laboratory experiments of Steele and Yentsch (1960) showed that the sinking rate of phytoplankton cells increased as the culture aged, and that the rate could be reduced by transferring the cells to nutrient-rich media or by darkening the culture (Figs. 15 and 16). Thus the general physio-

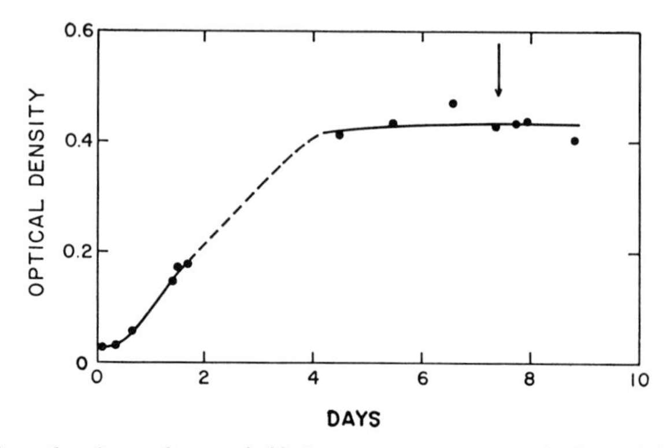

FIG. 15. Growth of a culture of *Skeletonema costatum* as indicated by increase in optical density of culture.

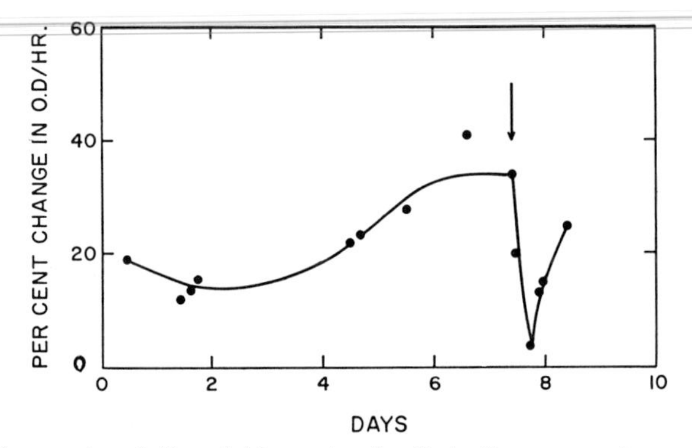

FIG. 16. Changes in relative sinking rate of cells in the same culture as Fig. 17, as indicated by reduction in the optical density (O.D.) of the supernatant in a cuvette standing undisturbed for 1 hour. At the point indicated by the arrow, fresh nutrient medium was added and the culture was transferred to darkness. (From Steele and Yentsch, 1960.)

logical condition of the cell may considerably influence its buoyancy, as shown previously by Gross and Zeuthen (1948) in the diatom *Ditylum brightwellii*. This would provide a possible explanation for seasonal variation in the vertical distribution of cellular chlorophyll to be found in the sea. During early spring, when nutrients are abundant, the cell populations undergo rapid growth near the surface, and tend to remain there because of adequate nutrition and a low sinking rate. In late summer, when nutrients are depleted, the rate of settling increases and the plants tend to accumulate in the lower regions of the euphotic zone, where sufficient nutrients are still present and further sinking is retarded. A renewal of vertical mixing during fall and winter disturbs this accumulation and redistributes the plants throughout the euphotic zone.

REFERENCES

Atkins, W. R. G. (1928). Seasonal variations in phosphate and silicate content of sea water during 1926 and 1927 in relation to the phytoplankton crop. *J. Marine Biol. Assoc. United Kingdom* **15,** 191–205.

Banse, K. (1957). Ergebnisse eines hydrographisch-produktionsbiologischen Längsschnittes durch die Ostee im Sommer 1956. Die Verteilung von Sauerstoff, Phosphat und suspendierter Substanz. *Kiel. Meeresforsch.* **13,** Heft 2, 186–201.

Barker, H. A. (1935). The culture and physiology of the marine dinoflagellates. *Arch. Mikrobiol.* **6,** 157–181.

Beklemishev, C. W., Petrikova, M. N., and Semina, H. J. (1961). [On the cause of the buoyancy of plankton diatoms] *Trudy Inst. Okeanol. Akad. Nauk S.S.S.R.* **51,** 33–36. (In Russian.)

Braarud, T. (1951). Salinity as an ecological factor in marine phytoplankton. *Physiol. Plantarum* **5,** 28–34.

Cooper, L. H. N. (1937a). On the ratio of nitrogen to phosphorus in the sea. *J. Marine Biol. Assoc. United Kingdom* **22,** 177–182.

Cooper, L. H. N. (1937b). The nitrogen cycle in the sea. *J. Marine Biol. Assoc. United Kingdom* **22,** 183–204.

Cooper, L. H. N. (1938). Redefinition of the anomaly of the nitrate-phosphate ratio. *J. Marine Biol. Assoc. United Kingdom* **23,** 179.

Doty, M. S., and Oguri, M. (1957). Evidence for a photosynthetic daily periodicity. *Limnol. and Oceanog.* **2,** 37–40.

Dugdale, R. C., Menzel, D. W., and Ryther, J. H. (1961). Nitrogen fixation in the Sargasso Sea. *Deep-Sea Research* **7,** 297–300.

English, T. S. (1959). Primary production in the central North Polar Sea, Drifting Station Alpha, 1957–1958, *Proc. Intern. Oceanog. Congr., New York, August–September, 1959* pp. 338–389.

Fleming. R. H. (1940). The composition of plankton and units for reporting populations and production. *Proc. Pacific Sci. Congr., Pacific Sci. Assoc., 6th Congr., 1939* Part 3, pp. 535–540.

Fogg, G. E., and Wolfe, M. (1954). The nitrogen metabolism of the Blue-Green Algae (Myxophyceae). *In* "Autotrophic Microorganisms." *Symposium Soc. Gen. Microbiol.* **4,** 99–125.

Goldberg, E. D. (1952). Iron assimilation by marine diatoms. *Biol. Bull.* **102,** 243.

Gross, F., and Zeuthen, E. (1948). The buoyancy of plankton diatoms: a problem of cell physiology. *Proc. Roy. Soc.* **B135,** 382.

Harris, E. (1959). The nitrogen cycle in Long Island Sound. *Bull. Bingham Oceanog. Collection Yale Univ.* **17,** 31–63.

Harris, E., and Riley, G. A. (1956). Oceanography of Long Island Sound. VIII. Chemical composition of the plankton. *Bull. Bingham Oceanog. Collection Yale Univ.* **15,** 315–323.

Harvey, H. W. (1926). Nitrate in the sea. *J. Marine Biol. Assoc. United Kingdom* **14,** 71–88.

Harvey, H. W. (1937). The supply of iron to diatoms. *J. Marine Biol. Assoc. United Kingdom* **22,** 208–219.

Harvey, H. W. (1940). Nitrogen and phosphorus required for the growth of phytoplankton. *J. Marine Biol. Assoc. United Kingdom* **24,** 115–123.

Harvey, H. W. (1953). Synthesis of organic nitrogen and chlorophyll by *Nitzschia closterium. J. Marine Biol. Assoc. United Kingdom* **31,** 477–478.

Harvey, H. W. (1955). "The Chemistry and Fertility of Sea Waters," 224 pp. Cambridge Univ. Press, London and New York.

Holmes, R. W. (1957). Solar radiation, submarine daylight, and photosynthesis. *In* "Treatise on Marine Ecology and Paleoecology" (J. W. Hedgpeth, ed.), Vol. I, pp. 109–128. *Geol. Soc. Am. Mem.* **No. 67.**

Hutchinson, G. E. (1957). "A Treatise on Limnology," Vol. 1: Geography; Physics and Chemistry, 1013 pp. Wiley, New York.

Jenkin, P. M. (1937). Oxygen production by the diatom *Coscinodiscus excentricus* in relation to submarine illumination in the English Channel. *J. Marine Biol. Assoc. United Kingdom* **22,** 301–342.

Jerlov, N. G. (1951). Optical studies of ocean waters. Repts. Swed. Deep Sea Expedition, 1947/48, Vol. III, No. 1, pp. 1–59.

Kalle, K. (1938). Zum Problem der Meereswasserfarbe. *Deut. Seewarte, Hamburg. Ann. Hydrog. Maritimen Meteorol.* **66,** 1–13.

Ketchum, B. H. (1939a). The absorption of phosphate and nitrate by illuminated cultures of *Nitzschia closterium. Am. J. Botany* **26,** 399–407.

Ketchum, B. H. (1939b). The development and restoration of deficiencies in the phosphorus and nitrogen composition of unicellular plants. *J. Cellular Comp. Physiol.* **13,** 373–381.

Ketchum, B. H., Ryther, J. H., Yentsch, C. S., and Corwin, N. (1958a). Productivity in relation to nutrients. *J. conseil, Conseil permanent intern. exploration mer. Rappt. et proc. verb.* **144,** 132–140.

Ketchum, B. H., Vaccaro, R. F., and Corwin, N. (1958b). The annual cycle of phosphorus and nitrogen in New England coastal waters. *J. Marine Research* **17,** 282–301.

Krey, J., Banse, K., and Hagmeier, E. (1957). Über die Bestimmung von Eiweis im Plankton mittels der Biuretreaktion. *Kiel. Meeresforsch.* **13,** 35–40.

McLachlan, J., and Yentsch. C. S. (1959). Observations on the growth of *Dunaliella euchlora* in culture. *Biol. Bull.* **116,** 461–471.

McLeod, G. C. (1958). Delayed light action spectra of several algae in visible and ultra-violet light. *J. Gen. Physiol.* **42,** 243–250.

Munk, W. H., and Riley, G. A. (1952). Absorption of nutrients by aquatic plants. *J. Marine Research* **11,** 215.

Nordli, E. (1957). Experimental studies on the ecology of *Ceratia. Oikos* **8,** 200–265.

Phifer, L. D., and Thompson, T. G. (1937). Seasonal variations in the surface waters of San Juan Channel during the five year period Jan. 1931–Dec. 30, 1935. *J. Marine Research* **1,** 34–59.

Redfield, A. C. (1934). On the proportions of organic derivatives in sea water and their relation to the composition of plankton. *In* James Johnstone Mem. Vol., pp. 176–192. Univ. of Liverpool, England.

Redfield, A. C., Smith, H. P., and Ketchum, B. H. (1937). The cycle of organic phosphorus in the Gulf of Maine. *Biol. Bull.* **73,** 421–443.

Richards, F. A. (1958). Dissolved silicate and related properties of some western North Atlantic and Caribbean waters. *J. Marine Research.* **17,** 449–465.

Riley, G. A. (1956). The oceanography of Long Island Sound. II. Physical oceanography. *Bull. Bingham Oceanog. Collection Yale Univ.* **15,** 15–47.

Ryther, J. H. (1954a). The ecology of phytoplankton blooms in Moriches Bay and Great South Bay, Long Island, New York. *Biol. Bull.* **106,** 198–209.

Ryther, J. H. (1954b). The ratio of photosynthesis to respiration in marine plankton algae and its effect upon the measurement of productivity. *Deep-Sea Research* **2,** 134.

Ryther, J. H. (1956). Photosynthesis in the ocean as a function of light intensity. *Limnol. and Oceanog.* **1,** 61.

Ryther, J. H., and Guillard, R. R. L. (1959). Enrichment experiments as a means of studying nutrients limiting to phytoplankton production. *Deep-Sea Research* **6,** 65–69.

Ryther, J. H., and Menzel, D. W. (1959). Light adaptation of marine phytoplankton. *Limnol. and Oceanog.* **4,** 492–497.

Ryther, J. H., and Yentsch, C. S. (1957). The estimation of phytoplankton production in the ocean from chlorophyll and light data. *Limnol. and Oceanog.* **2,** 281–286.

Ryther, J. H., and Yentsch, C. S. (1958). Primary production of continental shelf waters off New York. *Limnol. and Oceanog.* **3,** 327–335.

Ryther, J. H., Yentsch, C. S., Hulburt, E. M., and Vaccaro, R. F. (1958). The dynamics of a diatom bloom. *Biol. Bull.* **115,** 257–268.

Shimada, B. M. (1958). Diurnal fluctuation in photosynthetic rate and chlorophyll *a* content of phytoplankton from eastern Pacific waters. *Limnol. and Oceanog.* **3,** 336–339.

Steele, J. H., and Yentsch, C. S. (1960). The vertical distribution of chlorophyll. *J. Marine Biol. Assoc. United Kingdom* **39,** 217–226.

Steemann Nielsen, E. (1952). The use of radioactive carbon (C^{14}) for measuring radioactive production in the sea. *J. conseil, Conseil permanent intern. exploration mer.* **18,** 117.

Steemann Nielsen, E. (1955). The production of antibiotics by plankton algae and its effects upon bacterial activities in the sea. Papers in Marine Biol. and Oceanog. to honor Henry Bryant Bigelow. *Deep-Sea Research. Suppl.* **3,** 281–286.

Steemann Nielsen, E., and Aabye Jensen, E. (1957). Primary oceanic production. The autotrophic production of organic matter in the oceans. *Galathea Rept.* **1,** 49.

Steemann Nielsen, E., and Al Kholy, A. A. (1956). Use of C^{14} technique in measuring photosynthesis of phosphorus or nitrogen deficient algae. *Physiol. Plantarum* **9,** 144.

Steemann Nielsen, E., and Hansen, V. K. (1959a). Light adaptation in marine phytoplankton populations and its interrelation with temperature. *Physiol. Plantarum* **12,** 370.

Steemann Nielsen, E., and Hansen, V. K. (1959b). Measurements with the carbon 14 technique of rates of respiration in natural populations of phytoplankton. *Deep-Sea Research* **5,** 222–233.

Vaccaro, R. F., and Ryther, J. H. (1960). Marine phytoplankton and the distribution of nitrite in the sea. *J. conseil, Conseil permanent intern. exploration mer.* **25,** 260–271.

Yentsch, C. S. (1960). The influence of phytoplankton pigments on the color of sea water. *Deep-Sea Research* **7,** 1–9.

Yentsch, C. S., and Ryther, J. H. (1957). Short-term variations in phytoplankton chlorophyll and their significance. *Limnol. and Oceanog.* **2,** 140–142.

Yentsch, C. S., and Ryther, J. H. (1959). Absorption curves of acetone extracts of deep water particulate matter. *Deep-Sea Research* **6,** 72–74.

Yentsch, C. S., and Scagel, R. F. (1958). Diurnal study of phytoplankton pigments. An *in situ* study in East Sound, Washington. *J. Marine Research* **17,** 567–584.

Yentsch, C. S., and Vaccaro, R. F. (1958). Phytoplankton nitrogen in the oceans. *Limnol. and Oceanog.* **3,** 443–448.

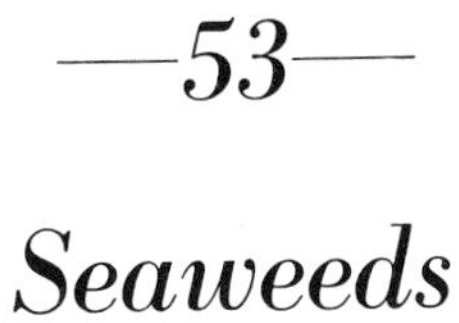

Seaweeds

R. BIEBL

Institute for Plant Physiology, University of Vienna, Austria

I. Introduction

Ecology is the study of the relationships of organisms to their environment. Among the objectives of marine algal ecology are: (i) the description of the zonation of the algal flora in various coastal areas; (ii) the determination and measurement of the chief ecological factors in the various habitats; and (iii) the experimental investigation of morphological and physiological adaptations which enable algae to occupy these particular habitats. Only the last consideration will be discussed in this chapter.

The literature on the ecology of marine algae has been reviewed by Oltmanns (1905, 1923), Feldmann (1951), Chapman (1957), Doty (1957), and Gessner (1955, 1959); see also Heim (1959).

II. Ecological Factors and Physiological Adaptation

A. General Observations

Since intertidal algae are periodically exposed at low tide, they must be able to withstand desiccation, and must possess a wide osmotic tolerance of dilution of the surrounding medium by rain and of increased salt concentrations resulting from evaporation. They must tolerate diurnal and seasonal changes of air temperature more extreme than those of the sublittoral habitat, and considerably higher light intensities. In addition they must withstand the mechanical stresses imposed by the force of the waves (Johnson and Skutch, 1928a; Feldmann, 1937), and, in northern latitudes, the denuding effects of ice in winter. Littoral zonation is affected by all of these factors, whereas as a rule in the sublittoral zone only light is of major importance as an ecological factor, apart from limitations imposed by various substrates.

Algae growing in tide pools, though submerged even at ebb tide, may still be subjected to great changes in salinity, dissolved oxygen and carbon dioxide, and temperature.

Detached portions of certain marine algae, particularly species of fucoids, can survive and proliferate in shallow bays or lagoons, on salt marshes, or in such oceanic regions as the Sargasso Sea (Parr, 1939). Under such conditions they often adopt an unusual form of thallus, and generally remain sterile; the physiological bases of this sterility have not been explained.

In general, it is found that sublittoral algae are less tolerant of extreme conditions than are the seaweeds of the shore, and that tolerance limits of the various species are wider as one ascends through the littoral zones towards dry land. Experimental evidence is presented in this article.

B. Emersion: Resistance to Desiccation

The ability of certain intertidal algae to resist desiccation over a period of days has long been recognized. Berthold (1882) observed that *Bangia fuscopurpurea* and *Porphyra leucosticta* revived in sea water after being dried for 8 to 14 days in air. Muenscher (1915) found that *Fucus evanescens, Gloiopeltis furcata*, and *Porphyra perforata* survived after 1 or 2 days' air-drying, but that the sublittoral algae *Nereocystis luetkeana* and *Alaria valida* died after only 1 hour of exposure. Isaac (1933, 1935) reported similar findings for the comparative resistance of the intertidal *Pelvetia canaliculata* and the sublittoral *Laminaria digitata.*

Intertidal and sublittoral algae of similar morphological structure exhibit no differences in their rate of drying, their desiccation curves essentially resembling those of gelatin or agar. Differences in their resistance to desiccation must, therefore, result primarily from differences within the protoplasm itself. As shown in Table I, algae of the intertidal zone at Helgoland tolerated desiccation for 14 hours in air of 83 to 86% relative humidity (r.h.), while sublittoral algae succumbed even in moist air at 98.8% r.h. In a similar experiment carried out at Pacific Grove, California, *Porphyra perforata* survived after 14 hours at 60.6% r.h. but not at 48.3% r.h.; *Enteromorpha clathrata* from a salt pond survived at 83.9% r.h. but not at 60.6% r.h.; *Ulva rigida*, though from the same tide level as the *Porphyra*,

TABLE I

Survival of Marine Algae under Desiccating Conditions[a]

Habitat	Species	Survival after 14 hours in air (% relative humidity) 100 — 95 — 90 — 85 — 83.9
Intertidal zone	*Porphyra laciniata*	
	Rhodochorton floridulum	
	Elachista fucicola	
	Ulva lactuca	
	Enteromorpha linza	
	Cladophora rupestris	
	Cladophora gracilis	
Low-tide level	*Polysiphonia nigrescens*	
	Membranoptera alata	
	Ptilota elegans	
	Dictyota dichotoma	
Sublittoral region	*Plocamium coccineum*	
	Antithamnion plumula	
	Trailiella intricata	
	Halarachnion ligulatum	

[a] From Biebl (1938a). [For *Trailiella* read *Trailliella.*]

tolerated only 90% r.h. and was almost completely dead after 14 hours at 83.9% r.h. (Biebl, 1956a).

Some intertidal algae appear to require periodic desiccation, or at least emersion. Fischer (1929) observed that plants of *Pelvetia canaliculata* and *Fucus spiralis* soon died when kept constantly submerged, possibly as a consequence of anoxia during darkness. Similarly *Bangia fuscopurpurea*, which in Naples grows in the spray zone, began to die after 9 days when continuously immersed in sea water, and after 21 days was completely dead. In contrast, thalli of the same species and of *Urospora penicilliformis*, dried on a glass plate and exposed to room air for 21 days, were found be still alive when returned to sea water (Biebl, 1939a).

The photosynthetic efficiency of intertidal algae while exposed to the air depends largely upon their water content (Stocker and Holdheide, 1937). The thick, mucilaginous Fucaceae can retain water longer (Zaneveld, 1937), and can therefore photosynthesize longer, than the thin thalli of

Porphyra, *Ulva*, or *Enteromorpha* which, even under a cloudy sky, may dry out after not more than 1 to 2 hours and reach the compensation point where net photosynthesis falls to zero. This length of time corresponds to only one-fourth or one-fifth of the period for which they are exposed. Despite so short a period of photosynthesis during emergence, *Porphyra* and *Enteromorpha*, on account of their relatively large surface area, have a net photosynthetic efficiency comparable to that of the Fucaceae with which they are associated.

Montfort (1937) examined the photosynthetic efficiency of specimens of three species of *Fucus* collected at the same site and dried in air for 5 hours. *F. platycarpus*, generally found in the upper littoral zone, retained 97% of its efficiency; *F. vesiculosus*, which normally occupies a somewhat lower zone, retained 72%; while *F. serratus*, which typically is located near low-water mark, retained only 42% as compared with controls which had not been dried, and took several hours to recover. This indicates that such differences are genetically controlled, and are not determined merely by the habitat.

C. Salinity: Resistance to Osmotic Changes

Some marine algae can tolerate a wide range of sea-water concentrations without suffering damage; others are adapted to much narrower ranges. (See reviews by Oltmanns, 1923; Fritsch, 1945; Blinks, 1951; Feldmann, 1951; Schwenke, 1960). Algae which grow on ships plying between rivers and the sea, and estuarine species in tidal waters, must be capable of tolerating salinities between 0 and 3.5%, and rapid changes within a few hours from one to the other. Studies of several such species on freighters in San Francisco Harbor were reported by Osterhout (1906).

Intertidal algae frequently experience no less abrupt alterations in salt content during the ebb tide, when evaporation concentrates the sea water trapped around them or rain showers dilute it. In most seas, sublittoral algae live in a milieu of essentially constant salt content, though in certain areas, e.g., in the Baltic Sea, even in the depths there may be fluctuations of salinity between 1 and 3% (see Kändler, 1951).

The first physiological observations on hypertonic death and the osmotic resistance of marine algae were those of Höfler (1930, 1931), obtained in the course of a comparative study of the protoplasmic properties of different taxonomic groups of plants. Similar studies, though from an ecological viewpoint, were subsequently carried out at Plymouth, England (Biebl, 1937, 1952a, b), Helgoland, Germany (Biebl, 1938a), Naples, Italy (Biebl, 1939b), Pacific Grove, California (Biebl, 1956a), and Roscoff, France (Biebl, 1958); all gave concordant results. Most intertidal algae could survive in a concentration range of 0.1 to 3.0 times that of sea water (in some cases tolerating even distilled water or 4.0 times the sea-water concentration for 24 hours). Algae from tidal pools or near low-water mark, however, exhibited smaller osmotic ranges, in this respect tending to resemble sublittoral algae, few of which can tolerate concentrations outside the range of 0.5 to 1.5 times that of sea water.

The limits of the hypertonic resistance of sublittoral Red Algae usually coincide with the osmotic value of the cells, i.e., with the concentration which causes plasmolysis, since plasmolysis of such algae results in permanent damage to the cells (Höfler, 1930, 1931; Kylin, 1938). Intertidal Red Algae, on the other hand, seem to have several possible means of avoiding plasmolytic damage in hypertonic concentrations of sea water (Biebl, 1939b). (i) They may possess small cells with very dense contents, and cell walls which swell considerably in concentrated sea water, thereby preventing plasmolysis by following the cell contents as they shrink on the withdrawal of water (e.g., *Porphyra*). (ii) They may possess so high an intracellular osmotic value that under normal conditions harmful plasmolysis can never occur (e.g., *Polysiphonia urceolata* or *Callithamnion tetragonum* var. *brachiatum*, which only begin to plasmolyze in 2.1 times the concentration of sea water). (iii) They may in some unknown way survive periods of plasmolysis without possessing an especially high internal osmotic value, e.g., *Ceramium ciliatum.* (iv) They may have both a high internal osmotic value and great tolerance to plasmolysis (e.g., *Rhodochorton floridulum* at Helgoland or *Polysiphonia pulvinata* at Naples).

Most thin intertidal algae, especially Green and Brown Algae, appear to

TABLE II

CHANGES OF RELATIVE INTERNAL OSMOTIC VALUES OF *Enteromorpha clathrata* IMMERSED IN VARIOUS CONCENTRATIONS OF SEA WATER[a]

Days after transfer	*Relative osmotic value of medium*					
	0.0	0.5	1.0	2.0	3.0	4.0
0	2.9	2.9	2.9	2.9	2.9	2.9
1	0.8	2.1–2.2	2.1–2.2	2.9	*depl.*	*pl.*
2	0.6	2.1–2.2	2.1–2.2	2.8	~3.6	*pl.*
3	0.6	2.0–2.1	2.1–2.2	2.8	~3.6	*pl.*
5	0.6	1.9–2.0	2.1–2.2	2.8	~3.6	*pl.*
7	*dead*	1.9	2.1–2.2	2.8	—	—

[a] KEY: 1.0 = equivalent to sea water; *depl.* = deplasmolyzed; *pl.* = plasmolyzed.

possess the additional physiological property of accumulating salts against a diffusion gradient, thereby preventing the onset of plasmolysis, or even reversing plasmolysis already in progress. Such an osmoregulatory mechanism was shown, for example, in the Brown Algae *Pylaiella littoralis* and *Elachistea fucicola* (Biebl, 1938b). It was especially well demonstrated in the following experiment using the chlorophyte *Enteromorpha clathrata* collected from salt ponds at Moss Landing near Monterey (Biebl, 1956a). The alga grew as dense mats floating on water with a salinity equivalent to 1.5 times sea water. Samples were transferred to water of various concentrations, ranging from zero (distilled water) to 6 times that of sea water (of salinity exceeding 20%). Portions were removed at intervals, and the internal osmotic pressure of the cells determined by the method of incipient plasmolysis. Though the highest concentrations tested killed almost all the cells within 24 hours, lower concentrations over a wide range were tolerated for several days. The results are summarized in Table II. How the algae control their osmotic value to balance or exceed that of the internal medium —whether by passive diffusion alone or by active salt uptake involving the expenditure of metabolic energy—remains unknown.

Hoffman (1943, 1959) and Schwenke (1958a) showed that sublittoral algae from the Gulf of Kiel (salinity about 2%) are more tolerant of hypotonic conditions than seaweeds from the English Channel at Plymouth (salinity 3.5%). The osmotic resistance range of *Membranoptera alata*, for example, was found to be 1.4 to 6.5% in Plymouth (Biebl, 1937) and 0.2 to 6.0% in Kiel (Schwenke, 1958a). Whether this is due to individual adaptation or to the development of physiological races must be decided by transplantation experiments.

The resistance of marine algae to hypotonic solutions increases with the calcium content of the medium (Schwenke, 1958b; Eppley and Cyrus, 1960), possibly correlated with the fact that calcium deficiency may cause a rapid loss of potassium from the cells (Eppley, 1958). Characteristic differences between the responses of littoral and sublittoral algae to media of various osmotic strength are revealed not only by their viability and susceptibility to plasmolysis, but also by changes in their respective rates of respiration and photosynthesis. Thus Montfort (1931) found that certain sublittoral algae are immediately and irreversibly damaged by diluted sea water or by fresh water, but that intertidal algae under similar conditions exhibit merely a reversible inhibition of their photosynthetic activity. The photosynthesis of typical intertidal algae, such as species of *Enteromorpha*, is scarcely inhibited by a temporary dilution of the sea water. Similarly, many intertidal algae show no marked alteration of the respiratory rate in diluted sea water, whereas algae from habitats below the low-water level exhibit temporarily increased rates of respiration which can be taken to indicate damage (Hoffmann, 1929).

D. Temperature: Resistance to Cold and Heat

1. Littoral and Sublittoral Algae

Observations on the survival of algae are easily made, and a number of comparative investigations have been reported. For the purposes of comparison, temperature tolerance experiments should be of uniform duration, since the point of death is dependent upon both the temperature and the time of exposure. *Ceramium tenuissimum,* for example, lives for 300 to 400 minutes at 28°C., 75 to 85 minutes at 33°C., and only 7 to 10 minutes at 38° (Ayres, 1916). A duration of 12 hours was selected by Biebl (1938a, 1958) as a practical and ecologically relevant period for such studies.

Five intertidal algae from Naples and eight from Roscoff were all able to survive the lowest temperature tested, −8°C., at which they were frozen hard in ice. Similarly, all ten of the intertidal algae investigated by Kylin (1917) on the Swedish south coast survived after freezing at −17δC. for 10 hours. Only three of the ten species (*Cladophora rupestris, Pylaiella littoralis,* and *Chondrus crispus*) were killed at −18 to −20δC. In the Arctic, *Fucus vesiculosus* survives temperatures of −40°C. for many months (Kanwisher, 1957). Experiments by Parker (1960) indicated that the cold tolerance of this species may vary seasonally, from about −30°C. in summer to below −45°C. in winter. Kanwisher pointed out that resistance to frost is often correlated with resistance to desiccation, since freezing may result in the formation of ice crystals associated with the physical abstraction of water from the protoplasm (Levitt, 1956, 1958). For example, when *Ulva pertusa* is frozen to −10°C., at which temperature it survives for at least 24 hours, frost plasmolysis can be observed, but not intracellular freezing (Terumoto, 1960a). In temperate zones, intertidal algae can usually withstand direct desiccation in air as well as physiological desiccation in hypertonic media or at temperatures low enough for ice formation. However, it may be noted that the freshwater "lake-ball alga," *Aegagropila sauteri,* is quite sensitive to desiccation, though it can tolerate −20°C. for as long as 24 hours, and survives even lower temperatures if treated with agents such as ethylene glycol, which inhibit the formation of intracellular ice (Terumoto, 1959, 1960b).

Sublittoral algae are more sensitive. Thus of thirteen species investigated at Naples of thirty at Roscoff, not a single specimen survived freezing. The annual water-temperature ranges at these two locations are 13 to 26°C. and 10 to 16°C., respectively. About half of the species survived cooling to −2°C. without ice formation, while the remainder died at 1 to 3°C. A few exceptions will be discussed below.

The upper temperature limit for the survival of intertidal algae from Naples was about 35°C. in six species, and about 30°C. in one other. By contrast, at Roscoff only the extremely resistant *Enteromorpha intestinalis* survived a 12-hour period at 35°C., while all the other seven species tested died at temperatures above 30°C. Differences between the sublittoral algae in the two areas were less marked. Of the thirteen species tested from Naples, eight survived at 27°C. and six at 30°C.; among the thirty species at Roscoff, twenty-one survived at no higher temperatures than 27°C. and the remaining nine tolerated 30°C. All the sublittoral algae investigated at both locations were killed when subjected to 35°C. for 12 hours.

Intertidal algae are subjected not only to a range of water temperatures, but also to the combined effects of desiccation, heating, and insolation, which necessitate a wider physiological tolerance. For example, dry rocks at Naples and Helgoland, at temperatures of 38 to 40°C., bore still viable thalli of *Bangia fuscopurpurea* and *Urospora penicilliformis* (Biebl, 1939a, b). Dried plants of *Bangia fuscopurpurea* were shown experimentally to survive in a relative humidity of 17.2% at 42°C. for 24 hours, although the same species in sea water could not tolerate even 35°C. for a similar length of time.

In order to investigate the effects of temperature shocks, such as occur when algae are resubmerged by the returning tide, Montfort *et al.* (1955, 1957) subjected several littoral and sublittoral Baltic algae, grown at 14°C. for 4 months, to 32°C. for 3 hours and then determined their rates of respiration and photosynthesis. Although the intertidal algae *Chaetomorpha linum* and *Fucus vesiculosus* suffered only a slight reduction in their photosynthetic activity, algae from 8-meters depth, (*Laminaria saccharina, Furcellaria fastigiata, Delesseria sanguinea,* and *Rhodomela subfusca*) showed depressions, lasting for many days, far below the compensation point.

Montfort *et al.* also attempted to adapt seaweeds to cold by maintaining them for 4 to 7 months at −1 to +1°C. However, in most cases the thermal tolerance of the algae was apparently unchanged, being evidently dependent entirely on hereditary factors; only in a few species, such as *Delesseria sanguinea* and *Laminaria saccharina,* could a slight shift of about 2°C. in the lethal temperature be detected.

Schwenke (1959) carried out similar temperature stress and adaptation experiments by transferring specimens of Baltic algae from sea water at 2° to 20°C. and back again within a 24-hour period. Small differences of resistance were noted. For example, *Delesseria sanguinea,* a cold-boreal alga, tolerated three such changes without damage and was still living 4 weeks later, whereas in *Phycodrys rubens,* a subarctic form, 10 to 75% of the cells were dead after the second change of water temperature, and 25 to 90% after the third change. *Phycodrys rubens* also proved to be the more sensitive in 12-hour heating experiments.

2. Temperature Resistance and Climatic Zones

The differences in the thermal tolerance of littoral algae at Naples and Roscoff might be considered examples of adaptation to climatic conditions. All but one of the Neapolitan intertidal algae examined by Biebl tolerated a 12-hour period in sea water at 35°C. However, of the intertidal algae from the more northerly situated Roscoff, this temperature was survived only by *Enteromorpha compressa*; the rest died at temperatures above 30°C.

The temperature ranges between −2° to +3°C. and 27° to 30°C. seem to delimit the tolerance of most sublittoral algae of the temperate zone, since all the species from Naples and Roscoff survived exposures within this range. However, in Plymouth the sublittoral alga *Sphondylothamnion multifidum* (Biebl, 1937) and in Roscoff certain other deep-water algae such as *Sphondylothamnion multifidum* var. *piliferum*, *Drachiella spectabilis* (Ernst and Feldmann, 1957), *Rhodophyllis divaricata*, and *Cryptopleura ramosa*, though capable of tolerating a temperature of 27°C., were partially or completely killed at 5°C. (Biebl, 1958). On the other hand, a second group of deep-water algae from Roscoff (*Dictyopteris membranacea*, *Desmarestia ligulata*, *Callophyllis laciniata*, *Calliblepharis ciliata*, *Plocamium coccineum*, *Polyneura hilliae* and *Heterosiphonia plumosa*) survived a temperature of −2°C. but died within 12 hours in sea water at 27°C.

The first group probably comprises northerly vanguards of a southern, cold-sensitive flora, while the second group may be regarded as representatives of a more heat-sensitive flora which have migrated southwards from their native cold seas. The range of 3° to 5°C. has long been recognized as a limit of cold resistance of tropical land plants (Molisch, 1897), but sublittoral algae from the warm seas around Puerto Rico died when cooled and maintained for 12 hours at 6δ to 14δC. (Biebl, unpublished).

E. Light

During low tide, algae of the tidal zone are exposed to full sunlight, while those of the sublittoral zone receive a light intensity that decreases with depth and turbidity of the water. In clear coastal waters, algae have been dredged from depths of 100 meters, whereas in turbid waters the light intensity may be sufficient for growth only in the uppermost few meters. Differences in resistance to damage by light can be expected among the algae of these two habitats, just as between sun- and shade-loving angiosperms.

With increasing depth the light alters not only in intensity but also in quality (Atkins, 1926, 1939, 1945; Poole and Atkins, 1926, 1928; Seybold,

1934; Clarke, 1936; Levring, 1947; Levring and Fish, 1956; Jerlov, 1951; see Yentsch, Chapter 52, this treatise). The water acts as a light filter, red light being completely absorbed in the upper layers, so that in the depths a blue-green twilight prevails. On physical grounds, therefore, Red Algae would be expected to be the most suited to deeper water since the prevalent light consists essentially of the wavelengths most strongly absorbed by the algae themselves, whereas Green Algae, which utilize mainly the longer wavelengths of light for photosynthesis, might be best suited to the upper sublittoral and the littoral zones. Detailed investigations of light absorption and light utilization for photosynthesis in variously colored marine algae by Seybold (1934), Montfort (1938, 1940, 1942), Seybold and Weissweiler (1942), Levring (1947, 1960), *et al.*, have tended to confirm this. A comparison of the light absorption and photosynthetic action spectra of marine algal thalli shows clearly that Green Algae effectively utilize light absorbed throughout virtually the entire red-absorption band of chlorophyll (Fig. 1), whereas Red Algae exhibit their highest photosynthetic rates in the spectral regions absorbed by the phycobilins, phycoerythrin and phycocyanin (Fig. 2) (Haxo and Blinks, 1950).

Engelmann (1883, 1884) and Gaidukov (1902, 1904) pointed out a

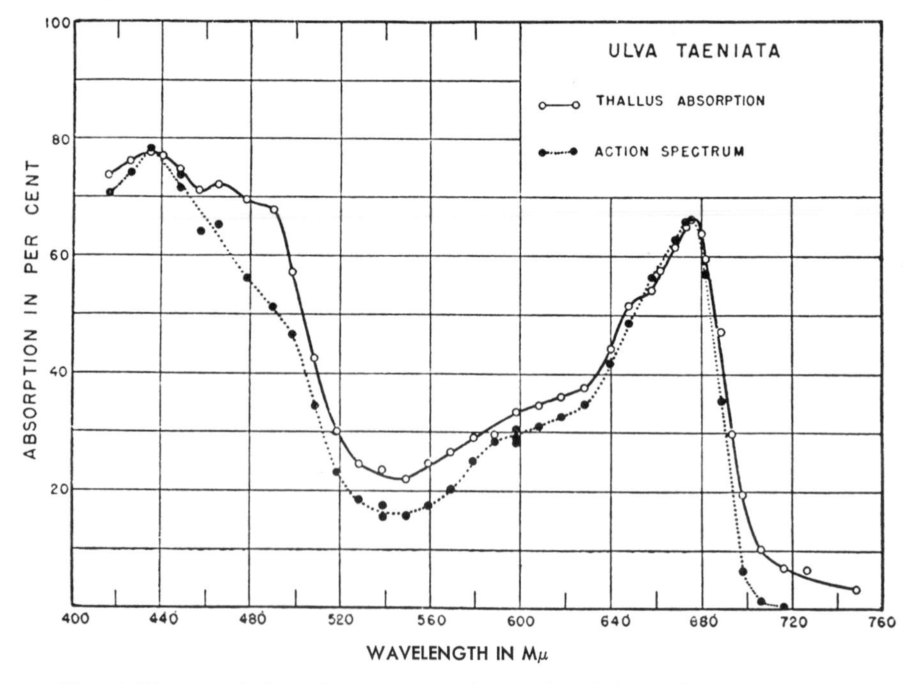

FIG. 1. Photosynthetic action spectrum of a species of Green Algae, *Ulva taeniata*. From Haxo and Blinks (1950; Fig. 8).

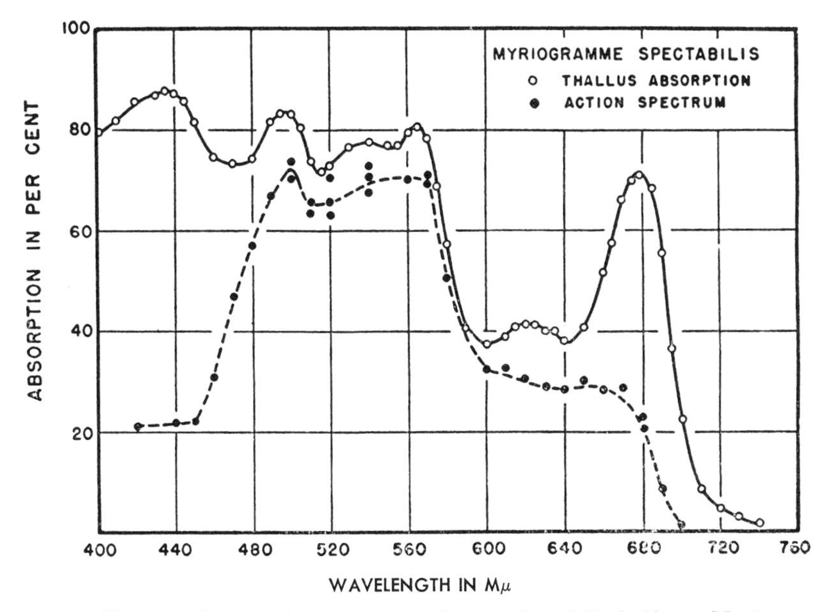

FIG. 2. Photosynthetic action spectrum of a species of Red Algae, *Myriogramme spectabilis*. From Haxo and Blinks (1950; Fig. 22).

tendency for the marine littoral flora to exhibit a descending sequence of predominantly Green, Brown, and Red Algae, indicating a correlation with the *quality* of the ambient light. Thus in deeper water, where blue and violet light prevails, the red pigments of the Rhodophyta appear adapted for its efficient absorption. Exceptions are not rare, however: many Red Algae grow in the tidal range, while many Green Algae can be found surprisingly deep in the sublittoral zone. On the other hand, Berthold (1882) and Oltmanns (1892) considered the *intensity* of light to be the chief controlling factor in the zonation of the algae at various depths. These two opposing views were reconciled by Harder's demonstration (1923) of the obvious fact that both wavelength and intensity of light are important. This has been confirmed by many workers (Ehrke, 1932; Montfort, 1933, 1934, 1938; Seybold, 1934, 1936; Levring, 1947; *et al.*). According to Montfort (1936), a sudden increase in light intensity causes "bright-light" algae to exhibit an increase in photosynthetic efficiency, whereas the efficiency of light utilization by "weak-light" algae is reduced by such treatment.

Observations by Dellow and Cassie (1955) on littoral zonation in caves indicated the ability of certain marine algae to grow in very dim light. Daytime illuminance as low as 5 to 250 lux permitted growth of *Hildenbrandia crouanii*, *Rhodochorton rothii* (?), and *Cladophora* sp.

The adaptation of marine algae to the various levels of light intensity in their habitat can also be demonstrated by determination of their tolerance to light of high intensity. Experiments in Plymouth showed that algae of the tidal zone and tide pools (*Porphyra laciniata, Cladophora utricularis, Plumaria elegans, Griffithsia flosculosa,* etc.) suffered no damage after 2 hours of direct exposure to sunlight, whereas sublittoral algae (*Dictyota dichotoma, Neomonospora pedicellata, Polyneura hilliae, Antithamnion cruciatum*) were killed by the same exposure (Biebl, 1952a). Similarly, at Pacific Grove the three intertidal algae investigated (*Ulva lobata, Cladophora trichotoma* and *Porphyra perforata*) survived after direct exposure to sunlight for 5 hours (519 kilolux-hours). However, among the sublittoral algae tested, *Pterochondria woodii, Microcladia californica,* and *Callophyllis marginifructa* were partially or completely killed by 1 hour of direct exposure to sunlight (97 kilolux-hours), *Polyneura latissima* by 1.5 hours (149 kilolux-hours), and *Botryoglossum farlowianum* and *Cryptopleura violacea* by 2 hours (205 kilolux-hours) (Biebl, 1956b). Experiments with various ranges of filtered sunlight showed that the damaging effects were primarily attributable to light of the shorter wavelengths. However, even radiation of wavelengths longer than 550 mμ, which passes a Schott filter OG_2, partially killed *Pterochondria woodii* and *Callophyllis marginifructa* during a 5-hour exposure period, and produced inflation of the plastids and other morphological evidence of damage in *Botryoglossum farlowianum.*

Although Red, Green, and Brown Algae exhibit characteristic absorption spectra in the 300 to 400 mμ range, below 290 mμ the absorption curves are essentially flat (Biebl, 1952a). The sensitivity of various marine algae to short-wave ultraviolet light, below 320 mμ, shows no regular correlation with their color or habitat (Biebl, 1952b; see Godward, Chapter 34, this treatise).

F. Hydrogen-Ion Concentration

As a result of photosynthesis of algae in tide pools, the pH of such small bodies of water may rise to 10 (Atkins, 1919–1922; Gail, 1918; Johnson and Skutch, 1928b; Feldmann and Davy de Virville, 1933; Davy de Virville, 1934–1935; Biebl, 1937; Doty, 1957). Indeed Moore *et al.* (1923) utilized the acidity of the water as a quantitative measurement for the photosynthetic activity of tide-pool algae. Kylin (1927), who investigated the resistance of a wide variety of marine algae in sea water with the pH adjusted to values between 3.6 and 10.0, found that the majority could survive for at least 1 to 3 days within a pH range of 6.8 to 9.6.

G. Nutrient Salts

Experimental investigations of the nutrition of attached marine algae have shown that the availability of nutrients is an important ecological factor (see Blinks, 1951; Henkel, 1952; Kinne-Diettrich, 1955; Schwenke, 1960). Under natural conditions, phosphorus or nitrogen is often limiting, and mineralization following autolysis of dead algae plays an important role in the regeneration of these elements. Hoffmann (1953, 1956; see also Hoffmann and Reinhardt, 1952) found that from dead thalli of *Fucus vesiculosus*, *Ulva lactuca* and *Laminaria saccharina*, as much as 30% of the organic phosphorus was remineralized within 12 hours.

References

Atkins, W. R. G. (1919–1922). The influence upon algal cells of an alteration in the hydrogen ion concentration of sea water. *J. Marine Biol. Assoc. United Kingdom* **12,** 789–791.

Atkins, W. R. G. (1926). Quantitative considerations of some factors concerned in plant growth in water. I. II. *J. conseil, Conseil permanent intern. exploration mer* **1,** 99–126, 197–226.

Atkins, W. R. G. (1939). Illumination in algal habitats. *Botan. Notiser*, pp. 145–147.

Atkins, W. R. G. (1945). Daylight and its penetration into the sea. *Trans. Illum. Eng. Soc. (London)* **10,** 1–12.

Ayres, A. H. (1916). The temperature coefficient of the duration of life of *Ceramium tenuissimum*. *Botan. Gaz.* **62,** 65–69.

Berthold, G. (1882). Über die Verteilung der Algen im Golf von Neapel. *Mitt. Zool. Sta. Neapel* **3,** 393–536.

Biebl, R. (1937). Ökologische und zellphysiologische Studien an Rotalgen der englischen Südküste. *Botan. Centr. Beih.* **A57,** 381–424.

Biebl, R. (1938a). Trockenresistenz und osmotische Empfindlichkeit der Meeresalgen verschieden tiefer Standorte. *Jahrb. wiss. Botan.* **86,** 350–386.

Biebl, R. (1938b). Zur Frage der Salzpermeabilität bei Braunalgen. *Protoplasma* **31,** 518–623.

Biebl, R. (1939a). Über Temperaturresistenz von Meeresalgen verschiedener Klimazonen und verschieden tiefer Standorte. *Jahrb. wiss. Botan.* **88,** 389–420.

Biebl, R. (1939b). Protoplasmatische Ökologie der Meeresalgen. *Ber. deut. botan. Ges.* **57,** 78–90.

Biebl, R. (1952a). Resistenz der Meeresalgen gegen sichtbares Licht und gegen kurzwellige UV-Strahlen. *Protoplasma* **44,** 353–377.

Biebl, R. (1952b). Ecological and non-environmental constitutional resistance of the protoplasm of marine algae. *J. Marine Biol. Assoc. United Kingdom* **31,** 307–315.

Biebl, R. (1956a). Zellphysiologisch-ökologische Untersuchungen an *Enteromorpha clathrata* (Roth) Greville. *Ber. deut. botan. Ges.* **69,** 75–86.

Biebl, R. (1956b). Lichtresistenz von Meeresalgen. *Protoplasma* **46,** 75–86.

Biebl, R. (1958). Temperatur- und osmotische Resistenz von Meeresalgen der bretonischen Küste. *Protoplasma* **50,** 217–242.

Blinks, L. R. (1951). Physiology and biochemistry of algae. *In* "Manual of Phycology—An Introduction to the Algae and their Biology" (G. M. Smith, ed.), pp. 263–291. Chronica Botanica, Waltham, Massachusetts.

Chapman, V. J. (1957). Marine algal ecology. *Botan. Rev.* **23,** 320–350.

Clarke, G. L. (1936). Light penetration in the western North Atlantic and its application to biological problems. *Rappts. et proc.-verb., J. conseil, Conseil permanent intern. exploration mer* **101,** Part 2.

Davy de Virville, A. (1934–1935). Recherches ecologiques sur la flore des flaques du littoral de l'Océan Atlantique et de la Manche. *Rev. gén. botan.* **46,** 705–721; **47,** 26–43, 96–114, 160–177, 230–243, 308–323.

Dellow, V., and Cassie, R. M. (1955). Littoral zonation in two caves in the Aukland district. *Trans. Roy. Soc. New Zealand* **83,** 321–331.

Doty, M. S. (1957). Rocky intertidal surfaces. *In* "Treatise on Marine Ecology and Paleoecology" (J. W. Hedgpeth, ed.) *Geol. Soc. Am. Mem.* **No. 67,** Vol. I, Chapter 18, pp. 535–585.

Ehrke, G. (1932). Über die Assimilation komplementär gefärbter Meeresalgen im Lichte von verschiedenen Wellenlängen. *Planta* **17,** 650–665.

Engelmann, T. W. (1883). Farbe und Assimilation. *Botan. Zentr.* **41,** 1–29.

Engelmann, T. W. (1884). Untersuchungen über die quantitativen Beziehungen zwischen Absorption des Lichtes und Assimilation in Pflanzenzellen. *Botan. Zentr.* **42,** 82–95.

Eppley, R. W. (1958). Potassium-dependent sodium extrusion by cells of *Porphyra perforata*, a red marine alga. *J. Gen. Physiol.* **42,** 281–288.

Eppley, R. W., and Cyrus, C. C. (1960). Cation regulation and survival of red alga, *Porphyra perforata*, in diluted and concentrated sea water. *Biol. Bull.* **118,** 55–65.

Ernst, J., and Feldmann, J. (1957). Une nouvelle Delessériacée des Côtes de Bretagne: *Drachiella spectabilis* nov. gen. nov. spec. *Rev. gén. botan.* **64,** 3–15.

Feldmann, J. (1937). Recherches sur la végétation marine de la Méditerranée. La côte des Albères. *Rev. Algol.* **10,** 1–339.

Feldmann, J. (1951). Ecology of marine algae. *In* "Manual of Phycology—an Introduction to the Algae and their Biology" (G. M. Smith, ed.), pp. 313–334. Chronica Botanica, Waltham, Massachusetts.

Feldmann, J., and Davy de Virville, A. (1933). Les conditions physiques et la végétation des flaques littorales de la côte des Albères. *Rev. gén. botan.* **45,** 621–654.

Fischer, E. (1929). Recherches de bionomie et d'océanographie littorale sur la Rance et le littoral de la Manche. *Ann. inst. océanog. (Paris)* **5,** 203–429.

Fritsch, F. E. (1945). "The Structure and Reproduction of the Algae," Vol. II. Cambridge Univ. Press, London and New York.

Gaidukov, N. (1902). Über den Einfluss farbigen Lichtes auf die Färbung lebender Oscillatorien. *Abhandl. preuss. Akad. Wiss., Math-naturw. Kl.*, pp. 1–36.

Gaidukov, N. (1904). Zur Farbanalyse der Algen. *Ber. deut. botan. Ges.* **22,** 23.

Gail, F. W. (1918). Some experiments with *Fucus* to determine the factors controlling its vertical distribution. *Publ. Puget Sound Biol. Sta. Univ. Wash.* **2,** 139–152.

Gessner, F. (1955). "Hydrobotanik: Die physiologischen Grundlagen der Pflanzenverbreitung im Wasser," Vol. I: Energiehaushalt. VEB Deutscher Verlag der Wissenschaften, Berlin.

Gessner, F. (1959). "Hydrobotanik: Die physiologischen Grundlagen der Pflanzenverbreitung im Wasser," Vol. II: Stoffhaushalt. VEB Deutscher Verlag der Wissenschaften, Berlin.

Harder, R. (1923). Über die Bedeutung von Lichtintensität und Wellenlänge für die Assimilation farbiger Algen. *Z. Botan.* **15,** 305–355.

Haxo, F. T., and Blinks, L. R. (1950). Photosynthetic action spectra of marine algae. *J. Gen. Physiol.* **33,** 389–442.

Heim, R. *Ed.* (1959). "Ecologie des Algues Marines, Dinard 1957." *Colloq. intern. centre natl. recherche sci. (Paris)* **81,** 276 pp.

Henkel, R. (1952). Ernährungsphysiologische Untersuchungen an Meeresalgen, insbesonders an *Bangia pumila. Kiel. Meeresforsch.* **8,** 192–211.

Höfler, K. (1930). Das Plasmolyse-Verhalten der Rotalgen. *Z. Botan.* **23,** 370–388.

Höfler, K. (1931). Hypotonietod und osmotische Resistenz einiger Rotalgen. *Österr. Botan. Z.* **80,** 51–71.
Hoffmann, C. (1929). Die Atmung der Meeresalgen und ihre Beziehung zum Salzgehalt. *Jahrb. wiss. Botan.* **71,** 214–268.
Hoffmann, C. (1943). Der Salzgehalt des Seewassers als Lebensfaktor mariner Pflanzen. *Kiel. Blätter* pp. 160–176.
Hoffmann, C. (1953). Weitere Beiträge zur Kenntnis der Remineralisation des Phosphors bei Meeresalgen. *Planta* **42,** 156–176.
Hoffmann, C. (1956). Untersuchungen über die Remineralisation des Phosphors im Plankton. *Kiel. Meeresforsch.* **12,** 25–36.
Hoffmann, C. (1959). Études écologiques et physiologiques de quelques algues de la mer Baltique. *In* "Ecologie des Algues marines, Dinard 1957." *Colloq. intern. centre natl. recherche sci. (Paris)* **81,** 205–218.
Hoffmann, C., and Reinhardt, M. (1952). Zur Frage der Remineralisation des Phosphors bei Benthosalgen. *Kiel. Meeresforsch.* **8,** 135–144.
Isaac, W. E. (1933). Some observations and experiments on the drought resistance of *Pelvetia canaliculata. Ann. Botany (London)* **47,** 343–348.
Isaac, W. E. (1935). Preliminary study of the water loss of *Laminaria digitata* during intertidal exposure. *Ann. Botany (London)* **49,** 109–117.
Jerlov, N. G. (1951). Optical studies of ocean waters. Repts. Swed. Deep-Sea Exped., 1947/1948, Vol. III: Physics and Chemistry, Fasc. 1, pp. 1–59.
Johnson, D. S., and Skutch, A. F. (1928a). The littoral vegetation on a headland of Mt. Desert Island, Maine. I. Submersible or strictly littoral vegetation. *Ecology* **9,** 188–215.
Johnson, D. S., and Skutch, A. F. (1928b). The littoral vegetation on a headland of Mt. Desert Island, Maine. II. Tide-pools and the environment and classification of submersible plant communities. *Ecology* **9,** 307–338.
Kändler, R. (1951). Der Einfluss der Wetterlage auf die Salzgehaltsschichtung im Übergangsgebiet zwischen Nord- und Ostee. *Deut. Hydrograph. Z.* **4,** (4/5/6), 150–160.
Kanwisher, J. (1957). Freezing and drying in intertidal algae. *Biol. Bull.* **113,** 275–285.
Kinne-Diettrich, E. (1955). Beiträge zur Kenntnis der Ernährungsphysiologie mariner Blaualgen. *Kiel. Meeresforsch.* **11,** 34–47.
Kylin, H. (1917). Über die Kälteresistenz der Meeresalgen. *Ber. deut. botan. Ges.* **35,** 370–384.
Kylin, H. (1927). Über den Einfluss der Wasserstoffionenkonzentration auf einige Meeresalgen. *Botan. Notiser,* pp. 243–254.
Kylin, H. (1938). Über den osmotischen Druck und die osmotische Resistenz einiger Meeresalgen. *Svensk Botan. Tidskr.* **32,** 238–248.
Levitt, J. (1956). "The Hardiness of Plants." Academic Press, New York.
Levitt, J. (1958). Frost, drought and heat resistance. *In* "Protoplasmatologia—Handbuch der Protoplasmaforschung" (L. V. Heilbrunn and F. Weber, eds.), Vol. VIII, pp. 1–87. Springer, Vienna.
Levring, T. (1947). Submarine daylight and the photosynthesis of marine algae. *Göteborgs Kgl. Vetenskaps-Vitterhets-Samhäll. Handl. Ser.* **B5,** 1–90.
Levring, T. (1960). Submarines Licht und die Algenvegetation. *Botan. Marina* **1,** 67–73.
Levring, T., and Fish, G. R. (1956). The penetration of light in some tropical East African waters. *Oikos* **7,** 98–109.

Molisch, H. (1897). "Untersuchungen über das Erfrieren der Pflanzen." Fischer, Jena, Germany.

Montfort, C. (1931). Assimilation und Stoffgewinn der Meeresalgen bei Aussüssung und Rückversalzung. *Ber. deut. botan. Ges.* **49,** (50–(58).

Montfort, C. (1933). Über Lichtempfindlichkeit und Leistungen roter Tiefseealgen und Grottenflorideen an freier Meeresoberfläche. *Protoplasma* **19,** 385–413.

Montfort, C. (1934). Farbe und Stoffgewinn im Meer. *Jahrb. wiss. Botan.* **79,** 493–592.

Montfort, C. (1936). Umwelt, Erbgut und physiologische Gestalt. I. Lichttod und Starklichtresistenz bei Assimilationsgeweben. *Jahrb. wiss. Botan.* **84,** 1–57.

Montfort, C. (1937). Die Trockenresistenz der Gezeitenpflanzen und die Frage der Übereinstimmung von Standort und Vegetation. *Ber. deut. botan. Ges.* **55,** (85)–(95).

Montfort, C. (1938). Funktionstypen des Assimilationsapparates und das Problem der gelben Blattfarbstoffe. I. Aufdeckung und vergleichende Analyse der Carotinoidwirkungen. *Kiel. Meeresforsch.* **2,** 301–344.

Montfort, C. (1940). Die Photosynthese brauner Zellen im Zusammenwirken von Chlorophyll und Carotinoiden. *Z. physik. Chem.* (*Leipzig*) **A186,** 57.

Montfort, C. (1942). Vergleichende Untersuchungen zur quantitativen Auswertung von Absorptionskurven für Fragen der Lichtenergiebilanz. *Botan. Arch.* **43,** 322.

Montfort, C., Ried, A., and Ried, J. (1955). Die Wirkung kurzfristiger warmer Bäder auf Atmung und Photosynthese im Vergleich von eurythermen und kaltstenothermen Meeresalgen. *Beitr. Biol. Pflanz.* **31,** 349–375.

Montfort, C. Ried, A., and Ried, J. (1957). Abstufungen der funktionellen Wärmeresistenz bei Meeresalgen in ihren Beziehungen zur Umwelt und Erbgut. *Biol. Zentr.* **76,** 257–289.

Moore, B., Whitley, E., and Webster, T. A. (1923). Studies of photosynthesis in marine algae. *Trans. Liverpool Biol. Soc.* **37,** 38–51.

Muenscher, W. L. G. (1915). Ability of seaweeds to withstand desiccation. *Publ. Puget Sound Biol. Sta., Univ. Wash.* **1,** 19–23.

Oltmanns, F. (1892). Über die Kultur- und Lebensbedingungen der Meeresalgen. *Jahrb. wiss. Botan.* **23,**

Oltmanns, F. (1905). "Morphologie und Biologie der Algen." 1st ed., Vol. II. Fischer, Jena, Germany.

Oltmanns, F. (1923). "Morphologie und Biologie der Algen." 2nd ed., Vol. III. Fischer, Jena, Germany.

Osterhout, W. J. V. (1906). The resistance of certain marine algae to changes in osmotic pressure and temperature. *Univ. Calif.* (*Berkeley*) *Publ., Publ. Botany* **2,** 227–228.

Parker, J. (1960). Seasonal changes in cold-hardiness of *Fucus vesiculosus*. *Biol. Bull.* **119,** 474–478.

Parr, A. E. (1939). Quantitative observations on the pelagic *Sargassum* vegetation of the western North Atlantic. *Bull. Bingham Oceanog. Collection* **VI,** 1–94.

Poole, H. H., and Atkins, W. R. G. (1926). On the penetration of light into sea-water. *J. Marine Biol. Assoc. United Kingdom* **41,** 177–198.

Poole, H. H., and Atkins, W. R. G. (1928). Further photo-electric measurements of the penetration of light into sea-water. *J. Marine Biol. Assoc. United Kingdom* **15,** 455–483.

Schwenke, H. (1958a). Über die Salzgehaltsresistenz einiger Rotalgen der Kieler Bucht. *Kiel. Meeresforsch.* **14,** 11–22.

Schwenke, H. (1958b). Über einige zellphysiologische Faktoren der Hypotonieresistenz mariner Rotalgen. *Kiel. Meeresforsch.* **14,** 130–150.

Schwenke, H. (1959). Untersuchungen zur Temperaturresistenz mariner Algen der westlichen Ostee. I. Das Resistenzverhalten von Tiefenrotalgen bei ökologischen und nichtökologischen Temperaturen. *Kiel. Meeresforsch.* **15,** 34–50.

Schwenke, H. (1960). Neuere Erkenntnisse über die Beziehungen zwischen den Lebensfunktionen mariner Pflanzen und dem Salzgehalt des Meer- und Brackwassers. *Kiel. Meeresforsch.* **16,** 28–47.

Seybold, A. (1934). Über die Lichtenergiebilanz submerser Wasserpflanzen, vornehmlich der Meeresalgen. *Jahrb. wiss. Botan.* **79,** 593–654.

Seybold, A. (1936). Über den Lichtfaktor photophysiologischer Prozesse. *Jahrb. wiss. Botan.* **82,** 741–795.

Seybold, A., and Weissweiler, A. (1942). Weitere spektrophotometrische Messungen an Laubblättern und an Chlorophylllösungen sowie an Meeresalgen. *Botan. Arch.* **44,** 102-153.

Stocker, O., and Holdheide, W. (1937). Die Assimilation Helgoländer Gezeitenalgen während der Ebbezeit. *Z. Botan.* **32,** 1–59.

Terumoto, I. (1959). Frost and drought injury in lake-balls. *Low Temp. Sci., Ser.* **B17,** 1–7.

Terumoto, I. (1960a). Frost-resistance in a marine alga, *Ulva pertusa* Kjellman. *Low Temp. Sci. Ser.* **B18,** 35–38.

Terumoto, I. (1960b). Effect of protective agents against freezing injury in lake ball. *Low Temp. Sci. Ser.* **B18,** 43–50.

Zaneveld, J. S. (1937). The littoral zonation of some Fucaceae in relation to desiccation. *J. Ecol.* **25,** 431–468.

—54—

Lichens

VERNON AHMADJIAN

Department of Biology, Clark University, Worcester, Massachusetts

I. Introduction

By means of cultural techniques it has now been established that the algal symbionts of lichens, hereafter referred to as the "phycobionts" (Scott, 1957), are independent organisms quite comparable and related to similar free-living algae. There is no good evidence for a genetic relationship between the alga and fungus in a lichen thallus, as postulated by Schmidt (1953). Some thirty genera of algae have been reported as occurring in lichen associations (Ahmadjian, 1958), but a considerable number of phycobionts belong to the chlorophyte genus *Trebouxia* (cf. Des Abbayes, 1951).

Physiological races of lichen algae, differing for the most part in their rates of growth, temperature optima, and nutrient preferences, have been reported (Chodat, 1913; Thomas, 1939; Ahmadjian, 1959a). However, there is no evidence that each lichen species has its own specific phycobiont or that the various races of algae result from the influence of the fungal partner. In some cases, the same species of lichen from different locations may embody different phycobionts. For example, from two samples of the marine lichen *Lichina confinis* from Norway and Spain it was possible to establish cultures of *Calothrix pulvinata*, whereas a third sample from southern Sweden yielded *Calothrix crustacea* (Ahmadjian, unpublished). Lichen algae should be viewed as autonomous organisms, unwilling partners in the lichen association.

Most published accounts of the physiological nature of the lichen association are speculative. Rarely have modern techniques been employed in studies of this relationship. For example, the use of radio-isotopes would enable one to follow the transfer of metabolites from the algal to the fungal partner. "Symbiosis" and "mutualism" have been the most popular terms applied to the lichen association. Leonian (1936) suggested that the fungal partners derive growth substances from the phycobionts. Quispel (1943–1945) postulated an exchange of nutrilites or metabolites between the two partners, the alga receiving from the fungus ascorbic acid or other substances which may promote photosynthesis, while the fungus receives from the phycobiont certain indispensable metabolites, possibly vitamins such as thiamine and biotin, which have been shown to be necessary for the growth of some lichen fungi (Zehnder, 1949; Hale, 1958; Ahmadjian, 1961). The successful resynthesis of isolated components of *Acarospora fuscata* clearly showed a symbiotic relation between the two partners; indirect evidence revealed that the mycobiont received carbon and nitrogen compounds from the phycobiont, which it supplied with mineral salts (Ahmadjian, unpublished). However, other evidence suggests that a lichen association should be regarded as a case of parasitism rather than of symbiosis or mutualism. The many reports of fungal haustoria penetrating the cells of the phycobiont (Geitler, 1955; Schiman, 1957; Ahmadjian, 1959a) support this view. Electron micrographs of *Cladonia cristatella* and *Lecidea* sp., representative of the two largest lichen genera, show haustoria penetrating the cell walls and making direct contact with the protoplast membrane of the phycobiont (Moore and McAlear, 1960), features which have not been revealed by light microscopy. Proponents of the parasitism theory also stress the common occurrence of normally free-living algae as lichen phycobionts. It would be unwise to assert that all lichenized associations are either entirely symbiotic or entirely parasitic. There are undoubtedly examples of each condition, and of both in varying degrees, depending on the species concerned and on the nature and location of the substratum.

Many lichens are characterized by the production of special pigments, such as depsides, depsidones, and other phenolic derivatives (see Hale, 1961). There has apparently been no record of a depside or depsidone produced by an isolated mycobiont (Hale, 1961), but other tannin-like Substances are formed both by fungi and by algae (see Ogino, Chapter 26, this treatise).

II. Analysis and Resynthesis of Lichen Associations

Despite their discouragingly slow growth, the separate cultivation of the individual lichen components is easily accomplished in the laboratory. Single-celled Green phycobionts of many lichens can often be directly isolated from the crushed thallus by means of a micropipette (Warén, 1918–1919; Jaag, 1929). Filamentous Green or Blue-green phycobionts can be isolated by transferring small fragments of the lichen thallus into an illuminated mineral solution, and allowing them to outgrow the heterotrophic fungal partner. Although each of the lichen components can be grown separately under suitable cultural conditions, most of the numerous experimental attempts to establish lichen synthesis in the laboratory were clearly unsuccessful or inconclusive (Hérisset, 1946; Zehnder, 1949; Mish, 1953; Tomaselli, 1958), while three early reports of apparently successful synthesis (Stahl, 1877; Bonnier, 1889; Thomas, 1939) are open to doubt since none has been duplicated (Ahmadjian, 1959c). It has been shown that the lichenized condition is unstable in a medium containing organic compounds; under such conditions the components of the composite plant tend to dissociate and develop independently (Thomas, 1939; Ahmadjian, 1959b). Initial stages of lichen synthesis were achieved under laboratory conditions only on a completely inorganic agar medium, in which the fungal component was forced, as it were, by starvation into physiological union with its phycobiont (Thomas, 1939; Ahmadjian, 1959b). On a washed agar substrate without supplements of any kind, lichen tissue consisting of a pseudoparenchyma with some evidence of initial thallus differentiation was formed by isolated components of *Acarospora fuscata* (Ahmadjian, unpublished). The lichenized association in this case was evidently dependent on the balanced growth of its fungal and algal components.

III. Physiology of the Phycobionts

Bond and Scott (1955), using isotopic nitrogen, showed that nitrogen fixation by the Blue-green phycobiont of *Collema granosum* occurs within the lichen thallus. Henriksson (1951, 1957) demonstrated that a species of *Nostoc* isolated from *Collema tenax* is capable of fixing atmospheric nitrogen, and in pure culture excretes into the medium various amounts of nitrogenous compounds depending on the cultural conditions. She concluded that the fungal partner of these lichens is normally dependent on organic substances excreted by the phycobiont during its growth (Henriksson, 1958). In culture, however, the mycobiont of this lichen has apparently a lethal action on its customary phycobiont, as well as on other algae (Henriksson, 1958; Ahmadjian and Henriksson, 1959).

There is general agreement that lichen algae grow better in media supplemented with organic compounds, and in such media many, mostly species of *Trebouxia*, can develop in complete darkness, though their growth is slower and their color may be paler than that of cells grown in light. Hexose sugars, especially glucose and fructose, appear to be the most readily assimilable sources for these algae. For most phycobionts, particularly when cultured in darkness, inorganic sources of nitrogen are generally inferior to organic compounds, and result in slower growth. The best source of nitrogen varies; some phycobionts favor asparagine, glycine, or alanine, others show preference for peptone, and the form and color of colonies on agar tend to vary accordingly (Warén, 1918–1919). Quispel (1943–1945) found that the addition of lichen fungus extract, nicotinic acid, or yeast extract, has a stimulatory effect on the growth of *Trebouxia* phycobionts. Zehnder (1949), on the other hand, reported that synthetic vitamins did not influence the growth of the phycobionts he investigated, which indeed may produce vitamin B_1 in excess. The growth of various phycobionts in culture has been shown to be somewhat inhibited by yeast extract (0.5%; *Trebouxia* spp.; Ahmadjian, unpublished), by indoleacetic acid (1.25 mg./liter) and by streptomycin (10–25 mg./liter) (Zehnder, 1949, 1951).

Many lichen algae (*Trebouxia*) can tolerate a wide range of pH. Quispel (1943–1945) demonstrated good growth of the algae between pH 4.0 and 7.4, while Mish (1953) reported a tolerance between pH 4.1 and 9.1 with an optimum at 7.4.

The optimum growth temperature of *Trebouxia* phycobionts varies widely from 4° to 24°C., depending on the algal strain, its position in the lichen thallus (Thomas, 1939), and the climatic conditions under which it naturally grows. The temperature tolerance range of these algae in the dried condition is extremely wide. Mish (1953) demonstrated that a *Trebouxia* phycobiont from *Umbilicaria papulosa* could survive temperatures as low as −197°C. and could be preserved alive by lyophilization. Lange (1953) showed that some phycobionts in the dry state could withstand temperatures of 100°C. Many lichen algae can likewise withstand prolonged desiccation, some as long as 8 months (Ahmadjian, unpublished).

The study of the effect of light on the growth of lichen phycobionts is an interesting field, still relatively untouched. Many phycobionts are components of lichens with a dense and heavily pigmented cortex of fungal hyphae, which must greatly reduce the intensity of light reaching the algal cells within the lichen thallus. Ahmadjian (1959a) showed that certain phycobionts from such heavily pigmented lichens are unable to survive under light intensities which are suboptimal for many other lichen algae. Sustained growth of the phycobiont of *Buellia punctata* on mineral medium was light-dependent only at intensities up to about 500 lux (Lewin, personal communication).

References

Ahmadjian, V. (1958). A guide for the identification of algae occurring as lichen symbionts. *Botan. Notiser* **111,** 632–644.

Ahmadjian, V. (1959a). Experimental observations on the algal genus *Trebouxia* de Puymaly. *Svensk Botan. Tidskr.* **53,** 71–80.

Ahmadjian, V. (1959b). A contribution toward lichen synthesis. *Mycologia* **51,** 56–60.

Ahmadjian, V. (1959c). The taxonomy and physiology of lichen algae and problems of lichen synthesis. Ph.D. Dissertation, Harvard University, Cambridge, Massachusetts.

Ahmadjian, V., and Henriksson, E. (1959). Parasitic relationship between two culturally isolated and unrelated lichen components. *Science* **130,** 1251.

Ahmadjian, V. (1961). Studies on lichenized fungi. *The Bryologist* **64,** 168–179.

Bond, G., and Scott, G. D. (1955). An examination of some symbiotic systems for fixation of nitrogen. *Ann. Botany (London)* **19,** 67–77.

Bonnier, G. (1889). Recherches sur la synthèse des lichens. *Ann. soc. nat.* [7] **9,** 1–34.

Chodat, R. (1913). Monographie d'algues en culture pure. *Matér. la flore cryptogam. Suisse* **4** (2), 1–266.

Des Abbayes, H. (1951). "Traité de lichénologie," 217 pp. Paul Lechevalier, Paris, France.

Geitler, L. (1955). Gehäufte Haustorien bei einer Collematacee. *Österr. Botan. Z.* **102,** 317–321.

Hale, M. E. (1958). Vitamin requirements of three lichen fungi. *Bull. Torrey Botan. Club* **85,** 182–187.

Hale, M. E. (1961). "Lichen Handbook," 178 pp. Smithsonian Institution Pub. 4434, Washington, D. C.

Henriksson, E. (1951). Nitrogen fixation by a bacteria-free, symbiotic *Nostoc* strain isolated from *Collema. Physiol. Plantarum* **4,** 542–545.

Henriksson, E. (1957). Studies in the physiology of the lichen *Collema.* I. The production of extracellular nitrogenous substances by the algal partner under various conditions. *Physiol. Plantarum* **10,** 943–948.

Henriksson, E. (1958). Studies in the physiology of the lichen *Collema.* II. A preliminary report on the isolated fungal partner with special regard to its behavior when growing together with the symbiotic alga. *Svensk Botan. Tidskr.* **52,** 391–396.

Hérisset, A. (1946). Démonstration expérimentale du rôle du *Trentepohlia umbrina* (Kg.) Born. dans la synthèse des Graphidèes corticoles. *Compt. rend. Acad. sci.* **222,** 100–102.

Jaag, O. (1929). Recherches expérimentales sur les gonidies des lichens appartenant aux genres *Parmelia* et *Cladonia. Bull. soc. botan. Genève* **21,** 1–119.

Lange, O. L. (1953). Hitze- und Trockenresistenz der Flechten in Beziehung zu ihrer Verbreitung. *Flora (Jena)* **140,** 39–97.

Leonian, L. H. (1936). Effect of auxins from some green algae upon *Phytophthora cactorum. Botan. Gaz.* **97,** 854–859.

Mish, L. B. (1953). Biological studies of symbiosis between the alga and fungus in the rock lichen *Umbilicaria papulosa.* Ph.D. Dissertation, Harvard University, Cambridge, Massachusetts.

Moore, R. T., and J. H. McAlear. (1960). Fine structure of mycota. 2. Demonstration of the haustoria of lichens. *Mycologia* **52,** 805–807.

Quispel, A. (1943–1945). The mutual relations between algae and fungi in lichens. *Rec. trav. botan. néerl,* **40,** 413–541.

Schiman, H. (1957). Beiträge zur Lebensgeschichte homoeomerer und heteromerer Cyanophyceen-Flechten. *Öster. botan. Z.* **104,** 409–453.

Schmidt, A. (1953). Essai d'une biologie de l'holophyte des lichens. *Mem. muséum natl. hist. nat., sér botan.* **B3,** 1–159.

Scott, G. D. (1957). Lichen terminology. *Nature* **179,** 486–487.

Stahl, E. (1877). Beiträge aur Entwicklungsgeschichte der Flechte. Heft II. Über die Bedeutung der Hymenialgonidien. Leipzig, 32 pp.

Thomas, E. A. (1939). Über die Biologie von Flechtenbildern. *Beitr. Kryptogamenfl. Schweiz.* **9,** 6–208.

Tomaselli, R. (1958). "Esperienze di simbiosi 'innaturali' di lichni 'in vitro'," pp. 1–28. Tipografia Valbonesi, Forli, Italy.

Warén, H. (1918–1919). Reinkulturen von Flechtengonidien. *Ovfers. Finska Vetenskaps-Soc. Förhand.* **61,** 1–79.

Zehnder, A. (1949). Über den Einfluss von Wuchsstoffen auf Flechtenbildner. *Ber. Schweiz. botan. Ges.* **59,** 201–267.

Zehnder, A. (1951). Über den Einfluss antibiotischer Stoffe auf das Wachstum von Grünalgen. *Experientia* **7,** 100–102.

—55—

Endozoic Algae

JOHN J. A. McLAUGHLIN AND PAUL A. ZAHL

Haskins Laboratories, New York City, New York

I. Introduction

Unicellular symbiotic algae inhabiting the protoplasm of certain protozoa and the tissues and organs of a variety of metazoa are traditionally called "zoochlorellae" and "zooxanthellae"; see reviews by Caullery (1950); Buchner (1953); Yonge 1957). Originally these terms were employed in the taxonomic sense, but at present they are used only in descriptive reference to the color of the symbionts—respectively green, and light yellow to deep brown. Their variety and occurrence have been reviewed by Fritsch (1935, 1952). The following discussion relates chiefly to experimental studies on the algae themselves, which, though adapted to live and multiply in animal tissue, can nevertheless be cultured axenically.

II. Zoochlorellae

Whereas zoochlorellae are found mainly in the tissues of freshwater invertebrates, a few animal species of brackish-water or marine habitats possess them. In most cases, they are assumed to be strains or races of the genus *Chlorella* (Chlorophyta), but in *Paulinella chromatophora* the symbiont is a sausage-shaped cyanophyte.

Among the most interesting examples of animals associated with zoochlorellae is the turbellarian *Convoluta roscoffensis* (Keeble and Gamble, 1907). Initial investigation labeled the alga as a *Chlorella*, but when flagellate stages were observed it was assigned to the genus *Carteria* (also Chlorophyta). The symbiont probably enters the host by ingestion, and, as in the case of all intra-tissue invaders, presumably possesses some method of avoiding digestion by the host. However, reduced algal growth or lack of nutritional factors may induce host animals to digest their algae. By what mechanism the alga thus becomes susceptible to digestion is unknown.

Pringsheim (1915) was the first to cultivate the alga-inhabited *Paramecium bursaria* in a sterile inorganic medium. Siegel (1960) demonstrated that its algae, when isolated in axenic culture, exhibit host-strain differences, and that cross-infection of host strains is difficult but possible, either in the course of normal conjugation or by reinfection of desymbiotized host cells. Strain differences of the zoochlorellae were also manifested *in vitro*.

III. Zooxanthellae

Marine animals harboring zooxanthellae include certain protozoa, sponges, coelenterates, worms, rotifers, echinoderms, mollusks, and ascidians. Examination of the algal pigments extracted from various coelenterates led investigators to consider some of the zooxanthellae as dinoflagellates; others were classed as Blue-green or as Green Algae. Except for a few cases, the taxonomy of zooxanthellae has remained unsettled because we lack experimental studies of the algae in culture.

IV. Axenic Culture of Symbiotic Algae

Much of the confusion attending the taxonomy of symbiotic algae is due at least in part to the failure, until recently, to elicit motile forms from the vegetative cells in cultures. This was first accomplished by Kawaguti (1944), who isolated a *Gymnodinium* from the coral *Acropora corymbosa*. The authors (Zahl and McLaughlin, 1957, 1959; McLaughlin and Zahl, 1957, 1959; Freudenthal, 1962) obtained axenic cultures from sea anemones and corals by homogenizing or tissue-stripping the animals, treating a centrifugally concentrated mass of the algal cells with mixtures of antibiotics, and isolating the cells in synthetic or enriched sea-water media. A medium has been defined which supports the growth of symbionts from the coelenterates *Cassiopeia*, *Anthopleura*, *Condylactis*, and *Plexaurella*. These zooxanthellae are all dinoflagellate species.

Siegel's work (1960) on the zoochlorellae of *Paramecium* was confined to isolation of the symbiont on a mineral agar medium. No nutritional investigation of these algae has been reported. Strains of *Zoochlorella* isolated from various freshwater sponges and ciliates do not require organic growth factors when grown in pure culture (Lewin, 1961).

The following nutritional and physiological data on gymnodinioids in axenic culture have been obtained. (i) Vegetatively reproducing, though non-motile, cells can tolerate bactericidal concentrations of antibiotics in a mixture containing penicillin K (200 mg./liter), streptomycin (40 mg./liter), chloramphenicol (20 mg./liter), and oxytetracycline (2 mg./liter). (ii) Free-living motile stages can be elicited in synthetic media under alternating light and darkness. (iii) Growth of the vegetative forms is rapid in liquid media or on semi-solid slants. (iv) Growth is good at temperatures from 18 to 30°C. (v) Cultures on agar slants can withstand −5°C. for 48 hours. (vi) Either ammonium salts or nitrate can serve as a suitable nitrogen source. (vii) Organic nitrogen and phosphate sources, such as amino acids, nucleotides, and glycerophosphates are readily utilized. (viii) Salinity tolerances for good growth fall between 1.2 and 3.6% NaCl. (ix) Production of O_2 *in vitro* is appreciable in light at 6000–8000 lux.

V. Role of Symbiotic Algae

By the use of $C^{14}O_2$, the direct transfer of soluble photosynthetic products from alga to host was shown likely in the sea-anemone *Anthopleura elegantissima* and in the freshwater hydroid *Chlorohydra viridissima* (Muscatine and Hand, 1958; Lenhoff and Zimmermann, 1959). Some workers postulate that calcium deposition in corals, like other physiological activities of the host, may be considerably influenced by the symbionts. The reciprocal use of the host's products by the plants has led many investigators to regard CO_2, ammonia, and other animal catabolites as essential to the symbiont. Most investigators consider that the evolution of oxygen by the algal cell is useful to the host tissues, concluding that, though not absolutely essential, it may be as beneficial to the host metabolism as is carbon dioxide removal.

References

Buchner, P. (1953). "Endosymbiose der Tiere mit pflanzlichen Mikroorganismen." Berkhauser, Basel, Switzerland, and Stuttgart, Germany.

Caullery, M. (1950). Le parasitisme et la symbiose. *In* "Encyclopédie Scientifique," 2nd ed. Dion, Paris.

Freudenthal, H. D. (1962). *Symbiodinium* gen. nov. and *Symbiodinium microadriaticum* sp. nov., a zoozanthella; taxonomy, life cycle, and morphology. *J. Protozool.* **9**, 45–52.

Fritsch, F. E. (1935). "The Structure and Reproduction of The Algae," Vol. I. Cambridge Univ. Press, London and New York.

Fritsch, F. E. (1952). Algae in association with heterotrophic or holozoic organisms. *Proc. Roy. Soc.* **B139**, 185–192.

Kawaguti, S. (1944). On the physiology of reef corals. VII. Zooxanthella of the reef corals is *Gymnodinium* sp., Dinoflagellata; its culture *in vitro*. *Palao Trop. Biol. Sta. Studies* **2**, 675–679.

Keeble, F., and Gamble, F. W. (1907). The origin and nature of the green cells of *Convoluta roscoffensis. Quart. J. Microscop. Sci.* **51,** 167–219.

Lenhoff, H. M., and Zimmermann, K. F. (1959). Biochemical studies of symbiosis in *Chlorohydra viridissima. Anat. Record* **134,** 599.

Lewin, R. A. (1961). Phytoflagellates and algae. *In* "Handbuch der Pflangenphysiologie" (W. Ruhland, ed.), **14,** 401–417.

McLaughlin, J. J. A., and Zahl, P. A. (1957). Studies in marine biology. II. *In vitro* culture of zooxanthellae. *Proc. Soc. Exptl. Biol. Med.* **95,** 115–120.

McLaughlin, J. J. A., and Zahl, P. A. (1959). Studies in marine biology. III. Axenic zooxanthellae from various invertebrate hosts. *Ann. N. Y. Acad. Sci.* **77,** 55–72.

Muscatine, L., and Hand, C. (1958). Direct evidence for the transfer of materials from symbiotic algae to the tissues of a coelenterate. *Proc. Natl. Acad. Sci. U. S.* **44,** 1259–1263.

Pringsheim, E. G. (1915). Die Kultur von *Paramecium bursaria. Biol. Zentr.* **35,** 375–379.

Siegel, R. W. (1960). Hereditary endosymbiosis in *Paramecium bursaria. J. Exptl. Cell Research* **19,** 239–252.

Yonge, C. M. (1957). Symbiosis. *In* "Treatise on Marine Ecology and Paleoecology" (J. W. Hedgpeth, ed.), Vol. I, pp. 429–442. *Geol. Soc. Am. Mem.* **No. 67.**

Zahl, P. A., and McLaughlin, J. J. A. (1957). Isolation and cultivation of zooxanthellae. *Nature* **180,** 199–200.

Zahl, P. A., and McLaughlin, J. J. A. (1959). Studies in marine biology. IV. On the role of algal cells in the tissues of marine invertebrates. *J. Protozool.* **6,** 344–352.

—APPENDIX A—

Classification of Algae

PAUL C. SILVA

Department of Botany, University of California, Berkeley, California

The term "algae" embraces an assemblage of organisms which exhibit great diversity of structure, reproduction, and metabolism. Algae include all oxygen-evolving photosynthetic organisms except bryophytes (mosses and liverworts) and vascular plants. Many colorless organisms are also referable to the algae on the basis of their similarity to photosynthetic forms with respect to structure, life history, and the nature of cell-wall and storage products. In size and complexity the algae range from single cells a few microns in diameter to the giant kelps, which may reach 60 meters in length.

The following table of classification is presented in order to indicate taxonomic relationships among the various genera mentioned in this book. Taxonomy is based on degree of similarity or dissimilarity, and relationships are proposed and tested in accordance with the principle of correlation of multiple factors or characters. While morphological characters are usually thought to be of primary (and classical) importance in taxonomy, it should be noted that two biochemical characters (pigmentation and food reserves) are fundamental in segregating this heterogeneous array of algae into major groups or divisions. The table of classification may serve the physiologist and biochemist in at least two ways: (i) it may help to orient him in regard to the organisms with which he is working; and (ii) it may suggest other suitable organisms for comparative studies. Because correlations between metabolic and morphological characters suggest themselves at every turn, the taxonomist eagerly awaits new information on comparative biochemistry and physiology, especially for organisms whose taxonomic relationships are not clear.

The reader is cautioned that this table of classification is incomplete, as it includes only those organisms mentioned in this book. The omission of entire groups, such as the chloromonads, indicates lacunae in our knowledge of algal metabolism, which we hope will eventually be filled. The system of nomenclature that has evolved through the years is reasonably effective for expressing taxonomic relationships, but unfortunately its technical aspects are not readily comprehended, even by taxonomists. It is not surprising, therefore, that physiologists and biochemists are bewildered by various name changes. Some changes reflect differences in taxonomic opinion or greater refinement of classification, while others are accommodations to rules governing the orderliness of the nomenclatural system itself. For the purposes of this book, changes of specific epithet within the same genus are considered of small import and therefore have not been noted except in the case of certain classical species (e.g., see Note 9).* Similarly, changes of generic name that are caused by the segregation of a new genus from an existing genus, by the merging of two genera or by the technical renaming of an existing genus, should create no difficulty, and such changes can readily be followed in the table (e.g., see Notes 5, 6, and 7, respectively). On the other hand, it is of the utmost importance to be cognizant of name changes stemming from misidentification or misinterpretation, lest a certain body of information be attributed to the wrong organism (e.g., see Notes 2 and 22).

The reader should also bear in mind that a generic name does not necessarily encompass organisms of a similar metabolic nature. Although certain genera are based on sufficient morphological criteria to assure us that we are dealing with closely related forms, others (e.g., *Chlorella*) can be defined by so few structural features that they doubtless include a wide array of organisms with diverse metabolic characteristics. In order to be reproducible and capable of meaningful comparison, studies with algae in such genera as *Chlorella* should be made on clearly designated strains. Wherever possible, living or preserved material of the organism studied should be kept in order that its identity may be later checked if necessary or desirable. In this regard algal culture collections, such as those at Cambridge University and Indiana University, can be of great importance and usefulness.

Division Cyanophyta (Blue-green Algae)

ORDER CHROOCOCCALES[1]

Family Chroococcaceae: *Anacystis*,[2] *Aphanocapsa*, *Chroococcus*, *Coccochloris*, *Gloeocapsa*, *Gloeothece*, *Microcystis* (=*Diplocystis*), *Synechococcus*, *Synechocystis*

Family Entophysalidaceae: *Chlorogloea*

ORDER NOSTOCALES

Family Oscillatoriaceae: *Arthrospira*, *Katagnymene*, *Lyngbya*, *Oscillatoria*, *Phormidium*,[2] *Schizothrix*, *Spirulina*, *Symploca*, *Trichodesmium*

Family Beggiatoaceae: *Beggiatoa*

Family Thiotrichaceae: *Leucothrix*, *Thiothrix*

---

* All author's notes (1–50) are included at the end of this Appendix.

Family Nostocaceae: *Anabaena, Aphanizomenon, Cylindrospermum, Isocystis, Nostoc*
Family Scytonemataceae: *Plectonema, Scytonema, Tolypothrix*
Family Rivulariaceae: *Calothrix, Gloeotrichia, Rivularia*
ORDER STIGONEMATALES
Family Mastigocladaceae: *Mastigocladus*

Division Rhodophyta (Red Algae)

Class Bangiophyceae[3]
ORDER PORPHYRIDIALES
Family Porphyridiaceae: *Porphyridium*
ORDER GONIOTRICHALES
Family Goniotrichaceae: *Goniotrichum*
ORDER BANGIALES
Family Erythropeltidaceae: *Smithora*[4]
Family Bangiaceae: *Bangia, Porphyra*

Class Florideophyceae[3]
ORDER NEMALIONALES
Family Acrochaetiaceae: *Chromastrum,*[5] *Rhodochorton*
Family Batrachospermaceae: *Batrachospermum, Tuomeya*
Family Lemaneaceae: *Lemanea*
Family Helminthocladiaceae: *Helminthocladia, Liagora, Nemalion*
ORDER BONNEMAISONIALES
Family Chaetangiaceae: *Galaxaura*
Family Bonnemaisoniaceae: *Bonnemaisonia* (incl. *Trailliella*[6])
ORDER GELIDIALES
Family Gelidiaceae: *Acanthopeltis, Gelidium, Pterocladia*
ORDER CRYPTONEMIALES
Family Dumontiaceae: *Constantinea, Dilsea, Dumontia, Neodilsea*
Family Hildenbrandiaceae: *Hildenbrandia*
Family Corallinaceae: *Amphiroa, Bossiella*[7] (=*Bossea*), *Calliarthron, Corallina, Goniolithon, Lithophyllum* (incl. *Tenarea tortuosa*[8]), *Lithothamnium, Melobesia*
Family Endocladiaceae: *Gloiopeltis*
Family Tichocarpaceae: *Tichocarpus*
Family Cryptonemiaceae: *Grateloupia, Phyllymenia* (incl. *Cyrtymenia*)
Family Kallymeniaceae: *Callophyllis*
ORDER GIGARTINALES
Family Gracilariaceae: *Gracilaria*[9]
Family Plocamiaceae: *Plocamium*
Family Sphaerococcaceae: *Caulacanthus*
Family Furcellariaceae: *Furcellaria, Halarachnion*
Family Solieriaceae: *Eucheuma*
Family Rhodophyllidaceae: *Calliblepharis, Cystoclonium, Rhodophyllis*
Family Hypneaceae: *Hypnea*
Family Phyllophoraceae: *Ahnfeltia, Phyllophora*
Family Gigartinaceae: *Chondrus, Gigartina, Iridaea* (=*Iridophycus*), *Rhodoglossum*

ORDER RHODYMENIALES
Family Rhodymeniaceae: *Halosaccion, Rhodymenia*
Family Lomentariaceae: *Coeloseira*
ORDER CERAMIALES
Family Ceramiaceae: *Antithamnion, Bornetia, Callithamnion, Campylaephora, Ceramium, Corynospora* (=*Neomonospora*), *Griffithsia, Microcladia, Plumaria, Ptilota, Sphondylothamnion*
Family Dasyaceae: *Heterosiphonia*
Family Delesseriaceae: *Acrosorium, Botryoglossum, Cryptopleura, Delesseria, Drachiella, Membranoptera, Myriogramme, Nitophyllum, Phycodrys, Polyneura*
Family Rhodomelaceae: *Bostrychia, Chondria, Digenea, Laurencia, Odonthalia, Polysiphonia, Pterochondria, Rhodomela,*[9] *Rytiphlaea, Symphyocladia*

Division Cryptophyta[10]

Class Cryptophyceae
ORDER CRYPTOMONADALES
Family Cryptomonadaceae: *Chilomonas, Cryptomonas, Rhodomonas*
Family Senniaceae: *Hemiselmis, Sennia*
[Genus of uncertain position: *Cyanidium*[11]]

Division Pyrrophyta (predominantly dinoflagellates)

Class Desmophyceae[12]
ORDER PROROCENTRALES
Family Prorocentraceae: *Prorocentrum*
Class Dinophyceae
ORDER GYMNODINIALES
Family Pronoctilucaceae: *Oxyrrhis*
Family Gymnodiniaceae: *Amphidinium, Gymnodinium*
Family Noctilucaceae: *Noctiluca*
ORDER PERIDINIALES
Family Glenodiniaceae: *Glenodinium*
Family Peridiniaceae: *Peridinium*
Family Gonyaulacaceae: *Gonyaulax*

Division Bacillariophyta (diatoms)

Class Centrobacillariophyceae[13]
ORDER EUPODISCALES[14]
Family Coscinodiscaceae: *Coscinodiscus* (including *Ethmodiscus*), *Cyclotella, Melosira, Skeletonema, Thalassiosira*
ORDER RHIZOSOLENIALES[15]
Family Rhizosoleniaceae: *Lauderia*
ORDER BIDDULPHIALES[16]
Family Biddulphiaceae: *Biddulphia, Ditylum, Lithodesmium*
Class Pennatibacillariophyceae[17]
ORDER FRAGILARIALES[18]
Family Tabellariaceae: *Licmophora, Tabellaria*
Family Fragilariaceae: *Asterionella, Fragilaria, Opephora*

ORDER EUNOTIALES[19]
Family Eunotiaceae: *Actinella* (incl. *Desmogonium*)
ORDER ACHNANTHALES[20]
Family Achnanthaceae: *Achnanthes, Cocconeis*
ORDER NAVICULALES[21]
Family Naviculaceae: *Amphipleura, Anomoeoneis, Caloneis, Gyrosigma, Navicula, Pinnularia, Pleurosigma*
Family Cymbellaceae: *Amphora, Cymbella, Gomphonema*
Family Amphiproraceae: *Amphiprora*
ORDER PHAEODACTYLALES[22]
Family Phaeodactylaceae: *Phaeodactylum*
ORDER BACILLARIALES[23]
Family Nitzschiaceae: *Bacillaria, Cylindrotheca, Hantzschia, Nitzschia*[22]
ORDER SURIRELLALES[24]
Family Surirellaceae: *Cymatopleura, Surirella*

Division Phaeophyta (Brown Algae)

ORDER ECTOCARPALES
Family Ectocarpaceae: *Acinetospora* (incl. *Heterospora*), *Ectocarpus, Feldmannia,*[25] *Pylaiella, Spongonema, Waerniella*
Family Ralfsiaceae: *Nemoderma*
ORDER SPHACELARIALES
Family Sphacelariaceae: *Sphacelaria*
Family Stypocaulaceae: *Halopteris* (incl. *Stypocaulon*)
Family Cladostephaceae: *Cladostephus*
ORDER DICTYOTALES
Family Dictyotaceae: *Dictyopteris, Dictyota, Padina*
ORDER CHORDARIALES
Family Myrionemataceae: *Pleurocladia*
Family Elachistaceae: *Elachista*
Family Chordariaceae: *Chordaria, Heterochordaria*
Family Spermatochnaceae: *Nemacystus*
ORDER SPOROCHNALES
Family Sporochnaceae: *Sporochnus*
ORDER DESMARESTIALES
Family Arthrocladiaceae: *Arthrocladia*
Family Desmarestiaceae: *Desmarestia*
ORDER DICTYOSIPHONALES
Family Punctariaceae: *Myelophycus*
Family Scytosiphonaceae: *Scytosiphon*
Family Dictyosiphonaceae: *Coilodesme*
ORDER LAMINARIALES
Family Chordaceae: *Chorda*
Family Laminariaceae: *Costaria, Laminaria*
Family Lessoniaceae: *Macrocystis, Nereocystis, Pelagophycus*
Family Alariaceae: *Alaria, Ecklonia, Eisenia, Undaria*
ORDER FUCALES
Family Notheiaceae: *Hormosira*

Family Fucaceae: *Ascophyllum, Fucus, Pelvetia*
Family Himanthaliaceae: *Himanthalia*
Family Cystoseiraceae: *Cystoseira, Halidrys, Myagropsis* (incl. *Cystophyllum*)[26]
Family Sargassaceae: *Coccophora, Sargassum* (incl. *Hizikia*)

Division Chrysophyta[27]

ORDER PHAEOTHAMNIALES
Family Phaeothamniaceae: *Apistonema*
[Genus of uncertain position: *Pleurochrysis*[28]]
ORDER OCHROMONADALES
Family Ochromonadaceae: *Anthophysa, Ochromonas, Siderodendron, Siphomonas, Uroglena*
Family Dinobryaceae: *Dinobryon*
Family Synuraceae: *Mallomonas, Microglena, Synura*
Family Coccolithophoraceae: *Calyptrosphaera, Coccolithus, Hymenomonas, Syracosphaera, Zygosphaera*
Family Phaeocystidaceae: *Phaeocystis*
ORDER ISOCHRYSIDALES
Family Isochrysidaceae: *Isochrysis*
Family Prymnesiaceae: *Prymnesium*
ORDER THALLOCHRYSIDALES
Family Thallochrysidaceae: *Thallochrysis*
ORDER CHROMULINALES
Family Chromulinaceae: *Chromulina* (incl. *Chrysomonas*)
Family Chrysocapsaceae: *Gloeochrysis, Phaeaster* (incl. *Monochrysis*)
Family Hydruraceae: *Hydrurus*

Division Xanthophyta[29]

ORDER CHLORAMOEBALES[30]
Family Chloramoebaceae: *Olisthodiscus*
ORDER MISCHOCOCCALES[31]
Family Pleurochloridaceae: *Botrydiopsis, Polyedriella, Vischeria*
Family Chlorobotrydaceae: *Monodus*
[Genus of uncertain position: *Halosphaera*[32]]
ORDER TRIBONEMATALES[33]
Family Heterotrichaceae: *Bumilleria*
Family Tribonemataceae: *Tribonema*
ORDER VAUCHERIALES[34]
Family Botrydiaceae: *Botrydium*
Family Vaucheriaceae: *Vaucheria*[34]

Division Euglenophyta

ORDER EUGLENALES
Family Euglenaceae: *Euglena, Khawkinea, Lepocinclis, Phacus, Trachelomonas*
Family Astasiaceae: *Astasia*
Family Peranemataceae: *Peranema*

Division Chlorophyta (Green Algae)

[Genus of uncertain position: *Chlorochytridion*[35]]

ORDER VOLVOCALES

Suborder Volvocineae

Family Polyblepharidaceae: *Dunaliella, Polytomella, Pyramimonas*

Family Chlamydomonadaceae: *Brachiomonas, Chlamydomonas* (incl. *Carteria ovata*[36]), *Hypnomonas, Platymonas, Polytoma*

Family Haematococcaceae: *Haematococcus* (incl. *Sphaerella lacustris* and *Balticola*[37]), *Stephanosphaera*

Family Volvocaceae: *Eudorina, Gonium, Pandorina, Volvox*

Suborder Tetrasporineae

Family Palmellaceae: *Gloeocystis*

Family Tetrasporaceae: *Tetraspora*

ORDER CHLOROCOCCALES

Family Chlorococcaceae: *Chlorococcum, Trebouxia* (incl. *Cystococcus cohaerens*)

Family Protosiphonaceae: *Protosiphon*

Family Hydrodictyaceae: *Hydrodictyon, Pediastrum, Tetraedron*[38]

Family Chlorellaceae: *Chlorella, Palmellococcus, Prototheca, Zoochlorella*

Family Oocystaceae: *Oocystis, Rhopalocystis,*[39] *Scotiella*

Family Botryococcaceae: *Botryococcus*

Family Coelastraceae: *Coelastrum*

Family Selenastraceae: *Ankistrodesmus* (incl. *Rhaphidium*), *Selenastrum*

Family Dictyosphaeriaceae: *Dictyosphaerium*

Family Scenedesmaceae: *Crucigenia, Scenedesmus*

ORDER SIPHONOCLADALES

Family Valoniaceae: *Dictyosphaeria, Valonia*

Family Siphonocladaceae: *Siphonocladus*

ORDER CODIALES[40]

Family Bryopsidaceae: *Bryopsis*

Family Codiaceae: *Codium*

ORDER DERBESIALES[41]

Family Derbesiaceae: *Derbesia* (incl. *Halicystis*[42])

ORDER CAULERPALES[43]

Family Udoteaceae: *Chlorodesmis, Halimeda, Penicillus, Rhipocephalus, Udotea*

Family Caulerpaceae: *Caulerpa*

ORDER DASYCLADALES

Family Dasycladaceae: *Acetabularia*[44] (incl. *Acicularia schenkii*), *Cymopolia, Dasycladus, Halicoryne, Neomeris*

ORDER CHLOROSPHAERALES[45]

Family Coccomyxaceae: *Coccomyxa, Nannochloris*

Family Chlorosphaeraceae: *Chlorosarcina*

ORDER ULOTRICHALES

Family Ulotrichaceae: *Chlorhormidium*[46] (=*Hormidium*), *Raphidonema, Stichococcus, Ulothrix*

Family Protococcaceae: *Protococcus*[47] (=*Pleurococcus*; incl. *Chlorococcum botryoides*)

Family Microsporaceae: *Microspora*
Family Chaetophoraceae: *Chaetophora, Draparnaldia, Microthamnion, Stigeoclonium*
Family Coleochaetaceae: *Coleochaete*
Family Trentepohliaceae: *Phycopeltis, Trentepohlia*
Family Sphaeropleaceae: *Sphaeroplea*
ORDER ULVALES
Family Monostromaceae: *Monostroma*
Family Ulvaceae: *Enteromorpha, Ulva*
ORDER SCHIZOGONIALES
Family Schizogoniaceae: *Prasiola*
ORDER CLADOPHORALES
Family Cladophoraceae: *Acrosiphonia*,[48] *Chaetomorpha, Cladophora* (incl. *Aegagropila*), *Rhizoclonium, Spongomorpha*,[48] *Urospora*[48]
ORDER OEDOGONIALES
Family Oedogoniaceae: *Oedogonium*
ORDER ZYGNEMATALES
Family Zygnemataceae: *Mougeotia, Spirogyra, Temnogametum, Zygnema, Zygogonium* (probably incl. *Pleurodiscus*)[49]
Family Mesotaeniaceae: *Ancylonema, Mesotaenium*
Family Desmidiaceae: *Closterium, Cosmarium, Euastrum, Micrasterias, Oocardium, Penium, Pleurotaenium, Staurastrum, Tetmemorus*

Division Charophyta (stoneworts)

ORDER CHARALES
Family Characeae: *Chara, Nitella, Nitellopsis* (incl. *Tolypellopsis*)

Apochromatic Groups of Uncertain Position

Protomastigineae

Family Trypanosomaceae: *Strigomonas*
Family Bicosoecaceae: *Bicosoeca*
Family Choanogastraceae: *Choanogaster*[50]

Notes

[1] Drouet and Daily (*Butler Univ. Studies* **12,** 1956) have made an exhaustive taxonomic study of the coccoid Cyanophyta. However, because the nomenclature of this group is notoriously involved, no attempt has been made to bring the names used in the present book into line with Drouet and Daily's classification.

[2] The formation of short but multicellular filaments by the alga extensively cultured and investigated under the name "*Anacystis nidulans*" indicates that this strain should not be assigned to the Chroococcales. Drouet (personal communication) has tentatively identified it as *Phormidium mucicola* (Nostocales).

[3] Class Bangiophyceae Cronquist (*Botan. Rev.* **26,** 437, 1960); class Florideophyceae Cronquist (*op. cit.* p. 438).

[4] *Smithora* Hollenberg (*Pacific Naturalist* **1,** No. 8, 1959) is based on *Porphyra naiadum.*

[5] *Chromastrum,* which was segregated from the *Acrochaetium-Rhodochorton*

assemblage by Papenfuss (*Univ. Calif. (Berkeley) Publs. Botan.* **18,** 320, 1945), includes *Rhodochorton floridulum.*

[6] *Trailliella intricata* has been shown to be the sporophytic phase of *Bonnemaisonia hamifera* (cf. Feldmann and Feldmann, *Ann. Sci. Nat. Bot.* [11] **3,** 75, 1942; Koch, *Arch. Mikrobiol.* **14,** 635, 1950).

[7] *Bossiella* Silva (*Madroño* **14,** 46, 1957) = *Bossea* Manza, a later homonym of *Bossea* Reichenbach.

[8] The plant generally known as *Tenarea tortuosa* was shown by Huvé (*Bull. soc. botan. France* **104,** 132, 1957) to be referable to *Lithophyllum.*

[9] *Gracilaria verrucosa* is the correct name for the plant generally known as *G. confervoides* (cf. Papenfuss, *Hydrobiologia* **2,** 195, 1950). *Rhodomela confervoides* is the correct name for the plant generally known as *R. subfusca* (cf. Silva, *Univ. Calif. (Berkeley) Publs. Botan.* **25,** 269, 1952).

[10] Division Cryptophyta Dougherty et Allen (*in* M. B. Allen, ed., "Comparative Biochemistry of Photoreactive Systems," p. 142, Fig. 2. Academic Press, New York, 1960).

[11] *Cyanidium caldarium* has been variously referred to the Cyanophyta (as *Pluto*), the Rhodophyta (as *Rhodococcus*), and the Chlorophyta (as "acid *Chlorella*"). The presence of starch, zeaxanthin, and a phycocyanin indicates a possible relationship with the cryptomonads. However, the absence of chlorophyll *c*, the difference in type of phycocyanin, and the presence of β-carotene rather than α-carotene suggests that if *Cyanidium* is assigned to this group it should be placed in a class coordinate with the cryptomonads rather than in the class Cryptophyceae. (Cf. M. B. Allen, ed., "Comparative Biochemistry of Photoreactive Systems," p. 39. Academic Press, New York, 1960, and Lewin, *Phycol. News Bull.* **14,** 6, 1961.)

[12] Class Desmophyceae Fott (*Preslia* **32,** 143, 1960).

[13] Class Centrobacillariophyceae nom. et stat. nov. = order Centrales Karsten (*in* A. Engler, ed., "Die natürlichen Pflanzenfamilien," 2nd ed., Vol. 2, p. 201. Engelmann, Leipzig, 1928).

[14] Order Eupodiscales Bessey (*Univ. Nebraska Studies* **7,** 284, 1907).

[15] Order Rhizosoleniales nom. nov. = Soleniales Krieger (*in* A. Engler, ed., "Syllabus der Pflanzenfamilien," 12th ed., Vol. 1, p. 83. Borntraeger, Berlin, 1954). The name Soleniales will be untenable according to the Montreal code of botanical nomenclature (to be published shortly) because it is based on the stem of an illegitimate family name, Soleniaceae.

[16] Order Biddulphiales Krieger (*in* A. Engler, ed., "Syllabus der Pflanzenfamilien," 12th ed., Vol. 1, p. 83. Borntraeger, Berlin, 1954).

[17] Class Pennatibacillariophyceae nom. et stat. nov. = order Pennales Karsten (*in* A. Engler, ed., "Die natürlichen Pflanzenfamilien, 2nd ed., Vol. 2, p. 202. Engelmann, Leipzig, 1928).

[18] Order Fragilariales nom. et stat. nov. = suborder Fragilarioideae West (G. S. West, "A Treatise on the British Freshwater Algae," p. 280. Cambridge Univ. Press, London and New York, 1904) = order Araphidales Krieger (*in* A. Engler, ed., "Syllabus der Pflanzenfamilien," 12th ed., Vol. 1, p. 83. Borntraeger, Berlin, 1954). The ordinal name Araphidales is untenable because it is not based on the name of a family.

[19] Order Eunotiales nom. nov. = Raphidioidales Krieger (*in* A. Engler, ed., "Syllabus der Pflanzenfamilien," 12th ed., Vol. 1, p. 84. Borntraeger, Berlin, 1954). The ordinal name Raphidioidales is untenable because it is not based on the name of a family.

[20] Order Achnanthales nom. et stat. nov. = suborder Achnanthoideae West ("A Treatise on the British Freshwater Algae," p. 280. Cambridge Univ. Press, London and New York, 1904) = order Monoraphidales Krieger (*in* A. Engler, ed., "Syllabus der Pflanzenfamilien," 12th ed., Vol. 1, p. 84. Borntraeger, Berlin, 1954). The ordinal name Monoraphidales is untenable because it is not based on the name of a family.

[21] Order Naviculales Bessey (*Univ. Nebraska Studies* **7,** 284, 1907).

[22] Order Phaeodactylales nom. et stat. nov. = suborder Phaeodactylineae J. Lewin (*J. Gen. Microbiol.* **18,** 428, 1958). *Nitzschia* (*Nitzschiella*) *closterium* and the biradiate form of *Phaeodactylum tricornutum* (frequently misidentified as "*Nitzschia closterium* forma *minutissima*" or simply "*Nitzschia closterium*") have both been used extensively in physiological and biochemical studies, resulting in much confusion (cf. Lewin, *op. cit.*).

[23] Order Bacillariales Hendey (*Discovery Rept.* **16,** 200, 1937).

[24] Order Surirellales nom. et stat. nov. = suborder Surirelloideae West ("A Treatise on the British Freshwater Algae," p. 280. Cambridge Univ. Press, London and New York, 1904).

[25] *Feldmannia,* which was segregated from *Ectocarpus* by Hamel ("Phéophycées de France," p. xi. Paris, 1931–1939), includes *E. globifer.*

[26] Fensholt (*Am. J. Botan.* **42,** 305, 1955) has referred *Cystophyllum hakodatense* and *C. crassipes* to *Cystoseira* and *C. sisymbrioides* to *Myagropsis myagroides.*

[27] The taxonomy of the Chrysophyta is in a notable state of flux. The present author has arbitrarily followed the classification proposed by Bourrelly (*Rev. alg., Mém. hors-sér.* **1,** 1957) without attempting to bring the nomenclature into accord with the Botanical Code. (For other taxonomic treatments of this group, see Papenfuss, Classification of the algae. *In* "A Century of Progress in the Natural Sciences," p. 147. California Acad. Sci., San Francisco, 1955; and Fott, "Algenkunde." Fischer, Jena, 1959.)

[28] For a discussion of the taxonomic relationships of *Pleurochrysis,* see Bourrelly (*Rev. alg., Mém. hors-sér.* **1,** 127, 230, 1957).

[29] Division Xanthophyta Polyansky et Hollerbach (*Opredelitel' Presnovodnykh Vodoroslei S.S.S.R.* **4,** 11, 1951).

[30] Order Chloramoebales nom. nov. = Heterochloridales Pascher (*Hedwigia* **53,** 10, 1912). Heterochloridaceae cannot serve as a stem for an ordinal name because it is an illegitimate synonym (of Chloramoebaceae).

[31] Order Mischococcales Fott ("Algenkunde," p. 131. Fischer, Jena, 1959). Heterococcales Pascher (*Hedwigia* **53,** 14, 1912) is untenable as an ordinal name because it is not based on the name of a family.

[32] For a discussion of the taxonomic relationships of *Halosphaera,* see Pascher (*in* L. Rabenhorst, "Kryptogamenflora von Deutschland, Österreich, und der Schweiz," 2nd ed., Vol. 11, p. 910. Akademische Verlagsges., Leipzig, 1939).

[33] Order Tribonematales Pascher (*in* Rabenhorst, *op. cit.* p. 915). Heterotrichales Pascher (*Hedwigia* **53,** 18, 1912) is untenable as an ordinal name because at the time it was proposed, the family Heterotrichaceae had not yet been established.

[34] Order Vaucheriales Bohlin (Utkast till de gröna algernas. Thesis, p. 14. Univ. of Lund, Sweden, 1901). Heterosiphonales Pascher (*Hedwigia* **53,** 21, 1912) is not only a junior synonym, but it is untenable as an ordinal name because it is not based on the name of a family. *Vaucheria,* on

morphological and biochemical grounds, has been removed to the Xanthophyta from the Chlorophyta, with which it was usually classified prior to 1950.

[35] For discussions of the morphology and possible taxonomic relationships of *Chlorochytridion tuberculatum* (as *Pedinomonas tuberculata*), see Vischer (*Verhandl. Intern. Ver. Limnol.* **10,** 504, 1949) and Butcher (An introductory account of the smaller algae of British coastal waters, Part I. *Gt. Brit., Ministry Agr. Fish. & Food, Fish. Invest.* [IV], **1**(2), pp. 35–36).

[36] *Carteria ovata* is the diploid motile stage of *Chlamydomonas variabilis* (cf. Behlau, *Beitr. Biol. Pflanzen* **26,** 221, 1939).

[37] For a discussion of the taxonomy and nomenclature of *Haematococcus*, see Pocock (*Trans. Roy. Soc. S. Afr.* **36,** 5, 1960).

[38] Although this genus is usually assigned to the Oocystaceae, Starr (*Am. J. Botan.* **41,** 17, 1954) observed the production of zoospores in *Tetraedron bitridens* and therefore transferred this species to the Hydrodictyaceae.

[39] *Rhopalocystis* Schussnig (*Österr. botan. Z.* **102,** 444, 1955).

[40] Order Codiales Feldmann (*Congr. intern. botan., 8e Congr., Paris,* **1954,** *Rappt. et communs.* **17,** 97, 1954). Siphonales is not tenable as an ordinal name because it is not based on a legitimate family name.

[41] Order Derbesiales Feldmann (*Congr. intern. botan., 8e Congr., Paris,* **1954,** *Rappts. et communs.* **17,** 97, 1954).

[42] *Halicystis ovalis* and *H. parvula* have been shown to be the gametophytic phase of *Derbesia marina* and *D. tenuissima,* respectively.

[43] Order Caulerpales Feldmann (*Compt. rend. acad. sci.* **222,** 753, 1946).

[44] *Acetabularia acetabulum* is the correct name for the plant generally known as *A. mediterranea* (cf. Silva, *Univ. Calif.* (*Berkeley*) *Publs. Botan.* **25,** 255, 1952).

[45] Order Chlorosphaerales Herndon (*Am. J. Botan.* **45,** 307, 1958).

[46] *Chlorhormidium* Fott (*Preslia* **32,** 149, 1960) = *Hormidium* Kützing emend. Klebs, not *Hormidium* Lindley (a genus of orchids).

[47] The nomenclature of the alga which forms the ever-present coating on trunks of trees, etc., is almost inextricably involved. Technically, neither of the names commonly applied to this plant, *Protococcus* or *Pleurococcus,* can be used in this sense. (Cf. Drouet and Daily, *Butler Univ. Studies* **12,** 144, 145, 1956.)

[48] Some authors consider *Acrosiphonia* a section of *Spongomorpha.* Jónsson (*Compt. rend. acad. sci.* **248,** 835, 1959) showed that the sporophytic phase of *A. spinescens* is *Codiolum petrocelidis,* a unicellular form hitherto assigned to the Chlorococcales. *Spongomorpha* and *Urospora* have also been shown to alternate with unicellular phases. Largely on the basis of this alternation of heteromorphic generations and special features of cell-wall structure, Jónsson (*op. cit.* p. 1567) established the family Acrosiphoniaceae to include *Acrosiphonia, Spongomorpha,* and *Urospora.* For criticism of this treatment, see Archer and Burrows (*Brit. Phycol. Bull.* **2,** 31, 1960).

[49] For discussions of *Pleurodiscus purpureus,* see Skuja (*Rev. alg.* **6,** 137, 1932) and Tiffany (*Brittonia* **2,** 170, 1936).

[50] *Choanogaster* Pochmann (*Arch. Protistenk.* **103,** 526, 1959).

—APPENDIX B—

Uptake of Radioactive Wastes by Algae

R. W. EPPLEY

Bioastronautics Laboratory, Northrop Corporation, Hawthorne, California

Algae are among the bases of aquatic food chains, some of which lead to man; there is therefore considerable practical importance in their uptake of radioactive fission products from nuclear reactors and from bomb "fall-out." Several studies have indicated that both ionic and particulate forms of radionuclides and both passive and active mechanisms of uptake may be involved. In natural waters the uptake of a given isotope in relation to other fission products may be quite different from that on land. For example, planktonic algae were found to accumulate Ce^{144} to a much greater extent than Cs^{137}, whereas for land plants the reverse is true (Rice and Willis, 1959).

Observations of surface adsorption include those on the uptake of Sr^{90} and Y^{90} by various seaweeds (Spooner, 1949), of Ce^{144} and Sr^{90} by species of marine phytoplankton (Rice, 1956; Rice and Willis, 1959), and of Ru^{106} by *Porphyra laciniata* (Jones, 1960). Once adsorbed, such radioisotopes may be more or less firmly bound or readily eluted. For example, nitrosyl ruthenium apparently can form nonexchangeable complexes involving colloidal iron on the surfaces of diatom cells, although in the absence of iron this does not occur (Jones, 1960). Although Y^{90} uptake by *Carteria* seems to be limited to surface adsorption, Sr^{90} is accumulated chiefly by living cells (Rice, 1956). Other examples of biological accumulation include those of Co^{60} by *Rhodymenia palmata* (Scott and Ericson, 1954), Zn^{65} by a variety of benthonic algae (Bachman and Odum, 1960), and Cs^{137} by *Chlorella* and *Euglena* (Williams, 1960). In the latter cases, as in the study of *Carteria* mentioned above, absorption was shown to vary with the physiological age of the cells. In the case of Sr isotopes, the extent and mechanism of uptake is further complicated in those algae which deposit calcareous skeletons into which $SrCO_3$ may be incorporated (see Bowen, 1956). Without specification of the possible mechanism involved, "coefficients of accumulation" of radioisotopes of Ca, Co, Fe, Rb, S, Zn, and Zr were determined for *Scenedesmus quadricauda*, *Cladophora glomerata*, and *Spirogyra* sp. by Gileva (1960). Polikarpov (1961) compared the rates of uptake of Sr^{90}, Cs^{137}, and Ce^{144} by the marine alga *Cystoseira barbata*, and the readiness with which these isotopes are liberated into seawater when the thalli disintegrate.

References

Bachmann, R. W., and Odum, E. P. (1960). Uptake of Zn^{65} and primary productivity in marine benthic algae. *Limnol. Oceanog.* **5**, 349–355.

Bowen, H. J. M. (1956). Strontium and barium in sea water and marine organisms. *J. Marine Biol. Assoc. United Kingdom* **35**, 451–459.

Gileva, E. A. (1960). [Coefficients of radio-isotope accumulation by fresh-water plants.] (In Russian.) *Doklady Akad. Nauk S.S.S.R.* **132**, 948–949.

Jones, R. F. (1960). The accumulation of nitrosyl ruthenium by fine particles and marine organisms. *Limnol Oceanog.* **5**, 312–325.

Polikarpov, G. G. (1961). [The role of detritus formation in the migration of Sr^{90}, Cs^{137}, and Ce^{144}.] *Doklady Akad. Nauk S.S.S.R.* **136**, 921–923. (In Russian.)

Rice, T. R. (1956). The accumulation and exchange of strontium by marine planktonic algae. *Limnol. Oceanog.* **1**, 123–138.

Rice, T. R., and Willis, V. A. (1959). Uptake, accumulation and loss of radioactive cerium-44 by marine planktonic algae. *Limnol. Oceanog.* **4**, 277–290.

Scott, R., and Ericson, L. E. (1954). Some aspects of cobalt metabolism by *Rhodymenia palmata* with particular reference to vitamin B_{12} content. *J. Exptl. Botany* **6**, 348–361.

Spooner, G. M. (1949). Observations on the absorption of radioactive strontium and yttrium by marine algae. *J. Marine Biol. Assoc. United Kingdom* **28**, 587–625.

Williams, L. G. (1960). Uptake of cesium-137 by cells and detritus of *Euglena* and *Chlorella*. *Limnol. Oceanog.* **5**, 301–311.

—APPENDIX C—

Antibiotics from Algae

RALPH A. LEWIN

Scripps Institution of Oceanography, University of California, La Jolla, California

In addition to the extracellular metabolites produced by algae, some of which may possess antibiotic properties (as mentioned by Fogg in Chapter 30 of this treatise), there have been a few reports of antibiotic activity in extracts of seaweeds, variously prepared in water or organic solvents. Undoubtedly in certain instances the enthusiasm of the authors for publication of their findings was not justified by subsequent biological tests of such substances. To date none has proved of economic value, and none has been unequivocally isolated and identified, though active components of *Symphyocladia gracilis* and *Rhodomela larix* have been partially identified as brominated phenolic compounds (Saito and Nakamura, 1951; Mautner *et al.*, 1953; Saito and Sameshima, 1955). In some cases the effect may have been attributable to nonspecific inhibition by such substances as tannins (see Ogino, Chapter 26, this treatise), organic acids such as those in high concentration in the cell sap of *Desmarestia* spp. (see Schiff, Chapter 14), or toxic organic bases such as choline and trimethylamine, isolated from *Gonyaulax catenella* by Riegel *et al.* (1949).

The following reference list comprises most of the articles on this subject published to date.

References

Allen, M. B., and Dawson, E. Y. (1960). Production of antibacterial substances by benthic tropical marine algae. *J. Bacteriol.* **79,** 459–460.

Burkholder, P. R., Burkholder, L. M., and Almodóvar, L. R. (1960). Antibiotic activity of some marine algae of Puerto Rico. *Botan. Marina* **2,** 149–154.

Chesters, C. G. C., and Scott, J. A. (1956). Production of antibiotic substances by seaweeds. *Proc. Intern. Seaweed Symposium, 2nd Symposium, Trondhjem, 1955* pp. 49–53.

Kamimoto, K. (1955). Studies on the antibacterial substances extracted from seaweeds. I. On the effect of the extracts from seaweeds against the growth of some pathogenic organisms. *Nippon Saikungaku Zasshi* **10,** 897–902.

Kamimoto, K. (1956). Studies on the antibacterial substances extracted from seaweeds. II. Effect of extracts from seaweeds on the growth of some acid-fast bacteria. *Nippon Saikingaku Zasshi* **11,** 307–313.

McCutcheon, R. S., Arigoni, L., and Fischer, L. (1949). Phytochemical investigation of *Cymathere triplicata, Hedophyllum sessile* and *Egregia Menziesii. J. Am. Pharm. Assoc.* **38,** 196–200.

Mautner, H. C., Gardner, G. M., and Pratt, R. (1953). Antibiotic activity of seaweed extracts. II. *Rhodomela larix. J. Am. Pharm. Assoc.* **42,** 294–296.

Pratt, R., Mautner, H., Gardner, G. M., Yi-Hsien Sha, and Dufrenoy, J. (1951). Report on antibiotic activity of seaweed extracts. *J. Am. Pharm. Assoc.* **40,** 575–579.

Riegel, B., Stanger, D. W., Wikholm, D. M., Mold, J. D., and Summer, H. (1949). Paralytic shellfish poison. V. The primary source of the poison, the marine plankton organism, *Gonyaulax catenella. J. Biol. Chem.* **177,** 7–11.

Roos, H. (1957). Untersuchungen über das Vorkommen antimikrobieller Substanzen in Meeresalgen, *Kiel. Meeresforsch.* **13,** 41–58.

Saito, K., and Nakamura, Y. (1951). Sargalin and related phenols from marine algae and their medical functions. (Preliminary report.) *J. Chem. Soc. Japan, Pure Chem. Sect.* **72,** 992–993.

Saito, K., and Sameshima, J. (1955). Studies on antibiotic action of algae extracts. *J. Agr. Chem. Soc. Japan* **29,** 427–430.

Vacca, D. D., and Walsh, R. A. (1954). The antibiotic activity obtained from an extract of *Ascophyllum nodosum. J. Am. Pharm. Assoc.* **43,** 24–26.

Author Index

Numbers in italic indicate the page on which the full reference is listed.

Subject Index

(Organic anions are generally listed under the name of the appropriate acid. For species, genera, and all higher categories, see the Taxonomic Index.)

Taxonomic Index

The taxonomic position of each alga referred to in the text is given in Appendix A, pp. 827–837. Genera and species are italicized; higher categories are not. Authorities are not given. Higher plants and animals are not listed unless specifically named. Organisms which are not algae, but which are referred to in the text, are listed here in parentheses.